Problem Books in Mathematics

Series Editor:

Peter Winkler
Department of Mathematics
Dartmouth College
Hanover, NH
USA

More information about this series at http://www.springer.com/series/714

Antonio Caminha Muniz Neto

An Excursion through Elementary Mathematics, Volume III

Discrete Mathematics and Polynomial Algebra

 Springer

Antonio Caminha Muniz Neto
Universidade Federal do Ceará
Fortaleza, Ceará, Brazil

ISSN 0941-3502 ISSN 2197-8506 (electronic)
Problem Books in Mathematics
ISBN 978-3-030-08590-2 ISBN 978-3-319-77977-5 (eBook)
https://doi.org/10.1007/978-3-319-77977-5

Printed on acid-free paper

This Springer imprint is published by the registered company Springer International Publishing AG part
of Springer Nature.
The registered company address is: Gewerbestrasse 11, 6330 Cham, Switzerland

E horas sem conta passo, mudo,
O olhar atento,
A trabalhar, longe de tudo
O pensamento.

Porque o escrever - tanta perícia,
Tanta requer,
Que ofício tal.., nem há notícia
De outro qualquer.

Profissão de Fé (excerto) Olavo Bilac

Preface

This is the final volume of a series of three volumes (the other ones being [9] and [8]) devoted to the mathematics of mathematical olympiads. Generally speaking, they are somewhat expanded versions of a collection of six volumes, first published in Portuguese by the Brazilian Mathematical Society in 2012 and currently in its second edition.

The material collected here and in the other two volumes is based on course notes that evolved over the years since 1991, when I first began coaching students of Fortaleza to the Brazilian Mathematical Olympiad and to the International Mathematical Olympiad. Some ten years ago, preliminary versions of the Portuguese texts also served as textbooks for several editions of summer courses delivered at UFC to math teachers of the Cape Verde Republic.

All volumes were carefully planned to be a balanced mixture of a smooth and self-contained introduction to the fascinating world of mathematical competitions, as well as to serve as textbooks for students and instructors involved with math clubs for gifted high school students.

Upon writing the books, I have stuck myself to an invaluable advice of the eminent Hungarian-American mathematician George Pólya, who used to say that one cannot learn mathematics without *getting one's hands dirty*. That's why, in several points throughout the text, I left to the reader the task of checking minor aspects of more general developments. These appear either as small omitted details in proofs or as subsidiary extensions of the theory. In this last case, I sometimes refer the reader to specific problems along the book, which are marked with an * and whose solutions are considered to be an essential part of the text. In general, in each section I collect a list of problems, carefully chosen in the direction of applying the material and ideas presented in the text. Dozens of them are taken from former editions of mathematical competitions and range from the almost immediate to real challenging ones. Regardless of their level of difficulty, generous hints, or even complete solutions, are provided to virtually all of them.

As a quick look through the Contents pages readily shows, this time we concentrate on combinatorics, number theory, and polynomials. Although the chapters' names quickly link them to one of these three major themes, whenever possible or desirable later chapters revisit or complement material covered in earlier ones. We now describe, a bit more specifically, what is covered within each major topic.

Chapters 1 through 5 are devoted to the study of basic combinatorial techniques and structures. We start by reviewing the elementary counting strategies, emphasizing the construction of bijections and the use of recursive arguments throughout. We then go through a bunch of more sophisticated tools, as the inclusion-exclusion principle and double counting, the use of equivalence relations, metrics on finite sets, and generating functions. Turning our attention to the existence of configurations, the pigeonhole principle of Dirichlet and invariants associated with algorithmic problems now play the central role. Our tour through combinatorics finishes by studying some graph theory, all the way from the basic definitions to the classical theorems of Euler (on Eulerian paths), Cayley (on the number of labeled trees), and Turán (on complete subgraphs of a given graph), to name just a few ones.

We then turn to elementary number theory, which is the object of Chaps. 6–12. We begin, of course, by introducing the basic concepts and properties concerned with the divisibility relation and exploring the notion of greatest common divisor and prime numbers. Then we turn to diophantine equations, presenting Fermat's descent method and solving the famous Pell's equation. Before driving through a systematic study of congruences, we make an interlude to discuss the basics of multiplicative arithmetic functions and the distribution of primes, these two chapters being almost entirely independent of the rest of the book. From this point until Chap. 12, we focus on the congruence relation and its consequences, from the very beginnings to the finite field $\mathbb{Z}_p$, primitive roots, Gauss' quadratic reciprocity law, and Fermat's characterization of integers that can be written as the sum of two squares. All of the above material is, here more than anywhere else in the book, illustrated with lots of interesting and challenging examples and problems taken from several math competitions around the world.

The last nine chapters are devoted to the study of complex numbers and polynomials. Apart from what is usually present in high school classes—as the basics of complex numbers and the notion of degree, the division algorithm, and the concept of root for polynomials—we discuss several nonstandard topics. We begin by highlighting the use of complex numbers and polynomials as *tags* in certain combinatorial problems and presenting a complete proof of the fundamental theorem of algebra, accompanied with several applications. Then, we study the famous theorem of Newton on symmetric polynomials and the equally famous Newton's inequalities. The next theme concerns interpolation of polynomials, when particular attention is placed on Lagrange's interpolation theorem. Such a result is used to solve linear systems of Vandermonde with no linear algebra, which in turn allows us to, later, analyze an important particular class of linear recurrence relations. The book continues with the study of factorization of polynomials over $\mathbb{Q}$, $\mathbb{Z}$, and $\mathbb{Z}_p$, together with several interesting problems on irreducibility. Algebraic and transcendental numbers then make their appearance; among other topics, we present a simple proof of the fact that the set of algebraic numbers forms a field and discuss the rudiments of cyclotomic polynomials and transcendental numbers.

The final chapter develops the most basic aspects of complex power series, which are then used, disguised as complex generating functions, to solve general linear recurrence relations.

Several people and institutions contributed throughout the years for my efforts of turning a bunch of handwritten notes into these books. The State of Ceará Mathematical Olympiad, created by the Mathematics Department of the Federal University of Ceará (UFC) back in 1980 and now in its 37th edition, has since then motivated hundreds of youngsters of Fortaleza to deepen their studies of mathematics. I was one such student in the late 1980s, and my involvement with this competition and with the Brazilian Mathematical Olympiad a few years later had a decisive influence on my choice of career. Throughout the 1990s, I had the honor of coaching several brilliant students of Fortaleza to the Brazilian Mathematical Olympiad. Some of them entered Brazilian teams to the IMO or other international competitions, and their doubts, comments, and criticisms were of great help in shaping my view on mathematical competitions. In this sense, sincere thanks go to João Luiz de A. A. Falcão, Roney Rodger S. de Castro, Marcelo M. de Oliveira, Marcondes C. França Jr., Marcelo C. de Souza, Eduardo C. Balreira, Breno de A. A. Falcão, Fabrício S. Benevides, Rui F. Vigelis, Daniel P. Sobreira, Samuel B. Feitosa, Davi Máximo A. Nogueira, and Yuri G. Lima.

Professor João Lucas Barbosa, upon inviting me to write the textbooks to the Amílcar Cabral Educational Cooperation Project with Cape Verde Republic, had unconsciously provided me with the motivation to complete the Portuguese version of these books. The continuous support of Professor Hilário Alencar, president of the Brazilian Mathematical Society when the Portuguese edition was first published, was also of great importance for me. Special thanks go to my colleagues—professors Samuel B. Feitosa and Fernanda E. C. Camargo—who read the entire English version and helped me improve it in a number of ways. If it weren't for my editor at Springer-Verlag, Mr. Robinson dos Santos, I almost surely would not have had the courage to embrace the task of translating more that 1500 pages from Portuguese into English. I acknowledge all the staff of Springer involved with this project in his name.

Finally, and mostly, I would like to express my deepest gratitude to my parents Antonio and Rosemary, my wife Monica, and our kids Gabriel and Isabela. From early childhood, my parents have always called my attention to the importance of a solid education, having done all they could for me and my brothers to attend the best possible schools. My wife and kids fulfilled our home with the harmony and softness I needed to get to endure on several months of work while translating this book.

Fortaleza, Brazil
December 2017

Antonio Caminha Muniz Neto

Contents

Chapter 1
Elementary Counting Techniques

Check for
updates

In this first chapter, we develop the usual elementary tools for counting the number
of distinct configurations corresponding to a certain combinatorial situation, without
needing to list them one by one. As the reader will see, the essential ideas are the
construction of bijections and the use of recursive arguments.

Although we develop all material from scratch, the reader is expected to have
some previous experience with elementary counting techniques, and is warned that
the material collected here can be somewhat terse at places.

1.1 The Bijective Principle

In all that follows, we assume that the reader has a relative acquaintance with sets
and elementary operations on them. Given $n \in \mathbb{N}$, we let I_n denote the set

$$I_n = \{j \in \mathbb{N};\ 1 \leq j \leq n\}$$

of natural numbers from 1 to n.

A set A is **finite** if $A = \emptyset$ or if there exists a bijection $f : I_n \to A$, for some
$n \in \mathbb{N}$. If $A \neq \emptyset$ is finite and $f : I_n \to A$ is a bijection, then letting $a_j = f(j)$ we
write $A = \{a_1, \ldots, a_n\}$ and say that n is the **number of elements** of A (Problem 1
shows that this is a well defined notion). Also in this case, we write

$$|A| = n \ \ \text{or} \ \ \#A = n$$

to mean that A has n elements. For the sake of completeness, we say that $\emptyset$ has 0
elements and write $|\emptyset| = 0$.

© Springer International Publishing AG, part of Springer Nature 2018
A. Caminha Muniz Neto, *An Excursion through Elementary Mathematics, Volume III*,
Problem Books in Mathematics, https://doi.org/10.1007/978-3-319-77977-5_1

The elementary theory of counting (configurations) has its foundations in the following simple proposition, to which we will systematically refer as the **bijective principle**.

Proposition 1.1 *If A and B are nonempty finite sets, then* $|A| = |B|$ *if and only if there exists a bijection* $f : A \to B$.

Proof First of all, assume that there exists a bijection $f : A \to B$. If $|A| = n$, we can take a bijection $g : I_n \to A$, so that $f \circ g : I_n \to B$ is also a bijection. Hence, $|B| = n$.

Conversely, suppose that $|A| = |B| = n$, with bijections $g : I_n \to A$ and $h : I_n \to B$. Then $h \circ g^{-1} : A \to B$ is a bijection from A to B. $\qquad\qquad\square$

The following consequence of the bijective principle is sometimes referred to as the **additive principle** of counting. Before we state it, we recall that two sets A and B are said to be **disjoint** if $A \cap B = \emptyset$.

Proposition 1.2 *If A and B are finite disjoint sets, then*

$$|A \cup B| = |A| + |B|.$$

Proof Exercise (see Problem 2). $\qquad\qquad\square$

A typical application of the former proposition consists in counting the number of different ways of choosing exactly one object out of two possible distinct kinds, such that there is a finite number of possibilities for each kind of object. In the statement of the proposition, the *kinds* correspond to the disjoint sets A and B, whereas the possibilities for each kind correspond to the elements of A and B.

An easy induction allows us to generalize the additive principle for n finite pairwise disjoint sets. This is the content of the coming

Corollary 1.3 *If* $A_1, A_2, \ldots, A_n$ *are finite pairwise disjoint sets, then*

$$\# \bigcup_{j=1}^{n} A_j = \sum_{j=1}^{n} |A_j|.$$

Proof Exercise (see Problem 3). $\qquad\qquad\square$

For what comes next, given two sets A and B, we let $A \setminus B$ denote the set

$$A \setminus B = \{x \in A; \ x \notin B\}$$

and say that $A \setminus B$ is the **difference** between A and B, in this order. If $B \subset A$, and if no danger of confusion arises, we shall sometimes refer to $A \setminus B$ as the **complement** of B in A, in which case we denote it as B^c, instead of $A \setminus B$.

The simple formula of the next corollary, which counts the number of elements of the complement of a subset of a finite set, is an additional consequence of the additive principle.

Corollary 1.4 *If A is a finite set and $B \subset A$, then*

$$|B| = |A| - |A \setminus B|.$$

Proof Since $A = B \cup (A \setminus B)$, a disjoint union, the additive principle gives

$$|A| = |B \cup (A \setminus B)| = |B| + |A \setminus B|.$$

$\square$

The general philosophy behind the use of the previous corollary in problems of counting is this: suppose we wish to count the number of elements of a certain finite set B, but do not know how to do it directly. An interesting strategy is to search for a finite set $A \supset B$ such that we know how to count both $|A|$ and $|A \setminus B|$. Then, we apply the formula of the corollary to compute the desired number of elements of B. Some concrete examples of this situation will be found in what is to come.

Our next result generalizes Proposition 1.1 and Corollary 1.4, computing the number of elements of a union of two finite sets. Formula (1.1) below is known as the **principle of inclusion-exclusion** for two finite sets, and will be generalized in Sect. 2.1 (cf. Theorem 2.1).

Proposition 1.5 *If A and B are finite sets, then*

$$|A \cup B| = |A| + |B| - |A \cap B|. \tag{1.1}$$

Proof Since A and $B \setminus A$ are finite, disjoint and such that $A \cup B = A \cup (B \setminus A)$, the additive principle gives

$$|A \cup B| = |A \cup (B \setminus A)| = |A| + |B \setminus A|.$$

On the other hand, we also have the disjoint union

$$B = (B \setminus A) \cup (A \cap B),$$

so that, again from the additive principle, $|B| = |B \setminus A| + |A \cap B|$. Then, $|B \setminus A| = |B| - |A \cap B|$, and once we plug this formula into the above relation for $|A \cup B|$ we get

$$|A \cup B| = |A| + (|B| - |A \cap B|).$$

$\square$

For what comes next, recall that the **cartesian product** of sets A and B is the set $A \times B$ whose elements are the ordered pairs (a, b) with $a \in A$ and $b \in B$. In mathematical symbols,

$$A \times B = \{(a, b); a \in A, b \in B\}.$$

It is also worth recalling that ordered pairs possess the following important property: if $(a, b), (c, d) \in A \times B$, then

$$(a, b) = (c, d) \Leftrightarrow a = c \text{ and } b = d.$$

In this respect, see Problem 5.

The coming result, together with its subsequent corollary, are known as the **multiplicative principle** or as the **fundamental principle of counting**.

Proposition 1.6 *If A and B are nonempty finite sets, then $A \times B$ is also finite, with*

$$|A \times B| = |A| \cdot |B|.$$

Proof Writing $B = \{y_1, \ldots, y_m\} = \bigcup_{j=1}^{m}\{y_j\}$, it follows from Problem 6 that

$$A \times B = A \times \left(\bigcup_{j=1}^{m}\{y_j\} \right) = \bigcup_{j=1}^{m}(A \times \{y_j\}),$$

a disjoint union. Hence, Corollary 1.3 gives

$$|A \times B| = \left| \bigcup_{j=1}^{m}(A \times \{y_j\}) \right| = \sum_{j=1}^{m}|A \times \{y_j\}|. \tag{1.2}$$

Now, since $f : A \to A \times \{y_j\}$ given by $f(x) = (x, y_j)$ is a bijection (with inverse $g : A \times \{y_j\} \to A$ given by $g(x, y_j) = x$), the bijective principle guarantees that $|A| = |A \times \{y_j\}|$ for $1 \leq j \leq m$. Hence, it follows from (1.2) that

$$|A \times B| = \sum_{j=1}^{m}|A| = |A| \cdot m = |A| \cdot |B|.$$

$\square$

In applications, we ought to invoke this version of the fundamental principle of counting whenever we need to choose two objects *simultaneously*, such that one of the objects is of one of two possible kinds and the other object is of the other kind (and each kind comprises a finite number of possibilities). In the statement of the proposition, the two possible kinds correspond to the sets A and B, whereas the possibilities for each kind correspond to the elements of A and B.

It is time we look at a concrete example.

Example 1.7 How many natural numbers have two distinct nonzero algarisms, both less that or equal to 5? How many of them are such that the first algarism is smaller that the second one?

Solution Let $A = \{1, 2, 3, 4, 5\}$. The first part of the problem is clearly equivalent to counting how many ordered pairs $(a, b) \in A \times A$ are such that $a \neq b$; the second part requires that $a < b$.

Let us start counting the number of ordered pairs $(a, b) \in A \times A$, without additional restrictions. Since this is the same as counting the number of elements of $A \times A$, the multiplicative principle gives $5 \times 5 = 25$ possible pairs (a, b). In order to count how many of them arc such that $a \neq b$, let us use Corollary 1.4: the number of such pairs is obtained by discounting, from the total number of pairs, those for which $a = b$. Since there are 5 pairs (a, a) with $a \in A$, we conclude that there are $25 - 5 = 20$ pairs (a, b) with $a \neq b$.

For what is left, note that (a, b) is an ordered pair such that $a < b$ if and only if (b, a) is an ordered pair for which $b > a$; in other words, the correspondence $(a, b) \mapsto (b, a)$ is a bijection between the set whose elements are the ordered pairs $(a, b) \in A \times A$ such that $a < b$ and that formed by the ordered pairs $(a', b') \in A \times A$ for which $a' > b'$. Hence, these sets have the same number of elements (namely, ordered pairs); since they are disjoint and their union equals the set of pairs $(a, b) \in A \times A$ such that $a \neq b$, it follows from Proposition 1.2 that the desired number of ordered pairs is $\frac{20}{2} = 10$. □

For the subsequent discussion, we need to extend the concept of cartesian product to an arbitrary finite number of finite nonempty sets. To this end, given finite nonempty sets $A_1, A_2, \ldots, A_n$, let's define their cartesian product $A_1 \times A_2 \times \cdots \times A_n$ as the set of sequences (to which we shall sometimes refer to as **n-tuples**) $(a_1, a_2, \ldots, a_n)$, such that $a_1 \in A_1, a_2 \in A_2, \ldots, a_n \in A_n$.[1]

Corollary 1.8 *If $A_1, A_2, \ldots, A_n$ are finite nonempty sets, then*

$$|A_1 \times A_2 \times \cdots \times A_n| = \prod_{j=1}^{n} |A_j|.$$

Proof Exercise (see Problem 7). □

The next corollary brings an important elaboration of the multiplicative principle.

Corollary 1.9 *Let $A_1, A_2, \ldots, A_k$ be finite nonempty sets with $|A_1| = n_1, |A_2| = n_2, \ldots, |A_k| = n_k$. Then, there are exactly $n_1 n_2 \ldots n_k$ sequences $(a_1, a_2, \ldots, a_k)$ with $a_j \in A_j$ for $1 \leq j \leq k$.*

---

[1]In view of this definition, in principle we have two distinct definitions for the elements of $A \times B$: on the one hand, they consist of the ordered pairs (a, b) such that $a \in A$ and $b \in B$; on the other, they are sequences (a, b) for which $a \in A$ and $b \in B$. Since the ordered pair (a, b) is defined by $(a, b) = \{\{a\}, \{a, b\}\}$ (cf. Problem 5) and the sequence (a, b) (with $a \in A$ and $b \in B$) is the function $f : \{1, 2\} \to A \cup B$ such that $f(1) := a \in A$ and $f(2) := b \in B$, we come to the conclusion that, although we have been using the same notation, they are distinct mathematical objects. However, for our purposes the identification of the ordered pair (a, b) to the sequence (a, b) is totally harmless and will be done, from now on, without further comments.

Proof As we have told above, a sequence $(a_1, a_2, \ldots, a_k)$ such that $a_1 \in A_1, a_2 \in A_2, \ldots, a_k \in A_k$ is nothing but an element of the cartesian product $A_1 \times A_2 \times \cdots \times A_k$. Hence, the number of such sequences equals the number of elements of $A_1 \times A_2 \times \cdots \times A_k$, which in turn equals, by the previous corollary, $|A_1||A_2|\ldots|A_k| = n_1 n_2 \ldots n_k$.   □

In words, the above corollary furnishes a method for counting in how many (distinct) ways we can choose k objects *in order* but *independently*, with the first object being of type 1, the second being of type 2, $\ldots$, the k-th being of type k and having at our disposal one or more possibilities for the choice of each type of object. Here again, the distinct types of objects correspond to the sets $A_1, \ldots, A_k$, whereas the numbers of possibilities for each type of object correspond to the elements of the sets under consideration.

As a special case of the situation of Corollary 1.9, we can count in how many ways it is possible to choose, in order but independently, k elements out of a set of n elements, with repeated choices being allowed. More precisely, we have the following

Corollary 1.10 *If $|A| = n$, then there are exactly n^k sequences of k terms, all chosen from the elements of A.*

Proof Make $A_1 = \cdots = A_k = A$ in Corollary 1.9.   □

In other words, we say that the above corollary counts how many **arrangements with repetition** of k objects, chosen out of a set with n objects, there exist.

Yet another way of looking at the result of Corollary 1.10 is to start by recalling that a sequence of k terms, all belonging to I_n, is simply a function $f : I_k \to I_n$; hence, what we established in that corollary was the fact that the number of distinct functions $f : I_k \to I_n$ is exactly n^k. Later, we shall compute how many of such functions are injective and how many are surjective.

The coming example applies the circle of ideas above to a concrete situation. For its solution, the reader might find it helpful to recall the *criterion of divisibility* by 3, which will be derived in Problem 1, page 162: *the remainder of a natural number n upon division by 3 equals that of the sum of its algarisms.*

Example 1.11 Compute the quantity of natural numbers n, of ten algarisms and satisfying the following conditions:

(a) n does not end in 0.
(b) n is divisible by 3.

Solution If $n = (a_1 a_2 \ldots a_9 a_{10})$ the decimal representation of n, then $m = (a_1 a_2 \ldots a_9)$ is a natural number of nine algarisms and $a_{10} \in \{1, 2, 3, \ldots, 9\}$. Now, for a fixed $m = (a_1 a_2 \ldots a_9)$, the criterion of divisibility by 3 assures that each of the numbers $(a_1 a_2 \ldots a_9 1)$, $(a_1 a_2 \ldots a_9 4)$ and $(a_1 a_2 \ldots a_9 7)$ leave the same remainder upon division by 3, the same happening with the numbers $(a_1 a_2 \ldots a_9 2)$, $(a_1 a_2 \ldots a_9 5)$ and $(a_1 a_2 \ldots a_9 8)$, as well as with the numbers $(a_1 a_2 \ldots a_9 3)$, $(a_1 a_2 \ldots a_9 6)$ and $(a_1 a_2 \ldots a_9 9)$. Moreover, letting r_1, r_2 and r_3 denote such common remainders, then r_1, r_2 and r_3 are pairwise distinct.

Hence, for each natural number $m = (a_1 a_2 \ldots a_9)$ of nine algarisms, we have exactly three natural numbers $n = (a_1 a_2 \ldots a_9 a_{10})$ which are divisible by 3. Since $a_1 \in \{1, 2, \ldots, 9\}$ and $a_2, \ldots, a_9 \in \{0, 1, 2, \ldots, 9\}$, Corollary 1.9 guarantees that the total of numbers we want to compute is $9 \times 10^8 \times 3 = 27 \times 10^8$. □

Problems: Sect. 1.1

1. * Prove that the notion of number of elements of a nonempty finite set is a well defined concept. More precisely, prove that there exists a bijection $f : I_m \to I_n$ if and only if $m = n$.
2. * Prove the additive principle of counting from the definition for the number of elements of a finite set. More precisely, prove that if A and B are nonempty, finite disjoint sets, with $|A| = m$ and $|B| = n$, then there exists a bijection $f : I_{m+n} \to A \cup B$.
3. * Prove Corollary 1.3.
4. * Let A and B be nonempty finite sets, with $|A| = |B|$. Prove that a function $f : A \to B$ is injective if and only if it is surjective.
5. * Given sets A and B and elements $a \in A$, $b \in B$, we formally define the **ordered pair** (a, b) by letting[2]

$$(a, b) = \{\{a\}, \{a, b\}\}.$$

   Use this definition to show that, for $a, c \in A$ and $b, d \in B$, one has $(a, b) = (c, d) \Leftrightarrow a = c$ and $b = d$.
6. * Given nonempty sets $A, B_1, \ldots, B_n$, prove that

$$A \times \left(\bigcup_{j=1}^{n} B_j \right) = \bigcup_{j=1}^{n} (A \times B_j).$$

   Moreover, if $B_1, \ldots, B_n$ are pairwise disjoint, prove that $A \times B_1, \ldots, A \times B_n$ are also pairwise disjoint.
7. * Prove Corollary 1.8.

   For the next problem, we define an **ordered partition** of a set A as a sequence $(A_1, \ldots, A_k)$ of subsets of A satisfying the following conditions: (i) $A = A_1 \cup \ldots \cup A_k$; (ii) $A_1, \ldots, A_k$ are pairwise disjoint. In this case, we use to say that $A_1, \ldots, A_k$ (in this order) form an *ordered partition of A into k of its subsets*.

8. Let $n, k \in \mathbb{N}$ and A be a finite set with n elements. Show that there exist exactly k^n ordered partitions of A into k subsets $A_1, \ldots, A_k$.

---

[2]Such a definition is due to Kazimierz Kuratowski, Polish mathematician of the twentieth century.

9. In the cartesian plane, a pawn moves according to the following rule: being at
   the point (a, b), he can go to one of the points $(a + 1, b + 1)$, $(a + 1, b - 1)$,
   $(a - 1, b + 1)$ or $(a - 1, b - 1)$. If the pawn starts at the point $(0, 0)$, how many
   distinct positions can he occupy after his first n moves?

10. (AIME—adapted) Given $k, n \in \mathbb{N}$, do the following items:

    (a) For a given integer $1 \leq j \leq n$, prove that there are exactly $(n - j + 1)^k$
        sequences formed by k elements of I_n (possibly with repetitions) such that
        the smallest term of the sequence is greater than or equal to j.
    (b) For a given integer $1 \leq j \leq n$, prove that there are exactly $(n - j + 1)^k -
        (n - j)^k$ sequences formed by k elements of I_n (possibly with repetitions),
        such that the smallest term of the sequence equals j.
    (c) Prove that the sum of the smallest terms of all n^k sequences of k elements
        of I_n (possibly with repetitions) equals $1^k + 2^k + \cdots + n^k$.

11. (France) Let $k \in \mathbb{N}$, $A = \{1, 2, 3, \ldots, 2^k\}$ and X be a subset of A satisfying
    the following condition: if $x \in X$, then $2x \notin X$. Find, with proof, the greatest
    possible number of elements of X.

1.2   More Bijections

In this section, with the elementary results of the previous section at our disposal, we
present some instances of deeper applications of the bijective principle to establish
the equality of the numbers of elements of two finite sets.

We start by computing, in two different ways, the number of subsets of a finite
set. If A is any set, we let $\mathcal{P}(A)$ denote the **power set** of A, i.e., the *family*[3]

$$\mathcal{P}(A) = \{B; \ B \subset A\}.$$

Given finite sets A and B, both with n elements, a bijection $f : A \to B$ naturally
induces a bijection $\tilde{f} : \mathcal{P}(A) \to \mathcal{P}(B)$, defined for $C \subset A$ by

$$\tilde{f}(C) = \{f(x); \ x \in C\}.$$

In words, given a subset C of A, we let $\tilde{f}(C)$ be the subset of B whose elements are
the images of the elements of C by f. In particular, note that

$$\tilde{f}(\emptyset) = \{f(x); \ x \in \emptyset\} = \emptyset,$$

since it is impossible to choose $x \in \emptyset$.

---

[3]In Set Theory, a **family** is a set whose elements are also sets.

Hence, as a corollary to the bijective principle, if A and B are nonempty finite disjoint sets, then

$$|A| = |B| \Rightarrow |\mathcal{P}(A)| = |\mathcal{P}(B)|. \tag{1.3}$$

We shall also need another concept, which will be useful in future discussions.

Definition 1.12 Let A be a nonempty set. For $B \subset A$, the **characteristic function** of B (with respect to A) is the function $\chi_B : A \to \{0, 1\}$, defined by

$$\chi_B(x) = \begin{cases} 0, & \text{if } x \notin B \\ 1, & \text{if } x \in B \end{cases}. \tag{1.4}$$

For example, if $A = \{a, b, c, d, e, f, g\}$ and $B = \{c, f, g\}$, then χ_B is the function from A to $\{0, 1\}$ such that

$$\chi_B(a) = \chi_B(b) = \chi_B(d) = \chi_B(e) = 0 \text{ and } \chi_B(c) = \chi_B(f) = \chi_B(g) = 1.$$

For what follows, if $A = \{a_1, \ldots, a_n\}$ is a set with n elements, whenever convenient we can associate to A the sequence $(a_1, \ldots, a_n)$; whenever we do that, we shall say that $(a_1, \ldots, a_n)$ is an **ordering** of (the elements of) A or, sometimes, the **natural ordering** of A.

If $B \subset A$, a (combinatorially) more interesting way of looking at the characteristic function χ_B of B with respect to A is to consider it as a sequence of 0's and 1's, with the positions of the 1's (with respect to the natural ordering of A) corresponding to the elements of B. For instance, let $A = \{a, b, c, d, e, f, g\}$, furnished with the ordering induced from the **lexicographical** (i.e., *alphabetical*) **order**; if $B = \{c, f, g\}$, then the characteristic function of B corresponds to the sequence $(0, 0, 1, 0, 0, 1, 1)$.

More generally, let $A = \{a_1, \ldots, a_n\}$, furnished with the natural ordering. For $B \subset A$, the **characteristic sequence** of B in A is the sequence $s_B = (\alpha_1, \ldots, \alpha_n)$, such that

$$\alpha_j = \begin{cases} 0, & \text{if } a_j \notin B \\ 1, & \text{if } a_j \in B \end{cases}. \tag{1.5}$$

We can finally compute the number of subsets of a set with n elements.

Theorem 1.13 *A set with n elements has exactly 2^n subsets.*

Proof Let $A = \{a_1, \ldots, a_n\}$ be a set with n elements, furnished with its natural ordering, and let S be the set of sequences of n terms, formed by the elements of the set $\{0, 1\}$. By Corollary 1.10, we have $|S| = 2^n$. We will show that $|\mathcal{P}(A)| = 2^n$, and to this end it suffices to show that the function

$$f : \mathcal{P}(A) \longrightarrow S$$
$$B \longmapsto s_B$$

(i.e., the function which associates to each subset of A its characteristic sequence with respect to A) is a bijection. For what is left to do, check that the function $g : S \to \mathcal{P}(A)$, given by

$$g(\alpha_1, \ldots, \alpha_n) = \{a_j \in A; \; \alpha_j = 1\},$$

is the inverse of f. $\qquad\qquad\qquad\qquad\qquad\qquad\qquad\qquad\qquad\qquad\qquad\qquad$ □

It is instructive to note that we can give another proof of the previous result, this time relying more directly on the bijective principle. Indeed, by (1.3) it suffices to show that the set $A = \{0, 1, \ldots, n-1\}$ has precisely 2^n subsets. To this end, we let

$$f : \mathcal{P}(A) \to \{0, 1, 2, \ldots, 2^n - 1\}$$

be given by $f(\emptyset) = 0$ and, for $\emptyset \neq B = \{a_1, \ldots, a_k\} \subset A$, with $0 \leq a_1 < \cdots < a_k \leq n - 1$,

$$f(B) = 2^{a_1} + \cdots + 2^{a_k}.$$

Since

$$1 \leq 2^{a_1} + \cdots + 2^{a_k} \leq 2^0 + 2^1 + \cdots + 2^{n-1} = 2^n - 1,$$

f is well defined. To conclude that it is a bijection, recall Example 4.12 of [8], which shows that every integer $1 \leq m \leq 2^n - 1$ can be written, in a unique way up to the order of the summands, as a sum of distinct powers of 2 (obviously, none of these can be bigger that 2^{n-1}); such a way of writing m is called its *binary representation*. Thus, the uniqueness of the binary representation is equivalent to the injectivity of f, whereas the existence of such a representation is equivalent to the surjectivity of f.

Notice how the above proof highlights the strength of the bijective principle in counting problems. Let us see two more examples.

Example 1.14 Let n be a natural number of the form $4k + 1$ or $4k + 2$, for some nonnegative integer k. Prove that I_n contains exactly 2^{n-1} subsets with even sum of elements (and under the convention that the sum of the elements of the empty set is 0).

Proof Let $\mathcal{F}_0$ and $\mathcal{F}_1$ be the families of the subsets of I_n with sum of elements respectively even and odd. Since $\mathcal{F}_0$ and $\mathcal{F}_1$ are disjoint and such that $\mathcal{F}_0 \cup \mathcal{F}_1 = \mathcal{P}(A)$, it follows from the additive principle and Theorem 1.13 that

$$|\mathcal{F}_0| + |\mathcal{F}_1| = |\mathcal{P}(A)| = 2^n.$$

Thus, if we show that $|\mathcal{F}_0| = |\mathcal{F}_1|$, we will get $|\mathcal{F}_0| = |\mathcal{F}_1| = 2^{n-1}$.

To what is left to do, let's construct a bijection between $\mathcal{F}_0$ and $\mathcal{F}_1$. To this end, consider the function

$$f : \mathcal{P}(A) \longrightarrow \mathcal{P}(A)$$
$$B \longmapsto B^c \ ,$$

where $B^c = A \setminus B$ denotes the *complement* of B in A. Since $(B^c)^c = A \setminus (A \setminus B) = B$, we have $f(f(B)) = B$ for every $B \subset A$; hence, $f \circ f = \mathrm{Id}_{\mathcal{P}(A)}$, so that f is a bijection (cf. Example 6.40 of [8]). Now, the idea is to show that, if $n = 4k + 1$ or $n = 4k + 2$, then f induces a bijection between $\mathcal{F}_0$ and $\mathcal{F}_1$.

Given $B \subset A$, we have

$$\sum_{x \in B} x + \sum_{x \in B^c} x = \sum_{x \in A} x = \sum_{x=1}^{n} x = \frac{n(n+1)}{2}$$

$$= (4k+1)(2k+1) \text{ or } (2k+1)(4k+3),$$

according to whether $n = 4k + 1$ or $n = 4k + 2$. In any case,

$$\sum_{x \in B} x + \sum_{x \in B^c} x$$

is an odd number, so that the sums of the elements of B and of B^c have distinct parities; in symbols,

$$B \in \mathcal{F}_0 \Leftrightarrow B^c \in \mathcal{F}_1.$$

Therefore, the restriction g of f to $\mathcal{F}_0$ applies $\mathcal{F}_0$ into $\mathcal{F}_1$, whereas the restriction h of f to $\mathcal{F}_1$ applies $\mathcal{F}_1$ into $\mathcal{F}_0$. However, since g and h are clearly inverses of each other (for, $f = f^{-1}$), it follows that g and h are also bijections, so that $|\mathcal{F}_0| = |\mathcal{F}_1|$. □

The bijective principle is particularly useful to understand the properties of the *partitions* of a natural number n. Here and in all that follows, a **partition** of the natural number n is a way of writing n as a sum of (one or more) not necessarily distinct natural summands. For example, the distinct partitions of 4 are

$$4 = 1 + 3 = 1 + 1 + 2 = 1 + 1 + 1 + 1 = 2 + 2.$$

The coming example is due to L. Euler.

Example 1.15 (Euler) Prove that the number of partitions of a natural n in odd summands equals the number of partitions of n in distinct summands.

Proof Letting $\mathcal{I}_n$ denote the set of partitions of n in odd summands and $\mathcal{D}_n$ the set of partitions of n in distinct summands, it suffices to construct a bijection $f : \mathcal{I}_n \rightarrow$

$\mathcal{D}_n$. To understand how to define f, consider the following partition of 51 in odd summands:

$$1 + 1 + 3 + 3 + 3 + 3 + 3 + 3 + 5 + 5 + 7 + 7 + 7.$$

Then, we can also write

$$\begin{aligned} 51 &= 2 \cdot 1 + 6 \cdot 3 + 2 \cdot 5 + 3 \cdot 7 \\ &= 2^1 \cdot 1 + (2^2 + 2^1) \cdot 3 + 2^1 \cdot 5 + (2^1 + 2^0) \cdot 7 \\ &= 2^1 \cdot 1 + 2^2 \cdot 3 + 2^1 \cdot 3 + 2^1 \cdot 5 + 2^1 \cdot 7 + 2^0 \cdot 7 \\ &= 2 + 12 + 6 + 10 + 14 + 7, \end{aligned}$$

which is a partition of 51 in distinct summands. In what follows, we shall make the above particular case into a general argument to get the desired bijection.

A partition $P \in \mathcal{I}_n$ is

$$n = \underbrace{1 + \cdots + 1}_{a_1} + \underbrace{3 + \cdots + 3}_{a_3} + \underbrace{5 + \cdots + 5}_{a_5} + \underbrace{7 + \cdots + 7}_{a_7} + \cdots,$$

where $a_1, a_3, a_5, \ldots \geq 0$ and, from some natural number k on, the number a_{2k+1} of summands equal to $2k + 1$ is 0. Then, we can write

$$n = a_1 \cdot 1 + a_3 \cdot 3 + a_5 \cdot 5 + a_7 \cdot 7 + \cdots.$$

From here, in order to get a partition $f(P)$ of n into distinct summands, substitute each positive coefficient a_{2k+1} by its binary representation (cf. Example 4.12 de [8]) and, if 2^l is one of the summands of such a representation, let $2^l(2k + 1)$ be one of the summands in $f(P)$ (you can easily notice that this is exactly what we did in the particular case of the partitions of 51).

By systematically proceeding this way, we claim that we get a partition of n into distinct summands. Indeed, any two summands of $f(P)$ can be written as

$$2^{l_1}(2k_1 + 1) \quad \text{and} \quad 2^{l_2}(2k_2 + 1);$$

if $k_1 \neq k_2$, then such summands are clearly different from one another; if $k_1 = k_2$, then we must have $l_1 \neq l_2$, for 2^{l_1} and 2^{l_2} are two summands of the binary representation of $a_{2k_1+1} = a_{2k_2+1}$. In any case, f is well defined.

To prove that f is a bijection, let's define a function $g : \mathcal{D}_n \to \mathcal{I}_n$ (which will be the inverse of f) in the following way: let Q be the partition

$$\begin{aligned} n = {}&(m_{11} + m_{12} + \cdots + m_{1t_1}) + (m_{31} + m_{32} + \cdots + m_{3t_3}) \\ &+ (m_{51} + m_{52} + \cdots + m_{5t_5}) + \cdots \end{aligned}$$

of n into distinct summands, where we have grouped, within each pair of parentheses, all summands with a single *odd part*. More precisely, what we are saying is that $m_{2k+1,i} = 2^{l_i}(2k+1)$, for some nonnegative integer l_i. Then,

$$m_{2k+1,1} + \cdots + m_{2k+1,t_k} = 2^{l_1}(2k+1) + \cdots + 2^{l_{t_k}}(2k+1)$$

$$= (2^{l_1} + \cdots + 2^{l_{t_k}})(2k+1),$$

so that, letting $a_{2k+1} = 2^{l_1} + \cdots + 2^{l_{t_k}}$, we get

$$n = a_1 \cdot 1 + a_3 \cdot 3 + a_5 \cdot 5 + a_7 \cdot 7 + \cdots.$$

We thus define $g(Q)$ to be the partition

$$n = \underbrace{1 + \cdots + 1}_{a_1} + \underbrace{3 + \cdots + 3}_{a_3} + \underbrace{5 + \cdots + 5}_{a_5} + \underbrace{7 + \cdots + 7}_{a_7} + \cdots,$$

so that $g(Q) \in \mathcal{I}_n$.

Finally, it is immediate to verify that f and g are inverses of each other. $\square$

Problems: Sect. 1.2

1. Compute the number of subsets of the set I_n which contain n.
2. * (APMO) We are given a natural number n and a finite nonempty set A. Show that there are exactly $(2^n - 1)^{|A|}$ sequences $(A_1, A_2, \ldots, A_n)$, formed by subsets of A such that $A = A_1 \cup A_2 \cup \ldots \cup A_n$.
3. (Soviet Union) We are given n straight lines in the plane, in *general position*, i.e., such that no two of them are parallel and no three of them pass through the same point. Compute the number of regions into which these lines divide the plane.

    For the next problem, we say that a family $\mathcal{F}$ of subsets of I_n is an **intersecting system** if, for every distinct $A, B \in \mathcal{F}$, we have $A \cap B \neq \emptyset$.

4. (Bulgaria—adapted) Let $\mathcal{F}$ be an intersecting system in I_n.

    (a) Give an example of such an $\mathcal{F}$ in which $|\mathcal{F}| = 2^{n-1}$.
    (b) Prove that $|\mathcal{F}| \leq 2^{n-1}$, for any such $\mathcal{F}$.

5. * Given $n \in \mathbb{N}$, let $\mathcal{I}_n$ be the family of subsets of I_n with odd numbers of elements. If $A = \{x_1, x_2, \ldots, x_{2k+1}\} \in \mathcal{I}_n$, with $x_1 < x_2 < \cdots < x_{2k+1}$, we say that x_{k+1} is the *central element* of A, and write $x_{k+1} = c(A)$. In this respect, do the following items:

    (a) For $1 \leq x_1 < x_2 < \cdots < x_{2k+1} \leq n$, show that the correspondence

    $$\{x_1, x_2, \ldots, x_{2k+1}\} \mapsto \{n+1 - x_{2k+1}, \ldots, n+1 - x_2, n+1 - x_1\}$$

establishes a bijection between the sets in $\mathcal{I}_n$ having central elements respectively equal to i and $n + 1 - i$.

(b) In Problem 5, page 20, we will show that $|\mathcal{I}_n| = 2^{n-1}$. Use the result of item (a), together with this fact, to show that

$$\sum_{A \in \mathcal{I}_n} c(A) = (n + 1) \cdot 2^{n-2}.$$

6. (TT) Given $n \in \mathbb{N}$, we define the *diversity* of a partition of n as the number of distinct summands in it. Also, let $p(n)$ stand for the number of distinct partitions of n and $q(n)$ for the sum of the diversities of all of the $p(n)$ partitions of n. Prove that:

   (a) $q(n) = 1 + p(1) + p(2) + \cdots + p(n - 1)$.
   (b) $q(n) \le \sqrt{2n}\, p(n)$.

1.3 Recursion

The general philosophy behind *recursive counting* is the following: we wish to count the number a_n of elements of a set A_n, which is declared by means of some specific rule(s) concerning the natural number n. We do this in a two-step process: firstly, we use the declaration of A_n to get a **recurrence relation** for the sequence $(a_n)_{n \ge 1}$, i.e., a relation of the form

$$a_n = F(a_1, \ldots, a_{n-1}, n), \tag{1.6}$$

where F is some function of n variables; secondly, we use the recurrence relation, together with algebraic and/or analytical arguments, to compute or estimate a_n in terms of n.

In this section we concentrate our efforts in the first step above, by examining in detail some specific examples ranging in difficulty from almost trivial to real challenging. As for the second step, for the time being we assume that the reader is familiar with the solution of linear recurrence relations of order at most three and with constant coefficients, as presented in Chapter 3 of [8]. Chapters 3 and 21 will develop more powerful tools for the treatment of more general recurrence relations.

We start with a recursive counting for the number os subsets of a finite set.

Example 1.16 If A is a set with n elements, then A has exactly 2^n subsets.

Proof Let a_n be the number of subsets of A (here, we are implicitly using (1.3) when we write $|A|$ as a function of n). For a fixed $x \in A$, there are two kinds of subsets of A: those which contain x and those which do not.

If B is a subset of A containing x, then $B \setminus \{x\}$ is a subset of $A \setminus \{x\}$; conversely, if B' is a subset of $A \setminus \{x\}$, then $B' \cup \{x\}$ is a subset of A containing x. Since such correspondences are clearly inverses of each other, we conclude that there are as many subsets of A containing x as there are subsets of $A \setminus \{x\}$; however, since $|A \setminus \{x\}| = n - 1$, it follows that there are exactly a_{n-1} of such subsets of A. On the other hand, since the subsets of A not containing x are precisely the subsets of $A \setminus \{x\}$, we have exactly a_{n-1} subsets of this second kind.

The argument of the previous paragraph clearly gives $a_n = 2a_{n-1}$ for every $n > 1$. Since we obviously have $a_1 = 2$, the formula for the general term of a GP gives

$$a_n = a_1 \cdot 2^{n-1} = 2^n.$$

$\square$

Our next example provides a recursive approach to Problem 3, page 13.

Example 1.17 (Soviet Union) We are given n straight lines in the plane, in *general position*, i.e., such that no two of them are parallel and no three of them pass through the same point. Compute the number of regions into which these lines divide the plane.

Solution Let a_k be the number of regions into which the plane gets divided by some k straight lines in general position. We trace out one more straight line, say r, such that the $k + 1$ resulting lines are also in general position. Since r intersects the k other lines in k distinct points, we conclude that r gets divided in $k + 1$ intervals by these k points. In turn, each one of these $k + 1$ intervals corresponds to exactly one region, from the a_k regions we had, that r splits into two new regions. Therefore, when we trace out line r we extinguish $k + 1$ older regions and generate $2(k + 1)$ new regions. Thus, if a_{k+1} denotes the total number of regions we get after tracing out line r, we conclude that

$$a_{k+1} = a_k - (k + 1) + 2(k + 1) = a_k + (k + 1).$$

Finally, since $a_1 = 2$, the formula for telescoping sums (formula (3.15) of [8]) gives

$$a_n = a_1 + \sum_{k=1}^{n-1}(a_{k+1} - a_k) = 2 + \sum_{k=1}^{n-1}(k + 1)$$

$$= 1 + \sum_{k=1}^{n} k = 1 + \frac{n(n + 1)}{2}.$$

$\square$

The coming example is perhaps the most celebrated of all elementary applications of recursive reasoning, being known as the **Fibonacci problem**.[4]

Example 1.18 A couple of rabbits starts generating descendants when is at least 2 months old. When this is so, it generates a new couple of baby rabbits per month. If we start with a couple of just-born baby rabbits at month one, how many couples will we have after 12 months?

Solution Let F_n denote the total number of couples of rabbits after n months, so that $F_1 = 1$ and $F_2 = 1$ (the first couple of descendants will be born only after 3 months, so will be counted just in F_3).

Now, let's look at the F_{k+2} couples of rabbits we will have after $k + 2$ months. There are two possibilities for such a couple: either it already existed after $k + 1$ months, or else it was born at the $(k + 2)$-th month. The first possibility amounts, by definition, to a total of F_{k+1} couples. As for the second one, note that the couple of rabbits under consideration descends from one of the couples that already existed after k months; conversely, each couple that already existed after k months generates a new couple of descendants that will be born at the $(k + 2)$-th month, so that this gives an additional number of F_k couples of rabbits after $k + 2$ months.

Hence, the total numbers of couples of rabbits after k, $k + 1$ and $k + 2$ months are related by the recursive relation

$$F_{k+2} = F_{k+1} + F_k,$$

for every $k \geq 1$. Thus,

$$F_3 = F_2 + F_1 = 2, \quad F_4 = F_3 + F_2 = 3, \ldots, \quad F_{12} = 144.$$

□

In the notations of the previous example, we recall that $(F_n)_{n \geq 1}$ is known as the **Fibonacci sequence**.

Example 1.19 Using the letters A, B and C, compute how many words of ten letters (not necessarily with a meaning) one can form, so that there are no consecutive consonants.

Solution For $n \geq 1$, let a_n be the number of *words* of n letters A, B, C satisfying the given condition on consonants. For $k \geq 3$, a word of k letters can finish in A, B or C. If it finishes in A, the $k - 1$ remaining letters form any of the a_{k-1} word of $k - 1$ letters and without consecutive consonants; if it finishes in B or C, then

---

[4]After Leonardo di Pisa, also known as Fibonacci, Italian mathematician of the eleventh century. Apart from its own contributions to Mathematics—as the problem we are presently describing— one of the greatest merits of Fibonacci was to help revive, in Middle Age Europe, the Mathematics of Classical Antiquity; in particular, Fibonacci's famous book *Liber Abaci* introduced, in Western Civilization, the Hindu-Arabic algarisms and numbering system.

the next to last letter must necessarily be an A (since we cannot have consecutive consonants), and the $k - 2$ initial letters form any of the a_{k-2} words of $k - 2$ letters having no consecutive consonants.

The reasoning of the previous paragraph gives us the recurrence relation

$$a_k = a_{k-1} + 2a_{k-2}, \ \forall \ k \geq 3.$$

Since $a_1 = 3$ and $a_2 = 5$ (the possible words of two letters and with no consecutive consonants are AA, AB, AC, BA and CA), we successively compute $a_3 = 11$, $a_4 = 21$, $a_5 = 43$, $a_6 = 85$, $a_7 = 171$, $a_8 = 341$, $a_9 = 683$ and $a_{10} = 1365$. $\square$

As the two last examples suggest, there are several combinatorial situations, of distinctive recursive characters, that give rise to the algebraic problem of computing, in terms of $n \in \mathbb{N}$, the n-th term of a sequence $(a_n)_{n \geq 1}$ that satisfies a certain recurrence relation. The coming example, of professor Emanuel Carneiro, generates a third order linear recurrence relation. In this respect, perhaps the reader might find it helpful to pause for a moment and review the content of Section 3.3 of [8]. Alternatively, he/she can take a quick look at Problem 5, page 79, or at the material of Chap. 21 (especially Theorem 21.22.)

Example 1.20 For each set $A = \{x_1, x_2, \ldots, x_m\}$ of real numbers, with $x_1 < x_2 < \cdots < x_m$, let

$$\Delta(A) = \sum_{k \leq j \leq m} x_j - \sum_{1 < j < k} x_j,$$

where $k = \frac{m+1}{2}$. Given an integer $n > 1$, show that

$$\sum_{\emptyset \neq A \subset I_n} \Delta(A) = (n^2 + n + 2) \cdot 2^{n-3}.$$

Solution Let $d_n = \sum_{\emptyset \neq A \subset I_n} \Delta(A)$. By examining some particular cases we easily get $d_1 = 1$, $d_2 = 4$ and $d_3 = 14$. In general,

$$d_{n+1} = \sum_{\emptyset \neq A \subset I_{n+1}} \Delta(A)$$

$$= \sum_{\substack{A \subset I_{n+1} \\ n+1 \in A}} \Delta(A) + \sum_{\emptyset \neq A \subset I_n} \Delta(A) \tag{1.7}$$

$$= \sum_{\substack{A \subset I_{n+1} \\ n+1 \in A}} \Delta(A) + d_n.$$

To analyse the last sum above, let $A' = A \setminus \{n+1\} \subset I_n$, for each $A \subset I_{n+1}$ such that $n+1 \in A$. Writing $\mathcal{P}_n$ and $\mathcal{I}_n$ for the families of subsets of I_n with even and odd numbers of elements, respectively, we have

$$\sum_{\substack{A \subset I_{n+1} \\ n+1 \in A}} \Delta(A) = \sum_{A' \in \mathcal{P}_n} \Delta(A' \cup \{n+1\}) + \sum_{A' \in \mathcal{I}_n} \Delta(A' \cup \{n+1\}). \tag{1.8}$$

In Problem 5 (see also Problem 7, page 29) we shall prove that $|\mathcal{P}_n| = |\mathcal{I}_n| = 2^{n-1}$. Assuming this fact for the time being, let's separately consider the two sums at the right hand side of (1.8):

(a) If $A' \in \mathcal{P}_n$, it easily follows from the definition of $\Delta(\cdot)$ that

$$\Delta(A' \cup \{n+1\}) = (n+1) + \Delta(A'),$$

where $\Delta(\emptyset) = 0$. However, since $|\mathcal{P}_n| = 2^{n-1}$, we get

$$\sum_{A' \in \mathcal{P}_n} \Delta(A' \cup \{n+1\}) = \sum_{\emptyset \neq A' \in \mathcal{P}_n} \Delta(A') + 2^{n-1}(n+1).$$

(b) If $A' \in \mathcal{I}_n$, say $A' = \{x_1, x_2, \ldots, x_{2k+1}\}$, with $x_1 < x_2 < \cdots < x_{2k+1}$, then

$$\begin{aligned}
\Delta(A' \cup \{n+1\}) &= \Delta(\{x_1, x_2, \ldots, x_{2k+1}, n+1\}) \\
&= (x_{k+2} + \cdots + x_{2k+1} + (n+1)) \\
&\quad - (x_1 + x_2 + \cdots + x_{k+1}) \\
&= (n+1) + (x_{k+1} + \cdots + x_{2k+1}) \\
&\quad - (x_1 + \cdots + x_k) - 2x_{k+1} \\
&= (n+1) + \Delta(A') - 2x_{k+1}.
\end{aligned}$$

Note that $x_{k+1} \in I_n$ is the *central element* of $A' \in \mathcal{I}_n$. Hence, letting $c(A')$ denote the central element of $A' \in \mathcal{I}_n$, we obtain

$$\begin{aligned}
\sum_{A' \in \mathcal{I}_n} \Delta(A' \cup \{n+1\}) &= \sum_{A' \in \mathcal{I}_n} ((n+1) + \Delta(A') - 2c(A')) \\
&= (n+1) \cdot 2^{n-1} + \sum_{A' \in \mathcal{I}_n} \Delta(A') - 2 \sum_{A' \in \mathcal{I}_n} c(A').
\end{aligned}$$

In Problem 5, page 13 (see also Problem 18, page 31), we saw that

$$\sum_{A' \in \mathcal{I}_n} c(A') = (n+1) \cdot 2^{n-2}.$$

Therefore,

$$\sum_{\substack{A' \in \mathcal{I}_n}} \Delta(A' \cup \{n+1\}) = \sum_{\substack{A' \in \mathcal{I}_n}} \Delta(A').$$

Substituting the results of items (a) and (b) successively in (1.8) and (1.7), we arrive at

$$d_{n+1} = d_n + \sum_{\substack{A \subset I_{n+1} \\ n+1 \in A}} \Delta(A)$$

$$= d_n + (n+1) \cdot 2^{n-1} + \sum_{\emptyset \neq A' \in \mathcal{P}_n} \Delta(A') + \sum_{A' \in \mathcal{I}_n} \Delta(A')$$

$$= d_n + (n+1) \cdot 2^{n-1} + \sum_{\emptyset \neq A' \subset I_n} \Delta(A')$$

$$= 2d_n + (n+1) \cdot 2^{n-1}.$$

Now, since the sequence $n \mapsto \frac{d_{n+1} - 2d_n}{2^{n-1}} = n+1$ is an arithmetic progression, it follows that

$$\frac{d_{n+3} - 2d_{n+2}}{2^{n+1}} + \frac{d_{n+1} - 2d_n}{2^{n-1}} = 2 \cdot \frac{d_{n+2} - 2d_{n+1}}{2^{n+1}}.$$

After clearing out denominators, such an equality gives us the linear recurrence relation

$$d_{n+3} - 6d_{n+2} + 12d_{n+1} - 8d_n = 0.$$

Finally, upon solving such a linear recurrence (cf. Problems 6 and 7, page 393—see also Example 3.20 of [8]) we get

$$d_n = (n^2 + n + 2) \cdot 2^{n-3}.$$

$\square$

Problems: Sect. 1.3

1. For each $n \in \mathbb{N}$, let a_n be the number of distinct ways of writing n as a sum of summands equal to 1, 3 or 4. Prove that $a_{n+4} = a_{n+3} + a_{n+1} + a_n$, for every $n \geq 1$.
2. Show that every convex n-gon has exactly $\frac{n(n-3)}{2}$ diagonals.

Fig. 1.1 The tower of Hanoi
game

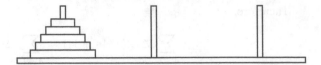

3. A $2 \times n$ checkerboard must be tiled with dominoes (each of which is supposed to have the shape of a 1×2 rectangle). If a_n denotes the number of such distinct tilings, do the following items:

   (a) Prove that $a_{k+2} = a_{k+1} + a_k$, for every integer $k \geq 1$.
   (b) Compute a_n as a function of n.

4. Let $n > 1$ distinct circles be given in the plane, such that any two of them have a common chord and no three of them pass through a single point. Compute, as a function of n, the number of regions into which the plane gets divided by the circles.

5. * We are given $n \in \mathbb{N}$ and a set A with n elements.

   (a) Let a_n and b_n respectively denote the totals of subsets of A with even and odd numbers of elements. Use a recursive argument to show that $a_n = a_{n-1} + b_{n-1} = b_n$.
   (b) Conclude that A has exactly 2^{n-1} subsets with an even number of elements and 2^{n-1} subsets with an odd number of elements.

   The coming problem is a particular case of **Kaplansky's first lemma**, which will be dealt with, in general form, in Example 1.28.

6. Three nonconsecutive chairs are to be chosen out of a row of ten chairs. How many are the ways of doing this?

7. The **Tower of Hanoi game** consists of three parallel equal rods, vertically attached to the surface of a table, together with a pile of n disks of pairwise unequal sizes, all having a hole in the center through which they can slide along the rods. Figure 1.1 shows the initial configuration of the game for $n = 5$, with all of the disks piled at the leftmost rod, from the smallest disk at the top to the largest at the bottom of the pile. The purpose of the game is to move all disks to the rightmost rod, possibly with the help of the central rod and subjected to the following rule: at no moment a disk can be placed above a smaller one, in any rod. For general n, let a_n denote the last number of movements needed to finish the game. Show that:

   (a) $a_{k+1} = 2a_k + 1$, for every $k \in \mathbb{N}$.
   (b) $a_n = 2^n - 1$, for every $n \in \mathbb{N}$.

8. * Given $k, n \in \mathbb{N}$, the **Stirling number of second kind**[5] $S(n, k)$ is defined as the number of ways of partitioning a set of n elements into k nonempty, pairwise disjoint subsets. For example, $S(3, 2) = 3$, since

---

[5]After James Stirling, Scottish mathematician of the eighteenth century. With respect to Stirling numbers of second kind, see also Problem 4, page 57.

$$\{1, 2\} \cup \{3\}, \quad \{1, 3\} \cup \{2\} \text{ and } \{2, 3\} \cup \{1\}$$

are the only ways of partitioning $\{1, 2, 3\}$ into two nonempty disjoint subsets.

(a) Verify that $S(n, 1) = S(n, n) = 1$ and $S(n, k) = 0$ if $n < k$.
(b) Prove that, for $1 \le k < n$, one has the recurrence relation

$$S(n + 1, k + 1) = S(n, k) + (k + 1)S(n, k + 1).$$

(c) Use items (a) and (b) to compute the number of ways of distributing 7 distinct coins into three equal 3 boxes, such that no box stays empty.

The coming problem will be posed two other times along these notes: one in Problem 8 and the other in Problem 8, page 92.

9. Compute, in terms of $n \in \mathbb{N}$, the number of sequences of n terms, all of which equal to 0, 1, 2 or 3 and having an even number of 0's.
10. A *flag with n strips* consists of n consecutive horizontal strips, each of which is colored red, white or blue and such that any two adjacent strips have different colors.

(a) Calculate how many are the distinct flags with n strips.
(b) If a_n denotes the total number of flags with n strips and such that the first and last strips have different colors, show that $a_{n+1} + a_n = 3 \cdot 2^n$ for every $n \ge 1$.
(c) Compute a_n as a function of n, for every $n \ge 1$.

11. (Bulgaria) A finite nonempty set A of positive integers is said to be *selfish* if $|A| \in A$. A selfish set A is *minimal* if A does not contain a selfish set different from itself. Do the following items:

(a) Prove that the number of minimal selfish subsets of I_n that do not contain n coincides with the number of minimal selfish subsets of I_{n-1}.
(b) Given $m \in \mathbb{Z}$ and $X \subset \mathbb{Z}$, let $X+m$ denote the set $X+m = \{x+m; x \in X\}$. Show that, for every integer $n > 2$, the correspondences

$$A \mapsto (A \setminus \{n\}) - 1 \text{ and } B \mapsto (B + 1) \cup \{n\}$$

are inverse bijections between the family of minimal selfish subsets of I_n that do not contain n and that of minimal selfish subsets of I_{n-2}.
(c) Show that I_n has exactly F_n minimal selfish subsets of I_n, where F_n denotes the n-th Fibonacci number.

For the next two problems, given a nonempty finite set A of real numbers, we let $\sigma(A)$ and $\pi(A)$ respectively denote the sum and the product of the elements of A, with the convention that $\sigma(A) = \pi(A) = a$ if $A = \{a\}$.

12. Given a nonempty finite set $X \subset \mathbb{N}$, let a_X denote the sum

$$a_X = \sum_{\emptyset \neq A \subset X} \frac{1}{\pi(A)}.$$

Do the following items:

(a) If $m \in \mathbb{N} \setminus X$ and $Y = X \cup \{m\}$, show that $a_Y = \left(\frac{m+1}{m}\right) a_X + \frac{1}{m}$.

(b) Given $n \in \mathbb{N}$, write a_n to denote a_{I_n}. Show that

$$a_{n+1} = \left(\frac{n+2}{n+1}\right) a_n + \frac{1}{n+1}.$$

(c) Show that $a_n = n$ for every $n \in \mathbb{N}$.

13. (USA—adapted) For a nonempty finite set $X \subset \mathbb{N}$, let b_X denote the sum

$$b_X = \sum_{\emptyset \neq A \subset X} \frac{\sigma(A)}{\pi(A)}.$$

(a) If $m \in \mathbb{N} \setminus X$ and $Y = X \cup \{m\}$, show that

$$b_Y = b_X + \sum_{\emptyset \neq A \subset X} \frac{m + \sigma(A)}{m\pi(A)} + 1 = \left(\frac{m+1}{m}\right) b_X + a_X + 1,$$

where a_X is the sum defined in the previous problem.

(b) Given $n \in \mathbb{N}$, write b_n instead of b_{I_n}. Show that

$$b_{n+1} = \left(\frac{n+2}{n+1}\right) b_n + (n+1).$$

(c) Deduce that, for $n \in \mathbb{N}$, one has

$$b_n = (n+1)^2 - (n+1)\left(1 + \frac{1}{2} + \cdots + \frac{1}{n}\right) - 1.$$

14. (France—adapted) We say that a set X of natural numbers is *good* provided it satisfies the following property: for every natural number x, if $x \in X$ then $2x \notin X$. For each natural k, let $A_k = \{1, 2, 3, \ldots, 2^k\}$ and b_k be the greatest number of elements which a good subset of A_k can have.

(a) Show that $b_k = b_{k-2} + 2^{k-1}$ for every $k \geq 2$.

(b) For each natural n, compute b_n as a function of n.

1.4 Arrangements, Combinations and Permutations

As an application of the ideas of the previous section, we use recursive arguments to solve three specific, though very important, counting problems, namely, those of counting the number of *arrangements without repetitions*, of *permutations* and of *combinations*. For what follows, we recall that $I_n = \{1, 2, \ldots, n\}$ for every $n \in \mathbb{N}$.

We start by counting, in the coming result, the number of injective functions between two nonempty finite sets.

Proposition 1.21 *Let $n, k \in \mathbb{N}$. If A is a set with n elements, then there are exactly $n(n-1)\ldots(n-k+1)$ injective functions $f : I_k \to A$.*

Proof Let B be a set with m elements and let a_{km} denote the number of injective functions $f : I_k \to B$.

Evidently, $a_{1n} = n$ (since there are exactly n possible choices for $f(1) \in A$) and, in this case, the formula in the statement of the proposition is true.

From now on, assume that $k > 1$. For a fixed $x \in A$, if $f : I_k \to A$ is an injective function such that $f(k) = x$, then the restriction of f to I_{k-1} is an injective function $\tilde{f} : I_{k-1} \to A \setminus \{x\}$. Conversely, given an injective function $\tilde{f} : I_{k-1} \to A \setminus \{x\}$, we extend $\tilde{f}$ to an injective function $f : I_k \to A$ by letting $f(k) = x$.

Since such operations of restriction and extension of injective functions are clearly inverses of each other, we conclude that there are as many injective functions $f : I_k \to A$ with $f(k) = x$ as there are injective functions $\tilde{f} : I_{k-1} \to A \setminus \{x\}$. Thus, the number of such functions is exactly $a_{k-1,n-1}$. However, since there are n possible choices for $x \in A$ (for, $|A| = n$), we get the recurrence relation

$$a_{kn} = na_{k-1,n-1}, \tag{1.9}$$

which is valid for every natural $n > 1$.

Now, observe that $a_{k1} = 0$ for every $k > 1$ (for, if $|A| = 1$, there is no way of choosing distinct $f(1)$ and $f(2)$ in A). By induction, assume that $a_{kn} = 0$ whenever $k > n$ and $1 \leq n < m$, where $m > 1$ is a natural number. Given $k, m \in \mathbb{N}$ such that $k > m$, we have $k - 1 > m - 1$, so that, by induction hypothesis, $a_{k-1,m-1} = 0$. But then, (1.9) gives $a_{km} = ma_{k-1,m-1} = 0$. Therefore, it follows by induction that $a_{kn} = 0$ for every $k, n \in \mathbb{N}$ such that $k > n$. Yet in this case we have $n-k+1 \leq 0$, so that $n(n-1)\ldots(n-k+1) = 0$ and the formula of the statement of the proposition is valid.

Suppose, from now on, that $k \leq n$. Then, again from (1.9), we have

$$a_{kn} = na_{k-1,n-1} = n(n-1)a_{k-2,n-2}$$

$$= \cdots$$

$$= n(n-1)\ldots(n-k+2)a_{1,n-k+1}$$

$$= n(n-1)\ldots(n-k+2)(n-k+1).$$

$\square$

Since an injective function $f : I_k \to A$ is simply a sequence of k pairwise distinct terms, all chosen from the elements of A, we can rephrase the above proposition according to the following

Corollary 1.22 *If* $|A| = n$, *then there are exactly* $n(n-1)\ldots(n-k+1)$ *sequences formed by* k *pairwise distinct elements of* A.

We can apply this corollary to a given combinatorial situation whenever it asks us to count how many are the *ordered choices of* k *pairwise distinct elements* of a given set A. For this reason, from now on we shall say that the previous corollary counts the number of **arrangements without repetition** of k pairwise distinct elements, chosen from a given set of n elements.

Example 1.23 How many are the natural numbers of three pairwise distinct algarisms?

Solution To choose a natural number with three pairwise distinct algarisms is the same as to choose a sequence (a, b, c) such that $a, b, c \in \{0, 1, 2, \ldots, 9\}$ are pairwise distinct and $a \neq 0$. Corollary 1.4 assures that, in order to count the number of possible such sequences, it suffices to count the number of sequences (a, b, c), with pairwise distinct $a, b, c \in \{0, 1, 2, \ldots, 9\}$, and then to *discount* the number of such sequences of the form $(0, b, c)$.

By the previous corollary, the number of sequences (a, b, c) with $a, b, c \in \{0, 1, 2, \ldots, 9\}$ pairwise distinct but without the restriction $a \neq 0$ is $10 \times 9 \times 8 = 720$. On the other hand, the number of such sequences of the form $(0, b, c)$ is $9 \times 8 = 72$. Hence, the desired result is $720 - 72 = 648$.   □

The counting of arrangements without repetition has a very important consequence, based on the following

Definition 1.24 A **permutation** of the elements of a nonempty set A is a bijection $f : A \to A$.

If A is finite and nonempty, it is not difficult to prove (see Problem 4, page 7) that a function $f : A \to A$ is bijective if and only if it is injective. Hence, letting $k = n$ in the formula of Proposition 1.21, we immediately get our next result.

Corollary 1.25 *If* $|A| = n$, *then there are exactly* $n!$ *permutations of the elements of* A.

Given a finite set A with n elements and a natural number k such that $0 \leq k \leq n$, the coming result uses a recursive argument to compute the number of subsets of A with k elements each. In order to properly state it, recall (cf. Section 4.2 of [8]) that, for nonnegative integers n and k with $0 \leq k \leq n$, one defines the binomial number $\binom{n}{k}$ by letting

$$\binom{n}{k} = \frac{n!}{k!(n-k)!}.$$

Moreover, for $1 \leq k \leq n$ such numbers satisfy the recurrence relation

$$\binom{n}{k} = \binom{n-1}{k} + \binom{n-1}{k-1},$$

which is known as *Stifel's relation* and allows us to prove that $\binom{n}{k} \in \mathbb{N}$ for every n and k as above.

Proposition 1.26 *If A is a finite set with n elements and $0 \leq k \leq n$, then A possesses exactly $\binom{n}{k}$ subsets of k elements.*

Proof If $k = 0$ there is nothing to do, for $\emptyset$ is the only subset of A having 0 elements and $\binom{n}{0} = 1$. Thus, let $1 \leq k \leq n$ and C_k^n be the number of subsets of A with k elements.

For a fixed $x \in A$, there are two kinds of subsets of A with k elements: those which do not contain x and those which do contain x. The former ones are precisely the k-element subsets of $A \setminus \{x\}$; since $|A \setminus \{x\}| = n - 1$, there are exactly C_k^{n-1} of these subsets of k elements of A.

On the other hand, if $B \subset A$ has k elements and $x \in B$, then $B \setminus \{x\} \subset A \setminus \{x\}$ has $k-1$ elements; conversely, if $B' \subset A \setminus \{x\}$ has $k-1$ elements, then $B' \cup \{x\} \subset A$ has k elements, one of which is x. Since such correspondences are clearly inverses of each other, we conclude that there are as many k-element subsets of A containing x, as there are $k - 1$ element subsets of $A \setminus \{x\}$; thus, there are exactly C_{k-1}^{n-1} such k-element subsets of A.

Taking these two contributions into account, we obtain for $1 \leq k \leq n$ the recurrence relation

$$C_k^n = C_k^{n-1} + C_{k-1}^{n-1},$$

which is identical to Stifel's relation for the binomial numbers $\binom{n}{k}$. Finally, since $C_0^n = 1 = \binom{n}{0}$ and $C_1^n = n = \binom{n}{1}$ (for A has exactly n subsets of 1 element each— the sets $\{x\}$, with $x \in A$), an easy induction gives $C_k^n = \binom{n}{k}$ for $0 \leq k \leq n$. $\square$

In words, the previous proposition computes how many are the *unordered choices* of k distinct elements of a set having n elements; one uses to say that such choices are the **combinations of n objects, taking k at a time**. Also thanks to the former proposition, one uses to refer to the binomial number $\binom{n}{k}$ as "*n chooses k*".

Example 1.27 When all diagonals of a certain convex octagon have been drawn, one noticed that there were no three of them passing through a single interior point of the octagon. How many points in the interior of the octagon are intersection points of two of its diagonals?

Solution Firstly, note that the condition on the diagonals of the octagon guarantees that each one of the points of intersection we wish to count is determined by a single pair of diagonals. Hence, it suffices to count how many pairs of diagonals of the octagon intersect in its interior. To this end, note that each 4-element subset of

the set of vertices of the octagon determines exactly two diagonals which intersect
in the interior of it; conversely, if two diagonals of the octagon do intersect in its
interior, then the set of their endpoints is a 4-element subset of the set of vertices of
the octagon. Therefore, the bijective principle assures that there are as many pairs of
diagonals intersecting in the interior of the octagon as there are 4-element subsets
of its set of vertices. By Proposition 1.26, then, the number of points of intersection
we wish to count equals $\binom{8}{4} = 70$. $\square$

Our next example brings a beautiful application of combinations, due to the
American mathematician of the twentieth century Irving Kaplansky and known as
Kaplansky's first lemma.[6]

Example 1.28 (Kaplansky) Given $n, k \in \mathbb{N}$, with $n \geq 2k - 1$, show that the number
k-element subsets of I_n without consecutive elements equals $\binom{n-k+1}{k}$.

Proof We first note that if $A \subset I_n$ has k elements, no two of which being
consecutive, then $n \geq 2k - 1$, for between any two elements of A there is at least
one element of I_n not belonging to A. Hence, the assumption $n \geq 2k - 1$ made
in the statement is natural and gives $n - k + 1 \geq k$, so that the binomial number
$\binom{n-k+1}{k}$ is well defined.

Now, recall that the characteristic function of A in I_n is a sequence of n terms,
with exactly k of them being equal to 1 and the other $n - k$ being equal to 0;
moreover, our assumption on A assures that any two 1's in such a sequence are
nonconsecutive. Since the characteristic function determines A, it suffices to count
the number of such sequences.

To what is left to do, let's first consider a sequence of $2(n - k) + 1$ terms, in
which the $n - k$ terms with even indices are equal to 0:

$$\underbrace{(\_, 0, \_, 0, \_, \ldots, \_, 0, \_)}_{2(n-k)+1}.$$

If we are to have exactly k terms equal to 1, no two of which being consecutive, it is
necessary and sufficient to choose, from the set of $n - k + 1$ positions corresponding
to odd indices, the k positions in which we want to put the 1's (as we previously
pointed out, this is possible because $n - k + 1 \geq k$). According to Proposition 1.26,
there are precisely $\binom{n-k+1}{k}$ possible choices for such positions. Finally, once we
erase the not chosen positions, we are left with the characteristic sequence of a set
satisfying the conditions of the problem. Conversely, it is immediate that every such
characteristic sequence can be obtained as above, so that the answer to our problem
is $\binom{n-k+1}{k}$. $\square$

Now, we are given natural numbers k and n and nonnegative integers $n_1, \ldots, n_k$
satisfying $n = n_1 + \cdots + n_k$. We also have k distinct types of objects, say $a_1, \ldots, a_k$,

---

[6]There is also a *Kaplansky's second lemma*, which will be the object of Problem 12. Kaplansky
devised these two results in order to solve Lucas' problem—cf. Problem 10, page 41.

and would like to count how many sequences of n terms have exactly n_1 terms equal to $a_1, \ldots, n_k$ terms equal to a_k. Yet in another way, we say that such a sequence is a **permutation with repeated elements** of n objects, each of which being of one of the types $a_1, \ldots, a_k$ and with n_1 objects of type $a_1, \ldots, n_k$ objects of type a_k. Let's now look at an equivalent problem.

If $f : I_n \to \{a_1, \ldots, a_k\}$ is a sequence satisfying the conditions of the previous paragraph (i.e., with n_1 terms equal to $a_1, \ldots, n_k$ terms equal to a_k), then, taking inverse images, we obtain

$$I_n = f^{-1}(a_1) \cup \ldots \cup f^{-1}(a_k).$$

At the right hand side above, the sets $f^{-1}(a_j)$, $1 \leq j \leq k$, are pairwise disjoint and such that $|f^{-1}(a_j)| = n_j$ for $1 \leq j \leq k$; we say that the union at the right hand side is the **partition** of I_n **induced** by f. Conversely, associated to a partition of I_n of the form $I_n = A_1 \cup \ldots \cup A_k$, with $|A_j| = n_j$ for $1 \leq j \leq k$, we have the sequence $f : I_n \to \{a_1, \ldots, a_k\}$ such that $f(A_j) = \{a_j\}$ for $1 \leq j \leq k$. Therefore, the bijective principle guarantees that a problem equivalent to the one of the previous paragraph is that of counting the number of distinct partitions of a set of n elements into k pairwise disjoint subsets $A_1, \ldots, A_k$, under the restriction that $|A_1| = n_1, \ldots, |A_k| = n_k$.

This being said, the coming result solves both counting problems described above.

Proposition 1.29 *Let $n, k \in \mathbb{N}$ and $n = n_1 + \cdots + n_k$ be a partition of n in positive integers. If A is a set with n elements and $\binom{n}{n_1, \ldots, n_k}$ denotes the number of partitions of A into sets $A_1, \ldots, A_k$, with $|A_j| = n_j$ for $1 \leq j \leq k$, then*

$$\binom{n}{n_1, \ldots, n_k} = \frac{n!}{n_1! \ldots n_k!}. \tag{1.10}$$

Proof We shall make induction on $k \in \mathbb{N}$. Firstly, the case $k = 1$ is trivial, for in this case $n_1 = n$ and the only subset of A with n elements is A itself. As induction hypothesis, let $l \geq 2$ be a natural number and assume that (1.10) is true for $1 \leq k < l$ and every partition of n into k positive integers $n_1, \ldots, n_k$. Then, fix a partition $n = n_1 + \cdots + n_l$ of n in positive integers.

The bijective principle assures that, for each subset A_l of A, with $|A_l| = n_l$, there are as many partitions of A as in the statement of the proposition as there are partitions

$$A \setminus A_l = A_1 \cup \ldots \cup A_{l-1},$$

of $A \setminus A_l$ into $l - 1$ subsets $A_1, \ldots, A_{l-1}$, with $|A_j| = n_j$ for $1 \leq j \leq l - 1$.

Since $|A \setminus A_l| = n - n_l$, the total number of such partitions of $A \setminus A_l$ is, by definition, equal to $\binom{n-n_l}{n_1, \ldots, n_{l-1}}$; however, since there are exactly $\binom{n}{n_l}$ ways of choosing the subset A_l of A, the fundamental principle of counting gives the recurrence relation

$$\binom{n}{n_1, \ldots, n_l} = \binom{n - n_l}{n_1, \ldots, n_{l-1}}\binom{n}{n_l}.$$

Finally, by applying the induction hypothesis to $\binom{n-n_l}{n_1,\ldots,n_{l-1}}$, we conclude that

$$\binom{n}{n_1, \ldots, n_l} = \frac{(n - n_l)!}{n_1! \ldots n_{l-1}!} \cdot \binom{n}{n_l} = \frac{n!}{n_1! \ldots n_{l-1}! n_l!}.$$

$\square$

Remark 1.30 In the notations of the proposition above, it is worth noting that

$$\binom{n}{k, n - k} = \binom{n}{k},$$

where the right hand side denotes the usual binomial number. Such and equality reflects the fact that choosing a k-element subset out of a set with n elements is the same as partitioning such a set into two subsets, one with k elements and the other with $n - k$ elements.

Problems: Sect. 1.4

1. If $n \in \mathbb{N}$ and A is a set with n elements, prove that there are exactly $n^k - n(n - 1)\ldots(n - k + 1)$ sequences of k terms, all taken out of A but not pairwise distinct.
2. * Do the following items:

    (a) Use a combinatorial argument to prove the formula for binomial expansion:

    $$(x + y)^n = \sum_{k=0}^{n} \binom{n}{k} x^{n-k} y^k. \qquad (1.11)$$

    (b) Generalize item (a), proving combinatorially the **multinomial expansion formula**:

    $$(x_1 + \cdots + x_k)^n = \sum_{\substack{n_1,\ldots,n_k \in \mathbb{Z}_+ \\ n_1 + \cdots + n_k = n}} \binom{n}{n_1, \ldots, n_k} x_1^{n_1} \ldots x_k^{n_k}, \qquad (1.12)$$

    where, for $n_1, \ldots, n_k \in \mathbb{Z}_+$ such that $n = n_1 + \cdots + n_k$, the number $\binom{n}{n_1,\ldots,n_k}$ is defined as in (1.10).

3. Given $n, k \in \mathbb{N}$, use the result of Proposition 1.29, together with item (b) of the previous problem, to compute the number of ordered partitions of a set A, with $|A| = n$, into k subsets $A_1, \ldots, A_k$ (cf. Problem 8, page 7).

4. Prove that the Stirling number of second kind $S(n, k)$ (cf. Problem 8, page 20) can be computed with the aid of the following formula:

$$S(n, k) = \frac{1}{k!} \sum_{\substack{n_1, \ldots, n_k \in \mathbb{N} \\ n_1 + \cdots + n_k = n}} \binom{n}{n_1, \ldots, n_k}.$$

5. (IMO—shortlist) Given a permutation $(a_1, a_2, \ldots, a_n)$ of I_n, we say that a_j is a *local maximum* if a_j is greater than its neighbors (for $j = 1$, this means that $a_1 > a_2$; for $j = n$, that $a_n > a_{n-1}$). Compute the number of permutations of I_n having exactly one local maximum.

6. (OCM) We are given $n \geq 6$ points on a circle Γ, and draw all $\binom{n}{2}$ chords joining two of them. We further assume that no three of such chords pass through a single point of the corresponding open disk D bounded by Γ. Compute the number of distinct triangles satisfying the two following properties:

   i. their vertices belong to D.
   ii. their sides lie on three of the $\binom{n}{2}$ drawn chords.

7. * Use combinations to show that a set with n elements has exactly 2^{n-1} subsets with an even number of elements and 2^{n-1} subsets with an odd number of elements.

8. Compute, in terms of $n \in \mathbb{N}$, the number of sequences of n terms, all of which equal to 0, 1, 2 or 3 and having an even number of 0's.

9. * Given $n, k \in \mathbb{N}$, show that the equation $x_1 + \cdots + x_k = n$ has exactly $\binom{n+k-1}{k-1}$ nonnegative integer solutions.

10. * Given $n, k \in \mathbb{N}$, show that the equation $x_1 + \cdots + x_k = n$ has exactly $\binom{n-1}{k-1}$ positive integer solutions.

11. The purpose of this problem is to give another proof of Kaplansky's first lemma (cf. Example 1.28). To this end, do the following items:

   (a) If $A = \{a_1, \ldots, a_k\} \subset I_n$ is a set without consecutive elements, let $x_1 = a_1 - 1$, $x_{k+1} = n - a_k$ and $x_j = a_j - a_{j-1} - 2$ for $2 \leq j \leq k$. Show that $x_1, \ldots, x_{k+1}$ solves, in nonnegative integers, equation $x_1 + \cdots + x_{k+1} = n - 2k + 1$.

   (b) If $x_1, \ldots, x_{k+1}$ is a solution of equation $x_1 + \cdots + x_{k+1} = n - 2k + 1$ in nonnegative integers, show how to use it to get a set $A = \{a_1, \ldots, a_k\} \subset I_n$ without consecutive elements.

   (c) Obtain Kaplansky's first lemma from the results of items (a) and (b) and from Problem 9.

12. * Prove **Kaplansky's second lemma**: given $n, k \in \mathbb{N}$, with $n \geq 2k$, the number of k-element subsets of I_n without consecutive elements, and considering 1 and n to be consecutive, is equal to

$$\frac{n}{n-k}\binom{n-k}{k}.$$

For the coming problem, we define a **multiset** as an ordered pair (A, f), where A is a set and f is a function from A into $\mathbb{N}$. If $x \in A$, we say that x is an *element* of multiset (A, f); in such a case, $f(x) \in \mathbb{N}$ is the *frequency* of x as an element of (A, f). A multiset (A, f) is *finite* if A is a finite set. In this case, the *number of elements* of the multiset is defined as the sum of the frequencies of the elements of A, seen as elements of (A, f); in symbols,

$$|(A, f)| = \sum_{x \in A} f(x). \qquad (1.13)$$

Informally, we denote a multiset (A, f) simply by A_f (or even A, whenever f is understood and there is no danger of confusion with the set A itself); we declare A_f between double curly braces, instead of simple ones (which are reserved for declaring sets), with each $x \in A$ being repeated exactly $f(x)$ times; for instance, if $A = \{a, b, c\}$ and $f : A \to \mathbb{N}$ is given by $f(a) = 2$, $f(b) = 3$ and $f(c) = 1$, we declare the multiset A_f by writing

$$A_f = \{\{a, a, b, b, b, c\}\}.$$

13. Let $n, k \in \mathbb{N}$ be given, with $n \geq k$, and let A be a set with k elements. Compute the number of multisets (A, f) of n elements.

14. For each integer $n > 1$, prove that there are exactly $\binom{2n}{n}$ subsets of I_{2n} with equal quantities of even and odd elements.

15. We are given natural numbers m and n, with $m > n$. If

$$\mathcal{F} = \{(A_1, \ldots, A_n); \ A_1, \ldots, A_n \subset I_m\},$$

compute, in terms of m and n, the value of the sum

$$\sum_{(A_1, \ldots, A_n) \in \mathcal{F}} |A_1 \cup \ldots \cup A_n|.$$

16. To each permutation $\sigma = (a_1, a_2, \ldots, a_n)$ of I_n, let

$$S_\sigma = (a_1 - a_2)^2 + (a_2 - a_3)^2 + \cdots + (a_{n-1} - a_n)^2.$$

Show that

$$\frac{1}{n!} \sum_\sigma S_\sigma = \binom{n+1}{3},$$

where, in the left hand side above, σ varies over all $n!$ permutations of I_n.

17. (Saint Petersburg—adapted) We are give $2n + 1$ points on a circle, such that no two of them are the endpoints of a diameter. Prove that, among all triangles having three of the given points as their vertices, at most

$$\frac{1}{6}n(n + 1)(2n + 1)$$

of them are acute.[7]

18. * The purpose of this problem is to use the material of this section to present another solution to Problem 5, page 13. To this end, do the following items:

(a) Fixed $1 \leq i \leq n$, show that i is the central element of exactly

$$\sum_j \binom{i - 1}{j}\binom{n - i}{j} = \binom{n - 1}{i - 1}$$

sets $A \in \mathcal{I}_n$, where the above sum extends to all indices j satisfying $0 \leq j \leq \min\{i - 1, n - i\}$.

(b) Conclude that

$$\sum_{A \in \mathcal{I}_n} c(A) = \sum_{i=1}^{n} \binom{n - 1}{i - 1}i = (n + 1) \cdot 2^{n-2}.$$

19. (IMO shortlist—adapted) Let[8] $n \in \mathbb{N}$ and $f : I_n \to I_n$ be such that $f(f(x)) = x$ for every $x \in I_n$.

(a) Show that f is a bijection and that, if f has exactly k fixed points, then $n - k$ must be even.

(b) Conversely, given $0 \leq k \leq n$ such that $n - k$ is even, show that for each choice of elements $x_1, \ldots, x_k \in I_n$ and for each partition $\{\{a_1, b_1\}, \ldots, \{a_l, b_l\}\}$ of the remaining $n - k$ elements of I_n in subsets of two elements each, there exists a single $f : I_n \to I_n$ having $x_1, \ldots, x_k$ as its fixed points and satisfying $f(a_i) = b_i$ for $1 \leq i \leq l$ and $f(f(x)) = x$ for each $x \in I_n$.

(c) Conclude that the number of functions $f : I_n \to I_n$ such that $f(f(x)) = x$ for each $x \in I_n$ is given by

$$\sum_{\substack{0 \leq k \leq n \\ n-k \text{ even}}} \binom{n}{k} \cdot \binom{n - k}{2, \ldots, 2}_{(n-k)/2} = \sum_{\substack{0 \leq k \leq n \\ n-k \text{ even}}} \frac{n!}{k! \cdot 2^{\frac{n-k}{2}}}.$$

---

[7] For another proof to this problem, see Example 4.18.

[8] For a generalization of this problem, see Problem 8, page 58.

20. * (BMO—adapted)[9] Let $n > 2$ be an integer and $\mathcal{F}$ be a family of 3-element subsets of I_n, such that, if $A, B \in \mathcal{F}$ and $A \neq B$, then $|A \cap B| \leq 1$. Prove that

$$|\mathcal{F}| \leq \frac{1}{6}(n^2 - n).$$

21. (Romania) Let $k, m \in \mathbb{N}$ be given, with $m \geq k$, and let A be a set of m elements. Choose pairwise distinct subsets $A_1, A_2, \ldots, A_n$ of A, such that $|A_i| \geq k$ for $1 \leq i \leq n$ and $|A_i \cap A_j| \leq k - 1$ for $1 \leq i < j \leq n$. Prove that $n \leq \binom{m}{k}$.

22. Let $A = \{a_1, a_2, \ldots, a_n\}$, and let $P_1, P_2, \ldots, P_n$ be pairwise distinct 2-element subsets of A, such that

$$P_i \cap P_j \neq \emptyset \Leftrightarrow \exists\, 1 \leq k \leq n;\ P_k = \{a_i, a_j\}.$$

Prove that each element of A belongs to exactly two of the sets $P_1, P_2, \ldots, P_n$.

---

[9]For the other half of the original problem, see Example 2.29.

Chapter 2
More Counting Techniques

In this chapter we study a few more elaborate counting techniques. We start by discussing the *inclusion-exclusion principle*, which, roughly speaking, is a formula for counting the number of elements of a finite union of finite sets. The presentation continues with the notion of *double counting* for, counting a certain number of configurations in two distinct ways, to infer some hidden result. Then, a brief discussion of equivalence relations and their role in counting problems follows. Among other interesting results, we illustrate it by proving a famous theorem of B. Bollobás, on extremal set theory. The chapter ends with a glimpse on the use of the language of *metric spaces* in certain specific counting problems.

2.1 The Inclusion-Exclusion Principle

In this section we extend the formula of Proposition 1.5 for the number of elements of the union of two finite sets. The following result is known as the **inclusion-exclusion principle**, and is usually attributed to the French mathematician of the eighteenth century Abraham de Moivre.

Theorem 2.1 (de Moivre) *If A_1, A_2, ..., A_n are finite sets, then*

$$|A_1 \cup A_2 \cup \ldots \cup A_n| = \sum_{k=1}^{n}(-1)^{k-1} \sum_{i_1 < \cdots < i_k} |A_{i_1} \cap \ldots \cap A_{i_k}|, \qquad (2.1)$$

with the last sum above extending over all sequences $(i_1, i_2, \ldots, i_k)$ of integers such that $1 \le i_1 < i_2 < \cdots < i_k \le n$.

Proof Fix an $x \in A_1 \cup A_2 \cup \ldots \cup A_n$ and suppose that x belongs to exactly l of the sets A_1, A_2, ..., A_n. We shall show that the expression at the right hand side of (2.1) counts x exactly once.

© Springer International Publishing AG, part of Springer Nature 2018
A. Caminha Muniz Neto, *An Excursion through Elementary Mathematics, Volume III*,
Problem Books in Mathematics, https://doi.org/10.1007/978-3-319-77977-5_2

Note that, for a fixed $1 \le k \le n$, x is counted in the sum

$$\sum_{i_1 < \cdots < i_k} |A_{i_1} \cap \ldots \cap A_{i_k}|$$

as many times as the number of sequences $(i_1, \ldots, i_k)$ such that $1 \le i_1 < \cdots < i_k \le n$ and $x \in A_{i_1} \cap \ldots \cap A_{i_k}$.

Since x belongs to exactly l of the given sets, we can restrict ourselves to the case in which $k \le l$; indeed, if $k > l$, then x will not belong to at least one of the sets $A_{i_1}, \ldots, A_{i_k}$, so that $x \notin A_{i_1} \cap \ldots \cap A_{i_k}$.

For $k \le l$, the number of sequences $(i_1, \ldots, i_k)$ as above equals the number of ways of choosing k of the l sets containing x. Therefore, there are exactly $\binom{l}{k}$ such sequences, and we conclude that x is counted at the right hand side of (2.1) exactly

$$\sum_{k=1}^{l} (-1)^{k-1} \binom{l}{k}$$

times. However, with the aid of the binomial expansion we get

$$\sum_{k=1}^{l} (-1)^{k-1} \binom{l}{k} = \binom{l}{0} - \sum_{k=0}^{l} (-1)^{k} \binom{l}{k} = 1 - (1-1)^l = 1,$$

as wished. □

Example 2.2 Let $n \ge 3$ be a natural number. Show that there does not exist sets A_1, $A_2, \ldots, A_n$, with n elements each and satisfying the following conditions:

(a) Any two of the A_i's have exactly two elements in common.
(b) Any three of the A_i's have no elements in common.

Solution By the sake of contradiction, assume that there does exist sets $A_1, A_2, \ldots, A_n$ satisfying the stated conditions. If we show that $|A_1 \cup A_2 \cup \ldots \cup A_n| = n$, we would have $A_i \subset A_1 \cup A_2 \cup \ldots \cup A_n$ and $|A_i| = |A_1 \cup A_2 \cup \ldots \cup A_n|$ for $1 \le i \le n$, so that $A_i = A_1 \cup A_2 \cup \ldots \cup A_n$ for $1 \le i \le n$. In particular, this would give $A_1 = A_2 = \cdots = A_n$, and item (b) will give us a contradiction.

To what is left, let's compute the number of elements of $A_1 \cup A_2 \cup \ldots \cup A_n$ with the aid of the inclusion-exclusion principle:

$$|A_1 \cup \ldots \cup A_n| = \sum_{k=1}^{n} (-1)^{k-1} \sum_{i_1 < \cdots < i_k} |A_{i_1} \cap \ldots \cap A_{i_k}|$$

$$= \sum_{i_1} |A_{i_1}| - \sum_{i_1 < i_2} |A_{i_1} \cap A_{i_2}|,$$

where, in the last equality above, we used the fact that $A_{i_1} \cap A_{i_2} \cap A_{i_3} = \emptyset$ for $1 \leq i_1 < i_2 < i_3 \leq n$. Now, note that

$$\sum_{i_1}^{n} |A_{i_1}| = \sum_{i=1}^{n} |A_i| = \sum_{i=1}^{n} n = n^2$$

and

$$\sum_{i_1 < i_2} |A_{i_1} \cap A_{i_2}| = \sum_{i_1 < i_2} 2 = 2 \binom{n}{2} = n(n-1),$$

for, there are exactly $\binom{n}{2} = \frac{n(n-1)}{2}$ possible choices of indices i_1 and i_2 satisfying $1 \leq i_1 < i_2 \leq n$. Hence, $A_1 \cup A_2 \cup \ldots \cup A_n$ has exactly $n^2 - n(n-1) = n$ elements. $\qquad\square$

Another proof of the inclusion-exclusion principle can be given by using characteristic functions of sets (cf. Definition 1.12). To this end, we shall need two preliminary results on such functions.

Lemma 2.3 *Let A be a nonempty set. If B and C are subsets of A, then:*

(a) $\chi_B = \chi_C \Leftrightarrow B = C$.
(b) $\chi_{B \cap C} = \chi_B \chi_C$.
(c) $\chi_{B^c} = 1 - \chi_B$.
(d) $\chi_{B \cup C} = \chi_B + \chi_C - \chi_B \chi_C$.

Proof Exercise (see Problem 1). $\qquad\square$

Proposition 2.4 *Let A be a finite set and $A_1, A_2, \ldots, A_n$ be subsets of A. If $\chi, \chi_j : A \rightarrow \{0, 1\}$ respectively denote the characteristic functions of $A_1 \cup \ldots \cup A_n$ and A_j in A, then*

$$\chi(x) = 1 - \prod_{j=1}^{n} (1 - \chi_j(x)) \tag{2.2}$$

for every $x \in A$.

Proof For $x \in A$, we have

$$\chi(x) = 1 \Leftrightarrow x \in A_1 \cup \ldots \cup A_n$$

$$\Leftrightarrow \exists\, 1 \leq j \leq n;\ x \in A_j$$

$$\Leftrightarrow \exists\, 1 \leq j \leq n;\ \chi_j(x) = 1$$

$$\Leftrightarrow \prod_{j=1}^{n} (1 - \chi_j(x)) = 0$$

$$\Leftrightarrow 1 - \prod_{j=1}^{n} (1 - \chi_j(x)) = 1.$$

Therefore, both sides of (2.2) coincide at every $x \in A$, as wished. $\qquad\square$

Now, observe that if $B \subset A$ is finite, then

$$|B| = \sum_{x \in A} \chi_B(x). \tag{2.3}$$

On the other hand, in the notations of the former proposition, we have

$$\chi(x) = 1 - \prod_{j=1}^{n}(1 - \chi_j(x))$$

$$= 1 - \left(1 + \sum_{1 \leq k \leq n}\sum_{i_1 < \cdots < i_k}(-1)^k \chi_{i_1}(x)\ldots\chi_{i_k}(x)\right)$$

$$= \sum_{1 \leq k \leq n}\sum_{i_1 < \cdots < i_k}(-1)^{k-1}\chi_{i_1\ldots i_k}(x),$$

where $\chi_{i_1\ldots i_k}$ denotes the characteristic function of $A_{i_1} \cap \ldots \cap A_{i_k}$ and we used item (b) of Lemma 2.3 in the last equality. Hence, the above and (2.3) give

$$|A_1 \cup \ldots \cup A_n| = \sum_{x \in A}\chi(x)$$

$$= \sum_{x \in A}\sum_{1 \leq k \leq n}\sum_{i_1 < \cdots < i_k}(-1)^{k-1}\chi_{i_1\ldots i_k}(x)$$

$$= \sum_{1 \leq k \leq n}\sum_{i_1 < \cdots < i_k}(-1)^{k-1}\sum_{x \in A}\chi_{i_1\ldots i_k}(x)$$

$$= \sum_{1 \leq k \leq n}\sum_{i_1 < \cdots < i_k}(-1)^{k-1}|A_{i_1} \cap \ldots \cap A_{i_k}|.$$

We now present two classical applications of the inclusion-exclusion principle. For the first one, we say that a permutation $(a_1, a_2, \ldots, a_n)$ of I_n is a **derangement** if $a_i \neq i$ for $1 \leq i \leq n$. This is the same as saying that the function $f : I_n \to I_n$, such that $f(i) = a_i$ for $1 \leq i \leq n$, has no fixed points. The coming example computes how many are these permutations of I_n (in this respect, see also Problem 6, page 79).

Example 2.5 Given an integer $n > 1$, there are precisely

$$d_n = n!\sum_{j=0}^{n}\frac{(-1)^j}{j!} \tag{2.4}$$

derangements of I_n.

Proof Let $\mathcal{D}_n$ be the set of derangements of I_n and, for $1 \leq i \leq n$, let $\mathcal{C}_i$ be the set of those permutations $(a_1, \ldots, a_n)$ of I_n for which $a_i = i$. Letting $\mathcal{P}_n$ denote the set of all permutations of I_n, we clearly have

$$\mathcal{P}_n \setminus \mathcal{D}_n = \mathcal{C}_1 \cup \mathcal{C}_2 \cup \ldots \cup \mathcal{C}_n.$$

Hence, the inclusion-exclusion principle gives

$$|\mathcal{P}_n \setminus \mathcal{D}_n| = |\mathcal{C}_1 \cup \ldots \cup \mathcal{C}_n|$$

$$= \sum_{k=1}^{n} (-1)^{k-1} \sum_{i_1 < \cdots < i_k} |\mathcal{C}_{i_1} \cap \ldots \cap \mathcal{C}_{i_k}|. \tag{2.5}$$

Since $d_n = |\mathcal{D}_n|$, we have

$$|\mathcal{P}_n \setminus \mathcal{D}_n| = |\mathcal{P}_n| - |\mathcal{D}_n| = n! - d_n.$$

On the other hand, since

$$(a_1, \ldots, a_n) \in \mathcal{C}_{i_1} \cap \ldots \cap \mathcal{C}_{i_k} \Leftrightarrow a_{i_1} = i_1, \ldots, a_{i_k} = i_k,$$

in order to count how many permutations $(a_1, \ldots, a_n)$ of I_n belong to $\mathcal{C}_{i_1} \cap \ldots \cap \mathcal{C}_{i_k}$, it suffices to count how many are the permutations of $I_n \setminus \{i_1, \ldots, i_k\}$. Since this is a set of $n - k$ elements, it follows that

$$|\mathcal{C}_{i_1} \cap \ldots \cap \mathcal{C}_{i_k}| = |\mathcal{P}_{n-k}| = (n - k)!.$$

Substituting both these equalities in (2.5), we get

$$n! - d_n = \sum_{k=1}^{n} (-1)^{k-1} \sum_{i_1 < \cdots < i_k} (n - k)!$$

$$= \sum_{k=1}^{n} (-1)^{k-1} \binom{n}{k} (n - k)!$$

$$= n! \sum_{k=1}^{n} \frac{(-1)^{k-1}}{k!},$$

where, in the next to last equality, we used the fact that there are exactly $\binom{n}{k}$ ways of choosing integers $1 \leq i_1 < \cdots < i_k \leq n$. Therefore,

$$d_n = n! - n! \sum_{k=1}^{n} \frac{(-1)^{k-1}}{k!} = n! \sum_{k=0}^{n} \frac{(-1)^k}{k!}.$$

$\square$

Example 2.6 We are given $m, n \in \mathbb{N}$ and finite sets A and B, such that $|A| = m$ and $|B| = n$. Prove that the number of surjective functions $f : A \to B$ equals

$$\sum_{k=0}^{n}(-1)^k \binom{n}{k}(n - k)^m.$$

Proof Let $\mathcal{F}_{m,n}$ denote the set of functions $f : A \to B$, and $\mathcal{S}_{m,n}$ the subset of $\mathcal{F}_{m,n}$ consisting of the surjective ones. Also, let $B = \{b_1, \ldots, b_n\}$ and, for $1 \leq i \leq n$, $\mathcal{F}_i$ be the set of functions $f : A \to B$ such that $b_i \notin \mathrm{Im}\,(f)$. Then,

$$\mathcal{F}_{m,n} \setminus \mathcal{S}_{m,n} = \mathcal{F}_1 \cup \ldots \cup \mathcal{F}_n,$$

and the inclusion-exclusion principle gives

$$|\mathcal{F}_{m,n} \setminus \mathcal{S}_{m,n}| = |\mathcal{F}_1 \cup \ldots \cup \mathcal{F}_n|$$

$$= \sum_{k=1}^{n}(-1)^{k-1} \sum_{i_1 < \cdots < i_k} |\mathcal{F}_{i_1} \cap \ldots \cap \mathcal{F}_{i_k}|. \tag{2.6}$$

Letting $s_{m,n} = |\mathcal{S}_{m,n}|$, we have

$$|\mathcal{F}_{m,n} \setminus \mathcal{S}_{m,n}| = |\mathcal{F}_{m,n}| - |\mathcal{S}_{m,n}| = n^m - s_{m,n},$$

where in the last equality we used the discussion in the last paragraph of Sect. 1.1. On the other hand, for a given function $f : A \to B$, we obviously have

$$f \in \mathcal{F}_{i_1} \cap \ldots \cap \mathcal{F}_{i_k} \Leftrightarrow b_{i_1}, \ldots, b_{i_k} \notin \mathrm{Im}\,(f);$$

in turn, this is equivalent to the fact that f can be seen as a function from A into $B \setminus \{b_{i_1}, \ldots, b_{i_k}\}$. However, since $|B \setminus \{b_{i_1}, \ldots, b_{i_k}\}| = n - k$ (and invoking once more the discussion in the last paragraph of Sect. 1.1), we get

$$|\mathcal{F}_{i_1} \cap \ldots \cap \mathcal{F}_{i_k}| = |\mathcal{F}_{m,n-k}| = (n - k)^m.$$

Substituting these relations in (2.6), we obtain

$$n^m - s_{m,n} = \sum_{k=1}^{n}(-1)^{k-1} \sum_{i_1 < \cdots < i_k} (n - k)^m$$

$$= \sum_{k=1}^{n}(-1)^{k-1} \binom{n}{k}(n - k)^m,$$

where, in the last equality, we used the fact that there are $\binom{n}{k}$ ways of choosing integers $1 \leq i_1 < \cdots < i_k \leq n$. Hence,

$$s_{m,n} = n^m - \sum_{k=1}^{n}(-1)^{k-1}\binom{n}{k}(n-k)^m = \sum_{k=0}^{n}(-1)^k\binom{n}{k}(n-k)^m.$$

$\square$

We shall reobtain the results of the two examples above, by other means, in Chap. 3 (cf. Problem 7, page 91, and Problem 9, page 92).

Example 2.7 (Romania) Compute the number of ways of coloring the vertices of a regular dodecagon using two colors, in such a way that no monochromatic set of vertices is the set of vertices of a regular polygon.

Solution A regular polygon having its vertices among those of the dodecagon is either an equilateral triangle, a square, a regular hexagon or the regular dodecagon itself. Since alternating vertices of a regular hexagon form an equilateral triangle, we conclude that it suffices to avoid equilateral triangles and squares. Let red (R) and blue (B) be the used colors.

There are four equilateral triangles whose vertices are among those of the given regular dodecagon, and such equilateral triangles have pairwise disjoint sets of vertices. Since each one of them can be colored in exactly six distinct and non monochromatic ways, the number of distinct colorings of the vertices of the dodecagon without monochromatic equilateral triangles equals $6^4 = 1296$.

Let's calculate how many of these 1296 colorings contain at least one monochromatic square. To this end assume, without loss of generality, that such a coloring contains a red square. Each one of its vertices is a vertex of exactly one of the four possible equilateral triangles. Since such a vertex is red, the other two vertices of the corresponding equilateral triangle could be colored BB, BR or RB. By the fundamental principle of counting, exactly $2 \cdot 3 \cdot 3^4 = 486$ (2 colors, 3 squares and 3 possible colorings for the other two vertices of each one of the four equilateral triangles) of the 1296 colorings without monochromatic equilateral triangles contain a monochromatic square.

Let's now compute how many of the 1296 colorings without monochromatic equilateral triangles contain at least two monochromatic squares. We do this considering two separate cases, according to whether the two monochromatic squares have equal or different colors. For each equilateral triangle, exactly one of its vertices is not a vertex of any of the two monochromatic squares. Assuming (without loss of generality) them to be red, we conclude that there is exactly one possible color (blue) for the third vertex. Assuming them to be one red and the other one blue, we conclude that there are two possible colors (either red or blue) for the third vertex. Since there are $\binom{3}{2} = 3$ ways of choosing two of the three squares, the number of colorings with at least two monochromatic squares equals

$$\binom{3}{2} \cdot (2 \cdot 1^4 + 2 \cdot 2^4) = 102$$

(note that if both squares have the same color, then there are two possibilities: they are both blue or both red; if they have different colors, there are also two possibilities: BR ou RB).

Finally, let's calculate how many of the 1296 colorings without monochromatic equilateral triangles have three monochromatic squares. The nonexistence of monochromatic equilateral triangles forbids the three squares to have a single color. Hence, there are six possible distinct colorings for their sets of vertices: BBR, BRB, RBB, RRB, RBR and BRR.

Therefore, by the inclusion-exclusion principle, the number of colorings we wish to count is

$$1296 - 486 + 102 - 6 = 906.$$

□

Problems: Sect. 2.1

1. * Prove Lemma 2.3.
2. How many natural numbers from 1 to 1000 have a prime factor less than 10?
3. * Given $m, n \in \mathbb{N}$, with $m < n$, prove that

$$\sum_{k=0}^{n-1} (-1)^k \binom{n}{k} (n-k)^m = 0.$$

4. Given $n \in \mathbb{N}$, compute the number of ways of forming a line with n couples, in such a way that no wife is a neighbor of her husband and vice-versa.
5. (England) How many are the permutations $(a_1, a_2, \ldots, a_6)$ of I_6 such that, for $1 \le j \le 5$, the sequence $(a_1, \ldots, a_j)$ is not a permutation of I_j? Explain your answer.

   The next two problems admit from the reader a certain degree of acquaintance with the decomposition of a natural number $n > 1$ as a product of powers of distinct primes, as well as with the notion of greatest common divisor of two nonzero integers. Specifically for Problem 7, the reader may find it useful to recall the following fact: if $n > 1$ is natural but not prime, then n admits a prime divisor which is less than or equal to $\sqrt{n}$. The necessary material is collected in Chap. 6. Alternatively, the reader can safely skip these problems without loss of continuity.

6. Let $n > 1$ be a natural number and $n = p_1^{\alpha_1} \ldots p_k^{\alpha_k}$ its factorisation as a product of powers of distinct primes $p_1 < \cdots < p_k$. Prove that there are exactly

$$n\left(1 - \frac{1}{p_1}\right)\cdots\left(1 - \frac{1}{p_k}\right) \tag{2.7}$$

integers $1 \leq m \leq n$ such that $\gcd(m, n) = 1$.

For the next problem, we let $\lfloor x \rfloor$ stand for the **integer part** of $x \in \mathbb{R}$, i.e., the greatest integer which is less than or equal to x. Thus, we have (for instance) $\lfloor 1 \rfloor = 1$, $\lfloor \sqrt{2} \rfloor = 1$ and $\lfloor \pi \rfloor = 3$.

7. Let $n > 1$ be natural and $p_1 < p_2 < \cdots < p_k$ be the prime numbers less than or equal to $\sqrt{n}$. Prove that the quantity of prime numbers less than or equal to n is given by the formula

$$n - \sum_{j=1}^{k}(-1)^{j-1} \sum_{1 \leq i_1 < \cdots < i_j \leq k} \left\lfloor \frac{n}{p_{i_1}\cdots p_{i_j}} \right\rfloor.$$

8. For $n \in \mathbb{N}$, we let $p(n)$ denote the number of distinct partitions of n in naturals summands. Also, for $1 \leq k \leq n$ we write $p_k(n)$ to denote the number of partitions of n into exactly k distinct natural summands. Prove that

$$p(n) = \sum_{1 \leq k < \sqrt{2n}}(-1)^{k-1} \sum_{\frac{k^2}{2} < l < n} p_k(l)p(n - l).$$

9. (Mongolia) Given $k, m, n \in \mathbb{N}$, let $p = \min\left\{n, \frac{m}{k+1}\right\}$. Show that equation $x_1 + x_2 + \cdots + x_n = m$ has exactly

$$\sum_{j=0}^{p}(-1)^j\binom{n}{j}\binom{m - j(k + 1) + n - 1}{n - 1}$$

solutions $(x_1, x_2, \ldots, x_n)$ with $x_t \in \mathbb{Z}$ and $0 \leq x_t \leq k$, for $1 \leq t \leq n$.

10. **Lucas' problem**[1] aims at counting the number of ways in which n couples can sit in $2n$ chairs around a circle, such that no wife sits beside her husband and no two people of the same sex sit on adjacent chairs. The purpose of this problem is to present I. Kaplansky's solution to this problem. To this end, do the following items:

   (a) Conclude that there are $2(n!)$ ways of choosing the chairs of the husbands. Once we are done with (a), conclude that the number of desired configurations is equal to $2(n!)a_n$, where a_n counts the number of ways of distributing the wives in the n left chairs, in such a way that no wife sits beside her husband.

---

[1] After Édouard Lucas, French mathematician of the nineteenth century.

(b) Label the chairs of the husbands, counterclockwise, as $h_1, h_2, \ldots, h_n$, and those to be occupied by the wives (in some allowed order) as $1, 2, \ldots, n$, with chair 1 situated between h_1 and h_2, chair 2 situated between h_2 and h_3, and so on. For $1 \le i \le n$, let w_i denote the wife of h_i, and A_{2i-1} (resp. A_{2i}) denote the set of ways of distributing the wives such that w_i sits in chair $i - 1$ (resp. chair i. Here, chair 0 is the same as chair n). Conclude that

$$a_n = n! - |A_1 \cup A_2 \cup A_3 \cup A_4 \cup \ldots \cup A_{2n-1} \cup A_{2n}|.$$

(c) If $\{i_1, i_2, \ldots, i_k\} \subset \{1, 2, \ldots, 2n\}$ has two consecutive elements (with $2n$ and 1 seen as consecutive), show that $A_{i_1} \cap A_{i_2} \cap \ldots \cap A_{i_k} = \emptyset$.

(d) If $\{i_1 < i_2 < \ldots < i_k\} \subset \{1, 2, \ldots, 2n\}$ has no consecutive elements (also with $2n$ and 1 seen as consecutive), show that

$$|A_{i_1} \cap A_{i_2} \cap \ldots \cap A_{i_k}| = (n - k)!.$$

(e) Apply Kaplansky's second lemma (cf. Problem 12, page 29) to conclude that there are exactly

$$\frac{2n}{2n - k} \binom{2n - k}{k}$$

ways of composing an intersection of the form $A_{i_1} \cap A_{i_2} \cap \ldots \cap A_{i_k}$, with $\{i_1 < i_2 < \ldots < i_k\} \subset \{1, 2, \ldots, 2n\}$ having no consecutive elements (and $2n$ and 1 seen as consecutive).

(f) Conclude that

$$a_n = \sum_{k=0}^{n} (-1)^k \frac{2n}{2n - k} \binom{2n - k}{k} (n - k)!.$$

2.2 Double Counting

Double counting is nothing but the bijective principle in action once again, albeit under a different perspective. The central idea is that, if we count a certain number of configurations in two different ways and without making mistakes, then the two results thus obtained must coincide. Sometimes this naive approach generates interesting conclusions.

The best way to master double counting is by looking at some examples, and we start with one that allows us to reobtain an important result, already derived by other means.

Example 2.8 If n and k are integers such that $0 \leq k \leq n$, then every set with n elements has exactly $\binom{n}{k}$ subsets with k elements.

Proof Note first that it suffices to consider the case $1 \leq k \leq n$.

Under this assumption, fix a set A with n elements and consider the problem of counting how many sequences of k terms, chosen from the elements of A, can be formed. By Corollary 1.22, the answer is $n(n-1)\ldots(n-k+1)$.

On the other hand, by permuting the elements of each k-element subset of A we generate $k!$ different sequences of k terms; moreover, it is immediate to see that the sequences we get this way are all sequences of k terms chosen from A, and they are all distinct. Hence, if (as in the previous chapter) C_k^n denotes the number of k-element subsets of A, we conclude that there are exactly $k!C_k^n$ sequences of k terms, all chosen from the elements of A.

Now, since $n(n-1)\ldots(n-k+1)$ and $k!C_k^n$ are answers to the same counting problem, we conclude that

$$k!C_k^n = n(n-1)\ldots(n-k+1)$$

or, which is the same,

$$
\begin{aligned}
C_k^n &= \frac{n(n-1)\ldots(n-k+1)}{k!} \\
&= \frac{n(n-1)\ldots(n-k+1)\cdot(n-k)!}{k!(n-k)!} \\
&= \binom{n}{k}.
\end{aligned}
$$

$\square$

The result of the next example is known as **Lagrange's identity**, after the French mathematician of the eighteenth century Joseph L. Lagrange.

Example 2.9 Prove that, for every natural number n, one has

$$\binom{2n}{n} = \binom{n}{0}^2 + \binom{n}{1}^2 + \binom{n}{2}^2 + \cdots + \binom{n}{n}^2.$$

Proof Consider a group of $2n$ people, composed by n men and n women. We shall compute the number of ways of choosing exactly n of the $2n$ people, by using two different approaches.

On the one hand, the answer obviously is $\binom{2n}{n}$. On the other, for $0 \leq k \leq n$, the fundamental principle of counting guarantees that we can choose k men and $n-k$ women in exactly $\binom{n}{k}\binom{n}{n-k}$ different ways; adding such possibilities as k varies from 0 to n and using the fact that $\binom{n}{k} = \binom{n}{n-k}$, we get the sum

$$\sum_{k=0}^{n} \binom{n}{k}\binom{n}{n-k} = \sum_{k=0}^{n} \binom{n}{k}^2$$

as another answer to the posed problem. Hence, the stated equality actually take place.   □

Yet concerning the previous example, a nontrivial difficulty in the implementation of the double counting argument relied in looking at both sides of the desired equality as different ways of counting the number of configurations relative to a single combinatorial situation. There is no general rule on how to find out such a counting problem; just experience and careful thinking will help. Therefore, let's go on and discuss another (and tougher) example.

Example 2.10 Let m and n be positive integers. Prove that

$$\sum_{j} \binom{m}{j}\binom{n}{j}2^j = \sum_{k} \binom{m}{k}\binom{n+k}{m},$$

with the above sums ranging over all possible integer values of the indices j and k.

Proof Let A and B be disjoint sets, with $|A| = m$ and $|B| = n$. Let's calculate the number of ordered pairs (X, Y), with $X \subset A$ and $Y \subset B \cup X$ such that $|Y| = m$.

If $|X| = k$, there are $\binom{m}{k}$ ways of choosing X; for each of them, the disjointness of A and B gives $|B \cup X| = n + k$, so that there are $\binom{n+k}{m}$ ways of choosing Y. Therefore, by successively applying the multiplicative and additive principles, we conclude that the number of such ordered pairs is

$$\sum_{k} \binom{m}{k}\binom{n+k}{m}.$$

We now choose Y first. More precisely, we compute how many ordered pairs (X, Y) satisfy the given conditions and are such that $|Y \cap B| = j$. There are $\binom{n}{j}$ ways of choosing j elements from B to put in Y; once they are chosen, we are left to choosing $m - j$ elements for Y, which ought to be elements of A. Then, there are $\binom{m}{m-j} = \binom{m}{j}$ ways of choosing these $m - j$ elements of A, and the condition $Y \subset B \cup X$ forces such elements to belong to X. Apart from these $m - j$ elements, the remaining elements of X can form an arbitrary subset of the set formed by the remaining j elements of A, so that we can choose such a subset in exactly 2^j different ways. Hence, the number of pairs (X, Y) is given by

$$\sum_{j} \binom{n}{j}\binom{m}{j}2^j.$$

Finally, since both of the above countings solve the same problem, the only possibility is to have the stated equality. □

The coming example illustrates how one can use double counting arguments to get inequalities.

Example 2.11 (IMO) Let n and k be positive integers and S be a set of n points in the plane satisfying the following conditions:

(a) no three points of S are collinear.
(b) For every point $P \in S$, there are at least other k points of S equidistant from P.

Prove that $k < \frac{1}{2} + \sqrt{2n}$.

Proof Let

$$\mathcal{F} = \{(A, B, C); \ A, B, C \in S \text{ and } \overline{AB} = \overline{AC}\}.$$

We shall count $|\mathcal{F}|$ in two different ways.

Firstly, for each $A \in S$ there are at least k points of S equidistant from A and such points generate at least $k(k-1)$ pairs (B, C) such that $(A, B, C) \in \mathcal{F}$. Since there are n possible choices for $A \in S$, we conclude that

$$|\mathcal{F}| \geq nk(k-1).$$

Now, for fixed $B, C \in S$, we claim that there are at most two points A in S such that $(A, B, C) \in \mathcal{F}$. Indeed, since any such A belongs to the perpendicular bisector of BC, and condition (a) gives at most two of them. However, since there are $n(n-1)$ possible choices for (B, C), we conclude that

$$|\mathcal{F}| \leq 2n(n-1).$$

Thus,

$$nk(k-1) \leq 2n(n-1)$$

and, hence,

$$k \leq \frac{1 + \sqrt{8n-7}}{2} < \frac{1 + \sqrt{8n}}{2} = \frac{1}{2} + \sqrt{2n}.$$

□

We finish this section by showing how to use a double counting argument to obtain formula (2.4) for the number of derangements of I_n. Prior to that, however, we need to establish the following auxiliary result, which will also be useful in Sect. 18.1.

Lemma 2.12 *If* $a_0, a_1, \ldots, a_n$ *and* $b_0, b_1, \ldots, b_n$ *are real numbers such that* $a_k = \sum_{j=0}^{k} \binom{k}{j} b_j$ *for* $0 \leq k \leq n$, *then*

$$b_k = \sum_{j=0}^{k} (-1)^{k+j} \binom{k}{j} a_j \qquad (2.8)$$

for $0 \leq k \leq n$.

Proof Let's make induction on $0 \leq k \leq n$, noting that the result is obvious if $k = 0$. As induction hypothesis, assume (2.8) to be true when $k = l - 1$, for some $1 \leq l \leq n$. For $k = l$, we get

$$a_l = \sum_{j=0}^{l} \binom{l}{j} b_j = \sum_{j=0}^{l-1} \binom{l}{j} b_j + b_l,$$

so that, by applying the induction hypothesis and the result of Problem 1, we obtain

$$b_l = a_l - \sum_{j=0}^{l-1} \binom{l}{j} b_j$$

$$= a_l - \sum_{j=0}^{l-1} \binom{l}{j} \sum_{i=0}^{j} (-1)^{j+i} \binom{j}{i} a_i$$

$$= a_l - \sum_{j=0}^{l-1} \sum_{i=0}^{j} (-1)^{j+i} \binom{l}{j} \binom{j}{i} a_i$$

$$= a_l - \sum_{i=0}^{l-1} \sum_{j=i}^{l-1} (-1)^{j+i} \binom{l}{j} \binom{j}{i} a_i.$$

If we now observe that

$$\binom{l}{j} \binom{j}{i} = \binom{l}{i} \binom{l-i}{l-j},$$

the above computations give

$$b_l = a_l - \sum_{i=0}^{l-1} \sum_{j=i}^{l-1} (-1)^{j+i} \binom{l}{i} \binom{l-i}{l-j} a_i$$

$$= a_l - \sum_{i=0}^{l-1} (-1)^i \binom{l}{i} a_i \sum_{j=i}^{l-1} (-1)^j \binom{l-i}{l-j}$$

$$= a_l - \sum_{i=0}^{l-1} (-1)^i \binom{l}{i} a_i \sum_{t=1}^{l-i} (-1)^{l-t} \binom{l-i}{t}.$$

Finally, since

$$\sum_{t=1}^{l-i}(-1)^t\binom{l-i}{t} = \sum_{t=0}^{l-i}(-1)^t\binom{l-i}{t} - 1 = (1-1)^{l-i} - 1 = -1,$$

we get

$$b_l = a_l + \sum_{i=0}^{l-1}(-1)^{l+i}\binom{l}{i}a_i = \sum_{i=0}^{l}(-1)^{l+i}\binom{l}{i}a_i.$$

$\square$

We present the promised example next.

Example 2.13 If d_n stands for the number of derangements of I_n, then

$$k! = \sum_{j=0}^{k}\binom{k}{j}d_j. \tag{2.9}$$

In particular, $d_k = k!\sum_{j=0}^{k}\frac{(-1)^j}{j!}$.

Proof As in Example 2.5, let $\mathcal{P}_k$ denote the set of all permutations of I_k and let's count $|\mathcal{P}_k|$ in two different ways.

Firstly, Corollary 1.25 gives $|\mathcal{P}_k| = k!$.

Secondly, for $0 \le j \le k$, let $\mathcal{F}_j$ denote the subset of $\mathcal{P}_k$ formed by those permutations of I_k having exactly j fixed points. Then,

$$\mathcal{P}_k = \mathcal{F}_0 \cup \mathcal{F}_1 \cup \ldots \cup F_k;$$

however, since the union at the right hand side above is disjoint, we obtain

$$|\mathcal{P}_k| = |\mathcal{F}_0| + |\mathcal{F}_1| + \cdots + |F_k|. \tag{2.10}$$

Now, for $f \in \mathcal{F}_j$, we can choose its j fixed points in exactly $\binom{k}{j}$ ways; on the other hand, the restriction of f to the subset of I_k formed by the other $k - j$ points clearly forms a derangement of I_{k-j}. Hence, the fundamental principle of counting gives

$$|\mathcal{F}_j| = \binom{k}{j}d_{k-j} = \binom{k}{k-j}d_{k-j}.$$

It then follows from (2.10) that

$$|\mathcal{P}_k| = \sum_{j=0}^{k} |\mathcal{F}_j| = \sum_{j=0}^{k} \binom{k}{k-j} d_{k-j} = \sum_{j=0}^{k} \binom{k}{j} d_j,$$

and it suffices to compare both expressions above for $|\mathcal{P}_k|$ to get (2.9).

For what is left, we apply the previous lemma to (2.9) (with $a_k = k!$ and $b_k = d_k$) to get

$$d_k = \sum_{j=0}^{k} (-1)^{k+j} \binom{k}{j} j! = \sum_{j=0}^{k} (-1)^{k+j} \frac{k!}{(k-j)!}$$

$$= k! \sum_{j=0}^{k} \frac{(-1)^{k-j}}{(k-j)!} = k! \sum_{j=0}^{k} \frac{(-1)^j}{j!}.$$

□

Problems: Sect. 2.2

1. * We are given $n \in \mathbb{N}$ and, for $0 \le i, j \le n$, a real number a_{ij}. Show that

$$\sum_{j=0}^{n} \sum_{i=0}^{j} a_{ij} = \sum_{i=0}^{n} \sum_{j=i}^{n} a_{ij}.$$

2. Use double counting to prove that, for each natural number n, one has

$$\binom{n}{0} + \binom{n}{1} + \binom{n}{2} + \cdots + \binom{n}{n} = 2^n.$$

3. Prove **Vandermonde's identity**[2]: given $m, n, p \in \mathbb{N}$, one has

$$\sum_{k} \binom{m}{k} \binom{n}{p-k} = \binom{m+n}{p},$$

with the sum at the left hand side above ranging through all possible nonnegative integer values of k.

---

[2]After Alexandre-Théóphile Vandermonde, French mathematician of the eighteenth century. For another proof of Vandermonde's identity, see Problem 3, page 355.

4. For each integer $n > 1$, prove that the number of subsets of I_{2n} with equal quantities of odd and even elements is equal to $\binom{2n}{n}$.

5. A battalion commander asked volunteers to compose 11 patrols, all of which formed by the same number of men. If each man entered exactly two patrols and each two patrols had exactly one man in common, compute the numbers of volunteers and of members of each patrol.[3]

6. (IMO) Given $n, k \in \mathbb{N}$, let $p_n(k)$ denote the number of permutations of I_n having exactly k fixed points. Prove that $\sum_{k=0}^{n} k p_n(k) = n!$.

7. (TT) The set of natural numbers is partitioned into m infinite and nonconstant arithmetic progressions, of common ratios $d_1, d_2, \ldots, d_m$. Prove that

$$\frac{1}{d_1} + \frac{1}{d_2} + \cdots + \frac{1}{d_m} = 1.$$

The result of the coming problem is known as **Sperner's lemma**, after the German mathematician of the twentieth century Emanuel Sperner.

8. A triangle ABC is partitioned into a finite number of smaller triangles, in such a way that any two of them are either disjoint, have exactly a common vertex or exactly a common side. Such a partition generates a certain number of points lying in the interior or on the sides of ABC and which are vertices of smaller triangles. We call them *vertices of the partition*. Randomly label the points lying on side AB (and different from A or B) as "A" or "B"; accordingly, randomly label the points lying on side AC (and different from A or C) as "A" or "C", and those lying on side BC (also different from B or C) as "B" or "C". Then, label each of the remaining vertices of the partition (all lying in the interior of ABC) as A, B or C, also randomly. Prove that at least one of the smaller triangles will be labelled as ABC.

9. (Russia) A senate has 30 senators, any two of which are either friends or enemies. It is also known that each senator has exactly six enemies. If each three senators form a commission, compute the number of those commissions such that its members are either pairwise friends or pairwise enemies.

10. We are given a convex n-gon in the plane. Prove that it is possible to choose $n - 2$ points in the plane such that every triangle formed by three vertices of the given n-gon has exactly one of the chosen points in its interior.

2.3 Equivalence Relations and Counting

Let A be a fixed nonempty set, and recall that a **relation** in A is a subset of the cartesian product $A \times A$.

---

[3]For another approach to this problem, see Problem 21, page 134.

We say that a relation $\mathcal{R}$ in A is **reflexive** if $a\mathcal{R}a$ for every $a \in A$; **symmetric** if, whenever $a, b \in A$, we have that $a\mathcal{R}b \Rightarrow b\mathcal{R}a$; **transitive** if, whenever $a, b, c \in A$, we have that $a\mathcal{R}b$ and $b\mathcal{R}c$ imply $a\mathcal{R}c$. Finally, an **equivalence relation** in A is a relation $\mathcal{R}$ in A which is simultaneously reflexive, symmetric and transitive. In Mathematics, equivalence relations are customarily denoted by $\sim$, instead of $\mathcal{R}$ (i.e., by writing $a \sim b$ instead of $a\mathcal{R}b$); from now on, we shall stick to this usage without further comments.

Examples 2.14

(a) Let $\mathcal{F}$ be a partition of a nonempty set A. The relation $\sim$ in A, defined by

$$a \sim b \Leftrightarrow \exists\, B \in \mathcal{F};\ a, b \in B,$$

is of equivalence. We say that such an equivalence relation is the one **induced** by the partition $\mathcal{F}$.
(b) Let A and B be nonempty sets and $f : A \to B$ be a given function. Relation $\sim$ in A, defined by

$$a \sim b \Leftrightarrow f(a) = f(b)$$

is of equivalence. We say that such a relation is **induced** by f.

The following definition is central to all that follows.

Definition 2.15 Let A be a nonempty set and $\sim$ be an equivalence relation in A. For $a \in A$, the **equivalence class** of a is the subset $\overline{a}$ of A defined by

$$\overline{a} = \{x \in A;\ x \sim a\}. \tag{2.11}$$

If $\sim$ is an equivalence relation in a nonempty set A and $a \in A$, then the equivalence class of $\overline{a}$ is clearly nonempty, for $a \in \overline{a}$. Our first result analyses the equivalence classes of a generic equivalence relation.

Proposition 2.16 *Let $\sim$ be an equivalence relation in a nonempty set A. For $a, b \in A$, one has*

$$a \sim b \Leftrightarrow \overline{a} \cap \overline{b} \neq \emptyset \Leftrightarrow \overline{a} = \overline{b}.$$

In particular, the equivalence classes of $\sim$ form a partition of A.

Proof We shall show that $a \sim b \Rightarrow \overline{a} \cap \overline{b} \neq \emptyset \Rightarrow \overline{a} = \overline{b} \Rightarrow a \sim b$:

- If $a \sim b$, then $a \in \overline{b}$ and, hence, $a \in \overline{a} \cap \overline{b}$; therefore, $\overline{a} \cap \overline{b} \neq \emptyset$.
- If $c \in \overline{a} \cap \overline{b}$ (i.e., if $\overline{a} \cap \overline{b} \neq \emptyset$), then $c \sim a$ and $c \sim b$; hence, the symmetry of $\sim$ guarantees that $a \sim c$ and $c \sim b$ and, thus, its transitivity implies $a \sim b$. We shall now show that $\overline{a} \subset \overline{b}$ (the converse inclusion being totally analogous): if $x \in \overline{a}$, then $x \sim a$; however, since $a \sim b$, it follows once more from the transitivity of $\sim$ that $x \sim b$, i.e., $x \in \overline{b}$.

• If $\bar{a} = \bar{b}$, then $a \in \bar{a}$ implies $a \in \bar{b}$. But this is the same as saying that $a \sim b$.

□

If $\sim$ is an equivalence relation in a nonempty set A, it follows immediately from the previous proposition that the family $A/\sim$, defined by

$$A/\sim = \{\bar{a}; \ a \in A\} \tag{2.12}$$

(i.e., the family of the equivalence classes of the elements of A with respect to $\sim$), is a partition of A. Such a family is called the **quotient set** of A by $\sim$, and the function

$$\pi : A \longrightarrow A/\sim$$
$$a \longmapsto \bar{a} \tag{2.13}$$

is the **projection function** of A on $A/\sim$.

Given an equivalence relation $\sim$ in a nonempty set A, a subset B of A is said to be a **system of distinct representatives** (we abbreviate **SDR**) for $\sim$ if the following conditions are satisfied:

(a) For all $a, b \in B$, either $a = b$ or $a \not\sim b$.
(b) For every $a \in A$, there exists $b \in B$ such that $a \sim b$.

In words, a SDR for an equivalence relation $\sim$ in A is a subset of A formed by pairwise nonrelated elements such that every element of A is related to exactly one of them. Yet in another way, if B is a SDR for the equivalence relation $\sim$ in A, then

$$A/\sim = \{\bar{x}; \ x \in B\}. \tag{2.14}$$

The following proposition provides an effective way of constructing SDR for equivalence relations.

Proposition 2.17 *Let $\sim$ be an equivalence relation in a nonempty set A. If $f : A/\sim \to A$ is a function such that $f(\bar{a}) \in \bar{a}$ for every $a \in A$, then $Im(f)$ is a SDR for $\sim$.*

Proof Let $B = Im\, f$. If $a, b \in B$, say $a = f(\bar{\alpha})$ and $b = f(\bar{\beta})$, with $\alpha, \beta \in A$, then the hypothesis on f assures that $a \in \bar{\alpha}$ and $b \in \bar{\beta}$. If $a \sim b$, then, since $\alpha \sim a$ and $b \sim \beta$, we have by transitivity that $\alpha \sim \beta$; thus, $\bar{\alpha} = \bar{\beta}$ and, hence,

$$a = f(\bar{\alpha}) = f(\bar{\beta}) = b.$$

In order to verify the validity of the second condition in the definition of a SDR, take $a \in A$ and let $b = f(\bar{a})$. Then $b \in Im\,(f)$ and the definition of f gives $b \in \bar{a}$, i.e., $a \sim b$.   □

In the notations of the previous proposition, we say that f is a **choice function** for the equivalence relation $\sim$. In words, the role of f is to choose one element out of each equivalence class in $A/\sim$.

Concerning the role of equivalence relations in counting problems, we have finally arrived at the result we are interested in.

Proposition 2.18 *Let $\sim$ be an equivalence relation in a finite nonempty set A. If all of the equivalence classes of A with respect to $\sim$ have the same number k of elements, then there are exactly $|A|/k$ distinct equivalence classes.*

Proof If $\{a_1, \ldots, a_m\}$ is a SDR with respect to $\sim$, then $A = \bigcup_{j=1}^{m} \bar{a}_j$, a disjoint union. Hence,

$$|A| = \sum_{j=1}^{m} |\bar{a}_j| = \sum_{j=1}^{m} k = km,$$

so that $m = |A|/k$. $\square$

As a first example of application of the previous proposition, we shall now reobtain the formula for the number of k-element subsets of a given set A, of n elements (the following reasoning is a slight variation of that in Example 2.8, the major change being in the adopted viewpoint). To this end, let $1 \le k \le n$ and $S_k(A)$ be the set of sequences of k distinct terms, all belonging to A. We already know that

$$|S_k(A)| = n(n-1)\ldots(n-k+1) = \frac{n!}{(n-k)!}.$$

Let $\sim$ be defined in $S_k(A)$ by

$$(a_1, \ldots, a_k) \sim (b_1, \ldots, b_k) \Leftrightarrow \{a_1, \ldots, a_k\} = \{b_1, \ldots, b_k\}. \qquad (2.15)$$

It's immediate to check that $\sim$ is of equivalence, and such that the equivalence class of a sequence $(a_1, \ldots, a_k)$ is the set of all of its permutations. Hence, such an equivalence class has precisely $k!$ elements, and Proposition 2.18 assures that the number of distinct equivalence relations is equal to

$$\frac{|S_k(A)|}{k!} = \frac{n!}{k!(n-k)!} = \binom{n}{k}.$$

Finally, note that (2.15) allows us to identify the equivalence classes with the k-element subsets of A, so that there are exactly $\binom{n}{k}$ of such subsets.

Another interesting situation is the following: in the set $S(A)$ of the permutations of a finite nonempty set A, we define an equivalence relation $\sim$ in the following way:

$$(a_1, \ldots, a_n) \sim (b_1, \ldots, b_n)$$

$$\Updownarrow \qquad\qquad (2.16)$$

$$\exists\, 1 \leq k \leq n; \; b_j = \begin{cases} a_{j+k}, & \text{if } 1 \leq j \leq n-k \\ a_{j+k-n}, & \text{if } n-k < j \leq n \end{cases}.$$

In other words, $(a_1, \ldots, a_n) \sim (b_1, \ldots, b_n)$ means that, for some $1 \leq k \leq n$, one has

$$(b_1, \ldots, b_n) = (a_{k+1}, a_{k+2}, \ldots, a_{n-1}, a_n, a_1, \ldots, a_k).$$

It's immediate to verify that $\sim$ is actually of equivalence in $S(A)$. Moreover, for each sequence $(a_1, \ldots, a_n) \in S(A)$, its equivalence class $\overline{(a_1, \ldots, a_n)}$ contains exactly n sequences; more precisely,

$$\overline{(a_1, \ldots, a_n)} = \{(a_1, a_2 \ldots, a_{n-1}, a_n), (a_2, a_3, \ldots, a_n, a_1)$$

$$(a_3, \ldots, a_n, a_1, a_2), \ldots, (a_n, a_1, a_2, \ldots, a_{n-1})\}.$$

With respect to the equivalence relation just described, the equivalence class of a sequence $(a_1, \ldots, a_n)$ is said to be the **circular permutation** of the sequence $(a_1, \ldots, a_n)$. In particular, it follows from the general discussion on equivalence relations that all of the sequences

$$(a_1, a_2 \ldots, a_{n-1}, a_n), (a_2, a_3, \ldots, a_n, a_1), \ldots, (a_n, a_1, a_2, \ldots, a_{n-1})$$

define the same circular permutation.

With the concept of circular permutation at our disposal, we now have the following result.

Proposition 2.19 *The number of distinct circular permutations of a set of n elements is $(n-1)!$.*

Proof If A is a set having n elements, we wish to count the number of equivalence classes of $S(A)$ with respect to the equivalence relation just described. Since $|S(A)| = n!$ and each equivalence class has exactly n elements, it follows from Proposition 2.18 that there are precisely $\frac{n!}{n} = (n-1)!$ distinct equivalence classes (i.e., circular permutations). $\qquad\qquad\qquad\qquad\qquad\qquad \square$

For what comes next, we say that two sets A and B are **incomparable with respect to inclusion**[4] if $A \not\subset B$ and $B \not\subset A$. The result below brings a deeper application of the ideas of this section, and is due to the Hungarian mathematician Béla Bollobás. Our discussion follows [25].

---

[4]This terminology comes from the fact that the inclusion relation is a *partial order* in the family of all subsets of a given set. In this respect, see the discussion at Sect. 4.3.

Theorem 2.20 (Bollobás)   *If $\mathcal{F}$ is a family of subsets of I_n, pairwise incomparable with respect to inclusion, then*

$$\sum_{A \in \mathcal{F}} \binom{n}{|A|}^{-1} \leq 1.$$

Proof Let $\mathcal{P}$ be the set of permutations $(a_1, \ldots, a_n)$ of I_n such that, for some $1 \leq k \leq n$, we have $\{a_1, \ldots, a_k\} \in \mathcal{F}$. Being a set of permutations of I_n, it is immediate that $|\mathcal{P}| \leq n!$.

We consider in $\mathcal{P}$ the relation $\sim$, given by

$$(a_1, \ldots, a_n) \sim (b_1, \ldots, b_n) \Leftrightarrow \{a_1, \ldots, a_k\} = \{b_1, \ldots, b_k\} \in \mathcal{F},$$

for some integer $1 \leq k \leq n$. We claim that $\sim$ is of equivalence. Indeed, it is clear that $\sim$ is reflexive and symmetric. In order to show that it is transitive, take a third permutation $(c_1, \ldots, c_n)$ in $\mathcal{P}$ and assume that $\{a_1, \ldots, a_k\} = \{b_1, \ldots, b_k\} \in \mathcal{F}$ and $\{b_1, \ldots, b_l\} = \{c_1, \ldots, c_l\} \in \mathcal{F}$, for some integers $1 \leq k, l \leq n$. Without loss of generality, we can also assume that $k \leq l$. However, if $k < l$, then

$$\{a_1, \ldots, a_k\} = \{b_1, \ldots, b_k\} \subset \{b_1, \ldots, b_l\} = \{c_1, \ldots, c_l\} \in \mathcal{F},$$

a strict inclusion; this would contradict the fact that the sets in $\mathcal{F}$ are pairwise incomparable with respect to inclusion. Therefore, $k = l$ and, thus,

$$\{a_1, \ldots, a_k\} = \{b_1, \ldots, b_k\} = \{c_1, \ldots, c_k\} \in \mathcal{F},$$

i.e., $(a_1, \ldots, a_n) \sim (c_1, \ldots, c_n)$.

We now fix a permutation $(a_1, \ldots, a_n) \in \mathcal{P}$, with $\{a_1, \ldots, a_k\} = A \in \mathcal{F}$. Then, $(b_1, \ldots, b_n) \sim (a_1, \ldots, a_n)$ if and only if $\{b_1, \ldots, b_k\} = A$, i.e., if and only if $(b_1, \ldots, b_k)$ is a permutation of the elements of A (and, hence, $(b_{k+1}, \ldots, b_n)$ is a permutation of the elements of $I_n \setminus A$); since we can permute the elements of A and of $I_n \setminus A$ in exactly $|A|!$ and $(n - |A|)!$ ways, respectively, we conclude that there are exactly $|A|!(n - |A|)!$ permutations $(b_1, \ldots, b_n)$ which are equivalent to the fixed permutation $(a_1, \ldots, a_n) \in \mathcal{P}$. Yet in another way, if A is the set in $\mathcal{F}$ associated to $(a_1, \ldots, a_n) \in \mathcal{P}$, then the equivalence class of $(a_1, \ldots, a_n)$ in $\mathcal{P}$ contains exactly $|A|!(n - |A|)!$ permutations.

Finally, by Proposition 2.16, the quotient set $\mathcal{P}/\sim$ of equivalence classes of $\mathcal{P}$ with respect to $\sim$ forms a partition of $\mathcal{P}$. However, since distinct equivalence classes are associated to distinct sets $A \in \mathcal{F}$ and every set $A \in \mathcal{F}$ gives rise to an equivalence class, it follows that

$$n! \geq |\mathcal{P}| = \sum_{\overline{(a_1, \ldots, a_n)} \in \mathcal{P}/\sim} \#\overline{(a_1, \ldots, a_n)} = \sum_{A \in \mathcal{F}} |A|!(n - |A|!).$$

This is the same as

$$\sum_{A \in \mathcal{F}} \frac{|A|!(n - |A|!)}{n!} \leq 1,$$

which in turn is equivalent to the desired inequality. □

For our next example we first need to recall some simple facts concerning the notion of divisibility of integers. Further details can be found in Chap. 6.

Given $a, b \in \mathbb{Z}$, with $b \neq 0$, we say that b *divides* a, or that a *is a multiple of* b, if $\frac{a}{b}$ is an integer. Letting $\frac{a}{b} = c$, this is the same as saying that $a = bc$, with $c \in \mathbb{Z}$. For instance, 7 divides 28, for $\frac{28}{7} = 4 \in \mathbb{Z}$. If b divides a, we write $b \mid a$; otherwise, we write $b \nmid a$ and say that b *does not divide* a. If a is not a multiple of b, the *division algorithm*[5] shows that there exist unique integers q and r such that $a = bq + r$ and $0 \leq r < |b|$.

We now introduce a whole class of equivalence relations in $\mathbb{Z}$, which will reveal itself to be of paramount importance to the material of Chaps. 10–20.

For a given integer $n > 1$, we say that two integers a and b are *congruent modulo* n if n divides $a - b$. If a and b are congruent modulo n, we write $a \equiv b \pmod{n}$; otherwise, we write $a \not\equiv b \pmod{n}$. For instance, we have $1 \equiv -5 \pmod{3}$, since $3 \mid (1 - (-5))$, but $-3 \not\equiv 7 \pmod{4}$, since $4 \nmid (-3 - 7)$. Note that if $a = nq + r$, with $0 \leq r < n$, then $n \mid (a - r)$, so that $a \equiv r \pmod{n}$. Therefore, every integer is congruent, modulo n, to precisely one of $0, 1, \ldots, n - 1$.

Example 2.21 Let $n > 1$ be a given integer. The relation $\sim$, defined in $\mathbb{Z}$ by

$$a \sim b \Leftrightarrow a \equiv b \pmod{n},$$

is an equivalence relation in $\mathbb{Z}$, which is known as the relation of **congruence modulo** n.

Proof Reflexivity of $\sim$ follows from

$$a \sim a \Leftrightarrow a \equiv a \pmod{n} \Leftrightarrow n \mid (a - a),$$

which is clearly true. On the other hand, since

$$a \sim b \Leftrightarrow a \equiv b \pmod{n} \Leftrightarrow n \mid (a - b),$$

and $n \mid (a - b) \Leftrightarrow n \mid (b - a)$, we have that

$$a \sim b \Rightarrow n \mid (b - a) \Rightarrow n \mid (b - a)$$
$$\Rightarrow b \equiv a \pmod{n} \Rightarrow b \sim a.$$

---

[5]For a formal definition of an *algorithm*, see the footnote at page 116.

Finally, for the transitivity of $\sim$, if $a \sim b$ and $b \sim c$, then $n \mid (a - b)$ and $n \mid (b - c)$, i.e., $\frac{a-b}{n}$ and $\frac{b-c}{n}$ are both integers; hence, the number

$$\frac{a - c}{n} = \frac{a - b}{n} + \frac{b - c}{n}$$

is also an integer, which in turn guarantees that $n \mid (a - c)$. But this is the same as saying that $a \equiv c \,(\mathrm{mod}\, n)$, so that $a \sim c$.   □

For the coming example, given an integer $m > 1$, a set $A \subset I_m$ and another integer $0 \le r \le m - 1$, we let $A + r$ denote the set

$$A + r = \{a + r \,(\mathrm{mod}\, m);\ a \in A\}.$$

Thus,

$$A + r = \{x + r;\ x \in A \text{ and } 1 \le x \le m - r\} \cup$$

$$\cup\ \{x + r - m;\ x \in A \text{ and } m - r < x \le m\}.$$

For instance, if $m = 7$, $A = \{1, 3, 5\}$ and $r = 4$, then

$$A + r = \{1 + 4, 3 + 4, 5 + 4 \,(\mathrm{mod}\, 7)\} = \{5, 7, 2\}.$$

Example 2.22 (Austrian-Polish) Let k and n be given natural numbers, such that $3 \nmid k$ and $1 \le k \le 3n$. Prove that the set I_{3n} has exactly $\frac{1}{3}\binom{3n}{k}$ subsets of k elements, such that the sum of its elements is a multiple of 3.

Solution Let $\mathcal{F}$ be the family of k-element subsets of I_{3n}. Consider a relation $\sim$ in $\mathcal{F}$ be defining $A \sim B$ if and only if there exists an integer $0 \le j \le 2$ such that $B = A + j$ modulo 3. It is immediate to check (cf. Problem 2) that $\sim$ is an equivalence relation, and that the equivalence class of a subset A of I_n contains exactly the three sets A, $A + 1$ and $A + 2$ (these last two modulo 3).

On the other hand, letting $S = \sum_{x \in A} x$, we obviously have

$$\sum_{x \in A+1} x \equiv S + k \,(\mathrm{mod}\, 3) \quad \text{and} \quad \sum_{x \in A+2} x \equiv S + 2k \,(\mathrm{mod}\, 3).$$

However, since $3 \nmid k$, we conclude that the three numbers S, $S + k$ and $S + 2k$ are pairwise incongruent modulo 3. But since every integer is congruent to precisely one of the integers 0, 1 or 2, modulo 3, we conclude that exactly one of S, $S + k$ and $S + 2k$ is divisible by 3. This is the same as saying that exactly one of the sets A, $A + 1$ and $A + 2$ has sum of elements equal to a multiple of 3. Hence, there are as many k-element subsets of I_{3n} whose sum of elements is divisible by 3 as there are equivalence classes of $\mathcal{F}$ with respect to $\sim$. But since each such equivalence class contains three elements, Proposition 2.18 assures that the number of distinct equivalence classes is precisely $\frac{1}{3}\binom{3n}{k}$.   □

Problems: Sect. 2.3

1. In each of the items of Example 2.14, prove that the given relations are, indeed, of equivalence. Also, identify the corresponding equivalence classes.
2. * Given $k, m \in \mathbb{N}$, define a relation $\sim$ in the family of k-element subsets of I_m by letting $A \sim B$ if and only if there exists $0 \leq j \leq 2$ for which $B = A + j$ modulo 3. Prove that $\sim$ is an equivalence relation.
3. Show that the number of distinct ways of partitioning a set of ab elements into a sets of b elements each is equal to $\frac{(ab)!}{a!(b!)^a}$.
4. Given $n, k \in \mathbb{N}$, with $1 \leq k \leq n$, let $S(n, k)$ denote the corresponding Stirling number of second kind (cf. Problem 8, page 20) and let A be a finite set with n elements. Do the following items:

   (a) Show that every equivalence relation on A, with exactly k equivalence classes, can be obtained from a surjective function $f : A \to B$, with $|B| = k$, as prescribed by item (b) of Example 2.14.
   (b) Show that the number of equivalence relations on A having exactly k equivalence classes is equal to $S(n, k)$.
   (c) Conclude, from the previous items, that

   $$S(n, k) = \frac{1}{k!} \sum_{j=0}^{k} (-1)^j \binom{k}{j} (k - j)^n.$$

   For the next problem, recall that a natural number $p > 1$ is said to be **prime** if 1 and p are the only positive divisors of p. We also recall (see Sects. 6.2 and 6.3 for details) that if $a, b \in \mathbb{N}$ are such that $p \mid (ab)$ but $p \nmid a$, then $p \mid b$.

5. We are given a prime number p and m pairwise distinct colors. Prove that there are exactly $\frac{m^p - m}{p} + m$ distinct necklaces, each of which formed by p beads, each bead being of one of the m given colors (there may well be beads of repeated colors). Then, use this counting to establish **Fermat's little theorem**[6]: if p is prime and m is a natural number which is not divisible by p, then $m^{p-1} \equiv 1 \pmod{p}$.
6. (IMO shortlist) Given $n \in \mathbb{N}$, we say that a sequence $(x_1, x_2, \ldots, x_n)$, with $x_j \in \{0, 1\}$ for $1 \leq j \leq n$, is *aperiodic* if there does not exist a positive divisor d of n such that the sequence is formed by the juxtaposition of $\frac{n}{d}$ copies of the subsequence $(x_1, \ldots, x_d)$. If a_n denotes the number of aperiodic sequences of size n, prove that $n \mid a_n$.
7. Prove the following theorem of Sperner: if $\mathcal{F}$ is a family of subsets of I_n, pairwise uncomparable with respect to inclusion, then

---

[6]After Pierre S. de Fermat, French mathematician of the seventeenth century. For a different proof of this result, see Sect. 10.2.

$$|\mathcal{F}| \le \binom{n}{\lfloor n/2 \rfloor},$$

with equality if $\mathcal{F} = \{A \subset I_n; \ |A| = \lfloor n/2 \rfloor\}$.

8. This problem generalizes Problem 19, page 31. To this end, let $n \in \mathbb{N}$, p be prime and $f : I_n \to I_n$ be such that $f^{(p)} = \mathrm{Id}$, with $f^{(p)}$ standing for the composition of f with itself, p times.

   (a) Show that f is a bijection and, if $\sim$ is the relation defined in I_n by

   $$a \sim b \Leftrightarrow b = f^{(j)}(a), \ \exists j \in \mathbb{Z},$$

   then $\sim$ is an equivalence relation.
   (b) Each equivalence class of I_n with respect to $\sim$ has either 1 or p elements. In particular, if f has exactly k fixed points, then $p \mid (n - k)$.
   (c) Conversely, given $0 \le k \le n$ such that $n - k = pl$, with $l \in \mathbb{Z}_+$, show that for each choice of elements $x_1, \dots, x_k \in I_n$ and for each partition

   $$I_n \setminus \{x_1, \dots, x_k\} = A_1 \cup \dots \cup A_l$$

   of $I_n \setminus \{x_1, \dots, x_k\}$ into pairwise disjoint p-element sets, there are exactly $((p - 1)!)^l$ functions f as in the statement, having precisely $x_1, \dots, x_k$ as their fixed points and satisfying $f(A_i) = A_i$ for $1 \le i \le l$.
   (d) Conclude that the number of functions $f : I_n \to I_n$ such that $f^{(p)} = \mathrm{Id}$ is given by

   $$\sum_{\substack{0 \le k \le n \\ p \mid (n-k)}} \binom{n}{k} \cdot \underbrace{\binom{n-k}{p, \dots, p}}_{(n-k)/p} ((p-1)!)^{\frac{n-k}{p}} = \sum_{\substack{0 \le k \le n \\ p \mid (n-k)}} \frac{n!}{k! \cdot p^{\frac{n-k}{p}}}.$$

   For the coming problem, recall that a family $\mathcal{F}$ of subsets of I_n is an *intersecting system* of I_n if, for all $A, B \in \mathcal{F}$, we have $A \cap B \ne \emptyset$.

9. The purpose of this problem is to prove the famous **Erdös-Ko-Rado theorem**[7]: for positive integers k and n, with $1 \le k \le n$, if $\mathcal{F}$ is an intersecting system formed by k-element subsets of I_n, then

   $$|\mathcal{F}| \le \begin{cases} \binom{n}{k}, & \text{if } k > n/2 \\ \binom{n-1}{k-1}, & \text{if } k \le n/2 \end{cases}.$$

   To this end, do the following items:

---

[7] After the Hungarian, Chinese and German twentieth century mathematicians Paul Erdös, Chao Ko and Richard Rado, respectively.

(a) If $k > \frac{n}{2}$, show that $|\mathcal{F}| \leq \binom{n}{k}$.

(b) If $k \leq \frac{n}{2}$, give an example of an intersecting system $\mathcal{F}$ of I_n, formed by k-element subsets of I_n and such that $|\mathcal{F}| = \binom{n-1}{k-1}$.

(c) Also in the case $k \leq \frac{n}{2}$, fix an arbitrary intersecting system $\mathcal{F}$ of I_n, formed by k-element subsets of I_n. Let S_n denote the set of permutations $\sigma : I_n \rightarrow I_n$, and C_n denote the set of circular permutations of I_n, so that $|S_n| = n!$, $|C_n| = (n-1)!$ and, for $\sigma \in S_n$, we have $\overline{\sigma} \in C_n$, with $|\overline{\sigma}| = n$ (here, and as in the text, $\overline{\sigma}$ stands for the equivalence class of σ with respect to the equivalence relation defining circular permutations). Define $A \in \mathcal{F}$ and $\overline{\sigma} \in C_n$ as being *compatible* provided $A = \{a_1, \ldots, a_k\}$, for some permutation $(a_1, \ldots, a_k, a_{k+1}, \ldots, a_n) \in \overline{\sigma}$.

    i. For $\sigma \in S_n$, show that at most k sets in $\mathcal{F}$ are compatible with $\overline{\sigma}$. Then, conclude that there are at most $(n-1)!k$ pairs $(A, \overline{\sigma})$ with $A \in \mathcal{F}, \sigma \in S_n$ and A and $\overline{\sigma}$ compatible to each other.

    ii. For a fixed $A \in \mathcal{F}$, show that there are exactly $k!(n-k)!$ elements $\overline{\sigma} \in C_n$ such that A and $\overline{\sigma}$ are compatible. From this, conclude that there are exactly $k!(n-k)!|\mathcal{F}|$ pairs $(A, \overline{\sigma})$ such that $A \in \mathcal{F}, \sigma \in S_n$ and A and $\overline{\sigma}$ are compatible.

    iii. Complete the proof of Erdös-Ko-Rado theorem.

2.4 Counting with Metrics

In this section, we present a counting technique with a distinguished *geometric* flavor. To this end, we shall need the following important definition, which is actually a concept pervasive to all of Mathematics.

Definition 2.23 If X is a nonempty set, a **metric** in X is a function $d : X \times X \rightarrow \mathbb{R}$ satisfying, for all $x, y, z \in X$, the following conditions:

(a) $d(x, y) = d(y, x)$.

(b) $d(x, y) \geq 0$ and $d(x, y) = 0 \Leftrightarrow x = y$.

(c) $d(x, z) \leq d(x, y) + d(y, z)$.

The prototype of metric is the distance function in Euclidean space. More precisely, letting E^3 denote the three dimensional Euclidean space and $d : E^3 \times E^3 \rightarrow \mathbb{R}$ be given by

$$d(P, Q) = \overline{PQ}, \ \forall \ P, Q \in E^3,$$

then conditions (a), (b) and (c) above are clearly satisfied, condition (c) following from the triangle inequality.

Thanks to the geometric example of the previous paragraph, from now on we shall refer ourselves to condition (c) of Definition 2.23 as the **triangle inequality**.

Another piece of terminology borrowed from that example is isolated in the next definition.

Definition 2.24 Let d be a metric in the nonempty set X. Given $x \in X$ and $r > 0$, the **ball**[8] of **center** x and **radius** r is the subset $B(x; r)$ of X, given by

$$B(x; r) = \{y \in X;\ d(y, x) \le r\}.$$

Back to the former example of metric, given $P \in E^3$ and $r > 0$ we have

$$B(P; r) = \{Q \in E^3;\ \overline{PQ} \le r\}$$

$$= \text{solid sphere of center } P \text{ and radius } r.$$

Thus, at least at a heuristic level, this example justifies the geometric intuition attached to the name *ball*.

In what concerns Combinatorics, our primary interest relies in metrics defined on *finite* nonempty sets. To understand why this is so, let us assume that X is such a set and d is a metric on it. Further, suppose that $x_1, \ldots, x_k \in X$ are such that the union of the balls centered at those points and with some positive radius r *cover* X, i.e., that

$$B(x_1; r) \cup \ldots \cup B(x_k; r) = X. \tag{2.17}$$

If we have an upper bound on the number of elements of any ball of radius r in X, say

$$|B(x; r)| \le c(r), \ \forall\, x \in X,$$

with $c(r) > 0$ depending only on r (and not on the chosen $x \in X$), then (2.17) gives us

$$|X| = |B(x_1; r) \cup \ldots \cup B(x_k; r)|$$

$$\le |B(x_1; r)| + \cdots + |B(x_k; r)|$$

$$\le k \cdot c(r).$$

Therefore,

$$k \ge \frac{1}{c(r)} |X|. \tag{2.18}$$

---

[8]The reader with previous acquaintance with the theory of Metric Spaces will notice that what we call a *ball* is generally known in the literature as a *closed ball*. Nevertheless, in order to ease the writing we will stick to this slightly different terminology. We believe that, as far as these notes are concerned, this will be a harmless practice.

Thus, an *upper* bound on the number of elements of any ball of radius r automatically gives a *lower* bound on the number of balls of radius r needed to cover X.

With such a general reasoning at our disposal, let us take a look at a first relevant example.

Example 2.25 Given $n \in \mathbb{N}$, let $\mathcal{S}_n$ be the set of sequences of n terms, all of which equal to 0 or 1, i.e.,

$$\mathcal{S}_n = \{a = (a_1, a_2, \ldots, a_n); \ a_i = 0 \text{ or } 1, \ \forall \ 1 \leq i \leq n\}.$$

Let $d : \mathcal{S}_n \times \mathcal{S}_n \to \mathbb{R}$ be the function given, for $a = (a_1, \ldots, a_n)$ and $b = (b_1, \ldots, b_n)$ in $\mathcal{S}_n$, by

$$d(a, b) = \#\{1 \leq i \leq n; \ a_i \neq b_i\}.$$

Then, d is a metric on $\mathcal{S}_n$, called the **Hamming metric**[9] on $\mathcal{S}_n$.

Proof We clearly have $d(a, b) \geq 0$ and $d(a, b) = 0$ if and only if $a = b$. It is also clear that $d(a, b) = d(b, a)$. In order to verify the triangle inequality, note that if $a, b \in \mathcal{S}_n$, then

$$d(a, b) = \sum_{i=1}^{n} |a_i - b_i|.$$

Therefore, for $a, b, c \in \mathcal{S}_n$, the triangle inequality for real numbers gives

$$d(a, c) = \sum_{i=1}^{n} |a_i - c_i| \leq \sum_{i=1}^{n} (|a_i - b_i| + |b_i - c_i|)$$

$$= \sum_{i=1}^{n} |a_i - b_i| + \sum_{i=1}^{n} |b_i - c_i|$$

$$= d(a, b) + d(b, c).$$

$\square$

In the notations of the previous example, given $a \in \mathcal{S}_n$ and $k \in \mathbb{N}$, we have

$$B(a; k) = \{b \in \mathcal{S}_n; \ d(a, b) \leq k\}$$

$$= \{b \in \mathcal{S}_n; \ a_i \neq b_i \text{ for at most } k \text{ indices } 1 \leq i \leq n\}.$$

---

[9]After the American mathematician of the twentieth century Richard Hamming, who used such a concept to study coding problems in Computer Science and Telecommunications Engineering. The famous *Hamming codes* were also named after him.

However, since there are exactly $\binom{n}{j}$ ways of choosing j indices $1 \le i \le n$ in which the sequences a and b are to differ, we conclude that

$$|B(a; k)| = \binom{n}{0} + \binom{n}{1} + \cdots + \binom{n}{k}. \tag{2.19}$$

The above being said, the discussion that precedes the former example will give the coming one.

Example 2.26 Let $n \in \mathbb{N}$ and $\mathcal{S}_n$ be the set of sequences of n terms, all of which equal to 0 or 1. Show that if we choose less than

$$\frac{2^n}{\binom{n}{0} + \binom{n}{1} + \cdots + \binom{n}{k}}$$

sequences in $\mathcal{S}_n$, then there will always be at least one sequence in $\mathcal{S}_n$ which differs from each of the chosen ones in at least $k + 1$ distinct positions.

Proof Choose l distinct sequences in $\mathcal{S}_n$, say $x_1, \ldots, x_l$, with

$$l < \frac{2^n}{\binom{n}{0} + \binom{n}{1} + \cdots + \binom{n}{k}}.$$

Then, (2.19) and computations analogous to those which led to (2.18) furnish

$$|B(x_1; k) \cup \ldots \cup B(x_l; k)| \le |B(x_1; k)| + \cdots + |B(x_l; k)|$$

$$= \left(\binom{n}{0} + \binom{n}{1} + \cdots + \binom{n}{k} \right) l < 2^n.$$

Now, the fundamental principle of counting gives $|\mathcal{S}_n| = 2^n$. Therefore, we conclude from the last inequality above that the union of balls $B(x_1; k) \cup \ldots \cup B(x_l; k)$ does not cover $\mathcal{S}_n$. Hence, there exists at least one sequence $x \in \mathcal{S}_n$ such that $x \notin B(x_i; k)$, for $1 \le i \le l$. But this is the same as saying that

$$d(x, x_i) > k, \ \forall \ 1 \le i \le l,$$

and the rest follows from the definition of the Hamming metric. $\qquad\square$

Prior to presenting another relevant example, recall that, given sets X and Y, one defines their **symmetric difference** as the set $X \triangle Y$, given by

$$X \triangle Y = (X \cup Y) \setminus (X \cap Y).$$

In words, $X \triangle Y$ is the set formed by the elements which belong to *exactly one* of the sets X and Y.

It follows promptly from the definition that $X \triangle \emptyset = X$ and $X \triangle Y = Y \triangle X$; moreover, the reader can easily verify that

$$X \triangle Y = (X \setminus Y) \cup (Y \setminus X),$$

for all sets X and Y.

The following lemma gathers together other properties of the symmetric difference which will be of our interest.

Lemma 2.27 *If X, Y and Z are any sets, then:*

(a) $X \triangle Y = \emptyset \Leftrightarrow X = Y$.
(b) $(X \triangle Y) \triangle Z = X \triangle (Y \triangle Z)$.
(c) $X \triangle Y = Z \Leftrightarrow X = Y \triangle Z$.

Proof

(a) Since we clearly have $X \triangle X = \emptyset$, it suffices to show that $X \triangle Y = \emptyset \Rightarrow X = Y$. To this end, observe that $X \triangle Y = \emptyset \Rightarrow X \cup Y = X \cap Y$ and, hence,

$$X \subset X \cup Y = X \cap Y \subset Y;$$

similarly, $Y \subset X$.

(b) It is not difficult to verify that

$$(X \triangle Y) \triangle Z = (X \cup Y \cup Z) \setminus ((X \cap Y) \cup (X \cap Z) \cup (Y \cap Z)),$$

and analogously to $X \triangle (Y \triangle Z)$. We leave as an exercise to the reader the task of checking the details.

(c) Let us show one implication, the other being completely analogous. If $X \triangle Y = Z$, it follows from item (b) that

$$X = \emptyset \triangle X = (Y \triangle Y) \triangle X = Y \triangle (Y \triangle X) = Y \triangle Z.$$

$\square$

With the above results at our disposal, we can finally present another important example of metric in a family of finite sets, known as the **metric of symmetric difference**.

Example 2.28 If $\mathcal{F}$ is a nonempty family of finite sets, then the function $d : \mathcal{F} \times \mathcal{F} \to \mathbb{R}$, given by $d(X, Y) = |X \triangle Y|$, is a metric in $\mathcal{F}$.

Proof Since $X \triangle Y = Y \triangle X$, we have

$$d(X, Y) = |X \triangle Y| = |Y \triangle X| = d(Y, X)$$

for all $X, Y \in \mathcal{F}$. It is also clear that $d(X, Y) \geq 0$ and $d(X, Y) = 0$ if and only if $X \triangle Y = \emptyset$; however, by item (a) of the previous lemma, this last equality takes place if and only if $X = Y$.

We are left to showing that $d(X, Y) \leq d(X, Z) + d(Z, Y)$ for all $X, Y, Z \in \mathcal{F}$, or, which is the same, that $|X \triangle Y| \leq |X \triangle Z| + |Z \triangle Y|$. To this end, since $|A \cup B| \leq |A| + |B|$ for finite sets A and B, it is enough that we establish the inclusion

$$X \triangle Y \subset (X \triangle Z) \cup (Z \triangle Y).$$

In turn, such an inclusion follows at once from the equality

$$(X \triangle Z) \cup (Z \triangle Y) = (X \cup Y \cup Z) \setminus (X \cap Y \cap Z),$$

whose proof we leave as an exercise to the reader.   □

Fix $m \in \mathbb{N}$ and consider the metric of symmetric difference in $\mathcal{F} = \mathcal{P}(m)$, the family of subsets of I_m. If $X \subset I_m$ is given, let us compute the number of subsets of I_m in the ball $B(X; k)$, where $k \geq 0$ is a given integer. To this end, start by noticing that

$$Y \in B(X; k) \Leftrightarrow |X \triangle Y| \leq k;$$

therefore, in order to count the number of elements in $B(X; k)$, it suffices to count how many are the subsets $Y \subset I_m$ for which $|X \triangle Y| \leq k$. For a fixed $Z \subset I_m$, item (c) of Lemma 2.27 guarantees the existence of a single $Y \subset I_m$ such that $Z = X \triangle Y$. Hence,

$$\{Y \in \mathcal{P}(m); \ |X \triangle Y| \leq k\} = \{Z \in \mathcal{P}(m); \ |Z| \leq k\}.$$

Finally, since for each $0 \leq j \leq k$ there are exactly $\binom{m}{j}$ ways of choosing $Z \subset I_m$ such that $|Z| = j$, it follows from the equality of families above that

$$|B(X; k)| = \#\{Z \in \mathcal{P}(m); \ |Z| \leq k\}$$
$$= \binom{m}{0} + \binom{m}{1} + \cdots + \binom{m}{k}. \tag{2.20}$$

For the coming example, let $\mathcal{P}_k(n)$ be the family of k-element subsets of I_n and d be the metric of symmetric difference in $\mathcal{P}_k(n)$. For distinct $X, Y \in \mathcal{P}_k(n)$, we have

$$d(X, Y) = |X \triangle Y| = |X \cup Y| - |X \cap Y|$$
$$= |X| + |Y| - 2|X \cap Y| \tag{2.21}$$
$$= 2(k - |X \cap Y|),$$

an even number.

Example 2.29 (BMO—Adapted) Let $n > 2$ be a given integer and $\mathcal{F}$ be a maximal family of 3-element subsets of I_n satisfying the following condition: for any $A, B \in \mathcal{F}$, with $A \neq B$, we have $|A \cap B| \leq 1$. Prove that

$$|\mathcal{F}| > \frac{1}{18}(n^2 - n).$$

Proof Let d be the metric of symmetric difference in $\mathcal{P}_3(n)$. It follows from (2.21) that, for $X, Y \in \mathcal{F}$,

$$|X \cap Y| \leq 1 \Leftrightarrow d(X, Y) \geq 4.$$

We now claim that if $\mathcal{F} = \{A_1, \ldots, A_k\}$, then

$$\mathcal{P}_3(n) = B(A_1; 2) \cup \ldots \cup B(A_k; 2). \tag{2.22}$$

Indeed, if this was not the case, then there would exist $A \in \mathcal{P}_3(n) \setminus (B(A_1; 2) \cup \ldots \cup B(A_k; 2))$, i.e., such that $d(A; A_i) \geq 4$ for $1 \leq i \leq k$. By what we did above, this would be the same as having $|A \cap A_i| \leq 1$ for $1 \leq i \leq k$. In turn, this would assure that $\mathcal{F}' = \{A, A_1, \ldots, A_k\}$ would be a family strictly containing $\mathcal{F}$ and also satisfying the stated conditions. This contradicts the maximality of $\mathcal{F}$.

Thus, it follows from (2.22) that

$$\binom{n}{3} = |\mathcal{P}_3(n)| = |B(A_1; 2) \cup \ldots \cup B(A_k; 2)| \tag{2.23}$$
$$\leq |B(A_1; 2)| + \cdots + |B(A_k; 2)|.$$

Now, given $A = \{a, b, c\} \in \mathcal{P}_3(n)$, let us compute the number of sets in $B(A; 2) \subset \mathcal{P}_3(n)$: recall that, for $A' \in \mathcal{P}_3(n)$, the distance $d(A', A)$ is even; therefore, (2.21) gives

$$d(A', A) \leq 2 \Leftrightarrow d(A', A) = 0 \text{ or } 2$$
$$\Leftrightarrow A' = A \text{ or } |A' \cap A| = 2$$
$$\Leftrightarrow A' = \{a, b, c\}, \{a, b, x\}, \{a, c, x\} \text{ or } \{b, c, x\},$$

for some $x \in I_n \setminus \{a, b, c\}$. We then conclude that

$$|B(A; 2)| = 1 + 3(n - 3) = 3n - 8.$$

We finish by using (2.23) and the last computation above to obtain

$$k \geq \frac{1}{3n - 8}\binom{n}{3} = \frac{n(n - 1)(n - 2)}{6(3n - 8)}$$
$$> \frac{n(n - 1)(n - 2)}{6(3n - 6)}$$
$$= \frac{n(n - 1)}{18}.$$

$\square$

Problems: Sect. 2.4

1. Prove that the Hamming metric is a particular case of the metric of symmetric difference.
2. (Ireland) Let $n \geq 11$ be an odd integer and $\mathcal{T}$ be the set of n-tuples $(x_1, \ldots, x_n)$ such that $x_i \in \{0, 1\}$ for $1 \leq i \leq n$. Given $x = (x_1, \ldots, x_n)$ and $y = (y_1, \ldots, y_n)$ in $\mathcal{T}$, let

$$f(x, y) = \#\{1 \leq i \leq n; \ x_i \neq y_i\}.$$

Suppose that there exists $\mathcal{S} \subset \mathcal{T}$ satisfying the following conditions:

(a) $|\mathcal{S}| = 2^{\frac{n+1}{2}}$.
(b) For each $x \in \mathcal{T}$, there exists a single $y \in \mathcal{S}$ for which $f(x, y) \leq 3$.

Prove that $n = 23$.
3. * Given an integer $n > 3$, let $f(n)$ be the largest possible number of 4-element subsets of I_n, such that any two of them have at most two elements in common. Prove that

$$\frac{1}{16}\binom{n}{3} < f(n) \leq \frac{1}{3}\binom{n}{3}.$$

4. * Let $n \geq 2$ be an integer and $A_1, \ldots, A_k \subset I_n$ be such that, for $1 \leq i < j \leq k$, we have $|A_i \triangle A_j| \geq 2r + 1$, where r is a given positive integer. Prove that

$$k \leq \frac{2^n}{\binom{n}{0} + \binom{n}{1} + \cdots + \binom{n}{r}}.$$

5. (India) Out of a committee of n members, 1997 distinct subcommittees are composed. It is known that, for any two of these subcommittees, there are at least five people participating of exactly one of them. Prove that $n \geq 19$.
6. (Brazil) During a summer camp, Arnold and Beatrice talked to each other using smoke clouds, sometimes big ones, sometimes small ones. Every day before breakfast, Arnold sent to Beatrice a sequence of 24 clouds. As Beatrice not always manages to distinguish a small cloud from a big one, they composed a dictionary before entering the camp. This dictionary collected n sequences of 24 sizes of clouds (for instance, one such sequence could be SBSBSBSBSBSBBS-BSBSBSBSBS, where B stands for *big* and S stands for *small*). For each of the n sequences, the dictionary indicated its meaning. Moreover, in order to prevent misinterpretation, Arnold and Beatrice avoided including similar sequences in the dictionary; more precisely, each two sequences in the dictionary always differed in at least 8 of their 24 positions. Show that $n \leq 4096$.

Chapter 3
Generating Functions

In this chapter we briefly discuss the powerful method of *generating functions*, i.e., the use of *power series* in the analysis of counting problems that are inaccessible by the more elementary techniques presented so far.

Unlike other approaches to this subject (e.g., that of the classic [41]), the material collected here makes heavy use of the methods of Differential and Integral Calculus, in the level of [3] or [8], for instance. Although working with *convergent* power series *apparently* requires far more background than using *formal* power series, it is actually much more complicated to rigorously justify the correctness of some algebraic manipulations with this last type of series. Furthermore, we believe that, in a first acquaintance with generating functions, the reader with some Calculus background feels much more comfortable in working in the realm of convergent power series. Anyhow, for the reader's convenience Sect. 3.2 briefly reviews the basic facts on the convergence of power series needed for our subsequent discussions.

For a comprehensive exposition on the uses of generating functions in Combinatorics, we refer the interested reader to the above mentioned wonderful book of professor H. Wilf.

3.1 Introduction

By definition, the (**ordinary**) **generating function** of a sequence $(a_n)_{n \geq 0}$ of real numbers[1] is the power series

$$\sum_{n \geq 0} a_n x^n.$$

---

[1]We shall sometimes refer to the generating function of a sequence $(a_n)_{n \geq 1}$ of real numbers, but this should be no source of confusion.

Among all generating functions, one of the simplest ones is, in a certain sense, that of the (sequence formed by the) nonnegative integers: $\sum_{n\geq 0} x^n$. As the reader probably knows from earlier studies (cf. Section 7.4 of [8], for instance), such a power series converges to $\frac{1}{1-x}$ when $|x| < 1$. Thus, for $|x| < 1$, we have

$$\frac{1}{1-x} = \sum_{n\geq 0} x^n, \tag{3.1}$$

and say that the series at the right hand side is a **geometric series**.

In order to illustrate the kind of use we have in mind for generating functions, we consider in this section two preliminary examples. Their role is to give the reader an easy glimpse on how the method works.

The first example revisits Problem 7, page 49, and employs some elementary facts on continuous functions, which can be reviewed in Chapter 8 of [8], for instance.

Example 3.1 (TT) The set of natural numbers is partitioned into m infinite and nonconstant arithmetic progressions, of common differences $d_1, d_2, \ldots, d_m$. Prove that

$$\frac{1}{d_1} + \frac{1}{d_2} + \cdots + \frac{1}{d_m} = 1.$$

Proof Let $f(x) = \sum_{n\geq 1} x^n$ be the generating function of the natural numbers. Letting a_i be the initial term of the arithmetic progression of common difference d_i, the stated condition assures that, for $|x| < 1$, we have

$$f(x) = \sum_{i=1}^{m} (x^{a_i} + x^{a_i+d_i} + x^{a_i+2d_i} + \cdots).$$

Hence, for $|x| < 1$, it follows from (3.1) (with x^{d_i} in place of x, in each series in the sum at the right hand side above) that

$$\frac{x}{1-x} = \sum_{i=1}^{m} \frac{x^{a_i}}{1-x^{d_i}}.$$

Multiplying both sides of this last equality by $1 - x$ and doing some algebra at the right hand side, we get

$$x = \sum_{i=1}^{m} \frac{x^{a_i}}{1 + x + x^2 + \cdots + x^{d_i-1}}. \tag{3.2}$$

Now, note that both sides of the equality above define continuous functions in the closed interval $[0, 1]$, and our reasoning so far assures that such functions coincide in the half-open interval $[0, 1)$. Therefore (and it is precisely here where we use the basic theory of continuous functions), they coincide also for $x = 1$, and evaluation of (3.2) at $x = 1$ gives

$$1 = \sum_{i=1}^{m} \frac{1}{d_i}.$$

$\square$

Our next example uses generating functions to obtain a positional formula for the n-th Fibonacci number.

Example 3.2 Let $(F_n)_{n \geq 1}$ be the Fibonacci sequence and $f(x) = \sum_{n \geq 1} F_n x^n$ be its generating function. For the time being, assume that the series defining f converges on an open interval of the form $(-r, r)$, for some real number $r > 0$ (we shall see, in the paragraph following Proposition 3.7, that this is indeed the case). Then, we can write

$$f(x) = F_1 x + F_2 x^2 + \sum_{n \geq 3} F_n x^n$$

$$= x + x^2 + \sum_{n \geq 3} (F_{n-1} + F_{n-2}) x^n$$

$$= x + x^2 + x \sum_{n \geq 3} F_{n-1} x^{n-1} + x^2 \sum_{n \geq 3} F_{n-2} x^{n-2}$$

$$= x + x^2 + x(f(x) - F_1 x) + x^2 f(x)$$

$$= x + (x + x^2) f(x),$$

thus getting

$$f(x) = \frac{x}{1 - x - x^2}$$

for $x \in (-r, r)$.

Now, writing $1 - x - x^2 = (1 - \alpha x)(1 - \beta x)$, with $\alpha, \beta \in \mathbb{R}$, we have $\alpha + \beta = 1$, $\alpha \beta = -1$ and

$$f(x) = \frac{x}{1 - x - x^2} = \frac{x}{(1 - \alpha x)(1 - \beta x)}.$$

By imposing, without loss of generality, that $\alpha > \beta$, we get $\alpha = \frac{1+\sqrt{5}}{2}$ and $\beta = \frac{1-\sqrt{5}}{2}$, so that $\alpha - \beta = \sqrt{5}$. Thus,

$$f(x) = \frac{1}{\sqrt{5}} \left(\frac{1}{1 - \alpha x} - \frac{1}{1 - \beta x} \right).$$

By applying (3.1) (with αx and βx in place of x) to expand the right hand side in power series, we get

$$f(x) = \frac{1}{\sqrt{5}} \left\{ \sum_{n \geq 0} (\alpha x)^n - \sum_{n \geq 0} (\beta x)^n \right\}$$

$$= \sum_{n \geq 1} \left(\frac{\alpha^n - \beta^n}{\sqrt{5}} \right) x^n.$$

It is important to point out that the discussion at the beginning of this section guarantees that the geometric series involved in the above computations converge for every $x \in \mathbb{R}$ for which $|\alpha x|, |\beta x| < 1$, i.e., for $|x| < \frac{1}{|\alpha|}$. Hence, letting $s = \min \left\{ r, \frac{1}{|\alpha|} \right\} > 0$, it follows that

$$\sum_{n \geq 1} F_n x^n = \sum_{n \geq 1} \left(\frac{\alpha^n - \beta^n}{\sqrt{5}} \right) x^n \tag{3.3}$$

for $|x| < s$. This way, if the power series representing f is unique, there will be no alternative than to infer, from the last equality above, that

$$F_n = \frac{\alpha^n - \beta^n}{\sqrt{5}} \tag{3.4}$$

for every $n \geq 1$.

Finally, the uniqueness result we hinted to above is indeed true, and will be explicitly stated in Corollary 3.6. This validates the last step above and shows that (3.4) does hold.

Problems: Sect. 3.1

1. In the notations of Example 3.1, if $a_1, a_2, \ldots, a_m$ are the initial terms of the progressions, prove that

$$\frac{a_1}{d_1} + \frac{a_2}{d_2} + \cdots + \frac{a_m}{d_m} = \frac{m + 1}{2}.$$

2. The Lucas sequence $(L_n)_{n\geq 1}$ is defined by $L_1 = 1$, $L_2 = 3$ and $L_{k+2} = L_{k+1} + L_k$, for every integer $n \geq 1$. Prove that:

   (a) $L_n = \alpha^n + \beta^n$ for every $n \in \mathbb{N}$, where $\alpha = \frac{1+\sqrt{5}}{2}$ and $\beta = \frac{1-\sqrt{5}}{2}$.
   (b) $L_{2n} = L_n^2 + 2(-1)^{n-1}$ for every $n \in \mathbb{N}$.
   (c) $L_n^2 - 5F_n^2 = 4(-1)^n$ for every $n \in \mathbb{N}$, where $(F_n)_{n\geq 1}$ denotes the Fibonacci sequence.

3.2 Power Series

In order to justify and extend arguments like those of the two examples of the previous section to more general situations, we shall review in this section some facts on the convergence of general *real power series*. Since a thorough exposition of the theory of real power series belongs more properly to the realm of basic real analysis, we have chosen to be somewhat sketchy, referring the interested reader to Chapter 11 of [8] for a more detailed account of the theory. Later, in Chap. 21, extensions to complex power series will be discussed and applied.

Let $r > 0$ be a real number. We say that a function $f : (-r, r) \to \mathbb{R}$ admits the (power series) expansion

$$f(x) = \sum_{n\geq 0} a_n x^n$$

in the interval $(-r, r)$, or that the **power series** $\sum_{n\geq 0} a_n x^n$ **converges** to f in the interval $(-r, r)$, if

$$f(x) = \lim_{m\to+\infty} \sum_{n=0}^{m} a_n x^n$$

for every $x \in (-r, r)$.

A relevant example is that of the geometric series (3.1). As a corollary we note that, for a given $\alpha \neq 0$, we have

$$\frac{1}{1 - \alpha x} = \sum_{n\geq 0} \alpha^n x^n \tag{3.5}$$

when $|x| < \frac{1}{|\alpha|}$. Indeed, if $|x| < \frac{1}{|\alpha|}$, then $|\alpha x| < 1$, so that it suffices to apply (3.1) to αx in place of x.

We shall frequently need to combine the power series expansions of two given functions to obtain that of another function, obtained from those by means of the arithmetic operations of addition, subtraction or multiplication of functions. In this respect, the very definition of convergence of power series assures that, if $f(x) = \sum_{n\geq0} a_n x^n$ and $g(x) = \sum_{n\geq0} b_n x^n$ along $(-r, r)$, then

$$(f \pm g)(x) = \sum_{n\geq0} (a_n \pm b_n) x^n$$

in $(-r, r)$. Concerning multiplication, we have the following result.

Proposition 3.3 *If the functions* $f, g : (-r, r) \to \mathbb{R}$ *have power series expansions* $f(x) = \sum_{k\geq0} a_k x^k$ *and* $g(x) = \sum_{l\geq0} b_l x^l$, *then the function* $fg : (-r, r) \to \mathbb{R}$ *has power series expansion* $(fg)(x) = \sum_{n\geq0} c_n x^n$, *with*

$$c_n = \sum_{k+l=n} a_k b_l = \sum_{k=0}^{n} a_k b_{n-k}. \tag{3.6}$$

Thus, for $|x| < 1$ and letting $a_k = 1$ for every $k \geq 0$, we get from (3.1) that

$$\frac{1}{(1-x)^2} = \frac{1}{1-x} \cdot \frac{1}{1-x} = \left(\sum_{k\geq0} a_k x^k \right) \left(\sum_{l\geq0} a_l x^l \right)$$

$$= \sum_{n\geq0} \left(\sum_{k=0}^{n} a_k a_{n-k} \right) x^n = \sum_{n\geq0} \left(\sum_{k=0}^{n} 1 \right) x^n$$

$$= \sum_{n\geq0} (n+1) x^n.$$

Moreover, such a reasoning can be easily generalized to give the power series expansion for the function $x \mapsto \frac{1}{(1-x)^k}$, for $|x| < 1$ (cf. Problems 1 and 2, page 78).

For the purposes we have in mind, the main result on power series is the one collected in the coming theorem.

Theorem 3.4 *Let* $r > 0$ *be such that the power series* $\sum_{n\geq0} a_n x^n$ *converges at every* $x \in (-r, r)$. *If* $f : (-r, r) \to \mathbb{R}$ *is given by* $f(x) = \sum_{n\geq0} a_n x^n$, *then* f *is differentiable, with* $f'(x) = \sum_{n\geq1} n a_n x^{n-1}$ *for every* $x \in (-r, r)$.

In words, the previous theorem guarantees that we can *termwise differentiate* a convergent power series, obtaining, as result, the power series corresponding to the derivative of the function defined by the original series.

Example 3.5 This example shows how to use the previous result to reobtain, for $|x| < 1$, the power series expansion of the function $x \mapsto \frac{1}{(1-x)^2}$. Since $\frac{1}{1-x} = \sum_{n \geq 0} x^n$ for $|x| < 1$, differentiating both sides of such an equality we get

$$\frac{1}{(1-x)^2} = \sum_{n \geq 1} nx^{n-1} = \sum_{n \geq 0} (n+1)x^n,$$

and this is also valid for $x \in (-1, 1)$.

A very important corollary of Theorem 3.4 is the uniqueness of the power series representing a certain function.

Corollary 3.6 *If $f : (-r, r) \to \mathbb{R}$ admits a power series representation in the interval $(-r, r)$, then f is infinitely differentiable. Moreover, its power series representation is unique and is given by*

$$f(x) = \sum_{n \geq 0} a_n x^n,$$

with $a_n = \frac{1}{n!} f^{(n)}(0)$, where $f^{(n)}(0)$ denotes the value of the n-th derivative of f at 0.

Proof Firstly, note that $f(0) = a_0$. Now, the previous theorem gives us $f'(x) = \sum_{n \geq 1} na_n x^{n-1}$ in $(-r, r)$; in particular, this gives $f'(0) = a_1$. Again by the former result, f' is differentiable and we get $f''(x) = \sum_{n \geq 2} n(n-1)a_n x^{n-2}$, so that $f''(0) = 2a_2$. Proceeding inductively this way, we easily finish the proof. $\square$

The actual importance of Theorem 3.4 will completely reveal itself only if we have at our disposal some criterion which allows one to establish the convergence of a given power series $\sum_{n \geq 0} a_n x^n$ in some open interval centered at 0. For our purposes, the coming result, which is a particular case of the comparison test for series, will suffice.

Proposition 3.7 *Let $(a_n)_{n \geq 0}$ be a sequence of real numbers. If there exist positive reals c and M such that $|a_n| \leq cM^n$ for every $n \geq 0$, then the power series $\sum_{n \geq 0} a_n x^n$ converges on the interval $\left(-\frac{1}{M}, \frac{1}{M}\right)$.*

Proof Since

$$|a_n x^n| = |a_n||x|^n \leq cM^n |x|^n = c|Mx|^n$$

and (cf. (3.5)) the geometric series $\sum_{n \geq 0} |Mx|^n$ converges when $|x| < \frac{1}{M}$, the comparison test for series assures that $\sum_{n \geq 0} a_n x^n$ converges when $|x| < \frac{1}{M}$. $\square$

With the above proposition at hand, we are in position to complete the analysis of Example 3.2. Indeed, letting $(F_n)_{n \geq 1}$ denote the Fibonacci sequence, an easy induction shows that $F_n \leq 2^n$ for every $n \geq 1$. Hence, by the previous result, the corresponding generating function, $\sum_{n\geq 1} F_n x^n$, converges whenever $x \in \left(-\frac{1}{2}, \frac{1}{2}\right)$. In turn, this assures that all computations made at that example are actually valid whenever $|x| < \min\left\{\frac{1}{2}, \frac{1}{|\alpha|}\right\} = \frac{1}{2}$. In particular, we can use Corollary 3.6 to conclude, from (3.3), that (3.4) does hold.

As Problem 5 will show, generating functions can be used to solve more general **second order linear recurrence relations**. Actually, when we have complex numbers at our disposal, we will completely analyse a much more general version of this result, in Chap. 21.

Example 3.8 As a relevant application of the previous proposition in conjunction with Theorem 3.4, let us show that

$$e^x = \sum_{n\geq 0} \frac{1}{n!} x^n \tag{3.7}$$

for every $x \in \mathbb{R}$.

We first claim that, given $A > 0$, there exist $\alpha > 0$ and $n_0 \in \mathbb{N}$ such that $n! > \alpha A^n$ for every $n > n_0$. Indeed, fixed a natural number $n_0 > A$ we have, for $n > n_0$ also natural,

$$n! = n_0! \underbrace{(n_0 + 1) \ldots (n - 1)n}_{n - n_0} > n_0! A^{n-n_0} = \frac{n_0!}{A^{n_0}} A^n.$$

It thus suffices to take $\alpha = \frac{n_0!}{A^{n_0}}$.

Now, given $A > 0$, take $\alpha > 0$ and $n_0 \in \mathbb{N}$ as above. Letting $a_n = \frac{1}{n!}, c = \frac{1}{\alpha}$ and $M = \frac{1}{A}$, we conclude that $0 < a_n < cM^n$ for every natural number $n > n_0$, and the previous proposition assures that the power series

$$\sum_{n\geq 0} \frac{1}{n!} x^n = \sum_{n=0}^{n_0} \frac{1}{n!} x^n + \sum_{n > n_0} \frac{1}{n!} x^n$$

defines a function $f : (-A, A) \to \mathbb{R}$ (note that $\frac{1}{M} = A$).

However, since $A > 0$ was arbitrarily chosen, we conclude that $f(x) = \sum_{n\geq 0} \frac{1}{n!} x^n$ is actually defined in the whole real line $\mathbb{R}$. Moreover, Theorem 3.4 guarantees that f is differentiable and such that

$$f'(x) = \sum_{n\geq 1} n \cdot \frac{1}{n!} x^{n-1} = \sum_{n\geq 0} \frac{1}{n!} x^n = f(x)$$

for every $x \in \mathbb{R}$. In turn, this gives $\frac{d}{dx}\left(e^{-x} f(x)\right) = 0$, so that $x \mapsto e^{-x} f(x)$ is constant in $\mathbb{R}$. Evaluating at $x = 0$ then yields $f(x) = e^x$ for every $x \in \mathbb{R}$.

For future use, we notice that changing x by ax in the equality of the previous example, we get

$$e^{ax} = \sum_{n \geq 0} \frac{a^n}{n!} x^n \tag{3.8}$$

for every $x \in \mathbb{R}$.

A variation of the argument used in Example 3.8 allows us to expand $f(x) = (1 + x)^\alpha$ in power series when $|x| < 1$ (and for a fixed real number $\alpha \neq 0$).

To this end, given $\alpha \in \mathbb{R}$ and $n \in \mathbb{Z} \setminus \{0\}$, we begin by defining the **generalized binomial number** $\binom{\alpha}{n}$ by letting $\binom{\alpha}{0} = 1$ and, for $n \geq 1$,

$$\binom{\alpha}{n} = \frac{\alpha(\alpha - 1)(\alpha - 2) \ldots (\alpha - n + 1)}{n!}. \tag{3.9}$$

Lemma 3.9 *Given $\alpha \in \mathbb{R}$ and $n \in \mathbb{N}$, we have:*

(a) $\binom{\alpha}{n} = \binom{\alpha-1}{n} + \binom{\alpha-1}{n-1}$.

(b) $\frac{n}{\alpha} \binom{\alpha}{n} = \binom{\alpha-1}{n-1}$, *for every $\alpha \neq 0$.*

(c) $\left| \binom{\alpha}{n} \right| \leq 1$ *whenever $|\alpha| \leq 1$.*

Proof

(a) Is an easy computation:

$$\binom{\alpha}{n} - \binom{\alpha-1}{n} = \frac{1}{n!} \alpha(\alpha - 1)(\alpha - 2) \ldots (\alpha - n + 1)$$

$$- \frac{1}{n!}(\alpha - 1)(\alpha - 2) \ldots (\alpha - n)$$

$$= \frac{1}{n!}(\alpha - 1)(\alpha - 2) \ldots (\alpha - n + 1)(\alpha - (\alpha - n))$$

$$= \frac{1}{(n-1)!}(\alpha - 1)(\alpha - 2) \ldots (\alpha - n + 1)$$

$$= \binom{\alpha - 1}{n - 1}.$$

(b) Follows immediately from (3.9).

(c) If $|\alpha| \leq 1$, it follows from (3.9) and the triangle inequality that

$$\left| \binom{\alpha}{n} \right| \leq \frac{|\alpha|(|\alpha| + 1)(|\alpha| + 2) \ldots (|\alpha| + n - 1)}{n!} \leq \frac{1 \cdot 2 \cdot \ldots \cdot n}{n!} = 1.$$

$\square$

The coming result is usually referred to as the **binomial series expansion** or **Newton's binomial theorem**, and is due to Sir Isaac Newton.[2]

Theorem 3.10 (Newton) *For $\alpha \neq 0$ and $|x| < 1$, we have*

$$(1+x)^\alpha = \sum_{n \geq 0} \binom{\alpha}{n} x^n. \tag{3.10}$$

Proof Let us firstly consider the case $0 < |\alpha| \leq 1$. Since $\left|\binom{\alpha}{n} x^n\right| \leq |x|^n$ (by item (c) of the previous lemma) and $\sum_{n \geq 0} |x|^n$ converges when $|x| < 1$, Proposition 3.7 assures that $\sum_{n \geq 0} \binom{\alpha}{n} x^n$ converges when $|x| < 1$. Hence, thanks to Theorem 3.4 the function $f : (-1, 1) \to \mathbb{R}$, given by $f(x) = \sum_{n \geq 0} \binom{\alpha}{n} x^n$, is differentiable, with

$$f'(x) = \sum_{n \geq 1} n \binom{\alpha}{n} x^{n-1} = \sum_{n \geq 1} \alpha \binom{\alpha - 1}{n - 1} x^{n-1} \tag{3.11}$$

(note that we have used item (c) of the previous lemma in the last equality above). The above expression for $f'(x)$, together with item (a) of the referred lemma, gives

$$(1+x)f'(x) = \alpha(1+x) \sum_{n \geq 1} \binom{\alpha - 1}{n - 1} x^{n-1}$$

$$= \alpha \left(\sum_{n \geq 1} \binom{\alpha - 1}{n - 1} x^{n-1} + \sum_{n \geq 1} \binom{\alpha - 1}{n - 1} x^n \right)$$

$$= \alpha \left(1 + \sum_{n \geq 2} \binom{\alpha - 1}{n - 1} x^{n-1} + \sum_{n \geq 2} \binom{\alpha - 1}{n - 2} x^{n-1} \right)$$

$$= \alpha \left(1 + \sum_{n \geq 2} \binom{\alpha}{n - 1} x^{n-1} \right)$$

$$= \alpha \sum_{n \geq 0} \binom{\alpha}{n} x^n = \alpha f(x).$$

Therefore, letting $g(x) = (1 + x)^{-\alpha} f(x)$, we get for $|x| < 1$ that

$$g'(x) = -\alpha(1+x)^{-\alpha-1} f(x) + (1+x)^{-\alpha} f'(x)$$

$$= (1+x)^{-\alpha-1} \{-\alpha f(x) + (1+x) f'(x)\} = 0.$$

---

[2]As quoted in [8], Newton is considered to be one of the greatest scientists ever, being difficult to properly address the scope of his contributions to the development of science.

In turn, this means that g is constant in the interval $(-1, 1)$. However, since $g(0) = 1$, we get $(1 + x)^{-\alpha} f(x) = 1$ for $|x| < 1$, as wished.

For the general case, assume that $(1 + x)^{\alpha} = \sum_{n \geq 0} \binom{\alpha}{n} x^n$ for some $\alpha \neq 0$ and every $x \in (-1, 1)$. We shall show that analogous formulas hold for $\alpha - 1$ and $\alpha + 1$ (and every $|x| < 1$):

(a) For $\alpha - 1$: Theorem 3.4 gives, for $|x| < 1$,

$$(1 + x)^{\alpha - 1} = \frac{1}{\alpha} \cdot \frac{d}{dx}(1 + x)^{\alpha} = \frac{1}{\alpha} \sum_{n \geq 1} n \binom{\alpha}{n} x^{n-1}$$

$$= \sum_{n > 1} \binom{\alpha - 1}{n - 1} x^{n-1} = \sum_{n \geq 0} \binom{\alpha - 1}{n} x^n.$$

(b) For $\alpha + 1$: we have

$$(1 + x)^{\alpha + 1} = (1 + x) \sum_{n \geq 0} \binom{\alpha}{n} x^n = \sum_{n \geq 0} \binom{\alpha}{n} x^n + \sum_{n \geq 0} \binom{\alpha}{n} x^{n+1}$$

$$= 1 + \sum_{n \geq 1} \left(\binom{\alpha}{n} + \binom{\alpha}{n - 1} \right) x^n = 1 + \sum_{n \geq 1} \binom{\alpha + 1}{n} x^n$$

$$= \sum_{n \geq 0} \binom{\alpha + 1}{n} x^n.$$

A straightforward inductive argument now shows that (3.10) holds for every $\alpha \neq 0$ and every $|x| < 1$. □

Corollary 3.11 *For $\alpha, \beta \neq 0$, we have*

$$(1 + \beta x)^{\alpha} = \sum_{n \geq 0} \binom{\alpha}{n} (\beta x)^n,$$

for every $x \in \mathbb{R}$ satisfying $|x| < \frac{1}{|\beta|}$.

Proof It suffices to apply (3.10) for βx in place of x, observing that $|\beta x| < 1 \Leftrightarrow |x| < \frac{1}{|\beta|}$. □

As an example of application of the former corollary, for $|x| < \frac{1}{2}$ we have

$$(1 - 2x)^{-1/2} = \sum_{n \geq 0} \binom{-1/2}{n} (-2x)^n,$$

with

$$\binom{-1/2}{n} = \frac{1}{n!}\left(-\frac{1}{2}\right)\left(-\frac{1}{2}-1\right)\cdots\left(-\frac{1}{2}-n+1\right)$$

$$= \frac{(-1)^n}{n!}\cdot\frac{1}{2}\cdot\frac{3}{2}\cdot\frac{5}{2}\cdots\frac{2n-1}{2}$$

$$= \frac{(-1)^n}{n!}\cdot\frac{(2n)!}{2^n\cdot2\cdot4\cdots(2n)}$$

$$= \frac{(-1)^n}{n!}\frac{(2n)!}{2^n\cdot2^n n!}$$

$$= \frac{(-1)^n}{4^n}\binom{2n}{n}.$$

Hence, for $|x| < \frac{1}{2}$ we get

$$(1-2x)^{-1/2} = \sum_{n\geq0}\frac{1}{2^n}\binom{2n}{n}x^n, \tag{3.12}$$

an equality that will be quite useful in the analysis of Example 3.14.

Problems: Sect. 3.2

1. * Generalize the result of Proposition 3.3, showing that if $f : (-r, r) \to \mathbb{R}$ has power series expansion $f(x) = \sum_{n\geq0} a_n x^n$, then, for $k \in \mathbb{N}$, we have $f(x)^k = \sum_{n\geq0} c_n x^n$, where

$$c_n = \sum a_{i_1} a_{i_2} \ldots a_{i_k}$$

and the sum extends over all k-tuples $(i_1, \ldots, i_k)$ of nonnegative integers satisfying $i_1 + \cdots + i_k = n$.

2. * Given $k \in \mathbb{N}$ and $|x| < 1$, prove that

$$\frac{1}{(1-x)^k} = \sum_{n\geq0}\binom{k+n-1}{n}x^n.$$

3. * The purpose of this problem is to show that the function $f : (-1, 1) \to \mathbb{R}$, given by $f(x) = \log(1+x)$, admits a power series expansion, as well as to get such an expansion. To this end, do the following items:

(a) If $f(x) = \sum_{n \geq 0} a_n x^n$, use Corollary 3.6 to conclude that $a_0 = 0$ and $a_n = \frac{(-1)^{n-1}}{n}$ for $n \geq 1$.

(b) Use Proposition 3.7, in conjunction with Theorem 3.4, to conclude that the power series $\sum_{n \geq 1} \frac{(-1)^{n-1}}{n} x^n$ defines a differentiable function $g : (-1, 1) \to \mathbb{R}$.

(c) Show that $g'(x) = \frac{1}{1+x}$ for every $x \in (-1, 1)$; then, conclude that $f(x) = g(x)$ for every $x \in (-1, 1)$.

4. * Let $(a_n)_{n \geq 0}$ be a sequence of real numbers such that $|a_n| \leq cM^n$, for some $c, M > 0$ and every $n \geq 0$.

(a) Use Proposition 3.7 to show that, for every $x \in \left(-\frac{1}{M}, \frac{1}{M}\right)$, the power series $\sum_{n \geq 0} \frac{a_n}{n+1} x^{n+1}$ converges.

(b) If $f, F : \left(-\frac{1}{M}, \frac{1}{M}\right) \to \mathbb{R}$ are given by $f(x) = \sum_{n \geq 0} a_n x^n$ and $F(x) = \sum_{n \geq 0} \frac{a_n}{n+1} x^{n+1}$, Theorem 3.4 assures that $F' = f$. Use this fact, together with the Fundamental Theorem of Calculus, to show that

$$\int_0^x \sum_{n \geq 0} a_n t^n \, dt = \sum_{n \geq 0} \frac{a_n}{n+1} x^{n+1},$$

i.e., that we can compute the integral of a function defined by a power series by *termwise integrating* the corresponding series.

5. Given $u, v \in \mathbb{R}$, with $v \neq 0$, let $(a_n)_{n \geq 1}$ be such that $a_{k+2} = ua_{k+1} + va_k$ for every $k \geq 1$. Moreover, assume that the equation $x^2 - ux - v = 0$ has real roots[3] α and β.

(a) If $\alpha \neq \beta$, show that $a_n = A\alpha^{n-1} + B\beta^{n-1}$ for every $n \geq 1$, where A and B are the solutions of the linear system $\begin{cases} A + B = a_1 \\ \alpha A + \beta B = a_2 \end{cases}$.

(b) If $\alpha = \beta$, show that $a_n = (A + Bn)\alpha^{n-1}$ for $n \geq 1$, where A and B are the solutions of the linear system $\begin{cases} A + B = a_1 \\ (A + 2B)\alpha = a_2 \end{cases}$.

6. * For $x \in \mathbb{R} \setminus \mathbb{Z}$, we define the **integer closest to** x, denoted $[x]$, as the single integer n such that $|x - n| < \frac{1}{2}$. In symbols,

$$n = [x] \Leftrightarrow |x - n| < \frac{1}{2}.$$

If $n > 1$ is an integer and d_n denotes the number of derangements of I_n, prove that $d_n = \left[\frac{n!}{e}\right]$.

---

[3]Actually, as will be seen in Chap. 21—cf. also Problem 6, page 393—such a restriction is unnecessary.

3.3   More Examples

In this section, we look at some recursive problems through the eyes of generating functions. Such examples were chosen with a twofold purpose: on the one hand, to illustrate the diversity of situations to which the method developed in the previous two sections can be applied; on the other, to work out some techniques which are themselves applicable to other relevant settings.

Example 3.12 Compute, with the aid of generating functions, the number of nonnegative integer solutions to the equation

$$a_1 + a_2 + \cdots + a_k = m,$$

where k and m are given natural numbers.

Solution Note that $a_1 + a_2 + \cdots + a_k = m$ if and only if $x^{a_1} x^{a_2} \ldots x^{a_k} = x^m$. Therefore, there are as many nonnegative integer solutions $(a_1, a_2, \ldots, a_k)$ of the given equation as there are ways of getting a summand x^m in the product

$$f(x) = \underbrace{(1 + x + x^2 + \cdots)(1 + x + x^2 + \cdots) \ldots (1 + x + x^2 + \cdots)}_{k}$$

(yet in another way, taking x^{a_1} in the first factor, x^{a_2} in the second, ..., x^{a_k} in the k-th factor, we get x^m in the product). However, for $|x| < 1$ we have

$$f(x) = \left(\frac{1}{1-x} \right)^k = \frac{1}{(1-x)^k} = \sum_{n \geq 0} \binom{k+n-1}{n} x^n,$$

where, in the last equality, we applied the result of Problem 2 of the previous section. Hence, the answer to our problem is the coefficient of x^m in the above series, i.e., $\binom{k+m-1}{m}$ (in turn, note that this agrees with the result of Problem 9, page 29.   □

The method of generating functions is particularly useful in the study of sequences $(a_n)_{n \geq 0}$ satisfying recurrence relations more general than those treated in the previous section. This is the case, for instance, for the sequence $(a_n)_{n \geq 0}$ such that $a_0 = 1$, $a_1 = -1$ and $a_n = -\frac{a_{n-1}}{n} + 2a_{n-2}$ for $n \geq 2$, since the coefficients of the recurrence relation are not constant. In this case, the idea is to consider the corresponding generating function, say $f(x) = \sum_{n \geq 0} a_n x^n$, and to follow the sequence of steps delineated below:

I. To use the initial values and the satisfied recurrence relation to conclude that the generating function converges on some interval of the form $(-r, r)$, with $r > 0$.

II. Again with the aid of the initial values and the given recurrence relation, to perform appropriate operations with the series expansion of f in order to get a *closed form* (i.e., a *formula*) for $f(x)$.

III. To expand in power series the formula for $f(x)$ obtained in the previous step.

IV. To use the uniqueness of power series representation, given by Corollary 3.6, to conclude that a_n is equal to the coefficient of x^n in the power series expansion of step III.

Note that the steps just described are precisely those which were followed in Example 3.2, the first having been concluded right after Proposition 3.7. We shall run through them once more, this time to solve the next example.

Example 3.13 Let $(a_n)_{n \geq 0}$ be the sequence given by $a_0 = 1$, $a_1 = -1$ and

$$a_n = -\frac{a_{n-1}}{n} + 2a_{n-2},$$

for every $n \geq 2$. Compute a_n as a function of n.

Solution Let $f(x) = \sum_{n \geq 0} a_n x^n$ be the generating function corresponding to the given sequence.

Step I: we shall try to apply the comparison test for power series, as given by Proposition 3.7. More precisely, assuming that $|a_{n-2}| \leq \alpha^{n-2}$ and $|a_{n-1}| \leq \alpha^{n-1}$, it comes from the triangle inequality, together with the given recurrence relation, that

$$|a_n| < \frac{|a_{n-1}|}{n} + 2|a_{n-2}| \leq \frac{\alpha^{n-1}}{n} + 2\alpha^{n-2}.$$

Therefore, we shall have $|a_n| \leq \alpha^n$ provided $\frac{\alpha^{n-1}}{n} + 2\alpha^{n-2} \leq \alpha^n$, or, which is the same,

$$\frac{\alpha}{n} + 2 \leq \alpha^2.$$

Since this last inequality is true for $\alpha = 2$ and every $n \geq 2$, and since $|a_0| \leq 2^0$ and $|a_1| \leq 2^1$, it follows by induction that $|a_n| \leq 2^n$ for every $n \geq 1$. Hence, Proposition 3.7 guarantees that f is defined and is differentiable in the open interval $\left(-\frac{1}{2}, \frac{1}{2} \right)$.

Step II: writing the given recurrence relation as $na_n = -a_{n-1} + 2na_{n-2}$ for $n \geq 2$, we have

$$f'(x) = \sum_{n \geq 1} n a_n x^{n-1} = a_1 + \sum_{n \geq 2} n a_n x^{n-1}$$

$$= a_1 + \sum_{n \geq 2} (-a_{n-1} + 2n a_{n-2}) x^{n-1}$$

$$= a_1 - \sum_{n \geq 2} a_{n-1} x^{n-1} + 2x \sum_{n \geq 2} n a_{n-2} x^{n-2}$$

$$= a_1 - (f(x) - a_0) + 2x \left(\sum_{n \geq 2} (n-2) a_{n-2} x^{n-2} + 2 \sum_{n \geq 2} a_{n-2} x^{n-2} \right)$$

$$= a_1 + a_0 - f(x) + 2x(f'(x) + 2f(x)).$$

However, since $a_1 + a_0 = 0$, we get $f'(x) = (4x - 1)f(x) + 2xf'(x)$ or, yet,

$$(2x - 1)f'(x) = -(4x - 1)f(x).$$

In order to *integrate* (i.e., to find the solutions of) the above differential equation, note first that f is positive in some interval $(-r, r)$, for some $0 < r \leq \frac{1}{2}$ (this comes from the fact that $f(0) = a_0 = 1 > 0$ and f, being differentiable, is continuous, hence has the same sign as $f(0)$ in a suitable neighborhood of 0). Thus, for $|x| < r$ we can write

$$\frac{f'(x)}{f(x)} = -\frac{4x - 1}{2x - 1} = -2 - \frac{1}{2x - 1}$$

and then, for $|x| < r \leq \frac{1}{2}$,

$$\log f(x) = \log f(t) \Big|_0^x = \int_0^x \frac{f'(t)}{f(t)} dt$$

$$= -\int_0^x \left(2 + \frac{1}{2t - 1} \right) dt$$

$$= -2x - \frac{1}{2} \log(1 - 2x).$$

Hence, for $|x| < r \leq \frac{1}{2}$ we have

$$f(x) = e^{-2x}(1 - 2x)^{-1/2}. \tag{3.13}$$

Step III: firstly, recall that the power series expansion of e^{-2x} is given by letting $a = -2$ in (3.8), and is valid in the whole real line:

$$e^{-2x} = \sum_{k \geq 0} \frac{(-2)^k}{k!} x^k.$$

Therefore, for f given as in (3.13) and $|x| < r \le \frac{1}{2}$, it follows from (3.12) and Proposition 3.3 that

$$f(x) = \left(\sum_{k \ge 0} \frac{(-2)^k}{k!} x^k \right) \left(\sum_{l \ge 0} \frac{1}{2^l} \binom{2l}{l} x^l \right)$$

$$= \sum_{n \ge 0} \left(\sum_{k+l=n} \frac{(-2)^k}{k!} \cdot \frac{1}{2^l} \binom{2l}{l} \right) x^n$$

$$= \sum_{n \ge 0} \left(\sum_{l=0}^{n} \frac{(-2)^{n-l}}{(n-l)!} \cdot \frac{1}{2^l} \binom{2l}{l} \right) x^n$$

$$= \sum_{n \ge 0} \left(2^n \sum_{l=0}^{n} \frac{(-1)^{n-l}}{4^l (n-l)!} \binom{2l}{l} \right) x^n .$$

Step IV: comparing the previous expansion with $f(x) = \sum_{n \ge 0} a_n x^n$, we get

$$a_n = 2^n \sum_{l=0}^{n} \frac{(-1)^{n-l}}{4^l (n-l)!} \binom{2l}{l} .$$

$\square$

In order to apply the circle of ideas discussed so far to counting problems of recursive character, apart from the four steps previously listed one has to execute a further one, namely: given a combinatorial problem in which one wants to compute, as a function of $n \in \mathbb{N}$, the total number a_n of possible configurations, one has to compute the initial values a_n and to get a recurrence relation satisfied by the sequence $(a_n)_{n \ge 1}$. Let us carefully examine a relevant example.

Example 3.14 Compute, as a function of $n \ge 3$, the number of distinct ways of partitioning a convex n-gon into triangles, by drawing diagonals which do not intersect in the interior of the polygon.

Solution Letting a_n be the number of partitions we wish to compute, it is clear that $a_3 = 1$. To obtain a recurrence relation valid for $n \ge 4$, label the vertices of the n-gon as $A_1, A_2, \ldots, A_n$, consecutively and in the counterclockwise sense. There are two possibilities:

(i) No diagonal departs from A_1: hence, diagonal $A_2 A_n$ must be drawn, so that we are left to partitioning the convex $(n-1)$-gon $A_2 A_3 \ldots A_n$, with no diagonals intersecting in its interior. The number of such partitions is, of course, a_{n-1}.

(ii) Diagonal $A_1 A_k$ is drawn, for some $3 \le k \le n-1$, but no diagonal $A_1 A_j$, with $j < k$, is drawn: then, the partition of $A_1 A_2 \ldots A_n$ induces partitions with the stated property in the convex polygons $A_1 A_2 \ldots A_k$ and $A_1 A_k A_{k+1} \ldots A_n$; moreover, in the partition of $A_1 A_2 \ldots A_k$, no diagonal departs from A_1.

Conversely, given any partition of $A_1 A_2 \ldots A_n$ with no diagonals intersecting in its interior but some diagonal departing from A_1, there exists a unique $3 \leq k \leq n-1$ such that the partition under consideration comes from exactly one pair of partitions of $A_1 A_2 \ldots A_k$ and $A_1 A_k A_{k+1} \ldots A_n$ as in the previous paragraph.

Now, if $k = 3$, there is exactly 1 way of partitioning $A_1 A_2 A_3$; if $3 < k \leq n - 1$, then arguing as in item (i) we conclude that there are exactly a_{k-1} ways of partitioning $A_1 A_2 \ldots A_k$ with no diagonal departing from A_1. A simple way of unifying these possibilities is to set $a_2 = 1$, and we shall make this convention henceforth.

Since there are a_{n-k+2} ways of partitioning $A_1 A_k A_{k+1} \ldots A_n$ as desired, success applications of the multiplicative and additive principles give exactly $\sum_{k=3}^{n-1} a_{k-1} a_{n-k+2}$ partitions of type (ii) for $A_1 A_2 \ldots A_n$.

Finally, taking both cases above into account (and recalling that $a_2 = 2$), we obtain the following recurrence relation for a_n, with $n \geq 4$:

$$
\begin{aligned}
a_n &= a_{n-1} + \sum_{k=3}^{n-1} a_{k-1} a_{n-k+2} \\
&= \sum_{k=3}^{n} a_{k-1} a_{n-k+2} \\
&= \sum_{k=2}^{n-1} a_k a_{n-k+1}.
\end{aligned}
$$

For what is to come, it is worth noting that the last expression above also makes sense for $n = 3$, and that if we set $a_1 = 0$ we can write it as

$$
a_n = \sum_{k=1}^{n} a_k a_{n-k+1}, \ \forall \, n \geq 3. \tag{3.14}
$$

As far as we know, there is no simple way to establish the convergence, in some open interval centered at 0, of the generating function $\sum_{n \geq 1} a_n x^n$ corresponding to the sequence $(a_n)_{n \geq 1}$. Nevertheless, it is not difficult for the reader to convince himself/herself that the fulfillment of this step is unnecessary *if* we obtain, in the end, a positional formula for a_n for which $\sum_{n \geq 1} a_n x^n$ does converge in some open interval centered at 0. We shall have more to say on this in a little while.

Back to the analysis of (3.14), note that the right hand side pretty much resembles formula (3.6) for the coefficients of the product of two convergent power series. Therefore, setting $f(x) = \sum_{n \geq 1} a_n x^n$, we are led to compute $f(x)^2$:

$$
\begin{aligned}
f(x)^2 &= \left(\sum_{k \geq 1} a_k x^k \right) \left(\sum_{l \geq 1} a_l x^l \right) = \sum_{n \geq 2} \left(\sum_{k+l=n} a_k a_l \right) x^n \\
&= \sum_{n \geq 2} \left(\sum_{k=1}^{n-1} a_k a_{n-k} \right) x^n = \sum_{n \geq 4} \left(\sum_{k=1}^{n-1} a_k a_{n-k} \right) x^n.
\end{aligned}
$$

Now, (3.14) gives $a_{n-1} = \sum_{k=1}^{n-1} a_k a_{n-k}$, so that the computations above give

$$f(x)^2 = \sum_{n \geq 4} a_{n-1} x^n = x \sum_{n \geq 3} a_n x^n = x(f(x) - x^2).$$

Putting $g(x) = \sum_{n \geq 0} a_{n+2} x^n$, we get $f(x) = x^2 g(x)$, and it follows from the equality above that

$$x^4 g(x)^2 = x(x^2 g(x) - x^2) = x^3(g(x) - 1),$$

i.e., $xg(x)^2 - g(x) + 1 = 0$. Thus,

$$g(x) = \frac{1 \pm \sqrt{1 - 4x}}{2x} = \frac{2}{1 \mp \sqrt{1 - 4x}},$$

and the fact that $g(0) = a_3 = 1$ gives

$$g(x) = \frac{1 - \sqrt{1 - 4x}}{2x} = \frac{2}{1 + \sqrt{1 - 4x}}.$$

(Recall that we are assuming that the power series defining f—and, hence, g—converges in some open interval of the form $(-r, r)$, for a certain $r > 0$; therefore, for the square roots above to represent real numbers we may assume—by diminishing r, if necessary—that $r < \frac{1}{4}$.)

Finally, we get from (3.12) and Lemma 3.9 that

$$
\begin{aligned}
g(x) &= \frac{1}{2x}\left(1 - \sum_{n \geq 0} \binom{1/2}{n}(-4x)^n\right) \\
&= -\frac{1}{2x} \sum_{n \geq 1} \frac{1}{2n}\binom{-1/2}{n-1}(-4x)^n \\
&= -\sum_{n \geq 1} \frac{1}{4n} \cdot \frac{(-1)^{n-1}}{4^{n-1}}\binom{2(n-1)}{n-1}(-4)^n x^{n-1} \\
&= \sum_{n \geq 1} \frac{1}{n}\binom{2(n-1)}{n-1} x^{n-1} = \sum_{n \geq 0} \frac{1}{n+1}\binom{2n}{n} x^n.
\end{aligned}
$$

Hence, letting $C_n = \frac{1}{n+1}\binom{2n}{n}$ for $n \geq 0$, it follows from $x^2 g(x) = f(x) = \sum_{n \geq 1} a_n x^n$ that

$$a_n = \frac{1}{n-1}\binom{2(n-2)}{n-2} = C_{n-2}$$

for $n \geq 3$.

At this point, rigorously speaking we do not know whether or not the above formula does solve the problem under consideration, for, in order to obtain it, we assumed the convergence of the power series $\sum_{n\geq 1} a_n x^n$ in some open interval centered at 0. Yet in another way, since we cannot guarantee *a priori* the convergence of $\sum_{n\geq 1} a_n x^n$ in some interval of the form $(-r, r)$, with $r > 0$, the whole process above is nothing but a heuristic reasoning that made us infer that C_{n-2} is, most likely, the correct value for a_n. Nevertheless, in order to conclude it suffices to show that, letting $a_1 = 0$, $a_2 = 1$ and $a_n = C_{n-2}$ for $n \geq 3$, we have $a_n = \sum_{k=1}^n a_k a_{n-k+1}$ for every $n \geq 3$.

For what is left to do, let us start by showing that the power series $\sum_{n\geq 0} C_n x^n$ converges in some open interval centered at 0, which is quite easy: since

$$\frac{1}{n+1}\binom{2n}{n} \leq \binom{2n}{n} \leq 2^{2n} = 4^n$$

for $n \geq 0$, it suffices to apply Proposition 3.7 to conclude that the series converges in the interval $\left(-\frac{1}{4}, \frac{1}{4}\right)$. Thus, letting

$$g(x) = \sum_{n\geq 0} C_n x^n = \sum_{n\geq 0} \frac{1}{n+1}\binom{2n}{n} x^n$$

and reverting the previous steps, we conclude that

$$g(x) = \frac{1 - \sqrt{1 - 4x}}{2x}$$

and, hence, that $xg(x)^2 - g(x) + 1 = 0$. Therefore, letting $f(x) = x^2 g(x)$, we get

$$f(x) = \sum_{n\geq 2} C_{n+2} x^n = \sum_{n\geq 1} a_n x^n$$

and $f(x)^2 = x(f(x) - x^2)$. Hence,

$$\left(\sum_{n\geq 1} a_n x^n\right)^2 = x\left(\sum_{n\geq 1} a_n x^n - x^2\right),$$

an equality that, once expanded, furnishes for $n \geq 3$ the desired recurrence relation,

$$a_n = \sum_{k=1}^n a_k a_{n-k+1}.$$

Thus, this shows that the sequence $(a_n)_{n\geq 1}$ such that $a_n = C_{n-2}$ for $n \geq 3$ is, indeed, the only solution to our problem. $\square$

Yet with respect to the previous example, we say that

$$C_n = \frac{1}{n+1}\binom{2n}{n}, \tag{3.15}$$

$n \geq 0$, is the n-th **Catalan number.**[4] Since $a_n = C_{n-2}$, $n \geq 3$, satisfies the recurrence relation (3.14), it is immediate to verify that the sequence $(C_n)_{n\geq 0}$ satisfies the recurrence relation

$$C_n = \sum_{k=0}^{n-1} C_k C_{n-k-1}, \tag{3.16}$$

which is then known as **Catalan's recurrence relation**. Problem 5 brings another combinatorial situation modeled by Catalan's recurrence.

We close this chapter by using generating functions to present an alternative proof to Example 5.27 of [8]. In order to do so, we shall need to recall two analytical facts, collected below.

Lemma 3.15 *The series* $\sum_{n\geq 1} \frac{1}{n^2}$ *and* $\sum_{n\geq 1} \frac{(-1)^{n-1}}{n^2}$ *converge. Moreover,*

$$\sum_{n\geq 1} \frac{(-1)^{n-1}}{n^2} = \frac{1}{2} \sum_{n\geq 1} \frac{1}{n^2}.$$

Proof The convergence of the series is quite standard, and can be found in most Calculus or Introductory Analysis texts (see, for instance, Example 7.39 and Proposition 7.48 of [8]). For what is left to do, it suffices to observe that, if $s = \sum_{n\geq 1} \frac{1}{n^2}$, then

$$\sum_{n\geq 1} \frac{(-1)^{n-1}}{n^2} = \sum_{n\geq 1} \frac{1}{n^2} - 2\sum_{k\geq 1} \frac{1}{(2k)^2} = s - \frac{1}{2}\sum_{k\geq 1} \frac{1}{k^2} = s - \frac{s}{2} = \frac{s}{2}.$$

□

The proof of the second result we shall need is considerably more refined than those of the other Calculus results used so far. Therefore, we shall limit ourselves to state it, referring the interested reader to Theorem 8.2 of [33] or (for a guided approach) to problem 7 of Section 11.3 of [8].

Lemma 3.16 *Assume that the power series* $\sum_{n\geq 0} a_n x^n$ *converges on the interval* $(-1, 1)$. *If the number series* $\sum_{n\geq 0} a_n$ *also converges, then*

---

[4]After Eugène Catalan, Belgian mathematician of the nineteenth century.

$$\lim_{x \to 1-} \sum_{n \geq 0} a_n x^n = \sum_{n \geq 0} a_n.$$

We can finally present the promised example. Apart from some details, we follow Chapter 17 of [23].

Example 3.17 Let $n > 1$ be an integer and $A = \{a_1, a_2, \ldots, a_n\}$ be a set of n positive integers, such that the sums of the elements of any two of its nonempty subsets are distinct. Prove that

$$\frac{1}{a_1} + \frac{1}{a_2} + \cdots + \frac{1}{a_n} < 2.$$

Proof We first claim that it is enough to prove that $\sum_{i=1}^{n} \frac{1}{a_i} \leq 2$. Indeed, this being the case, let $A' = A \cup \{2^N\}$, with $2^N > a_1 + \cdots + a_n$. The choice of N guarantees that any two nonempty subsets of A' also have distinct sums of elements, so that

$$\sum_{i=1}^{n} \frac{1}{a_i} < \sum_{i=1}^{n} \frac{1}{a_i} + \frac{1}{2^N} \leq 2.$$

For what is left to do, if $0 < x < 1$, then, in the notations of the paragraph that precedes Problem 12, page 22, the stated conditions assure that

$$\prod_{i=1}^{k}(1 + x^{a_i}) = 1 + \sum_{\emptyset \neq S \subset A} x^{\sigma(S)} \leq 1 + \sum_{j \geq 1} x^j = \frac{1}{1 - x}.$$

Taking natural logarithms and invoking the result of Problem 3, page 78, we successively get

$$\sum_{i=1}^{k} \log(1 + x^{a_i}) \leq -\log(1 - x)$$

and

$$\sum_{i=1}^{k} \left(x^{a_i} - \frac{x^{2a_i}}{2} + \frac{x^{3a_i}}{3} - \cdots \right) \leq x + \frac{x^2}{2} + \frac{x^3}{3} + \cdots,$$

for $0 < x < 1$. Dividing both sides above by x, we get the inequality

$$\sum_{i=1}^{k} \left(x^{a_i - 1} - \frac{x^{2a_i - 1}}{2} + \frac{x^{3a_i - 1}}{3} - \cdots \right) \leq 1 + \frac{x}{2} + \frac{x^2}{3} + \cdots,$$

which is valid on all of $[0, 1)$.

Now, fix $x \in (0, 1)$ and integrate the last inequality above on the interval $(0, x)$ to get

$$\sum_{i=1}^{k} \int_0^x \left(t^{a_i-1} - \frac{t^{2a_i-1}}{2} + \frac{t^{3a_i-1}}{3} - \cdots \right) dt \leq \int_0^x \left(1 + \frac{t}{2} + \frac{t^2}{3} + \cdots \right) dt$$

or, according to Problem 4, page 79,

$$\sum_{i=1}^{k} \frac{1}{a_i} \left(x - \frac{x^{2a_i}}{2^2} + \frac{x^{3a_i}}{3^2} - \cdots \right) \leq x + \frac{x^2}{2^2} + \frac{x^3}{3^2} + \cdots .$$

At this point, the simultaneous application of the first part of Lemma 3.15 and of Lemma 3.16 guarantee that, as $x \to 1-$, the limits of both left and right hand sides above do exist and can be computed termwise. Hence, by taking those limits, we conclude that

$$\sum_{i=1}^{k} \frac{1}{a_i} \left(1 - \frac{1}{2^2} + \frac{1}{3^2} - \cdots \right) \leq 1 + \frac{1}{2^2} + \frac{1}{3^2} + \cdots .$$

Finally, the second part of Lemma 3.15 furnishes

$$\sum_{i=1}^{k} \frac{1}{a_i} < \frac{\sum_{j\geq 1} \frac{1}{j^2}}{\sum_{j\geq 1} \frac{(-1)^{j-1}}{j^2}} = 2.$$

$\square$

Problems: Sect. 3.3

1. Use generating functions to solve the recurrence $a_1 = 2$ and $a_{k+1} = a_k + (k+1)$, for $k \geq 1$.
2. The sequence $(a_n)_{n\geq 0}$ is given by $a_0 = 1$ and $a_{n+1} = 2a_n + n$ for $n \geq 0$. In order to compute a_n in terms of n, do the following items:

   (a) Assuming that $a_n \leq \alpha^n$, conclude that $a_{n+1} \leq \alpha^{n+1}$, provided $\alpha^n (\alpha - 2) \geq n$. Then, show that $a_n \leq 3^n$ for every $n \geq 0$.
   (b) Show that the generating function of $(a_n)_{n\geq 0}$ converges in the interval $\left(-\frac{1}{3}, \frac{1}{3} \right)$ and is given by $f(x) = \frac{1-2x+2x^2}{(1-x)^2(1-2x)}$.
   (c) Find real constants A, B and C such that

$$\frac{1 - 2x + 2x^2}{(1 - x)^2(1 - 2x)} = \frac{A}{(1 - x)^2} + \frac{B}{1 - x} + \frac{C}{1 - 2x}.$$

(d) Expand each summand of the right hand side above as a power series to conclude that $a_n = 2^{n+1} - (n + 1)$ for $n \geq 0$.

3. Given $k, m \in \mathbb{N}$, use generating functions to compute the number of integer solutions of the equation $a_1 + a_2 + \cdots + a_k = m$, such that $a_i \geq 1$ for $1 \leq i \leq k$.
4. Use generating functions to compute the number of nonnegative integer solutions of the equation $a_1 + a_2 + a_3 + a_4 = 20$, satisfying $a_1 \geq 2$ and $a_3 \leq 7$.
5. A particle moves on the cartesian plane in such a way that from point (a, b) it can go to either $(a + 1, b)$ or $(a, b + 1)$. Given $n \in \mathbb{N}$, let a_n be the number of distinct ways the particle has to go from $A_0(0, 0)$ to $A_n(n, n)$, without ever touching a point (x, y) situated above the bisector of odd quadrants (i.e., one such point for which $y > x$). In this respect, do the following items:

(a) Let $A_k(k, k)$, with $0 \leq k < n$. Prove that there are exactly $a_k a_{n-1-k}$ distinct trajectories for the particle in which A_k is the last point (before A_n) on the line $y = x$.
(b) Conclude that $a_n = \sum_{k=0}^{n-1} a_k a_{n-1-k}$ and, hence, that $a_n = C_n$ for $n \geq 1$, where C_n is the n-th Catalan number.

For the coming problem, the reader may find it convenient to read again the paragraph that precedes Example 1.15.

6. For $n \in \mathbb{N}$, we let a_n denote the number of partitions of n in natural summands, none of which exceeds 3. The purpose of this problem is to compute a_n as a function of n, and to this end do the following items:

(a) Show that, for $|x| < 1$, one has

$$\sum_{n \geq 1} a_n x^n = \frac{1}{(1 - x)(1 - x^2)(1 - x^3)}.$$

(b) Find $a, b, c, d \in \mathbb{R}$ for which

$$\frac{1}{(1 - x)(1 - x^2)(1 - x^3)} = \frac{a}{(1 - x)^3} + \frac{b}{(1 - x)^2} + \frac{c}{1 - x^2} + \frac{d}{1 - x^3}.$$

(c) Conclude that

$$a_n = \begin{cases} \frac{1}{6}\binom{n+2}{2} + \frac{1}{4}(n + 1) + \frac{7}{12}, & \text{if } 6 \mid n \\ \frac{1}{6}\binom{n+2}{2} + \frac{1}{4}(n + 1) + \frac{1}{4}, & \text{if } 2 \mid n \text{ but } 3 \nmid n \\ \frac{1}{6}\binom{n+2}{2} + \frac{1}{4}(n + 1) + \frac{1}{3}, & \text{if } 2 \nmid n \text{ but } 3 \mid n \\ \frac{1}{6}\binom{n+2}{2} + \frac{1}{4}(n + 1), & \text{if } 2, 3 \nmid n \end{cases}.$$

Let a_n count a certain number of configurations in terms of $n \in \mathbb{N}$. Up to this point, we have tried to explicitly compute a_n with the aid of the ordinary generating function $\sum_{n\geq 0} a_n x^n$ of the sequence $(a_n)_{n\geq 0}$. Nevertheless, there are some instances in which is more adequate to invoke the **exponential generating function** of $(a_n)_{n\geq 0}$, i.e., the power series

$$\sum_{n\geq 0} \frac{a_n}{n!} x^n .$$

The remaining problems present a few such situations.

7. As in Example 2.5, let d_n denote the number of distinct derangements of I_n. The purpose of this problem is to use the exponential generating functions $\sum_{n\geq 2} \frac{d_n}{n!} x^n$ to compute d_n as a function of n. To this end, do the following items:

   (a) Given integers $n \geq 3$ and $1 \leq k \leq n - 1$, let $(a_1, \ldots, a_n)$ be a derangement of $\{1, 2, \ldots, n\}$ for which $a_n = k$.

       i. If $a_k = n$, prove that there are exactly d_{n-2} such permutations.
       ii. If $a_k \neq n$, prove that there are exactly d_{n-1} such permutations.

   (b) Conclude from (a) that $d_n = (n-1)(d_{n-1} + d_{n-2})$ for $n \geq 3$.
   (c) Imposing that $d_k \leq k! \alpha^k$ for every $k < n$ and some $\alpha > 0$, show that $d_n \leq n! \alpha^n$ if $\alpha \geq 2$. Then, conclude that the series $\sum_{n\geq 2} \frac{d_n}{n!} x^n$ converges on the interval $(-\frac{1}{2}, \frac{1}{2})$ and, hence, defines a differentiable function $f : (-\frac{1}{2}, \frac{1}{2}) \to \mathbb{R}$.
   (d) Use the result of (b) to show that $(1-x)f'(x) = x(f(x) + 1)$.
   (e) Since $f(0) = \frac{1}{2}$, by taking an adequate real number $0 < r < \frac{1}{2}$, if needed, we can assume that $f(x) > 0$ in $(-r, r)$. Conclude that, in $(-r, r)$, one has

   $$\frac{d}{dx} \log(f(x) + 1) = -1 + \frac{1}{1-x}.$$

   (f) Deduce from (c) that $f(x) = -1 + \frac{1}{1-x} \cdot e^{-x}$.
   (g) Use (3.1), (3.8) and Proposition 3.3 to obtain

   $$f(x) = \sum_{n\geq 1} \left(\sum_{l=0}^{n} \frac{(-1)^l}{l!} \right) x^n .$$

   (h) Finally, compute d_n as a function of n with the aid of (g) and Corollary 3.6.

In the previous problem, the convenience of the use of the exponential generating function of $(d_n)_{n\geq 2}$ was mostly due (cf. item (c)) to the fact that $\sum_{n\geq 2} \frac{d_n}{n!} x^n$ converges at some interval centered at 0, whereas $\sum_{n\geq 2} d_n x^n$ does not. Nevertheless, in the coming two problems, the involved combinatorics will compel us to use exponential, instead of ordinary, generating functions.

8. We want to compute, in terms of $n \in \mathbb{N}$, the number s_n of sequences of n terms, all of which equal to 0, 1, 2 or 3 and having an even number of 0's. To this end, do the following items:

(a) Show that

$$s_n = \sum \frac{n!}{(2a_0)! a_1! a_2! a_3!},$$

the sum extending over all nonnegative values of a_0, a_1, a_2, a_3 such that $2a_0 + a_1 + a_2 + a_3 = n$.

(b) Set $s_0 = 1$ and conclude that $\sum_{n \geq 0} \frac{s_n}{n!} x^n$ is the exponential generating function of

$$f(x) = \left(1 + \frac{x^2}{2!} + \frac{x^4}{4!} + \cdots\right)\left(1 + \frac{x}{1!} + \frac{x^2}{2!} + \cdots\right)^3.$$

(c) Show that $f(x) = \frac{1}{2}(e^{4x} + e^{2x})$, and hence that $s_n = \frac{1}{2}(4^n + 2^n)$.

9. One can also use exponential generating functions to compute the number of surjective functions between two finite sets (cf. Example 2.6). To this end, given $m, n \in \mathbb{N}$ and finite sets A and B such that $|A| = m$ and $|B| = n$, do the following items:

(a) Show that the number of surjections $f : A \rightarrow B$ is equal to the number of ways of distributing m distinct objects among n distinct boxes, in such a way that no box remains empty.

(b) Conclude that the number of surjections $f : A \rightarrow B$ is given by

$$\sum \frac{m!}{a_1! a_2! \dots a_n!},$$

the above sum extending over all n-tuples $(a_1, \dots, a_n)$ of positive integers such that $a_1 + \cdots + a_n = m$.

(c) If $g(x) = \left(x + \frac{x^2}{2!} + \frac{x^3}{3!} + \cdots\right)^n$, use the result of Problem 1, page 78, to show that the term corresponding to x^m in the expansion of g is given by

$$\sum \frac{x^{a_1}}{a_1!} \cdot \frac{x^{a_2}}{a_2!} \cdots \frac{x^{a_n}}{a_n!},$$

the above sum extending over all n-tuples $(a_1, \dots, a_n)$ of positive integers such that $a_1 + \cdots + a_n = m$. Conclude that the number of desired surjective functions is equal to the coefficient of $\frac{x^m}{m!}$ in the expansion of g.

(d) Use (3.8) to show that

$$g(x) = (e^x - 1)^n = \sum_{k=0}^{n} \binom{n}{k} (-1)^k e^{(n-k)x}.$$

Then, expanding $e^{(n-k)x}$ in power series with the aid of (3.8), conclude from item (c) that the number of surjections we want to compute is equal to

$$\sum_{k=0}^{n} (-1)^k \binom{n}{k} (n-k)^m.$$

Chapter 4
Existence of Configurations

Generally speaking, Combinatorics deals with two different kinds of problems: those in which we want to count the number of distinct ways of making a certain choice and those in which we want to make sure that some configuration does appear. In order to solve a problem of the first kind above, we employ, among others, the counting techniques discussed in the previous chapters. For the second kind of problem, up to this moment we have not developed any idea that could be systematically used. It is our purpose in this chapter to remedy this state of things. To this end, we start by discussing the famous *pigeonhole's principle* of Dirichlet, along with several interesting examples. Then we move on to some applications of the principle of mathematical induction to the existence of configurations. The chapter continues with the study of partial order relations, exploring Mirsky's theorem on the relation between chains and anti-chains. We close the chapter by explaining how, in some situations, the search for an adequate invariant or a semi-invariant can give the final outcome of certain seemingly random algorithms.

4.1 Pigeonhole's Principle

In its simplest version, **pigeonhole's principle**, also known as **Dirichlet's principle**,[1] can be stated as follows.

Proposition 4.1 (Pigeonhole's Principle I) *If we distribute n pigeons into n − 1 cages (the* pigeonholes*), then at least one cage will necessarily contain at least two pigeons.*

---

[1] After the German mathematician of the nineteenth century Gustav L. Dirichlet. Actually, we shall have to wait until Sect. 7.2 (cf. Lemma 7.7) to appreciate the original problem that motivated Dirichlet to introduce the pigeonhole principle.

Proof By contraposition, if each one of the $n - 1$ cages was to contain at most one pigeon, then we would have at most $n - 1$ pigeons.   □

As naive as it seems, the pigeonhole principle allows us to establish astonishing consequences, as shown by the coming examples. In all that follows, we shall always assume that the relation of getting acquainted to someone else is *symmetric*, i.e., if A knows B, then B also knows A.

Example 4.2 There are n guests in a party. Show that we can always find two of them who, within the party, are acquainted with the same number of people.

Proof Firstly, note that each of the n guests is acquainted with at least 0 and at most $n - 1$ other ones within the party. Therefore, there are two cases to consider:

(i) Each guest knows at least one other person in the party: take $n - 1$ rooms, numbered from 1 to $n - 1$, and put in room i all of the guests (if any) who know exactly i other guests. Since we have $n - 1$ rooms (the pigeonholes) and n guests (the pigeons), Dirichlet's principle assures that at least one room will contain at least two people. By the way we have allocated the guests in the rooms, these two guests know, within the party, the same number of other people.

(ii) There exists at least one "guest" who, actually, has no acquaintances within the party: then, no guest knows all of the other $n - 1$ people, so that, in this case, we can number the rooms from 0 to $n - 2$ and reason as in (i). Once more, pigeonhole's principle guarantees that at least one of the rooms will contain at least two people. Also as in (i), these two people know, within the party, the same number of other people.

                                                                                □

Example 4.3 (IMO Shortlist) Each subset of the set $\{1, 2, \ldots, 10\}$ is painted with one, our of n possible colors. Find the greatest possible value of n for which one can always find two distinct and nonempty sets $A, B \subset \{1, 2, \ldots, 10\}$, such that A, B and $A \cup B$ are all painted with the same color.

Solution Let $X = \{1, 2, \ldots, 10\}$. Given ten colors $C_1, C_2, \ldots, C_{10}$, let us paint the k-element subsets of X with color C_k, for $1 \le k \le 10$. This way, given two distinct and nonempty subsets A and B of X, there are two possibilities: either A and B have different numbers of elements, and hence were painted with different colors, or A and B have the same number of elements; in this second case, $A \cup B$ has more elements than A and B, and thus was painted with a color different from that of A and B. Therefore, we conclude that $n \le 9$.

Now, take only nine colors, and let $A_i = \{1, \ldots, i\}$ for $1 \le i \le 10$. Since

$$A_1 \subset A_2 \subset \cdots \subset A_{10}$$

is a *chain* of ten distinct and nonempty subsets of X (the pigeons) but we have only nine colors (the pigeonholes), Dirichlet's principle assures the existence of indices $1 \leq i < j \leq 10$ such that A_i and A_j are painted with a single color. Since $A_i \cup A_j = A_j$, it suffices to let $A_i = A$ and $A_j = B$ to fulfill the stated conditions. Thus, the largest possible value of n is 9.   □

There are several interesting generalizations of the pigeonhole principle, and the coming result collects the one who is probably the most important of them. For what is to come, recall (cf. the paragraph preceding Problem 7, page 41) that the **integer part** of a real number x, denoted $\lfloor x \rfloor$, is the greatest integer which is less than or equal to x. Thus, if $\lfloor x \rfloor = n \in \mathbb{Z}$, then $n \leq x < n + 1$. It is frequently useful to write down such inequalities as

$$\lfloor x \rfloor \leq x < \lfloor x \rfloor + 1. \tag{4.1}$$

Proposition 4.4 (Pigeonhole's Principle II) *If n pigeons are distributed among k different cages, then at least one cage will contain at least $\left\lfloor \frac{n-1}{k} \right\rfloor + 1$ pigeons.*

Proof Arguing once more by contraposition, let us suppose that each of the k cages contains at most $\left\lfloor \frac{n-1}{k} \right\rfloor$ pigeons. Then, altogether we have at most $k \left\lfloor \frac{n-1}{k} \right\rfloor$ pigeons, and it suffices to note that, by the left inequality in (4.1),

$$k \left\lfloor \frac{n-1}{k} \right\rfloor \leq k \left(\frac{n-1}{k} \right) = n - 1 < n.$$

□

Let us examine a few other examples.

Example 4.5 Prove that, in any group of twenty people, at least three of them were born in the same day of the week.

Proof Let the twenty people be the pigeons and the 7 days of the week be the cages. Then, associate a person to a day of the week if he/she was born in that day. The second version of the pigeonhole principle guarantees that at least one cage will contain at least $\left\lfloor \frac{20-1}{7} \right\rfloor + 1 = 3$ people. Thus, these three people were born in the same day of the week.   □

Example 4.6 (Leningrad[2]) Each 1×1 square of a 5×41 chessboard is painted either red or blue. Prove that it is possible to choose three lines and three columns of the chessboard so that the nine 1×1 squares into which they intersect are painted with the same color.

Proof Since we have used only two colors, the second version of the pigeonhole principle implies that, in each column (of five 1×1 squares), one of the two colors must occur at least $\lfloor \frac{5-1}{2} \rfloor + 1 = 3$ times. Call such a color *dominant* and, for each

---

[2]Former name of the Russian city of Saint Petersburg.

of the 41 columns, take note of its dominant color. Since one out of the two colors dominates in each of the 41 columns, by invoking once more the second version of the pigeonhole principle we conclude that one of the two colors is the dominant one in at least $\lfloor \frac{41-1}{2} \rfloor + 1 = 21$ columns. Assume, without loss of generality, that such a color is red, and call a column as *red* if red is its dominant color. Then, choose 21 red columns and, in each of them, choose three 1×1 red squares.

Notice that there are exactly $\binom{5}{2} = 10$ possible ways of choosing three of the five 1×1 squares of each column. Therefore, out of the 21 red columns (and once again from the second version of the pigeonhole principle), the three 1×1 chosen red squares occupy exactly the same positions in at least $\lfloor \frac{21-1}{10} \rfloor + 1 = 3$ columns.

In other words, we have at least three columns having 1×1 red squares in the same set of three lines, and this is exactly what we were asked to prove.   □

The next example uses an obvious generalization of the pigeonhole principle for the case in which we have a finite number of cages and an infinite number of pigeons.

Example 4.7 (Bosnia) A set A of natural numbers is said to be *good* if, for some $n \in \mathbb{N}$, the equation $x - y = n$ has infinitely many solutions (x, y) with $x, y \in A$. If $\mathbb{N} = A_1 \cup A_2 \cup \cdots \cup A_{100}$, prove that at least one of the sets A_i is good.

Proof By the first version of Dirichlet's principle, out of each set of 101 natural numbers (the pigeons), at least two will belong to a single one of the 100 sets A_1, $A_2, \ldots, A_{100}$ (the cages). In particular, for each $k \in \mathbb{N}$, at least two of the 101 natural numbers $101(k - 1) + 1, 101(k - 1) + 2, \ldots, 101k$, which we shall call x_k and y_k, belong to a single one of the 100 sets $A_1, A_2, \ldots, A_{100}$.

Assume, without loss of generality, that $x_k > y_k$ for each k. This way, we obtain infinitely many pairs $x_1 > y_1, x_2 > y_2, \ldots$ of natural numbers satisfying the following conditions:

(i) the two numbers in each pair are in one of the sets $A_1, A_2, \ldots, A_{100}$;
(ii) $1 \leq x_k - y_k \leq 100$ for each $k \geq 1$;
(iii) $i > j \Rightarrow x_i > y_i > x_j > y_j$.

Looking at the integers $k \geq 1$ as the pigeons and at the 100 sets $A_1, A_2, \ldots, A_{100}$ as the cages, we conclude (by the obvious generalization of the pigeonhole principle alluded to above) that there exist an index $1 \leq m \leq 100$ and an infinite set $B \subset \mathbb{N}$ such that

$$i \in B \Rightarrow x_i, y_i \in A_m.$$

We claim that A_m is a good set. Indeed, since $x_i - y_i \in \{1, 2, 3, \ldots, 100\}$ for each $i \in B$, a further application of the obvious generalization of the pigeonhole principle guarantees the existence of an integer $1 \leq n \leq 100$ and an infinite set $C \subset B$ such that

$$i \in C \Rightarrow x_i - y_i = n.$$

However, since $i \in C \Rightarrow x_i, y_i \in A_m$, we conclude from (iii) that the equation $x - y = n$ admits infinitely many solutions in A_m. □

As our next application of Dirichlet's principle, we shall prove a particular case of a famous result of the Hungarian mathematicians Paul Erdős and Esther Szekeres (cf. [15]), which is known in mathematical literature as **Erdős-Szekeres theorem** (for another proof of it, see Sect. 4.3).

Example 4.8 (Erdős-Szekeres) Out of any sequence of $n^2 + 1$ distinct real numbers, one can always find a monotone subsequence of at least $n + 1$ terms.

Proof Let $A = \{x_1, x_2, \dots, x_{n^2+1}\}$ be the set whose elements are the $n^2 + 1$ terms of the given sequence, and assume that the given sequence does not possess any increasing subsequence with at least $n + 1$ terms. Then, every increasing subsequence starting at a certain $x \in A$ has at most n terms. Define

$$f : A \to \{1, 2, \dots, n\}$$

by letting $f(x) = $ *size of the largest increasing subsequence of the given sequence, starting at x.* Since A has $n^2 + 1$ elements, the pigeonhole principle guarantees that we should have at least

$$\left\lfloor \frac{(n^2 + 1) - 1}{n} \right\rfloor + 1 = n + 1$$

elements of A with the same image, say $x_{i_1}, x_{i_2}, \dots, x_{i_{n+1}}$, with $i_1 < i_2 < \cdots < i_{n+1}$. If $x_{i_j} < x_{i_l}$ for some $1 \le j < l \le n + 1$, we would have $f(x_{i_j}) > f(x_{i_l})$, for we could enlarge an increasing sequence starting at x_{i_l} by placing x_{i_j} before it. But since $f(x_{i_j}) > f(x_{i_l})$ is impossible, we conclude that

$$x_{i_1} > x_{i_2} > \cdots > x_{i_{n+1}}.$$

Thus, we have found a decreasing subsequence of $n + 1$ terms. □

We end this section by presenting four additional examples which illustrate the wide versatility of applications of Dirichlet's principle.

Example 4.9 (South Korea) For each positive integer m, prove that there exist integers a and b such that $|a|, |b| \le m$ and

$$0 < a + b\sqrt{2} \le \frac{1 + \sqrt{2}}{m + 2}.$$

Proof Let A be the set of positive reals of the form $a + b\sqrt{2}$, with $0 \le a, b \le m$ and a and b not both equal to 0. Since $x_1 + y_1\sqrt{2} = x_2 + y_2\sqrt{2}$, with $x_1, x_2, y_1, y_2 \in \mathbb{Z}$, implies $x_1 = x_2$ and $y_1 = y_2$, we conclude, with the aid of the fundamental principle of counting, that A has exactly $(m + 1)^2 - 1 = m^2 + 2m$ elements, the largest one being $m(1 + \sqrt{2})$.

Let $\alpha = \frac{1+\sqrt{2}}{m+2}$, and $I_j = ((j-1)\alpha, j\alpha]$, with $1 \le j \le m^2 + 2m$. The union of the I_j's is the interval $(0, m(1 + \sqrt{2})]$, which contains A. If one of the elements of A belongs to I_1, there is nothing left to do. Otherwise, there will be left $m^2 + 2m - 1$ intervals to contain all $m^2 + 2m$ elements of A, and the Dirichlet principle assures that at least one of these intervals, say I_k, contains at least two elements of A, say $x_1 + y_1\sqrt{2}$ and $x_2 + y_2\sqrt{2}$, with $x_1 + y_1\sqrt{2} < x_2 + y_2\sqrt{2}$. Now, since I_k has length α, we get

$$0 < (x_2 - x_1) + (y_2 - y_1)\sqrt{2} = (x_2 + y_2\sqrt{2}) - (x_1 + y_1\sqrt{2}) < \alpha.$$

Finally, noticing that $|x_2 - x_1|, |y_2 - y_1| \le m$, we conclude that it suffices to let $a = x_2 - x_1$ and $b = y_2 - y_1$.   □

The next example uses the existence part of the Fundamental Theorem of Arithmetic (Theorem 6.43), together with the notation of congruence modulo 2 (cf. Example 2.21).

Example 4.10 (IMO) Let A be a set of 1985 positive integers, none of which has a prime divisor larger than 26. Show that it is possible to find four elements of A whose product is a fourth perfect power.

Proof The given conditions assure that every $x \in A$ is of the form $x = 2^{\alpha_{1x}} 3^{\alpha_{2x}} \ldots 23^{\alpha_{9x}}$, with $\alpha_1, \alpha_2, \ldots, \alpha_9 \in \mathbb{Z}_+$. Modulo 2, the number of distinct 9-tuples $(\alpha_1, \ldots, \alpha_9)$ is equal to $2^9 = 512$ (for each exponent α_{ix} is either odd or even). Since $1985 > 512$, pigeonhole's principle assures that we can choose $x_1, y_1 \in A$ such that $\alpha_{ix_1} \equiv \alpha_{iy_1} \pmod{2}$ for $1 \le i \le 9$. Hence, $\alpha_{ix_1} + \alpha_{iy_1} \equiv 0 \pmod{2}$ for $1 \le i \le 9$, so that $x_1 y_1 = z_1^2$, for some $z_1 \in \mathbb{N}$. Since $1985 - 2 > 512$, by invoking Dirichlet's principle once more, we can choose $x_2, y_2 \in A \setminus \{x_1, y_1\}$ such that $x_2 y_2 = z_2^2$, for some $z_2 \in \mathbb{N}$. Continuing this way, we obtain pairwise distinct elements $x_1, y_1, x_2, y_2, \ldots, x_{513}, y_{513} \in A$, with $x_j y_j = z_j^2$ for some $z_j \in \mathbb{N}$ and every $1 \le i \le 513$.

Now, since the prime factorisations of $z_1, z_2, \ldots, z_{513}$ also contain only primes less than 26 and $513 > 512$, an additional application of the pigeonhole principle guarantees the existence of indices $1 \le k < l \le 513$ such that $z_k z_l = w^2$, for some $w \in \mathbb{N}$. Hence,

$$x_k y_k x_l y_l = z_k^2 z_l^2 = (z_k z_l)^2 = (w^2)^2 = w^4.$$

□

Example 4.11 (IMO) Let $x_1, x_2, \ldots, x_n$ be real numbers for which $x_1^2 + x_2^2 + \cdots + x_n^2 = 1$. Prove that, for every integer $k \ge 2$, there exist integers $a_1, a_2, \ldots, a_n$ such that at least one of which is nonzero, $|a_i| \le k - 1$ for $1 \le i \le n$ and

$$|a_1 x_1 + a_2 x_2 + \cdots + a_n x_n| \le \frac{(k-1)\sqrt{n}}{k^n - 1}.$$

Proof Firstly, if $(a_1, a_2, \ldots, a_n)$ is one of the k^n sequences of integers satisfying $0 \leq a_i \leq k - 1$ for $1 \leq i \leq n$, then Cauchy's inequality (cf. Section 5.2 of [8], for instance) gives

$$\left| \sum_{j=1}^{n} a_j x_j \right| \leq \left(\sum_{j=1}^{n} a_j^2 \right)^{1/2} \left(\sum_{j=1}^{n} x_j^2 \right)^{1/2} \leq (k - 1)\sqrt{n}.$$

Now, subdivide interval $[0, (k - 1)\sqrt{n}]$ into $k^n - 1$ disjoint subintervals, each of which having length $\frac{(k-1)\sqrt{n}}{k^n-1}$. Since there exist k^n distinct sequences $(a_1, a_2, \ldots, a_n)$ of integers, pigeonhole's principle guarantees that at least two of such sequences, say $(b_1, b_2, \ldots, b_n)$ and $(c_1, c_2, \ldots, c_n)$, are such that both of the sums $\sum_{i=1}^{n} b_i x_i$ and $\sum_{i=1}^{n} c_i x_i$ belong to a single one of those $k^n - 1$ subintervals. Hence,

$$\left| \sum_{j=1}^{n} (b_j - c_j) x_j \right| = \left| \sum_{j=1}^{n} b_j x_j - \sum_{j=1}^{n} c_j x_j \right| \leq \frac{(k - 1)\sqrt{n}}{k^n - 1},$$

and it suffices to let $a_i = b_i - c_i$ for $1 \leq i \leq n$. $\qquad\square$

Our final example is a problem with two parts, in which only the easiest one makes use of Dirichlet's principle. Nevertheless, insight from such a use gives a valuable clue for an effective approach to the second part.

Example 4.12 (EKMC) Let $k \geq 3$ and n be integers such that $n > \binom{k}{3}$. Prove that if a_i, b_i, c_i $(1 \leq i \leq n)$ are $3n$ distinct real numbers, then one can find at least $k + 1$ distinct numbers among the $3n$ sums $a_i + b_i, a_i + c_i, b_i + c_i$. Show also that this claim is not necessarily true if $n = \binom{k}{3}$.

Proof By contradiction, assume that we have at most k distinct numbers among the $3n$ sums $a_i + b_i, a_i + c_i, b_i + c_i$, say $x_1, x_2, \ldots, x_k$.

Since $a_i + b_i, a_i + c_i$ and $b_i + c_i$ are pairwise distinct, $\binom{k}{3}$ is the number of 3-element subsets of $\{x_1, x_2, \ldots, x_k\}$ and $n > \binom{k}{3}$, we conclude that there exist $1 \leq i < j \leq n$ for which

$$\{a_i + b_i, a_i + c_i, b_i + c_i\} = \{a_j + b_j, a_j + c_j, b_j + c_j\}.$$

But it is not difficult to verify that such an equality would imply that at least one element of $\{a_i, b_i, c_i\}$ would be equal to at least one element of $\{a_j, b_j, c_j\}$, which is an absurd.

For a counterexample when $n = \binom{k}{3}$, we want to have only k distinct numbers among the $3n$ numbers $a_i + b_i, a_i + c_i, b_i + c_i$, even though the $3n$ numbers a_i, b_i, c_i are still pairwise distinct. To this end, first note that, by arbitrarily choosing k

distinct real numbers $x_1, x_2, \ldots, x_k$, we can form $n = \binom{k}{3}$ subsets of $\{x_1, x_2, \ldots, x_k\}$, of three elements each, and choose an arbitrary bijection between the sets

$$\left\{ a_i + b_i, a_i + c_i, b_i + c_i;\ 1 \le i \le n = \binom{k}{3} \right\}$$

and

$$\{ x_r + x_s + x_t;\ 1 \le r < s < t \le k \}.$$

Once we have done that, we would have, for $1 \le i < j \le n$,

$$a_i + b_i = x_{\alpha_i}, \quad a_i + c_i = x_{\beta_i}, \quad b_i + c_i = x_{\gamma_i}$$

and

$$a_j + b_j = x_{\alpha_j}, \quad a_j + c_j = x_{\beta_j}, \quad b_j + c_j = x_{\gamma_j},$$

where $1 \le \alpha_i, \alpha_j, \beta_i, \beta_j, \gamma_i, \gamma_j \le k$ are such that $\alpha_i, \beta_i, \gamma_i$ are pairwise distinct, the same happening with $\alpha_j, \beta_j, \gamma_j$.

Facing each of the two 3-tuples of equalities above as a linear system of equations, we can compute a_i, b_i, c_i (resp. a_j, b_j, c_j) in terms of $x_{\alpha_i}, x_{\beta_i}, x_{\gamma_i}$ (resp. $x_{\alpha_j}, x_{\beta_j}, x_{\gamma_j}$), getting three distinct numbers in each case (for $x_{\alpha_i}, x_{\beta_i}, x_{\gamma_i}$—resp. $x_{\alpha_j}, x_{\beta_j}, x_{\gamma_j}$—are pairwise distinct).

Therefore, the central point to the construction of a counterexample is that we ought to have

$$\{ a_i, b_i, c_i \} \cap \{ a_j, b_j, c_j \} = \emptyset$$

or, which is the same,

$$\{ x_{\alpha_i} + x_{\beta_i} - x_{\gamma_i}, x_{\alpha_i} + x_{\gamma_i} - x_{\beta_i}, x_{\beta_i} + x_{\gamma_i} - x_{\alpha_i} \}$$

and

$$\{ x_{\alpha_j} + x_{\beta_j} - x_{\gamma_j}, x_{\alpha_j} + x_{\gamma_j} - x_{\beta_j}, x_{\beta_j} + x_{\gamma_j} - x_{\alpha_j} \}$$

to be disjoint.

It is thus enough to show that it is possible to choose $x_1, x_2, \ldots, x_k$ in such a way that

$$x_r + x_s - x_t \ne x_u + x_v - x_w$$

whenever r, s, t and u, v, w are two 3-tuples of distinct integers of the interval $[1, k]$, with $\{r, s, t\} \neq \{u, v, w\}$. Since we may have $t \in \{u, v\}$ or $w \in \{r, s\}$, a possible choice is to take

$$x_1 = 2^1, \quad x_2 = 2^3, \quad x_3 = 2^5, \ldots, \quad x_k = 2^{2k-1}.$$

The uniqueness of binary representation of naturals (cf. Section 4.1 of [8], for instance) assures that the above conditions will be satisfied, even if we have $t \in \{u, v\}$ or $w \in \{r, s\}$. □

Problems: Sect. 4.1

1. We mark at random five points inside a square of side length 2. Show that it is always possible to find two of these five points whose distance from one another is at most $\sqrt{2}$.

2. Show that if we choose 800 points inside a cube of edge length 10, at least one of the line segments determined by two of these points will have length less than 2.

3. Let $x \in \mathbb{R}$ and n be an integer greater than 1. Show that, among the numbers x, $2x, \ldots, (n-1)x$, there is at least one whose distance from an integer is at most $\frac{1}{n}$.

    The coming problem brings yet another useful version of the pigeonhole principle. In words, it says that at least one of a number of given pigeonholes will contain at least as much pigeons as their average value.

4. * We are given n pigeonholes and a positive integer m. We place a_1 pigeons in the first pigeonhole, a_2 pigeons in the second, $\ldots$, a_n pigeons in the n-th. If

$$\frac{a_1 + a_2 + \cdots + a_n}{n} \geq (\text{resp. } >) \, m,$$

    prove that at least one of the pigeonholes will contain at least, m (resp. $m + 1$) pigeons.

5. The natural numbers from 1 to 15 are distributed around a circle. Show that it is always possible to find five consecutive numbers whose sum is greater than or equal to 40.

6. We are given two equal disks A and B, each of which is divided into 200 equal sectors, which, in turn, are painted either black or white. In disk A, there are 100 white sectors and 100 black sectors, in an unknown order. In disk B, we do not know how many sectors are white. Prove that it is possible to place disk A directly above disk B in such a way that at least 100 sectors of disk A lie above sectors of disk B, whose colors match theirs.

The next eight problems use elementary facts on the notion of divisibility of integers. Before trying to solve them, the reader might find it convenient to run through the statements of the definitions and results of Chap. 6.

7. Prove that in any set of 52 integers, there exist at least two whose sum or difference is a multiple of 100.

8. Prove that every natural number has a nonzero multiple whose decimal representation contains solely the digits 0 and 1.

9. (TT) Prove that every natural number has a multiple whose sum of digits is odd.

10. Show[3] that, if a and n are relatively prime natural numbers, then at least one of the numbers $a, a^2, \ldots, a^{n-1}$ leaves remainder 1 upon division by n.

11. We are given a set $A = \{a_1, a_2, \ldots, a_n\}$, of $n > 1$ positive integers. Prove that there exist natural numbers k and l such that $1 \le k \le l \le n$ and the sum $a_k + a_{k+1} + \cdots + a_l$ is a multiple of n.

12. Show that, by arbitrarily choosing $n+1$ elements from I_{2n}, we shall necessarily choose two such that one divides the other.

13. (Iran) Let A be a set of 33 natural numbers, each of which has all of its prime factors among the prime numbers 2, 3, 5, 7, 11. Prove that there exist two distinct elements of A whose product is a perfect square.

14. (Russia) We have chosen 15 pairwise relatively prime elements from the set $\{1, 2, \ldots, 1998\}$. Prove that at least one of the 15 chosen elements is necessarily a prime number.

15. Given a set A of ten positive integers of two digits each, prove that it is always possible to obtain two nonempty disjoint subsets of A having the same sum of elements.

16. (India) A set of real numbers is *free of sums* provided no two of its elements (not necessarily distinct) are such that their sum equals another element of the set. Find the greatest number of elements that a free of sums set may have, knowing that it is a subset of $A = \{1, 2, 3, \ldots, 2n + 1\}$.

17. Show that, by choosing $n + 1$ elements from the set $\{1, 2, \ldots, 3n\}$, we have necessarily chosen two elements x and y such that $xy + 1$ or $4xy + 1$ is a perfect square.

18. For each positive integer n, find the least positive integer $f(n)$ such that, for every partition of the set $\{1, 2, \ldots, f(n)\}$ into n sets, there exist integers $a \ge 0$ and $1 \le x \le y$ for which $a + x, a + y$ and $a + x + y$ all belong to a single one of the n sets of the partition.

19. (USA)

   (a) Each 1×1 square of a 3×7 chessboard is painted either white or black. Prove that, independently of the way we have painted the squares, there will always be four squares of the same color, which are the corner squares of a rectangle with sides parallel to those of the chessboard.

---

[3]Let $\varphi(n)$ denote expression (2.7), deduced at Problem 6, page 40, and let $a \in \mathbb{Z}$ be relatively prime with n. In Sect. 10.2, we shall prove a theorem of L. Euler which assures that $a^{\varphi(n)}$ always leave remainder 1 upon division by n (in the language of Sect. 2.3, $a^{\varphi(n)} \equiv 1 \pmod{n}$).

  (b) Exhibit a way of painting the squares of a 4×6 chessboard so that the
      condition of item (a) is never satisfied.

20. (IMO) Seventeen people discuss one of three possible subjects by letter. More
    precisely, each two of the seventeen people exchange a letter in which exactly
    one of the three possible subjects is touched upon. Prove that there are three of
    them who exchange letters on the same subject.

4.2   Induction and Existence of Configurations

In this section, we examine some examples that illustrate how mathematical
induction can be used, in combinatorial situations, to establish the existence of
configurations possessing certain properties. We assume from the reader a thorough
acquaintance with the principle of induction, referring to Section 4.1 of [8] for
an elementary exposition of the necessary background. In particular, most often
we shall not write formal inductive proofs, i.e., sometimes initial cases will not be
checked and induction hypotheses will not be explicitly stated.

Example 4.13 (Germany) Given $n \in \mathbb{N}$, prove that one can assemble a square of
side length 2^n by using one square piece of side length 1 and several *L-triminoes*,
i.e., pieces of the shape below, where each small square has side length equal to 1:

Proof We shall use induction on n to show that each square of side length 2^n can be
assembled as required, with the square piece of side length 1 occupying a corner of
the $2^n \times 2^n$ square.

  For $n = 1$, the assembling is immediate, as shown in the figure below:

  By induction hypothesis, assume that every square of side length 2^k can be
assembled as prescribed. For the inductive step, take a $2^{k+1} \times 2^{k+1}$ square and
divide it into four squares of side length 2^k, as shown in the figure below. Then,
place an L-trimino in the center of the larger square, so that it occupies one square

of side length 1 in three of the four squares of side length 2^k. Finally, place the 1×1 square piece in the corner of the $2^{k+1} \times 2^{k+1}$ square that belongs to the square of side length 2^k which was not intersected by the L-trimino already placed.

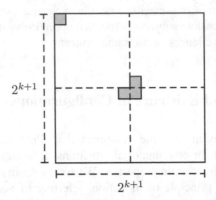

The placement of these two pieces has the effect of letting each of the four squares of side length 2^k with exactly one corner occupied by a square piece of side length 1. Hence, induction hypothesis assures that we can finish the assembling of each one of these four squares by using L-triminoes only, and taking these lets us conclude that the same is true for the $2^{k+1} \times 2^{k+1}$ square (with the square piece of side length 1 placed in one of its corners).   □

As one should expect, the coming example shows that, sometimes, strong induction is in force.

Example 4.14 (Leningrad) List all nonempty subsets of the set $\{1, 2, \ldots, n\}$ which do not contain consecutive elements. For each one of them, compute the square of the product of its elements. Prove that the sum of all of these numbers is equal to $(n + 1)! - 1$.

Proof As was anticipated above, we shall make strong induction on $n \geq 1$.

If $n = 1$, the only nonempty subset of $\{1\}$ is itself, so that the required sum is $1 = 2! - 1$.

By induction hypothesis, suppose that, for $1 \leq n \leq k$, the desired sum is equal to $(n + 1)! - 1$. Now, notice that the nonempty subsets of $\{1, 2, \ldots, k + 1\}$ without consecutive elements can be divided in two categories:

(i) the nonempty subsets of $\{1, 2, \ldots, k\}$ without consecutive elements;
(ii) the sets of the form $A \cup \{k+1\}$, where A is a subset of $\{1, 2, \ldots, k-1\}$ without consecutive elements.

The induction hypothesis guarantees that the sum of the squares of the products of the elements of the sets of type (i) equals $(k + 1)! - 1$, while that of the sets of type (ii) equals $(k + 1)^2 (k! - 1) + (k + 1)^2$ (the summand $(k + 1)^2$ corresponding to the set $\{k + 1\} = \emptyset \cup \{k + 1\}$).

Hence, the sum of the squares of the products of the elements of all desired subsets of $\{1, 2, \ldots, k + 1\}$ equals

$$(k + 1)! - 1 + (k + 1)^2(k! - 1) + (k + 1)^2 =$$
$$= (k + 1)! - 1 + (k + 1)^2 \cdot k!$$
$$= (k + 1)! - 1 + (k + 1)!(k + 1)$$
$$= (k + 1)!(k + 2) - 1$$
$$= (k + 2)! - 1.$$

□

We now give two examples that show that an inductive argument can compose just *part* of the analysis of a certain combinatorial situation.

Example 4.15 (Russia) Let $n > 1$ be an odd integer. In open field, n children are so positioned that, for each one of them, the distances to the other $n - 1$ children are pairwise distinct. Each child has a water pistol and, at the sound of a whistle, fires at the child closest to him/her. Show that at least one child will remain dry.

Proof Firstly, let us look at the case $n = 3$. Let A, B and C be the children and assume, without loss of generality, that $AB < BC < AC$. Then, A fires at B and B at A, so that C remains dry.

Now, assume that whenever $2k - 1$ children ($k > 1$) are positioned in open field as prescribed in the statement of the problem, at least one of them remains dry. We then consider $2k + 1$ children, also positioned as required by the problem. Since the distances between pairs of children are pairwise distinct, there exist two children, say A and B, such that the distance between them is the smallest one of all distances between two of the given children. Thus, A fires at B and vice-versa. Discarding A and B, we are left with $2k - 1$ children, and there are two distinct possibilities:

(i) One of these $2k - 1$ children fires at A or B: in this case, at most $2k - 2$ shots are fired towards some of the remaining $2k - 1$ children (except A and B). By the pigeonhole principle, at least one of them remains dry.
(ii) None of the remaining $2k - 1$ children fires at A or B: by induction hypothesis, at least one of these $2k - 1$ children remains dry.

□

Example 4.16 (IMO Shortlist) We wish to write one of the numbers 0, 1 or 2 in each entry of a table of 19 lines and 86 columns, in such a way that the following conditions are satisfied:

(a) Each column has exactly k zeros.
(b) For any two chosen columns, there exists a line such that the two entries of this line situated at the chosen columns are, in some order, equal to 1 and 2.

Find all values of k for which this is possible.

Solution We say that a set of m lines *distinguishes* n columns if, for any choice of two of these n columns, it is possible to choose one of the m lines in such a way that the entries of this line situated in the two chosen columns are equal to 1 and 2, in some order.

We claim that m lines distinguish at most 2^m columns. Indeed, it is clear that a single line can distinguish at most two columns. Now, assume that j lines distinguish at most 2^j columns. Then, since a single line distinguishes at most two columns, $j + 1$ lines will distinguish at most two groups of 2^j columns (the distinction made by the $(j + 1)$-th line between the two groups of 2^j columns is a permutation of the pattern

$$\underbrace{11\ldots 1}_{2^j}\underbrace{22\ldots 2}_{2^j}.$$

Hence, $j + 1$ lines will distinguish at most $2^j + 2^j = 2^{j+1}$ columns.

Back to the given table, since $86 > 2^6$, the above claim assures that we need at least 7 lines to distinguish all of the 86 columns. On the other hand, 7 lines will suffice, for $86 < 2^7$. Actually, it suffices to choose 86 distinct sequences of 7 terms, each of which equal to 1 or 2, and fulfill the first 7 lines with the terms of these 86 sequences. The remaining lines can be randomly fulfilled.

Therefore, we must have $0 \le k \le 19 - 7 = 12$.   □

For the coming example, recall that, given $n \in \mathbb{N}$, we let $I_n = \{1, 2, \ldots, n\}$. Also recall that, given sets A and B, their *symmetric difference* $A \triangle B$ is given by

$$A \triangle B = (A \cup B) \setminus (A \cap B).$$

Example 4.17 (Argentina) Given $n \in \mathbb{N}$, let $\mathcal{P}_n$ be the family of all subsets of I_n. If $f : \mathcal{P}_n \to I_n$ is any function, show that there exist distinct $A, B \in \mathcal{P}_n$ and such that

$$f(A) = f(B) = \max(A \triangle B).$$

Proof For $n = 1$, we have $f : \{\emptyset, \{1\}\} \to \{1\}$ and $\emptyset \triangle \{1\} = \{1\}$, so that

$$f(\emptyset) = f(\{1\}) = 1 = \max(\{1\}).$$

Suppose that the statement is true for $n = k$, and take a function $f : \mathcal{P}_{k+1} \to I_{k+1}$. There are two cases to consider:

(i) f maps all sets in $\mathcal{P}_k$ to elements of I_k: in this case, it suffices to apply the induction hypothesis to the restriction of f to I_k.
(ii) There exists $A \in \mathcal{P}_k$ such that $f(A) = k + 1$: let $\mathcal{F} \subset \mathcal{P}_{k+1}$ be the family of the elements of $\mathcal{P}_{k+1}$ containing $k + 1$, i.e.,

$$\mathcal{F} = \{X \cup \{k + 1\}; \ X \in \mathcal{P}_k\}.$$

We look at two subcases:

- there exists $B \in \mathcal{F}$ such that $f(B) = k + 1$: then, $k + 1 \in B \setminus A$, so that $A \neq B$, $k + 1 \in A \triangle B$ and

$$f(A) = f(B) = k + 1 = \max(A \triangle B).$$

- f maps $\mathcal{F}$ into I_k: let $g : \mathcal{P}_k \to I_k$ be given by setting $g(X) = f(X \cup \{k+1\})$, for every $X \in \mathcal{P}_k$. By induction hypothesis, there exist distinct A', $B' \in \mathcal{P}_k$, such that

$$g(A') = g(B') = \max(A' \triangle B').$$

Since $(A' \cup \{k+1\}) \triangle (B' \cup \{k+1\}) = A' \triangle B'$, we conclude that $f(A' \cup \{k+1\})$ and $f(B' \cup \{k + 1\})$ are both equal to $\max\left((A' \cup \{k + 1\}) \triangle (B' \cup \{k + 1\})\right)$, with $A' \cup \{k + 1\} \neq B' \cup \{k + 1\}$.

□

We finish this section by revisiting Problem 17, page 31, this time with the aid of mathematical induction.

Example 4.18 (Saint Petersburg—Adapted) We are given $2n + 1$ points on a circle, such that no two of them are the endpoints of a diameter. Prove that, among all triangles having three of the given points as their vertices, at most

$$\frac{1}{6}n(n + 1)(2n + 1)$$

of them are acute.

Proof For $n = 1$ there is nothing to do, since three points on a circle determine exactly one triangle.

By induction hypothesis, assume that the problem is true when $n = k - 1$, for some integer $k \geq 2$. Then, let $n = k$, so that we have $2k + 1$ points on the circle. Out of them, take all pairs of points such that the chord joining them leaves k points in one arc and $k - 1$ points in the other arc of the circle. Among all such pairs of points, let A and B be two such that AB is at minimum distance from the center O of the circle (if there is more than one pair of points with this property, choose any one of them).

Now, there are three kinds of acute triangles having vertices at three of the $2k + 1$ given points:

(i) Those having A and B as two of their vertices: since O must belong to the interior of such a triangle, and AB leaves k points at one side and $k - 1$ points at the other side, there are at most k such triangles.

(ii) Those having A or B, but not both of them, as one of their vertices: let C and D be the other two vertices of such an acute triangle (so that the third one is either A or B). We consider two subcases:

- C and D are at the same side of AB: if ACD was to be acute, then AC or AD would leave B and O at different half-planes. In turn, this would assure that the distance from AC or AD to O would be less than the distance of AB to O, thus contradicting our choice of A and B. Therefore, ACD is not acute and, accordingly, the same is true of BCD.
- C and D are in opposite sides with respect to AB: since $ACBD$ is convex, at most one of the triangles ACD or BCD contains the center O, and hence at most one of such triangles is acute. On the other hand, by the choice of A and B the set $\{C, D\}$ can be chosen in exactly $k(k - 1)$ ways, and this is the largest possible number of acute triangles of this kind.

(iii) Those having neither A nor B as one of their vertices: by induction hypothesis, there are at most $\frac{1}{6}(k - 1)k(2k - 1)$ such triangles.

Adding the contributions of the three cases above, we conclude that the number of acute triangles having its vertices at three of the $2k + 1$ given points is at most

$$k + k(k - 1) + \frac{1}{6}(k - 1)k(2k - 1) = \frac{1}{6}k(k + 1)(2k + 1).$$

$\square$

Problems: Sect. 4.2

1. (India) Prove that, for every $n \geq 6$, every square can be partitioned into n other squares.
2. (Brazil) Given a natural number $n > 1$, we write a real number of modulus less than 1 in each cell of an $n \times n$ chessboard, in such a way that the sum of the numbers written in the four cells of any 2×2 square is equal to 0. If n is odd, show that the sum of the numbers written in the n^2 cells is less than n.
3. (TT) Point O is situated in the interior of the convex polygon $A_1A_2 \ldots A_n$. Consider all of the angles $\angle A_i O A_j$, with distinct $1 \leq i, j \leq n$. Prove that at least $n - 1$ of them are not acute.
4. (TT) In a convex polygon P some diagonals were drawn, such that no two of them intersect in the interior of P. Show that there are at least two vertices of P such that none of the traced diagonals is incident to none of these two vertices.
5. (Hungary) In an $n \times n$ chessboard we have a certain number of towers (at most one tower for each 1×1 square). A tower can move from one square to another if and only if these two squares belong to a single line or column of the board and there are no other towers in between. On the other hand, if two towers belong to a single line or column and there are no other towers in between, then we say that any of these towers can *attack* the other. Prove that one can paint the towers with one of three colors in such a way that no tower can attack another tower of the same color.

6. We are given n distinct lines in the plane. Prove that it is possible to paint the regions into which these lines divide the plane of black or white, so that if two regions share a common edge, then they have distinct colors.

7. (TT) Each square of a chessboard is painted either red or blue. Prove that the squares of one of these two colors are such that a chess queen can move through all of them (possibly passing more than once through one or more of the 1×1 squares of this color), without visiting a single square of the other color. We recall that a chess queen can move an arbitrary number of squares along any line, column or diagonal of the chessboard.

8. We are given n points in the plane, not all of which being collinear. Prove that:

   (a) There exists at least one line that passes through exactly two of these n points.

   (b) The n points determine at least n distinct lines.

   The result of item (a) is known as the **Sylvester-Gallai theorem**,[4] while that of item (b) is due to Erdös and de Bruijn.[5]

9. In the plane we are given $2n$ points in *general position*, i.e., such that no three of them are collinear. If n of the given points are blue and the other n are red, prove that it always possible to draw n line segments satisfying the following conditions:

   (a) Each line segment joins a blue point to a red point.

   (b) They do not intersect each other.

10. (OCS) Prove that, given a positive integer n, there exists another positive integer k_n with the following property: given any set of k_n points in space, no four of them being coplanar, and associating an integer number from 1 to n to each line segment joining two of these k_n points, one necessarily gets a triangle having vertices at three of the k_n points and such that the numbers associated to its sides are all equal.

4.3 Partially Ordered Sets

This short section discusses a few facts on *partially ordered sets*, focusing on the theorems of Mirsky[6] (cf. [30]) and Dilworth[7] (cf. [14]). As a corollary of Mirsky's theorem, we obtain another proof of Erdös-Szekeres theorem. Dilworth's theorem, in turn, will be applied to Graph Theory in Problem 25, page 134.

---

[4]James Sylvester, English mathematician, and Tibor Gallai, Hungarian mathematician, both of the twentieth century.

[5]Nicolaas de Bruijn, Dutch mathematician of the twentieth century.

[6]After Leon Mirsky, Russian mathematician of the twentieth century.

[7]After Robert Dilworth, American mathematician of the twentieth century.

We start by presenting the concept of partially ordered set, together with some examples. In the coming definition, $\preceq$ denotes a relation on the set A, i.e., a subset of the cartesian product $A \times A$.

Definition 4.19 We say that a nonempty set A is **partially ordered** by $\preceq$, or that $(A, \preceq)$ is a **partially ordered set**, if the following conditions are satisfied, for all $a, b \in A$:

(a) (**Reflexivity**) $a \preceq a$.
(b) (**Antisymmetry**) $a \preceq b$ and $b \preceq a \Rightarrow a = b$.
(c) (**Transitivity**) $a \preceq b$ and $b \preceq c \Rightarrow a \preceq c$.

In this case, we also say that $\preceq$ is a **partial order relation** in A.

In the notations of the previous definition, $\preceq$ must be thought of as a way of *comparing* elements of A. This being said, note that we are not demanding that $\preceq$ provides a way of comparing any two elements of A; in other words, it may well happen that there exist $a, b \in A$ such that $a \npreceq b$ and $b \npreceq a$ (i.e., such that $a \preceq b$ and $b \preceq a$ are both false). If this is so, then we say that a and b are *incomparable* (via $\preceq$); otherwise, a and b are said to be *comparable*.

If $\preceq$ is a partial order relation on A for which any two elements of A are comparable, we say that $\preceq$ is a **total order relation**, or simply a total order. We also say that $(A, \preceq)$ is a **totally ordered set**.

It is time that we take a look at a few relevant examples of order relations. We refer to Problem 1 for details.

Example 4.20

(a) The set of real numbers is totally ordered by the order relation $\preceq$, defined by $a \preceq b \Leftrightarrow a \leq b$. In this case, we say that $\leq$ is the *usual order* in $\mathbb{R}$.
(b) Let X be a nonempty set and $\mathcal{F} = \mathcal{P}(X)$ be the family of all subsets of X. For $Y, Z \in \mathcal{F}$ (i.e., $Y, Z \subset X$), we define $Y \preceq Z \Leftrightarrow Y \subset Z$. It is immediate to verify that $(\mathcal{F}, \preceq)$ is a partially ordered set, which is usually denoted simply by writing $(\mathcal{P}(X), \subset)$. In this case, we say that $\subset$ is the **inclusion relation** in X.
(c) Given $a, b \in \mathbb{N}$, recall that $a \mid b$ (one reads "a *divides* b") if there exists $c \in \mathbb{N}$ such that $b = ac$. Letting $\preceq$ be defined on $\mathbb{N}$ by $a \preceq b \Leftrightarrow a \mid b$, it is also straightforward that $(\mathbb{N}, \preceq)$ is a partially ordered set. In view of this example, one frequently says that $\mathbb{N}$ is partially ordered by the **divisibility relation**.
(d) Let A be a nonempty set partially ordered by $\preceq$. If $B \subset A$ is nonempty, the restriction of $\preceq$ to elements of B turns it into another partially ordered set. In this case, we say that B is partially ordered by the order relation **induced** from A.

If $(A, \preceq)$ is a partially ordered set and $a, b \in A$, we write $a \prec b$ to mean that $a \preceq b$ and $a \neq b$. As we indicated before, we write $a \npreceq b$ to mean that $a \preceq b$ is false. For what we have in mind, the coming definition is central.

Definition 4.21 Let A be a nonempty set ordered by $\preceq$. With respect to $\preceq$, a nonempty subset B of A is:

(a) A **chain** if, for every $a, b \in B$, we have either $a \preceq b$ or $b \preceq a$.
(b) An **antichain** if, for every $a, b \in B$, we have $a \npreceq b$ and $b \npreceq a$.

In words (and in the above notations), B is a chain if, furnished with the order relation induced from A, is a totally ordered set. On the other hand, B is an antichain if any two of its elements are incomparable by $\preceq$.

Example 4.22 Let $\mathbb{N}$ be partially ordered by the divisibility relation. Letting $P = \{2, 3, 5, 7, 11, \ldots\}$ denote the set of prime numbers, it will be proved in Sect. 6.3 that P is infinite. Thus, P is an infinite antichain on $\mathbb{N}$, whereas $\{2, 2 \cdot 3, 2 \cdot 3 \cdot 5, \ldots\}$ is an infinite chain.

Example 4.23 If $X = \{a_1, a_2, \ldots, a_m\}$ is a set with m elements, then, ordering $\mathcal{F} = \mathcal{P}(X)$ with respect to inclusion, the family

$$\mathcal{G} = \{\emptyset, \{a_1\}, \{a_1, a_2\}, \{a_1, a_2, a_3\} \ldots, \{a_1, a_2, \ldots, a_m\}\}$$

is a chain. Actually, $\mathcal{F}$ does not contain longer chains. In order to prove this assume, by the sake of contradiction, that

$$\mathcal{C} = \{A_1 \subset A_2 \subset \ldots \subset A_{m+2}\}$$

is a chain on $\mathcal{F}$, formed by $m + 2$ distinct subsets of X. Then, since $|A_i| \in \{0, 1, 2, \ldots, m\}$ for $1 \leq i \leq m + 2$, the pigeonhole principle assures that there exist $1 \leq i < j \leq m + 2$ for which $|A_i| = |A_j|$. In turn, together with $A_i \subset A_j$, this gives $A_i = A_j$, which is an absurd.

On the other hand, it is immediate to verify that, for each integer $0 \leq k \leq m$, the family

$$\mathcal{F}_k = \{Y \subset X;\ |Y| = k\}$$

is an antichain in $\mathcal{F}$ (note that $\mathcal{F}_0 = \{\emptyset\}$). Furthermore, note that

$$\mathcal{F} = \mathcal{F}_0 \cup \mathcal{F}_1 \cup \ldots \cup \mathcal{F}_m.$$

We are finally in position to state and prove Mirsky's theorem, which is a direct generalization of the situation described in the above example. For the proof of it, if $(A, \preceq)$ is a partially ordered set and $B = \{a_1 \preceq a_2 \preceq \ldots \preceq a_m\}$ is a chain in A formed by m distinct elements, then we say that B has **length** m and a_m is its **maximum element**.

Theorem 4.24 (Mirsky) *Let $(A, \preceq)$ be a partially ordered set. If A does not possess chains of length $n + 1$, then A can be written as the union of at most n antichains.*

Proof Let us make induction on n, noticing firstly that, if A does not possess chains of length 2, then A is itself an antichain, and the theorem is trivially true. By induction hypothesis, suppose that each partially ordered set which does not posses chains of length k can be written as a union of at most $k - 1$ antichains.

Let $(A, \preceq)$ be a partially ordered set which does not possess chains of length $k + 1$. Define B as the subset of A formed by the maximum elements of the chains of maximum length in A. We first claim that B is an antichain with respect to $\preceq$. By contradiction, suppose that there existed $x, y \in B$ such that $x \prec y$. Since $x \in B$, we can find a chain C in A, of maximum length and whose maximum element is x; but then, $C \cup \{y\}$ would be a chain in A, with maximum element y and longer than C, which is an absurd.

Now, note that $(A \setminus B, \preceq)$ does not have chains of length k. Indeed, if C was such a chain in $(A \setminus B, \preceq)$, then, since A does not possess chains of length $k + 1$, the chain C would be one of maximum length in A. Therefore, the maximum element of C would belong to B, which is a contradiction.

By applying the induction hypothesis to $A \setminus B$, we conclude that it can be written as the union of at most $k - 1$ antichains. Finally, adjoining B to such a collection of antichains, we write A as the union of at most $(k - 1) + 1 = k$ antichains. $\square$

As we mentioned before, we now obtain the Erdös-Szekeres theorem as a corollary of Mirsky's theorem. Actually, the following version generalizes that of Example 4.8.

Theorem 4.25 (Erdös-Szekeres) *If $a, b \in \mathbb{N}$ and $n = ab + 1$, then every sequence $(x_1, \ldots, x_n)$, of distinct real numbers, either possess an increasing subsequence of $a + 1$ terms or a decreasing subsequence of $b + 1$ terms.*

Proof Define $A = \{(i, x_i); 1 \le i \le n\}$ and let $\preceq$ be the relation in A given by

$$(i, x_i) \preceq (j, x_j) \Leftrightarrow (i = j) \text{ or } (i < j \text{ and } x_i < x_j).$$

It is immediate to verify that $\preceq$ is a partial order in A, and that

$$(i_1, x_{i_1}) \prec (i_2, x_{i_2}) \prec \cdots \prec (i_k, x_{i_k})$$

is a chain in A if and only if $(x_{i_1}, \ldots, x_{i_k})$ an increasing subsequence of $(x_1, \ldots, x_n)$. Accordingly, for $1 \le i_1 < i_2 < \cdots < i_k \le n$,

$$\{(i_1, x_{i_1}), (i_2, x_{i_2}), \ldots, (i_k, x_{i_k})\}$$

is an antichain in A if and only if $(x_{i_1}, \ldots, x_{i_k})$ is a decreasing subsequence in $(x_1, \ldots, x_n)$. Indeed, since $i_r < i_s$ for $1 \le r < s \le k$ and all of the x_i's are distinct, we get

$$(i_r, x_{i_r}) \not\preceq (i_s, x_{i_s}) \Leftrightarrow x_{i_r} < x_{i_s} \text{ is false} \Leftrightarrow x_{i_r} > x_{i_s}.$$

Now, assume that $(x_1, \ldots, x_n)$ does not contain an increasing sequence of $a + 1$ terms. Then, A does not have chains of length $a+1$, so that Mirsky's theorem assures that A can be written as the union of $l \leq a$ antichains, say $A_1, \ldots, A_l$. However, since

$$ab + 1 = n = |A| = \left| \bigcup_{i=1}^{l} A_i \right| \leq \sum_{i=1}^{l} |A_i|,$$

it follows from the pigeonhole principle that, for some index $1 \leq i \leq l$, we must have

$$|A_i| \geq \left\lfloor \frac{ab}{l} \right\rfloor + 1 \geq b + 1.$$

$\square$

Problems: Sect. 4.3

1. * Check that all of the relations defined in Example 4.20 are, indeed, order relations.
2. Let A be a set of $n^2 + 1$ positive integers. Prove that there exists a subset B of A, with $n + 1$ elements and satisfying one of the following conditions:

   (a) For all $a, b \in B$, either $a \mid b$ or $b \mid a$.
   (b) For all $a, b \in B$, one has $a \nmid b$ and $b \nmid a$.

3. State and prove a generalization of the Erdös-Szekeres theorem for a partially ordered finite set.
4. Let be given $m, n \in \mathbb{N}$ and an $m \times \left(n^{2^m} + 1\right)$ table of pairwise distinct real numbers. Show that it is possible to choose $n + 1$ columns of the table such that the $m \times (n + 1)$ table composed by the chosen columns satisfies the following condition: the numbers of each line, written from left to right, form an increasing or decreasing sequence.
5. In a finite partially ordered set, let A and B be two chains having the greatest possible lengths. If $A \cap B \neq \emptyset$, prove that A and B have equal lengths.

   For the next problem, if $(A, \preceq)$ is a partially ordered set, we say that $a \in A$ is a **maximal element** of A if the following condition is satisfied:

$$a \preceq x, \ x \in A \Rightarrow a = x.$$

6. * The purpose of this problem is to **Dilworth's theorem** on partially ordered sets: *if a finite partially ordered set $(A, \preceq)$ does not contain antichains of $n + 1$ elements, then A can be written as the union of n pairwise disjoint chains.* To this end, and arguing by induction on $|A|$, do the following items (cf. [19]):

(a) Let a be a maximal element of A and assume, by induction hypothesis, that $A \setminus \{a\}$ contains an antichain B_0 of k elements and can be written as the union of k pairwise disjoint chains $A_1, \ldots, A_k$. Conclude that $A_i \cap B_0 \neq \emptyset$ for $1 \leq i \leq k$.

(b) For $1 \leq i \leq k$, let $x_i \in A_i$ be the maximum element of A_i belonging to an antichain of k elements in $A \setminus \{a\}$. If $B = \{x_1, x_2, \ldots, x_k\}$, show that B is an antichain.

(c) Assume that $x_i \preceq a$ for some $1 \leq i \leq k$.

    i. Let $C = \{a\} \cup \{y \in A_i;\ y \preceq x_i\}$. Show that $A \setminus C$ contains an antichain of $k-1$ elements but does not contain an antichain of k elements.

    ii. Apply the induction hypothesis to $A \setminus C$ and, then, conclude that A can be written as the union of k pairwise disjoint chains.

(d) Assume that $x_i \npreceq a$ for every $1 \leq i \leq k$. Show that:

    i. $B \cup \{a\}$ is an antichain of $k+1$ elements.

    ii. A can be written as the union of $k+1$ pairwise disjoint chains.

4.4 Invariants

In the combinatorial situations we discuss here, we generally consider *algorithms*[8] involving somewhat random choices of entries, and try to study the set of possible outcomes after a certain number of iterations. In this sense, a frequently fruitful approach consists in trying to associate an **invariant** to the algorithm, i.e., a mathematical object that behaves in a predictable way along each iteration of it, independently of the chosen inputs.

Since there are no general rules that teach us which invariant should be associated to each particular algorithmic situation, we shall limit ourselves here to examine some interesting examples, in order to help the reader to grasp the general idea and to prepare him/her for the posed problems.

Example 4.26 There are several $+$ and $-$ signs written on a blackboard. At each second we can replace two signs by a single one, according to the following rule: if the two erased signs are equal, we replace them by a $+$ sign, whereas if they are different, we replace them by a $-$ sign. Show that, when just one sign is left on the blackboard, it will not depend on the order we made the replacements.

---

[8]Formally, an **algorithm** is a finite sequence of well defined operations that, once performed on some (more or less) arbitrary set of data (called the **input** of the algorithm), furnish a definite result, called the **outcome** or **output** of the algorithm. On the other hand, each such performance of the algorithm is generally referred to as an **iteration** of it. We shall encounter algorithms several times along these notes.

Solution By the sign rule for products, the product of the signs written on the blackboard is an invariant for the described replacement algorithm. Thus, the final sign on the blackboard will be equal to the result of the multiplication of all of the $+$ and $-$ signs initially written and, hence, does not depend on the order of the replacements.   □

Example 4.27 (Bulgaria) There are 2000 white balls in a box. Also, there is a sufficient supply of white, green and red balls outside the box. The following replacement operations are allowed with the balls lying inside the box:

(a) Two white balls by a green one.
(b) Two red balls by a green one.
(c) Two green balls by a white and a red ones.
(d) One white ball and one green ball by a red one.
(e) One green ball and one red ball by a white one.

After a finite number of these replacement operations, only three balls were left in the box. Prove that at least one of them is green. Does there exist a finite number of operations leaving just one ball in the box?

Solution Assume that, in a certain moment, we have x white balls, y green balls and z red balls in the box. Let us look at how the quantity $x + 2y + 3z$ behaves when performing one of the allowed operations:

- (a): $x + 2y + 3z$ does not change;
- (b): $x + 2y + 3z$ changes to $x + 2(y + 1) + 3(z - 2) = x + 2y + 3z - 4$;
- (c): $x + 2y + 3z$ changes to $(x + 1) + 2(y - 2) + 3(z + 1) = x + 2y + 3z$;
- (d): $x + 2y + 3z$ changes to $(x - 1) + 2(y - 1) + 3(z + 1) = x + 2y + 3z$;
- (e): $x + 2y + 3z$ changes to $(x + 1) + 2(y - 1) + 3(z - 1) = x + 2y + 3z - 4$.

According to the analysis above, the remainder of the division of $x + 2y + 3z$ by 4 is an invariant for the described replacement algorithm. Since we began with $x = 2000$ and $y = z = 0$, we conclude that $x + 2y + 3z$ is always a multiple of 4.

Now, note that with only three balls in the box, we have $x + y + z = 3$ and $x + 2y + 3z = 4k$, for some $k \in \mathbb{N}$. If $y = 0$, then $x + z = 3$ and $x + 3z = 4k$; but, then, $4k = x + 3z = x + 3(3 - x) = 9 - 2x$, which is an absurd. Therefore, we must have $y \geq 1$, so that there will be at least one green ball in the box.

On the other hand, if at some moment we had only one ball in the box, we should have $x + y + z = 1$ and $x + 2y + 3z = 4k$. The first equation shows that exactly one of x, y and z is equal 1, while the other two are equal to 0; but these possibilities are clearly inconsistent with the second equation. Therefore, we will never reach a situation in which just one ball is left in the box.   □

In the following, we analyse more sophisticated examples.

Example 4.28 (Soviet Union) At first, we have $n > 1$ real numbers written in a blackboard. An allowed operation is to erase two of them, say a and b, and to write $\frac{a+b}{4}$ in their place. Thus, after each operation, the total quantity of numbers written in the blackboard is one less than it was before the operation has been executed.

If all of the n initial numbers were equal to 1, prove that, when just one number is left on the blackboard, it will necessarily be greater than or equal to $\frac{1}{n}$.

Proof Suppose that, at some moment, we have the numbers $x_1, x_2, \ldots, x_{k+1}$ written on the blackboard. At this moment, associate the number

$$f(x_1, \ldots, x_{k+1}) = \frac{1}{x_1} + \frac{1}{x_2} + \cdots + \frac{1}{x_{k+1}}$$

to the blackboard. Without loss of generality, assume that we erase x_k and x_{k+1}, writing $x'_k = \frac{x_k + x_{k+1}}{4}$ in place of them. Since

$$\frac{4}{x_k + x_{k+1}} \leq \frac{1}{x_k} + \frac{1}{x_{k+1}}$$

(cf. Example 5.1 (b) of [8], for instance) it is clear that

$$f(x_1, \ldots, x_{k-1}, x'_k) \leq f(x_1, \ldots, x_k, x_{k+1}).$$

Hence, $f(x_1, \ldots, x_k)$ is a *semi-invariant* for our problem (i.e., the order relation between the values of $f(x_1, \ldots, x_k)$, computed before and after each operation, is an invariant for the problem—in our case, such values for a nonincreasing sequence). Thus, when a single number x will be written on the blackboard, we ought to have

$$n = f(1, 1, \ldots, 1) \geq f(x) = \frac{1}{x}.$$

In turn, this is the same as having $x \geq \frac{1}{n}$.   □

Example 4.29 (Putnam) Let A be the total number of distinct sequences $(a_1, a_2, \ldots, a_{10})$ of positive integers such that

$$\frac{1}{a_1} + \frac{1}{a_2} + \cdots + \frac{1}{a_{10}} = 1.$$

Decide whether A is even or odd. Justify your answer.

Solution Firstly, let us prove that the equation

$$\frac{1}{x_1} + \frac{1}{x_2} + \cdots + \frac{1}{x_{10}} = 1$$

has an even number of solutions $x_1 = a_1, \ldots, x_{10} = a_{10}$ in positive integers, satisfying the additional condition $a_1 \neq a_2$. Indeed, if $x_1 = a_1, x_2 = a_2, x_3 = a_3, \ldots, x_{10} = a_{10}$ is such a solution, then $x_1 = a_2, x_2 = a_1, x_3 = a_3, \ldots, x_{10} = a_{10}$ is another one, and vice-versa. Therefore, we can group the above solutions in pairs, so that there is an even number of them.

Thanks to the above argument, in order to compute the parity of the number of solutions of the given equation in positive integers, it suffices to compute the parity of the number of such solutions satisfying the additional condition $a_1 = a_2$ (here is where the idea of invariance plays a role: the parity of the number of solutions in positive integers does not vary if we impose that $a_1 = a_2$).

Analogously, it suffices to compute the parity of the number of solutions in positive integers satisfying the conditions $a_3 = a_4$, $a_5 = a_6$, $a_7 = a_8$ and $a_9 = a_{10}$. In turn, this being the case, we get

$$\frac{1}{a_1} + \frac{1}{a_3} + \frac{1}{a_5} + \frac{1}{a_7} + \frac{1}{a_9} = \frac{1}{2}. \tag{4.2}$$

Following the same reasoning as above, we conclude that there is an even number of solutions of (4.2) in positive integers and such that $a_1 \neq a_3$, as well as an even number of such solutions in which $a_5 \neq a_7$. Therefore, in order to compute the parity of the total number of solutions in positive integers, we can assume that $a_1 = a_3$ and $a_5 = a_7$. The given equation now reduces to

$$\frac{2}{a_1} + \frac{2}{a_5} + \frac{1}{a_9} = \frac{1}{2}. \tag{4.3}$$

By repeating the same reasoning once more, we conclude that the parity of the total number of solutions is the same as that of the total number of solutions of (4.3) in which $a_1 = a_5$, i.e., the parity of the total number of solutions of

$$\frac{4}{a_1} + \frac{1}{a_9} = \frac{1}{2}.$$

The last equation above is equivalent to

$$a_9 = \frac{2a_1}{a_1 - 8} = 2 + \frac{16}{a_1 - 8}, \tag{4.4}$$

so that $a_1 - 8 = 1, 2, 4, 8$ or 16.

Hence, we have exactly five positive integer solutions for (4.4), and the invariance of parity assures that A is odd. $\qquad \square$

Example 4.30 (Belarus) A cube lies on an infinite grid of 1×1 squares in such a way that one of its faces (say $\mathcal{F}$) is situated exactly above one of the squares of the grid. The cube begins to roll on the grid, passing from one position to the next by rotating over one of its edges. At some moment, the cube stops exactly on the same 1×1 square where it began to move. Then, one notices that the face of the cube in contact with the grid is $\mathcal{F}$ again. Is it possible that, this time, $\mathcal{F}$ is rotated by 90° with respect to its initial orientation?

Proof It is not possible! In order to prove this, let us paint the vertices of the grid alternately black and white, as in ordinary chessboard. When the cube is about to start rolling, $\mathcal{F}$ has two black and two white vertices; then, paint the remaining four vertices of the cube also black and white, in such a way that each of its edges joins two vertices of distinct colors. It is easy to check that, as the cube rolls on the grid from one position to the next, black (resp. white) vertices of it are always in contact with black (resp. white) vertices of the grid. (Here is the invariance!) However, if at some future moment $\mathcal{F}$ revisits its original position rotated by 90° with respect to the initial orientation, then we would have two black vertices of $\mathcal{F}$ exactly above two white vertices of the grid. This contradiction proves our initial claim.   □

Our last example uses a semi-invariant reasoning to construct an optimal configuration.

Example 4.31 (IMO Shortlist) 155 birds are standing on a circle of center O. Two birds A and B are said to be *mutually visible* if and only if $A\widehat{O}B \leq 10°$. Assume that more than one bird can stand on a single point. Compute the smallest possible number of pairs of mutually visible birds.

Solution Assume that at least one bird stands at a point A, and at least another one stands at a point B, where A and B are distinct points of the circle such that $A\widehat{O}B \leq 10°$. Suppose, further, that h birds are visible from A but not from B, whereas k birds are visible from B but not from A. If $h \leq k$ and we let all birds in B to fly to A, then the total number of pairs of mutually visible birds (obviously) does not increase. By repeating such an operation several times, we obtain at the end a configuration in which two birds are mutually visible if and only if they are placed at a single point. Moreover, since the total number of pairs of mutually visible birds did not increase along the performed operations (here we have the semi-invariant!), in order to minimize this number, it suffices to consider such configurations. Also, since $\frac{1}{36} \cdot 360° = 10°$, a further consequence of this reasoning and the pigeonhole principle is that we ought to have birds in at most 35 distinct points of the circle.

Now, take 35 points along the circle, labelled $1, 2, 3, \ldots, 35$, and place $x_i \geq 0$ birds at the point i. Our problem reduces to minimizing

$$\sum_{i=1}^{35} \binom{x_i}{2} = \frac{1}{2} \sum_{i=1}^{35} x_i(x_i - 1),$$

under the constraint $\sum_{i=1}^{35} x_i = 155$. It thus suffices to minimize $\sum_{i=1}^{35} x_i^2$, under the condition $\sum_{i=1}^{35} x_i = 155$. To this end, first note that if $1 \leq i < j \leq 35$ are such that $x_i - x_j > 1$, then

$$(x_i^2 + x_j^2) - ((x_i - 1)^2 + (x_j + 1)^2) = 2(x_i - x_j - 1) > 0.$$

Therefore, a minimizing placement of the birds must be such that the values of the x_i's must differ by at most 1, from where we can let $x_1 = \ldots = x_k = a$ and $x_{k+1} = \ldots = x_{35} = a + 1$. This way, we get

$$\sum_{i=1}^{35} x_i^2 = ka^2 + (35 - k)(a + 1)^2 = 35(a + 1)^2 - (2a + 1)k.$$

Finally, since $ka + (35 - k)(a + 1) = 155$ or, which is the same, $35a - k = 120$, it is immediate to check that the minimum possible value of $\sum_{i=1}^{35} x_i^2$ is obtained for $a = 4$ and $k = 20$. Hence, the above discussion assures that the minimum possible value of mutually visible birds is

$$\binom{5}{2} \cdot 15 + \binom{4}{2} \cdot 20 = 270.$$

□

Remark 4.32 Yet with respect to the solution above, let $n \in \mathbb{N}$ be given and $x_1, \ldots, x_n$ be nonnegative integer variables. The argument just presented, of minimizing $\sum_{i=1}^{n} x_i^2$ subjected to the constraint $\sum_{i=1}^{n} x_i = m$, is an example of **discrete optimization** and is quite important in Combinatorics. It will make its appearance again, in disguised form, in Problems 26 and 27, page 135, as well as in Problem 3, page 142.

Problems: Sect. 4.4

1. (OCM) One has six equal bottles, five of them containing 2l of water each and the sixth one containing 1l. At any time, an allowed operation is to choose two bottles and split the total amount of water they contain into two equal parts (by pouring water from one bottle to the other). Is it possible to reach the situation in which every bottle has the same amount of water?

2. (Soviet Union) At the beginning of a class, a teacher writes a second degree trinomial on the blackboard. At later moments, an allowed operation with the trinomial written on the blackboard, say $ax^2 + bx + c$, consists in replacing it by $cx^2 + bx + a$ (if $c \neq 0$) or by any trinomial of the form $a(x + t)^2 + b(x + t) + c$, with $t \in \mathbb{R}$ chosen at random. If the trinomial initially written was $x^2 - x - 1$, is there any sequence of allowed operations that makes the trinomial $x^2 + 3x - 1$ appear in the blackboard?

3. We cut out two 1×1 squares situated in opposite corners of an 8×8 chessboard. Is it possible to cover the 62 remaining squares with 31 rectangles 2×1? Justify your answer!

4. (Argentina) Firstly, numbers 1, 2, 3, ..., 1998 are written on a blackboard. An allowed operation is to replace two of the numbers written, say a and b, by $|a - b|$. After 1997 operation, there is just one number written on the blackboard. Prove that, regardless of the order in which the replacements have been made, this last number will always be odd.

5. A Mathematics teacher posed the following activity to his students: one of them would write on the blackboard a list of six integer numbers; then, a second student would choose three of these six numbers, say x, y and z, replacing them respectively by $x - y - z$, $3x - 3y - 2z$ and $4x - 2y + 4z$ (thus getting, immediately below the former one—and by repeating the three numbers that have not been chosen—a new list of six integer numbers on the blackboard). Such a procedure would, then, be repeated with this new list of six numbers and another student. After the class, the blackboard was partially erased, there remaining only two lists os numbers:

$$1 \ 2 \ 3 \ 4 \ 5 \ 6 \text{ and } 3 \ 7 \ 2 \ 15 \ 8 \ 8$$

(we don't know how many lists there were between these two left ones). Prove that, at some replacement, a student made a mistake.

6. (TT) Ten coins are placed around a circle, all of them showing *heads* (so that *tails* cannot be seen at the beginning). Two moves are allowed:

   (a) To turn down four consecutively placed coins.
   (b) To turn down four coins placed as in *XXOXX* (here, X represents a coin to be turned down, whereas O represents one that should stay as it is).

   Is it possible to have all ten coins showing tails after a finite sequence of moves?

7. (Brazil) In the parliament of Terra Brasilis, each member has at most three enemies. Prove that the parliament can be divided into two houses, in such a way that, within his/her house, each member has at most one enemy.

8. (Leningrad) To begin with, numbers $1, 2, 3, \ldots, n$ are written on a blackboard. At any subsequent time, if there are at least two numbers on the blackboard, an allowed operation consists of erasing two of them, say a and b, and writing $a + b + ab$ instead (so that, after the operation is performed, we have one less number on the blackboard). After such an operation is carried out $n - 1$ times, we shall have only one number written on the blackboard. Prove that it will not depend on the order the operations were executed.

9. At each vertex of a square we have a certain quantity of cards (possibly none). An allowed operation is to remove a certain quantity of cards from one vertex and to put twice this quantity of cards in one of the two vertices adjacent to that one (to this end, there is a sufficiently large supply of cards outside the square, all of them ready to be used). Assume that we begin with just one card in one vertex, the other three vertices being empty. Is it possible, after a certain number of steps, to have exactly 1, 9, 8 and 9 cards in the vertices of the square, in clockwise order?

   The next problem revisits Problem 9, page 111, with the aid of invariants.

10. In the plane we are given $2n$ points in *general position*, i.e., such that no three of them are collinear. If n of these points are blue and the other n are red, prove that it is always possible to draw n line segments satisfying the following conditions:

(a) Each line segment joins a blue point to a red point.

(b) They do not intersect each other.

11. A cube of edge length n is partitioned into n^3 unit cubes. Two unit cubes are said to be *adjacent* if they share a common face. Students A and B play the following game on the bigger cube: A starts in a unit cube of his/her choice, then moving to an adjacent unit cube. Then B, departing from this last unit cube, moves to a third one, adjacent to the one he departed from, and so on. A and B play alternately, without ever revisiting a unit cube visited before. If the first who cannot play loses the game, and if A and B play with the best possible strategies, who will be the winner?

12. (IMO shortlist)

(a) Decide whether or not it is possible to number the unit squares of an 8×8 chessboard from 1 to 64 in such a way that, for every set of four unit squares of one of the shapes below, the sum of the numbers written on them is a multiple of 4:

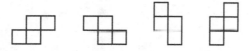

(b) Do the same for the sets of four unit squares of one of the following shapes:

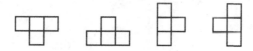

13. (Leningrad) All 120 faces of twenty equal cubes are painted black or white, with three faces of each color per cube. Prove that the cubes can be placed on an opaque table in such a way that the faces touching it form the board of a 6×6 square and the number of black visible faces equals the number of white visible ones.

For the coming problem the reader will find it useful to recall the following fact, to be proved later (cf. Proposition 6.33), on the greatest common divisor and least common multiple of two natural numbers: given $a, b \in \mathbb{N}$, one has

$$ab = \gcd(a, b) \operatorname{lcm}(a, b).$$

14. (TT) Several positive integers are written on a blackboard. An allowed operation is to erase two distinct numbers and to write, in their places, their gcd and lcm. Prove that, after we have repeated such an operation a certain finite number of times the numbers written on the blackboard do not change any more, no matter which additional operations we perform.

Chapter 5
A Glimpse on Graph Theory

We begin this chapter by considering the following three combinatorial problems:

Problem 1 in a party with 100 guests, there is always an even number of people who know, within the party, an odd number of other people.

Problem 2 a country has 10 cities and 37 two-way roads, such that each road connects two distinct cities and each pair of cities is connected by at most one road. Under such conditions, is it always possible to use the roads to travel from one city to another (perhaps passing through at least one other city)?

Problem 3 a regular polygon of 100 sides is drawn on the plane, together with all of its diagonals. Suppose we choose at random 2501 of its sides or diagonals, and paint them red. Is it necessarily true that a red triangle, having vertices at three of the vertices of the polygon, will be formed?

In spite of the seemingly lack of correlation between the above problems, in all of them we have certain sets of "*objects*" (guests, cities, vertices), together with relations between them (knowing each other, being connected by a road, being connected by a red segment, according to the case). These common features will allow us to analyse the three problems within the same abstract context, namely, *Graph Theory*. There, we concentrate on the fact that two objects may or may not be related, without paying actual attention to the particular kinds of objects or relationships between them.

It is the purpose of this chapter to develop the most basic aspects of Graph Theory, together with some interesting applications. Along the way, among others we shall prove Euler's theorem on the characterization of the existence of Eulerian paths, Cayley's theorem on the number of labelled trees and Turán's extremal theorem on the existence of cliques. Since we will barely touch the surface of Graph Theory, we refer the interested reader to [13, 22] or [40] for more comprehensive introductions to this subject.

5.1 Basic Concepts

In this chapter, unless stated otherwise, all sets under consideration will be finite. Given a nonempty set A, we shall denote by $\mathcal{P}_2(A)$ the family of the subsets of A having two elements:

$$\mathcal{P}_2(A) = \{B \subset A; \ |B| = 2\}.$$

In particular,

$$\#\mathcal{P}_2(A) = \binom{|A|}{2}.$$

Definition 5.1 A (**simple**) **graph** is a pair $G = (V; E)$, where V is a finite nonempty set and $E \subset \mathcal{P}_2(V)$. The elements of the set V are called the **vertices** of the graph G, whereas the elements of the family E are called the **edges** of G.

If there is any possibility of confusion, we may write $V(G)$ and $E(G)$ to denote the sets of vertices and edges of a graph G.

If $G = (V; E)$ is a graph and u and v are two of its vertices, we shall say that u and v are **adjacent** (or **neighbors**) provided $\{u, v\} \in E$; in this case, we shall also say that the edge $\{u, v\}$ is **incident** to the vertices u and v. Whenever there is no danger of confusion, we shall denote the edge $\{u, v\}$ simply by uv or vu. If u and v are not adjacent, we shall say that they are **non adjacent** vertices of G.

For most purposes, it is quite convenient to represent a graph $G = (V; E)$ by means of a diagram in which the elements of V are depicted as points or tiny circles in the plane and the edges of G as *arcs* joining the corresponding vertices. Well understood, the diagram thus obtained has no geometric meaning, and its purpose is nothing but to schematically represent the adjacency relations between pairs of vertices of G. For instance, if

$$G = (\{a, b, c, d\}; \{\{a, b\}, \{a, c\}, \{a, d\}, \{b, c\}\}),$$

then G may be represented by any one of the two diagrams shown in Fig. 5.1, for both of them embody the same adjacency relations.

We shall have more to say on this when we study the notion of *isomorphism* for graphs.

Fig. 5.1 Two representations of a graph

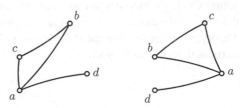

Example 5.2 Given a set V with n elements and a graph $G = (V; E)$, Definition 5.1 allows two extreme cases, namely, those in which $E = \emptyset$ or $E = \mathcal{P}_2(V)$. In the first case, we say that $G = (V; \emptyset)$ is the **trivial graph** on n vertices, which can be represented by a set of n points in the plane, with no arcs. In the second case, G is the **complete graph** (or a **clique**) on n vertices, which can also be represented by a set of n points in the plane, with any two of them being adjacent. From now on, we shall denote a complete graph on n vertices by K_n; in particular, note that K_n has exactly $\binom{n}{2}$ edges.

Back to the notion of adjacency, fixed a vertex u of $G = (V; E)$, we shall denote by $N_G(u)$ the set of vertices adjacent to u:

$$N_G(u) = \{v \in V; \ uv \in E\}.$$

The following definition is central to all that follows.

Definition 5.3 Let $G = (V; E)$ be a graph. For $u \in V$, the **degree** of u, denoted $d_G(u)$ is the number of vertices of G which are adjacent to u:

$$d_G(u) = |N_G(u)|.$$

Remark 5.4 Whenever the graph $G = (V; E)$ is understood, for $u \in V$ we shall denote the degree of u and the set of neighbors of u simply by $d(u)$ and $N(u)$, respectively.

We are finally in position to state and prove the most basic result of Graph Theory, which is due to Euler.[1]

Theorem 5.5 (Euler) *In every graph $G = (V; E)$, the sum of the degrees of the vertices is equal to twice the number of edges. In symbols:*

$$2|E| = \sum_{u \in V} d_G(u). \tag{5.1}$$

Proof We use double counting. It suffices to observe that, if $\epsilon = \{u, v\}$ is an edge of G, then ϵ is counted exactly twice in the right hand side of (5.1): once in the summand $d_G(u)$ and another time in the summand $d_G(v)$. Hence, the right hand side of (5.1) has to be equal to $2|E|$, since this number also counts each edge of G exactly twice. □

The coming corollary, also due to Euler, is the most important consequence of the previous theorem.

Corollary 5.6 (Euler) *In any graph, the number of vertices of odd degree is even.*

---

[1] Actually, as we shall see later, Euler is usually credited as being the founder of Graph Theory.

Proof Let $G = (V; E)$ be a graph and, for an integer $k \geq 0$, let $v_k(G)$ denote the number of vertices of G with degree k. Since the sum at the right hand side of (5.1) has exactly $v_k(G)$ summands equal to k, we have (again by double counting) the equality

$$\sum_{u \in V} d_G(u) = \sum_{k \geq 0} k v_k(G). \qquad (5.2)$$

It thus follows from Euler's theorem that

$$2|E| = \sum_{u \in V} d_G(u) = \sum_{k \geq 0} k v_k(G)$$

$$= \sum_{j \geq 0} (2j + 1) v_{2j+1}(G) + \sum_{j \geq 1} 2j v_{2j}(G)$$

$$= \sum_{j \geq 0} v_{2j+1}(G) + \sum_{j \geq 1} 2j (v_{2j}(G) + v_{2j+1}(G)).$$

Therefore,

$$\sum_{j \geq 0} v_{2j+1}(G) = 2|E| - \sum_{j \geq 1} 2j (v_{2j}(G) + v_{2j+1}(G)),$$

which is an even number.   □

Example 5.7 Prior to a committee's reunion, some of its ten members shook hands. Is it possible that the numbers of handshakes have been, in some order, equal to 1, 1, 1, 3, 3, 3, 4, 6, 7 and 8?

Proof Look at the committee's members as the vertices of a graph, with two vertices being adjacent if the corresponding people shook hands. Then, if the situation just described could have occurred, the degrees of the vertices of the graph would have been, in some order, equal to 1, 1, 1, 3, 3, 3, 4, 6, 7 and 8. But this would contradict the previous corollary, since we would have an odd number of vertices of odd degree. Therefore, such a situation cannot have occurred.   □

The following definition brings the appropriate notion of equivalence for graphs.

Definition 5.8 Graphs $G = (V_1; E_1)$ and $H = (V_2; E_2)$ are said to be **isomorphic** if there exists a bijection $f : V_1 \to V_2$ which preserves incidence, i.e., such that, for every two distinct vertices u and v of G, we have

$$\{u, v\} \in E_1 \Leftrightarrow \{f(u), f(v)\} \in E_2.$$

In this case, we denote $G_1 \simeq G_2$.

It is clear from the previous definition that two isomorphic graphs have equal numbers of vertices. On the other hand, Problem 4 guarantees that, for each $n \in \mathbb{N}$, the notion of isomorphism of graphs induces an equivalence relation (cf. Sect. 2.3) in the set of graphs of n vertices.

The coming proposition gives other necessary conditions for two graphs to be isomorphic.

Proposition 5.9 *Two isomorphic graphs have equal numbers of vertices of a certain degree. In particular, they have equal numbers of edges.*

Proof Let $G = (V_1; E_1)$ and $H = (V_2; E_2)$ be isomorphic graphs and $f : V_1 \rightarrow V_2$ be an incidence-preserving bijection. If u is a vertex of G having degree $k > 0$, such that $N_G(u) = \{u_1, \ldots, u_k\}$, then it is immediate to see that $N_H(f(u)) = \{f(u_1), \ldots, f(u_k)\}$. In particular, $f(u)$ is a vertex of H having degree k. An analogous reasoning assures that if u has degree 0, then the same holds for $f(u)$.

Since the argument of the last paragraph is symmetric with respect to G and H (here we are using the fact that the notion of isomorphism of graphs is an equivalence relation), we conclude that G and H have, for every integer $k \geq 0$, equal quantities of vertices of degree k.

For what is left, we apply Euler's theorem twice: since $d_G(u) = d_H(f(u))$ for every $u \in V_1$ and f is a bijection between the sets of vertices of G and H, we obtain

$$2|E_1| = \sum_{u \in V_1} d_G(u) = \sum_{u \in V_1} d_H(f(u)) = \sum_{f(u) \in V_2} d_H(f(u)) = 2|E_2|.$$

Therefore, $|E_1| = |E_2|$.   □

Example 5.10 If $G = (V; E)$ is a graph of n vertices we can assume, whenever needed, that $V = I_n$. Indeed, since $|V| = n$, we can choose a bijection $f : V \rightarrow I_n$ and let $H = (I_n; F)$ be defined by setting, for $i \neq j$ in I_n,

$$\{i, j\} \in F \Leftrightarrow \{f^{-1}(i), f^{-1}(j)\} \in E.$$

The previous example allows us to introduce a quite useful *algebraic representation* of a graph, namely, its *adjacency matrix*. In this respect, we assume the reader is acquainted with the most basic concepts on matrices, which can be revisited at Chapter 2 of [4], for instance.

Definition 5.11 Given a graph $G = (V; E)$, with $|V| = n$, assume, according to the previous example, that $V = I_n$. The **adjacency matrix** of G is the $n \times n$ matrix $\text{Adj}(G) = (a_{ij})$ such that

$$a_{ij} = \begin{cases} 1, & \text{if } i \neq j \text{ and } \{i, j\} \in E \\ 0, & \text{else} \end{cases}.$$

Lemma 5.12 *The adjacency matrix of a graph is symmetric, with zeros in the main diagonal.*

Proof In the notations of the previous definition we have, for $i \neq j$ in I_n, that

$$a_{ij} = 1 \Leftrightarrow \{i, j\} \in E \Leftrightarrow \{j, i\} \in E \Leftrightarrow a_{ji} = 1;$$

hence, $\mathrm{Adj}(G)$ is symmetric. The rest is immediate from the definition.   □

Remark 5.13 Given a graph $G = (V; E)$, we shall sometimes say that G is **labelled**, an allusion to the fact that its vertices have *names*, or *labels*. The notion of isomorphism of graphs allows us to introduce the concept of *unlabelled* graph. More precisely, if V is a set with n elements and $\mathcal{G}$ is the set of graphs having V as set of vertices, then the restriction of the relation of graph isomorphism to $\mathcal{G}$ induces an equivalence relation in $\mathcal{G}$. A **unlabelled graph** with n vertices is an equivalence class in $\mathcal{G}$, with respect to the isomorphism relation. Unless we explicitly say otherwise, in these notes we shall consider only labelled graphs, generally omitting the adjective "*labelled*".

We finish this section by examining the important notion of *subgraph* of a graph.

Definition 5.14 A graph H is a **subgraph** of a graph G if $V(H) \subset V(G)$ and $E(H) \subset E(G)$. The graph H is a **spanning subgraph** of G if H is a subgraph of G for which $V(H) = V(G)$.

We now show how to build two fundamental examples of subgraphs of a given graph.

Example 5.15 Given a graph $G = (V; E)$ and $\epsilon \in E$, the subgraph of G obtained by the **edge excision** of ϵ is the graph $H = (V; E \backslash \epsilon)$. From now on, we shall denote such a subgraph of G simply by $G - \epsilon$. In words, $G - \epsilon$ is the subgraph obtained from G by *erasing* the edge ϵ. Note, further, that $G - \epsilon$ is a spanning subgraph of G.

Example 5.16 Given a graph $G = (V; E)$ and $u \in V$, the subgraph of G obtained by **vertex excision** of u is the graph $H = (V \setminus \{u\}; E')$, where

$$E' = E \setminus \{\epsilon \in E; \, \epsilon \text{ is incident with } u\}.$$

From now on, we shall denote such a subgraph of G simply by $G - u$. In words, $G - u$ is the subgraph of G obtained by *erasing* vertex u, together with all of the edges of G incident to it. Note, further, that $G - u$ has one vertex less and $d_G(u)$ edges less than G.

Corollary 5.6 gave a necessary condition for a list of nonnegative integers to be the list of degrees of some simple graph. Although there is no simple sufficient condition to decide whether this is so, we finish this section by presenting Havel-Hakimi's[2] simple algorithm, that allows one to quickly decide it for any given list of nonnegative integers. We follow [22].

Theorem 5.17 (Havel-Hakimi) *If s is a positive integer, then the nonnegative integers*

$$s \geq t_1 \geq t_2 \geq \ldots \geq t_s \geq d_1 \geq \ldots \geq d_n$$

are the degrees of the vertices of some simple graph if and only if so are

$$t_1 - 1, t_2 - 1, \ldots, t_s - 1, d_1, \ldots, d_n.$$

Proof Let G be a graph whose vertices have degrees $s \geq t_1 \geq t_2 \geq \ldots \geq t_s \geq d_1 \geq \ldots \geq d_n$. Since s is positive and the vertex u of degree s has exactly s neighbors, we conclude that $t_1, t_2, \ldots, t_s \geq 1$ (note that this follows from the fact that we have listed the degrees in descending order). For $1 \leq i \leq s$, let v_i be the vertex of degree t_i and, for $1 \leq j \leq n$, let w_j be the vertex of degree d_j. We distinguish two cases:

(i) $N_G(u) = \{v_1, \ldots, v_s\}$: then the vertices of $H := G - u$ have degrees $t_1 - 1, t_2 - 1, \ldots, t_s - 1, d_1, \ldots, d_n$.

(ii) $N_G(u) \neq \{v_1, \ldots, v_s\}$: then, for some $1 \leq i \leq s$ and some $1 \leq j \leq n$, we have u nonadjacent to v_i and adjacent to w_j. In particular, $t_i \geq d_j$, and we consider two subcases (make some drawings to follow the reasoning):

- $t_i = d_j$: let G' be the graph obtained from G by exchanging v_i and w_j.
- $t_i > d_j$: then v_i has a neighbor x which is not a neighbor of w_j. Let G' be the graph obtained from G by erasing the edges $\{u, w_j\}$, $\{v_i, x\}$ and adding the edges $\{u, v_i\}$, $\{w_j, x\}$.

After having performed either i. or ii., we are left with a graph G' whose list of degrees is equal to that of G, and such that $N_{G'}(u) \cap \{v_1, \ldots, v_s\}$ strictly contains $N_G(u) \cap \{v_1, \ldots, v_s\}$. If $N_{G'}(u) = \{v_1, \ldots, v_s\}$, we return to i. Otherwise, we repeat ii. until we reach such a situation. In the end of this process, we will find a graph H having degrees $t_1 - 1, t_2 - 1, \ldots, t_s - 1, d_1, \ldots, d_n$.

Conversely, assume that $(t_1 - 1, t_2 - 1, \ldots, t_s - 1, d_1, \ldots, d_n)$ is the sequence of degrees of some simple graph H, and let u_i be a vertex in H with degree $t_i - 1$. Form a graph G by adding a vertex to H and making it adjacent to $u_1, \ldots, u_s$. Then, the sequence of degrees of the vertices of G is precisely $(s, t_1, t_2, \ldots, t_s, d_1, \ldots, d_n)$. $\square$

---

[2]After Václav J. Havel, Czech mathematician, and Seifollah L. Hakimi, Iranian-American mathematician, both of the twentieth century.

Problems: Sect. 5.1

1. Solve *Problem 1* at the beginning of this chapter.
2. Prove that every graph has at least two vertices with the same degree.
3. Given graphs $G_1 = (V_1; E_1)$ and $G_2 = (V_2; E_2)$, with $|V_1| = |V_2|$, prove that G_1 and G_2 are isomorphic if and only if $\text{Adj}(G_2)$ can be obtained from $\text{Adj}(G_1)$ by means of a permutation of rows.
4. * Prove that graph isomorphism is an equivalence relation in the family of graphs.
5. Compute the number of (labelled) graphs with vertices $u_1, \ldots, u_n$.
6. Prove that there does not exist a graph of seven vertices, with degrees 1, 1, 2, 3, 4, 5 and 6, in some order.
7. Give an example of a graph of eight vertices, of degrees 1, 1, 2, 3, 4, 4, 5 and 6 in some order.
8. Given $n \in \mathbb{N}$, show that there exists a simple graph of $2n$ vertices, whose degrees are $n, n, n - 1, n - 1, \ldots, 3, 3, 2, 2, 1, 1$.
9. * If $G = (V; E)$ is a graph, its **complement** is the graph $\overline{G} = (V; E^c)$, where E^c denotes the complement of E in $\mathcal{P}_2(V)$. If $|V| = n$, prove that:

   (a) $d_G(u) + d_{\overline{G}}(u) = n - 1$, for every $u \in V$.
   (b) $|E| + |E^c| = \binom{n}{2}$.

10. A graph is **self-complementary** if it is isomorphic to its complement. If G is a self-complementary graph with n vertices, prove that n leaves remainder 0 or 1 upon division by 4.
11. * Generalize Examples 5.15 and 5.16. More precisely, given a graph $G = (V; E)$ and $\emptyset \neq B \subsetneq V$, $\emptyset \neq A \subset E$, define the subgraphs $G - B$ and $G - A$ of G, respectively obtained by the excision of the vertices in B and of the edges in A.
12. * Given a graph G and a subset A of $V(G)$, the subgraph of G **induced** by A is the graph $G_{|A} = (A; E')$, where

$$E' = \{\{u, v\} \in E; \ u, v \in A\}.$$

    Prove that:

    (a) For $u \in V(G)$, one has $G - u = G_{|V(G)\setminus\{u\}}$.
    (b) $G_{|A} = G - A^c$, where A^c denotes the complement of A in $V(G)$ and $G - A^c$ is defined as in the previous problem.

13. Let $G = (V; E)$ be a graph. We choose two nonadjacent vertices, say u and v, and add edge $\{u, v\}$ to E, thus obtaining a new graph $H = (V; F)$. Prove that the difference between the numbers of odd degree vertices in H and in G is -2, 0 or 2. Then, use this fact to give another proof of Corollary 5.6.

    For the coming problem, the reader may find it convenient to recall the discussion on congruence modulo n, in Sect. 2.3.

14. * Given integers $1 \leq n < m$, let $G(m, n) = G(I_m; E)$ be the graph such that, for $1 \leq i, j \leq m$,

$$\{i, j\} \in E \Leftrightarrow j \equiv i \pm n \pmod{m}.$$

   (a) Represent $G(m, n)$ when $m = 6$ and n varies from 1 to 5.
   (b) If $m = 2n$, prove that each vertex has degree 1 and $|E| = n$. Then, represent the graph thus obtained.
   (c) If $m \neq 2n$, prove that each vertex has degree 2 and $|E| = m$.
   (d) Prove that $G(m, n) \simeq G(m, m - n)$.

15. * In the notations of the previous problem, represent the graph obtained from $G(5, 1)$ and $G(5, 2)$ by joining, for $1 \leq i \leq 5$, vertex i of $G(5, 1)$ to vertex $2i$ of $G(5, 2)$. The graph thus obtained is known as **Petersen graph**.[3]

   For the next problem, one says that a nonempty subset A of the set of vertices of a graph G is **independent** if any two vertices in A are nonadjacent in G.

16. * A graph $G = (V; E)$ is **bipartite** if one can write $V = V_1 \cup V_2$, with V_1 and V_2 being disjoint, nonempty and independent sets of vertices. Prove that $|E| \leq |V_1| \cdot |V_2|$, with equality if and only if each vertex of V_1 is adjacent to some vertex of V_2, and vice-versa.

17. * In the notations of the previous problem, a bipartite graph $G = (V_1 \cup V_2; E)$ is said to be **complete** if $|E| = |V_1| \cdot |V_2|$. If $H = (W_1 \cup W_2; F)$ is another complete bipartite graph, prove that

$$G \simeq H \Leftrightarrow |V_1| = |W_1| \text{ and } |V_2| = |W_2|, \text{ or vice-versa.}$$

   Thanks to such a result, if $|V_1| = m$ and $|V_2| = n$, we write simply $K_{m,n}$ to denote the complete bipartite graph $G = (V_1 \cup V_2; E)$.

18. * In the notations of Problem 16, let $V_1 = \{1, \ldots, m\}$ and $V_2 = \{m+1, \ldots, m+n\}$. Prove that the adjacency matrix of G has the form

$$\text{Adj}(G) = \begin{bmatrix} 0 & A \\ A^{\top} & 0 \end{bmatrix},$$

   where A is an $m \times n$ matrix and $A^{\top}$ denotes the transpose of A.

19. A bipartite graph $G = (V_1 \cup V_2; E)$, with independent sets of vertices V_1 and V_2, has 16 vertices of degree 5 and some (at least one) vertices of degree 8. If all of the vertices of degree 8 belong to V_1, compute how many vertices of degree 8 the graph G can have.

20. * A graph G is r-**regular** if all vertices of G have degree r.

---

[3] After Julius P. C. Petersen, Danish mathematician of the nineteenth century.

(a) If G is an r-regular graph with n vertices, prove that $2 \mid nr$ and that G has exactly $\frac{nr}{2}$ edges.

(b) Given natural numbers n and r, such that $r < n$ and nr is even, show how to construct an r-regular graph with n vertices.

21. A battalion commander asked volunteers to compose 11 patrols, all of which formed by the same number of men. If each men entered exactly two patrols and each two patrols had exactly one men in common, compute the numbers of volunteers and of members of each patrol.[4]

22. Let $G = (V_1 \cup V_2; E)$ be a bipartite graph, with $|V_1| = m$ and $|V_2| = n$. It is known that all vertices in V_1 have degree g, all vertices in V_2 have degree 2 and $|N(u) \cap N(v)| = 1$, for all distinct vertices $u, v \in V_1$.

(a) Compute g and n in terms of m.

(b) For the values of g and n computed as in (a), show that there is a bipartite graph satisfying the stated conditions.

23. (Japan) A total of x students entered a Mathematics competition in which $2y$ problems were proposed. It is known that each student solved y problems and each problem was solved by the same number of students; moreover, each two students solved exactly three problems in common. Compute the values of all possible pairs (x, y) and, for each such pair, exhibit a configuration showing which problems were solved by each student.

24. (Hungary) Prove that it is impossible to place the numbers $1, 2, 3, \dots, 13$ around a circle so that, for any two neighboring numbers x and y, we have $3 \le |x-y| \le 5$.

25. Let $G = (V; E)$ be a graph. A subset V' of V is a **vertex cover** of G if every edge of G is incident to least one vertex in V'; a vertex cover V' is **minimal** if G has no vertex cover with less elements than V'. A **matching** in G is a subset E' of E formed by pairwise disjoint edges; a matching E' is **maximal** if G has no matching with more edges than E'. The purpose of this problem is to prove **König's theorem**[5]: in every bipartite graph, the number of vertices in a minimal vertex cover is equal to the number of edges in a maximal matching. To this end, let $G = (V_1 \cup V_2; E)$ be bipartite, with independent sets of vertices V_1 and V_2, and do the following items:

(a) Let $\preceq$ be the relation on $V_1 \cup V_2$ given by $u \preceq v \Leftrightarrow u \in V_1, v \in V_2$. Show that $\preceq$ is a partial order on $V_1 \cup V_2$.

(b) With respect to $\preceq$, show that:

 i. An antichain is the same as the complement of a vertex cover in G.

 ii. A collection of disjoint chains of lengths greater than 1 is the same as a matching in G.

---

[4]For another approach to this problem, see Problem 5, page 49.

[5]After Dénes König, Hungarian mathematician of the twentieth century.

Fig. 5.2 A transitive (left) or cyclic (right) K_3

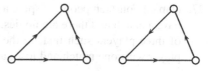

(c) Apply Dilworth's theorem (cf. Problem 6, page 115) to prove König's theorem.

For the coming problem, we define a **directed graph**, or **digraph** as a graph G such that we associate an *orientation* to each one of its edges. Informally, this can be thought of as an *arrow* along the edge; formally, each edge of a digraph is an *ordered pair* (instead of a set) (u, v), for some distinct $u, v \in V(G)$, such that if (u, v) is an edge then (v, u) is not. Given a vertex u of a digraph G, we define the **out degree** $d_G^+(u)$ of u as the number of arrows *departing from* u; analogously, the **in degree** of u, denoted $d_G^-(u)$, is the number of arrows *arriving at* u. Hence,

$$d_G(u) = d_G^+(u) + d_G^-(u)$$

is the degree of u in G, thought of as an ordinary graph (i.e., the graph obtained by removing the orientations of the edges of the digraph).

If a digraph contains a K_3, then such a K_3 is said to be **transitive** or **cyclic**, provided it satisfies the situation depicted at left or right, respectively, in Fig. 5.2. A digraph G is **complete**, or a **tournament** if it is a complete graph when we remove the orientations of its edges.

26. * If G is a complete digraph of n vertices, prove that:

(a) $\sum_{u \in V} d^+(u) = \binom{n}{2}$, where V is the set of vertices of G and we write $d^+(u)$ in place of $d_G^+(u)$.

(b) The average out degree of G is $\overline{d}^+ = \frac{n-1}{2}$.

(c) The number of transitive K_3's in G is $\sum_{u \in V} \binom{d^+(u)}{2}$.

(d) The number of cyclic K_3's in G is given by

$$\binom{n}{3} + \frac{1}{2}\binom{n}{2} - \frac{1}{2}\sum_{u \in V} d^+(u)^2.$$

(e) The number of cyclic K_3's in G is also given by

$$\frac{n(n^2 - 1)}{24} - \sum_{u \in V}(d^+(u) - \overline{d}^+)^2.$$

Then, conclude that this is at most

$$\begin{cases} n(n^2 - 1)/24, & \text{if } n \text{ is odd} \\ n(n^2 - 4)/24, & \text{if } n \text{ is even} \end{cases}.$$

27. (Japan) Fourteen people dispute a chess tournament in which each two players
have a match and there are no ties. Compute the largest possible number of sets
of three players such that, in the three matches involving two of them, each
player wins one match and loses another one.

5.2 Paths, Walks and Cycles

In this section we study the problem of deciding whether a graph is *connected*, i.e.,
has only one *piece*. As a byproduct, we introduce the concepts that name the section
and prove several important results related to them. In particular, we give a necessary
and sufficient condition for the existence of an Eulerian walk in a connected graph,
and characterize bipartite graphs in terms of cycles.

Definition 5.18 A **walk** of **length** $k \geq 1$ in a graph G is a sequence

$$\mathcal{P} = (u_0, u_1, \ldots, u_k) \tag{5.3}$$

of (not necessarily distinct) vertices of G, such that u_{i-1} is adjacent to u_i, for $1 \leq i \leq k$. A walk $\mathcal{P}$ as above is said to be **closed** if $u_0 = u_k$.

The graphs we wish to qualify are those for which there is a walk between any
two vertices. Since these graphs are quite important for our subsequent discussion,
we isolate them in the coming

Definition 5.19 A graph is **connected** if there exists a walk between any two of its
vertices. A non connected graph is said to be **disconnected**.

The coming proposition furnishes a simple (sufficient, albeit obviously not
necessary) criterion for a graph to be connected. For its statement, we need to
introduce a concept which will also be useful in other future context: given a graph
$G = (V; E)$, its **minimum degree** is the nonnegative integer $\delta(G)$ given by

$$\delta(G) = \min\{d_G(u); \ u \in V(G)\}. \tag{5.4}$$

Proposition 5.20 *If G is a graph of n vertices for which $\delta(G) \geq \lfloor \frac{n}{2} \rfloor$, then G is connected.*

Proof If $n = 1$, there is nothing to do. If $n = 2$, then $\delta(G) = 1$ and, hence, the two
vertices of G are adjacent. Suppose, then, that $n > 2$, and let u and v be distinct
vertices of G; we shall show that there is a walk between them. If u and v are
adjacent, there is nothing to do. Otherwise, since

$$dG(u) + d_G(v) \geq 2\lfloor n/2 \rfloor > n - 2 = \#(V(G) \setminus \{u, v\}),$$

the pigeonhole principle guarantees the existence of a vertex $w \neq u, v$, adjacent to both u and v. Therefore, the walk (u, w, v) joins u and v. □

For a fixed graph G, it is immediate to verify that the relation $\sim$ in $V(G)$, defined by letting

$$u \sim v \Leftrightarrow u = v \text{ or there exists a walk in } G \text{ joining } u \text{ and } v, \tag{5.5}$$

is of equivalence. Also, note that an equivalence class of $V(G)$ with respect to such a relation is, in particular, a set of vertices of G.

For the next definition, the reader may find it convenient to review the notion of *subgraph induced by a set of vertices*, in Problem 12, page 132.

Definition 5.21 The **connected components** of a graph G are the subgraphs $G_{|A}$ of G, induced by the equivalence classes A of $V(G)$ with respect to the equivalence relation (5.5).

If A is an equivalence class of vertices of G with respect to (5.5), then, in particular, there is a walk between any two distinct vertices in A, so that $G_{|A}$ is a connected graph. On the other hand, if H is a connected subgraph of G containing $G_{|A}$, then $H = G_{|A}$. Indeed, if $u \in V(H)$ and $v \in A \subset V(H)$, then the connectedness of H assures the existence of a walk in H (and hence in G) joining u and v; therefore, $u \in A$, since A is an equivalence class of vertices of G with respect to the existence of walks. Thus, we have $A \subset V(H) \subset A$, from where it follows that $H = G_{|A}$. The above argument allows us to assert that the connected components of a graph are precisely its *maximal connected subgraphs*.

Back to the analysis of the walks in a graph, we collect yet another relevant piece of terminology, which goes back to Euler. In this sense, if $\mathcal{P}$ is a walk in a graph G as in (5.3), we say that $\{u_{i-1}, u_i\}$, for $1 \leq i \leq k$, are the edges of G *traversed* by $\mathcal{P}$.

Definition 5.22 An **Eulerian walk** in a connected graph is a closed walk which traverses each edge of the graph exactly once. A connected graph is **Eulerian** if it contains an Eulerian walk.

Let G be an Eulerian graph and $\mathcal{P}$ be an Eulerian walk in G, as in (5.3). For an arbitrary vertex u of G, if $k \geq 1$ is the number of occurrences of u in $\mathcal{P}$, we claim that $d_G(u) = 2k$, an even number. This is obvious if all such occurrences of u are of the form $u = u_i$, with $1 \leq i < k$ (for in this case u_{i-1} and u_{i+1} are distinct vertices of G, both adjacent to u); on the other hand, if $u_0 = u$, note that $u_k = u$ too, for $\mathcal{P}$ is closed. Therefore, a necessary condition for a connected graph G to be Eulerian is that all vertices of G are of even degree. Theorem 5.24 below guarantees that such condition is also sufficient for the existence of an Eulerian walk. Before we discuss it, we need an important auxiliary result.

Lemma 5.23 *If G is a connected and nontrivial graph in which every vertex has even degree, then G contains a closed walk which does not traverse the same edge twice.*

Proof Choose a vertex u_0 of G and, starting at u_0, construct a walk in G by performing the following operation as many times as possible: being at a vertex u_{k-1} of G, with $k \geq 1$, choose a vertex u_k of G, adjacent to u_{k-1} and such that the edge $\{u_{k-1}, u_k\}$ has not been chosen yet.

In order to show that the algorithm described above does generate a closed walk in G, assume that, after a certain number of (at least one) operations, we find ourselves in a vertex u_k of G. There are two possibilities:

 (i) $u_k = u_0$: there is nothing to do, for the closed walk is $(u_0, u_1, \ldots, u_k)$.
(ii) $u_k \neq u_0$: suppose that the vertex u_k has appeared (with smaller indices) other l times along the performed operations; this used exactly $2l + 1$ of the edges incident to u_k (two edges for each of the l previous occurrence of u_k, and an extra edge to reach u_k from u_{k-1}). However, since u_k has even degree, there exists at least one edge incident to u_k and which has not been used yet; hence, the algorithm does not stop at u_k.

Finally, since the number of edges in G is finite, the above discussion assures that the algorithm stops at u_0 and, when it does so, we have succeeded in constructing a closed walk in G which does not traverse the same edge twice. $\square$

We are now in position to prove the result of Euler alluded to above.

Theorem 5.24 (Euler) *A connected and nontrivial graph G is Eulerian if and only if all of its vertices have even degree.*

Proof We have already seen, in the paragraph preceding Lemma 5.23, that the condition that all vertices of G have even degree is indeed necessary for the existence of an Eulerian walk. In order to establish the sufficiency, let us make induction on the number of edges of G.

Since G is nontrivial, if all of its vertices have even degree, then it has at least three edges; moreover, if G has exactly three edges, then it is immediate to prove that it is isomorphic to K_3, which is clearly Eulerian.

Now, assume that G has $n > 3$ edges, and that the theorem is true for all connected graphs with less than n edges and such that all vertices have even degree. By the previous lemma, we can take in G a closed walk $\mathcal{P} = (u_0, u_1, \ldots, u_{k-1}, u_0)$, that does not traverse the same edge twice. Let A be the set of edges in $\mathcal{P}$, and $H_1, \ldots, H_l$ be the nontrivial connected components of $G - A$ (cf. Problem 11, page 132). If a vertex u of H_i occurs $j \geq 0$ times in $\mathcal{P}$, then

$$d_G(u) = d_{H_i}(u) + 2j,$$

so that $d_{H_i}(u)$ is even. Since this is true for all indices $1 \leq i \leq l$ and all vertices $u \in V(H_i)$, we conclude that $H_1, \ldots, H_l$ do satisfy the hypotheses of the theorem.

Since each H_i has less edges than G, the induction hypothesis guarantees the existence of an Eulerian walk $\mathcal{P}_i$ for H_i, for $1 \leq i \leq l$. At this point, the idea is to *glue* the Eulerian walks $\mathcal{P}_1, \ldots, \mathcal{P}_l$ to $\mathcal{P}$ one by one, in a way to get an Eulerian walk for G. To this end, note that $\mathcal{P}$ and $\mathcal{P}_1$ have at least one vertex in common, say u. We may assume, without loss of generality, that $\mathcal{P}$ and $\mathcal{P}_1$ start and end at u (it suffices to change the starting points of both these closed walks, if necessary). Hence, following $\mathcal{P}$ and subsequently $\mathcal{P}_1$, we obtain a closed walk $\mathcal{P}_1'$ in G, which encompasses all of the edges of $\mathcal{P}$ and $\mathcal{P}_1$ and does not traverse one of them twice. It now suffices to repeat the above argument more $l-1$ times, gluing $\mathcal{P}_2$ to $\mathcal{P}_1'$ to get a closed walk $\mathcal{P}_2'$, and so on. The closed walk $\mathcal{P}_l'$, obtained at the end of this process will be an Eulerian walk in G. □

The above result, proved by L. Euler in 1735, marks the birth of Graph Theory. Euler came to it after having been inquired on the possibility of performing a walk along the streets of the prussian city of Königsberg (nowadays the Russian city of Kaliningrad), traversing exactly once each of the seven bridges the city had, at that time, over the Pregel river.

Taking into account that these bridges join two islands of the Pregel to the remaining parts of the city, Euler modelled the problem posed to him as the search for finding, in the graph of Fig. 5.3, a walk of the type we refer today as being *Eulerian*. Although such a graph is not simple (notice the *multiple edges* joining two vertices, a situation that, by the sake of simplicity, we did not allow in our presentation of the theory), a moment's reflection will easily convince the reader that the previous theorem remains true for *non simple connected graphs*. Therefore, this result (actually, the adaptation of the argument for the proof of the necessity part of it, given at the paragraph preceding Lemma 5.23) guarantees that the graph of Fig. 5.3 does not contain an Eulerian walk, for all of its vertices have odd degree.

Back to the general development of the theory, we finish this section by studying the concepts of *paths* and *cycles* in graphs, as well as looking at some relevant applications of them.

Definition 5.25 A **path** in a graph is a walk with distinct vertices.

Fig. 5.3 Graph representing the bridges of Königsberg

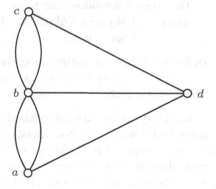

Even though every path is a walk, it is clear that the converse is not true in general, namely, that there are walks which are not paths. Nevertheless, it is not difficult to convince ourselves (cf. Problem 7) that, if a graph contains a walk between two of its vertices, then it also contains a path between them.

Definition 5.26 A k-**cycle** in a graph G is a closed walk

$$(u_0, u_1, \ldots, u_{k-1}, u_0)$$

in G, of length $k \geq 3$ and such that $(u_0, u_1, \ldots, u_{k-1})$ is a path.

The coming result uses the concept of cycle to characterize bipartite graphs (cf. Problem 16, page 133).

Proposition 5.27 *A graph G is bipartite if and only if every cycle in G has even length.*

Proof Firstly, let us assume that G is bipartite, with independent sets of vertices V_1 and V_2. If $C = (u_0, u_1, u_2, \ldots, u_{k-1}, u_0)$ is a cycle in G, then the independence of the sets V_1 and V_2 guarantees that vertices u_i and u_{i+1} alternate between the sets V_1 and V_2, for $0 \leq i \leq k - 1$ (with $u_k = u_0$). In particular, the path $(u_0, u_1, \ldots, u_{k-1})$ alternates $k - 1$ times between V_1 and V_2 for, starting from u_0, arrive at u_{k-1}. However, since u_{k-1} and u_0 are adjacent, they cannot belong to a single one of the sets V_1 or V_2 and, hence, $k - 1$ must be odd. Therefore, the length k of C must be even.

Conversely, assume that every cycle in G has even length. We use induction on the number of edges to show that G is bipartite. If G contains some cycle C, erase an edge $\epsilon = \{u, v\}$ of C. Since no further cycles are created, all of the cycles of $G - \epsilon$ are also of even length. Hence, by induction hypothesis, $G - \epsilon$ is bipartite, with independent sets of vertices V_1 and V_2, say. We shall show that V_1 and V_2 are also independent in G. To this end, note that the path $C - \epsilon$ in $G - \epsilon$ joins u to v and has odd length; since its vertices alternate between V_1 and V_2, it follows that $u \in V_1$ and $v \in V_2$, or vice-versa. Therefore, V_1 and V_2 are also independent in G, so that G is bipartite too.   □

The coming definition, due to the nineteenth century Irish astronomer, mathematician and physicist William R. Hamilton, is, in a certain sense, *dual* to the definition of Eulerian walk.

Definition 5.28 A **hamiltonian cycle** in a graph is a cycle that passes through all of the vertices of the graph. A connected graph is **hamiltonian** if it contains a hamiltonian cycle.

Not every connected graph is hamiltonian; for instance, any graph without cycles is not hamiltonian (for not-so-trivial examples, see Example 5.30 or Problem 21).

The question of the existence or not of hamiltonian cycles in graphs is much more difficult than the corresponding problem for Eulerian walks. Actually, up to now there does not exist a set of simple necessary and sufficient conditions for a graph to be hamiltonian. On the other hand, the coming result, due to the Hungarian mathematician of the twentieth century Gabriel Dirac, provides a simple sufficient condition for the existence of hamiltonian cycles in a graph. For its statement, the reader may find it helpful to review the definition of minimum degree of a graph, in (5.4).

Theorem 5.29 (Dirac) *If G is a graph with $n \geq 3$ vertices and such that $\delta(G) \geq \frac{n}{2}$, then G is hamiltonian.*

Proof Proposition 5.20 gives the connectedness of G. Now, let $\mathcal{P} = (u_0, \ldots, u_k)$ be a maximal path in G (i.e., one for which there is no other of greater length). If v is a neighbor of u_0 not belonging to $\mathcal{P}$, we can change $\mathcal{P}$ by $\mathcal{P}_1 = (v, u_0, \ldots, u_k)$, thus obtaining a path in G bigger than $\mathcal{P}$, which is a contradiction. Hence, $\mathcal{P}$ includes all of the neighbors of u_0 and, analogously, all of the neighbors of u_k. In turn, this assures that the length of $\mathcal{P}$ is at least $\frac{n}{2}$.

We first claim that there exists an index $0 \leq i \leq k - 1$ such that u_0 is adjacent to u_{i+1} and u_i is adjacent to u_k. In order to show this, let $A = \{0 \leq i \leq k - 1;\ u_{i+1} \in N_G(u_0)\}$ and $B = \{0 \leq i \leq k - 1;\ u_i \in N_G(u_k)\}$. Since all of the vertices in $N_G(u_0)$ and all of the vertices in $N_G(u_k)$ are in $\mathcal{P}$, it follows that $|A|, |B| \geq \frac{n}{2}$. Since $|A \cup B| \leq k$, we then get

$$|A \cap B| = |A| + |B| - |A \cup B| \geq n - k > 0,$$

for $\mathcal{P}$ is a path in G with $k + 1$ vertices, so that $k + 1 \leq n$.

Now, choose $0 \leq i \leq k - 1$ as in the above claim, and consider the cycle

$$\mathcal{C} = (u_0, u_{i+1}, u_{i+2}, \ldots, u_{k-1}, u_k, u_i, u_{i-1}, \ldots u_1, u_0).$$

We claim that $\mathcal{C}$ is hamiltonian. By contradiction, suppose that some vertex of G, say v, is not in $\mathcal{C}$. Since $\mathcal{C}$ has more than $\frac{n}{2}$ vertices and v has at least $\frac{n}{2}$ neighbors, pigeonhole's principle (applied as in the proof of Proposition 5.20) guarantees the existence of a vertex w in $\mathcal{C}$ such that w is adjacent to v. Renaming the vertices of $\mathcal{C}$, we can assume that

$$\mathcal{C} = (v_0, v_1, \ldots, v_k, v_0),$$

with $v_0 = w$. But then, $\mathcal{P}_1 = (v, v_0, v_1, \ldots, v_k)$ would be a path in G bigger than $\mathcal{P}$, which is impossible. □

The lower bound on $\delta(G)$ asked by Dirac's theorem is the best possible one, in the sense of the following example.

Example 5.30 Let $n > 1$ be odd and let G be the graph of n vertices formed by the union of two copies of $K_{(n+1)/2}$ with a single common vertex, say u (i.e., take two disjoint copies of $K_{(n+1)/2}$, choose one vertex out of each of them and make these two vertices coincide). One clearly has $\delta(G) = \frac{n-1}{2} < \frac{n}{2}$. Nevertheless, G does not contain a hamiltonian cycle, for every cycle in G containing u must be entirely contained in one of the copies of $K_{(n+1)/2}$ which originated G. Indeed, if C is such a cycle, then $C - u$ is a connected subgraph of $G - u$, which, in turn is the disjoint union of two copies of $K_{(n-1)/2}$; therefore, $C - u$ is contained in a single one of these two copies and, hence, is also contained in a single one of the two copied of $K_{(n+1)/2}$ from which we formed G.

Problems: Sect. 5.2

1. For this problem, recall that if $A = (a_{ij})_{m \times n}$ and $B = (b_{jk})_{n \times p}$ are $m \times n$ and $n \times p$ matrices with entries a_{ij} and b_{jk}, then the product matrix of A and B (in this order) is $AB = (c_{ik})_{m \times p}$ such that, for $1 \leq i \leq m$ and $1 \leq k \leq p$, one has

$$c_{ik} = \sum_{j=1}^{n} a_{ij} b_{jk}. \tag{5.6}$$

This being said, do the following items:

   (a) If M is the adjacency matrix of a graph G of vertices $v_1, \ldots, v_n$, use the above formula to prove that, for $i \neq j$, entry (i, j) of M^2 equals the number of distinct paths of length 2 in G, joining vertices v_i and v_j.
   (b) Generalize (5.6), showing that if $A_1, \ldots, A_k$ are $n \times n$, with $A_l = (a_{ij}^l)_{n \times n}$, then the product matrix of $A_1, \ldots, A_k$ (in this order) is $A_1 \ldots A_k = (b_{ij})_{n \times n}$, such that

$$b_{ij} = \sum_{l_1, \ldots, l_k} a_{il_1}^1 a_{l_1 l_2}^2 \ldots a_{l_{k-2} l_{k-1}}^{k-1} a_{l_{k-1} j}^k,$$

   with the indices $l_1, \ldots, l_k$ in the above sum varying over all possibilities $1 \leq l_1, \ldots, l_k \leq n$.
   (c) Generalize item (a) in the following way: if M is the adjacency matrix of a graph G of vertices $v_1, \ldots, v_n$ and $k \geq 1$ is an integer, show that entry (i, j) of the matrix M^k equals the number of distinct walks of length k in G, joining vertices v_i and v_j.

2. Given a graph G and a vertex u of G, let $H = (V; E)$, where V is the set of the vertices of all paths in G departing from u and E is the set formed by the edges of these paths. Prove that H is the connected component of G containing u.
3. Prove that every graph with n vertices and at least $\binom{n-1}{2} + 1$ edges is connected. Give an example of a graph with n vertices and $\binom{n-1}{2}$ edges which is not connected.
4. Solve *Problem 2*, posed at the beginning of this chapter.

For the coming problem, you shall need the following fact, to be established in Proposition 6.26: given nonzero integers m, n and c, there exist integers x and y such that $c = mx + ny$ if and only if $\gcd(m, n)$ divides c.

5. Let m and n be given naturals and $d = \gcd(m, n)$. Prove that the graph $G(m, n)$, described in Problem 14, page 132, has exactly $\gcd(m, n) = d$ connected components.

6. Show that if a graph G is disconnected, then its complement $\overline{G}$ (cf. Problem 9, page 132) contains a complete bipartite graph with the same set of vertices of G. In particular, conclude that either G or $\overline{G}$ is connected.

7. * If there is a walk between two vertices of a graph, prove that there is also a path between them.

8. Find an Eulerian walk in the graph depicted below:

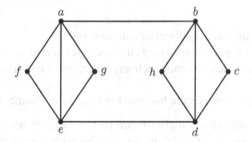

9. A **semi-Eulerian walk** in a connected graph G is a walk that starts and ends in distinct vertices of G and traverses each edge of G exactly once. Prove that G has a semi-Eulerian walk if and only if it has exactly two odd degree vertices.

10. Prove that, in every connected graph, two paths whose lengths are the largest possible ones have at least one common vertex.

11. A graph G has $n + 2$ vertices, $u_1, u_2, \ldots, u_n, v, w$, and satisfies the following conditions: v and w are both adjacent to $u_1, \ldots, u_n$, but not to one another; $u_1, \ldots, u_n$ are pairwise adjacent. For a fixed integer $2 \le k \le n + 1$, compute the number of paths of length k in G, joining v to w.

12. If all vertices of a graph have degrees at least 2, prove that the graph contains a cycle.

13. Prove that every 2-regular graph (cf. Problem 20, page 133) is a cycle.

14. (TT) During a meeting, each one of five mathematicians slept exactly twice. It is also known that, for each two of those five mathematicians, there was a moment in which both were sleeping. Prove that, at some moment, three of them were sleeping.

15. (IMO) Let G be a connected graph with n edges. Prove that it is possible to label the edges of G with the integers 1 to n, in such a way that the labels attributed to the edges incident at each vertex of degree greater than 1 are relatively prime.

16. * An edge ϵ of a graph G is said to be a **cutting edge** provided $G - \epsilon$ has more connected components than G. If G is a connected graph and ϵ is a cutting edge of G, prove that $G - \epsilon$ has exactly two connected components.

For the coming problem, given a digraph G (cf. Problem 26, page 135) and two of its vertices, say u and v, we define an (oriented) **walk** in G, from u to v, to be a sequence $(u_0, \ldots, u_k)$ of vertices of G such that $u_0 = u$, $u_k = v$ and the edge joining u_i and u_{i+1} is oriented from u_i to u_{i+1}, for $0 \le i < k$. In this case, we say that k is the length of the walk. A **path** is a walk with distinct vertices. Finally, recall that a *complete digraph* or *tournament* is a digraph that turns into a complete graph when we remove the orientation of its edges.

17. In a tournament G, show that there exists a vertex from which one can reach any other vertex with a path of length at most 2.

18. (TT) In Shvambrania there are n cities, each two of them connected by a road. These roads do not intersect; if necessary, viaducts are used for some of them to overpass others. An evil wizard wants to establish one way directions in each of the roads, so that, if someone leaves a certain city, he/she can no longer return. Prove that:

   (a) The wizard can actually establish such rules.
   (b) Whatever the way the wizard chooses to set his rules, there will always be a city from which one can reach any other, and a city from which one cannot leave.
   (c) The wizard can execute his intent in exactly $n!$ distinct ways.

19. (Brazil) Consider an $n \times n$ chessboard ($n > 1$ integer), and choose one of its n^2 unit squares, say c_0. A *path*[6] in the chessboard is a sequence $(c_0, c_1, \ldots, c_m)$ of distinct unit squares of the chessboard such that, for each $0 \le j < m$, squares c_j and c_{j+1} share a common edge. Such a path is said to be *optimal* if it contains all of the n^2 unit squares of the chessboard. Moreover, we say that a path as above contains an U if, for some $0 \le j \le m - 3$, the unit squares c_j and c_{j+3} share a common edge. Show that every optimal path contains an U.

20. (Tchecoslovaquia) Prove that, in every connected graph, there exists a vertex such that the average degree of its neighbors is greater than or equal to the average degree of all vertices of the graph.

21. The purpose of this problem is to prove that Petersen's graph (cf. Problem 15, page 133) is not hamiltonian. To this end assume, by contradiction, that such a graph possesses a hamiltonian cycle C. In the notations of Fig. 5.4, do the following items:

   (a) Prove that C contains either two or four of the edges $\{a, a'\}$, $\{b, b'\}$, $\{c, c'\}$, $\{d, d'\}$ and $\{e, e'\}$.
   (b) If four of the edges in (a) belong to C suppose, without loss of generality, that $\{b, b'\}$ is the one not belonging to C. Conclude that $\{a', b'\}$, $\{b', c'\}$, $\{b, e\}$ and $\{b, d\}$ are in C and use this to reach a contradiction.
   (c) If two of the edges of (a) are in C assume, without loss of generality, that $\{a, a'\}$ is one of them. Conclude that exactly one of the edges $\{a, c\}$ or

---

[6]Note that, in principle, this definition of *path* does not coincide with the one given in the text. Nevertheless, we have chosen to translate it according to the original.

Fig. 5.4 Petersen's graph

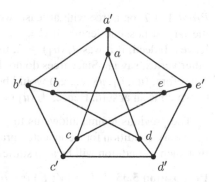

$\{a, d\}$ is in the cycle, say $\{a, c\}$. If $\{a, c\}$ is in $\mathcal{C}$, conclude that the other edge of (a) which is in $\mathcal{C}$ is $\{d, d'\}$, and reach a contradiction.

22. Let k and n be integers such that $2 < k < n$. If $f(n, k)$ denotes the largest possible number of K_k's in K_n, such that any two of these K_k's have at most $k - 2$ vertices in common, prove that

$$\frac{1}{k^2}\binom{n}{k-1} < f(n, k) \leq \frac{1}{k}\binom{n}{k}.$$

23. (Vietnam) Find the least positive integer k for which there exists a graph of 25 vertices such that each one of them is adjacent to exactly k others, and each two nonadjacent vertices are both adjacent to a third vertex.

24. (Austrian-Polish) A rectangular building is formed by two rows of fifteen square rooms (situated as the unit squares of two adjacent rows of an ordinary chessboard). Each room has one, two or three doors, that lead respectively to one, two or three adjacent rooms. It is known that the doors are distributed in such a way that it is always possible to go from one room to another without leaving the building. Compute how many are the distinct ways of distributing the doors along the building.

5.3 Trees

In this section we study connected graphs which do not have cycles, also said to be **acyclic**. The relevant definition is the one below.

Definition 5.31 A **tree** is a connected and acyclic graph.

If T is a tree, a **leaf** of T is a vertex of degree 1, whereas a **node** of T is a vertex of degree greater than 1. It is an important fact that every tree has leaves, as we now establish.

Lemma 5.32 *Every tree with more than one vertex has at least two leaves.*

Proof Let T be a tree with at least two vertices and consider, in T, a path $\mathcal{C}$ with the largest possible length. If $\mathcal{C} = (u_1, u_2, \ldots, u_n)$, we claim that u_1 and u_n are leaves. Indeed, if it was $d_T(u_1) \geq 2$, then u_1 would be adjacent to u_2 and to some other vertex, say u_0. Since trees do not have cycles, we must have $u_0 \neq u_3, \ldots, u_n$. Therefore, $(u_0, u_1, u_2 \ldots, u_n)$ would be a path in T with length greater than that of $\mathcal{C}$, which is an absurd. Hence, $d_T(u_1) = 1$ and, analogously, $d_T(u_n) = 1$. $\square$

The previous lemma allows us to prove the coming proposition, which gives a necessary condition for a connected graph to be a tree. Soon hereafter we will show that such a condition will be also sufficient.

Proposition 5.33 *If $T = (V; E)$ is a tree, then $|E| = |V| - 1$.*

Proof We make induction on $|V|$. Firstly, it is clear that every tree with two vertices has exactly one edge. Assume, by induction hypothesis, that every tree with $n > 1$ vertices has exactly $n - 1$ edges, and consider a tree T with $n + 1$ vertices. By the lemma, T has a leaf u; hence, $T - u$ is a tree with n vertices, and the induction hypothesis assures that $T - u$ has $n - 1$ edges. However, since T has exactly one edge more than $T - u$, it follows that T has exactly n edges. $\square$

Corollary 5.34 *If $G = (V; E)$ is a connected graph, then G contains at least one cycle if and only if $|E| \geq |V|$.*

Proof Suppose first that G contains at least one cycle. We shall show that $|E| \geq |V|$ by induction on $|E|$. If ϵ is the edge of a cycle in G, then $G - \epsilon$ is also connected and has $|V|$ vertices and $|E| - 1$ edges. Then, there are two possibilities:

- $G - \epsilon$ contains a cycle: it follows from the induction hypothesis that $|E| - 1 \geq |V|$;
- $G - \epsilon$ is acyclic: since it is still connected, $G - \epsilon$ is a tree. Therefore, the previous proposition guarantees that $|V| = (|E| - 1) + 1 = |E|$.

In any case, we concluded that $|E| \geq |V|$.

Conversely, if G is acyclic, then G is a tree, so that (again from the previous result) $|E| = |V| - 1 < |V|$. $\square$

Corollary 5.35 *Let $T = (V; E)$ be a connected graph. If $|E| = |V| - 1$, then T is a tree.*

Proof If T has at least one cycle, it would follow from the previous corollary that $|E| \geq |V|$, which is a contradiction. Thus, T is connected and acyclic, hence, a tree. $\square$

For what comes next, given a connected graph G, associate to each edge ϵ of G a **weight**, i.e., a positive real number $w(\epsilon)$ (to motivate this concept, think of the vertices of the graph as several cities of some region and of the weights as measures of travel expenses between adjacent cities). This way, G becomes a **weighted graph** and the **weight** $w(G)$ of G is defined by setting

$$w(G) = \sum_{\epsilon \in E} w(\epsilon),$$

with $E = E(G)$.

Given a connected graph G, a **spanning tree** of G is a spanning subgraph of G (cf. Definition 5.14) which is a tree. An interesting problem in the elementary theory of trees is the following one: given a weighted graph, find a spanning tree of least possible weight, said to be a **minimal spanning tree**.

We deal with this problem in the coming proposition, due to the American mathematician of the twentieth century Joseph Kruskal. Well understood, since the set of spanning trees of a graph is finite (and nonempty, as we shall see), there always exist minimal spanning trees in weighted graphs. The point of Kruskal's theorem is that the proof of it actually furnishes an *algorithm* for finding a minimal spanning tree. Such an algorithm is known in Computer Science as **Kruskal's algorithm** or **greedy** algorithm.

Proposition 5.36 (Kruskal) *If G is a weighted graph, then G has minimal spanning trees.*

Proof Form a subgraph T of G according to the following algorithm:

(i) Choose an edge of G with least possible weight and put it in T.
(ii) After having chosen a certain number of edges of G for T, take, among the edges not chosen yet, one of minimum weight and whose inclusion in T does not generate cycles.
(iii) Repeat step (ii) as long as possible.

Firstly, note that the steps above give, as result, a spanning tree T of G. Indeed, T is certainly acyclic; it is also connected (and thus a tree), for otherwise we could, due to the connectedness of G, execute the algorithm one further time. Moreover, if T was not spanning, there would exist in G a vertex u not belonging to T but adjacent to some vertex v of T. Without loss of generality, we may assume that u and v are such that $w(\{u, v\})$ is minimal, among all possible choices of u and v as above. Hence, the algorithm would continue with the inclusion of the edge $\{u, v\}$ in T, which is an absurd.

Now, we let S be another spanning tree of G, and will show that $w(T) \leq w(S)$.

Since S and T have the same number of edges (cf. Problem 3), there exists an edge ϵ of T such that ϵ does not belong to S. Again by Problem 3, the inclusion of ϵ in S generates a cycle C; but since T is a tree, there exists some edge ϵ' in C, distinct from ϵ and not belonging to T. In particular, ϵ' is in S.

We claim that $w(\epsilon) \leq w(\epsilon')$. Indeed, if $w(\epsilon) > w(\epsilon')$, then, upon executing Kruskal's algorithm, at the step we chose ϵ we would have chosen ϵ', for it would have a smaller weight. However, since ϵ' is not in T, this gives a contradiction.

Since $w(\epsilon) \leq w(\epsilon')$, changing ϵ' by ϵ in S, we obtain a new generating tree S_1, such that $w(S_1) \leq w(S)$ and S_1 and T have one more edge in common than S and T do. If T has k edges not belonging to S, then, repeating the above reasoning more $k - 1$ times, we obtain a sequence $S_1, S_2, \ldots, S_k$ of spanning trees such that $S_k = T$

and

$$w(S_k) \leq \cdots \leq w(S_1) \leq w(S).$$

Thus, $w(T) \leq w(S)$. □

We finish this section by counting the number of labelled trees (cf. Remark 5.13) with a given number of vertices and a given set of labels. This result is due to the nineteenth century English mathematician Arthur Cayley and, for the sake of simplicity, we omit the set of labels from its statement.

Theorem 5.37 (Cayley) *There are exactly n^{n-2} labelled trees of n vertices.*

Proof The given formula is trivially true for $n = 1$ and $n = 2$. Assume $n > 2$ and, without loss of generality, that the vertices of the trees will be labelled $1, 2, \ldots, n$. Let a_n be the total number of such trees and, for $1 \leq i \leq n$, let A_i stands for the set of them such that the vertex with label i is a leaf. Since it is impossible that all of the n vertices are leaves, the inclusion-exclusion principle (cf. Theorem 2.1) gives

$$a_n = |A_1 \cup \ldots \cup A_n|$$

$$= \sum_{k=1}^{n-1} \sum_{1 \leq i_1 < \cdots < i_k \leq n} (-1)^{k-1} |A_{i_1} \cap \ldots \cap A_{i_k}|. \tag{5.7}$$

Now, given integers $1 \leq k \leq n$ and $1 \leq i_1 < \cdots < i_k \leq n$, note that $A_{i_1} \cap \ldots \cap A_{i_k}$ is the set of trees in which vertices $i_1, i_2, \ldots, i_k$ are leaves. If T stands for one such tree, then

$$T_{i_1 i_2 \ldots i_k} := T - \{i_1, i_2, \ldots, i_k\}$$

is a tree with $n-k$ vertices, with labels in $I_n \setminus \{i_1, i_2, \ldots, i_k\}$. The key to count $|A_{i_1} \cap \ldots \cap A_{i_k}|$ is to look at how T can be reconstructed from $T_{i_1 i_2 \ldots i_k}$. To understand this, note that for a fixed $1 \leq j \leq k$ vertex i_j must be joined to one of the $n - k$ vertices of $T_{i_1 i_2 \ldots i_k}$; since the involved choices are independent, we conclude that there are exactly $(n - k)^k$ possible ways of choosing the vertices of $T_{i_1 i_2 \ldots i_k}$ to which vertices $i_1, i_2, \ldots, i_k$ will link themselves. Since there are a_{n-k} possibilities for $T_{i_1 i_2 \ldots i_k}$, we therefore conclude that

$$|A_{i_1} \cap \ldots \cap A_{i_k}| = (n - k)^k a_{n-k}. \tag{5.8}$$

In the above formula, note that the right hand side depends solely on k, but not on $i_1, i_2, \ldots, i_k$. Since there are exactly $\binom{n}{k}$ ways of choosing integers $1 \leq i_1 < \cdots < i_k \leq n$, Eqs. (5.7) and (5.8) give us

$$a_n = \sum_{k=1}^{n-1} (-1)^{k-1} \binom{n}{k} (n-k)^k a_{n-k}.$$

Now assume, by induction hypothesis, that $a_l = l^{l-2}$ for $1 \leq l < n$. Then, the recurrence relation above gives

$$a_n = \sum_{k=1}^{n-1} (-1)^{k-1} \binom{n}{k} (n-k)^{n-2}.$$

We finish the proof by applying the formula of Problem 3, page 40, with $m = n - 2$ to get $a_n = n^{n-2}$.   □

Problems: Sect. 5.3

1. * Prove that the following claims on a graph G are equivalent:

   (a) G is a tree.
   (b) Given arbitrarily two vertices of G, there is a single path in G connecting them.
   (c) G is connected and every edge of G is a cutting edge (cf. Problem 16, page 143).

   Then, conclude that, in a tree with n vertices, there are exactly $\binom{n}{2}$ distinct paths.

2. Let T be a tree with a node of degree k. Prove that T has at least k leaves.

3. * Let G be a connected graph.

   (a) Prove that any two spanning trees of G have the same number of edges.
   (b) If T is a spanning tree of G and ϵ is an edge of G that is not in T, prove that $T \cup \{\epsilon\}$ contains cycles.
   (c) If G has a single spanning tree, prove that G is a tree.
   (d) If T is a spanning tree of G and G has k edges not belonging to T, prove that G contains at least k distinct cycles.

4. Prove that the following algorithm also constructs minimal spanning trees of weighted graphs:

   i. Choose a vertex and, among all edges incident to it, the one with least possible weight.
   ii. After having chosen a certain number of vertices and edges, choose the edge of least weight, among all edges not yet chosen and that join an already chosen vertex to a not chosen yet one.
   iii. Repeat the previous item as many times as possible.

The above algorithm is due to the Dutch mathematician of the twentieth century Edsger Dijkstra, being known as **Dijkstra's algorithm**.

ab	ac	ad	ae	af	ag	bc	bg	cd	de	ef	fg
6	3	2	4	3	7	6	6	2	3	1	8

5. Let G be a graph with set of vertices $\{a, b, c, d, e, f, g\}$ and with edge weights given as in the following table: Find all possible spanning trees of minimal weight.

   For the coming problem, we shall use the fact, established in Problem 1, that a tree with n vertices contains exactly $\binom{n}{2}$ distinct paths.

6. Let T be a tree with n vertices and satisfying the following condition: it is possible to associate to each one of the $n - 1$ edges of T a natural weight, so that the $\binom{n}{2}$ distinct paths in T have, in some order, weights equal to $1, 2, \ldots, \binom{n}{2}$. The purpose of this problem is to show that, in this case, either n or $n - 2$ must be a perfect square. To this end, do the following items:

   (a) Show that the number of paths in T having odd weight is equal to:

$$
\begin{cases}
\frac{1}{2}\binom{n}{2}, & \text{if } \binom{n}{2} \text{ is even} \\[2mm]
\frac{1}{2}\left(\binom{n}{2} + 1\right), & \text{if } \binom{n}{2} \text{ is odd}
\end{cases}.
$$

   (b) Show that it is possible to paint the vertices of T either blue or red, in such a way that two adjacent vertices have distinct colors if and only if the edge joining them has odd weight.

   (c) If the coloring of item (b) generates x blue vertices and y red ones, show that the number of paths in T with odd weight is equal to xy.

   (d) If $\binom{n}{2}$ is even, show that $n = (x - y)^2$; if $\binom{n}{2}$ is odd, conclude that $n - 2 = (x - y)^2$.

5.4 Independent Sets and Cliques

This last section on graphs is devoted to a somewhat more detailed study of complete subgraphs of a given graph, and is partially based on the classnotes [16]. I kindly acknowledge professor Samuel B. Feitosa for having allowed me to use them.

   Given a graph G, recall that a nonempty subset A of $V(G)$ is said to be **independent** if any two vertices in A are nonadjacent. If G has n vertices, we can partition $V(G)$ in n sets (each set consisting of a single vertex) such that each one of them is independent. Then, there exists a least natural number k such that $V(G)$ can be partitioned into k independent sets. We name such a natural number according to the following

Definition 5.38 Given a graph G, its **chromatic number**, denoted $\mathcal{X}(G)$, is the least natural number k such that the set of vertices of G can be partitioned into k independent sets.

The reason for the word *chromatic* in the above definition is due to the fact that, if $\mathcal{X}(G) = k$ and $V(G) = A_1 \cup \ldots \cup A_k$, with $A_1, \ldots, A_k$ independent sets, then, painting the vertices in A_i with a color c_i (where $c_1, \ldots, c_k$ are distinct colors), we obtain a coloring of the vertices of G in which any two adjacent vertices have distinct colors; conversely, if k is the least possible number of distinct colors for which there exists a coloring of the vertices of G with k distinct colors and with no adjacent vertices of the same color, then the sets formed by the vertices of each color are k independent sets that partition $V(G)$.

Examples 5.39

(a) If G is a bipartite graph (cf. Problem 16, page 133) with at least one edge, it is immediate to see that $\mathcal{X}(G) = 2$.
(b) If G is a 2-cycle with n vertices, then it is clear that $\mathcal{X}(G) = 2$ if n is even (for we can color the vertices by alternating two colors along the cycle), but $\mathcal{X}(G) = 3$ if n is odd (for in this case we need a third color for the last vertex of the cycle).

Generalizing item (b) of the above example, the proof of the coming proposition describes a rather simple algorithm which guarantees that if no vertex of a graph has degree greater than k, then its chromatic number is at most $k + 1$. In all that follows, given a graph G we let

$$\Delta(G) = \max\{d_G(u); \; u \in V(G)\}$$

and say that $\Delta(G)$ is the **maximum degree** in G.

Proposition 5.40 *In every graph G, we have $\mathcal{X}(G) \leq \Delta(G) + 1$.*

Proof Let $\Delta(G) = n$ and fix $n + 1$ distinct colors. Start by painting an arbitrary vertex of G with one of the colors. Then, assume that, at some moment, $k < n$ vertices of G have been painted with the $n + 1$ given colors and in such a way that adjacent vertices have distinct colors. Choose a vertex u that has not been painted yet. Since $d_G(u) \leq \Delta(G) = n$, at most n of the given colors were used to paint vertices adjacent to u; hence, we are left with at least one of the $n + 1$ colors to paint u, so that it has a color distinct from those of all of its neighbors. $\square$

The following definition furnishes a lower bound for the chromatic number of a graph.

Definition 5.41 An *n*-**clique** in a graph G is a subgraph of G which is isomorphic to K_n. The **clique number** of G, denoted $\omega(G)$, is the greatest integer $n \geq 0$ such that G contains an *n*-clique.[7]

Since one needs n distinct colors to color the vertices of K_n in such a way that no two adjacent vertices have the same color, we have

$$\mathcal{X}(G) \geq \omega(G),$$

for every graph G. On the other hand, we can estimate $\omega(G)$ upon the intuition that, if a graph has lots of edges (when compared to its number of vertices), then $\omega(G)$ must be large. Let us take a look at a relevant

Example 5.42 Let G be a connected graph with $n > 1$ vertices. If G has more than $\lfloor \frac{n^2}{4} \rfloor$ edges, then $\omega(G) \geq 3$.

Proof By contraposition, assume that G does not contain a K_3.

Let u be a vertex of G with maximum degree, say k. We can assume that $k > 1$, for, otherwise, Euler's Theorem 5.5 would give us at most $\frac{n}{2}$ edges in G, and this would contradict the fact that $\frac{n}{2} < \lfloor \frac{n^2}{4} \rfloor + 1$ for $n > 1$.

Since G has no K_3, the k vertices of G adjacent to u form an independent set. Therefore, every edge of G which is not incident to u is incident to some of the $n - k - 1$ vertices of G not belonging to $N_G(u) \cup \{u\}$. However, since each one of these $n - k - 1$ vertices has degree less than or equal to k, we conclude that G has at most

$$k + (n - 1 - k)k = -k^2 + nk$$

edges.

In order to finish the proof, it suffices to note that

$$\mathbb{Z} \ni -k^2 + nk \leq \frac{n^2}{4},$$

so that we must have $-k^2 + nk \leq \lfloor \frac{n^2}{4} \rfloor$. $\square$

The coming problem brings an interesting application of the previous one.

Example 5.43 (China) Given distinct real numbers $x_1, x_2, \ldots, x_n$, prove that there exists at most $\lfloor \frac{n^2}{4} \rfloor$ ordered pairs (x_i, x_j) such that $1 < |x_i - x_j| < 2$.

---

[7]We call the reader's attention for not confusing notations $w(G)$ for the weight of a weighted graph and $\omega(G)$ for the clique number of a graph; the first uses the Latin letter w, whereas the second uses the Greek letter ω (*omega*).

Proof Construct a graph G having the x_i's as vertices, with x_i adjacent to x_j if and only if $1 < |x_i - x_j| < 2$. We claim that G does not contain K_3's. Indeed, if x_i, x_j and x_k were the vertices of a K_3 in G, with $x_i < x_j < x_k$, say, then

$$x_k - x_i = (x_k - x_j) + (x_j - x_i) > 1 + 1 = 2,$$

which is a contradiction. Therefore, by the previous example we conclude that G has at most $\lfloor \frac{n^2}{4} \rfloor$ edges. □

In order to generalize the reasoning of Example 5.42 from K_3's to K_l's, let us firstly construct a generic example of a graph with n vertices and without K_l's. To this end, we start by assuming that $l \le n$, for otherwise there is nothing to do; then, we use the division algorithm (cf. Proposition 6.7) to write

$$n = (l - 1)q + r,$$

with $0 \le r < l - 1$. Then, partition the set of n vertices into $l - 1$ subsets $A_1, A_2, \ldots,$ A_{l-1}, such that

$$|A_1| = \ldots = |A_r| = q + 1 \text{ and } |A_{r+1}| = \ldots = |A_{l-1}| = q;$$

that it is possible to get such a partition follows from the equality $r(q+1) + (l - r - 1)q = (l-1)q + r = n$. Continuing, impose that each A_i be an independent set, and also that each vertex of A_i is adjacent to each vertex of A_j, for all $1 \le i, j \le l - 1$, with $i \ne j$. The graph G thus obtained is called a **Turán graph**, after the Hungarian mathematician of the twentieth century Paul Turán, to whom we owe part of the material of this section.

The Turán graph just constructed does not contain any K_l, since a K_l contains l-cycles, whereas G does not contain any such cycle. On the other hand, by counting how many edges of K_n are not present in G, we conclude that it has exactly

$$T(n; l) := \binom{n}{2} - r\binom{q+1}{2} - (l - r - 1)\binom{q}{2}$$

edges. Substituting $q = \frac{n-r}{l-1}$ and doing some elementary algebra, we get

$$T(n; l) = \frac{n^2(l - 2) - r(l - 1 - r)}{2(l - 1)}, \tag{5.9}$$

even though $\omega(G) < l$.

The coming result (cf. [39]) assures that, if G is a graph with n vertices and such that $\omega(G) < l$, then G must have at most $T(n; l)$ edges.

Theorem 5.44 (Turán) *If a graph G of n vertices has more than $T(n; l)$ edges, then $\omega(G) \ge l$.*

Proof In the notations of the preceding discussion, let us make a proof by induction on q, assuming that r and l are fixed. If $q = 0$, then $n = r$ and it follows from (5.9) that

$$
\begin{aligned}
T(n; l) &= \frac{n^2(l-2) - n(l-1-n)}{2(l-1)} \\
&= \frac{n^2(l-1) - n(l-1)}{2(l-1)} \\
&= \frac{n^2 - n}{2} = \binom{n}{2}.
\end{aligned}
$$

Since a graph with n vertices cannot have more that $\binom{n}{2}$ edges, the statement of the theorem is *vacuously true*.[8]

Now, consider a graph G with n vertices, without K_l's and with the largest possible number of edges. Clearly, G contains a K_{l-1}, for otherwise we could add at least one more edge to G and it would still contain no K_l. Let H be a K_{l-1} in G, and A be the set of the remaining $n - l + 1$ vertices of G (i.e., those not belonging to H). There are three kinds of edges in G:

* The edges of H: overall, $\binom{l-1}{2}$ edges.
* The edges joining vertices in A to vertices in H: since G contains no K_l's, each vertex in A is adjacent to at most $l - 2$ vertices in H. This gives us at most $(n - l + 1)(l - 2)$ edges of this kind.
* The edges of the induced subgraph $G_{|A}$ (cf. Problem 12, page 132): since $G_{|A}$ does not contain K_l's too, and $n - l + 1 = (q - 1)(l - 1) + r$, we can apply the induction hypothesis to $G_{|A}$, so that $G_{|A}$ has at most $T(n - l + 1; l)$ edges.

Therefore, the graph G has at most

$$
T(n - l + 1; l) + (n - l + 1)(l - 2) + \binom{l-1}{2}
$$

edges.

Finally, substituting the formula for $T(n - l + 1; l)$ (obtained from (5.9)) and performing some elementary algebra once more, we get

$$
T(n - l + 1; l) + (n - l + 1)(l - 2) + \binom{l-1}{2} = T(n; l),
$$

so that G has at most $T(n; l)$ edges.   □

---

[8]In terms of Logic, an implication is true whenever it premise is false. See Chapter 1 of [34], for instance.

A nice way of grasping the details of the proof of Turán's theorem is to rework it in the particular case $(l - 1) \mid n$. In this respect, we refer the reader to Problems 2 and 5.

We finish this section by examining yet another invariant of a graph G, one which is closely related to $\omega(G)$.

Definition 5.45 Given a graph G, its **independence number**, denoted $\alpha(G)$, is the number of elements of a maximal independent set of vertices of G.

For the coming proposition, the reader may find it useful to recall the concept of complementary graph (cf. Problem 9, page 132).

Proposition 5.46 *Let a graph G and a nonempty subset A of the set of vertices of G be given. The subgraph of G induced by A is a clique in G if and only if A is an independent set in $\overline{G}$. In particular,*

$$\omega(G) = \alpha(\overline{G}) \quad \text{and} \quad \omega(\overline{G}) = \alpha(G).$$

Proof The first part follows immediately from the definitions. For the second part, it suffices to use the fact that $\overline{\overline{G}} = G$. □

Example 5.47 Let G be a graph with nine vertices. Suppose that, given any five of the vertices of G, there exist at least two edges with both endpoints among them. Compute the least possible number of edges G can have.

Proof We consider two cases separately:

(a) $\omega(\overline{G}) \geq 4$: fix a K_4 in $\overline{G}$ and let A be the set formed by the vertices of this K_4. Then, A is independent in G and, if u is one of the five remaining vertices, the stated condition assures that u is adjacent to at least two of the vertices in A. Thus, we conclude that G has at least $5 \cdot 2 = 10$ edges.

(b) $\omega(\overline{G}) \leq 3$: by Turán's theorem, $\overline{G}$ has at most $T(9; 4) = 27$ edges and, hence G has at least $\binom{9}{2} - 27 = 9$ edges.

For an example of such a G with 9 edges, take the union of three pairwise disjoint K_3's. Given five vertices in G, there are two possibilities: either three of these five vertices form one of the K_3's of G, or we choose the five vertices by taking two vertices of two of the K_3's of G, together with one vertex of the third K_3. In any case, there are at least two edges of G having both of its endpoints among the five vertices. □

Problems: Sect. 5.4

1. If G is a graph with at least six vertices, prove that $\omega(G) \geq 3$ or $\omega(\overline{G}) \geq 3$.
2. The purpose of this problem is to provide a slightly different proof of the result of Example 5.42 when G is a graph of $2q$ vertices, $q > 1$, and at least $q^2 + 1$

edges. To this end, fix two adjacent vertices u and v of G and do the following items:

(a) If at least $2q - 1$ edges depart from u or v to the remaining $2q - 2$ vertices, use the pigeonhole principle to guarantee that G has a K_3 of the form uvw.
(b) If at most $2q - 2$ edges depart from u or v to the remaining $2q - 2$ vertices, apply an appropriate induction hypothesis to $G - \{u, v\}$ (cf. Problem 11, page 132) to conclude that such a graph (and, hence, also G) contains a K_3.

3. Solve *Problem 3* at the beginning of this chapter.
4. Let G be a graph with 10 vertices and 26 edges. Show that G must contain at least five K_3's.
5. Let $l \geq 3$ and q be natural numbers, and G be a graph of $(l - 1)q$ vertices, with at least $\frac{1}{2}(l - 1)(l - 2)q^2$ edges.[9] Argue along the same lines of Problem 2 to show that G contains a K_l.
6. (TT) Twenty teams entered a soccer tournament. Find the minimum number of matches such a tournament must have so that, among any three teams, at least two have played against each other.
7. In a given a set of n people, it is known that out of any group of three of them at least two are acquainted with each other, whereas out of any group of four at least two are not acquainted with each other. Compute the largest possible value of n.
8. (TT) We are given 21 points on a circle. Prove that there exist at least 100 arcs, each of which having two of the given points as endpoints and whose measures are less than or equal to $120°$.
9. (Poland) Let n points on a circle with radius 1 be given. Show that there is at most $\frac{n^2}{3}$ line segments joining pairs of these points and having lengths greater than $\sqrt{2}$.
10. (USA) In a certain society, each pair of people is classified either as *amicable* or *hostile*. Members of amicable pairs are said to be *friends* and members of hostile pairs are said to be *enemies*. Suppose that the society has n people, q amicable pairs and that, among any three people, at least two form an hostile pair. Prove that there exists at least one member of the society whose set of enemies contains, at most, no more than $q\left(1 - \frac{4q}{n^2}\right)$ amicable pairs.
11. Let G be a graph with n vertices, such that $\omega(G) < k$, where $k \geq 2$. Prove that G contains at least $\lfloor\frac{n}{k-1}\rfloor$ vertices with degrees less than or equal to $\lfloor\frac{(k-2)n}{k-1}\rfloor$.
12. A set of 1001 people is such that each subset of 11 people contains at least two who know each other. Show that there exist at least 101 people in the set such that each one of them knows at least 100 other people within the set.
13. (China) Let A be a subset of $S = \{1, 2, 3, \ldots, 10^6\}$ having 101 elements. Prove that there exist $x_1, x_2, \ldots, x_{100}$ in S such that, for $1 \leq j \leq 100$, the sets $A_j := \{x + x_j; x \in A\}$ are pairwise disjoint.

---

[9]Note that this is precisely Turán's number $T(n; l)$ when $n = (l - 1)q$.

Chapter 6
Divisibility

This chapter is devoted to the elementary properties concerning the *relation of divisibility* in the set of integers, with a particular emphasis on the division algorithm, on the notion of greatest common divisor and the fundamental role played by prime numbers. In spite of the elementary character of the arguments we shall use, we will meet several interesting problems and results along the way, like Bézout's theorem on the characterization of the greatest common divisor of two integers and Euclid's theorem on the infinitude of primes.

6.1 The Division Algorithm

We start by isolating our central object of study, namely, the relation of divisibility between two integers.

Definition 6.1 Given $a, b \in \mathbb{Z}$, with $b \neq 0$, we say that b **divides** a, and write $b \mid a$, if there exists $c \in \mathbb{Z}$ such that $a = bc$. In case b doesn't divide a, we write $b \nmid a$.

Let b be a nonzero integer. If b divides a, we also say that b is a **divisor** of a, that a is **divisible** by b or that a is a **multiple** of b. If $b \mid a$ and $b > 0$, then b is a **positive divisor** of a. Note that every nonzero integer is a divisor of itself and of 0. Also, ± 1 divide every integer.

The coming example starts by formalizing two concepts we are certainly familiar with.

Example 6.2 An integer n is **even** provided it is a multiple of 2; otherwise, n is said to be **odd**. According to Definition 6.1, the even integers are precisely those which can be written in the form $2k$, for some $k \in \mathbb{Z}$, i.e., the integers

$$\ldots, -6, -4, -2, 0, 2, 4, 6, \ldots.$$

© Springer International Publishing AG, part of Springer Nature 2018
A. Caminha Muniz Neto, *An Excursion through Elementary Mathematics, Volume III*,
Problem Books in Mathematics, https://doi.org/10.1007/978-3-319-77977-5_6

The remaining integers, namely,

$$\ldots, -5, -3, -1, 1, 3, 5, \ldots,$$

are the odd ones. Thus, every odd integer is one plus an even integer, so that can be generically denoted by writing $2k + 1$, where $k \in \mathbb{Z}$.

The next example isolates an application of a simple algebraic identity which will be quite useful for us.

Example 6.3 Let a and b be two given integers and k be a natural number.

(a) If $a \neq b$, then $(a - b) \mid (a^k - b^k)$.
(b) If $a \neq -b$ and k is odd, then $(a + b) \mid (a^k + b^k)$.

Proof Standard basic algebra (cf. problem 2.1.18 of [8], for instance) gives

$$a^k - b^k = (a - b)(a^{k-1} + a^{k-2}b + \cdots + ab^{k-2} + b^{k-1})$$

and, for odd k,

$$a^k + b^k = (a + b)(a^{k-1} - a^{k-2}b + \cdots - ab^{k-2} + b^{k-1}).$$

Since both $a^{k-1} + a^{k-2}b + \cdots + ab^{k-2} + b^{k-1}$ and $a^{k-1} - a^{k-2}b + \cdots - ab^{k-2} + b^{k-1}$ are integers, it suffices to recall Definition 6.1.   □

Example 6.4 Prove that, for every k natural, $10^k - 1$ is divisible by 9.

Proof It suffices to apply item (a) of the previous example: $10^k - 1 = 10^k - 1^k$ is divisible by $10 - 1 = 9$. Alternatively, note that

$$10^k - 1 = \underbrace{99 \ldots 9}_{k \text{ digits}} = 9 \cdot \underbrace{11 \ldots 1}_{k \text{ digits}},$$

which is clearly a multiple of 9, or write, with the aid of the binomial formula,

$$10^k - 1 = (9 + 1)^k - 1 = \sum_{j=0}^{k} \binom{k}{j} 9^{k-j} - 1 = 9 \sum_{j=0}^{k-1} \binom{k}{j} 9^{k-1-j},$$

again a multiple of 9.   □

The following proposition establishes some basic properties of the relation of divisibility. The reader should make an effort to keep them in mind for future use.

Proposition 6.5 *Let a, b, c be nonzero integers and x, y be any integers.*

(a) *If $b \mid a$ and $a \mid b$, then $a = \pm b$.*
(b) *If $c \mid b$ and $b \mid a$, then $c \mid a$.*

(c) If $c \mid a$ and $c \mid b$, then $c \mid (ax + by)$.
(d) If $b \mid a$, then $|b| \le |a|$.
(e) If $c \mid b$, then $c \mid ab$.
(f) If $b \mid a$, then $bc \mid ac$.

Proof

(a) If a' and b' are integers such that $a = ba'$ and $b = ab'$, then $a = (ab')a' = a(a'b')$, so that $a'b' = 1$. Therefore, $a' = \pm 1$, and it follows that $a = \pm b$.
(b) If $a = ba'$ and $b = cb'$, with $a', b' \in \mathbb{Z}$, then $a = (cb')a' = c(a'b')$, with $a'b' \in \mathbb{Z}$. Thus, $c \mid a$.
(c) Let $a = ca'$ and $b = cb'$, with $a', b' \in \mathbb{Z}$. Then $ax + by = ca'x + cb'y = c(a'x + b'y)$, with $a'x + b'y \in \mathbb{Z}$. We conclude that $c \mid (ax + by)$.
(d) Let $a = ba'$, with $a' \in \mathbb{Z}$. Then, $a \ne 0 \Rightarrow a' \ne 0$, and hence $|a'| \ge 1$. This gives $|a| = |ba'| = |b||a'| \ge |b|$.
(e) If $b = cb'$, with $b' \in \mathbb{Z}$, then $ab = c(ab')$, with $ab' \in \mathbb{Z}$. Therefore, $c \mid ab$.
(f) If $a = ba'$, with $a' \in \mathbb{Z}$, then $ac = (bc)a'$, and hence $bc \mid ac$. $\qquad\square$

Remarks 6.6

  i. As a particular case of item (c) of the previous proposition, if $c \mid a$ and $c \mid b$, then $c \mid (a \pm b)$. We shall use this remark several times in what follows.
  ii. Item (c) of the previous proposition can be easily generalized to prove that, if $c \mid a_1, \ldots, a_n$, then $c \mid (a_1x_1 + \cdots + a_nx_n)$ for all $x_1, \ldots, x_n \in \mathbb{Z}$.
  iii. It follows from item (d) above that every nonzero integer has only a finite number of divisors.

We continue by recalling that the **integer part** $\lfloor x \rfloor$ of $x \in \mathbb{R}$ is defined by

$$\lfloor x \rfloor = \max\{n \in \mathbb{Z};\ n \le x\}. \tag{6.1}$$

In other words, given $n \in \mathbb{Z}$,

$$\lfloor x \rfloor = n \Leftrightarrow n \le x < n + 1. \tag{6.2}$$

For instance, since $1 < \sqrt{2} < 2$ we have $\lfloor \sqrt{2} \rfloor = 1$; on the other hand, since $-3 < -2.3 < -2$, then $\lfloor -2.3 \rfloor = -3$.

Our next result is usually referred to as the **division algorithm**[1] for integers, and is one of the cornerstones of the basic theory of divisibility.

Proposition 6.7 *Given $a, b \in \mathbb{Z}$, with $b \ne 0$, there exist unique $q, r \in \mathbb{Z}$ such that $a = bq + r$ and $0 \le r < |b|$. Moreover, if $b > 0$, then*

---

[1] At this point, the reader may wish to take another look at the definition of algorithm, at the footnote of page 116.

$$q = \left\lfloor \frac{a}{b} \right\rfloor \ \text{and} \ r = a - \left\lfloor \frac{a}{b} \right\rfloor b. \tag{6.3}$$

*Such integers q and r are respectively called the **quotient** and the **remainder** upon division of a by b.*

Proof Suppose first that $b > 0$, and let q be the greatest integer such that $bq \leq a$. Then $bq \leq a < b(q+1)$, so that $0 \leq a - bq < b$ and it suffices to define $r = a - bq$. Note also that, since

$$q \leq \frac{a}{b} < q + 1,$$

such an integer q is precisely $\left\lfloor \frac{a}{b} \right\rfloor$.

If $b < 0$, then $-b > 0$, so that there exist $q, r \in \mathbb{Z}$ for which $a = (-b)q' + r$, with $0 \leq r < -b$. Hence, $a = b(-q') + r$, with $0 \leq r < -b = |b|$. Make $q = -q'$.

Now, suppose that $a = bq + r = bq' + r'$, with $q, q', r, r' \in \mathbb{Z}$ and $0 \leq r, r' < |b|$. Then,

$$|r' - r| < |b| \ \text{and} \ b(q - q') = r' - r.$$

If $q \neq q'$, then $|q - q'| \geq 1$, so that

$$|b| \leq |b| \cdot |q - q'| = |r' - r| < |b|,$$

a contradiction. Therefore, $q = q'$ and hence $r = r'$. $\square$

The proof of the following corollary illustrates a typical application of the division algorithm. For its statement, recall that a nonnegative integer m is a **perfect square** provided $m = n^2$, for some nonnegative integer n.

Corollary 6.8 *Every perfect square leaves remainder:*

(a) 0 or 1 upon division by 3.
(b) 0 or 1 upon division by 4.
(c) 0, 1 or 4 upon division by 8.

Proof Let n be a natural number.

(a) By the division algorithm, n leaves remainder 0, 1 or 2 upon division by 3, so that $n = 3q$, $3q + 1$ or $3q + 2$, for some $q \in \mathbb{Z}$. Now, let's consider these three possibilities:

- If $n = 3q$, then $n^2 = 3 \cdot 3q^2$.
- If $n = 3q + 1$, then $n^2 = 3(3q^2 + 2q) + 1$.
- If $n = 3q + 2$, then $n^2 = 3(3q^2 + 4q + 1) + 1$.

Since $3q^2 \in \mathbb{Z}$, the uniqueness part of the division algorithm assures that, in the first case above, n^2 leaves remainder 0 upon division by 3. By the same argument, in the other two cases n^2 leaves remainder 1 upon division by 3.

(b) By the division algorithm, the remainder upon division of n by 4 is 0, 1, 2 or 3, so that $n = 4q, 4q + 1, 4q + 2$ or $4q + 3$, for some $q \in \mathbb{Z}$. Hence, we can certainly give to this item a proof that parallels that of (a). Nevertheless, let's see a simpler one: again by invoking the division algorithm (or Example 6.2), we can write $n = 2q$ or $2q + 1$, for some $q \in \mathbb{Z}$.

- If $n = 2q$, then $n^2 = 4q^2$.
- If $n = 2q + 1$, then $n^2 = 4(q^2 + q) + 1$.

Hence, upon division by 4, in the first case n^2 leaves remainder 0, whereas in the second it leaves remainder 1.

(c) Once more, we could give a proof of this fact as a direct application of the division algorithm, by writing $n = 8q + r$ for some $q \in \mathbb{Z}$ and some $r \in \{0, 1, 2, \dots, 7\}$. However, in order to avoid having to consider eight different cases, we proceed as in the proof of item (b):

- If $n = 2q$, then $n^2 = 4q^2$. Now, there are two possibilities:

  - If q^2 is even, say $q^2 = 2k$ for some $k \in \mathbb{N}$, then $n^2 = 8k$.
  - If q^2 is odd, say $q^2 = 2k + 1$ for some $k \in \mathbb{N}$, then $n^2 = 8k + 4$.

- If $n = 2q + 1$, then $n^2 = 4q(q + 1) + 1$. However, since one of the integers q and $q + 1$ is even, in any case the product $q(q + 1)$ is even, so that we can write $q(q + 1) = 2k$ for some $k \in \mathbb{N}$. Then, $n^2 = 4 \cdot 2k + 1 = 8k + 1$.

Finally, as in the proof of the previous items, the expressions for n^2 obtained above guarantee that its possible remainders upon division by 8 are exactly 0, 1 or 4. $\qquad\square$

The following elaboration of the division algorithm with play an important role in subsequent discussions.

Corollary 6.9 *Let a_1, a_2 and b be given integers, with b being nonzero. Then, $b \mid (a_1 - a_2)$ if and only if a_1 and a_2 leave equal remainders upon division by b.*

Proof Suppose first that $a_1 = bq_1 + r$ and $a_2 = bq_2 + r$, with $q_1, q_2, r \in \mathbb{Z}$. Then, $a_1 - a_2 = b(q_1 - q_2)$, i.e., $b \mid (a_1 - a_2)$.

Conversely, assume that $b \mid (a_1 - a_2)$ and let $a_1 = bq_1 + r_1$ and $a_2 = bq_2 + r_2$, with $q_1, q_2, r_1, r_2 \in \mathbb{Z}$ and $0 \le r_1, r_2 < |b|$. Then,

$$a_1 - a_2 = (q_1 - q_2)b + (r_1 - r_2)$$

and, since $b \mid (a_1 - a_2)$ and $b \mid (q_1 - q_2)b$, item (c) of Proposition 6.5 guarantees that b divides $r_1 - r_2 = (a_1 - a_2) - (q_1 - q_2)b$. On the other hand, inequalities $0 \le r_1, r_2 < |b|$ give $|r_1 - r_2| < |b|$, so that the only possibility for b dividing $r_1 - r_2$ is having $|r_1 - r_2| = 0$ or, which is the same, $r_1 = r_2$. $\qquad\square$

As an application of the previous result, the coming example computes the possible values of the last digit of a perfect square.

Example 6.10 Every perfect square ends in 0, 1, 4, 5, 6 or 9.

Proof First of all, notice that the remainder of a natural number upon division by 10 coincides with its last digit (i.e., the rightmost one, in the decimal representation of the number). Indeed, let $m = 10m' + a_0$, with $m' \in \mathbb{Z}_+$ and $0 \le a_0 \le 9$; since $10m'$ ends in 0, it follows that the decimal representation of the sum $10m' + a_0$ (which equals m) has a_0 as rightmost digit.

Now, let $n \in \mathbb{N}$ and $q, r \in \mathbb{Z}$ be such that $n = 10q + r$, with $0 \le r \le 9$. Then,

$$n^2 = (10q + r)^2 = 100q^2 + 20qr + r^2$$
$$= 10(10k^2 + 2kr) + r^2,$$

so that $10 \mid (n^2 - r^2)$, and Corollary 6.9 guarantees that n^2 and r^2 leave equal remainders upon division by 10. Hence, our initial remark assures that the last digit of n^2 equals that of r^2.

In order to finish the proof, it suffices to check which are the possible values of the last digit of r^2 when r varies from 0 to 9: $0^2 = 0$; 1^2 and 9^2 end in 1; 2^2 and 8^2 end in 4; 3^2 and 7^2 end in 9; 4^2 and 6^2 end in 6; 5^2 ends in 5. Thus, the possible values of the last digit of n^2 are 0, 1, 4, 5, 6 or 9. $\square$

Example 6.11 Let $n > 1$ be an integer. Show that any sequence of n consecutive integers contains exactly one multiple of n.

Proof Let $a + 1, a + 2, \ldots, a + n$ be a sequence of n consecutive integers, with $a = nq + r, 0 \le r < n$. Then, $0 < n - r \le n$, so that $a + (n - r)$ is one of the numbers of our sequence. Now, note that

$$a + (n - r) = (nq + r) + (n - r) = n(q + 1),$$

which is a multiple of n.

To what is left to do, we shall apply again the result of Corollary 6.9: the numbers $a + 1, a + 2, \ldots, a + n$ leave pairwise disjoint remainders upon division by n, since the modulus of the difference between any two of them is at least 1 and at most $n - 1$. Therefore, our sequence contains at most one multiple of n. $\square$

Problems: Sect. 6.1

1. Prove the **divisibility criteria** by 3 and by 9: the remainder of the division of a natural number by 3 (resp. by 9) equals the remainder of the division by 3 (resp. by 9) of the sum of the digits of its decimal representation. In particular, a natural number is a multiple of 3 (resp. of 9) if and only if the sum of its digits is so.

2. Find the remainder of the division of 10^k by 11.

3. Prove the **criterion of divisibility** by 11: if the natural number n has decimal representation $n = (a_k a_{k-1} \ldots a_1 a_0)_{10}$, then its remainder upon division by 11 equals that of the *alternating sum*

$$a_0 - a_1 + a_2 - \ldots + (-1)^{k-1} a_{k-1} + (-1)^k a_k$$

upon division by 11. In particular, $11 \mid n$ if and only if 11 divides the alternating sum of the digits of n.

4. (IMO) Find all natural numbers n of three digits, such that $11 \mid n$ and $\frac{n}{11}$ is equal to the sum of the squares of the digits of n.

5. (IMO) Find all $n \in \mathbb{N}$ such that the product of the digits of the decimal representation of n is equal to a $n^2 - 10n - 22$.

6. * Given a natural number n, prove that the product of any n consecutive integers is always divisible by $n!$.

7. (Hungary) For $n \in \mathbb{N}$, prove that $8^n - 3^n - 6^n + 1^n$ is a multiple of 10.

8. (Hungary) Prove that 5 divides $1^{99} + 2^{99} + 3^{99} + 4^{99} + 5^{99}$.

   For the next problem, given $m \in \mathbb{Z}$ we shall let $\mathbb{Z}m$ denote the set

$$n\mathbb{Z} = \{nq;\ q \in \mathbb{Z}\} = \{0, \pm n, \pm 2n, \pm 3n, \ldots\}.$$

9. Let S be a set of integers containing 0 and satisfying the following condition:

$$x, y \in S,\ x \neq y \Rightarrow x - y \in S.$$

   If $S \neq \{0\}$, prove that:

   (a) $S \cap \mathbb{N} \neq \emptyset$.
   (b) If $x, y \in S$, then $x + y \in S$.
   (c) If $n = \min(S \cap \mathbb{N})$, then $S = n\mathbb{Z}$.

10. * Compute the remainder of the division of $2^{64} + 1$ by $2^{32} + 1$. More generally, given integers a, m and n, all greater than 1 and such that $m > n$, compute the remainder of the division of $a^{2^m} + 1$ by $a^{2^n} + 1$.

    The coming problem generalizes the binary representation of natural numbers, established in Example 4.12 of [8].

11. * Let $m, a \in \mathbb{N}$, with $a > 1$. Show that there exist unique integers $k, m_0, m_1, \ldots, m_k$, with $k \geq 0$, $0 \leq m_j \leq a - 1$ for $0 \leq j \leq k$, $m_k > 0$ and

$$m = m_k a^k + m_{k-1} a^{k-1} + \cdots + m_1 a + m_0.$$

    The above expression for m is known as its **representation in base a**.

12. * For $x \in \mathbb{R}$, prove the following items:

    (a) $x - 1 < \lfloor x \rfloor \leq x$.
    (b) $\lfloor x + 1 \rfloor = \lfloor x \rfloor + 1$.

(c) $\lfloor x \rfloor + \lfloor -x \rfloor = \begin{cases} 0, & \text{if } x \in \mathbb{Z} \\ -1, & \text{if } x \notin \mathbb{Z} \end{cases}$.

13. * Let $x, y \in \mathbb{R}$ and $m, n \in \mathbb{Z}$, with $n > 0$. Prove the following items:

(a) $x \le y \Rightarrow \lfloor x \rfloor \le \lfloor y \rfloor$.

(b) $\lfloor nx \rfloor \ge n\lfloor x \rfloor$.

(c) $\left\lfloor \frac{m+1}{n} \right\rfloor \le \left\lfloor \frac{m}{n} \right\rfloor + 1$.

(d) $\left\lfloor \frac{\lfloor x \rfloor}{n} \right\rfloor = \left\lfloor \frac{x}{n} \right\rfloor$.

(e) $\lfloor x \rfloor + \lfloor y \rfloor \le \lfloor x + y \rfloor \le \lfloor x \rfloor + \lfloor y \rfloor + 1$.

(f) $\lfloor x + y \rfloor + \lfloor x \rfloor + \lfloor y \rfloor \le \lfloor 2x \rfloor + \lfloor 2y \rfloor$.

14. For $x \in \mathbb{R}$, prove that its **closest integer** is given by $\left\lfloor x + \frac{1}{2} \right\rfloor$.

15. (ORM) Find all $a, b \in \mathbb{N}$ such that

$$\left\lfloor \frac{a^2}{b} \right\rfloor + \left\lfloor \frac{b^2}{a} \right\rfloor + 1 = \left\lfloor \frac{a^2 + b^2}{ab} \right\rfloor + ab.$$

16. (Hungary) Prove that there are no odd integers x, y and z such that

$$(x + y)^2 + (y + z)^2 = (x + z)^2.$$

17. Let x, y and z be integers such that $x^2 + y^2 = z^2$. Prove that $6 \mid xy$.

   For the coming problem, recall that a **perfect cube** is an integer of the form q^3, for some $q \in \mathbb{Z}$.

18. (Brazil—adapted)

(a) Find all possible remainders upon the division of a perfect cube by 7.

(b) If x, y, z are integers such that $x^3 + y^3 - z^3$ is a multiple of 7, prove that at least one of x, y, z is a multiple of 7.

19. (Brazil) Find all natural numbers x, y, z, n such that $n^x + n^y = n^z$.

20. (Soviet Union) For $n > 3$ integer, show that $1! + 2! + 3! + \cdots + n!$ is never a perfect square.

21. (Brazil) Prove that there doesn't exist integers x and y such that $15x^2 - 7y^2 = 9$.

22. (France—adapted) For $k, m, n \in \mathbb{N}$, with $k > 2$, let $F_m = 2^{2^m} + 1$ be the m-th **Fermat's number**[2] and $G_n = \frac{n(n-1)}{2}(k - 2) + n$. For a fixed k, we want to find (if any) all $m, n \in \mathbb{N}$ such that $F_m = G_n$. To this end, do the following items:

(a) Prove that, if there exists such a pair (m, n), then there also exists a nonnegative integer r such that $n = 2^r + 1$.

---

[2]After Pierre S. de Fermat, French mathematician of the seventeenth century. As the perseverant reader will testify, along the rest of this book we shall have a lot more to say on the enormous legacy of Fermat to elementary Number Theory.

(b) Conclude from (a) that $k - 1 = 2^r t_1$ for some $t_1 \in \mathbb{N}$, so that $2^{2^m} + 1 = 2^{2r}(2^r t_1 + t_1 - 1)$.

(c) Use induction on $l \in \mathbb{N}$ to show that if we had $2^{2^m} + 1 = 2^{(l+1)r}(2^r t_l + t_l \mp 1)$ for some $t_l \in \mathbb{N}$, then $2^{2^m} + 1 = 2^{(l+2)r}(2^r t_{l+1} + t_{l+1} \pm 1)$.

(d) Conclude on the existence or not of such a pair (m, n).

23. (Russia) For each positive integer m, let $s(m)$ denote the sum of the digits of the decimal representation of m. Prove that there exist infinitely many natural numbers n for which $s(3^n) \geq s(3^{n+1})$.

6.2   GCD and LCM

We say that a nonzero integer d is a **common divisor** of given nonzero integers $a_1, \ldots, a_n$ if $d \mid a_1, \ldots, d \mid a_n$. Note that $a_1, \ldots, a_n$ always have common divisors: 1, for instance. Moreover, since every nonzero integer has only a finite number of divisors, $a_1, \ldots, a_n$ have only a finite number of common divisors. Thus, the following definition makes sense.

Definition 6.12 The **greatest common divisor** of the nonzero integers $a_1, a_2, \ldots, a_n$, denoted $\gcd(a_1, a_2, \ldots, a_n)$, is (as its name says) the greatest one of the common divisors of $a_1, a_2, \ldots, a_n$. Nonzero integers $a_1, a_2, \ldots, a_n$ are said to be **pairwise coprime** or **relatively prime**, if $\gcd(a_1, a_2, \ldots, a_n) = 1$.

The coming result is due to the French mathematician Étienne Bézout.[3] For its statement, given $n \in \mathbb{Z}$ we let $n\mathbb{Z}$ denote the set of integer multiples of n, namely,

$$n\mathbb{Z} = \{nx;\ x \in \mathbb{Z}\} = \{0, \pm n, \pm 2n, \pm 3n, \ldots\}.$$

Theorem 6.13 (Bézout) *Let $a_1, a_2, \ldots, a_n$ be nonzero integers. If*

$$S = \left\{ \sum_{i=1}^{n} a_i x_i;\ x_i \in \mathbb{Z}, \forall\, 1 \leq i \leq n \right\},$$

then $S = d\mathbb{Z}$, with $d = \gcd(a_1, a_2, \ldots, a_n)$. In particular, there exist $u_1, u_2, \ldots, u_n \in \mathbb{Z}$ such that

$$\gcd(a_1, a_2, \ldots, a_n) = a_1 u_1 + a_2 u_2 + \cdots + a_n u_n. \tag{6.4}$$

---

[3]French mathematician of the eighteenth century, Bézout is one of the precursors of the area of Mathematics known today as Algebraic Geometry. In spite of the coming result, his most famous theorem is probably the one which states that, if f and g are real polynomials in two indeterminates X, Y, with no common factors and with degrees respectively equal to m and n, then the number of points of intersection of the *curves* $f(x, y) = 0$ and $g(x, y) = 0$ in the cartesian plane is at most mn—for example, two distinct ellipses have at most $2 \cdot 2 = 4$ points in common. For an elementary exposition, we recommend [18] or [31].

Proof It's immediate to see that every multiple of an element of S belongs to S, i.e.,

$$u \in S \Rightarrow u\mathbb{Z} \subset S. \tag{6.5}$$

On the other hand, since d divides $a_1x_1 + a_2x_2 + \cdots + a_nx_n$ for all $x_1, x_2, \ldots, x_n \in \mathbb{Z}$, we certainly have $S \subset d\mathbb{Z}$.

In order to establish the opposite inclusion, note first that S does contain positive integers; indeed, by choosing $x_1 = a_1$ and $x_2 = \cdots = x_n = 0$, for instance, we conclude that

$$a_1^2 = a_1x_1 + a_2x_2 + \cdots + a_nx_n \in S.$$

Since S contains positive integers, there exists a least positive integer d' in S. If we show that $d' = d$, we will get $d \in S$, and (6.5) (with d in place of u) will give $d\mathbb{Z} \subset S$.

We first claim that $d' \mid a_1, \ldots, a_n$. Indeed, since $d' \in S$, there exist $u_1, u_2, \ldots, u_n \in \mathbb{Z}$ such that $d' = a_1u_1 + a_2u_2 + \cdots + a_nu_n$. Now, let $a_1 = d'q + r$, with $q, r \in \mathbb{Z}$ and $0 \leq r < d'$. Then,

$$
\begin{aligned}
r &= a_1 - d'q \\
&= a_1 - (a_1u_1 + a_2u_2 + \cdots + a_nu_n)q \\
&= a_1(1 - u_1q) + a_2(-u_2q) + \cdots + a_n(-u_nq),
\end{aligned}
$$

so that $r \in S$. If $0 < r < d'$, we would get a contradiction to the fact that d' is the least positive integer in S. Therefore, $r = 0$ and $d' \mid a_1$. Analogously, $d' \mid a_2, \ldots, a_n$.

Finally, since d' is a common divisor of $a_1, a_2, \ldots, a_n$, in order to get $d' = d$ is suffices to show that $d' \geq d$. To this end, if $a_1 = dq_1, a_2 = dq_2, \ldots, a_n = dq_n$, with $q_1, q_2, \ldots, q_n \in \mathbb{Z}$, then

$$
\begin{aligned}
d' &= a_1u_1 + a_2u_2 + \cdots + a_nu_n \\
&= dq_1u_1 + dq_2u_2 + \cdots + dq_nu_n \\
&= d(q_1u_1 + q_2u_2 + \cdots + q_nu_n),
\end{aligned}
$$

so that $0 < d \mid d'$. Thus, $d \leq d'$. $\qquad\square$

Remark 6.14 In the notations of Bézout's theorem, one uses to say, thanks to (6.4), that $\gcd(a_1, a_2, \ldots, a_n)$ is a **linear combination** of $a_1, a_2, \ldots, a_n$, with integer coefficients. For the time being, we shall concentrate on the *existence* of such a way of representing $\gcd(a_1, a_2, \ldots, a_n)$. However, an actual way of getting $u_1, u_2, \ldots, u_n \in \mathbb{Z}$ as in (6.4) will be an easy consequence of *Euclid's algorithm* (cf. Proposition 6.27), together with item (b) of Corollary 6.23.

In what follows, we collect several useful consequences of the previous theorem.

Corollary 6.15 *Let $a_1, a_2, \ldots, a_n$ be nonzero integers with* gcd *equal to d. If $d' \in$ $\mathbb{N}$, then $d' \mid a_1, a_2, \ldots, a_n$ if and only if $d' \mid d$.*

Proof Take integers $u_1, u_2, \ldots, u_n$ such that $d = a_1 u_1 + a_2 u_2 + \cdots + a_n u_n$. Since $d' \mid a_1, a_2, \ldots, a_n$, item i. of Remarks 6.6 guarantees that $d' \mid d$. The converse is immediate. $\qquad\square$

Corollary 6.16 *Let $a_1, a_2, \ldots, a_n$ be nonzero given integers with* gcd *equal to d.*

(a) *$d = 1$ if and only if there exist integers $u_1, u_2, \ldots, u_n$ such that $a_1 u_1 + a_2 u_2 + \cdots + a_n u_n = 1$.*
(b) *$\gcd\left(\frac{a_1}{d}, \frac{a_2}{d}, \ldots, \frac{a_n}{d}\right) = 1$.*

Proof

(a) If $d = 1$, the existence of such integers $u_1, u_2, \ldots, u_n$ follows from Bézout's theorem. Conversely, let $u_1, u_2, \ldots, u_n$ be a set of integers satisfying the stated conditions. Since $d \mid a_1, a_2, \ldots, a_n$, it follows once more from item i. of Remarks 6.6 that $d \mid (a_1 u_1 + \cdots + a_n u_n)$, i.e., $d \mid 1$. Thus, $d = 1$.
(b) Letting $d = a_1 u_1 + a_2 u_2 + \cdots + a_n u_n$, we have $\left(\frac{a_1}{d}\right) u_1 + \cdots + \left(\frac{a_n}{d}\right) u_n = 1$. Therefore, the desired result follows directly from item (a). $\qquad\square$

Example 6.17 (China) Let a, b, c and d be nonzero integers, such that $c + d \neq 0$ and $ad - bc = 1$. Prove that $\frac{a+b}{c+d}$ is an irreducible fraction.

Proof We wish to prove that $\gcd(a + b, c + d) = 1$. To this end, and according to the previous corollary, it suffices to find integers u and v such that

$$(a + b)u + (c + d)v = 1.$$

Since $ad - bc = 1$, one readily notices that $u = d$ and $v = -b$ is a possible choice. $\qquad\square$

Remark 6.18 Let $a_1, a_2, \ldots, a_n$ be nonzero integers with gcd equal to d. Letting $u_i = \frac{a_i}{d}$ for $1 \leq i \leq n$, item (b) of the previous corollary assures that $u_1, u_2, \ldots, u_n$ are relatively prime integers, for which $a_1 = du_1, a_2 = du_2, \ldots, a_n = du_n$.

The above remark gives a way of writing nonzero integers $a_1, a_2, \ldots, a_n$ which will reveal itself to be useful in a number of situations. The two coming examples stress this point.

Example 6.19 Every nonzero rational number can be written as an **irreducible fraction**, i.e., has a fractional representation of the form $\frac{a}{b}$, where a and b are nonzero relatively prime integers. Indeed, let r be a nonzero rational number and $r = \frac{m}{n}$ be a fractional representation of it. Then, m and n are nonzero integers and, letting $d = \gcd(m, n)$, we get $m = da$ and $n = db$, with $\gcd(a, b) = 1$. This gives

$$r = \frac{m}{n} = \frac{da}{db} = \frac{a}{b},$$

which gives the desired fractional representation of r.

Example 6.20 (Russia) Let a and b be natural numbers such that $\frac{a+1}{b} + \frac{b+1}{a} \in \mathbb{N}$. Prove that

$$\gcd(a, b) \le \sqrt{a + b}.$$

Proof Let $d = \gcd(a, b)$, so that there exist integers u and v such that $a = du$, $b = dv$ and $\gcd(u, v) = 1$. Then,

$$\frac{a+1}{b} + \frac{b+1}{a} = \frac{du+1}{dv} + \frac{dv+1}{du}$$
$$= \frac{u(du+1) + v(dv+1)}{duv}$$
$$= \frac{d(u^2 + v^2) + (u + v)}{duv} \in \mathbb{N}.$$

In turn, this assures that

$$uv \cdot \frac{d(u^2 + v^2) + (u + v)}{duv} = (u^2 + v^2) + \frac{u + v}{d}$$

is also a natural number. Hence, $d \mid (u + v)$, so that $d \le u + v$. But this gives

$$d^2 \le du + dv = a + b.$$

$\square$

Corollary 6.21 *Given nonzero integers $a_1, a_2, \ldots, a_n, k$, we have:*

(a) $\gcd(ka_1, ka_2, \ldots, ka_n) = |k| \gcd(a_1, a_2, \ldots, a_n)$.
(b) $\gcd(a_1, a_2, \ldots, a_n) = \gcd(\gcd(a_1, a_2, \ldots, a_{n-1}), a_n)$.

Proof

(a) Let $d = \gcd(a_1, \ldots, a_n)$ and $d' = \gcd(ka_1, \ldots, ka_n)$. Since d divides all of $a_1, \ldots, a_n$, we conclude that $|k|d$ divides all of $ka_1, \ldots, ka_n$, so that $|k|d$ is a positive common divisor of $ka_1, \ldots, ka_n$. However, since d' is the largest of such common divisors, it follows that $|k|d \le d'$. Conversely, since k divides d' and d' divides all of $ka_1, \ldots, ka_n$, we get that $\frac{d'}{|k|}$ is a positive integer which divides all of $a_1, \ldots, a_n$. This gives $\frac{d'}{|k|} \le d$ or, which is the same, $d' \le |k|d$. Therefore, $d' = |k|d$.

(b) Let $d = \gcd(a_1, \ldots, a_n)$ and $d' = \gcd(\gcd(a_1, a_2, \ldots, a_{n-1}), a_n)$. Since $d \mid$
    $a_1, \ldots, a_{n-1}$, we get from Corollary 6.15 that $d \mid \gcd(a_1, \ldots, a_{n-1})$. Then, d
    divides both $\gcd(a_1, \ldots, a_{n-1})$ and a_n, so that (once again from Corollary 6.15)
    d divides the gcd of these two numbers, that is, $d \mid d'$. A similar reasoning shows
    that $d' \mid d$, thus finishing the proof.

$\square$

We specialize our discussion to the greatest common divisor of two nonzero
integers. Given $a, b \in \mathbb{Z} \setminus \{0\}$, with $d = \gcd(a, b)$, Bézout's theorem guarantees
the existence of integer u and v such that $d = au + bv$. It is important to note that
such a way of writing $\gcd(a, b)$ is not unique; actually, if $t \in \mathbb{Z}$, then we also have
$d = a(u - tb) + b(v + ta)$. We shall have more to say about this in Proposition 6.26.

Later, we will establish a quite important and useful algorithm for effectively
finding the gcd of two nonzero given integers, namely, *Euclid's algorithm*. We start
by studying some properties of the gcd of two nonzero integers.

Proposition 6.22 *For any nonzero integers a, b and c, we have that:*

(a) If $c \mid ab$ and $\gcd(b, c) = 1$, then $c \mid a$.
(b) If $a + bc \neq 0$, then $\gcd(a + bc, b) = \gcd(a, b)$.
(c) If $\gcd(a, c) = 1$, then $\gcd(a, bc) = \gcd(a, b)$.
(d) If $c \mid b$ and $\gcd(u, b) = 1$, then $\gcd(a, c) = 1$.
(e) If $\gcd(b, c) = 1$, then $\gcd(a, bc) = \gcd(a, b)\gcd(a, c)$. In particular, if b and
    c are relatively prime and divide a, then bc divides a.

Proof

(a) Let $u, v \in \mathbb{Z}$ be such that $bu + cv = 1$. Multiplying both sides of this equality by
    a, we get $(ab)u + c(av) = a$. Finally, since $c \mid (ab)$, item (c) of Proposition 6.5
    shows that $c \mid a$.
(b) Let $d = \gcd(a + bc, b)$ and $d' = \gcd(a, b)$. Since $d' \mid a, b$, we have that
    $d' \mid a, a + bc$. Hence, by Corollary 6.15 we get $d' \mid d$. Conversely, since
    $d \mid (a + bc)$ and $d \mid b$, we conclude that $d \mid [(a + bc) - bc]$, i.e., $d \mid a$ and
    $d \mid b$. Again by Corollary 6.15, we obtain $d \mid d'$ and, thus, $d = d'$.
(c) Set $d = \gcd(a, b)$ and $d' = \gcd(a, bc)$. From $d \mid b$ it follows that $d \mid bc$.
    Therefore, $d \mid a$ and $d \mid bc$, so that $d \mid \gcd(a, bc) = d'$. In order to finish,
    let us show that $d' \mid d$: since $\gcd(a, c) = 1$, Bézout's theorem guarantees the
    existence of $u, v \in \mathbb{Z}$ such that $au + cv = 1$. This way, $a(bu) + (bc)v = b$, and
    such an equality, together with the fact that $d' \mid a$ and $d' \mid bc$, proves that $d' \mid b$.
    Then, $d' \mid a$ and $d' \mid b$, so that $d' \mid \gcd(a, b) = d$.
(d) Let $d \in \mathbb{Z}$ be such that $b = cd$, and choose $u, v \in \mathbb{Z}$ for which $au + bv = 1$.
    Then, $au + c(dv) = 1$, and item (a) of Corollary 6.16 gives $\gcd(a, c) = 1$.
(e) Once more, set $d = \gcd(a, b)$ and write $a = du$ and $b = dv$, with $u, v \in \mathbb{Z}$
    relatively prime. By successively applying Corollary 6.23 and the result of item
    (c), we obtain

$$\gcd(a, bc) = \gcd(du, dvc) = d\gcd(u, vc) = d\gcd(u, c).$$

However, $d \mid b$ and $\gcd(b, c) = 1$ imply (cf. item (d)) $\gcd(d, c) = 1$. Hence, one further application of item (c) gives us

$$\gcd(a, c) = \gcd(du, c) = \gcd(u, c).$$

Finally, taking together both relations above, we get

$$\gcd(a, bc) = d \cdot \gcd(u, c) = \gcd(a, b) \gcd(a, c).$$

The rest is immediate.

$\square$

Corollary 6.23 *Let a and b be nonzero integers and k, m and n be natural numbers.*

(a) *If $\gcd(a, b) = 1$, then $\gcd(a^m, b^n) = 1$.*
(b) *If $\gcd(a, b) = 1$ and $ab = k^n$, then there exist $u, v \in \mathbb{Z}$ such that $a = u^n$, $b = v^n$.*

Proof

(a) Item (c) of the previous proposition, with b^{n-1} in place of b and b in place of c, gives

$$\gcd(a, b^n) = \gcd(a, b^{n-1} \cdot b) = \gcd(a, b^{n-1}).$$

It follows by induction on n that $\gcd(a, b^n) = \gcd(a, b) = 1$. Analogously,

$$\gcd(a^m, b^n) = \gcd(a^{m-1} \cdot a, b^n) = \gcd(a^{m-1}, b^n)$$

and, by induction on m, we get $\gcd(a^m, b^n) = \gcd(a, b^n) = 1$.

Alternatively, since $\gcd(a, b) = 1$, Bézout's theorem guarantees the existence of $x, y \in \mathbb{Z}$ such that $ax + by = 1$. Hence, the formula for binomial expansion gives

$$1 = (ax + by)^n = \sum_{k=0}^{n-1} \binom{n}{k} (ax)^{n-k} (by)^k + (by)^n = aq + b^n y^n,$$

with $q = \sum_{k=0}^{n-1} \binom{n}{k} a^{n-k-1} x^{n-k} (by)^k$. Hence, item (a) of Corollary 6.16 shows that $\gcd(a, b^n) = 1$. Starting from this last relation and arguing in an entirely analogous way, we conclude that $\gcd(a^m, b^n) = 1$.

(b) Set $u = \gcd(a, k)$ and $v = \gcd(b, k)$. Since $\gcd(a, b) = 1$, from item (d) of the previous proposition we obtain

$$k = \gcd(k^n, k) = \gcd(ab, k) = \gcd(a, k) \cdot \gcd(b, k) = uv,$$

so that

$$ab = k^n = u^n v^n.$$

Now, $u \mid a$ and $\gcd(a, b) = 1$ give us (cf. item (d) of the former Proposition) $\gcd(u, b) = 1$; analogously, $\gcd(v, a) = 1$. It thus follows from item (a) that

$$\gcd(u^n, b) = 1 \text{ and } \gcd(v^n, a) = 1.$$

Finally,

$$\begin{cases} u^n \mid ab \text{ and } \gcd(u^n, b) = 1 \Rightarrow u^n \mid a \Rightarrow u^n \leq a \\ v^n \mid ab \text{ and } \gcd(v^n, a) = 1 \Rightarrow v^n \mid b \Rightarrow v^n \leq b \end{cases}.$$

However, since $ab = u^n v^n$, $u^n \leq a$ and $v^n \leq b$, the only possibility is to have $a = u^n$ and $b = v^n$.
$\square$

We are now in position to prove a result which was already used in [8]. We present it as the coming

Example 6.24 Given natural numbers n and k, with $k > 1$, either there exists $m \in \mathbb{N}$ such that $n = m^k$ or $\sqrt[k]{n}$ is an irrational number.

Proof Assume that $\sqrt[k]{n} = \frac{m}{p}$, with m and p relatively prime naturals (cf. Example 6.19). Then, $p^k n = m^k$, and condition $\gcd(m, p) = 1$, together with the results of the previous corollary, gives $\gcd(m^k, p^k) = 1$. On the other hand, $p^k n = m^k$ assures that $p^k \mid m^k$. Therefore, the only way of p^k and m^k being relatively prime is that $p^k = 1$. However, since $k > 1$, we must then have $p = 1$, so that $n = m^k$.
$\square$

Roughly speaking (for now, any attempt of giving a more precise definition would be useless), an equation in (more than one) integer variables is said to be a **diophantine equation**.[4] The analysis of general diophantine equation equations is generally a very difficult task, demanding quite sophisticated arguments to be accomplished. We shall meet (and solve) several important examples of Diophantine equations along the way. For the time being, let us take a look at a simple example.

Example 6.25 Prove that there does not exist $x, y \in \mathbb{N}$ such that $x^3 + 3 = 4y(y+1)$.

Proof On the contrary, assume that $x^3 + 3 = 4y(y + 1)$ for some $x, y \in \mathbb{N}$. Then,

$$x^3 + 4 = 4y(y + 1) + 1 = (2y + 1)^2$$

---

[4]After Diophantus of Alexandria, Greek mathematician of the third century BC, who is considered to be one of the founding fathers of Algebra and Number Theory. In his famous book *Aritmetica*, Diophantus tried to systematically study integer solutions of certain particular kinds of polynomial equations. These, in turn, passed to be deservedly known as *diophantine*.

and, hence,

$$x^3 = (2y+1)^2 - 2^2 = (2y-1)(2y+3).$$

Now, if $d = \gcd(2y-1, 2y+3)$, then d is odd and divides $(2y+3) - (2y-1) = 4$, so that $d = 1$. Therefore, the product of the relatively prime integers $2y - 1$ and $2y + 3$ is a perfect cube, and Corollary 6.23 assures that $2y - 1$ and $2y + 3$ must themselves be perfect cubes. However, one cannot find two perfect cubes whose difference is equal to 4, so that we have reached a contradiction. □

In spite of the *ad hoc* character of the solution of the previous example, the theory we have developed so far, on the gcd of two integers, allows us to completely solve the **diophantine linear equation** in two variables $ax + by = c$. Here, x and y are integer unknowns and a, b and c are given nonzero integer parameters.

Proposition 6.26 *Let a, b and c be nonzero given integers. The equation $ax + by = c$ admits integer solutions if and only if $\gcd(a, b) \mid c$. Moreover, if this is so and $d = \gcd(a, b)$ and $x = x_0$, $y = y_0$ is some integer solution, then the formulas*

$$x = x_0 + \frac{b}{d}t, \quad y = y_0 - \frac{a}{d}t, \tag{6.6}$$

$t \in \mathbb{Z}$, give all possible integer solutions. In particular, if $ab > 0$ and $x, y \in \mathbb{Z}$ is an integer solution, then we can assume that $x > 0 > y$ or $x < 0 < y$.

Proof Firstly, suppose that there exist $x, y \in \mathbb{Z}$ for which $ax + by = c$. Since $d \mid a$ and $d \mid b$, it follows that $d \mid (ax + by)$, i.e., $d \mid c$. Conversely, let $c = de$, with $e \in \mathbb{Z}$, and (by Bézout's theorem) u and v be integers such that $d = au + bv$. Then, $c = a \cdot eu + b \cdot ev$, so that the given equation admits the integer solution $x_0 = eu$, $y_0 = ev$.

For the second part, assume again that $d \mid c$, and let $x = x_0$, $y = y_0$ be some integer solution of the given equation. If $x = x_1$, $y = y_1$ is another integer solution, we must have $a(x_1 - x_0) = b(y_0 - y_1)$; cancelling d out of both sides of such an equality, we obtain

$$\frac{a}{d}(x_1 - x_0) = \frac{b}{d}(y_0 - y_1).$$

Thus, $\frac{b}{d} \mid \frac{a}{d}(x_1 - x_0)$ and, since $\gcd\left(\frac{a}{d}, \frac{b}{d}\right) = 1$, item (a) of the previous proposition guarantees that $\frac{b}{d} \mid (x_1 - x_0)$. Letting $x_1 - x_0 = \frac{b}{d}t$, we get $y_0 - y_1 = \frac{a}{d}t$, and the stated formulas easily follow. Conversely, it is immediate to verify that such formulas do provide integer solutions to the given equation.

For what is left, assume that $a, b > 0$ (the remaining case $a, b < 0$ can be dealt with in an analogous way), so that

$$x_0 + \frac{b}{d}t > 0 \Leftrightarrow t > -\frac{dx_0}{b} \text{ and } y_0 - \frac{a}{d}t < 0 \Leftrightarrow t > \frac{dy_0}{a}.$$

Therefore, by choosing $t \in \mathbb{Z}$ such that $t > -\frac{dx_0}{b}, \frac{dy_0}{a}$, and letting $x_1 = x_0 + \frac{b}{d}t$ and $y_1 = y_0 - \frac{a}{d}t$, we have $x_1 > 0$, $y_1 < 0$ and $ax_1 + by_1 = c$. Analogously, one can show that we can take an integer solution x, y of the given equation with $x < 0 < y$. $\qquad\qquad\qquad\qquad\qquad\qquad\qquad\qquad\qquad\qquad\qquad\qquad\qquad\qquad\quad\square$

Up to this moment, we only know that one *can* write the gcd of two nonzero integers a and b as a linear combination $au + bv$, for some $u, v \in \mathbb{Z}$. However, we still do not have at our disposal a method of *effectively* doing this. To remedy such a situation, we consider the following algorithm, usually attributed to Euclid himself.

Euclid's Algorithm

$$
\begin{array}{lll}
\text{Step 1} & : a = bq_1 + r_1 & 0 < r_1 < b \\
\text{Step 2} & : b = r_1 q_2 + r_2 & 0 < r_2 < r_1 \\
\text{Step 3} & : r_1 = r_2 q_3 + r_3 & 0 < r_3 < r_2 \\
& \quad\cdots & \cdots \\
\text{Step } j & : r_{j-2} = r_{j-1} q_j + r_j & 0 < r_j < r_{j-1} \\
\text{Step } j+1 & : r_{j-1} = r_j q_{j+1} + 0 &
\end{array}
$$

Note that the execution of the above algorithm actually stops after a finite number of steps, for, since $r_1, r_2, \ldots$ are integers satisfying $b > r_1 > r_2 > \cdots \geq 0$, there must exist a least index j such that r_j is the last nonzero remainder in the sequence of divisions above.

The importance of Euclid's algorithm relies upon the following result.

Proposition 6.27 *In the notations above, we have* $\gcd(a, b) = r_j$.

Proof Several applications of item (b) of Proposition 6.22 give

$$
\begin{aligned}
\gcd(a, b) &= \gcd(a - bq_1, b) = \gcd(r_1, b) \\
&= \gcd(r_1, b - r_1 q_2) = \gcd(r_1, r_2) \\
&= \gcd(r_1 - r_2 q_3, r_2) = \gcd(r_3, r_2) \\
&\quad\cdots \\
&= \gcd(r_{j-1}, r_j) = r_j,
\end{aligned}
$$

where, in the last equality above, we used the fact that $r_j \mid r_{j-1}$. $\qquad\qquad\square$

Euclid's algorithm provides a method for effectively finding a pair of integers u and v whose existence is assured by Bézout's theorem, i.e., such that $\gcd(a, b) = au + bv$. Let us see how to do this by looking at a numerical

Example 6.28 Use Euclid's algorithm to show that $\gcd(120, 84) = 12$. Then, solve the linear diophantine equation $120x + 84y = 12$.

Solution We start by executing Euclid's algorithm:

$$120 = 84 \cdot 1 + 36$$
$$84 = 36 \cdot 2 + 12$$
$$36 = 12 \cdot 3.$$

Since 12 is the last nonzero remainder, it follows from Proposition 6.27 that $\gcd(120, 84) = 12$.

We can now find integers u and v such that $12 = 120u + 84v$ by working *backwards* with the successive divisions that led us to the gcd:

$$12 = 84 \cdot 1 - 36 \cdot 2$$
$$= 84 \cdot 1 - (120 - 84 \cdot 1) \cdot 2$$
$$= 84 \cdot 1 - 120 \cdot 2 + 84 \cdot 2$$
$$= 84 \cdot 3 + 120(-2).$$

Finally, with the solution $x_0 = -2$, $y_0 = 3$ of $120x + 84y = 12$ in our hands, we can get all solutions by applying formulas (6.6):

$$x = -2 + (84/12)t = -2 + 7t;$$
$$y = 3 - (120/12)t = 3 - 10t.$$

□

Euclid's algorithm and the proof of its correctness (i.e., Proposition 6.27) are also useful for theoretical purposes, as shown by the coming example.

Example 6.29 Let a, m and n be given natural numbers, with $a > 1$. If $q, r \in \mathbb{Z}$ are such that $m = nq + r$, with $0 \leq r < n$, prove that

$$\gcd(a^m - 1, a^n - 1) = a^{\gcd(m,n)} - 1.$$

Proof Let us start by showing that, if $r = 0$, then $(a^n - 1) \mid (a^m - 1)$. To this end, it suffices to observe that $a^m - 1 = (a^n)^q - 1$, and to recall that we already know, from Example 6.3, that $a^n - 1$ divides $(a^n)^q - 1$.

We now prove that, if $r > 0$, then $\gcd(a^m - 1, a^n - 1) = \gcd(a^n - 1, a^r - 1)$. Indeed, letting $a^n = b$ whenever convenient, we get

$$a^m - 1 = a^{nq+r} - 1 = (a^{nq} - 1)a^r + (a^r - 1)$$
$$= ((a^n)^q - 1)a^r + (a^r - 1)$$
$$= (a^n - 1)(b^{q-1} + \cdots + b + 1)a^r + (a^r - 1).$$

Letting

$$c = (b^{q-1} + b^{q-2} + \cdots + b + 1)a^r$$
$$d = \gcd(a^m - 1, a^n - 1)$$
$$d' = \gcd(a^n - 1, a^r - 1)$$

we get

$$a^m - 1 = (a^n - 1)c + (a^r - 1).$$

Hence, item (b) of Proposition 6.22 gives

$$\gcd(a^m - 1, a^n - 1) = \gcd((a^n - 1)c + (a^r - 1), a^n - 1)$$
$$= \gcd(a^r - 1, a^n - 1).$$

For what is left to do, assume, without loss of generality, that $m \geq n$. If $m = n$, we are done. Assume, then, that $m > n$, and consider Euclid's algorithm for m and n:

$$
\begin{aligned}
m &= nq_1 + r_1 & 0 < r_1 < n; \\
n &= r_1 q_2 + r_2 & 0 < r_2 < r_1; \\
r_1 &= r_2 q_3 + r_3 & 0 < r_3 < r_2; \\
& \cdots & \cdots \\
r_{j-2} &= r_{j-1} q_j + r_j & 0 < r_j < r_{j-1}; \\
r_{j-1} &= r_j q_{j+1} + 0.
\end{aligned}
$$

Our previous discussion guarantee that

$$\gcd(m, n) = \gcd(n, r_1) = \gcd(r_1, r_2) = \cdots = \gcd(r_{j-1}, r_j) = r_j.$$

Therefore, by applying the above reasoning several times, we get

$$\gcd(a^m - 1, a^n - 1) = \gcd(a^n - 1, a^{r_1} - 1)$$
$$= \gcd(a^{r_1} - 1, a^{r_2} - 1)$$
$$\cdots$$
$$= \gcd(a^{r_{j-1}} - 1, a^{r_j} - 1)$$
$$= a^{r_j} - 1 = a^{\gcd(m,n)-1}.$$

$\square$

We finish this section by studying the *least common multiple* of a finite set of nonzero integers. Given nonzero integers $a_1, a_2, \ldots, a_n$, the positive integer

$|a_1 a_2 \ldots a_n|$ is a common multiple of $a_1, a_2, \ldots, a_n$. Therefore, there exists a least positive integer which is a common multiple of $a_1, a_2, \ldots, a_n$, and this gives sense to the following

Definition 6.30 Given nonzero integers $a_1, a_2, \ldots, a_n$, the **least common multiple** of $a_1, a_2, \ldots, a_n$, denoted lcm $(a_1, a_2, \ldots, a_n)$, is the smallest positive integer which is a common multiple of all of $a_1, a_2, \ldots, a_n$.

The coming results establish the basic properties of the lcm.

Proposition 6.31 *Let a_1, a_2, $\ldots$, a_n be nonzero integers with* lcm *equal to m. An integer M is a common multiple of $a_1, a_2, \ldots, a_n$ if and only if $m \mid M$.*

Proof Let M be a common multiple of $a_1, a_2, \ldots, a_n$ and write $M = mq + r$, with $q, r \in \mathbb{Z}$ and $0 \leq r < m$. Since $a_1 \mid m$ and $a_1 \mid M$, it follows from Proposition 6.5 that $a_1 \mid (M - mq)$, i.e., $a_1 \mid r$; analogously, $a_2, \ldots, a_n \mid r$, so that $0 \leq r < m$ is a common multiple of $a_1, a_2, \ldots, a_n$. However, since m is the least (positive) common multiple of $a_1, a_2, \ldots, a_n$, the only left possibility is that $r = 0$. Hence, $m \mid M$. $\square$

Lemma 6.32 *Let a_1, a_2, $\ldots$, a_n be nonzero integers and $k \in \mathbb{N}$. Then,* lcm $(ka_1, ka_2, \ldots, ka_n) = k \cdot$ lcm $(a_1, a_2, \ldots, a_n)$.

Proof If $M =$ lcm $(ka_1, ka_2, \ldots, ka_n)$ and $m =$ lcm $(a_1, a_2, \ldots, a_n)$, then km is a positive common multiple of $ka_1, ka_2, \ldots, ka_n$, so that $km \geq M$. On the other hand, since M is a (positive) common multiple of $ka_1, ka_2, \ldots, ka_n$, it follows that $k \mid M$ and M/k is a positive common multiple of $a_1, a_2, \ldots, a_n$; therefore, $M/k \geq m$ or, which is the same, $M \geq km$. We conclude that $M = km$, as desired. $\square$

Our next result relates the gcd and the lcm of two nonzero integers.

Proposition 6.33 *If a and b are nonzero integers, then*

$$\gcd(a, b) \cdot \text{lcm}\,(a, b) = |ab|.$$

Proof We shall begin by showing that $\gcd(a, b) = 1 \Rightarrow$ lcm $(a, b) = |ab|$. To this end, let m be a positive common multiple of a and b. Since $\gcd(a, b) = 1$, item (e) of Proposition 6.22 guarantees that $ab \mid m$. Hence, $m \geq |ab|$, so that $|ab|$ is the smallest positive common multiple of a and b, i.e., $|ab| =$ lcm (a, b).

For the general case, first note that lcm $(a, -b) =$ lcm (a, b); since we already have $\gcd(a, -b) = \gcd(a, b)$, we can assume, without loss of generality, that $a, b > 0$. Let $d = \gcd(a, b)$ and write $a = du$ and $b = dv$, so that u and v are positive and relatively prime integers. We wish to show that lcm $(du, dv)d = d^2 uv$ or, by the previous lemma, that lcm $(u, v) = uv$. But this is exactly what we have done in the paragraph above. $\square$

Example 6.34 (Japan) Find all pairs (a, b) of positive integers such that $a \geq b$ and

$$\text{lcm}\,(a, b) + \gcd(a, b) + a + b = ab.$$

Solution Set $d = \gcd(a, b)$, and let $u, v \in \mathbb{N}$ be such that $a = du$ and $b = dv$. Then, $a \geq b \Rightarrow u \geq v$ and

$$d \cdot \operatorname{lcm}(a, b) = \gcd(a, b) \cdot \operatorname{lcm}(a, b) = ab = d^2 uv,$$

so that $\operatorname{lcm}(a, b) = duv$. Substituting these expressions in the original equation, we conclude that it is equivalent to

$$u(v + 1) + (v + 1) = duv, \tag{6.7}$$

so that $u \mid (v + 1)$. Thus, $u \leq v + 1$ and, therefore, $u = v$ or $u = v + 1$. If $u = v$, it follows from $\gcd(u, v) = 1$ that $u = v = 1$, so that $d = 4$ and $a = b = 4$. If $u = v + 1$, it follows from (6.7) that

$$d = \frac{u(v + 1) + (v + 1)}{uv} = \frac{(v + 1)^2 + (v + 1)}{(v + 1)v}$$
$$= \frac{v^2 + 3v + 2}{v^2 + v} = 1 + \frac{2}{v},$$

which in turn implies $v = 1$ or $v = 2$. By separately examining each of these cases, we arrive at the remaining solutions: $a = 6$, $b = 3$ or $a = 6$, $b = 4$. $\qquad\square$

Problems: Sect. 6.2

1. (IMO) For $n \in \mathbb{N}$, prove that $\gcd(21n + 4, 14n + 3) = 1$.
2. (OIM) Find all positive integers m and n such that $2^m + 1 = n^2$.
3. Consider two infinite and nonconstant arithmetic progressions whose terms are positive integers. Prove that there exist infinitely many naturals which are terms of both of these sequences if and only if the gcd of their common differences divides the difference of their first terms.
4. * Given $a, m, n \in \mathbb{N}$, with $m \neq n$, prove that $\gcd(a^{2^n} + 1, a^{2^m} + 1) = 1$ or 2.
5. (Japan) Prove that $\gcd(n! + 1, (n + 1)! + 1) = 1$, for every $n \in \mathbb{N}$.
6. (TT) If a and b are natural numbers for which $ab \mid (a^2 + b^2)$, show that $a = b$.
7. (Brazil) We partition an $a \times b$ rectangle $(a, b \in \mathbb{N})$ into ab squares 1×1. Prove that each diagonal of the rectangle crosses the interior of exactly $a + b - \gcd(a, b)$ squares 1×1.
8. Let n and k be positive integers, with $n \geq k$. Prove that the gcd of the binomial numbers

$$\binom{n}{k}, \binom{n + 1}{k}, \dots, \binom{n + k}{k}$$

equals 1.

9. (Miklós-Schweitzer) Let $n > 1$ be a given integer and 2^k be the greatest power of 2 that divides it. Prove that the gcd of the binomial numbers

$$\binom{2n}{1}, \binom{2n}{3}, \binom{2n}{5}, \ldots, \binom{2n}{2n-1}$$

is equal to 2^{k+1}.

10. (USA—adapted) For a fixed $k \in \mathbb{N}$, show that

$$\max\{\gcd(n^2 + k, (n + 1)^2 + k); \ n \in \mathbb{N}\} = 4k + 1.$$

11. (Hungary) If a, b and c are positive integers such that $\frac{1}{a} + \frac{1}{b} = \frac{1}{c}$, prove that there exist positive integers q, u and v for which $\gcd(u, v) = 1$ and $a = qu(u + v)$, $b = qv(u + v)$ and $c = quv$.

12. (England) Let x and y be natural numbers such that $2xy$ divides $x^2 + y^2 - x$. Prove that x is a perfect square.

13. (USA) Of any set of ten consecutive natural numbers, show that one can always choose at least one element which is relatively prime to all of the nine remaining ones.

14. Let a, b and m be given naturals, with $\gcd(a, m) = 1$. Show that

$$\sum_{j=0}^{m-1} \left\lfloor \frac{aj + b}{m} \right\rfloor = \frac{1}{2}(a - 1)(m - 1) + b.$$

For the coming problem, recall (cf. [8], for instance) that the **fractionary part** of $x \in \mathbb{R}$ is the real number $\{x\}$ such that $\{x\} = x - \lfloor x \rfloor$. Thus, $\{x\} \in [0, 1)$ and $\{x\} = 0 \Leftrightarrow x \in \mathbb{Z}$.

15. (Japan) Let $n, r \in \mathbb{N}$ be such that $n > 1$ and $n \nmid r$. If $d = \gcd(n, r)$, prove that

$$\sum_{i=1}^{n-1} \left\{ \frac{ri}{n} \right\} = \frac{1}{2}(n - d).$$

16. (Putnam) Given $m, n \in \mathbb{N}$, with $m \geq n$, prove that

$$\frac{\gcd(m, n)}{m} \binom{m}{n} \in \mathbb{N}.$$

17. (Bulgaria) Let $(a_n)_{n \geq 1}$ be the sequence of positive integers defined by $a_1 = 2$ and $a_{n+1} = a_n^2 - a_n + 1$ for $n \in \mathbb{N}$. Show that any two terms of this sequence are relatively prime.

18. * As usual, let $(F_n)_{n \geq 1}$ be the Fibonacci sequence, i.e., the sequence given by $F_1 = F_2 = 1$ and $F_{k+2} = F_{k+1} + F_k$ for every $k \in \mathbb{N}$. Do the following items:

(a) Prove that $\gcd(F_n, F_{n+1}) = 1$ for every $n \in \mathbb{N}$.
(b) Prove that, for all $n, k \in \mathbb{N}$, with $n > 1$, one has $F_{n+k} = F_{n-1}F_k + F_n F_{k+1}$.
(c) Use item (b) to successively conclude that, for all $n, q, r \in \mathbb{N}$:

    (i) $\gcd(F_{nq}, F_n) = F_n$.
    (ii) $\gcd(F_{nq-1}, F_n) = 1$.
    (iii) $\gcd(F_{nq+r}, F_n) = \gcd(F_r, F_n)$.

(d) Show that $\gcd(F_n, F_m) = F_{\gcd(m,n)}$ for all $m, n \in \mathbb{N}$.

19. (Croatia) Let $(a_n)_{n \geq 1}$ be a sequence of integers such that $a_1 = 1$ and $a_{n+2} = a_2 a_{n+1} + a_n$ for every $n \in \mathbb{N}$. Prove that $\gcd(a_m, a_n) = a_{\gcd(m,n)}$.

20. Given relatively prime natural numbers a and b, establish the following results:

(a) Every natural number $n > ab$ can be written as $n = ax + by$, with $x, y \in \mathbb{N}$.
(b) The natural number ab cannot be written as in (a).
(c) Every integer $n > ab - a - b$ can be written in the form $n = ax + by$, for some nonnegative integers x and y.
(d) The integer $ab - a - b$ cannot be written as in (c).
(e) There are exactly $\frac{1}{2}(a - 1)(b - 1)$ nonnegative integers which cannot be written in the form $ax + by$, for some nonnegative integers x and y.

21. (IMO—adapted) Let a, b and c be pairwise relatively prime natural numbers.

(a) Show that there do not exist $x, y, z \in \mathbb{Z}_+$ such that $2abc - ab - bc - ca = xbc + yac + zab$.
(b) If $n \in \mathbb{N}$ satisfies $n > abc - a - bc$, use the result of the previous problem to establish the existence of $x, t \in \mathbb{Z}_+$ such that $n = xbc + ta$, with $0 \leq x \leq a - 1$.
(c) Show that the integer t of item (b) is greater than $bc - b - c$, and use the result of the previous problem once again to establish the existence of $y, z \in \mathbb{Z}_+$ for which $t = bz + cy$.
(d) If $n \in \mathbb{N}$ satisfies $n > 2abc - ab - bc - ca$, use the results of items (b) and (c) to show that there exist $x, y, z \in \mathbb{Z}_+$ such that $n = xbc + yac + zab$.

22. (IMO—shortlist) Generalize the result of the previous problem in the following way: if $a_1, a_2, \ldots, a_n$ are pairwise relatively prime natural numbers and

$$A = \left\{ \sum_{j=1}^{n} x_j a_1 \ldots \widehat{a}_j \ldots a_n; \ x_1, \ldots, x_n \in \mathbb{Z}_+ \right\}$$

(here, $\widehat{a}_j$ indicates that a_j is not present in the product $x_j a_1 \ldots \widehat{a}_j \ldots a_n$), then

$$\left(n - 1 - \sum_{i=1}^{n} \frac{1}{a_i} \right) \prod_{i=1}^{n} a_i$$

is the greatest natural number which does not belong to A.

23. It is interesting to ask how efficient Euclid's algorithm is, upon computing the gcd of two nonzero integers. The purpose of this problem is to shed light on this question, by proving a result of Lamé.[5] To this end, do the following items:

   (a) If F_k is the k-th Fibonacci number and $n \in \mathbb{N}$, prove that $F_{n+5} > 10F_n$. Then, deduce that $F_{5n+2} > 10^n$, so that the decimal representation of F_{5n+2} has at least $n + 1$ digits.

   (b) Given natural numbers $a > b$, assume that Euclid's algorithm takes n steps to compute $\gcd(a, b)$. Prove that $b \geq F_{n-1}$.

   (c) Prove **Lamé's theorem**: given natural numbers $a > b$, the number of steps needed in Euclid's algorithm to compute $\gcd(a, b)$ is at most five times the number of decimal digits of b.

24. (Brazil) Let $n > 1$ be an integer. Prove that

$$1 + \frac{1}{2} + \frac{1}{3} + \cdots + \frac{1}{n}$$

is never an integer.

6.3   Prime Numbers

An integer $p > 1$ is **prime** if its only positive divisors are 1 and p. An integer $a > 1$ which is not prime is said to be **composite**. We shall prove in a few moments (cf. Theorem 6.38) that the set of primes is infinite. For the time being, and for the reader's convenience, we list below all prime numbers less than 50; their *primalities* can be checked directly and without difficulty (see, also, the discussion below on the *Eratosthenes' sieve*):

$$2, 3, 5, 7, 11, 13, 17, 19, 23, 29, 31, 37, 41, 43, 47.$$

   Our purpose in this section is to study the elementary properties of prime numbers, paying special attention to their relation to composite numbers. We start with the following auxiliary result, which is known as **Euclid's lemma**.

Lemma 6.35 (Euclid) *Every integer $n > 1$ can be written as a product of finitely many primes, not necessarily distinct.*[6]

Proof Let us make induction on n. For $n = 2$ there is nothing to do, for 2 is prime. Now, let $m > 2$ be a given integer and assume that every integer n such that $2 \leq$

---

[5] After Gabriel Lamé, French mathematician of the nineteenth century.

[6] Here, for the sake of simplifying the writing, we say that a prime is a product of one prime. This is in accordance with the general definition of $\prod_{j=1}^{n} a_j$, for which $\prod_{j=1}^{1} a_j = a_1$.

$n < m$ can be written as a product of finitely many primes; we shall prove that the same is true for m itself, and to this end we distinguish two distinct cases:

(i) if m is prime, there is nothing to do.
(ii) If m is not prime, there exist integers a and b such that $m = ab$, with $1 < a, b < m$. By induction hypothesis, each of a and b can be written as a product of finitely many primes, say $a = p_1 \ldots p_k$ and $b = q_1 \cdots q_l$, with $k, l \geq 1$ and $p_1, \ldots, p_k, q_1, \ldots, q_l$ primes. Therefore,

$$m = ab = p_1 \ldots p_k q_1 \cdots q_l,$$

so that m is also a product of finitely many primes.

□

A useful consequence of the previous lemma is the following criterion for searching prime divisors of a composite numbers, due to the Greek mathematician of the third century BC Eratosthenes of Cyrene.[7]

Corollary 6.36 (Eratosthenes) *If an integer $n > 1$ is composite, then n has a prime divisor p such that $p \leq \sqrt{n}$.*

Proof Since n is composite, we can write $n = ab$, with a and b integers such that $1 < a \leq b$. Letting p be a prime divisor of a (which exists, by Euclid's lemma), it follows that $p \mid n$ and

$$p^2 \leq a^2 \leq ab = n.$$

Therefore, $p \leq \sqrt{n}$. □

The previous corollary can be successfully used to establish the primality of a small natural number, as shown by the coming example. The method involved is known in mathematical literature as **Eratosthenes' sieve**, and consists in dividing the natural number at scrutiny by each of the primes which are less that its square root. According to Eratosthenes' theorem, if none of such divisions is exact, then the number is prime.

Example 6.37 Use the previous result to prove that 641 is prime.

Proof Firstly, note that $25 < \sqrt{641} < 26$. Hence, if 641 is composite, then Eratosthenes theorem guarantees that 641 must possess a prime divisor $p \leq 25$, so that

$$p \in \{2, 3, 5, 7, 11, 13, 17, 19, 23\}.$$

---

[7]Apart from his work in Number Theory, another important achievement of Eratosthenes was an indirect measurement of the Earth's diameter, with an impressive degree of accuracy for his time.

Nevertheless, it is immediate to verify that, upon dividing 641 by each of the primes above, no division is exact (i.e., leaves zero remainder). Therefore, 641 is indeed a prime number.   □

A slight modification of the procedure above can be used to find prime numbers *recursively*. For example, imagine one wants to find all prime numbers less than or equal to 150. One starts by erasing, from the list of odd numbers from 3 to 150, those which are multiples of 3. After that, the least number greater than 3 and not erased is 5, which is then prime; thus, we perform a second step, by erasing all multiples of 5. This step leaves 7 as the least number not erased, so that 7 is prime; in the third step, we then erase all multiples of 7. Continuing to proceed this way, we get the following list as result:

$$
\begin{array}{cccccccccccccc}
3 & 5 & 7 & \not{9} & 11 & 13 & \not{15} & 17 & 19 & \not{21} & 23 & \not{25} & \not{27} & 29 \\
31 & \not{33} & \not{35} & 37 & \not{39} & 41 & 43 & \not{45} & 47 & \not{49} & \not{51} & 53 & \not{55} & \not{57} \\
59 & 61 & \not{63} & \not{65} & 67 & \not{69} & 71 & 73 & \not{75} & \not{77} & 79 & \not{81} & 83 & \not{85} \\
\not{87} & 89 & \not{91} & \not{93} & \not{95} & 97 & \not{99} & 101 & 103 & \not{105} & 107 & 109 & \not{111} & 113 \\
\not{115} & \not{117} & \not{119} & \not{121} & \not{123} & \not{125} & 127 & \not{129} & 131 & \not{133} & \not{135} & 137 & 139 & \not{141} \\
\not{143} & \not{145} & \not{147} & 149 & 151 & \not{153} & \not{155} & 157 & \not{159} & \not{161} & 163 & \not{165} & 167 & \not{169}
\end{array}
$$

The numbers which were not erased are prime.

In spite of the above, the importance of Eratosthenes' sieve lies on a theoretical, rather than practical, level. For instance, in order to decide whether or not 999997 is prime, we would have to divide it by all prime numbers less than $\sqrt{999997} \cong 1000$, which is not feasible without the aid of computer programming.[8]

Back to Euclid's Lemma 6.35, if we gather together equal primes we conclude that every integer $n > 1$ can be write in the form $n = p_1^{\alpha_1} \ldots p_k^{\alpha_k}$, with $p_1, \ldots, p_k$ pairwise distinct primes and $\alpha_1, \ldots, \alpha_k \in \mathbb{N}$. We shall see in Theorem 6.43 that, up to a permutation of factors, such a way of writing n is unique. However, before we do that, let us prove another result of Euclid, which shows that the set of primes is infinite.

Theorem 6.38 (Euclid)   *The set of prime numbers is infinite.*

Proof By induction on n, let us prove that if $\mathbb{N}$ contains n primes, then it contains $n + 1$ primes. Suppose, then, that $p_1, \ldots, p_n$ are pairwise distinct primes, and let

$$
m = p_1 \ldots p_n + 1.
$$

---

[8] Actually, the situation here is even more dramatic. With a little more work, it is possible to show that Eratosthenes' sieve is not a **polynomial algorithm**. This means that the quantity of arithmetic operations one needs to perform in order to compute the n-th prime with the aid of it grows exponentially with n. Thus, even with the aid of a powerful computer, Eratosthenes' sieve is not a feasible *primality algorithm* for large numbers.

By Euclid's Lemma 6.35, there exists a prime p such that $p \mid m$. If $p = p_i$ for some $1 \leq i \leq n$, then $p \mid p_1 \ldots p_n$. It follows from Proposition 6.5 that p divides the difference $m - p_1 \ldots p_n$, i.e., $p \mid 1$, which is an absurd. Therefore, p is a prime number distinct from all of the p_i's, so that we have at least $n + 1$ primes in $\mathbb{N}$. □

Remark 6.39 In spite of the above result, there are arbitrarily large gaps in the sequence of primes. Indeed, given an integer $k \geq 3$, for all of the consecutive and positive integers $a + 2, a + 3, \ldots, a + k$ to be composite it suffices that a be a common multiple of $2, 3, \ldots, k$. We shall have more to say on the distribution of primes along $\mathbb{N}$ in Chap. 9.

As the next example shows, small variations of the argument used in the proof of Euclid's theorem allow us to prove that several arithmetic progressions contain infinitely many primes among their terms.[9]

Example 6.40 Prove that there are infinitely many primes of the form $4k - 1$.

Proof Assume that there is only a finite number of primes of the form $4k - 1$, say $p_1 = 3$, $p_2 = 7$, $p_3 = 11$, $\ldots$, p_t, and consider the number

$$m = 4p_1 p_2 \ldots p_t - 1.$$

Then, $m > 1$ and, letting $m' = p_1 p_2 \ldots p_t$, we have $m = 4m' - 1$.

On the other hand, Lemma 6.35 guarantees the existence of odd primes $q_1, \ldots, q_l$ such that $m = q_1 \ldots q_l$. Now, observe that every odd prime is of one of the forms $4q' - 1$ or $4q' + 1$, for some $q' \in \mathbb{Z}$. If $q_i = 4q_i' + 1$ for $1 \leq i \leq t$, we would get

$$m = q_1 \ldots q_l = (4q_1' + 1) \ldots (4q_l' + 1) = 4q + 1,$$

for some $q \in \mathbb{N}$. But this would contradict the fact that $m = 4m' - 1$. Hence, there exists at least one index $1 \leq i \leq l$ such that $q_i = 4q_i' - 1$.

Finally, since $p_1, p_2, \ldots, p_t$ are all primes of the form $4k - 1$, we must have $q_i = p_j$ for some $1 \leq j \leq t$. However, since $q_i \mid m$, we conclude that p_j should be a divisor of $m = 4p_1 p_2 \ldots p_t - 1$, which is a contradiction. □

For the proof of Theorem 6.43 we shall need the following auxiliary result, which is important in itself.

Lemma 6.41 *If $a_1, \ldots, a_n \in \mathbb{N}$ and p is a prime such that $p \mid a_1 a_2 \ldots a_n$, then there exists $1 \leq i \leq n$ such that $p \mid a_i$. In particular, if $a_1, \ldots, a_n$ are also primes, then there exists $1 \leq i \leq n$ such that $p = a_i$.*

---

[9] Actually, it is possible to prove a much more general result, due to the Dirichlet, which states that every infinite and nonconstant arithmetic progression of naturals contains infinitely many primes, provided its first term and its common difference are relatively prime. We shall see a relevant particular case of Dirichlet's theorem at Sect. 20.2.

Proof We shall prove the case $n = 2$, the general case being totally analogous. Assume that $p \mid (ab)$ but $p \nmid a$, and let $d = \gcd(a, p)$. Since $d \mid p$, we have $d = 1$ or $d = p$; however, since $p \nmid a$ and $d \mid a$, we conclude that $d \neq p$, and hence $d = 1$. Thus, since $p \mid (ab)$ and $\gcd(a, p) = 1$, it follows from item (a) of Proposition 6.22 that $p \mid b$.

The particular case follows at once from the first part, together with the very definition of prime number. $\square$

Example 6.42 If p is a prime number, then p divides each of the binomial numbers $\binom{p}{k}$, for every integer $1 \leq k \leq p - 1$.

Proof We can restrict to the case of an odd p. Since

$$\binom{p}{k} = \frac{p!}{k!(p-k)!} = \frac{(k+1)\ldots(p-1)p}{(p-k)!}$$

is an integer, we have that $(p - k)!$ divides $(k + 1)\ldots(p - 1)p$. Now, since p is prime and $p \nmid 1, 2, \ldots, p - k$, the previous lemma guarantees that $p \nmid (p - k)!$, and hence $\gcd(p, (p - k)!) = 1$. Therefore, item (a) of Proposition 6.22 shows that $(p - k)!$ must divide $(k + 1)\ldots(p - 1)$. In turn, this fact allows us to write

$$\binom{p}{k} = \underbrace{\frac{(k+1)\ldots(p-1)}{(p-k)!}}_{\in \mathbb{N}} \cdot p,$$

so that $\binom{p}{k}$ is a multiple of p. $\square$

The coming result is the cornerstone of elementary Number Theory, and is known as the **Fundamental Theorem of Arithmetic**. Whenever convenient, we shall abbreviate it as **FTA**.

Theorem 6.43 *Every integer $n > 1$ can be written as a product of powers of distinct primes. Moreover, such a decomposition of n is unique in the following sense: if $n = p_1^{\alpha_1} \ldots p_k^{\alpha_k} = q_1^{\beta_1} \ldots q_l^{\beta_l}$, with $p_1 < \cdots < p_k$ and $q_1 < \cdots < q_l$ being primes and $\alpha_i, \beta_j \geq 1$ being integers, then $k = l$ and $p_i = q_i$, $\alpha_i = \beta_i$ for $1 \leq i \leq k$.*

Proof The existence was established in the paragraph preceding Theorem 6.38.

For the uniqueness, let the integer $n > 1$ have two decompositions as in the statement. Since $p_1 \mid n$, we have that $p_1 \mid q_1^{\beta_1} \ldots q_l^{\beta_l}$, and the previous lemma guarantees the existence of an index $1 \leq j \leq l$ such that $p_1 = q_j$. On the other hand, since $q_1 \mid n$, we have that $q_1 \mid p_1^{\alpha_1} \ldots p_k^{\alpha_k}$; again by the previous lemma, there exists an index $1 \leq i \leq k$ such that $q_1 = p_i$. Thus, $p_1 = q_j \geq q_1 = p_i \geq p_1$, from where we get $p_1 = q_1$ and, hence,

$$n = p_1^{\alpha_1} p_2^{\alpha_2} \ldots p_k^{\alpha_k} = p_1^{\beta_1} q_2^{\beta_2} \ldots q_l^{\beta_l}.$$

We shall now prove that $\alpha_1 = \beta_1$. By contradiction, assume that $\alpha_1 \neq \beta_1$. We have two different cases to consider:

(i) $\alpha_1 < \beta_1$: then $p_2^{\alpha_2} \ldots p_k^{\alpha_k} = p_1^{\beta_1 - \alpha_1} q_2^{\beta_2} \ldots q_l^{\beta_l}$, so that $p_1 \mid p_2^{\alpha_2} \ldots p_k^{\alpha_k}$. Arguing as above, there would exist an index $2 \leq i \leq k$ such that $p_1 = p_i$, which is an absurd.

(ii) $\alpha_1 > \beta_1$: then $p_1^{\alpha_1 - \beta_1} p_2^{\alpha_2} \ldots p_k^{\alpha_k} = q_2^{\beta_2} \ldots q_l^{\beta_l}$. Again as above, we would get $p_1 = q_j$ for some index $2 \leq j \leq l$, and this would give us the contradiction $p_1 = q_j > q_1 = p_1$.

Thus, $\alpha_1 = \beta_1$, and it follows that $p_2^{\alpha_2} \ldots p_k^{\alpha_k} = q_2^{\beta_2} \ldots q_l^{\beta_l}$.

Repeating the above reasoning several times (there is an induction here, which shall not be formalized for the sake of clearness), we successively conclude that $p_2 = q_2$ and $\alpha_2 = \beta_2$, $p_3 = q_3$ and $\alpha_3 = \beta_3$, etc. At the end, if $k < l$ we would get $1 = q_{k+1}^{\beta_{k+1}} \ldots q_l^{\beta_l}$, which is clearly impossible. Likewise, $k > l$ is also impossible, so that $k = l$ and there is nothing left to do. □

The FTA will be almost omnipresent along these notes. However, it is instructive to the reader that we pause for a moment to illustrate how one can profitably use such a result in concrete situations.

Example 6.44 The natural numbers x and y are such that $3x^2 + x = 4y^2 + y$. Prove that $x - y$ is a perfect square.

Proof Let p be a prime number and p^a, p^b and p^c be the greatest powers of p dividing x, y and $x - y$, respectively (here we are already invoking the FTA). Let us momentarily assume that $a \leq b$. Then, $p^{2a} \mid x^2$, y^2, and it follows from item (c) of Proposition 6.5 that $p^{2a} \mid (4y^2 - 3x^2)$. Since $x - y = 4y^2 - 3x^2$, we then have that $p^{2a} \mid (x - y)$, and the uniqueness part of the FTA gives $c \geq 2a$. On the other hand, writing

$$x^2 = (x - y) - 4(y^2 - x^2) = (x - y)[1 + 4(y + x)]$$

and recalling the way we chose c, we conclude that $p^c \mid x^2$. Therefore, once more from the uniqueness part of the FTA, we obtain $c \leq 2a$. Hence, $c = 2a$, which is an even number. If $a \geq b$ we conclude, quite similarly, that $c = 2b$. Finally, since the prime p was arbitrarily chosen, the FTA presents $x - y$ as the product of powers of primes with even exponents. Therefore, $x - y$ is a perfect square. □

The presentation of an integer $n > 1$ as the product of powers of distinct primes is said to be its **canonical factorisation** or **decomposition** in prime powers. The following corollaries collect some useful consequences of the existence of such a factorisation.

Corollary 6.45 *Let a, b, m and n be natural numbers for which $a^m = b^n$. If $\gcd(m, n) = 1$, then there exists $c \in \mathbb{N}$ such that $a = c^n$ and $b = c^m$.*

Proof It is clear that a and b are divisible by exactly the same primes, say $p_1 <$ $\ldots < p_k$. Let $a = p_1^{\alpha_1} \ldots p_k^{\alpha_k}$ and $b = p_1^{\beta_1} \ldots p_k^{\beta_k}$, with $\alpha_i, \beta_i \geq 1$, be the canonical decompositions of a and b. Then,

$$p_1^{m\alpha_1} \ldots p_k^{m\alpha_k} = a^m = b^n = p_1^{n\beta_1} \ldots p_k^{n\beta_k},$$

and the uniqueness part of the FTA gives

$$m\alpha_i = n\beta_i, \ \forall \ 1 \leq i \leq k. \tag{6.8}$$

Thus, $m \mid n\beta_i$ and, since m and n are relatively prime, item (a) of Proposition 6.22 allows us to conclude that $m \mid \beta_i$. Letting $u_i \in \mathbb{N}$ be such that $\beta_i = mu_i$, (6.8) gives $\alpha_i = nu_i$. Therefore, letting $c = p_1^{u_1} \ldots p_k^{u_k}$, we obtain

$$a = p_1^{nu_1} \ldots p_k^{nu_k} = c^n \text{ and } b = p_1^{mu_1} \ldots p_k^{mu_k} = c^m.$$

$\square$

Let us now learn how to obtain the canonical decompositions of the positive divisors of an integer greater than 1 and the gcd and lcm of two nonzero integers.

Corollary 6.46 *If* $n = p_1^{\beta_1} \ldots p_k^{\beta_k}$ *is the canonical decomposition of the integer* $n > 1$, *then the positive divisors of* n *are the natural numbers of the form* $p_1^{\alpha_1} \ldots p_k^{\alpha_k}$, *with* $0 \leq \alpha_i \leq \beta_i$ *for* $1 \leq i \leq k$. *In particular, letting* $d(n)$ *denote the number of positive divisors of* n, *then*

$$d(n) = \prod_{i=1}^{k} (\beta_i + 1). \tag{6.9}$$

Proof If $d > 1$ is a divisor of n and p is a prime that divides d, then p also divides n, so that p is equal to one of the primes $p_1, \ldots, p_k$. However, since this is valid for every prime divisor of d, it follows that $d = p_1^{\alpha_1} \ldots p_k^{\alpha_k}$, with $\alpha_i \geq 0$ for every i (we not necessarily have $\alpha_i \geq 1$, for it may well happen that $p_i \nmid d$ for one or more values of i). Now, if $d' \in \mathbb{N}$ is such that $n = dd'$, then, as above, we get $d' = p_1^{\alpha_1'} \ldots p_k^{\alpha_k'}$, with $\alpha_i' \geq 0$ for every i. Therefore,

$$p_1^{\alpha_1+\alpha_1'} \ldots p_k^{\alpha_k+\alpha_k'} = dd' = n = p_1^{\beta_1} \ldots p_k^{\beta_k},$$

and the uniqueness part of the FTA gives $\alpha_i + \alpha_i' = \beta_i$ for every i. In particular, $\alpha_i \leq \beta_i$ for every i.

Conversely, it is immediate to show that every natural number d of the form $p_1^{\alpha_1} \ldots p_k^{\alpha_k}$, with $0 \leq \alpha_i \leq \beta_i$ for $1 \leq i \leq k$, is a positive divisor of n.

For what is left to do, what we did above, again in conjunction with the uniqueness part of the FTA, assures that the number of positive divisors of n

coincides with the number of sequences $(\alpha_1, \ldots, \alpha_k)$ of integers such that $0 \le \alpha_i \le \beta_i$ for $1 \le i \le k$. Since there are $\beta_i + 1$ possibilities for α_i, formula (6.9) follows at once from the fundamental principle of counting (cf. Corollary 1.9). □

The following example presents a simple application of the former result.

Example 6.47 Prove that the natural number n is a perfect square if and only if $d(n)$ is odd.

Proof Firstly, since $1 = 1^2$ and $d(1) = 1$ is odd, we can assume that $n > 1$. Then, let $n > 1$ be an integer and $n = p_1^{\beta_1} \ldots p_k^{\beta_k}$ be its canonical factorisation into prime powers.

If n is a perfect square, say $n = m^2$, with $m \in \mathbb{N}$, then $m \mid n$, so that $m = p_1^{\alpha_1} \ldots p_k^{\alpha_k}$, with $0 \le \alpha_i \le \beta_i$ for $1 \le i \le k$. Therefore,

$$p_1^{\beta_1} \ldots p_k^{\beta_k} = n = m^2 = p_1^{2\alpha_1} \ldots p_k^{2\alpha_k},$$

and the uniqueness part of the FTA gives $\beta_i = 2\alpha_i$ for $1 \le i \le k$. Thus, the previous corollary gives

$$d(n) = (\beta_1 + 1) \ldots (\beta_k + 1) = (2\alpha_1 + 1) \ldots (2\alpha_k + 1),$$

which is and odd number.

Conversely, let $d(n) = (\beta_1 + 1) \ldots (\beta_k + 1)$ be odd. Then, $\beta_i + 1$ is odd for $1 \le i \le k$, which in turn gives β_i even for $1 \le i \le k$. Writing $\beta_i = 2\alpha_i$, we get

$$n = p_1^{2\alpha_1} \ldots p_k^{2\alpha_k} = \left(p_1^{\alpha_1} \ldots p_k^{\alpha_k}\right)^2 = m^2,$$

with $m = p_1^{\alpha_1} \ldots p_k^{\alpha_k}$. □

For the next corollary, it is useful to extend the canonical decomposition of an integer greater than 1 by allowing exponents equal to 0. For instance, we can write $48 = 2^4 \cdot 3$ and $270 = 2 \cdot 3^3 \cdot 5$ by using the prime numbers 2, 3 and 5 in both cases, so that

$$48 = 2^4 \cdot 3 \cdot 5^0 \quad \text{and} \quad 270 = 2 \cdot 3^3 \cdot 5.$$

Corollary 6.48 *Let* $a, b > 1$ *be natural numbers such that* $a = p_1^{\alpha_1} \ldots p_k^{\alpha_k}$ *and* $b = p_1^{\beta_1} \ldots p_k^{\beta_k}$, *with* $p_1 < \cdots < p_k$ *prime numbers and* $\alpha_i, \beta_i \ge 0$ *for* $1 \le i \le k$. *Then,*

$$\gcd(a, b) = \prod_{i=1}^{k} p_i^{\min\{\alpha_i, \beta_i\}} \quad \text{and} \quad \operatorname{lcm}(a, b) = \prod_{i=1}^{k} p_i^{\max\{\alpha_i, \beta_i\}}.$$

Proof Let us do the proof for the gcd of a and b (that for the lcm of a and b is totally analogous). Since $\min\{\alpha_i, \beta_i\} \leq \alpha_i, \beta_i$ for every i, the number $d = \prod_{i=1}^{k} p_i^{\min\{\alpha_i, \beta_i\}}$ divides both a and b. Now, let d' be a common positive divisor of a and b. Corollary 6.46 gives $d' = p_1^{\gamma_1} \ldots p_k^{\gamma_k}$, with $\gamma_i \geq 0$ for every i; moreover, since, $d' \mid a, b$, it also gives $\gamma_i \leq \alpha_i, \beta_i$ for every i. Hence, we have $\gamma_i \leq \min\{\alpha_i, \beta_i\}$ for every i, so that (once more from Corollary 6.46) $d' \mid d$. Therefore, $d = \gcd(a, b)$.

$\square$

Back to the last paragraph before the statement of the above corollary, we can now readily compute

$$\gcd(48, 270) = 2 \cdot 3 = 6 \quad \text{and} \quad \text{lcm}(48, 270) = 2^4 \cdot 3^3 \cdot 5 = 2160.$$

For what comes next, if a natural number n and a prime number p are such that $p \leq n$, then we obviously have that $p \mid n!$. The following result, due to A-M. Legendre,[10] teaches us how to compute the exponent of p in the canonical decomposition of $n!$. Accordingly, (6.10) is known as **Legendre's formula**.

Proposition 6.49 *Let $n > 1$ be natural and p be prime. If $e_p(n)$ denotes the exponent of p in the canonical factorisation of $n!$, then*

$$e_p(n) = \sum_{k \geq 1} \left\lfloor \frac{n}{p^k} \right\rfloor. \tag{6.10}$$

Proof Firstly, note that the above sum is actually finite, for $\lfloor \frac{n}{p^k} \rfloor = 0$ whenever $p^k > n$. Now, if $k \in \mathbb{N}$ and $p^k, 2p^k, \ldots, mp^k$ are the multiples of p^k which are less than or equal to n, then $mp^k \leq n < (m+1)p^k$ or, equivalently, $m \leq \frac{n}{p^k} < m+1$. Hence, m is the greatest integer which is less than or equal to $\frac{n}{p^k}$, i.e., $m = \lfloor \frac{n}{p^k} \rfloor$. In turn, this shows that for each integer $k \geq 1$ there are exactly

$$\left\lfloor \frac{n}{p^k} \right\rfloor - \left\lfloor \frac{n}{p^{k+1}} \right\rfloor$$

natural numbers which are less than or equal to n and are multiples of p^k but not of p^{k+1}. Each of these naturals contributes with exactly k factors p to $e_p(n)$, so that

$$e_p(n) = \sum_{k \geq 1} k \left(\left\lfloor \frac{n}{p^k} \right\rfloor - \left\lfloor \frac{n}{p^{k+1}} \right\rfloor \right) = \sum_{k \geq 1} \left\lfloor \frac{n}{p^k} \right\rfloor.$$

$\square$

The two coming examples bring interesting applications of Legendre's formula.

---

[10] After Adrien-Marie Legendre, French mathematician of the eighteenth and nineteenth centuries.

Example 6.50 (Soviet Union) In how many zeros does 1000! end?

Proof We have to find the greatest power of 10 that divides 1000!. To this end, and writing $1000! = 2^{e_2(1000)} \cdot 5^{e_5(1000)} m$, with $m \in \mathbb{N}$, it is enough to compute the smallest of $e_2(1000)$ and $e_5(1000)$. It is clear that the smallest of these is $e_5(1000)$, which can be promptly computed with the aid of Legendre's formula (and taking into account that $5^5 > 1000$):

$$e_5(1000) = \sum_{j=1}^{4} \left\lfloor \frac{1000}{5^j} \right\rfloor = 200 + 40 + 8 + 1 = 249.$$

Therefore, 1000! ends in 249 zeros. □

Example 6.51 (Yugoslavia) Prove that $2^n \nmid n!$, for each $n \in \mathbb{N}$.

Proof The result is clearly true for $n = 1$. For $n > 1$, Legendre's theorem assures that the greatest power of 2 dividing $n!$ has exponent $k = \sum_{j \geq 1} \left\lfloor \frac{n}{2^j} \right\rfloor$. However, $\lfloor x \rfloor \leq x$ for every x and $\left\lfloor \frac{n}{2^j} \right\rfloor = 0$ for every sufficiently large j; therefore, by means of the formula for the sum of the terms of a geometric series (cf. Proposition 7.38 of [8], for instance), we get

$$k = \sum_{j \geq 1} \left\lfloor \frac{n}{2^j} \right\rfloor < \sum_{j \geq 1} \frac{n}{2^j} = n.$$

Hence, $k < n$, so that $2^n \nmid n!$. □

Problems: Sect. 6.3

1. Let p and q be two consecutive odd primes. Prove that there exist integers $a, b, c > 1$, not necessarily distinct, such that $p + q = abc$.
2. (IMO—adapted) Let $k \in \mathbb{N}$ be such that $p = 3k + 2$ is prime. If m and n are natural numbers such that

$$\frac{m}{n} = 1 - \frac{1}{2} + \frac{1}{3} - \frac{1}{4} + \cdots - \frac{1}{2k} + \frac{1}{2k+1},$$

prove that $p \mid m$.
3. (Hungary) Given an odd prime p, find all $x, y \in \mathbb{N}$ such that $\frac{2}{p} = \frac{1}{x} + \frac{1}{y}$.
4. (IMO) Find all $n \in \mathbb{N}$ such that $n(n+1)(n+2)(n+3)$ has exactly three distinct prime divisors.
5. (Russia) Let $(a_n)_{n \geq 1}$ be a sequence of natural numbers such that $\gcd(a_i, a_j) = \gcd(i, j)$ whenever $i \neq j$. Prove that $a_n = n$ for every $n \geq 1$.

6. * We say that a natural n is **square free** if $n = 1$ or $n = p_1 \ldots p_k$, where $p_1 < \cdots < p_k$ are prime numbers. In this respect, do the following items:

   (a) Prove that every integer $n > 1$ can be uniquely written in the form $n = ab^2$, with $a, b \in \mathbb{N}$ and a being square free.
   (b) Use item (a) to show that, if there were exactly k prime numbers, then there would exist exactly 2^k square free integers and that the number of positive integers less than or equal to n would be at most $2^k \sqrt{n}$.
   (c) Use item (b) to give another proof of the fact that there are infinitely many primes.

7. Show that there are infinitely many primes of each of the forms $3k + 2$ and $6k + 5$.

8. (Yugoslavia) For each $n \in \mathbb{N}$, prove that the number $2^{2^n} - 1$ has at least n distinct prime factors.

9. If p is an odd prime, prove[11] that p divides $\binom{2p}{p} - 2$.

10. * Let a and n be positive integers, with $a > 1$. Do the following items:

    (a) If $a^n + 1$ is prime, show that n is a power of 2 (we shall see in Problem 7, page 252, that the converse is not true).[12]
    (b) If $n > 1$ and $a^n - 1$ is prime, show that n is itself prime and $a = 2$ (the converse is not true, as shown by the example $2^{11} - 1 = 23 \cdot 89$).

11. (IMO) Find all naturals $n > 6$ for which the natural numbers less than n and relatively prime with n form an arithmetic progression.

12. (IMO.) Let k, m and n be naturals such that $m + k + 1$ is a prime number greater than $n + 1$. If $c_s = s(s + 1)$ for each $s \in \mathbb{N}$, prove that

$$(c_{m+1} - c_k)(c_{m+2} - c_k) \ldots (c_{m+n} - c_k)$$

is divisible by $c_1 c_2 \cdots c_n$.

13. (Hungary) Let n be a given natural. Show that there are exactly $d(n^2)$ ordered pairs (u, v) with $u, v \in \mathbb{N}$ and such that $\mathrm{lcm}\,(u, v) = n$.

14. (IMO) Find all natural numbers a and b such that $a^{b^2} = b^a$.

15. (BMO) Find all $n \in \mathbb{N}$ such that $n = d_1^2 + d_2^2 + d_3^2 + d_4^2$, where $1 = d_1 < d_2 < d_3 < d_4$ are the smallest positive divisors of n.

16. (OIM) For each $n \in \mathbb{N}$, let $1 = d_1 < \cdots < d_k = n$ be the positive divisors of n. Find all n satisfying the following conditions:

    (a) $k \geq 15$.
    (b) $n = d_{13} + d_{14} + d_{15}$.
    (c) $(d_5 + 1)^3 = d_{15} + 1$.

---

[11] For a stronger version of this result, see Problem 15, page 310.

[12] Prime numbers of the form $2^{2^n} + 1$ are called **Fermat primes**. They enter decisively into Gauss' description of which regular polygons can be constructed with straightedge and compass (cf. [20], for instance).

17. (Japan)[13] For each integer $n > 1$, let $I(n)$ be the sum of the greatest odd divisors of the numbers $1, 2, \ldots, n$ and $T(n) = 1 + 2 + \cdots + n$. Prove that there exist infinitely many values of n for which $3I(n) = 2T(n)$.

18. (Australia) Find all naturals n for which $d(n) = \frac{n}{3}$, where $d(n)$ stands for the number of positive divisors of n.

19. (IMO) Let m and n be arbitrary nonnegative integers. Prove that

$$\frac{(2m)!(2n)!}{m!n!(m+n)!}$$

is always an integer.

20. For each positive integer n, let $a_n = \binom{2n}{n}$.

   (a) Show that the binomial number a_n is always even.
   (b) Prove that $4 \mid a_n$ if and only if n is not a power of 2.

21. (ORM) We say that a positive divisor d of a natural number n is *special* if $d < \sqrt{n}$ and there are no perfect squares between d and $\frac{n}{d}$. Prove that every natural n has at most one special divisor.

22. (Romania) Let n be a natural number whose binary representation has exactly k digits 1. Prove that 2^{n-k} divides $n!$.

23. Find all natural numbers a, b and k such that a and b are relatively prime and $(a + kb)(b + ka)$ is a power of a prime.

24. (Brazil) Show that there exists a set A of positive integers satisfying the following conditions:

   (a) A has 1000 elements.
   (b) The sum on any number of distinct elements of A (at least one such element) is not a perfect power of exponent greater than 1.

25. (Brazil—adapted) For each integer $n > 1$, let $p(n)$ be the greatest prime divisor of n. Do the following items:

   (a) If q is an odd prime, show that we cannot have $p(q^{2^k}) > p(q^{2^k} + 1)$ for every integer $k \geq 0$.
   (b) If $p(q^{2^k}) < p(q^{2^k} + 1)$ and k is minimum with such a property, show that $p(q^{2^k} - 1) < p(q^{2^k})$.
   (c) Show that there exist infinitely many naturals n satisfying $p(n - 1) < p(n) < p(n + 1)$.

26. Let $p_1 = 2$, $p_2 = 3$, $p_3 = 5$, $\ldots$ be the sequence of prime numbers. For each $k \in \mathbb{N}$, let $x_k = p_1 p_2 \ldots p_k + 1$ and $A = \{x_1, x_2, \ldots\}$. For a fixed $m \in \mathbb{N}$, prove that one can choose a subset B of A satisfying the following properties:

   (a) $|B| = m$.
   (b) Any two elements of B are relatively prime.
   (c) The sum of the elements of any subset of B having more than one element is not prime.

---

[13]For the converse to this problem, see Problem 5, page 226.

Chapter 7
Diophantine Equations

Our purpose in this chapter is to study some elementary diophantine equations. Among these we highlight Pythagoras' and Pell's equation, for which we characterize all solutions. We also present to the reader the important *Fermat's descent method*, which provides a frequently useful tool for showing that certain diophantine equations do not possess *nontrivial* solutions, in a way to be made precise. The aforementioned method is one of the major legacies of Pierre Simon de Fermat to Number Theory,[1] and will be frequently used hereafter.

7.1 Pythagorean Triples and Fermat's Descent

Let us start by studying the nonzero integer solutions of the equation $x^2 + y^2 = z^2$, which, thanks to Pythagoras' theorem of Euclidean Geometry, is known as **Pythagoras' equation**. After we find all such solutions, we shall see how to use that information to solve other interesting diophantine equations. The fundamental result is as follows.

Proposition 7.1 *The 3-tuples (x, y, z) of nonzero integers such that $x^2 + y^2 = z^2$ are given by:*

$$\begin{cases} x = 2uvd \\ y = (u^2 - v^2)d \\ z = (u^2 + v^2)d \end{cases} or \begin{cases} x = (u^2 - v^2)d \\ y = 2uvd \\ z = (u^2 + v^2)d, \end{cases} \tag{7.1}$$

where d, u and v are nonzero integers, with u and v of distinct parities and relatively prime.

---

[1]Other remarkable ones are Fermat's little theorem, which has already appeared in Problem 5, page 57, and will be proved by other methods in Sect. 10.2, and Fermat's theorem on sums of two squares, which will be the object of Sect. 12.4.

© Springer International Publishing AG, part of Springer Nature 2018

A. Caminha Muniz Neto, *An Excursion through Elementary Mathematics, Volume III*,
Problem Books in Mathematics, https://doi.org/10.1007/978-3-319-77977-5_7

Proof We can assume, without loss of generality, that $x, y, z > 0$. If $d = \gcd(x, y)$, then $d^2 \mid (x^2 + y^2)$, so that $d^2 \mid z^2$; therefore, $d \mid z$. Hence, there exist nonzero integers a, b and c for which $\gcd(a, b) = 1$ and $(x, y, z) = (da, db, dc)$. Moreover, since

$$x^2 + y^2 = z^2 \Leftrightarrow a^2 + b^2 = c^2,$$

we conclude that it suffices to find all nonzero integer solutions a, b, c of the above equation, subjected to the additional condition $\gcd(a, b) = 1$.

The requirement that $a^2 + b^2 = c^2$ and $\gcd(a, b) = 1$ easily give us (prove this!) $\gcd(a, c) = \gcd(b, c) = 1$. Now, recall (cf. Corollary 6.8) that the square of an integer t leaves remainder 0 or 1 upon division by 4, according to whether t is respectively even or odd. Therefore, if a and b are both odd, then $c^2 = a^2 + b^2$ will leave remainder 2 upon division by 4, which is a contradiction.

Since $\gcd(a, b) = 1$, we are left with two cases to consider: a is odd and b is even, or vice-versa. Let us look at the first case, the analysis of the second one being entirely analogous. Since a is odd, b is even and $c^2 = a^2 + b^2$, we conclude that c is also odd. Now, write

$$b^2 = (c - a)(c + a). \tag{7.2}$$

If $d' = \gcd(c - a, c + a)$, then d' divides both

$$(c + a) + (c - a) = 2c \text{ and } (c + a) - (c - a) = 2a$$

so that $d' \mid \gcd(2a, 2c) = 2$. However, since $c - a$ and $c + a$ are both even, we conclude that $d' = 2$, and we can write (7.2) as

$$\left(\frac{b}{2}\right)^2 = \left(\frac{c - a}{2}\right)\left(\frac{c + a}{2}\right),$$

with $\gcd\left(\frac{c-a}{2}, \frac{c+a}{2}\right) = 1$. Thus, item (b) of Corollary 6.23 assures the existence of relatively prime naturals u and v for which $c + a = 2u^2$ and $c - a = 2v^2$, so that

$$a = u^2 - v^2, \quad b = 2uv, \quad c = u^2 + v^2.$$

Furthermore, since $c = u^2 + v^2$ is odd, we see that u and v must have distinct parities.

Finally, direct substitution on the original equation shows that all 3-tuples as above are indeed solutions of Pythagoras' equation, and there is nothing left to do.
□

A 3-tuple (x, y, z) of positive integers such that $x^2 + y^2 = z^2$ is said to be a **pythagorean triple**, in allusion to the Greek mathematician and philosopher

Pythagoras of Samos and his famous theorem on right triangles. Indeed, and as we have already mentioned, such a triple (x, y, z) determines,[2] up to congruence, a single right triangle of legs of lengths x and y and hypotenuse of length z.

The coming example uses the previous result to find all integer solutions of another diophantine equation.

Example 7.2 Find all nonzero integer solutions of the equation $x^2 + y^2 = 2z^2$, with $x \neq \pm y$.

Solution In any such solution, x and y must be both even or both odd, for otherwise $x^2 + y^2$ would be odd, hence different from $2z^2$. Thus, taking $a = \frac{x+y}{2}$ and $b = \frac{x-y}{2}$, we have $a, b \in \mathbb{Z} \setminus \{0\}$ (since $x \neq \pm y$) and $x = a + b$, $y = a - b$. Substituting such expressions for x and y in the original equation, we conclude that

$$x^2 + y^2 = 2z^2 \Leftrightarrow a^2 + b^2 = z^2.$$

Since this last equation is that of Pythagoras, our previous result guarantees the existence of nonzero integers d, u and v, with u and v relatively prime and of distinct parities, such that

$$a = 2uvd, \quad b = (u^2 - v^2)d, \quad z = (u^2 + v^2)d$$

or

$$a = (u^2 - v^2)d, \quad b = 2uvd, \quad z = (u^2 + v^2)d.$$

Hence, the solutions (x, y, z) of the given equation, with $x \neq \pm y$, are of one of the forms below, where d, u and v satisfy the conditions prescribed above:

$$x = (u^2 - v^2 + 2uv)d, \quad y = (-u^2 + v^2 + 2uv)d, \quad z = (u^2 + v^2)d$$

or

$$x = (u^2 - v^2 + 2uv)d, \quad y = (u^2 - v^2 - 2uv)d, \quad z = (u^2 + v^2)d.$$

$\square$

The equations we have analysed up to this point are, in a certain sense, privileged, for they possess infinitely many nonzero integer solutions. Our next example will display a diophantine equation in x, y and z having $x = y = z = 0$ as its only integer solution. The proof we shall present is an application of **Fermat's descent method**, which can sometimes be profitably used to show that a diophantine equation does not have nonzero integer solutions.

---

[2]For two different proofs of this fact, see chapters 4 and 5 of [9], for instance.

Schematically, Fermat's method consists of the fulfillment of the following stages:

i Assume that the given equation has a solution in nonzero integers.

ii. From such an assumption, infer that it has a solution in nonzero integers which, in a certain sense (which will depend on the problem at hand) is *minimal*.

iii. Deduce the existence of yet another solution in nonzero integers which is *smaller* than the minimal one of item ii., thus arriving at a contradiction.

Example 7.3 Prove that the equation $3x^2 + y^2 = 2z^2$ does not have nonzero integer solutions.

Proof Firstly, observe that we cannot have exactly one of the integers x, y, z equal to 0. Hence, assume that the given equation possess a solution (x, y, z) with $x, y, z \in \mathbb{N}$ (for, if (x, y, z) is a solution, then so is $(\pm x, \pm y, \pm z)$). Among all such solutions (x, y, z), there must exist one for which z has the least possible value, say $x = a$, $y = b, z = c$. Let us work this solution.

If $3 \nmid b$, it follows from $3a^2 + b^2 = 2c^2$ that $3 \nmid c$. But then, Corollary 6.8 guarantees that b^2 and c^2 leave remainder 1 upon division by 3, and hence equality $3a^2 + b^2 = 2c^2$ gives us a contradiction. Therefore, $3 \mid b$, say $b = 3b_1$ for some $b_1 \in \mathbb{N}$, and it follows from

$$2c^2 = 3a^2 + b^2 = 3(a^2 + 3b_1^2)$$

that $3 \mid c$. Letting $c = 3c_1$, for some $c_1 \in \mathbb{N}$, the last equality above furnishes

$$6c_1^2 = a^2 + 3b_1^2,$$

so that $3 \mid a$, say $a = 3a_1$, with $a_1 \in \mathbb{N}$. Now, plugging $a = 3a_1$ in the last equality above, we get

$$2c_1^2 = 3a_1^2 + b_1^2,$$

so that (a_1, b_1, c_1) is another solution, in natural numbers, of the original equation. However, the relation $0 < c_1 = \frac{c}{3} < c$ is a contradiction, for we departed from a solution in natural numbers (a, b, c) for which the (natural) value c of z was the least possible one.

Since our original assumption (namely, that the given equation has *some* solution) leads to a contradiction, we are forced to conclude that the equation has no solution at all. $\square$

Back to Pythagoras' equation, a natural generalization of it, posed by Fermat, would be to study the more general diophantine equation below:

$$x^n + y^n = z^n. \tag{7.3}$$

Fermat himself claimed to have succeeded in using his descent method to prove that (7.3) has no nonzero integer solutions when $n > 2$, and for this reason such a result has eventually come to be known as **Fermat's last theorem**. Nevertheless, since his reasoning was never written down, some of the most brilliant minds of Mathematics tried to reproduce it, year after year. As astonishing as it may seen, apart from particular cases, all such attempts resulted unsuccessful, and this state of things forged the general perception that Fermat had probably made some mistake.

It took mankind almost 350 years to settle the issue. Indeed, it was only in the last decade of the twentieth century that the English mathematician Andrew Wiles entered History as the one to establish Fermat's conjecture. Wiles accomplishment came after several years of hard work and through two long and deep papers, the second of which in collaboration with his former student Richard Taylor. As is almost always the case with such profound results in modern Mathematics, Wiles and Taylor built on equally deep works of other prominent mathematicians, among the most notable of them we should mention the Japanese Yutaka Taniyama and Goro Shimura, and the German Gerhard Frey.[3]

Surprisingly enough, the descent method can be used to successfully approach the case $4 \mid n$ in **Fermat's equation** (7.3). We do this next.

Example 7.4 If the natural number n is a multiple of 4, then there does not exist nonzero integers x, y and z such that $x^n + y^n = z^n$.

Proof Let $n = 4k$, with $k \in \mathbb{N}$. From $x^n + y^n = z^n$, with $x, y, z \in \mathbb{Z} \setminus \{0\}$, we get

$$(x^k)^4 + (y^k)^4 = (z^{2k})^2,$$

so that (x^k, y^k, z^{2k}) is a solution of equation $a^4 + b^4 = c^2$ in nonzero integers. Thus, it suffices to show that this last equation does not admit nonzero integer solutions.

By the sake of contradiction, let us assume that there exist $a, b, c \in \mathbb{N}$ such that $a^4 + b^4 = c^2$. We can also suppose that a, b and c are chosen in such a way that there is no other solution (α, β, γ), with $\alpha, \beta, \gamma \in \mathbb{N}$ and $\gamma < c$ (as the reader surely expects, this is precisely the assumption that will give us a contradiction via the descent method).

Since (a^2, b^2, c) solves Pythagoras' equation, Proposition 7.1, together with the minimality of c, guarantees that $\gcd(a^2, b^2) = 1$, as well as that there exist relatively prime natural numbers u and v, of distinct parities and such that

$$a^2 = u^2 - v^2, \quad b^2 = 2uv, \quad c = u^2 + v^2$$

or

$$a^2 = 2uv, \quad b^2 = u^2 - v^2, \quad c = u^2 + v^2.$$

---

[3]For the reader interested in knowing a little more on this epic endeavor, we recommend first to run through the best seller of Singh [35], and then to turn to the marvelous book of professors Stewart and Tall [37].

Let us look at the first case above, the other one being totally analogous. In that case, a is odd and $a^2 + v^2 = u^2$. Therefore, by invoking once more the characterization of pythagorean triples (and the fact that $\gcd(u, v) = 1$) we conclude by the existence of relatively prime natural numbers p and q, of distinct parities and such that

$$a = p^2 - q^2, \quad v = 2pq, \quad u = p^2 + q^2.$$

Then,

$$b^2 = 2uv = 4pq(p^2 + q^2),$$

and since $\gcd(p, q) = 1$, it follows that both p and q are also relatively prime with $p^2 + q^2$. Hence, in order that $4pq(p^2 + q^2)$ is a perfect square, the only possibility is that all of p, q and $p^2 + q^2$ are also perfect squares, say

$$p = \alpha^2, \quad q = \beta^2, \quad p^2 + q^2 = \gamma^2,$$

with $\alpha, \beta, \gamma \in \mathbb{N}$. In turn, this finally gives

$$\alpha^4 + \beta^4 = p^2 + q^2 = \gamma^2,$$

with

$$c = u^2 + v^2 > u = p^2 + q^2 = \gamma^2 \geq \gamma,$$

thus contradicting the minimality of c.

Therefore, there are no nonzero solutions of $x^n + y^n = z^n$ when $4 \mid n$. □

The next section shall bring a nontrivial (and important) application of Fermat's descent. For the time being, let us close this section with a beautiful application of the characterization of pythagorean triples to a problem in Euclidean Geometry.

Example 7.5 (Crux—Adapted) Given a positive integer r, we want to compute the number of pairwise incongruent right triangles ABC, satisfying the following conditions:

(a) The radius of the circle inscribed in ABC equals r.
(b) The lengths of the sides of ABC are relatively prime integers.

Show that there are exactly 2^l such triangles, where l is the number of distinct prime factors of r.

Proof Let b and c be the legs and a be the hypotenuse of a right triangle whose inscribed circle has radius r. It is an easy exercise of Euclidean Geometry (which we leave to the reader) to show that $r = \frac{b+c-a}{2}$.

Since a, b and c are relatively prime and $b^2 + c^2 = a^2$, the characterization of pythagorean triples gives relatively prime integers $u > v > 0$, of distinct parities, for which

$$b = u^2 - v^2, \quad c = 2uv, \quad a = u^2 + v^2.$$

Thus, to count the number of triangles satisfying the given conditions is the same as to count the number of ordered paired (u, v) satisfying the above conditions. To this end, note that

$$r = \frac{b + c - a}{2} = v(u - v)$$

and that $\gcd(u, v) = 1 \Leftrightarrow \gcd(v, u - v) = 1$.

Letting $r = p_1^{\alpha_1} p_2^{\alpha_2} \cdots p_l^{\alpha_l}$, with $p_1 < p_2 < \cdots < p_l$ primes, we conclude that there are exactly 2^l distinct possibilities for v:

$$v = 1 \text{ or } v = p_{i_1}^{\alpha_{i_1}} \cdots p_{i_k}^{\alpha_{i_k}},$$

where $\{i_1, \ldots, i_k\}$ is a subset of $\{1, 2, \ldots, l\}$. For each such v, note that $u = v + \frac{r}{v}$ is completely determined and that $u = v + \frac{r}{v} > v$. We now have to check that u and v have distinct parities. To this end, assume, without loss of generality, that r is even (the analysis of the case of an odd r is quite similar). Then, v contains either all or none of the factors 2 of r. If v has all factors 2 of r, then $\frac{r}{v}$ is odd, and it follows that $u = v + \frac{r}{v}$ is also odd; if v is itself odd, then $\frac{r}{v}$ is even and $u = v + \frac{r}{v}$ is odd. In any event, u and v have, indeed, distinct parities.

In order to conclude that the number of right triangles we wish to count is precisely 2^l, it now suffices to show that, if (u_1, v_1) and (u_2, v_2) are two distinct choices for (u, v) as above, then the corresponding triangles are actually incongruent. To this end, assuming that $v_1 \neq v_2$ (the case $u_1 \neq u_2$ being totally analogous), we are going to show that the two corresponding right triangles have distinct hypotenuses. But this is straightforward:

$$u_1^2 + v_1^2 = u_2^2 + v_2^2 \Leftrightarrow \left(\frac{r}{v_1} + v_1\right)^2 + v_1^2 = \left(\frac{r}{v_2} + v_2\right)^2 + v_2^2$$

$$\Leftrightarrow \frac{r^2}{v_1^2} + 2v_1^2 = \frac{r^2}{v_2^2} + 2v_2^2$$

$$\Leftrightarrow r^2 \left(\frac{v_2^2 - v_1^2}{v_1^2 v_2^2}\right) = 2(v_2^2 - v_1^2)$$

$$\Leftrightarrow \frac{r}{v_1 v_2} = \sqrt{2},$$

which is an absurd. □

Problems: Sect. 7.1

1. (Brazil—adapted) Given $n, k \in \mathbb{N}$, prove that the equation $x^n + ky^n = z^{n+1}$ has infinitely many solutions in positive integers x, y and z.

2. Find all solutions, in positive integers, of the equation

$$(x + y)^2 + (y + z)^2 = (x + z)^2.$$

3. Show that the nonzero integer solutions of the equation $x^2 + 2y^2 = z^2$ are given by

$$x = \pm(u^2 - 2v^2)d, \quad y = \pm 2uvd, \quad z = \pm(u^2 + 2v^2)d,$$

where d, u and v are nonzero integers, with $\gcd(u, 2v) = 1$.

4. Show that none of the following equations have nonzero integer solutions:

   (a) $x^4 + 4y^4 = z^2$.
   (b) $x^4 + 2y^4 = z^2$.
   (c) $x^2 + y^2 = 3z^2$.
   (d) $x^3 + 5y^3 = 9z^3$.

5. Do the following items:

   (a) (Hungary) Show that there do not exist $x, y \in \mathbb{Q}$ such that $x^2 + xy + y^2 = 2$.
   (b) Find all $x, y \in \mathbb{Q}$ for which $x^2 + xy + y^2 = 1$.

6. Prove the following theorem of Euler: there does not exist nonzero integers w, x, y and z for which $w^2 + x^2 + y^2 = 7z^2$.

7. (Bulgaria) Prove that there does not exist rationals x, y and z such that

$$x^2 + y^2 + z^2 + 3(x + y + z) + 5 = 0.$$

8. Given[4] $a, b, c \in \mathbb{Q}$ such that $a + b\sqrt[3]{2} + c\sqrt[3]{4} \neq 0$, show that there exist $x, y, z \in \mathbb{Q}$ for which

$$\frac{1}{a + b\sqrt[3]{2} + c\sqrt[3]{4}} = x + y\sqrt[3]{2} + z\sqrt[3]{4}.$$

9. (IMO) Given $n \in \mathbb{N}$ and a circle of radius 1, show that we can choose n points $A_1, A_2, \ldots, A_n$ on it such that $\overline{A_i A_j}$ is rational, for any $1 \leq i < j \leq n$.

10. (IMO) Let a, b, c and d be integers such that $a > b > c > d > 0$. If

$$ac + bd = (b + d + a - c)(b + d - a + c),$$

prove that $ab + cd$ is not a prime number.

---

[4]For another approach to this problem, see Problem 9, page 486.

7.2 Pell's Equation

Recall (cf. Problem 6, page 190) that an integer greater than 1 is said to be **square free** if it is not divisible by the square of a prime number. In this section, we examine the integer solutions of equations of the form

$$x^2 - dy^2 = m, \tag{7.4}$$

where $d > 1$ is square free and m is a given integer. Such an equation is known as **Pell's equation**.[5]

Since d is a product of distinct primes, Example 6.24 guarantees that $\sqrt{d}$ is irrational. In particular, we conclude that the above equation has no nonzero integer solutions when $m = 0$. Indeed, if this was not the case, then we would have $x, y \neq 0$ and, hence, $\sqrt{d} = \frac{x}{y} \in \mathbb{Q}$. On the other hand, if $d, m < 0$, then Pell's equation obviously has no solutions, whereas if $d < 0 < m$, then (7.4) has at most a finite number of solutions. We also notice that, even if $d, m > 0$, (7.4) may have no solution at all, as Problem 1 shows.

From now on, we shall assume that $m = 1$. (The general case is partially dealt with at Problem 4; in this respect, see also [27]). Before we dive into a general analysis, let us give a simple example showing that, in this case, Pell's equation may have infinitely many solutions.

Example 7.6 Show that the equation $x^2 - 2y^2 = 1$ has infinitely many positive integer solutions.

Proof Note that $x = 3$, $y = 2$ is a solution. On the other hand, we can generate infinitely many integer solutions of this equation from one nonzero integer solution (a, b) in the following way: starting from $a^2 - 2b^2 = 1$, we write

$$(a + b\sqrt{2})(a - b\sqrt{2}) = 1$$

and, hence,

$$(a + b\sqrt{2})^2(a - b\sqrt{2})^2 = 1.$$

Expanding both binomials, we arrive at

$$(a^2 + 2b^2 + 2ab\sqrt{2})(a^2 + 2b^2 - 2ab\sqrt{2}) = 1$$

or, which is the same, at

$$(a^2 + 2b^2)^2 - 2(2ab)^2 = 1.$$

---

[5]After John Pell, English mathematician of the seventeenth century.

Therefore, $(a^2 + 2b^2, 2ab)$ is also an integer solution and, letting $a, b \in \mathbb{N}$, we have $a < a^2 + 2b^2$. Iterating the above argument, we obtain infinitely many solutions for the given equation. $\square$

The method used in the previous example can be easily generalized to show that the general Pell's equation $x^2 - dy^2 = 1$ admits infinitely many nonzero integer solutions, provided it admits at least one such solution. Moreover, as illustrated by Problems 4 and 7, for instance, small modifications in that approach also allow us to treat more general equations.

Even tough we can easily find infinitely many integer solutions of (7.4) in some particular cases, for the time being we do not know whether or not there are other integer solutions, not to say how to find all of them. In order to answer these questions, we shall need a preliminary result, due to Dirichlet, on approximation of irrational numbers by rational ones. For what comes next, recall (cf. the discussion preceding Problem 15, page 178) that, for $x \in \mathbb{R}$, the **fractionary part** of x is the real number $\{x\} \in [0, 1)$, given by $\{x\} = x - \lfloor x \rfloor$.

Lemma 7.7 (Dirichlet) *If α is an irrational number, then there exist infinitely many rationals $\frac{x}{y}$, with x and y nonzero, relatively prime integers such that*

$$\left| \alpha - \frac{x}{y} \right| < \frac{1}{y^2}. \tag{7.5}$$

Proof Let $n > 1$ be any integer and consider the $n + 1$ real numbers $\{j\alpha\} \in [0, 1)$, with $j = 0, 1, \ldots, n$. Since

$$[0, 1) = \left[0, \frac{1}{n} \right) \cup \left[\frac{1}{n}, \frac{2}{n} \right) \cup \cdots \cup \left[\frac{n-1}{n}, 1 \right),$$

a union of n disjoint half-open intervals, pigeonhole's principle guarantees the existence of indices $0 \le k < j \le n$ such that $\{j\alpha\}$ and $\{k\alpha\}$ belong to a single of these n half-open intervals.

Then, $|\{j\alpha\} - \{k\alpha\}| < \frac{1}{n}$ or, which is the same,

$$|(j - k)\alpha - (\lfloor j\alpha \rfloor - \lfloor k\alpha \rfloor)| < \frac{1}{n}.$$

It thus follows that

$$\left| \alpha - \frac{\lfloor j\alpha \rfloor - \lfloor k\alpha \rfloor}{j - k} \right| < \frac{1}{(j - k)n} \le \frac{1}{(j - k)^2}, \tag{7.6}$$

and letting $x = \lfloor j\alpha \rfloor - \lfloor k\alpha \rfloor$ and $y = j - k$, we have $0 < y \le n$ and $\left| \frac{x}{y} - \alpha \right| < \frac{1}{y^2}$. On the other hand, if $d = \gcd(x, y)$ and $x = dx_1, y = dy_1$, then

$$\left|\frac{x_1}{y_1} - \alpha\right| < \frac{1}{y^2} \leq \frac{1}{y_1^2},$$

and we can assume that $\gcd(x, y) = 1$.

In order to guarantee the existence of infinitely many such pairs, let x and y be nonzero, relatively prime integers such that $\left|\frac{x}{y} - \alpha\right| < \frac{1}{y^2}$. By choosing a natural number n_1 for which $\left|\frac{x}{y} - \alpha\right| > \frac{1}{n_1}$, and repeating the argument that led to (7.6) with n_1 in place of n, we obtain nonzero, relatively prime integers x_1 and y_1 such that $0 < y_1 \leq n_1$ and

$$\left|\frac{x_1}{y_1} - \alpha\right| < \frac{1}{n_1 y_1} \leq \frac{1}{y_1^2}.$$

Moreover, since

$$\left|\frac{x_1}{y_1} - \alpha\right| < \frac{1}{n_1 y_1} \leq \frac{1}{n_1} < \left|\frac{x}{y} - \alpha\right|,$$

we conclude that $\frac{x_1}{y_1} \neq \frac{x}{y}$.

Finally, by iterating the above reasoning we obtain infinitely many (x, y) of nonzero, relatively prime integers satisfying (7.5). □

As we shall now see, Dirichlet's lemma allows us to show that, for a fixed square free natural number $d > 1$, there exists at least one integer value of m for which (7.4) admits infinitely many integer solutions.

Lemma 7.8 *If $d > 1$ is a square free natural number, then there exists $m \in \mathbb{Z} \setminus \{0\}$ such that equation $x^2 - dy^2 = m$ has infinitely many integer solutions.*

Proof Since $\sqrt{d} \notin \mathbb{Q}$, Dirichlet's lemma assures that the set S of ordered pairs (x, y) of nonzero, relatively prime integers x, y satisfying $\left|\frac{x}{y} - \sqrt{d}\right| < \frac{1}{y^2}$ is infinite. However, if (x, y) is such a pair, then $|x - y\sqrt{d}| < \frac{1}{|y|}$, and triangle inequality gives us

$$|x^2 - dy^2| = |x - y\sqrt{d}||x + y\sqrt{d}| < \frac{1}{|y|}\left(|x - y\sqrt{d}| + 2|y|\sqrt{d}\right)$$

$$= \frac{1}{|y|}\left(\frac{1}{|y|} + 2|y|\sqrt{d}\right) < 2\sqrt{d} + 1.$$

Thus, for each $(x, y) \in S$, the set of integers $x^2 - dy^2$ is contained in the set of the nonzero integers of the interval $(-(2\sqrt{d} + 1), 2\sqrt{d} + 1)$. But since this is a finite set, there exists an integer $m \neq 0$, situated between $-(2\sqrt{d} + 1)$ and $2\sqrt{d} + 1$, that repeats itself an infinitely number of times among the values of $x^2 - dy^2$, for $(x, y) \in S$. Of course, this is the same as saying that, for this value of m, equation $x^2 - dy^2 = m$ admits infinitely many integer solutions. □

We are finally in position to characterize all integer solutions of (7.4) when $m = 1$; the key ingredients will be the previous lemma and Fermat's method of descent. We start by establishing the existence of solutions.

Proposition 7.9 *If $d > 1$ is a square free natural number, then equation $x^2 - dy^2 = 1$ admits at least one solution in positive integers x and y.*

Proof Take, by the previous lemma, $m \in \mathbb{Z} \setminus \{0\}$ for which the equation $x^2 - dy^2 = m$ has infinitely many integer solutions. Since there are only finitely many different remainders upon division by m, we can choose two such solution, say (x_1, y_1) and (x_2, y_2), such that $|x_1| \neq |x_2|$ and m divides $x_2 - x_1, y_2 - y_1$. Then,

$$(x_1 + y_1\sqrt{d})(x_2 - y_2\sqrt{d}) = (x_1 x_2 - dy_1 y_2) + (x_2 y_1 - x_1 y_2)\sqrt{d} \qquad (7.7)$$

and, by writing $x_2 - x_1 = mr$ and $y_2 - y_1 = ms$, with $r, s \in \mathbb{Z}$, we get

$$x_1 x_2 - dy_1 y_2 = x_1(x_2 - x_1) + (x_1^2 - dy_1^2) + (y_1 - y_2)dy_1$$
$$= m(rx_1 + 1 - sdy_1)$$

and

$$x_2 y_1 - x_1 y_2 = (x_2 - x_1)y_1 + x_1(y_1 - y_2) = m(r - s).$$

For the sake of simplicity, let $x_1 x_2 - dy_1 y_2 = mu$ and $x_2 y_1 - x_1 y_2 = mv$, with $u, v \in \mathbb{Z}$. It follows from (7.7) that

$$(x_1 + y_1\sqrt{d})(x_2 - y_2\sqrt{d}) = m(u + v\sqrt{d}) \qquad (7.8)$$

and, hence,

$$(x_1 - y_1\sqrt{d})(x_2 + y_2\sqrt{d}) = m(u - v\sqrt{d}).$$

By multiplying the left and right hand sides of these two relations, we arrive at

$$m^2 = (x_1^2 - dy_1^2)(x_2^2 - dy_2^2) = m^2(u^2 - dv^2),$$

so that $u^2 - dv^2 = 1$.

We are left to showing that $u, v \neq 0$. If $u = 0$, we would have $-dv^2 = 1$, which is clearly impossible. If $v = 0$ we would get $u = \pm 1$, and it would follow from (7.8) that

$$(x_1 + y_1\sqrt{d})(x_2 - y_2\sqrt{d}) = \pm m = \pm(x_2^2 - dy_2^2);$$

therefore,

$$x_1 + y_1\sqrt{d} = \pm(x_2 + y_2\sqrt{d}),$$

which in turn implies $|x_1| = |x_2|$, for $\sqrt{d} \notin \mathbb{Q}$. But this conclusion contradicts our choices of x_1 and x_2, and we are done. $\quad\square$

We are finally able to state and prove the result we have been seeking, on the characterization of all integer solutions of Pell's equation $x^2 - dy^2 = 1$, for a square free integer $d > 1$. Observe that the previous proposition guarantees that it has at least one solution.

Theorem 7.10 *If $d > 1$ is a square free integer, then the equation $x^2 - dy^2 = 1$ admits infinitely many solutions in positive integers x, y. More precisely, if $x = x_1$, $y = y_1$ is the solution in positive integers for which the sum $x + y\sqrt{d}$ is as small as possible, then the other positive integer solutions of the equation are given by the natural numbers x_n, y_n satisfying the equality*

$$x_n + y_n\sqrt{d} = (x_1 + y_1\sqrt{d})^n,$$

where $n \in \mathbb{N}$.

Proof Let $\alpha = x_1 + y_1\sqrt{d}$, with $x_1, y_1 \in \mathbb{N}$ being chosen as in the statement of the theorem.

Given $n \in \mathbb{N}$, the binomial expansion formula readily gives $x_n, y_n \in \mathbb{N}$ such that

$$(x_1 \pm y_1\sqrt{d})^n = x_n \pm y_n\sqrt{d}.$$

Thus,

$$
\begin{aligned}
1 = (x_1^2 - dy_1^2)^n &= (x_1 + y_1\sqrt{d})^n(x_1 - y_1\sqrt{d})^n \\
&= (x_n + y_n\sqrt{d})(x_n - y_n\sqrt{d}) = x_n^2 - dy_n^2,
\end{aligned}
$$

so that all of the ordered pairs (x_n, y_n) built as in the statement of the theorem are indeed solutions of the given equation.

Now, if (x, y) is any solution in positive integers, it suffices to show that there exists $n \in \mathbb{N}$ for which $x + y\sqrt{d} = \alpha^n$. For the sake of contradiction, assume that the converse is true. Since $\alpha > 1$, we have $\lim_{n \to +\infty} \alpha^n = +\infty$, so that there exists $n \in \mathbb{N}$ for which

$$\alpha^n < x + y\sqrt{d} < \alpha^{n+1}$$

or, which is the same,

$$1 < \frac{x + y\sqrt{d}}{\alpha^n} < \alpha. \tag{7.9}$$

However, relations $\alpha^n = x_n + y_n\sqrt{d}$ and $x_n^2 - dy_n^2 = 1$, would give us

$$\frac{x + y\sqrt{d}}{\alpha^n} = \frac{x + y\sqrt{d}}{x_n + y_n\sqrt{d}}$$

$$= (x_n - y_n\sqrt{d})(x + y\sqrt{d})$$

$$= (xx_n - dyy_n) + (x_ny - y_nx)\sqrt{d},$$

with

$$(xx_n - dyy_n)^2 - d(x_ny - y_nx)^2 = (x_n^2 - dy_n^2)(x^2 - dy^2) = 1.$$

Hence, $(xx_n - dyy_n, x_ny - y_nx)$ is an integer solution such that (by (7.9))

$$1 < (xx_n - dyy_n) + (x_ny - y_nx)\sqrt{d} = \frac{x + y\sqrt{d}}{\alpha^n} < \alpha.$$

Thus, if we show that $xx_n - dyy_n, x_ny - y_nx > 0$, we would arrive at a contradiction to the minimality of α (note that, here, we are invoking Fermat's descent method).

For what is left to do, letting $a = xx_n - dyy_n$ and $b = x_ny - y_nx$, we have

$$a + b\sqrt{d} = \frac{x + y\sqrt{d}}{\alpha^n} > 1 \text{ and } a^2 - db^2 = 1,$$

so that

$$a - b\sqrt{d} = \frac{1}{a + b\sqrt{d}} > 0.$$

Then, on the one hand,

$$2a = (a - b\sqrt{d}) + (a + b\sqrt{d}) > 0;$$

on the other,

$$a - b\sqrt{d} = \frac{1}{a + b\sqrt{d}} < 1$$

gives us $b\sqrt{d} > a - 1 \geq 0$, so that $b > 0$.   □

Example 7.11 Find all positive integer solutions of equation $x^2 - 2y^2 = 1$.

Solution Letting (x_1, y_1) be the solution in positive integers for which $x_1 + y_1\sqrt{2}$ is as small as possible, the previous result teaches us that the positive integer solutions of this equation are of the form (x_n, y_n), where x_n and y_n are such that

$$x_n + y_n\sqrt{2} = (x_1 + y_1\sqrt{2})^n,$$

for $n \in \mathbb{N}$.

Since the ordered pairs $(1, 1)$, $(1, 2)$, $(2, 1)$, $(2, 2)$, $(3, 1)$, $(1, 3)$, $(2, 3)$ and $(4, 1)$ are not solutions of the given equation but $(3, 2)$ is a solution, it is immediate to check that $x_1 = 3$ and $y_1 = 2$. Hence, the remaining solutions in positive integers are the ordered pairs (x_n, y_n) given by the equality

$$x_n + y_n\sqrt{2} = (3 + 2\sqrt{2})^n.$$

$\square$

Problems: Sect. 7.2

1. * If d and m leave remainder 3 upon division by 4, show that (7.4) does not possess any integer solutions.
2. With respect to Example 7.11, prove that the positive integer solutions (x_n, y_n) of $x^2 - 2y^2 = 1$ are recursively defined by $x_1 = 3$, $y_1 = 2$ and, for an integer $n \geq 1$,

$$x_{n+1} = 3x_n + 4y_n, \quad \text{and} \quad y_{n+1} = 2x_n + 3y_n.$$

3. Prove that the equation $x^2 - 2y^2 = -1$ has infinitely many integer solutions.
4. * Let $d, m \in \mathbb{N}$, with $d > 1$ being square free. If equation $x^2 - dy^2 = m$ has a solution (x_0, y_0) in positive integers, prove that it has infinitely many such solutions.
5. (Hungary) Show that there are infinitely many positive integers n for which $n^2 + (n + 1)^2$ is a perfect square.
6. (Turkey) Find infinitely many solutions, in positive integers, of the equation

$$y^2 + 1 = x(x + y).$$

7. Generalize the result of the previous problem in the following way: let $a, b, c \in \mathbb{Z}$ be such that $\Delta = b^2 - 4ac$ is greater than 1 and square free. If $n \in \mathbb{Z}$ is such that the equation

$$x^2 - \Delta y^2 = 4an$$

has at least one integer solution (x_0, y_0) for which $2a \mid (x_0 - by_0)$, show that the equation

$$ax^2 + bxy + cy^2 = n$$

has infinitely many integer solutions.

Chapter 8
Arithmetic Functions

This short chapter introduces an important class of functions, called *arithmetic multiplicative functions*, which play a prominent role in the elementary theory of numbers. Among the many arithmetic multiplicative functions we shall encounter here, two deserve all spotlights: the Euler function φ, which will reveal itself to be an indispensable tool for basically all further theoretical developments, and the Möbius function μ, which is essential to getting the celebrated *Möbius inversion formula* and its subsequent application to the Euler function.

In all that follows, we shall refer ourselves to a given function $f : \mathbb{N} \to \mathbb{R}$ as an **arithmetic function**.

Definition 8.1 An arithmetic function $f : \mathbb{N} \to \mathbb{R}$ is said to be **multiplicative** if, for all relatively prime $m, n \in \mathbb{N}$, one has $f(mn) = f(m)f(n)$.

Since 1 is relatively prime with itself, if $f : \mathbb{N} \to \mathbb{R}$ is an arithmetic multiplicative function, then $f(1) = f(1)^2$, so that $f(1) = 0$ or 1. In case $f(1) = 0$, we get

$$f(n) = f(n \cdot 1) = f(n)f(1) = 0, \ \forall \, n \in \mathbb{N},$$

i.e., f vanishes identically. Hence, from now on we shall implicitly assume that if $f : \mathbb{N} \to \mathbb{R}$ is an arithmetic multiplicative function, then $f(1) = 1$.

Note also that if $f : \mathbb{N} \to \mathbb{R}$ is an arithmetic multiplicative function and $n > 1$ is an integer with canonical decomposition $n = p_1^{\alpha_1} \dots p_k^{\alpha_k}$, then several applications of the definition give us

$$f(n) = f(p_1^{\alpha_1}) \dots f(p_k^{\alpha_k}). \tag{8.1}$$

In words, the above inequality says that, in order to compute the values $f(n)$, with $n \in \mathbb{N}$, it suffices to know how to compute the values $f(p^\alpha)$, with p prime and $\alpha \in \mathbb{N}$.

© Springer International Publishing AG, part of Springer Nature 2018
A. Caminha Muniz Neto, *An Excursion through Elementary Mathematics, Volume III*,
Problem Books in Mathematics, https://doi.org/10.1007/978-3-319-77977-5_8

Finally, if $f : \mathbb{N} \to \mathbb{R}$ is an arithmetic function such that $f(1) = 1$, then the equality $f(mn) = f(m)f(n)$ is always true when $m = 1$ or $n = 1$; therefore, in order to prove that such an f is multiplicative, we just have to consider the case of $m, n > 1$.

We shall repeatedly use the above remarks without further notice.

Example 8.2 We saw at Corollary 6.46 that if $n = p_1^{\alpha_1} \ldots p_k^{\alpha_k}$ is the canonical decomposition of a natural number $n > 1$ in primes, then

$$d(n) = (\alpha_1 + 1) \ldots (\alpha_k + 1)$$

is the number of positive divisors of n. We claim that the function $d : \mathbb{N} \to \mathbb{R}$ thus obtained is multiplicative. Indeed, if $m, n > 1$ are relatively prime integers with canonical decompositions $n = p_1^{\alpha_1} \ldots p_k^{\alpha_k}$ and $m = q_1^{\beta_1} \ldots q_l^{\beta_l}$, then $p_i \neq q_j$ for all $1 \leq i \leq k, 1 \leq j \leq l$. Therefore, the canonical decomposition of mn in primes is

$$mn = p_1^{\alpha_1} \ldots p_k^{\alpha_k} q_1^{\beta_1} \ldots q_l^{\beta_l}.$$

so that

$$d(mn) = (\alpha_1 + 1) \ldots (\alpha_k + 1)(\beta_1 + 1) \ldots (\beta_l + 1) = d(m)d(n).$$

For what comes next, given $n \in \mathbb{N}$ we shall write $D(n)$ to denote the set of positive divisors of n, so that $d(n) = |D(n)|$.

Lemma 8.3 *If m and n are relatively prime naturals, then the map*

$$f : D(m) \times D(n) \longrightarrow D(mn)$$
$$(x, y) \longmapsto xy$$

is a bijection.

Proof It follows from Example 8.2 and the fundamental principle of counting that

$$|D(mn)| = d(mn) = d(m)d(n) = |D(m)| \cdot |D(n)| = |D(m) \times D(n)|.$$

Hence, the domain and codomain of f have equal quantities of elements. Thus, in order to prove that f is a bijection, it suffices to establish its surjectivity. To this end, let us apply item (d) of Proposition 6.22: if $k \mid mn$, then the condition $\gcd(m, n) = 1$ guarantees that

$$k = \gcd(k, mn) = \gcd(k, m) \cdot \gcd(k, n).$$

Therefore, if we let $a = \gcd(k, m)$ and $b = \gcd(k, n)$, we get $a \in D(m), b \in D(n)$ and $f((a, b)) = ab = k$, so that k belongs to the image of f. But since $k \in D(mn)$ was arbitrarily chosen, we conclude that f is surjective. $\square$

Along the rest of this chapter, we shall write $\sum_{0<d|n} f(d)$ to denote the sum of the values $f(d)$, when d varies in $D(n)$. Although such an use of the letter d is conflictant with the notation for the function d of the previous example, this will be harmless in context.

The coming proposition establishes one of the most important properties of arithmetic multiplicative functions.

Proposition 8.4 *If $f : \mathbb{N} \to \mathbb{R}$ is an arithmetic multiplicative function, then so is the function $F : \mathbb{N} \to \mathbb{R}$, given by*

$$F(n) = \sum_{0<d|n} f(d).$$

Proof If $m, n \in \mathbb{N}$ are such that $\gcd(m, n) = 1$, then the previous lemma, together with the multiplicative character of f, gives

$$F(mn) = \sum_{0<d|mn} f(d) = \sum_{\substack{0<d_1|m \\ 0<d_2|n}} f(d_1 d_2) = \sum_{0<d_1|m} \sum_{0<d_2|n} f(d_1) f(d_2)$$

$$= \left(\sum_{0<d_1|m} f(d_1) \right) \left(\sum_{0<d_2|n} f(d_2) \right)$$

$$= F(m) F(n).$$

$\square$

Example 8.5 The function $f : \mathbb{N} \to \mathbb{R}$ such that $f(n) = n$ for every $n \in \mathbb{N}$ is obviously multiplicative. Hence, the previous proposition assures that so is the function $s : \mathbb{N} \to \mathbb{R}$, given by

$$s(n) = \sum_{0<d|n} f(d) = \sum_{0<d|n} d.$$

Since $s(n)$ adds the positive divisors of n, given a prime number p and $\alpha \in \mathbb{N}$, it follows from Corollary 6.46 that

$$s(p^\alpha) = \sum_{0<d|p^\alpha} d = \sum_{j=0}^{\alpha} p^j = \frac{p^{\alpha+1} - 1}{p - 1}.$$

Thus, if $n = p_1^{\alpha_1} \ldots p_k^{\alpha_k}$, with primes $p_1 < \cdots < p_k$ and natural numbers $\alpha_1, \ldots, \alpha_k$, relation (8.1) and the above computations give

$$s(n) = \prod_{j=1}^{k} s(p_j^{\alpha_j}) = \prod_{j=1}^{k} \left(\frac{p_j^{\alpha_j+1} - 1}{p_j - 1} \right). \tag{8.2}$$

For our next example, we recall (cf. Problem 6, page 190) that an integer $n > 1$ is *square free* if $n = p_1 \ldots p_k$, with $p_1 < \cdots < p_k$ being primes; equivalently, n is square free if there does not exist an integer $q > 1$ such that $q^2 \mid n$.

The coming definition presents one of the most important arithmetic multiplicative functions multiplicativas, the *Möbius function*.

Definition 8.6 The **Möbius function**[1] is the function $\mu : \mathbb{N} \to \mathbb{R}$ given by

$$\mu(n) = \begin{cases} 1, & \text{if } n = 1 \\ 0, & \text{if } q^2 \mid n, \text{ para algum inteiro } q > 1 \\ (-1)^k, & \text{if } n = p_1 \ldots p_k, \text{ com } p_1, \ldots, p_k \text{ distinct primes} \end{cases}.$$

In order to check that the Möbius function is indeed multiplicative, let $m, n > 1$ be relatively prime integers. Then, mn will be divisible by a perfect square greater than 1 if and only if either m or n do satisfy such a property; this being the case, it is immediate to see that

$$f(mn) = 0 = f(m)f(n).$$

On the other hand, if m and n are square free, say $n = p_1 \ldots p_k$ and $m = q_1 \ldots q_l$, with $p_1 < \cdots < p_k$ and $q_1 < \cdots < q_l$ being prime numbers, the condition $\gcd(m, n) = 1$ assures that $p_i \neq q_j$ for all i and j. Hence, $mn = p_1 \ldots p_k q_1 \ldots q_l$ is the canonical decomposition of mn in primes, so that

$$\mu(mn) = (-1)^{k+l} = (-1)^k (-1)^l = \mu(m)\mu(n).$$

The coming result brings a very important property of the Möbius function.

Proposition 8.7 *If* $\mu : \mathbb{N} \to \mathbb{R}$ *is the Möbius function, then*

$$\sum_{0 < d \mid n} \mu(d) = \begin{cases} 1, & \text{if } n = 1 \\ 0, & \text{if } n > 1 \end{cases}.$$

Proof Let $F : \mathbb{N} \to \mathbb{R}$ be the function given by

$$F(n) = \sum_{0 < d \mid n} \mu(d).$$

Proposition 8.4 guarantees that F is multiplicative, and we wish to show that $F(1) = 1$ and $F(n) = 0$ for $n > 1$. We consider three cases separately:

(i) $n = 1$: we have $F(1) = \sum_{0 < d \mid 1} \mu(d) = \mu(1) = 1$.

---

[1] After the German mathematician of the nineteenth century August Möbius.

(ii) $n = p^k$, with p prime and $k \geq 1$ integer: then

$$F(p^k) = \sum_{0<d\,|\,p^k} \mu(d) = \sum_{j=0}^{k} \mu(p^j) = \mu(1) + \mu(p) = 0.$$

(iii) $n = p_1^{\alpha_1} \ldots p_k^{\alpha_k}$, with $p_1 < \cdots < p_k$ being primes: since F is multiplicative, it follows from (8.1) and item (ii) that

$$F(n) = F(p_1^{\alpha_1}) \ldots F(p_k^{\alpha_k}) = 0.$$

$\square$

Theorem 8.8 is the reason behind the central role of Möbius function in the theory of arithmetic multiplicative functions. For its proof, notice that the function

$$\begin{aligned} f : D(n) &\longrightarrow D(n) \\ d &\longmapsto n/d \end{aligned} \tag{8.3}$$

is a bijection; indeed, this follows at once from the fact that $f \circ f = \mathrm{Id}_{D(n)}$, as the reader can promptly check.

In the notations of Proposition 8.4, formula (8.4) below, known as the **Möbius inversion formula**, teaches us how to recover function f from F (even if f is not multiplicative).

Theorem 8.8 (Möbius) *Let $f : \mathbb{N} \to \mathbb{R}$ be an arithmetic function (not necessarily multiplicative). If $F : \mathbb{N} \to \mathbb{R}$ is given by $F(n) = \sum_{0<d\,|\,n} f(d)$, then*

$$f(n) = \sum_{0<d\,|\,n} F\left(\frac{n}{d}\right) \mu(d) = \sum_{0<d\,|\,n} F(d)\mu\left(\frac{n}{d}\right). \tag{8.4}$$

Proof The second equality in (8.4) follows from the bijectivity of the function in (8.3). For the first equality, note first that

$$\sum_{0<d\,|\,n} F\left(\frac{n}{d}\right) \mu(d) = \sum_{0<d\,|\,n} \left(\mu(d) \sum_{0<d'\,|\,\frac{n}{d}} f(d') \right) = \sum_{0<d\,|\,n} \sum_{0<d'\,|\,\frac{n}{d}} \mu(d) f(d').$$

However, since $d \mid n$ and $d' \mid \frac{n}{d}$ if and only if $d' \mid n$ and $d \mid \frac{n}{d'}$, it follows from the above computations that

$$\sum_{0<d\,|\,n} F\left(\frac{n}{d}\right) \mu(d) = \sum_{0<d'\,|\,n} \sum_{0<d\,|\,\frac{n}{d'}} \mu(d) f(d') = \sum_{0<d'\,|\,n} \left(f(d') \sum_{0<d\,|\,\frac{n}{d'}} \mu(d) \right).$$

Now, if $\frac{n}{d'} > 1$ (i.e., if $d' < n$), then Proposition 8.7 furnishes

$$\sum_{0<d\mid\frac{n}{d'}} \mu(d) = 0.$$

Therefore, the next to last sum above reduces to the summand corresponding to $d' = n$, so that

$$\sum_{0<d'\mid n} \left(f(d') \sum_{0<d\mid\frac{n}{d'}} \mu(d) \right) = f(n) \sum_{0<d\mid 1} \mu(d) = f(n)\mu(1) = f(n).$$

□

Before we proceed with the development of the theory, we present an example which shows how Möbius inversion formula can be a useful tool in counting problems.

Example 8.9 Let $n \in \mathbb{N}$ be given. We say that a sequence $(x_1, x_2, \ldots, x_n)$, with $x_j \in \{0, 1\}$ for $1 \leq j \leq n$, is *aperiodic* if there does not exist a divisor $0 < d < n$ of n such that the sequence is formed by the juxtaposition of $\frac{n}{d}$ copies of the subsequence $(x_1, \ldots, x_d)$. Compute, in terms of n, the total number of aperiodic sequences of n terms.

Proof First of all, the fundamental principle of counting assures that there exist exactly 2^n sequences $(x_1, x_2, \ldots, x_n)$ with $x_j \in \{0, 1\}$ for $1 \leq j \leq n$.

On the other hand, for such a sequence $(x_1, x_2, \ldots, x_n)$, we define its *period* as the least positive divisor d of n for which the sequence is formed by the juxtaposition of n/d copies of the subsequence $(x_1, \ldots, x_d)$. In particular, an aperiodic sequence $(x_1, x_2, \ldots, x_n)$ has period n.

More generally, if the sequence $(x_1, x_2, \ldots, x_n)$ has period d, then $(x_1, \ldots, x_d)$ is necessarily aperiodic, and conversely. Hence, if a_k denotes the number of aperiodic sequences of k terms, we get

$$\sum_{0<d\mid n} a_d = 2^n.$$

If we now apply the Möbius inversion formula, we obtain

$$a_n = \sum_{0<d\mid n} \mu\left(\frac{n}{d}\right) 2^d,$$

as wished.   □

The theory developed so far allows us to introduce and study the main properties of another quite important arithmetic function, according to the following

Definition 8.10 Euler's φ function is the function $\varphi : \mathbb{N} \to \mathbb{N}$ given by

$$\varphi(n) = \#\{1 \le k \le n;\ \gcd(k, n) = 1\}.$$

In words, $\varphi(n)$ counts how many integers from 1 to n are relatively prime with n. In what comes next, among other properties we shall show that the function φ is multiplicative. We shall also use this result to compute $\varphi(n)$ in terms of the canonical decomposition of n in prime factors.[2] We start with a result which will be useful in other circumstances.

Proposition 8.11 *With respect to Euler's function $\varphi : \mathbb{N} \to \mathbb{N}$, we have*

$$\sum_{0 < d \mid n} \varphi(d) = \sum_{0 < d \mid n} \varphi \left(\frac{n}{d} \right) = n.$$

Proof The first equality comes from the bijectivity of function f in (8.3). For the second one, let $D(n) = \{1 = a_1 < a_2 < \cdots < a_t = n\}$. If $1 \le k \le n$, then $\gcd(k, n) \in D(n)$, i.e., $\gcd(k, n) = a_i$ for some $1 \le i \le t$. Therefore, letting $A_i = \{1 \le k \le n;\ \gcd(k, n) = a_i\}$, we get

$$I_n = A_1 \cup \ldots \cup A_t,$$

a disjoint union. Hence, $n = \sum_{i=1}^{t} |A_i|$. Now, note that

$$
\begin{aligned}
A_i &= \{1 \le k \le n;\ \gcd(k, n) = a_i\} \\
&= \{1 \le k/a_i \le n/a_i;\ k/a_i \in \mathbb{N} \text{ and } \gcd(k/a_i, n/a_i) = 1\} \\
&= \{1 \le l \le n/a_i;\ \gcd(l, n/a_i) = 1\},
\end{aligned}
$$

so that $|A_i| = \varphi \left(\frac{n}{a_i} \right)$. Thus,

$$n = \sum_{i=1}^{t} |A_i| = \sum_{i=1}^{t} \varphi \left(\frac{n}{a_i} \right) = \sum_{0 < d \mid n} \varphi \left(\frac{n}{d} \right).$$

$\square$

For the coming result, also due to Euler, we shall need the fact, left as an exercise to the reader (see Problem 1), that the product of two arithmetic multiplicative functions yields yet another such function.

Theorem 8.12 (Euler) *The Euler function $\varphi : \mathbb{N} \to \mathbb{N}$ is multiplicative.*

---

[2]Another approach was the object of Problem 6, page 40.

Proof Letting $G(n) = n$, the previous proposition gives us $\sum_{0<d|n} \varphi(d) = G(n)$. Therefore, it follows from Möbius inversion formula that

$$\varphi(n) = \sum_{0<d|n} \mu(d)G\left(\frac{n}{d}\right) = \sum_{0<d|n} \mu(d) \cdot \frac{n}{d} = n \sum_{0<d|n} \frac{\mu(d)}{d}. \tag{8.5}$$

However, since $f(d) = \frac{\mu(d)}{d}$ is a multiplicative function (verify this assertion!), Proposition 8.4 shows that function $F : \mathbb{N} \to \mathbb{R}$, defined by

$$F(n) = \sum_{0<d|n} \frac{\mu(d)}{d},$$

is also multiplicative. Hence, Problem 1 assures that $\varphi(n) = nF(n)$ is multiplicative, too. $\square$

As we have promised before, the coming corollary relates $\varphi(n)$ with the canonical decomposition of an integer $n > 1$ in prime factors.

Corollary 8.13 (Euler) *If the canonical decomposition of an integer $n > 1$ in primes is given by $n = p_1^{\alpha_1} \dots p_k^{\alpha_k}$, then*

$$\varphi(n) = n\left(1 - \frac{1}{p_1}\right)\dots\left(1 - \frac{1}{p_k}\right). \tag{8.6}$$

Proof We shall firstly compute the value of $\varphi(p^\alpha)$, with p prime and $\alpha \geq 1$ integer:

$$\begin{aligned}
\varphi(p^\alpha) &= \#\{1 \leq k \leq p^\alpha; \ \gcd(k, p^\alpha) = 1\} \\
&= \#\{1 \leq k \leq p^\alpha; \ \gcd(k, p) = 1\} \\
&= \#\left(\{1, 2, 3, \dots, p^\alpha\} \setminus \{p, 2p, 3p, \dots, p^{\alpha-1}p\}\right) \\
&= p^\alpha - p^{\alpha-1} \\
&= p^\alpha\left(1 - \frac{1}{p}\right).
\end{aligned}$$

Now, since φ is multiplicative, it follows from (8.1) that

$$\begin{aligned}
\varphi(n) &= \varphi(p_1^{\alpha_1})\dots\varphi(p_k^{\alpha_k}) \\
&= p_1^{\alpha_1}\left(1 - \frac{1}{p_1}\right)\dots p_k^{\alpha_k}\left(1 - \frac{1}{p_k}\right) \\
&= p_1^{\alpha_1}\dots p_k^{\alpha_k}\left(1 - \frac{1}{p_1}\right)\dots\left(1 - \frac{1}{p_k}\right) \\
&= n\left(1 - \frac{1}{p_1}\right)\dots\left(1 - \frac{1}{p_k}\right).
\end{aligned}$$

$\square$

The formula of the previous corollary allows us to deduce many interesting (and useful) properties of the Euler function. The coming example presents such a result.

Example 8.14 Given $m \in \mathbb{N}$, show that the equation $\varphi(n) = m$ has at most a finite number of solutions $n \in \mathbb{N}$.

Proof Firstly, if $n = p^{\alpha} r$, with $\alpha \geq 1$, p prime and $\gcd(p, r) = 1$, then the multiplicative character of the function φ gives

$$m = \varphi(n) = \varphi(p^{\alpha})\varphi(r) = p^{\alpha-1}(p-1)\varphi(r) \geq 2^{\alpha-1},$$

for $\varphi(r) \geq 1$ and $p \geq 2$. Hence, taking logarithms at base 2, we obtain

$$\alpha \leq 1 + \log_2 m.$$

Moreover, analogous computations furnish

$$m = p^{\alpha-1}(p-1)\varphi(r) \geq p - 1,$$

whence $p \leq m + 1$.

Now, if $n = p_1^{\alpha_1} \ldots p_k^{\alpha_k}$ is the canonical decomposition of n in primes and $\alpha = \max\{\alpha_1, \ldots, \alpha_k\}$, it follows from what we did above that $\alpha \leq 1 + \log_2 n$ and, hence,

$$n \leq (p_1 \ldots p_k)^{\alpha} \leq \left(\prod_{\substack{p \leq m+1 \\ p \text{ prime}}} p \right)^{1+\log_2 m}.$$

$\square$

Problems: Chap. 8

1. * Prove that the product of two arithmetic multiplicative functions is also an arithmetic multiplicative function.
2. Prove that, for every $n \in \mathbb{N}$, we have $\prod_{0<d|n} d = n^{d(n)/2}$.
3. Prove that $\frac{s(n)}{d(n)} \geq \sqrt{n}$, for every $n \in \mathbb{N}$.
4. (OCM) A teacher chose a positive integer n and, then, posed the following problems to two of his students: the first one should compute the number of ordered pairs (x, y) of positive integers satisfying the equation $\frac{1}{x} + \frac{1}{y} = \frac{1}{n}$; the second should do the same regarding the equation $\frac{1}{x} - \frac{1}{y} = \frac{1}{n}$. Knowing that the sum of the answers found by the students was 78, show that at least one of them made a mistake.
5. (Hungary—adapted) For $n \in \mathbb{N}$ and $0 \leq r \leq 3$, let $D_r(n)$ be the set of positive divisors of n leaving remainder r upon division by 4.

(a) If $\gcd(m, n) = 1$, prove that there exists a natural bijection

$$(D_1(m) \times D_1(n)) \cup (D_3(m) \times D_3(n)) \longrightarrow D_1(mn).$$

Do the same with respect to $(D_1(m) \times D_3(n)) \cup (D_3(m) \times D_1(n))$ and $D_3(mn)$.

(b) Prove that $|D_1(n)| \geq |D_3(n)|$.

6. A natural number $n > 1$ is said to be **perfect** if $s(n) = 2n$, where $s(n)$ denotes the sum of the positive divisors of n. Prove that:

(a) If $n > 1$ is perfect, then $\sum_{0 < d \mid n} \frac{1}{d} = 2$.
(b) If p is a prime number for which $2^p - 1$ is also prime, then $2^{p-1}(2^p - 1)$ is perfect.

7. The purpose of this problem is to establish the converse of the previous problem, thus proving the following theorem of Euler: if n is an even perfect number,[3] then there exists a prime number p such that $2^p - 1$ is prime and $n = 2^{p-1}(2^p - 1)$. To this end, let $n = 2^k q$ be an even perfect number, with $k, q \in \mathbb{N}$ and q being odd. Do the following items:

(a) Conclude that $(2^{k+1} - 1)s(q) = 2^{k+1}q$. Then, show that there exists $a \in \mathbb{N}$ for which $q = (2^{k+1} - 1)a$ and $s(q) = 2^{k+1}a$.
(b) If $a = 1$, show that $2^{k+1} - 1$ is prime, whence $k + 1 = p$, a prime number.
(c) If $a = 2^{k+1} - 1$, show that $s(q) \geq 1 + a + a^2 > (a+1)a = 2^{k+1}a$, which is a contradiction.
(d) If $a > 1$ and $a \neq 2^{k+1} - 1$, then q has at least four distinct positive divisors: $1, 2^{k+1} - 1, a$ and $(2^{k+1} - 1)a$. Use this fact to conclude that $s(q) > 2^{k+1}a$, thus arriving at a new contradiction.

8. A natural number n is *abundant* if $s(n) > 2n$, where $s(n)$ denotes the sum of the positive divisors of n. If $a \in \mathbb{N}$ is abundant, show that ab is also abundant, regardless of the value of $b \in \mathbb{N}$.

9. Let $f : \mathbb{N} \to \mathbb{R}$ be defined by $f(1) = 1$ and, for $n > 1$,

$$f(n) = \frac{(-1)^k}{p_1 p_2 \cdots p_k},$$

where $p_1, p_2, \ldots, p_k$ are the distinct prime divisors of n. Find all $n \in \mathbb{N}$ for which $\sum_{0 < d \mid n} f(d) = 0$.

10. If $f : \mathbb{N} \to \mathbb{R}$ is an arithmetic multiplicative function, prove that

$$\sum_{0 < d \mid n} \mu(d) f(d) = \prod_{\substack{p \text{ prime} \\ p \mid n}} (1 - f(p)).$$

Then, use the above result to establish the following items:

---

[3] Up to this day, no one knows whether or not there exist odd perfect numbers.

(a) $\sum_{0<d|n} d\mu(d) = \prod_{\substack{p \text{ prime} \\ p|n}} (1 - p)$.

(b) $\sum_{0<d|n} \mu(d)^2 = 2^k$, where k is the number of distinct prime divisors of n (note that $k = 0$ if $n = 1$).

11. Prove **Liouville's theorem**[4]: for each $n \in \mathbb{N}$, one has

$$\left(\sum_{0<j|n} d(j) \right)^2 = \sum_{0<j|n} d(j)^3.$$

12. * Let $f : \mathbb{N} \to \mathbb{R}$ be any function and $F : \mathbb{N} \to \mathbb{R}$ be given by $F(n) = \sum_{0<d|n} f(d)$. Prove that

$$\sum_{k=1}^{n} F(k) = \sum_{j=1}^{n} \left\lfloor \frac{n}{j} \right\rfloor f(j).$$

13. Let $f : \mathbb{N} \to \{-1, 1\}$ be given by $f(1) = 1$ and $f(n) = (-1)^{\alpha_1 + \cdots + \alpha_k}$, where $n = p_1^{\alpha_1} \cdots p_k^{\alpha_k}$ is the canonical decomposition of an integer $n > 1$ in primes. Prove that, for every integer $n \geq 1$, one has

$$\sum_{j=1}^{n} \left\lfloor \frac{n}{j} \right\rfloor f(j) = \lfloor \sqrt{n} \rfloor.$$

14. (Brazil) Prove that, for every natural number $n > 1$, we have

$$n \left(\frac{1}{2} + \frac{1}{3} + \cdots + \frac{1}{n} \right) < \sum_{k=1}^{n} d(k) \leq n \left(1 + \frac{1}{2} + \frac{1}{3} + \cdots + \frac{1}{n} \right).$$

15. Let F be an arithmetic multiplicative function and $f : \mathbb{N} \to \mathbb{R}$ be implicitly defined by the equality $F(n) = \sum_{0<d|n} f(d)$, for every $n \in \mathbb{N}$. Prove that f is also multiplicative.

16. The purpose of this problem is to give another proof of the multiplicative character of Euler's function φ. To this end, given relatively prime integers $m, n > 1$, arrange all natural numbers from 1 to mn as shown in the table

$$
\begin{array}{cccc}
1 & 2 \ldots & k \ldots & m \\
m + 1 & m + 2 \ldots & m + k \ldots & 2m \\
2m + 1 & 2m + 2 \ldots & 2m + k \ldots & 3m \\
& \cdots & & \cdots \\
(n-1)m + 1 & (n-1)m + 2 \ldots & (n-1)m + k \ldots & mn
\end{array}
$$

and do the following items:

---

[4]After Joseph Liouville, French mathematician of the nineteenth century. As we shall see in Sect. 20.4—cf. Theorem 20.31—Liouville was the one to give the very first example of a transcendental number.

(a) Prove that an entry of the table is relatively prime with mn if and only if it is relatively prime with both m and n.

(b) Show that, in any column, either all elements are prime with m or none of them satisfies such a property.

(c) Prove that there are exactly $\varphi(m)$ columns formed by integers relatively prime to m; moreover, each one of these columns contains precisely $\varphi(n)$ entries relatively prime with n.

(d) Conclude that $\varphi(mn) = \varphi(m)\varphi(n)$.

17. If $F(n) = \sum_{0<d|n} \frac{\varphi(d)}{d}$, compute $F(n)$ in terms of the canonical decomposition of n.

18. (OIM shortlist) For each $m \in \mathbb{N}$, let A_m be the set of ordered pairs (d, n) of integers such that d is a positive divisor of m, $1 \le n \le m$ and $\gcd(d, n) = 1$. Find all $m \in \mathbb{N}$ for which $|A_m| = 1993$.

19. * For an integer $n > 2$, prove the following items:

(a) If $P_n = \{1 \le k \le n; \gcd(k, n) = 1\}$, then the correspondence $k \mapsto n - k$ is a bijection of P_n.

(b) $\varphi(n) = 2l$, where l is the number of elements of P_n which are less than or equal to $\frac{n-1}{2}$; in particular, $\varphi(n)$ is even.

20. Given $m, n \in \mathbb{N}$, with $n > 2$, let $1 = a_1 < \cdots < a_k = n - 1$ be the positive integers prime with n and less than or equal to n, and $S_m(n) = \sum_{i=1}^{k} a_i^m$ be the sum of their m-th powers. Establish the following:

(a) If $k = 2l$, then $S_m(n) = \sum_{i=1}^{l}(a_i^m + (n - a_i)^m)$. Then, use this relation to conclude that $S_m(n)$ is even whenever n is itself even.

(b) $S_m(n) = \sum_{j=0}^{m}(-1)^j \binom{m}{j} n^{m-j} S_j(n)$.

(c) If m is odd, then $2S_m(n) = n \sum_{j=0}^{m-1}(-1)^j \binom{m}{j} n^{m-1-j} S_j(n)$. Then, show that $n \mid S_m(n)$ if m is odd.

21. In the notations of the statement of the previous problem, show that:

(a) Every $1 \le m \le n$ can be uniquely written as $m = \frac{n}{d} \cdot a$, with $a, d \in \mathbb{N}$ such that $d \mid n$ and $\gcd(a, d) = 1$.

(b) $\sum_{0<d|n} \frac{S_k(d)}{d^k} = \frac{1^k+2^k+\cdots+n^k}{n^k}$.

(c) $S_k(n) = n^k \sum_{0<d|n} \mu\left(\frac{n}{d}\right) \left(\frac{1^k+2^k+\cdots+d^k}{d^k}\right)$.

(d) $S_1(n) = \frac{1}{2}n\varphi(n)$.

(e) $S_2(n) = \frac{1}{3}n^2\varphi(n) + \frac{1}{6}n \prod_{\substack{p \text{ prime} \\ p|n}}(1 - p)$.

Chapter 9
Calculus and Number Theory

In this chapter, we assume that the reader is conversant with the rudiments of Calculus. More precisely, we shall assume from the reader familiarity with convergent sequences and series, as well as with the notions of limits and derivatives of functions. In this sense, the material of [8] covers all of what is necessary. Here, we shall present some basic examples and results on the distribution of prime numbers, having the classic book of professor G. Andrews [2] as guideline. We shall also discuss a rather interesting asymptotic result of Cesàro, on pairs of relatively prime natural numbers, following [29].

9.1 On the Distribution of Prime Numbers

One of the first results on prime numbers we have learned in this book was Euclid's theorem, which asserts that the set of primes is infinite. Nevertheless, as was already pointed out in Remark 6.39, there is an distinctive lack of uniformity on the way these numbers distribute along the naturals. It is, therefore, quite astonishing that a result such as Hadamard's **Prime Number Theorem**, quoted below, actually holds true. Before we state it, let us set some notation: for each positive real x, we shall write $\pi(x)$ to denote the number of prime numbers less than or equal to x.

Theorem 9.1 (Hadamard[1]) *In the above notations, one has*

$$\lim_{x \to +\infty} \frac{\pi(x)}{x/\log x} = 1, \tag{9.1}$$

where $\log : (0, +\infty) \to \mathbb{R}$ *stands for the natural logarithm function.*

---

[1] Jacques Hadamard, French mathematician of the nineteenth century.

© Springer International Publishing AG, part of Springer Nature 2018
A. Caminha Muniz Neto, *An Excursion through Elementary Mathematics, Volume III*,
Problem Books in Mathematics, https://doi.org/10.1007/978-3-319-77977-5_9

Although a proof of the above theorem is well beyond the scope of these notes, we refer the interested reader to the marvelous book of professor T. Apostol [5] for a self-contained approach. However, we stress that Hadamard's theorem assures that the functions $\pi(x)$ and $\frac{x}{\log x}$ are *asymptotically equal*, in the sense that limit (9.1) holds true.

A few years earlier than Hadamard, the Russian mathematician P. Chebyshev obtained a much simpler (yet rather interesting) result on prime distribution, guaranteeing the existence of positive constants c and C such that

$$c\frac{x}{\log x} \le \pi(x) \le C\frac{x}{\log x},$$

for every real $x \ge 2$.

A proof of the second one of the above inequalities will be the object of Problem 4; for a proof of the first one, we recommend to the reader the book of professor Andrews, already quoted above. On the other hand, it readily follows from the second one of the above inequalities that

$$\frac{\pi(x)}{x} \le \frac{C}{\log x}$$

for every real $x \ge 2$; in particular, one has

$$\lim_{x\to+\infty} \frac{\pi(x)}{x} = 0. \tag{9.2}$$

In words, (9.2) says that the total quantity of prime numbers less than or equal to x is an *infinitesimal* with respect to x (i.e., grows much slower than x itself) as $x \to +\infty$. In what follows, we shall give a direct prove of the validity of such a result. We start with the following preliminary

Proposition 9.2 *If k is a natural number, then*

$$\frac{\pi(x)}{x} < \frac{\varphi(k)}{k} + \frac{2k}{x},$$

where φ denotes Euler's function.

Proof For a real number $x > 0$, let $q, r \in \mathbb{Z}$ be such that $\lfloor x \rfloor = kq + r$, with $0 \le r < k$. Note that

$$\{1, 2, \ldots, \lfloor x \rfloor\} = \left(\bigcup_{j=0}^{q-1} \{kj + 1, kj + 2, \ldots, k(j + 1)\}\right) \cup$$

$$\cup \{kq + 1, kq + 2, \ldots, kq + r\}. \tag{9.3}$$

From 1 to k there is at most k prime numbers. On the other hand, if $j \geq 1$, then, for an element of $\{kj + 1, kj + 2, \ldots, k(j + 1)\}$ to be prime, it must necessarily be relatively prime with k. Hence, such a set contains at most $\varphi(k)$ prime numbers.

Now, since $q = \lfloor \frac{x}{k} \rfloor$, (9.3) allows us to write down the following upper estimate for $\pi(x)$:

$$\pi(x) \leq k + (q - 1)\varphi(k) + r < 2k + \left\lfloor \frac{x}{k} \right\rfloor \varphi(k) \leq 2k + \frac{x}{k}\varphi(k).$$

Finally, in order to get the stated result, it suffices to divide both sides of the last inequality above by x.   □

We shall also need the following technical

Lemma 9.3 *If $m > 1$ is an integer and $p_1, p_2, \ldots, p_n$ are the prime numbers less than or equal to m, then*

$$\sum_{j=1}^{m} \frac{1}{j} < \left[\left(1 - \frac{1}{p_1}\right)\left(1 - \frac{1}{p_2}\right) \cdots \left(1 - \frac{1}{p_n}\right) \right]^{-1}.$$

Proof Formula (3.1) for the sum of a convergent geometric series gives

$$\left[\left(1 - \frac{1}{p_1}\right)\left(1 - \frac{1}{p_2}\right) \cdots \left(1 - \frac{1}{p_n}\right) \right]^{-1} = \prod_{i=1}^{n} \left(1 + \frac{1}{p_i} + \frac{1}{p_i^2} + \cdots \right).$$

On the other hand, the choice of $p_1, p_2, \ldots, p_n$, together with the fundamental theorem of arithmetic, guarantees the validity of the inclusion

$$\left\{ 1, \frac{1}{2}, \ldots, \frac{1}{m} \right\} \subset \left\{ \frac{1}{p_1^{j_1} p_2^{j_2} \cdots p_n^{j_n}}; \ j_1, j_2, \ldots, j_n \geq 0 \right\}.$$

Therefore,

$$\sum_{j=1}^{m} \frac{1}{j} < \sum_{j_1, j_2, \ldots, j_n \geq 0} \frac{1}{p_1^{j_1} p_2^{j_2} \cdots p_n^{j_n}} = \prod_{i=1}^{n} \left(1 + \frac{1}{p_i} + \frac{1}{p_i^2} + \cdots \right),$$

and the desired result promptly follows.   □

The previous lemma, together with the divergence of the harmonic series (cf. Example 7.37 of [8], for instance), gives

$$\lim_{n \to +\infty} \left[\left(1 - \frac{1}{p_1}\right) \cdots \left(1 - \frac{1}{p_n}\right) \right]^{-1} = +\infty,$$

where p_j stands for the j-th prime. Therefore,

$$\lim_{n \to +\infty} \left(1 - \frac{1}{p_1}\right) \cdots \left(1 - \frac{1}{p_n}\right) = 0. \tag{9.4}$$

With the previous results at our disposal, we are finally in position to prove the promised weak version of Hadamard's theorem.

Theorem 9.4 $\lim_{x \to +\infty} \frac{\pi(x)}{x} = 0$.

Proof We have to prove that, given $\epsilon > 0$, there exists $x_0 > 0$ such that $x > x_0 \Rightarrow \frac{\pi(x)}{x} < \epsilon$. To this end, let $p_1, p_2, \ldots, p_n$ be the first n prime numbers, and $k = p_1 p_2 \ldots p_n$. Proposition 9.2, together with the formula for $\varphi(k)$, gives us

$$\frac{\pi(x)}{x} < \left(1 - \frac{1}{p_1}\right)\left(1 - \frac{1}{p_2}\right) \cdots \left(1 - \frac{1}{p_n}\right) + \frac{2p_1 p_2 \ldots p_n}{x}. \tag{9.5}$$

Now, (9.4) allows us to choose a natural number n such that

$$\left(1 - \frac{1}{p_1}\right)\left(1 - \frac{1}{p_2}\right) \cdots \left(1 - \frac{1}{p_n}\right) < \frac{\epsilon}{2}.$$

If we let $x_0 = \frac{4p_1 p_2 \ldots p_n}{\epsilon}$, then, for $x > x_0$, it follows from (9.5) that

$$\frac{\pi(x)}{x} < \frac{\epsilon}{2} + \frac{2p_1 p_2 \ldots p_n}{x} < \frac{\epsilon}{2} + \frac{2p_1 p_2 \ldots p_n}{x_0} = \epsilon,$$

as wished.   □

The arguments that led to (9.4) used the fact that the harmonic series diverges: $\sum_{n \geq 1} \frac{1}{n} = +\infty$. Heuristically, we can say that this is due to the fact that the sequence of naturals has *too many naturals* (actually, all of them). On the other hand, already in Lemma 3.15 we used the fact that the series of the inverses of perfect squares converges: $\sum_{n \geq 1} \frac{1}{n^2} < +\infty$. This time, and also from a heuristic point of view, we can attribute the convergence of the series to the fact that the sequence of perfect squares contains *too few naturals*.

In view of the above, a natural question poses itself in the context of prime numbers: if p_n stands for the n-th prime number, does the series $\sum_{n \geq 1} \frac{1}{p_n}$ have *too many* or *too few* naturals, in the above sense? Theorem 9.4 would encourage us to say that it does possess too few naturals.

At this point our intuition fails, for we shall prove below another theorem of the great L. Euler,[2] asserting the *divergence* of the series of the inverses of primes. To this end, we shall need the inequality below, which readily follows from (3.7):

$$e^x > 1 + x, \quad \forall \, x > 0. \tag{9.6}$$

---

[2]For a slightly different proof of such a result, see the problems of Section 10.9 of [8].

Theorem 9.5 (Euler) *If p_k denotes the k-th prime, then the series $\sum_{k\geq 1}\frac{1}{p_k}$ diverges.*

Proof If $m \in \mathbb{N}$ and $p_1, p_2, \ldots, p_k$ are the prime numbers less than or equal to m, it is immediate to see that

$$\left(\sum_{1\leq j<\sqrt{m}}\frac{1}{j^2}\right)\left(1+\sum_{1\leq i\leq k}\sum_{1\leq j_1<\cdots<j_i\leq k}\frac{1}{p_{j_1}p_{j_2}\cdots p_{j_i}}\right)\geq \sum_{j=1}^{m}\frac{1}{j}.$$

Now, observe that the second factor at the left hand side of the previous inequality is nothing but the sum of the inverses of the square free natural numbers less than or equal to m. Letting $\sum_{\substack{1\leq j\leq m\\ j \text{ sf}}}\frac{1}{j}$ denote this sum, we can write the above inequality as

$$\left(\sum_{1\leq j<\sqrt{m}}\frac{1}{j^2}\right)\left(\sum_{\substack{1\leq j\leq m\\ j \text{ sf}}}\frac{1}{j}\right)\geq \sum_{j=1}^{m}\frac{1}{j}.$$

Since $\sum_{j\geq 1}\frac{1}{j}$ diverges and $\sum_{j\geq 1}\frac{1}{j^2}$ converges, we conclude that

$$\sum_{\substack{j\geq 1\\ j \text{ sf}}}\frac{1}{j}$$

diverges. Finally, by the sake of contradiction, assume that the series of the inverses of primes converges to a certain real number a. Then, for every $n \in \mathbb{N}$, we get from (9.6) that

$$e^a > \exp\left(\sum_{\substack{p<n\\ p \text{ prime}}}\frac{1}{p}\right) = \prod_{\substack{p<n\\ p \text{ prime}}}\exp\left(\frac{1}{p}\right) > \prod_{\substack{p<n\\ p \text{ prime}}}\left(1+\frac{1}{p}\right) = \sum_{\substack{1\leq j<n\\ j \text{ sf}}}\frac{1}{j},$$

where $\exp : \mathbb{R} \to \mathbb{R}$ denotes the exponential function of basis e. But this is indeed a contradiction, for we already know that the sum of the inverses of the square free natural numbers diverges. □

Problems: Sect. 9.1

1. Prove that there exists a positive real x_0 for which $\pi(x) > \frac{\log x}{2x}$ for $x > x_0$.
2. For each integer $k \geq 1$, let a_k be the k-th natural number which is not a perfect square. Decide whether the series $\sum_{k\geq 1}\frac{1}{a_k}$ converges.

3. For each integer $k \geq 1$, let a_k be the k-th composite natural number. Decide whether the series $\sum_{k \geq 1} \frac{1}{a_k}$ converges.

4. The purpose of this problem is to show that

$$\pi(x) \leq (30 \log 2) \frac{x}{\log x}, \qquad (9.7)$$

for every real $x \geq 8$, where $\log : (0, +\infty) \to \mathbb{R}$ denotes the natural logarithm function. To this end, do the following items:

(a) For every $n \in \mathbb{N}$, prove that $\binom{2n}{n}$ is divisible by all prime numbers p satisfying $n < p \leq 2n$. Moreover, show that $\binom{2n}{n} < 2^{2n}$.

(b) Prove that, for a natural $n \geq 2$, one has $\pi(2n) < \pi(n) + (2 \log 2) \frac{n}{\log n}$.

(c) If $f : (0, +\infty) \to \mathbb{R}$ is given by $f(x) = \frac{x}{\log x}$, show that f is increasing in $(e, +\infty)$ and that $f\left(\frac{x+2}{2}\right) < \frac{15}{6} f(x)$ for $x \geq 8$.

(d) Use items (b) and (c) to conclude that $\pi(2n) < (32 \log 2) \frac{n}{\log n}$ for every integer $n \geq 2$.

(e) Deduce that (9.7) holds for every real $x \geq 8$.

5. The purpose of this problem is to establish the converse of Problem 17, page 191. To this end, given $n \in \mathbb{N}$, let $I(n)$ denote the sum of the greatest odd divisors of the numbers $1, 2, \ldots, n$ and do the following items:

(a) For $n \in \mathbb{N}$, let $\tau(n)$ be the greatest odd divisor of n and $i(n)$ the greatest odd integer less than or equal to n. Prove that

$$I(n) = (\tau(1) + \tau(3) + \cdots + \tau(i(n)))$$
$$+ (\tau(2) + \tau(4) + \cdots + \tau(2\lfloor n/2 \rfloor))$$
$$= \frac{1}{4}(i(n) + 1)^2 + I\lfloor n/2 \rfloor.$$

(b) In the notations of (a), show that

$$I(n) = \frac{1}{4} \sum_{k=0}^{t} (i(q_k) + 1)^2,$$

where 2^t is the greatest power of 2 which is less than or equal to n and, for $0 \leq k \leq t$, q_k is the quotient one gets upon dividing n by 2^k.

(c) Use the fact that $i(q_k) \leq \frac{n}{2^k}$ to get the estimate

$$I(n) \leq \frac{1}{4} \left(\frac{n^2}{3} \left(4 - \frac{1}{4^t} \right) + n \left(4 - \frac{1}{2^{t-1}} \right) + t + 1 \right).$$

(d) Use the fact that $t = \lfloor \log_2 n \rfloor \leq n$ to get the estimate

$$I(n) \leq \frac{4n^2 + 15n + 3}{12}.$$

(e) Use inequalities $\lfloor \log_2 n \rfloor \geq 0$, $2^{\lfloor \log_2 n \rfloor} \geq 2^{\log_2 n - 1} = \frac{n}{2}$ and

$$\tau(q_k) \geq q_k - 1 = \left\lfloor \frac{n}{2^k} \right\rfloor - 1 \geq \frac{n}{2^k} - 2$$

to show, in an analogous way, that

$$I(n) \geq \frac{4n^2 - 12n - 1}{12}.$$

(f) If $T(n) = 1 + 2 + \cdots + n$, conclude that

$$\frac{4n^2 - 12n - 1}{6n^2 + 6n} \leq \frac{I(n)}{T(n)} \leq \frac{4n^2 + 15n + 3}{6n^2 + 6n}.$$

Then, show that if $r \in \mathbb{Q} \setminus \{\frac{2}{3}\}$, then $\frac{I(n)}{T(n)} = r$ for at most a finite number of integer values of n.

9.2 Chebyshev's Theorem

The Prime Number Theorem guarantees that the difference $\pi(2x) - \pi(x)$ is asymptotically equal to

$$\frac{2x}{\log(2x)} - \frac{x}{\log x} = \frac{x}{\log x}\left(\frac{2\log x}{\log 2 + \log x} - 1\right)$$

$$= \frac{x}{\log x}\left(\frac{\log x - \log 2}{\log x + \log 2}\right).$$

However, since

$$\lim_{x \to +\infty} \frac{\log x - \log 2}{\log x + \log 2} = 1,$$

we conclude that $\pi(2x) - \pi(x)$ is asymptotically equal to $\frac{x}{\log x}$. It happens (for instance by L'Hôpital's rule) that

$$\lim_{x \to +\infty} \frac{x}{\log x} = +\infty,$$

so that the above reasoning assures that the number of primes between x and $2x$ grows unboundedly as $x \to +\infty$.

In this section we give a complete proof of a much weaker result, also due to Chebyshev, which shows that there is at least one prime number between n and $2n$, for every integer $n > 1$. For the proof of it, we shall need the two following auxiliary results.

Lemma 9.6 *Let $n, p \in \mathbb{N}$, with p prime, and let μ_p be the exponent of p in the factorisation of $\binom{2n}{n}$. Then,*

$$\mu_p = \sum_{j \geq 1} \left(\left\lfloor \frac{2n}{p^j} \right\rfloor - 2 \left\lfloor \frac{n}{p^j} \right\rfloor \right).$$

Moreover, if ν_p is the only integer such that $p^{\nu_p} \leq 2n < p^{\nu_p+1}$, then $\mu_p \leq \nu_p$.

Proof For the first part, since $\binom{2n}{n} = \frac{(2n)!}{(n!)^2}$, we readily conclude that μ_p is given by $p^{\mu_p} = \frac{p^{e_p(2n)}}{p^{2e_p(n)}}$, where $e_p(2n)$ and $e_p(n)$ are as in Legendre's formula (cf. Proposition 6.10). Hence,

$$\mu_p = e_p(2n) - 2e_p(n),$$

and it suffices to use (6.10) to get the stated formula for μ_p.

For the second part, it follows from the definition of ν_p that

$$j > \nu_p \Rightarrow p^j > 2n \Rightarrow \left\lfloor \frac{2n}{p^j} \right\rfloor - 2 \left\lfloor \frac{n}{p^j} \right\rfloor = 0.$$

On the other hand, for $j \geq 1$ we have

$$\left\lfloor \frac{2n}{p^j} \right\rfloor - 2 \left\lfloor \frac{n}{p^j} \right\rfloor < \frac{2n}{p^j} - 2 \left(\frac{n}{p^j} - 1 \right) = 2,$$

so that $\left\lfloor \frac{2n}{p^j} \right\rfloor - 2 \left\lfloor \frac{n}{p^j} \right\rfloor \leq 1$. Therefore, gathering together the two pieces of information above, we obtain

$$\mu_p = \sum_{j \geq 1} \left(\left\lfloor \frac{2n}{p^j} \right\rfloor - 2 \left\lfloor \frac{n}{p^j} \right\rfloor \right) = \sum_{j=1}^{\nu_p} \left(\left\lfloor \frac{2n}{p^j} \right\rfloor - 2 \left\lfloor \frac{n}{p^j} \right\rfloor \right) \leq \sum_{j=1}^{\nu_p} 1 = \nu_p.$$

$\square$

Lemma 9.7 *If $x \geq 2$ is a real number, then*

$$\prod_{\substack{p \leq x \\ p \ prime}} p < 4^x.$$

Proof The lemma is clearly true for $2 \leq x \leq 3$. On the other hand, if it is valid for $x = n$, where $n \geq 3$ is an odd integer, then it will also be valid for $n \leq y < n + 2$; indeed, in such a case $n + 1$ is even, so that

$$\prod_{\substack{p \leq y \\ p \text{ prime}}} p = \prod_{\substack{p \leq n \\ p \text{ prime}}} p < 4^n \leq 4^y.$$

It then suffices to show that the lemma is true for $x = n$, where $n \geq 3$ is an odd integer. To this end, we make induction on $n \geq 3$ odd, noticing that we already have the validity of the result for $n = 3$. By induction hypothesis, assume that it is also true for all odd integers less than a certain odd integer $n \geq 5$. Let $k = \frac{n \pm 1}{2}$, with the sign chosen in such a way that k is odd. Then, $k \geq 3$ and $n - k$ is even and such that

$$n - k = 2k \mp 1 - k \leq k + 1.$$

Hence, if p is a prime number for which $k < p \leq n$, then p is odd, $p \mid n!$, $p \nmid k!$ and $p \nmid (n - k)!$ (it cannot happen that $n - k = k + 1 = p$, since $n - k$ is even). Thus, we conclude that the product of all such primes divides

$$\binom{n}{k} = \frac{n!}{k!(n - k)!},$$

so that

$$\prod_{\substack{k < p \leq n \\ p \text{ prime}}} p \leq \binom{n}{k}. \tag{9.8}$$

Now, our choice of k assures that k and $n - k$ are distinct. However, since $\binom{n}{k} = \binom{n}{n-k}$ and such binomial numbers are summands in the binomial expansion of $2^n = (1 + 1)^n$, we get $\binom{n}{k} \leq 2^{n-1}$, and (9.8) furnishes

$$\prod_{\substack{k < p \leq n \\ p \text{ prime}}} p \leq 2^{n-1}. \tag{9.9}$$

Therefore, the induction hypothesis and (9.9) give

$$\prod_{\substack{p \leq n \\ p \text{ prime}}} p = \left(\prod_{\substack{p \leq k \\ p \text{ prime}}} p \right) \left(\prod_{\substack{k < p \leq n \\ p \text{ prime}}} p \right) < 4^k \cdot 2^{n-1} = 2^{2k+n-1} \leq 2^{2n} = 4^n.$$

$\square$

We have finally arrived at the desired result, which is known as **Chebyshev's theorem**.[3]

Theorem 9.8 (Chebyshev) *For each integer $n > 1$, there is at least one prime number between n and $2n$.*

Proof A general argument will prove the result when $n \geq 128$. For $n < 128$ take $p = 3$ if $n = 2$, $p = 5$ if $n = 3$ and:

$$p = 7 \text{ if } 4 \leq n \leq 6;$$
$$p = 13 \text{ if } 7 \leq n \leq 12;$$
$$p = 23 \text{ if } 13 \leq n \leq 22;$$
$$p = 43 \text{ if } 23 \leq n \leq 42;$$
$$p = 83 \text{ if } 43 \leq n \leq 82;$$
$$p = 131 \text{ if } 83 \leq n \leq 127.$$

Now, assume the result is false for some integer $n \geq 128$. In the notations of Lemma 9.6, such a supposition guarantees that

$$\binom{2n}{n} = \prod_{\substack{p \leq 2n \\ p \text{ prime}}} p^{\mu_p} = \prod_{\substack{p \leq n \\ p \text{ prime}}} p^{\mu_p}.$$

However, for any prime p such that $\frac{2n}{3} < p \leq n$, we have

$$p \geq 3, \quad p^2 > \frac{2}{3}np, \quad 1 \leq \frac{n}{p} < \frac{3}{2}, \quad \text{and } 2 \leq \frac{2n}{p} < 3,$$

so that $\frac{2n}{p^2} < \frac{3}{p} \leq 1$; it thus follows that

$$\mu_p = \sum_{j \geq 1} \left(\left\lfloor \frac{2n}{p^j} \right\rfloor - 2 \left\lfloor \frac{n}{p^j} \right\rfloor \right) = \left\lfloor \frac{2n}{p} \right\rfloor - 2 \left\lfloor \frac{n}{p} \right\rfloor = 2 - 2 = 0.$$

On the other hand, for any prime p such that $\sqrt{2n} < p \leq \frac{2n}{3}$, we have $p^2 > 2n$, and Lemma 9.6 gives $1 \leq \mu_p \leq \nu_p = 1$. Finally, for primes p such that $p \leq \sqrt{2n}$, we have, again by Lemma 9.6, that $p^{\mu_p} \leq p^{\nu_p} \leq 2n$.

---

[3] Pafnuty Chebyshev, Russian mathematician of the nineteenth century.

The estimates above give us (in what follows, p denotes a prime number)

$$\binom{2n}{n} = \left(\prod_{p \le \sqrt{2n}} p^{\mu_p} \right) \left(\prod_{\sqrt{2n} < p \le \frac{2n}{3}} p^{\mu_p} \right) \left(\prod_{\frac{2n}{3} < p \le n} p^{\mu_p} \right)$$

$$= \left(\prod_{p \le \sqrt{2n}} 2n \right) \left(\prod_{\sqrt{2n} < p \le \frac{2n}{3}} p \right)$$

$$< 4^{\frac{2n}{3}} \left(\prod_{p \le \sqrt{2n}} 2n \right),$$

where we used Lemma 9.7 in the inequality above.

Since 9 and the even integers greater than 2 are not primes, the condition $n \ge 128$ ($\Leftrightarrow \sqrt{2n} \ge 16$) assures that

$$\pi\left(\sqrt{2n} \right) \le \frac{\sqrt{2n} - 1}{2} - 1.$$

Substituting this last estimate into (9.10), we arrive at

$$\binom{2n}{n} < 4^{\frac{2n}{3}} (2n)^{\pi(\sqrt{2n})} < 4^{\frac{2n}{3}} (2n)^{\sqrt{\frac{n}{2}} - 1}. \tag{9.10}$$

Finally, note that $\binom{2n}{n}$ is the largest among all of the $2n + 1$ summands in the binomial expansion of $4^n = (1 + 1)^{2n}$. Since the first and last of these summands are both equal to 1, we get $2n\binom{2n}{n} > 4^n$ or, which is the same, $\binom{2n}{n} > \frac{4^n}{2n}$. In turn, such an estimate, combined with (9.10), furnishes

$$\frac{4^n}{2n} < 4^{\frac{2n}{3}} (2n)^{\sqrt{\frac{n}{2}} - 1},$$

which yields

$$2^{\frac{2n}{3}} < (2n)^{\sqrt{\frac{n}{2}}}.$$

Taking natural logarithms an dividing both sides by $\frac{\sqrt{2n}}{6}$, we conclude that the above inequality can be rewritten as

$$\sqrt{8n} \log 2 - 3 \log(2n) < 0. \tag{9.11}$$

Letting $f(x) = \sqrt{8x} \log 2 - 3 \log(2x)$, we have $f(128) = 8 \log 2 > 0$ and

$$f'(x) = \frac{\sqrt{2x} \log 2 - 3}{x} > 0$$

for $x \geq 128$. Thus, f is increasing for $x \geq 128$ and (9.11) is false, which gives us the desired contradiction.   □

The famous **Goldbach conjecture**[4] claims that every even number greater than 2 can be written as the sum of two (not necessarily distinct) primes. Although a serious account on the state of the art on Goldbach's conjecture is far beyond the scope of these notes, Chebyshev's theorem allows us to present a simple example on the possibility of writing natural numbers as sums of distinct primes.

Example 9.9 Let $\mathbb{P}$ stand for the set of prime numbers. Show that every natural number can be written as a sum of distinct elements of the set $\{1\} \cup \mathbb{P}$.

Proof Let $n > 1$ be a natural number and p be the greatest prime number less than or equal to n. If $n \geq 2p$, then Chebyshev's theorem would give a prime q between p and $2p$, which would contradict the maximality of p. Therefore, $p \leq n < 2p$ and, hence, $0 \leq n - p < p$. Arguing by induction, we conclude that $n - p$ can be written as a sum of distinct elements of the set $\{1\} \cup \mathbb{P}$, and writing $n = (n - p) + p$ we conclude that the same is true of n.   □

Problems: Sect. 9.2

1. If p_n is the n-th prime, prove that $p_{n+1} < 2p_n$.
2. * Prove that, for every integer $n > 1$, there exists a prime p whose exponent in the canonical decomposition of $n!$ is equal to 1.
3. (TT) Prove that, for every integer $n > 1$, the number $1!2! \cdots n!$ is not a perfect square.
4. Find all natural numbers m, n, x and y such that $m, n > 1$ and $(m!)^x = (n!)^y$.

   For the two coming problems, the reader will find it convenient to use the following *stronger version* of Chebyschev's theorem: for every integer $n \geq 6$, there are at least two primes between n and $2n$.
5. (OCS) Find all $m, n \in \mathbb{N}$ such that $(n - 1)!n! = m!$.
6. Find all natural numbers $n > 1$ such that every natural $1 < m < n$ which is relatively prime with n is indeed a prime number.

---

[4] After Christian Goldbach, German mathematician of the eighteenth century.

9.3   Cèsaro's Theorem

In 1881, E. Cesàro[5] proved that the odds that two randomly chosen natural numbers be relatively prime is equal to $\frac{6}{\pi^2}$. We finish this chapter by presenting an elementary proof of this fact. Our exposition follows a nonpublished manuscript of prof. Hudson N. Lima (cf. [29]). We start by recalling a few rather simple facts on the calculus of probabilities.

Let E be a finite nonempty set, said to be the **sampling space**, and P be a **probability distribution** in E, i.e., a function $P : E \to [0, 1]$ such that

$$\sum_{x \in E} P(x) = 1.$$

For each $x \in E$, we call $P(x)$ the **probability** of x; also, we say that the elements of E are **equiprobable** provided

$$P(x) = \frac{1}{|E|}, \ \forall \, x \in E.$$

More generally, an **event** in E is a subset X of E, and we let its probability be defined by

$$P(X) = \sum_{x \subset X} P(x).$$

Back to the problem we wish to analyse, let[6] $E = I_n \times I_n$ and assume that the elements of E are equiprobable, so that the probability of each one of them equals $\frac{1}{n^2}$. Given $n \in \mathbb{N}$, let P_n be the probability that a randomly chosen ordered pair $(a, b) \in E$ have relatively prime entries a and b. We want to show that

$$\lim_{n \to \infty} P_n = \frac{6}{\pi^2}.$$

We shall first of all find an adequate expression for P_n, and to this end we begin by observing that $P_n = \frac{|X|}{n^2}$, where

$$X = \{(a, b) \in I_n \times I_n; \ \gcd(a, b) = 1\}$$

is the event subjacent to P_n.

---

[5]Ernesto Cesàro, Italian mathematician of the nineteenth century.

[6]As in previous chapters, we set $I_n = \{1, 2, \ldots, n\}$.

Elementary counting gives

$$|X| = 2\#\{(a, b) \in I_n \times I_n;\ \gcd(a, b) = 1 \text{ and } a \leq b\}$$
$$- \#\{(a, a) \in I_n \times I_n;\ \gcd(a, a) = 1\}$$
$$= 2 \sum_{b=1}^{n} \#\{a \in I_n;\ \gcd(a, b) = 1 \text{ and } a \leq b\} - 1$$
$$= 2 \sum_{b=1}^{n} \varphi(b) - 1,$$

where φ denotes Euler's function. Hence,

$$P_n = \frac{2 \sum_{b=1}^{n} \varphi(b) - 1}{n^2}. \tag{9.12}$$

We now need the following auxiliary result.

Lemma 9.10 *If $f : \mathbb{N} \to \mathbb{R}$ is any function, then for $n \in \mathbb{N}$ we have:*

(a) $\sum_{k=1}^{n} \sum_{0 < d \mid k} f(d) = \sum_{k=1}^{n} f(k) \lfloor \frac{n}{k} \rfloor$.
(b) $\sum_{k=1}^{n} k \sum_{0 < d \mid k} f(d) = \frac{1}{2} \sum_{k=1}^{n} k f(k) \lfloor \frac{n}{k} \rfloor (\lfloor \frac{n}{k} \rfloor + 1)$.

Proof Item (a) is the content of Problem 12, page 219. Concerning (b), note firstly that the set of ordered pairs (d, k) of integers such that $1 \leq k \leq n$ and $0 < d \mid k$ coincides with the set of ordered pairs (d, k) of integers such that $1 \leq d \leq n$ and $k = ld$, for some integer $1 \leq l \leq n/d$. Hence, we can change the order of the involved sums, thus getting

$$\sum_{k=1}^{n} k \sum_{0 < d \mid k} f(d) = \sum_{d=1}^{n} \sum_{\substack{k=ld \\ 1 \leq l \leq \frac{n}{d}}} k f(d)$$
$$= \sum_{d=1}^{n} f(d) \left(d + 2d + \cdots + \left\lfloor \frac{n}{d} \right\rfloor d \right)$$
$$= \frac{1}{2} \sum_{d=1}^{n} d f(d) \left\lfloor \frac{n}{d} \right\rfloor \left(\left\lfloor \frac{n}{d} \right\rfloor + 1 \right).$$

$\square$

Now, recall that, according to (8.5),

$$\varphi(n) = n \sum_{0 < d \mid n} \frac{\mu(d)}{d},$$

where $\mu : \mathbb{N} \to \mathbb{R}$ stands for the Möbius function. Therefore, item (b) of the previous lemma furnishes

$$2 \sum_{k=1}^{n} \varphi(k) = 2 \sum_{k=1}^{n} k \sum_{0<d|k} \frac{\mu(d)}{d}$$

$$= \sum_{k=1}^{n} \mu(k) \left\lfloor \frac{n}{k} \right\rfloor \left(\left\lfloor \frac{n}{k} \right\rfloor + 1 \right) \tag{9.13}$$

$$= \sum_{k=1}^{n} \mu(k) \left\lfloor \frac{n}{k} \right\rfloor^2 + \sum_{k=1}^{n} \mu(k) \left\lfloor \frac{n}{k} \right\rfloor.$$

The second summand in the last sum above can be computed with the aid of item (a) of the previous lemma, together with the result of Proposition 8.7. Indeed, we have

$$\sum_{k=1}^{n} \mu(k) \left\lfloor \frac{n}{k} \right\rfloor = \sum_{k=1}^{n} \sum_{0<d|k} \mu(d) = 1,$$

since $\sum_{0<d|k} \mu(d)$ does not vanish only for $k = 1$. It follows from (9.12) and (9.13) that

$$P_n = \frac{1}{n^2} \sum_{k=1}^{n} \mu(k) \left\lfloor \frac{n}{k} \right\rfloor^2. \tag{9.14}$$

With the last formula above at our disposal, we can state and prove the coming proposition, which assures that the limit $\lim_{n\to\infty} P_n$ does exist and expresses it as another limit which, as we shall see, can be effectively computed. We begin by noticing that, since $|\mu(k)| \leq 1$ for every $k \in \mathbb{N}$ and $\sum_{k\geq 1} \frac{1}{k^2}$ is convergent, the series

$$\sum_{k\geq 1} \frac{\mu(k)}{k^2}$$

is absolutely convergent, hence convergent.

Proposition 9.11 *If P_n is as in (9.14), then $\lim_{n\to\infty} P_n$ exists and is such that*

$$\lim_{n\to\infty} P_n = \sum_{k\geq 1} \frac{\mu(k)}{k^2}.$$

Proof Firstly, from (9.14) we get

$$\left| P_n - \sum_{k=1}^{n} \frac{\mu(k)}{k^2} \right| = \left| \sum_{k=1}^{n} \mu(k) \left(\frac{1}{n^2} \left\lfloor \frac{n}{k} \right\rfloor^2 - \frac{1}{k^2} \right) \right|$$

$$\leq \sum_{k=1}^{n} \left| \frac{1}{k^2} - \frac{1}{n^2} \left\lfloor \frac{n}{k} \right\rfloor^2 \right|. \tag{9.15}$$

In order to estimate the last sum above, we claim that, given natural numbers n and k such that $1 \leq k \leq n$, we have

$$\left| \frac{1}{k^2} - \frac{1}{n^2} \left\lfloor \frac{n}{k} \right\rfloor^2 \right| < \frac{2}{nk} - \frac{1}{n^2}.$$

Indeed,

$$\frac{n}{k} - 1 < \left\lfloor \frac{n}{k} \right\rfloor \leq \frac{n}{k} \Rightarrow \frac{n^2}{k^2} - \frac{2n}{k} + 1 < \left\lfloor \frac{n}{k} \right\rfloor^2 \leq \frac{n^2}{k^2}$$

$$\Rightarrow \frac{1}{k^2} - \frac{2}{kn} + \frac{1}{n^2} < \frac{1}{n^2} \left\lfloor \frac{n}{k} \right\rfloor^2 \leq \frac{1}{k^2}$$

$$\Rightarrow 0 \leq \frac{1}{k^2} - \frac{1}{n^2} \left\lfloor \frac{n}{k} \right\rfloor^2 < \frac{2}{kn} - \frac{1}{n^2},$$

as wished.

Back to (9.15), we obtain from the above estimates that

$$\left| P_n - \sum_{k=1}^{n} \frac{\mu(k)}{k^2} \right| < \sum_{k=1}^{n} \left(\frac{2}{nk} - \frac{1}{n^2} \right) = \frac{2}{n} \sum_{k=1}^{n} \frac{1}{k} - \frac{1}{n}.$$

Now, from L'Hôpital's rule we get

$$\frac{2}{n} \sum_{k=1}^{n} \frac{1}{k} < \frac{2}{n} \left(1 + \int_1^n \frac{1}{t} dt \right) = \frac{2}{n} (\log n + 1) \to 0$$

as $n \to +\infty$. Hence,

$$\lim_{n \to +\infty} \left(\frac{2}{n} \sum_{k=1}^{n} \frac{1}{k} - \frac{1}{n} \right) = 0,$$

and our previous estimates assure that

$$\lim_{n \to +\infty} \left(P_n - \sum_{k=1}^{n} \frac{\mu(k)}{k^2} \right) = 0.$$

Finally, it follows from the above that

$$\lim_{n \to \infty} P_n = \lim_{n \to \infty} \left(P_n - \sum_{k=1}^{n} \frac{\mu(k)}{k^2} \right) + \sum_{k \geq 1} \frac{\mu(k)}{k^2} = \sum_{k \geq 1} \frac{\mu(k)}{k^2}.$$

□

The former result reduced the proof of Cesàro's theorem to showing that

$$\sum_{k \geq 1} \frac{\mu(k)}{k^2} = \frac{6}{\pi^2}.$$

To this end, we first need to show the coming result.

Theorem 9.12 $\displaystyle\sum_{k \geq 1} \frac{1}{k^2} = \frac{\pi^2}{6}.$

The proof we present is due to professors A. Yaglom and I. Yaglom (cf. Chapter 8 of [1]) and relies on some elementary results on complex numbers and roots of polynomials; Chaps. 13 and 16 cover all that is necessary. The classical alternative proof, relying upon the theory of *Fourier Series*, is delineated in problems 11 and 12 of Section 11.2 of [8] (alternatively, see Chapter 8 of [33] or Chapter 2 of [36]).

Proof The first of de Moivre's formulas (cf. Proposition 13.9) gives us

$$\sin(n\theta) = \text{Im}((\cos\theta + i\sin\theta)^n)$$

$$= \text{Im}\left\{ \sum_{j=0}^{n} \binom{n}{j} i^j (\cos\theta)^{n-j} (\sin\theta)^j \right\}$$

$$= \sum_{k=0}^{\lfloor \frac{n-1}{2} \rfloor} (-1)^k \binom{n}{2k+1} (\cos\theta)^{(n-2k-1)} (\sin\theta)^{(2k+1)}.$$

Therefore,

$$\frac{\sin(n\theta)}{(\sin\theta)^n} = \sum_{k=0}^{\lfloor \frac{n-1}{2} \rfloor} (-1)^k \binom{n}{2k+1} (\cot\theta)^{(n-2k-1)},$$

whenever $\sin\theta \neq 0$.

For $n = 2m + 1$, we have

$$\frac{\sin((2m + 1)\theta)}{(\sin \theta)^{(2m+1)}} = \sum_{k=0}^{m}(-1)^k \binom{2m + 1}{2k + 1}(\cot \theta)^{(2m-2k)}.$$

Thus, letting

$$f(x) = \sum_{k=0}^{m}(-1)^k \binom{2m + 1}{2k + 1}x^{m-k},$$

we have shown that

$$f((\cot \theta)^2) = \frac{\sin((2m + 1)\theta)}{(\sin \theta)^{(2m+1)}}. \tag{9.16}$$

Setting $\theta_k = (\frac{k}{2m+1})\pi$ for $k = 1, 2, \ldots, m$, it is immediate from (9.16) that $(\cot \theta_1)^2, (\cot \theta_2)^2, \ldots, (\cot \theta_m)^2$ are m distinct roots of f. However, since f has degree m, we conclude that those are all of its complex roots. Now, the Girard-Viète relations between coefficients and roots of a polynomial (cf. Proposition 16.6) give

$$(\cot \theta_1)^2 + (\cot \theta_2)^2 + \cdots + (\cot \theta_m)^2 = \frac{\binom{2m+1}{3}}{\binom{2m+1}{1}} = \frac{m(2m - 1)}{3}.$$

From this, basic Trigonometry allows us to compute

$$\begin{aligned}
(\csc \theta_1)^2 + \cdots + (\csc \theta_m)^2 &= (1 + (\cot \theta_1)^2) + \cdots + (1 + (\cot \theta_m)^2) \\
&= m + ((\cot \theta_1)^2 + \cdots + (\cot \theta_m)^2) \\
&= m + \frac{m(2m - 1)}{3} = \frac{2m(m + 1)}{3}.
\end{aligned}$$

Finally, since $\theta_j \in (0, \frac{\pi}{2})$ for $1 \le j \le m$, we have for $1 \le j \le m$ the inequalities

$$\sin \theta_j \le \theta_j \le \tan \theta_j$$

or, which is the same,

$$(\cot \theta_j)^2 \le \frac{1}{\theta_j^2} \le (\csc \theta_j)^2.$$

Therefore,

$$\frac{m(2m-1)}{3} = \sum_{j=1}^{m}(\cot\theta_j)^2 \le \sum_{j=1}^{m}\frac{1}{\theta_j^2} \le \sum_{j=1}^{m}(\csc\theta_j)^2 = \frac{2m(m+1)}{3}$$

and, taking the definition of the θ_j into account,

$$\frac{m(2m-1)}{3} \le \sum_{j=1}^{m}\frac{(2m+1)^2}{\pi^2 j^2} \le \frac{2m(m+1)}{3}.$$

Multiplying the last inequalities above by $\frac{\pi^2}{(2m+1)^2}$, letting $m \to +\infty$ and noticing that

$$\lim_{m\to\infty}\frac{m(2m-1)}{3(2m+1)^2} = \lim_{m\to\infty}\frac{2m(m+1)}{3(2m+1)^2} = \frac{1}{6},$$

the squeezing principle for limits of sequences gives us the desired result. □

Theorem 9.13 *If p_n stands for the n-th prime number, then*

$$\lim_{n\to\infty}\prod_{k=1}^{n}\left(1-\frac{1}{p_k^2}\right)^{-1} = \sum_{k\ge1}\frac{1}{k^2}.$$

Proof Since $p_n > n$ for every $n \in \mathbb{N}$, the fundamental theorem of arithmetic gives (upon expanding the right hand side below)

$$\sum_{k=1}^{n}\frac{1}{k^2} \le \left(1+\frac{1}{p_1^2}+\cdots+\frac{1}{p_1^{2l}}\right)\cdots\left(1+\frac{1}{p_n^2}+\cdots+\frac{1}{p_n^{2l}}\right)$$

for every integer $l \ge 1$. Moreover, also for every integer $l \ge 1$, we have

$$\left(1+\frac{1}{p_1^2}+\cdots+\frac{1}{p_1^{2l}}\right)\cdots\left(1+\frac{1}{p_n^2}+\cdots+\frac{1}{p_n^{2l}}\right) \le \sum_{k=1}^{p_1^l\cdots p_n^l}\frac{1}{k^2},$$

once more from the fundamental theorem of arithmetic, together with the fact that each summand obtained from the expansion of the products at the left hand side is also present at the right hand side.

Since

$$\sum_{k=1}^{p_1^l\cdots p_n^l}\frac{1}{k^2} \le \sum_{k\ge1}\frac{1}{k^2},$$

we obtain the set of inequalities

$$\sum_{k=1}^{n} \frac{1}{k^2} \leq \left(1 + \frac{1}{p_1^2} + \cdots + \frac{1}{p_1^{2l}}\right) \cdots \left(1 + \frac{1}{p_n^2} + \cdots + \frac{1}{p_n^{2l}}\right) \leq \sum_{k \geq 1} \frac{1}{k^2}.$$

Letting $l \to +\infty$ and observing that

$$\lim_{l \to \infty} \left(1 + \frac{1}{p_i^2} + \cdots + \frac{1}{p_i^{2l}}\right) = \sum_{j \geq 0} \frac{1}{p_i^{2j}} = \left(1 - \frac{1}{p_i^2}\right)^{-1},$$

we get

$$\sum_{k=1}^{n} \frac{1}{k^2} \leq \left(1 - \frac{1}{p_1^2}\right)^{-1} \cdots \left(1 - \frac{1}{p_n^2}\right)^{-1} \leq \sum_{k \geq 1} \frac{1}{k^2}$$

ou, which is the same,

$$\sum_{k=1}^{n} \frac{1}{k^2} \leq \prod_{k=1}^{n} \left(1 - \frac{1}{p_k^2}\right)^{-1} \leq \sum_{k \geq 1} \frac{1}{k^2}.$$

Finally, it now suffices to let $n \to +\infty$ and invoke the squeezing theorem. □

Theorem 9.14 *If p_n denotes the n-th prime number, then*

$$\lim_{n \to \infty} \prod_{k=1}^{n} \left(1 - \frac{1}{p_k^2}\right) = \sum_{k \geq 1} \frac{\mu(k)}{k^2}.$$

Proof In all that follows, we make the following convention on indices: $l < m \Rightarrow$ $i_l < i_m$. Given $n \in \mathbb{N}$, let:

- A_n be the set of natural numbers of the form $p_{i_1} p_{i_2} \cdots p_{i_{2s}}$, with $s \in \mathbb{N}$ and p_{i_1}, $p_{i_2}, \ldots, p_{i_{2s}}$ being prime numbers such that $p_{i_1} p_{i_2} \cdots p_{i_{2s}} \leq n$;
- $\tilde{A}_n$ be the set of natural numbers of the form $p_{i_1} p_{i_2} \cdots p_{i_{2s}}$, with $s \in \mathbb{N}$, p_{i_1} and $p_{i_2}, \ldots, p_{i_{2s}}$ being primes such that $i_1, i_2, \ldots, i_{2s} \leq n$;
- B_n be the set of naturals of the form $p_{i_1} p_{i_2} \cdots p_{i_{2s+1}}$, with $s \in \mathbb{N}$ and p_{i_1}, p_{i_2}, $\ldots, p_{i_{2s+1}}$ being prime numbers such that $p_{i_1} p_{i_2} \cdots p_{i_{2s+1}} \leq n$;
- $\tilde{B}_n$ be the set of naturals of the form $p_{i_1} p_{i_2} \cdots p_{i_{2s+1}}$, with $s \in \mathbb{N}$ and p_{i_1}, p_{i_2}, $\ldots, p_{i_{2s+1}}$ being prime numbers such that $i_1, i_2, \ldots, i_{2s+1} \leq n$;

Let also

$$a_n = 1 + \sum_{x \in A_n} \frac{1}{x^2}, \ \tilde{a}_n = 1 + \sum_{x \in \tilde{A}_n} \frac{1}{x^2}, \ b_n = \sum_{x \in B_n} \frac{1}{x^2} \ \text{and} \ \tilde{b}_n = \sum_{x \in \tilde{B}_n} \frac{1}{x^2}.$$

It is immediate to verify that

$$a_n - b_n = \sum_{k=1}^{n} \frac{\mu(k)}{k^2} \ \text{and} \ \tilde{a}_n - \tilde{b}_n = \prod_{k=1}^{n} \left(1 - \frac{1}{p_k^2}\right).$$

Moreover, the limits $\lim a_n$, $\lim b_n$, $\lim \tilde{a}_n$ and $\lim \tilde{b}_n$ all exist, for the corresponding sequences are nondecreasing and bounded above by $\sum_{k=1}^{\infty} \frac{1}{k^2}$. In particular, this reasoning assures the existence of the limit

$$\lim_{n \to \infty} \prod_{k=1}^{n} \left(1 - \frac{1}{p_k^2}\right).$$

Since $p_n > n$ for every $n \in \mathbb{N}$, we have $A_n \subset \tilde{A}_n \subset A_{p_1 p_2 \ldots p_n}$ and $B_n \subset \tilde{B}_n \subset B_{p_1 p_2 \ldots p_n}$. Therefore, the following inequalities are valid:

$$a_n \leq \tilde{a}_n \leq a_{p_1 p_2 \ldots p_n} \ \text{and} \ b_n \leq \tilde{b}_n \leq b_{p_1 p_2 \ldots p_n}.$$

Hence, once more from the squeezing theorem, we get

$$\lim_{n \to \infty} a_n = \lim_{n \to \infty} \tilde{a}_n \ \text{and} \ \lim_{n \to \infty} b_n = \lim_{n \to \infty} \tilde{b}_n,$$

so that

$$\lim_{n \to \infty} \prod_{k=1}^{n} \left(1 - \frac{1}{p_k^2}\right) = \lim_{n \to \infty} \tilde{a}_n - \lim_{n \to \infty} \tilde{b}_n = \lim_{n \to \infty} a_n - \lim_{n \to \infty} b_n$$

$$= \lim_{n \to \infty} \sum_{k=1}^{n} \frac{\mu(k)}{k^2} = \sum_{k \geq 1} \frac{\mu(k)}{k^2},$$

as wished. $\qquad\qquad\qquad\qquad\qquad\qquad\qquad\qquad\qquad\qquad\qquad\qquad\qquad\qquad$ □

We can finally gather together the results above to compute

$$\lim_{n\to\infty} P_n = \sum_{k\geq 1} \frac{\mu(k)}{k^2} = \lim_{n\to\infty} \prod_{k=1}^{n} \left(1 - \frac{1}{p_k^2}\right)$$

$$= \left[\lim_{n\to\infty} \prod_{k=1}^{n} \left(1 - \frac{1}{p_k^2}\right)^{-1}\right]^{-1}$$

$$= \left[\sum_{k\geq 1} \frac{1}{k^2}\right]^{-1} = \frac{6}{\pi^2}.$$

Chapter 10
The Relation of Congruence

In this chapter, we define and explore the most basic properties of the important relation of congruence modulo $n > 1$. Our central goal is to prove the famous *Fermat's little theorem*, as well as its generalization, due to Euler. The pervasiveness of these two results in elementary Number Theory owes a great deal to the fact that they form the starting point for a systematic study of the behavior of the remainders of powers of a natural number a upon division by a given natural number $n > 1$, relatively prime with a. We also present the no less famous *Chinese remainder theorem*, along with some interesting applications.

10.1 Basic Definitions and Properties

The central object of study in this section is the relation on $\mathbb{Z}$ defined as follows.

Definition 10.1 Let a, b and n be given integers, with $n > 1$. We say that a is **congruent** to b, modulo n, and denote $a \equiv b \,(\mathrm{mod}\, n)$, provided $n \mid (a - b)$. If a is not congruent to b modulo n, we write $a \not\equiv b \,(\mathrm{mod}\, n)$.

Examples 10.2 According to the above definition, we can write:

(a) $3 \equiv 5 \,(\mathrm{mod}\, 2)$, for $2 \mid (3 - 5)$.
(b) $-1 \equiv 11 \,(\mathrm{mod}\, 12)$, for $12 \mid (-1 - 11)$.
(c) $2 \equiv -1 \,(\mathrm{mod}\, 3)$, for $3 \mid (2 - (-1))$.
(d) $x \equiv -x \,(\mathrm{mod}\, 2)$, for $2 \mid (x - (-x))$.
(e) $1 \not\equiv 2 \,(\mathrm{mod}\, 3)$, for $3 \nmid (1 - 2)$.
(f) $20 \not\equiv 15 \,(\mathrm{mod}\, 7)$, for $7 \nmid (20 - 15)$.

What are we really looking at when we consider congruences modulo n? In order to answer such a question, let us observe what happens with the integers modulo 4, for instance:

$$4k \equiv 0 \,(\text{mod}\,4), \quad 4k+1 \equiv 1 \,(\text{mod}\,4),$$

$$4k+2 \equiv 2 \,(\text{mod}\,4) \quad \text{and} \quad 4k+3 \equiv 3 \,(\text{mod}\,4).$$

Thus, the list $\dots$, -5, -4, -3, -2, -1, 0, 1, 2, 3, 4, 5, $\dots$ of integers is equal, modulo 4, to

$$\dots, 3, 0, 1, 2, 3, 0, 1, 2, 3, 0, 1, \dots$$

(with 0 corresponding to 0, of course), and we readily see that every integer is congruent, modulo 4, to its remainder upon division by 4. This result still holds true in general, as shown by the following result.

Proposition 10.3 *Let a and n be given integers, with $n > 1$.*

(a) *If a leaves remainder r upon division by n, then $a \equiv r \,(\text{mod}\,n)$. In particular, every integer is congruent, modulo n, to exactly one of the numbers 0, 1, 2, $\dots$, $n-2$, $n-1$.*

(b) *$a \equiv b \,(\text{mod}\,n) \Leftrightarrow a$ and b leave equal remainders when divided by n.*

Proof

(a) Assume that a leaves remainder r upon division by n. Then, the division algorithm gives $a = qn + r$, for some integer q, and this gives that $n \mid (a - r)$. But this is the same as writing $a \equiv r \,(\text{mod}\,n)$. The rest is immediate.

(b) If $a \equiv b \,(\text{mod}\,n)$, then $n \mid (a - b)$ and Corollary 6.9 (with a and b in place of a_1 and a_2, and n in place of b) shows that a and b leave equal remainders when divided by n. Conversely, if a and b leave the same remainder r upon division by n, we can write $a = nq_1 + r$ and $b = nq_2 + r$, with $q_1, q_2 \in \mathbb{Z}$. Therefore, $a - b = n(q_1 - q_2)$, so that $n \mid (a - b)$. Hence, $a \equiv b \,(\text{mod}\,n)$.

□

Remark 10.4 The definition of congruence modulo n excludes modulus $n = 1$ for, otherwise, $a \equiv b \,(\text{mod}\,1)$ would be a synonym of $1 \mid (a - b)$, which is always true. Hence, any two integers would be indistinguishable modulo 1, so that such a relation would be useless, as far as divisibility is concerned.

Since congruence modulo n only sees remainders upon division by n, the reader may well be asking himself/herself what advantage(s) (if any) do we have in using it. As we shall see in a while, the first profit is computational in nature. Indeed, in the two coming propositions we prove some elementary properties of congruences that will allow us, for instance, to mechanically compute the remainder we get upon dividing 17^{2002} by 13. Such a task is not so easy to accomplish by using the methods we developed so far.

Proposition 10.5 *Given integers a, b, c and n, with $n > 1$, we have:*

(a) *$a \equiv a \,(\text{mod}\,n)$.*

(b) *$a \equiv b \,(\text{mod}\,n) \Rightarrow b \equiv a \,(\text{mod}\,n)$.*

(c) *$a \equiv b \,(\text{mod}\,n)$ and $b \equiv c \,(\text{mod}\,n) \Rightarrow a \equiv c \,(\text{mod}\,n)$.*

Proof Items (a) and (b) are immediate. Concerning (c), if $a \equiv b \pmod{n}$ and $b \equiv c \pmod{n}$, then $n \mid (a - b)$ and $n \mid (b - c)$, and item i. of Remark 6.6 assures that n also divides $a - c = (a - b) + (b - c)$. But this is the same as having $a \equiv c \pmod{n}$.

$\square$

Proposition 10.6 *Let a, b, c, d, m and n be given integers, with $m, n > 1$.*

(a) *If $a \equiv b \pmod{n}$ and $c \equiv d \pmod{n}$, then $a + c \equiv b + d \pmod{n}$ and $ac \equiv bd \pmod{n}$. In particular, $ac \equiv bc \pmod{n}$.*

(b) *If $a \equiv b \pmod{n}$, then $a^k \equiv b^k \pmod{n}$ for every $k \in \mathbb{N}$.*

(c) *If $c_0, c_1, \ldots, c_m \in \mathbb{Z}$ and $f(x) = c_m x^m + \cdots + c_1 x + c_0$ is a polynomial function, then*

$$a \equiv b \pmod{n} \Rightarrow f(a) \equiv f(b) \pmod{n}.$$

(d) *If $a \equiv b \pmod{n}$, then $\gcd(a, n) = \gcd(b, n)$.*

(e) *If $a + c \equiv b + c \pmod{n}$, then $a \equiv b \pmod{n}$.*

(f) *If $ac \equiv bc \pmod{n}$ and $\gcd(c, n) = d$, then $a \equiv b \pmod{\frac{n}{d}}$. In particular, if $\gcd(c, n) = 1$, then $a \equiv b \pmod{n}$.*

(g) *If $a \equiv b \pmod{mn}$, then $a \equiv b \pmod{m}$ and $a \equiv b \pmod{n}$.*

(h) *If $a \equiv b \pmod{n}$ and $a \equiv b \pmod{m}$, with $\gcd(m, n) = 1$, then $a \equiv b \pmod{mn}$.*

Proof

(a) Since $(a + c) - (b + d) = (a - b) + (c - d)$, $ac - bd = a(c - d) + (a - b)d$ and $n \mid (a - b)$, $n \mid (c - d)$, it follows from item (c) of Proposition 6.5 (see also item i. of Remark 6.6) that $n \mid [(a + c) - (b + d)]$ and $n \mid (ac - bd)$. But this is the same as having $a + c \equiv b + d \pmod{n}$ and $ac \equiv bd \pmod{n}$. Finally, the particular case follows from $c \equiv c \pmod{n}$.

(b) Letting $c = a$ and $d = b$ in the second part of item (a), we get $a^2 \equiv b^2 \pmod{n}$. On the other hand, had we already shown that $a^l \equiv b^l \pmod{n}$ for some $l \in \mathbb{N}$, then, once more from the second part of (a) (this time with $c = a^l$ and $d = b^l$), we obtain

$$a^{l+1} = a \cdot a^l \equiv b \cdot b^l = b^{l+1} \pmod{n}.$$

Item (b) follows, then, by induction on k.

(c) If $a \equiv b \pmod{n}$, then we get from items (a) and (b) that $c_k a^k \equiv c_k b^k \pmod{n}$, for $0 \le k \le m$. Hence, it follows from Problem 1 that

$$f(a) = \sum_{k=0}^{m} c_k a^k \equiv \sum_{k=0}^{m} c_k b^k = f(b) \pmod{n}.$$

(d) Since $a \equiv b \pmod{n}$, there exists $q \in \mathbb{Z}$ such that $a = b + nq$. We then wish to show that

$$\gcd(b + nq, n) = \gcd(b, n).$$

But this is immediate from item (b) of Proposition 6.22.

(e) If $a + c \equiv b + c \pmod{n}$, then n divides $(a + c) - (b + c) = a - b$, which is the same as $a \equiv b \pmod{n}$.

(f) Let $n = dn'$ and $c = dc'$, with c' and n' being relatively prime integers. From $ac \equiv bc \pmod{n}$ we get $(dn') \mid [dc'(a-b)]$ or, which is the same, $n' \mid c'(a-b)$. However, since $\gcd(n', c') = 1$, item (a) of Proposition 6.22 gives $n' \mid (a - b)$. Since $n' = \frac{n}{d}$, this is the same as $a \equiv b \pmod{\frac{n}{d}}$. The rest is immediate.

(g) If $a \equiv b \pmod{mn}$, then $mn \mid (a - b)$, so that $m \mid (a - b)$. This last relation is equivalent to $a \equiv b \pmod{m}$. Analogously, $a \equiv b \pmod{n}$.

(h) Since $m, n \mid (a - b)$ and $\gcd(m, n) = 1$, it follows from item (d) of Proposition 6.22 that $mn \mid (a - b)$, which is exactly what we wanted to prove.

$\square$

We are now in position to solve the example below.

Example 10.7 Compute the remainder we get upon dividing 17^{2002} by 13.

Solution Since $17 \equiv 4 \pmod{13}$ and $16 \equiv 3 \pmod{13}$, item (b) of the previous proposition gives

$$17^{2002} \equiv 4^{2002} = 16^{1001} \equiv 3^{1001} \pmod{13}.$$

Now, observing that $3^3 \equiv 1 \pmod{13}$ and applying items (a) and (b) of the same proposition, we obtain

$$3^{1001} = 3^2 \cdot 3^{999} = 9 \cdot (3^3)^{333} \equiv 9 \cdot 1^{333} = 9 \pmod{13}.$$

Then, Proposition 10.3 guarantees that 17^{2002} leaves remainder 9 when divided by 13.

$\square$

The elementary properties of congruence deduced in Proposition 10.6 allow us to establish the criterion of divisibility by 9 in a much simpler way that that hinted to in Problem 1, page 162. This is our next

Example 10.8 A natural number n and the sum of the algarisms of its decimal representation leave equal remainders upon division by 9.

Proof Letting $n = (a_k a_{k-1} \ldots a_1 a_0)_{10}$ be the decimal representation of the natural number n, we have

$$n = a_k 10^k + a_{k-1} 10^{k-1} + \cdots + a_1 10 + a_0.$$

Since $10 \equiv 1 \pmod 9$, item (c) of Proposition 10.6 (with $f(x) = a_k x^k + a_{k-1} x^{k-1} + \cdots + a_1 x + a_0$) gives

$$n = f(10) \equiv f(1) = a_k + a_{k-1} + \cdots + a_1 + a_0 \pmod 9.$$

The rest follows from item (b) of Proposition 10.3.   □

As an additional example of the computational simplification the congruence relation brings (and for future reference), we shall prove once again the results of Corollary 6.8 and Example 6.10. Before that, however, it is time we make the following simple remark.

As we already know, every integer is congruent to 0, 1, 2, 3, 4, 5 or 6, modulo 7; also, since

$$4 \equiv -3 \pmod 7, \quad 5 \equiv -2 \pmod 7 \quad \text{and} \quad 6 \equiv -1 \pmod 7,$$

we conclude that every integer is congruent, modulo 7, to one of $0, \pm 1, \pm 2$ or ± 3. On the other hand, every integer is congruent to 0, 1, 2, 3, 4, 5, 6 or 7, modulo 8; however, since

$$5 \equiv -3 \pmod 8, \quad 6 \equiv -2 \pmod 8 \quad \text{and} \quad 7 \equiv -1 \pmod 8,$$

it follows that every integer is congruent, modulo 8, to $0, \pm 1, \pm 2, \pm 3$ or 4. The advantage of replacing, modulo 7, the integers from 0 to 6 by $0, \pm 1, \pm 2, \pm 3$ lies on the fact that, if in some context one needs to raise the remainders upon division by 7 to some exponent k, then one is likely to have much less work by using $0, \pm 1, \pm 2, \pm 3$ in place of $0, 1, 2, 3, 4, 5, 6$, for $(-x)^k = \pm x^k$. By the same token, it is sometimes advantageous to replace, modulo 8, the usual remainders 0, 1, 2, 3, 4, 5, 6, 7 by $0, \pm 1, \pm 2, \pm 3, 4$.

Generalizing the discussion of the previous paragraph, it is not hard to verify that:

i. If $n = 2k$, then every integer is congruent, modulo n, to one of

$$0, \pm 1, \pm 2, \ldots, \pm(k-1), k.$$

ii. If $n = 2k + 1$, then every integer is congruent, modulo n, to one of

$$0, \pm 1, \pm 2, \ldots, \pm k.$$

We shall go one step further in formalizing the above discussion when we study the concept of *complete residue systems*, in Sect. 11.1. For the time being, we establish the above mentioned results.

Proposition 10.9 *For every* $a \in \mathbb{Z}$, *one has:*

(a) $a^2 \equiv 0, 1, 4, 5, 6$ *or* $9 \pmod{10}$.
(b) $a^2 \equiv 0$ *or* $1 \pmod 3$.
(c) $a^2 \equiv 0$ *or* $1 \pmod 4$.
(d) $a^2 \equiv 0, 1$ *or* $4 \pmod 8$.
(e) $a^4 \equiv 0$ *or* $1 \pmod{16}$.

Proof

(a) Modulo 10, we have $a \equiv 0, \pm 1, \pm 2, \pm 3, \pm 4$ or 5, so that

$$a^2 \equiv 0^2, (\pm 1)^2, (\pm 2)^2, (\pm 3)^2, (\pm 4)^2 \text{ or } 5^2 \pmod{10}$$

or, which is the same, $a^2 \equiv 0, 1, 4, 9, 6$ or $5 \pmod{10}$. However, since the last digit of a natural number is equal to the remainder we get upon dividing it by 10, it follows that the last digit of a^2 is equal to one of 0, 1, 4, 5, 6 or 9.
(b) We know that $a \equiv 0$ or $\pm 1 \pmod 3$, so that $a^2 \equiv 0^2$ or $(\pm 1)^2 \pmod 3$. Hence, $a^2 \equiv 0$ or $1 \pmod 3$.
(c) Since $a \equiv 0, \pm 1$ or $2 \pmod 4$, we get $a^2 \equiv 0^2, (\pm 1)^2$ or $2^2 \pmod 4$. Since $2^2 \equiv 0 \pmod 4$, it follows that $a^2 \equiv 0$ or $1 \pmod 4$.
(d) As in the previous items, we have $a \equiv 0, \pm 1, \pm 2, \pm 3$ or $4 \pmod 8$, and then

$$a^2 \equiv 0^2, (\pm 1)^2, (\pm 2)^2, (\pm 3)^2 \text{ or } 4^2 \pmod 8.$$

Now, $3^2 = 9 \equiv 1$ and $4^2 = 16 \equiv 0 \pmod 8$, so that $a^2 \equiv 0, 1$ or $4 \pmod 8$.
(e) By item (d), we have $a^2 = 8q + r$, with $q \in \mathbb{N}$ and $r = 0, 1$ or 4. Hence,

$$a^4 = (8q + r)^2 = 16(4q^2 + qr) + r^2 = 16q' + 0 \text{ or } 16q' + 1,$$

with $q' \in \mathbb{N}$.

$\square$

The coming examples show how we can use what we have learned up to this point about congruences to solve various interesting problems. Generally speaking, the solutions we shall present will be a mixture of a judicious use of one or more congruence relations, together with other ideas, each of which adapted to the situation at hand. The reader should make an effort to perceive exactly how each used tool helps in paving the way to the final solution, for this attitude will be of great help in preparing him/her to tackle the problems posed at the end of the section.

Example 10.10 (Italy) Find all $x, y \in \mathbb{N}$ for which $x^2 + 615 = 2^y$.

Solution Looking at the given equation modulo 3, we obtain

$$x^2 + 0 \equiv (-1)^y \pmod 3.$$

Now, by item (b) of Proposition 10.9, we have $x^2 \equiv 0$ or $1 \pmod 3$, so that the congruence above gives us the following possibilities:

$$0 \equiv (-1)^y \pmod 3 \text{ or } 1 \equiv (-1)^y \pmod 3.$$

The first one clearly never occurs. In what concerns the second, if y is odd, we get $1 \equiv -1 \pmod 3$, which never happens too. Hence, y must be even, say $y = 2z$, with $z \in \mathbb{N}$. The stated equation can then be written as

$$615 = 2^{2z} - x^2 = (2^z - x)(2^z + x).$$

Finally, since $2^z + x > 2^z - x$ and $615 = 3 \cdot 5 \cdot 41$, we are left to analysing the linear systems below:

$$\begin{cases} 2^z + x = 615 \\ 2^z - x = 1 \end{cases}, \quad \begin{cases} 2^z + x = 205 \\ 2^z - x = 3 \end{cases},$$

$$\begin{cases} 2^z + x = 123 \\ 2^z - x = 5 \end{cases} \text{ or } \begin{cases} 2^z + x = 41 \\ 2^z - x = 15 \end{cases}.$$

To this end, in each case we add both equations to obtain $2^{z+1} = 616, 208, 128$ or 56, respectively. However, since 2^{z+1} is a power of 2, the only allowed possibility is $2^{z+1} = 128$, so that

$$\begin{cases} 2^z + x = 123 \\ 2^z - x = 5 \end{cases}.$$

Then, $z = 6$ and $x = 59$, and the only solution to the original equation is $x = 59$ and $y = 2z = 12$. □

Example 10.11 (OIM) Find all $m, n \in \mathbb{N}$ such that $2^m + 1 = 3^n$.

Solution If $m = 1$, then clearly $n = 1$. If $m \geq 2$, then $2^m \equiv 0 \pmod 4$ and, looking at the equation modulo 4, we obtain

$$1 \equiv (-1)^n \pmod 4.$$

It follows that n must be even, say $n = 2u$, with $u \in \mathbb{N}$. Substituting this into the given equation, we get $2^m + 1 = 3^{2u}$ or, which is the same,

$$2^m = (3^u - 1)(3^u + 1).$$

In view of the Fundamental Theorem of Arithmetic, the equality above assures that both $3^u - 1$ and $3^u + 1$ must be powers of 2. To get further information, let $d = \gcd(3^u - 1, 3^u + 1)$ and observe that

$$d \mid [(3^u + 1) - (3^u - 1)],$$

i.e., $d \mid 2$; however, since $3^u - 1$ and $3^u + 1$ are both even numbers, we have $d = 2$. Now, the only way the gcd of two powers of 2 can be equal to 2 is when the smallest one is itself equal to 2, so that the only possibility is to have

$$\begin{cases} 3^u - 1 = 2 \\ 3^u + 1 = 2^{m-1} \end{cases}.$$

Thus, $u = 1$, $m = 3$ and, hence, $n = 2$. $\square$

Example 10.12 (OCM) Do the following items:

(a) Prove that there does exist $x, y, z \in \mathbb{N}$ such that $13x^4 + 3y^4 - z^4 = 2013$.
(b) Prove that there does not exist $x, y, z \in \mathbb{N}$ such that $13x^4 + 3y^4 - z^4 = 2014$.

Proof

(a) Letting $z = 2x$, we get $y^4 - x^4 = 671$ or, which is the same, $(y^2 - x^2)(y^2 + x^2) = 11 \cdot 61$. Hence, the Fundamental Theorem of Arithmetic gives us the possibilities

$$\begin{cases} y^2 - x^2 = 1 \\ y^2 + x^2 = 471 \end{cases} \text{ and } \begin{cases} y^2 - x^2 = 11 \\ y^2 + x^2 = 61 \end{cases},$$

so that $x = 5$, $y = 6$ and $z = 2x = 10$.
(b) By the sake of contradiction, assume that a solution does exist. Item (e) of Proposition 10.9 gives $a^4 \equiv 0$ or $1 \pmod 8$, for every $a \in \mathbb{Z}$. Therefore, $13x^4 + 3y^4 - z^4 \equiv 0, 2, 4, 5$ or $7 \pmod 8$. However, since $2014 \equiv 6 \pmod 8$, we have reached a contradiction.

$\square$

The reader will probably notice that the solutions of our last two examples are somewhat more terse than those of the previous ones. An effective training for the proposed problems is to fulfill the omitted details.

Example 10.13 (Romania) Let $m, n, p \in \mathbb{N}$, where p is an odd prime. If $\frac{7^m + p \cdot 2^n}{7^m - p \cdot 2^n} \in \mathbb{N}$, prove that such a number is itself prime.

Proof Let $a = \frac{7^m + p \cdot 2^n}{7^m - p \cdot 2^n}$. Since

$$a = 1 + \frac{p \cdot 2^{n+1}}{7^m - p \cdot 2^n} = -1 + \frac{2 \cdot 7^m}{7^m - p \cdot 2^n},$$

it follows that $(7^m - p \cdot 2^n) \mid \gcd(p \cdot 2^{n+1}, 2 \cdot 7^m)$. Now, there are two possibilities:

(i) $p = 7$: in this case,

$$(7^m - 7 \cdot 2^n) \mid \gcd(7 \cdot 2^{n+1}, 2 \cdot 7^m) = 14,$$

which gives $7^{m-1} - 2^n = 1$. Analysing such an equation modulo 3, we reach a contradiction.

(ii) $p \neq 7$: in this case, we get $\gcd(p \cdot 2^{n+1}, 2 \cdot 7^m) = 2$ and, hence, $(7^m - p \cdot 2^n) \mid 2$. Again, this implies $7^m - p \cdot 2^n = 1$ and, looking at such last equality also modulo 3, we obtain $1^m - p \cdot 2^n \equiv 1 \pmod 3$, so that $p = 3$. It follows that $7^m - 3 \cdot 2^n = 1$ and $a = 7^m + 3 \cdot 2^n$.

If $m = 1$, then $n = 1$ and, hence $a = 13$, a prime number. If $m > 1$, then $n > 1$ and

$$2^{n-1} = \frac{7^m - 1}{6} = 7^{m-1} + \cdots + 7 + 1.$$

It follows that m is even, say $m = 2k$. Thus, we get $49^k - 1 = 3 \cdot 2^n$ and distinguish two subcases:

- If $k = 1$, then $m = 2$, $n = 4$ and $a = 97$, once more a prime number.
- If $k > 1$ then, factoring $49^k - 1$, we get

$$49^{k-1} + \cdots + 49 + 1 = 2^{n-3}.$$

Such an equality gives k even and, writing $k = 2l$, we obtain

$$3 \cdot 2^n = 49^{2l} - 1 \equiv (-1)^{2l} - 1 \equiv 0 \pmod 5,$$

which is an absurd.

□

Example 10.14 (BMO) Let $(a_n)_{n \geq 1}$ be the sequence defined for $n \geq 1$ by $a_n = 2^n + 49$. Find all integers $n \geq 1$ for which $a_n = pq$ and $a_{n+1} = rs$, where p, q, r, s are primes such that $p < q, r < s$ and $q - p = s - r$.

Solution Let n be a natural number for which a_n and a_{n+1} satisfy the stated conditions. If $p, r > 3$, then we get $p, q, r, s \equiv 1$ or $5 \pmod 6$. However, since $5^2 \equiv 1 \pmod 6$, in any case we obtain $pq, rs \equiv 1$ or $5 \pmod 6$. On the other hand, if k denotes the odd element of $\{n, n+1\}$, then $2^k \equiv 2 \pmod 6$, whence $2^k + 49 \equiv 2 + 1 = 3 \pmod 6$, which, in turn, is an absurd. Thus, we must have $p \leq 3$ or $r \leq 3$, so that at least one of p, r must be equal to 3 (for a_n and a_{n+1} are both odd).

If $p \geq r$, then $q = s + p - r \geq s$, implying that $a_n = pq \geq rs = a_{n+1}$, a contradiction. Therefore, $p < r$, and hence $p = 3$ and $q = 3 + s - r$. We thus get $a_n = pq = 3(3 + s - r)$. On the other hand, we also have

$$2a_n = 2^{n+1} + 98 = (2^{n+1} + 49) + 49 = a_{n+1} + 49 = rs + 49,$$

so that $6(3 + s - r) = rs + 49$. This gives $r = 6 - \frac{67}{s+6}$, and hence $s = 61, r = 5$, $q = 59$. Then,

$$3 \cdot 59 = a_n = 2^n + 49 \Rightarrow 2^n = 128 \Rightarrow n = 7.$$

□

Problems: Sect. 10.1

1. * Generalize item (a) of Proposition 10.6, showing that if $m, n > 1$ are natural numbers and $a_1, \ldots, a_m, b_1, \ldots, b_m$ are integers such that $a_k \equiv b_k \pmod{n}$ for $1 \le k \le m$, then

$$\sum_{k=1}^{m} a_k \equiv \sum_{k=1}^{m} b_k \pmod{n} \text{ and } \prod_{k=1}^{m} a_k \equiv \prod_{k=1}^{m} b_k \pmod{n}.$$

2. Let $n = (a_k a_{k-1} \ldots a_1 a_0)_{10}$ be the decimal representation of the natural number. Prove that the remainder of n upon division by 11 is equal to that of

$$a_0 - a_1 + a_2 - a_3 + \cdots + (-1)^{k-1} a_{k-1} + (-1)^k a_k.$$

3. Find the remainder of the division of $1000^{55} + 55^{1000}$ by 7.
4. Show that $8^{100} + 3^{2001}$ is not a perfect square.
5. Find the remainder upon division of 73^{10} by 5.
6. Let $n \in \mathbb{N}$ be such that $n \equiv -1 \pmod{4}$. Prove that there exists a prime p such that $p \mid n$ and $p \equiv -1 \pmod{4}$.
7. * Use the fact that $2^4 + 5^4 = 641 = 2^7 \cdot 5 + 1$ to show that $2^{2^5} + 1$ is not a prime number.
8. (IMO) Find the smallest natural number n satisfying the two following conditions:

   (i) The last digit of the decimal representation of n is 6.
   (ii) If we erase this last digit 6 and write it immediately to the left of the first digit of n, we obtain the number $4n$.

9. Find all positive integers n such that $7 \mid (2^n + 3^n)$.
10. Find all prime numbers p and q for which $p^2 + 3pq + q^2$ is a perfect square.

    The coming three problems are concerned with the Fibonacci and Lucas sequences, so the reader may find it convenient to start by recalling (cf. Sects. 1.3 and 3.1) that the Fibonacci sequence $(F_n)_{n \ge 1}$ is given by $F_1 = F_2 = 1$ and $F_{k+2} = F_{k+1} + F_k$, whereas the Lucas sequence $(L_n)_{n \ge 1}$ is defined by $L_1 = 1, L_2 = 3$ and $L_{k+2} = L_{k+1} + L_k$, for every integer $n \ge 1$.

11. Prove the following properties of the Fibonacci and Lucas sequences:

    (a) $2 \mid F_n \Leftrightarrow 3 \mid n$ and $4 \mid F_n \Leftrightarrow 6 \mid n$.
    (b) $2 \mid L_n \Leftrightarrow 3 \mid n$ and $4 \mid L_n \Leftrightarrow n \equiv 3 \pmod{6}$.
    (c) $3 \mid L_n \Leftrightarrow n \equiv 2 \pmod{4}$.
    (d) $L_n \equiv 3 \pmod{4}$ if $2 \mid n$ and $3 \nmid n$.
    (e) $L_{n+12} \equiv L_n \pmod{8}$, for every $n \in \mathbb{N}$.

12. Also concerning the Fibonacci and Lucas sequences, prove that

$$\gcd(F_n, L_n) = \begin{cases} 1, & \text{if } 3 \nmid n \\ 2, & \text{if } 3 \mid n \end{cases}.$$

13. Prove the following properties of the Fibonacci and Lucas sequence, for every $m, n, k \in \mathbb{N}$ such that $2 \mid k$ and $3 \nmid k$.

(a) $2F_{m+n} = F_m L_n + F_n L_m$.

(b) $2L_{m+n} = 5F_m F_n + L_m L_n$.

(c) $F_{m+2kt} \equiv (-1)^t F_m \pmod{L_k}$, for every $t \in \mathbb{N}$.

(d) $L_{m+2k} \equiv (-1)^t L_m \pmod{L_k}$, for every $t \in \mathbb{N}$.

14. (USA) Prove that the equation $x_1^4 + x_2^4 + x_3^4 + \cdots + x_{14}^4 = 15,999$ does not have any integer solutions.

15. (Romania) Find all natural numbers m, n, k such that $2^m + 3^n = k^2$.

16. (Soviet Union) Find all solutions in integers $m, n, p > 1$ of the equation

$$1! + 2! + \cdots + n! = m^p.$$

17. (Bulgaria) Find all positive integers x, y and z for which $3^x + 4^y = 5^z$.

18. (Tchecoslovaquia) Find all positive integers x, y and p such that p is prime and $p^x - y^3 = 1$.

19. Find all $a, b, c \in \mathbb{N}$ such that a and b are even and $a^b + b^a = 2^c$.

20. (Hungary) Let a and b be given naturals and n be a nonnegative integer such that $a^n \mid b$. Prove that a^{n+1} divides $(a+1)^b - 1$.

21. (Brazil) Find all positive integer solutions of the equation $x^2 + 15^a = 2^b$.

22. (France) Compute the digit of units of the integer part of $\frac{10^{1992}}{10^{83}+7}$? Justify your answer.

23. (China—adapted) The objective of this problem is to solve equation $7^x - 3^y = 4$ in natural numbers, showing that $x = y = 1$ is its only solution. To this end, do the following items:

(a) Use modulo 8 to conclude that there are no solutions if x is even.

(b) If x is odd and $y > 1$, look at the equation modulo 9 to show that $x \equiv 2 \pmod 3$ and, hence, that $x \equiv 5 \pmod 6$.

(c) Writing $x = 6q + 5$, with $q \in \mathbb{N}$, conclude that $7^x \equiv \pm 2 \pmod{13}$.

(d) Show that every power of 3 is congruent to 1, 3 or 9, modulo 13.

(e) Assuming that $y > 1$, use items (c) and (d) to reach a contradiction.

24. (Bulgaria—adapted) The purpose of this problem is to show that the only solution of the equation $5^x 7^y + 4 = 3^z$ in nonnegative integers x, y and z is $x = 1$, $y = 0$ and $z = 2$. To this end, do the following items:

(a) If $x = 0$, the equation reduces to $7^y + 4 = 3^z$, so that $z \geq 2$. Use modulo 9 to show that, in this case, there are no solutions.

(b) From now on, assume $x \geq 1$. Use modulo 5 to conclude that $z = 2t$, for some $t \in \mathbb{N}$.

(c) Show that $\gcd(3^t - 2, 3^t + 2) = 1$. Then, conclude that: (i) $3^t - 2 = 1$ and $3^t + 2 = 5^x 7^y$, (ii) $3^t - 2 = 5^x$ and $3^t + 2 = 7^y$ or (iii) $3^t - 2 = 7^y$ and $3^t + 2 = 5^x$.

(d) In case (ii), use modulo 3 to show that there are no solutions.

(e) In case (iii), we have $5^x - 7^y = 4$. From this, use modulo 4 to conclude that y is even. Finally, if $x \geq 2$, use modulo 25 to get a contradiction.

25. (Miklós-Schweitzer—adapted) In order to solve equation $(x+1)^y - x^z = 1$ in integers $x, y, z > 1$, do the following items:

(a) Look at the equation modulo $x + 1$ to conclude that z is odd.
(b) Write $(x+1)^y = x^z + 1$ and factor the right hand side to show that x is even. Then, write $x^{z-1} = \frac{(x+1)^y - 1}{x}$ and factor the right hand side to conclude that y is also even.
(c) If $x = 2s$ and $y = 2t$, with $s, t \in \mathbb{N}$, show that $(x+1)^t - 1$ and $(x+1)^t + 1$ have gcd equal to 2. From this, use modulo x to show that $(x+1)^t - 1 = 2s^z$ and $(x+1)^t + 1 = 2^{z-1}$.
(d) Use inequality $2s^z < 2^{z-1}$ to show that $s = 1$. Then, successively get $x = 2, t = 1, y = 2$ and $z = 3$.

10.2   The Theorems of Euler and Fermat

The effective use of congruences to compute remainders is considerably simplified if we find exponents which make a certain power congruent to 1. For instance, knowing that $7^3 \equiv 1 \pmod 9$ it rests much easier to compute the remainder of 25^{1001} upon division by 9: since $25 \equiv 7 \pmod 9$, we have

$$25^{1001} \equiv 7^{1001} = (7^3)^{333} \cdot 7^2 \equiv 1^{333} \cdot 49 \equiv 4 \pmod 9.$$

In this direction, given relatively prime integers a and n, with $n > 1$, the purpose of this section is to find an exponent $k \in \mathbb{N}$ for which

$$a^k \equiv 1 \pmod n.$$

To this end, we shall first look at the case in which n is prime, thus proving one of the most important results of the elementary theory of congruences: **Fermat's little theorem**.[1]

Theorem 10.15 (Fermat) *Given $a, p \in \mathbb{Z}$, with p prime, we have $a^p \equiv a \pmod p$. In particular, if $\gcd(a, p) = 1$, then*

$$a^{p-1} \equiv 1 \pmod p. \tag{10.1}$$

Proof If $a^p \equiv a \pmod p$, then p divides $a^p - a = a(a^{p-1} - 1)$. Hence, if $\gcd(a, p) = 1$, then (10.1) follows from item (a) of Proposition 6.22.

---

[1] A combinatorial proof of Fermat's little theorem was the object of Problem 5, page 57.

It thus suffices to show that $a^p \equiv a \pmod{p}$, for every $a \in \mathbb{Z}$. If $p = 2$, there is nothing to do, for $a^2 - a = a(a-1)$, being the product of two consecutive integers, is always even. Let us now assume that $p > 2$ and prove the desired result, firstly for $a > 0$, by induction on a. For $a = 1$, there is nothing to do. Assume, by induction hypothesis, that the theorem holds for a certain natural value of a, i.e., assume that $k^p \equiv k \pmod{p}$, for some $k \in \mathbb{N}$. For $a = k + 1$, we have

$$(k + 1)^p - (k + 1) = (k^p - k) + \sum_{j=1}^{p-1} \binom{p}{j} k^{p-j}.$$

However, since $p \mid (k^p - k)$ (by induction hypothesis) and $p \mid \binom{p}{j}$ for $1 \leq j \leq p-1$ (by Example 6.42), it follows that p divides $(k + 1)^p - (k + 1)$, so that $(k + 1)^p \equiv (k + 1) \pmod{p}$.

Let us now look at the case $a \leq 0$ and p odd. If $a = 0$, once more there is nothing to do. If $a < 0$, then, since p is odd, what we did above gives

$$a^p = -(-a)^p \equiv -(-a) = a \pmod{p}.$$

$\square$

Before we extend Fermat's little theorem to a composite modulo, we collect a few examples that illustrate some applications of it.

Example 10.16 If p and q are distinct primes, prove that pq divides $p^{q-1} + q^{p-1} - 1$.

Proof Since p and q are distinct primes, we have $\gcd(p, q) = 1$. Hence, by Fermat's little theorem, q divides $p^{q-1} - 1$. However, since q obviously divides q^{p-1}, we conclude that q divides $q^{q-1} + (p^{q-1} - 1)$ as well. Analogously, p divides $p^{q-1} + (q^{p-1} - 1)$. Finally, since p and q both divide $p^{q-1} + q^{p-1} - 1$ and are relatively prime, item (d) of Proposition 6.22 guarantees that pq divides $p^{q-1} + q^{p-1} - 1$ too. $\square$

Example 10.17 (Romania) Let p and q be prime numbers, with $q \neq 5$. If $q \mid (2^p + 3^p)$, prove that $q > p$.

Proof Since $q \mid (2^p + 3^p)$, we clearly have $q \neq 2, 3$; since $q \neq 5$ by hypothesis, we thus have $q > 5$. Hence, we can assume that $p > 3$. If $q \leq p$, then $q - 1 < p$, so that $q - 1$ and p are relatively prime. In such a case, Bézout's theorem assures the existence of $x, y \in \mathbb{N}$ such that $px = (q - 1)y + 1$. Therefore, from $2^p \equiv -3^p \pmod{q}$ we successively get $(2^p)^x \equiv (-3^p)^x \pmod{q}$ and (since $-3^p = (-3)^p$)

$$2^{(q-1)y+1} \equiv (-3)^{(q-1)y+1} \pmod{q}.$$

Finally, since $q \neq 2, 3$, Fermat's little theorem gives $2^{q-1}, (-3)^{q-1} \equiv 1 \pmod{q}$; then, the above congruence reduces to $2 \equiv -3 \pmod{q}$, so that $q = 5$. Finally, such a conclusion is obviously a contradiction.   □

Example 10.18 (BMO) Let $p > 2$ be a prime number such that $3 \mid (p - 2)$, and let

$$S = \{y^2 - x^3 - 1; \; 0 \leq x, y < p \text{ and } x, y \in \mathbb{Z}\}.$$

Prove that S contains at most $p - 1$ multiples of p.

Proof If $0 \leq u, v \leq p - 1$ and $u^3 \equiv v^3 \pmod{p}$, we claim that $u = v$. To this end, start by noticing that $u^3 \equiv v^3 \equiv 0 \pmod{p}$ if and only if $u = v = 0$; on the other hand, for $1 \leq u, v \leq p - 1$, Fermat's little theorem, together with the fact that $p - 1 = 3k + 1$, give

$$u^{3k+1} \equiv v^{3k+1} \pmod{p}. \tag{10.2}$$

Now,

$$u^3 \equiv v^3 \pmod{p} \Rightarrow u^{3k} \equiv v^{3k} \pmod{p},$$

and it comes from (10.2) that

$$u^{3k}u \equiv v^{3k}v \equiv u^{3k}v \pmod{p}.$$

Finally, cancelling out u^3 from the above congruence, we obtain $u \equiv v \pmod{p}$; condition $1 \leq u, v \leq p - 1$ thus implies that $u = v$.

The above discussion guarantees that

$$\{x^3; \; 0 \leq x \leq p - 1\} = \{0, 1, \ldots, p - 1\}. \tag{10.3}$$

On the other hand, $y^2 - x^3 - 1 \in S$ is a multiple of p if and only if $y^2 \equiv x^3 - 1 \pmod{p}$. Hence, it follows from (10.3) that, for each $0 \leq y \leq p - 1$, there exists a single $0 \leq x \leq p - 1$ for which p divides $y^2 - x^3 - 1$.

The above reasoning would provide us with at most p multiples of p in S. Nevertheless, note that $0 = 1^2 - 0^3 - 1 = 3^3 - 2^3 - 1$ is represented twice in S, so that S actually contains no more that $p - 1$ multiples of p.   □

We generalize Fermat's little theorem with the following result of L. Euler, which brings Euler's function φ into prominence. For its statement, notice that if $a^k \equiv 1 \pmod{n}$, then item (d) of Proposition 10.6 assures that $\gcd(a^k, n) = \gcd(1, n) = 1$; in particular, $\gcd(a, n) = 1$.

Theorem 10.19 (Euler) *If a and n are relatively prime integers, with $n > 1$, then*

$$a^{\varphi(n)} \equiv 1 \pmod{n}. \tag{10.4}$$

Proof Let p be a prime factor of n and k be a natural number. We start by making induction on k to show that

$$a^{\varphi(p^k)} \equiv 1 \pmod{p^k}. \tag{10.5}$$

Case $k = 1$ reduces to Fermat's little theorem. Now, assume, by induction hypothesis, that $a^{\varphi(p^l)} \equiv 1 \pmod{p^l}$, for some natural l. Then, $a^{\varphi(p^l)} = p^l q + 1$ for some integer q, and we have

$$a^{\varphi(p^{l+1})} - 1 = a^{p \cdot \varphi(p^l)} - 1 = (a^{\varphi(p^l)})^p - 1$$

$$= (p^l q + 1)^p - 1 = \sum_{j=0}^{p} \binom{p}{j}(p^l q)^j - 1$$

$$= \binom{p}{1}p^l q + \sum_{j=2}^{p} \binom{p}{j} p^{jl} q^j$$

$$= p^{l+1} q + p^{2l} \sum_{j=2}^{p} \binom{p}{j} p^{(j-2)l} q^j.$$

$$\equiv 0 \pmod{p^{l+1}},$$

since $2l \geq l + 1$ for every integer $l \geq 1$.

In order to finish the proof, let $n = p_1^{\alpha_1} p_2^{\alpha_2} \ldots p_k^{\alpha_k}$ be the canonical decomposition of n into powers of distinct primes, so that (cf. Theorem 8.12)

$$\varphi(n) = \varphi(p_1^{\alpha_1})\varphi(p_2^{\alpha_2}) \ldots \varphi(p_k^{\alpha_k}).$$

Letting $m_j = \varphi(p_1^{\alpha_1}) \ldots \widehat{\varphi(p_j^{\alpha_j})} \ldots \varphi(p_k^{\alpha_k})$ (where the $\widehat{}$ above a factor means that it is omitted from the corresponding product) we have $\varphi(n) = \varphi(p_j^{\alpha_j})m_j$, and it follows from (10.5) that

$$a^{\varphi(n)} = \left(a^{\varphi(p_j^{\alpha_j})}\right)^{m_j} \equiv 1^{m_j} \equiv 1 \pmod{p_j^{\alpha_j}}.$$

However, since the powers $p_j^{\alpha_j}$ are pairwise relatively prime, item (h) of Proposition 10.6 guarantees that $a^{\varphi(n)} \equiv 1 \pmod{n}$. $\square$

As in the case of Fermat's little theorem, we pause to see, in a few examples, some interesting applications of Euler's theorem.

Example 10.20 (Brazil) Prove that there exists an integer $k > 2$ such that the number $1 \underbrace{99 \ldots 9}_{k} 1$ is a multiple of 1991.

Proof Notice firstly that

$$1\underbrace{99\ldots9}_{k}1 = 2 \cdot 10^{k+1} - 9.$$

Hence, we want to find an integer $k > 2$ such that

$$2 \cdot 10^{k+1} \equiv 9 \,(\mathrm{mod}\,1991).$$

To this end, since $2 \cdot 10^3 \equiv 9 \,(\mathrm{mod}\,1991)$, we have

$$2 \cdot 10^{k+1} \equiv 9 \,(\mathrm{mod}\,1991) \;\Leftrightarrow\; 2 \cdot 10^{k+1} \equiv 2 \cdot 10^3 \,(\mathrm{mod}\,1991)$$

$$\Leftrightarrow 10^{k-2} \equiv 1 \,(\mathrm{mod}\,1991).$$

For what is left to do, note that $1991 = 11 \cdot 181$ and 181 is prime (by Eratostenes' sieve, for instance). Hence, $\varphi(1991) = \varphi(11) \cdot \varphi(181) = 10 \cdot 180 = 1800$, and Euler's theorem gives $10^{1800} \equiv 1 \,(\mathrm{mod}\,1991)$. Therefore, it suffices to take $k - 2 = 1800$. $\qquad\square$

Example 10.21 (Romania) Let a and n be given natural numbers. Show that we can choose n pairwise relatively prime elements from the set

$$A = \{a^2 + a - 1, a^3 + a^2 - 1, a^4 + a^3 - 1, \ldots\}.$$

Proof By induction, let $k \in \mathbb{N}$ and assume that we have already established the existence of a subset $B_k \subset A$ such that $|B_k| = k$ and the elements of B_k are pairwise relatively prime. Let

$$m = \prod_{x \in B_k} x.$$

Since $m \equiv 1 \,(\mathrm{mod}\,a)$, we have $\gcd(a, m) = 1$. Therefore, Euler's theorem gives $a^{\varphi(m)} \equiv 1 \,(\mathrm{mod}\,m)$.

If $y = a^{\varphi(m)+1} + a^{\varphi(m)} - 1$, then $y \in A$. We claim that $\gcd(y, x) = 1$ for every $x \in B_k$. To this end, it suffices to show that $\gcd(y, m) = 1$. However, again from Euler's theorem, we get

$$y = a^{\varphi(m)+1} + a^{\varphi(m)} - 1 \equiv a + 1 - 1 = a \,(\mathrm{mod}\,m),$$

and item (d) of Proposition 10.6 guarantees that

$$\gcd(y, m) = \gcd(a, m) = 1.$$

Hence, the set $B_{k+1} = B_k \cup \{y\}$ is formed by $k + 1$ pairwise relatively prime elements of A, thus completing the induction step. $\qquad\square$

For the next example, note that if $a, k \in \mathbb{N}$, then the decimal representation of a^k has exactly $\lfloor k \log_{10} a \rfloor + 1$ digits. Indeed, letting m denote the number of algarisms of a^k, we have $10^{m-1} \le a^k < 10^m$ and, then,

$$m - 1 \le \log_{10} a^k < m;$$

hence, $m - 1 = \lfloor \log_{10} a^k \rfloor = \lfloor k \log_{10} a \rfloor$, as wished.

Example 10.22 Prove that there exists a power of 2 with 1000 consecutive zeros in its decimal representation.

Proof We wish to find $m \in \mathbb{N}$ such that the decimal representation of 2^m is of the form

$$2^m = (**\ldots*\underbrace{00\ldots0}_{1000}**\ldots*)_{10}.$$

Alternatively, letting A and B denote the natural numbers formed by the portions of the decimal expansion of 2^m situated respectively to the right and to the left of the sequence of the 1000 consecutive zeros, we want to find $m \in \mathbb{N}$ such that

$$2^m = B \cdot 10^{1000+a} + A,$$

with the decimal representation of A having at most a digits.

To this end, if we let $k = 1000 + a$ and $A = 2^k$, we will have

$$2^m = B \cdot 10^{1000+a} + A$$

if and only if the decimal representation of 2^k has at most $k - 1000$ algarisms and (dividing both sides of the last equality displayed above by 2^k) $2^{m-k} = B \cdot 5^k + 1$. In turn, this last condition will hold if and only if $2^{m-k} \equiv 1 \pmod{5^k}$.

We then have to show that it is possible to obtain $k, m \in \mathbb{N}$ such that: (i) $m > k$; (ii) the decimal representation of 2^k has at most $k - 1000$ algarisms; (iii) $2^{m-k} \equiv 1 \pmod{5^k}$. Let us start by analysing conditions (i) and (iii).

By Euler's theorem, we have $2^{\varphi(5^k)} \equiv 1 \pmod{5^k}$. Now, since $\varphi(5^k) = 4 \cdot 5^{k-1}$, there must exist $q \in \mathbb{N}$ such that $2^{4 \cdot 5^{k-1}} = 5^k q + 1$. Letting $m = k + 4 \cdot 5^{k-1}$, we have $m > k$ and $2^{m-k} \equiv 1 \pmod{5^k}$.

We are then left to showing that it is possible to find $k \in \mathbb{N}$ such that the decimal representation of 2^k has at most $k - 1000$ digits. To this end, the discussion preceding the statement of this example assures that the decimal representation of 2^k has exactly $\lfloor k \log_{10} 2 \rfloor + 1$, so that it suffices to ask that the following inequality holds:

$$\lfloor k \log_{10} 2 \rfloor + 1 \le k - 1000.$$

However, since

$$k - \lfloor k \log_{10} 2 \rfloor > k - k \log_{10} 2 = k \log_{10} 5,$$

it is enough to start by taking a natural number k for which $k \log_{10} 5 > 1001$. This is surely possible.   □

Problems: Sect. 10.2

1. (Slovenia) We are given several integers whose sum equals 1496. Is it possible that the sum of their seventh powers equals 1999?
2. (Australia) If p is a prime number, prove that

$$\underbrace{11 \ldots 11}_{p} \underbrace{22 \ldots 22}_{p} \ldots \underbrace{99 \ldots 9}_{p} - 123456789$$

   is divisible by p.
3. Given a sequence $(x_1, x_2, x_3, \ldots, x_{2n-2}, x_{2n-1}, x_{2n})$ of reals, an allowed operation is to replace it by the sequence

$$(x_{n+1}, x_1, x_{n+2}, x_2, \ldots, x_{2n-1}, x_{n-1}, x_{2n}, x_n).$$

   Assume that we start with the sequence $(1, 2, 3, \ldots, 2n-2, 2n-1, 2n)$, where $2n + 1$ is a prime number. Show that, after performing $2n$ allowed operations, all of the numbers will be back to their original positions.
4. (BMO) Prove that the equation $y^2 = x^5 - 4$ has no integer solutions.
5. (USA) Given a prime p, prove that there are infinitely many naturals n for which p divides $2^n - n$.
6. (Romania) Prove that there does not exist an integer $n > 1$ such that n divides $3^n - 2^n$.
7. (Bulgaria—adapted) Given prime numbers p and q, do the following items:

   (a) If $p \mid (5^p - 2^p)$, prove that $p = 3$.
   (b) If $p \geq q$ and $q \mid (5^p - 2^p)$, prove that $q = 3$.
   (c) Find all such p and q for which $pq \mid (5^p - 2^p)(5^q - 2^q)$.

8. (BMO) Prove that, for every given natural number n, there exists a natural number $m > n$ such that the decimal representation of 5^m is obtained from that of 5^n by adding a string of algarisms to the left of it.
9. (IMO) Prove that, for every integer $n > 1$, there exist pairwise distinct integers $k_1, k_2, \ldots, k_n > 1$ such that each two of the numbers $2^{k_1} - 3, 2^{k_2} - 3, \ldots, 2^{k_n} - 3$ are relatively prime.

10. Let $(a_n)_{n \geq 1}$ be the sequence of positive integers implicitly defined by the equality

$$\sum_{0 < d \mid n} a_d = 2^n,$$

which is valid for every $n \in \mathbb{N}$. Prove that $n \mid a_n$ for every $n \in \mathbb{N}$.

11. (Iran—adapted) Euler's theorem guarantees that, for every integer $n > 2$, each prime factor of n is also a prime factor of $2^{\varphi(n)} - 1$. Nevertheless, the converse claim is not true in general, namely, if $n > 3$ is odd, then $2^{\varphi(n)} - 1$ always has prime factors which are not prime factors of n. In order to prove this, do the following items:

(a) Let $p_1, \ldots, p_k$ be pairwise distinct odd primes, such that $k > 1$ or $p_1 \neq 3$. Show that there does not exist integers $\alpha_1, \ldots, \alpha_k \geq 0$, not all zero, such that

$$2^{(p_1-1)\cdots(p_k-1)} - 1 = p_1^{\alpha_1} \cdots p_k^{\alpha_k}.$$

(b) Conclude that if $n \neq 1, 3^k$ is and odd positive integer, then there exists a prime p such that $p \nmid n$ but $p \mid (2^{\varphi(n)} - 1)$.

(c) If $n = 3^k$, with $k \geq 2$, show that $2^{\varphi(n)} - 1 \equiv 0 \pmod 7$.

12. The purpose of this problem is to prove the following result, which is known as **Sophie Germain's theorem**[2]: if p is a prime number such that $2p + 1 = q$ is also prime, and if x, y and z are integers for which $x^p + y^p + z^p = 0$, then p divides at least one of x, y and z. To this end, show the items below:

(a) We can assume that x, y and z are pairwise relatively prime and that $p > 2$.

(b) If $x^p + y^p + z^p = 0$, then $p \mid (x + y + z)$.

(c) By contradiction, assume from now on that $p \nmid x, y, z$. If $r \neq p$ is a prime divisor of $y + z$, then

$$y^{p-1} - y^{p-2}z + \cdots - yz^{p-2} + z^{p-1} \equiv py^{p-1} \pmod r.$$

(d) Conclude from (c) that $r \nmid (y^{p-1} - y^{p-2}z + \cdots - yz^{p-2} + z^{p-1})$ and, hence, that $\gcd(y + z, y^{p-1} - y^{p-2}z + \cdots - yz^{p-2} + z^{p-1}) = 1$.

(e) Show that there exist $a, b, c, d \in \mathbb{Z}$ such that $y + z = a^p$, $z + x = b^p$, $x + y = c^p$ and $y^{p-1} - y^{p-2}z + \cdots - yz^{p-2} + z^{p-1} = d^p$.

(e) Since $p = \frac{q-1}{2}$, we have $x^{\frac{q-1}{2}} + y^{\frac{q-1}{2}} + z^{\frac{q-1}{2}} = 0$. Now, apply Fermat's little theorem to show that $q \mid xyz$.

---

[2] After Marie-Sophie Germain, French mathematician of the eighteenth and nineteenth centuries.

(f) If $q \mid x$, deduce that $b^p + c^p - a^p \equiv 0 \pmod{q}$. Then, substitute $p = \frac{q-1}{2}$ and use Fermat's little theorem once again to show that $q \mid abc$. Subsequently:

    i. If $q \mid b$ or $q \mid c$, conclude that $q \mid x, y$ or $q \mid x, z$, thus contradicting item (a). Therefore, $q \mid a = (y + z)$.

    ii. It follows from (c) that $d^p \equiv p y^{p-1} \equiv p c^{p(p-1)} \pmod{q}$. Substitute $p = \frac{q-1}{2}$ and apply Fermat's little theorem yet another time to show that $\pm 1 \equiv p \pmod{q}$, thus reaching a new contradiction.

13. (India—adapted) For each $x \in \mathbb{N}$, let $S(x)$ denote the set

$$S(x) = \{y \in \mathbb{N}; \ \varphi^{(k)}(y) = x, \ \text{for some } k \in \mathbb{N}\},$$

where φ stands for the Euler function, $\varphi^{(1)} = \varphi$ and, for each $k > 1$ natural, $\varphi^{(k)}$ is the composite of φ with itself, k times. Let also

$$T = \{2^a \cdot 3^b; \ a, b \in \mathbb{Z}_+, a \geq 1\}.$$

Do the items below to show that

$$x \in T \Leftrightarrow S(x) \text{ is infinite.}$$

(a) Prove that $x \in T \Rightarrow S(x)$ is infinite.

(b) For each $x \in \mathbb{N}$, let $u(x)$ denote the exponent of the greatest power of 2 that divides x. Show that $u(\varphi(x)) \geq u(x)$.

(c) If $x \notin T$, prove that $u(\varphi(x)) = u(x)$ if and only if $x = 2^m p^a$, where $m \in \mathbb{N}$ and $p \geq 7$ is a prime such that $p \equiv 3 \pmod{4}$.

(d) If $x \notin T$ is such that $u(\varphi^{(2)}(x)) = u(\varphi(x)) = u(x)$, prove that the natural number a if item (c) is equal to 1.

(e) Assuming that $S(x)$ is infinite, prove that there exist $a_1, a_2, a_3, \ldots \in \mathbb{N}$ such that

$$x = a_1, \ a_1 = \varphi(a_2), \ a_2 = \varphi(a_3), \ a_3 = \varphi(a_4), \ldots$$

(f) Continue assuming $S(x)$ to be infinite. Use the result of item (b) to prove that there exists $n \in \mathbb{N}$ for which

$$u(a_n) = u(a_{n+1}) = u(a_{n+2}) = \cdots .$$

(g) Conclude from item (d) that, with respect to the integer n of item (f), if $i \geq n + 2$ then there exists a prime $p_i \geq 7$ such that $p_i \equiv 3 \pmod{4}$ and $m_i \in \mathbb{N}$ such that $a_i = 2^{m_i} p_i$. Then, prove that $m_{n+2} = m_{n+3} = \cdots$ and, for every $i > 1$, $p_i = 2p_{i-1} + 1$.

(h) Prove that there is no infinite sequence $q_1, q_2, q_3, \ldots$ of primes such that
$q_i = 2q_{i-1} + 1$ for every $i > 1$.

(i) Conclude that $S(x)$ infinite $\Rightarrow x \in T$.

10.3   Linear Congruences and the Chinese Remainder Theorem

In this section, we study linear equations and systems of equations involving congruences.

We first consider **linear congruences**, i.e., congruences of the form

$$ax \equiv b \,(\text{mod}\,n), \qquad (10.6)$$

with a, b and n being given integers, such that $a \neq 0$ and $n > 1$, and we search for the **solutions** (or **roots**) $x \in \mathbb{Z}$ of the equation, i.e., the integers x for which (10.6) holds true.

As a particular case of the above situation, we say that a is **invertible** modulo n if the linear congruence (10.6) has a solution when $b = 1$, i.e., if there exists $x \in \mathbb{Z}$ such that

$$ax \equiv 1 \,(\text{mod}\,n).$$

If this is so, then such an integer x is said to be an **inverse** of a, modulo n. In this respect, we have the following important

Proposition 10.23 *An integer a is invertible modulo n if and only if $\gcd(a, n) = 1$. In this case, any two inverses of a, modulo n, are congruent modulo n.*

Proof If a is invertible modulo n, then there exists $x \in \mathbb{Z}$ such that $ax \equiv 1 \,(\text{mod}\,n)$. Then, there exists $y \in \mathbb{Z}$ for which $ax = ny+1$ or, which is the same, $xa+(-y)n = 1$. In view of such an equality, we know from Corollary 6.16 that $\gcd(a, n) = 1$.

Conversely, recall that Bézout's theorem guarantees that the gcd of any two integers can always be written as a linear combination of them. Hence, if $\gcd(a, n) = 1$ then there exist $x, y \in \mathbb{Z}$ such that $ax + ny = 1$, and it immediately follows from this that $ax \equiv 1 \,(\text{mod}\,n)$.

Finally, let both x and y be inverses of a, modulo n. Then

$$ax \equiv 1 \equiv ay \,(\text{mod}\,n),$$

so that $ax \equiv ay \,(\text{mod}\,n)$. However, since $\gcd(a, n) = 1$, Proposition 10.6 gives $x \equiv y \,(\text{mod}\,n)$. Thus, a has, modulo n, a single inverse.   □

Corollary 10.24 *If $a, p \in \mathbb{Z}$, with p prime, then a is invertible modulo p if and only if $p \nmid a$. Moreover, in this case the inverse of a modulo p is unique.*

Proof Immediate from the fact that $\gcd(a, p) = 1 \Leftrightarrow p \nmid a$. $\square$

Corollary 10.25 *Let $a, b, n \in \mathbb{Z}$, with $n > 1$. If a is invertible modulo n, then the linear congruence $ax \equiv b \pmod{n}$ always has a single solution modulo n, namely,*

$$x \equiv a^{\varphi(n)-1} b \pmod{n}.$$

In particular, every inverse of a, modulo n, is congruent to $a^{\varphi(n)-1}$, modulo n.

Proof If a is invertible modulo n, then $\gcd(a, n) = 1$. Hence, successively applying item (f) of Proposition 10.6 and Euler's theorem, we get

$$ax \equiv b \pmod{n} \Leftrightarrow a^{\varphi(n)} x \equiv a^{\varphi(n)-1} b \pmod{n} \Leftrightarrow x \equiv a^{\varphi(n)-1} b \pmod{n}.$$

$\square$

Since any two inverses of a modulo n are always congruent modulo n, from now on we shall say that a has a *single inverse* modulo n. Such an inverse will be denoted by a^{-1}, whenever there is no danger of confusion with the usual inverse $\frac{1}{a}$ of a with respect to the multiplication in $\mathbb{Q}$.

As an application of the notion of inversion modulo n, we shall now prove the famous *primality criterion* known as **Wilson's theorem**.[3]

Theorem 10.26 (Wilson) *A natural number p is prime if and only if*

$$(p - 1)! \equiv -1 \pmod{p}.$$

Proof If $p = mn$, with m and n integers and $1 < n < p$, then $(p - 1)! \equiv -1 \pmod{p}$ implies $(p - 1)! \equiv -1 \pmod{n}$. On the other hand, since $1 < n \leq p - 1$, we also have that $n \mid (p - 1)!$, i.e., $(p - 1)! \equiv 0 \pmod{n}$. Therefore, $0 \equiv (p - 1)! \equiv -1 \pmod{n}$, and hence $n \mid 1$, which is a contradiction.

Conversely, let p be prime and consider the function

$$f : \{1, 2, \ldots, p - 1\} \to \{1, 2, \ldots, p - 1\},$$

that associates to each $a \in \{1, 2, \ldots, p - 1\}$ its inverse $a^{-1} \in \{1, 2, \ldots, p - 1\}$, modulo p. Corollary 10.25 assures that f é uma bijeção, with $f(a) = b \Leftrightarrow f(b) = a$. Moreover,

---

[3] After John Wilson, English mathematician of the seventeenth century. Although it is a primality test, in practice Wilson's theorem is not so useful, for, according to Stirling's formula (cf. [8], for instance), $n!$ grows exponentially as $n \to +\infty$.

$$f(a) = a \Leftrightarrow a^2 \equiv 1 \ (\mathrm{mod}\ p)$$

$$\Leftrightarrow p \mid (a^2 - 1)$$

$$\Leftrightarrow p \mid (a - 1) \text{ or } p \mid (a + 1)$$

$$\Leftrightarrow a = 1 \text{ or } p - 1.$$

Hence, there exist $a_1, \ldots, a_{\frac{p-3}{2}} \in \{1, 2, \ldots, p - 1\}$ such that

$$\{1, 2, \ldots, p - 1\} = \{1, p - 1\} \cup \bigcup_{i=1}^{\frac{p-3}{2}} \{a_i, a_i^{-1}\}.$$

Finally, it follows from this, together with the fact that $a_i a_i^{-1} \equiv 1 \ (\mathrm{mod}\ p)$ for $1 \le i \le \frac{p-3}{2}$, that

$$(p - 1)! = (p - 1) \prod_{i=1}^{\frac{p-3}{2}} (a_i a_i^{-1}) \equiv -1 \ (\mathrm{mod}\ p).$$

□

Back to equations involving congruences, a natural generalization of the linear congruence (10.6) is obtained by considering solutions of a *system of linear congruences*. In this sense, the following result, known as the **Chinese remainder theorem**, examines the existence of solutions for such systems.

Theorem 10.27 *Let $m_1, m_2, \ldots, m_k$ be pairwise relatively prime naturals, all greater than 1. For arbitrarily given integers $a_1, a_2, \ldots, a_k$, the system of linear congruences*

$$\begin{cases} x \equiv a_1 \ (\mathrm{mod}\ m_1) \\ x \equiv a_2 \ (\mathrm{mod}\ m_2) \\ \quad \ldots \\ x \equiv a_k \ (\mathrm{mod}\ m_k) \end{cases} \tag{10.7}$$

admits a unique solution, modulo $m_1 m_2 \ldots m_k$. Yet in another way, there exists a single integer $0 \le y < m_1 m_2 \ldots m_k$ such that $x \in \mathbb{Z}$ satisfies the system above if and only if

$$x \equiv y \ (\mathrm{mod}\ m_1 m_2 \ldots m_k).$$

Proof Note first that if x_1 and x_2 are any two solutions of the above system, then

$$x_1 \equiv a_i \equiv x_2 \ (\mathrm{mod}\ m_i), \quad \forall\, 1 \le i \le k.$$

However, since $m_1, \ldots, m_k$ are pairwise relatively prime, it follows from Proposition 10.6 that

$$x_1 \equiv x_2 \pmod{m_1 m_2 \ldots m_k}.$$

Hence, if the system (10.7) has a solution at all, this will be unique, modulo $m_1 m_2 \ldots m_k$.

For the existence of a solution, for $1 \leq j \leq k$ let

$$y_j = \prod_{\substack{1 \leq i \leq k \\ i \neq j}} m_i,$$

so that $\gcd(y_j, m_j) = 1$. Let also b_j be the inverse of y_j modulo m_j and $x = \sum_{j=1}^{k} a_j b_j y_j$. For a fixed $1 \leq l \leq k$, we have $y_j \equiv 0 \pmod{m_l}$ whenever $j \neq l$; hence, modulo m_l we get

$$x \equiv a_l b_l y_l \equiv a_l \cdot 1 \equiv a_l \pmod{m_l}.$$

$\square$

Problem 5 offers an alternative proof for the existence of solutions for the system of linear congruences (10.7). In order to finish this section, let us see an example who shows how it is possible to apply the Chinese remainder theorem to obtain interesting results.[4]

Example 10.28 Given an integer $n > 1$, show that there exist n consecutive and composite natural numbers.

Proof Choose n distinct primes $p_1, p_2, \ldots, p_n$ and consider the system of linear congruences

$$\begin{cases} x \equiv -1 \pmod{p_1^2} \\ x \equiv -2 \pmod{p_2^2} \\ \ldots \\ x \equiv -n \pmod{p_n^2} \end{cases}.$$

Since $p_1, p_2, \ldots, p_n$ are pairwise relatively prime, the Chinese remainder theorem assures the existence of $m \in \mathbb{N}$ satisfying the above system. Hence, $p_j^2 \mid (m + j)$ for $1 \leq j \leq n$, so that $m + 1, m + 2, \ldots, m + n$ are all composite natural numbers. $\square$

---

[4]In this respect, see also Problems 12, page 309, and 4, page 314.

Problems: Sect. 10.3

1. Let a, b and n be given integers, with $n > 1$. With respect to the linear congruence $ax \equiv b \pmod{n}$, prove that:

   (a) There is a solution if and only if $\gcd(a, n) \mid b$.
   (b) If $\gcd(a, n) \mid b$, then the linear congruence has exactly $\gcd(a, n)$ pairwise incongruent solutions, modulo n.

2. Let $a_1, a_2, \ldots, a_k$, b and n be given integers, with $n > 1$. Prove that the congruence

$$a_1 x_1 + a_2 x_2 + \cdots + a_k x_k \equiv b \pmod{n}$$

   has solutions if and only if $\gcd(a_1, a_2, \ldots, a_k, n) \mid b$.

3. A group of soldiers was arranged as to form a rectangular array, with several rows. The commander observed that by placing 12 soldiers in each row, 7 soldiers would be left, while by placing 13 in each row, 5 would be left. Knowing that the total number of soldiers ranged between 600 and 700, find out how many were them.

4. In the hypotheses of Theorem 10.27, do the following items:

   (a) Show that solving (10.7) for $k = 2$ is equivalent to solving, in $u, v \in \mathbb{Z}$, the linear diophantine equation

$$m_1 u - m_2 v = a_2 - a_1.$$

   (b) Show that solving (10.7) is equivalent to finding out all $y \in \mathbb{Z}$ that solve the system of linear congruences

$$\begin{cases} m_1 y \equiv a_2 - a_1 \pmod{m_2} \\ m_1 y \equiv a_3 - a_1 \pmod{m_3} \\ \quad \cdots \\ m_1 y \equiv a_k - a_1 \pmod{m_k} \end{cases}.$$

   (c) Apply the procedure delineated along items (a) and (b) to find out all integer solutions of the system of congruences

$$\begin{cases} x \equiv 2 \pmod{5} \\ x \equiv 4 \pmod{11} \\ x \equiv 9 \pmod{13} \end{cases}.$$

5. In the notations of Theorem 10.27, if $y_j = \prod_{\substack{1 \le i \le k \\ i \ne j}} m_i$ for $1 \le j \le k$, prove that $x = \sum_{j=1}^{k} a_j y_j^{\varphi(m_j)}$ solves the system of linear congruences (10.7).

6. Let $p_1 = 2$, $p_2 = 3$, $p_3 = 5, \ldots$ be the sequence of primes. Show that there exists a natural number which leaves remainder $2^{p_{k-1}} - 1$ upon division by $2^{p_k} - 1$, for $2 \leq k \leq 100$.

7. Prove that there are arbitrarily long sequences of consecutive natural numbers such that none of them is a perfect power of exponent greater than 1.

8. Let $m = p_1^{\alpha_1} \ldots p_k^{\alpha_k}$ be the canonical decomposition of the integer $m > 1$ in powers of distinct primes, and f be a polynomial function with integer coefficients.

   (a) Prove that the congruence $f(x) \equiv 0 \pmod{m}$ has an integer solution if and only if each one of the congruences $f(x) \equiv 0 \pmod{p_i^{\alpha_i}}$, for $1 \leq i \leq k$, has itself an integer solution.[5]

   (b) Given $t \in \mathbb{N}$, let $N(t)$ denote the number of integer solutions, pairwise incongruent modulo t, for the congruence $f(x) \equiv 0 \pmod{t}$. Show that $N(m) = N(p_1^{\alpha_1}) \ldots N(p_k^{\alpha_k})$.

9. (IMO shortlist) Let be given a nonconstant arithmetic progression of natural numbers, whose initial term and common difference are relatively prime. If it contains a perfect square and a perfect cube among its terms, show that it also contains a perfect sixth power.[6]

---

[5]If p is prime and $\alpha \in \mathbb{N}$, a sufficient condition for the solvability of a congruence of the form $f(x) \equiv 0 \pmod{p^\alpha}$ will be presented in Problem 11, page 394.

[6]The hypotheses that the arithmetic progression is nonconstant and has relatively prime initial term and common ratio are actually unnecessary. They were assumed just to simplify the solution of the problem.

Chapter 11
Congruence Classes

In this chapter, we return to the point of view of Example 2.21, looking at congruence modulo n as an *equivalence relation*. As a byproduct of our discussion, a number of interesting applications will be presented, among which is an alternative, simpler proof of Euler's theorem. We will also introduce the quotient set $\mathbb{Z}_n$ and show that it can be furnished with operations of *addition* and *multiplication* quite similar to those of $\mathbb{Z}$. In particular, the case of $\mathbb{Z}_p$, with p prime, will be crucial to our future discussion of polynomials.

11.1 Systems of Residues

Given an integer $n > 1$, Proposition 10.5 assures that the relation $\sim$ in $\mathbb{Z}$, defined by

$$a \sim b \Leftrightarrow a \equiv b \,(\mathrm{mod}\, n),$$

is an equivalence relation, called the **congruence relation modulo n**.

For $a \in \mathbb{Z}$, we let the **congruence class of a, modulo n**, be defined as the equivalence class $\overline{a}$ of a with respect to the relation of congruence modulo n:

$$\begin{aligned}
\overline{a} &= \{x \in \mathbb{Z};\ x \equiv a \,(\mathrm{mod}\, n)\} \\
&= \{x \in \mathbb{Z};\ x = a + nq,\ \exists\, q \in \mathbb{Z}\} \\
&= \{\ldots, -2n + a, -n + a, a, n + a, 2n + a, \ldots\}.
\end{aligned} \qquad (11.1)$$

Given $a \in \mathbb{Z}$, we know that there exists a single integer $0 \le r < n$ (the remainder of the division of a by n) such that $a \equiv r \,(\mathrm{mod}\, n)$. Therefore, $\{0, 1, \ldots, n-1\}$ is

© Springer International Publishing AG, part of Springer Nature 2018
A. Caminha Muniz Neto, *An Excursion through Elementary Mathematics, Volume III*,
Problem Books in Mathematics, https://doi.org/10.1007/978-3-319-77977-5_11

a system of distinct representatives for the relation of congruence modulo n, and Proposition 2.16 together with (2.12) furnish the partition

$$\mathbb{Z} = \overline{0} \cup \overline{1} \cup \cdots \cup \overline{n-1} \qquad (11.2)$$

of $\mathbb{Z}$. More generally, we have the following important

Definition 11.1 A **complete residue system** modulo n, (abbreviated CRS) is a system of distinct representatives for the congruence relation modulo n.

It follows from (11.2) and (11.1) that every CRS modulo n is a set $\{a_0, a_1, \ldots, a_{n-1}\}$ of integers, such that $a_r \equiv r \pmod{n}$ for $0 \le r \le n-1$. Alternatively, since $\overline{0} = \overline{n}$, we can view a CRS modulo n a set $\{a_1, a_2, \ldots, a_n\}$ of integers, such that $a_r \equiv r \pmod{n}$ for $1 \le r \le n$. Yet in another way, a set of n integers is a CRS modulo n if and only if its elements are pairwise incongruent modulo n.

Example 11.2 Given an integer $n > 1$, the sets $\{0, 1, 2, \ldots, n-1\}$ and $\{1, 2, \ldots, n\}$ are obviously CRS modulo n. Another example is furnished by the set

$$\left\{ x \in \mathbb{Z}; \ -\frac{n}{2} \le x < \frac{n}{2} \right\}.$$

Indeed, note first that if $-\frac{n}{2} \le x < y < \frac{n}{2}$, then $x \not\equiv y \pmod{n}$, for $0 < y - x < n$. Now, it suffices to separately consider the cases n even and n odd, in order to show that the set displayed above has exactly n elements.

The coming proposition provides a useful way of constructing new CRS out of given ones.

Proposition 11.3

(a) *Let a, b and n be given integers, with $n > 1$ and a and n relatively prime. If the set $\{x_0, x_1, \ldots, x_{n-1}\}$ is a CRS modulo n, then the set $\{ax_0 + b, ax_1 + b, \ldots, ax_{n-1} + b\}$ is also a CRS modulo n.*

(b) *Let m and n be integers greater than 1 and relatively prime. If the sets $\{x_0, x_1, \ldots, x_{m-1}\}$ and $\{y_0, y_1, \ldots, y_{n-1}\}$ are CRS modulo m and n, respectively, then the set*

$$\{nx_i + my_j; \ 0 \le i < m, \ 0 \le j < n\}$$

is a CRS modulo mn.

Proof

(a) Since every CRS modulo n has n elements, it suffices to show that if $0 \le i, j < n$ and

$$ax_i + b \equiv ax_j + b \pmod{n},$$

then $i = j$. To this end, it follows from the above congruence that $ax_i \equiv ax_j \pmod{n}$. Now, since $\gcd(a, n) = 1$, item (f) of Proposition 10.6 guarantees that $x_i \equiv x_j \pmod{n}$. Finally, since $\{x_0, x_1, \ldots, x_{n-1}\}$ is a CRS modulo n, this last congruence gives us $i = j$.

(b) Analogously to the proof of item (a), it suffices to show that if $0 \le i, j < m$, $0 \le k, l < n$ and

$$nx_i + my_k \equiv nx_j + my_l \pmod{mn},$$

then $i = j$ and $k = l$. Assuming the validity of the above congruence, we have from item (g) of Proposition 10.6 that

$$nx_i + my_k \equiv nx_j + my_l \pmod{m} \quad \text{and} \quad nx_i + my_k \equiv nx_j + my_l \pmod{n}.$$

In turn, the last two congruences above readily give

$$nx_i \equiv nx_j \pmod{m} \quad \text{and} \quad my_k \equiv my_l \pmod{n}.$$

However, since m and n are relatively prime, by applying again item (f) of Proposition 10.6 we obtain

$$x_i \equiv x_j \pmod{m} \quad \text{and} \quad y_k \equiv y_l \pmod{n}.$$

Finally, since $\{x_0, x_1, \ldots, x_{m-1}\}$ and $\{y_0, y_1, \ldots, y_{n-1}\}$ are CRS modulo m and n, respectively, we get $i = j$ and $k = l$. □

The previous proposition furnishes a proof of a particular case of the chinese remainder theorem which sheds light in the proof of the general case presented in the last section. In this respect, see also Problems 2 and 3.

Example 11.4 Let m_1 and m_2 be relatively prime naturals greater than 1, and a_1 and a_2 be arbitrarily given integers. We will show that the system of linear congruences

$$\begin{cases} x \equiv a_1 \pmod{m_1} \\ x \equiv a_2 \pmod{m_2} \end{cases} \tag{11.3}$$

has a single solution, modulo $m_1 m_2$.

To this end, item (b) of the previous proposition assures that

$$\{m_1 u_1 + m_2 u_2; \ x_i \in \mathbb{Z} \text{ and } 1 \le u_i \le m_i \text{ for } i = 1, 2\}$$

is a CRS modulo $m_1 m_2$. If we set $x = m_1 u_1 + m_2 u_2$, for some such u_1 and u_2, then

$$x \equiv m_2 u_2 \pmod{m_1} \quad \text{and} \quad x \equiv m_1 u_1 \pmod{m_2}.$$

Therefore, by the transitivity of congruences, x will solve (11.3) if and only if we can choose $1 \le u_1 \le m_1$ and $1 \le u_2 \le m_2$ such that

$$m_2 u_2 \equiv a_1 \ (\mathrm{mod}\, m_1) \quad \text{and} \quad m_1 u_1 \equiv a_2 \ (\mathrm{mod}\, m_2).$$

Although Corollary 10.25 assures we can uniquely choose such u_1 and u_2, this is also an immediate consequence of item (a) of the previous proposition. Indeed, since $\gcd(m_1, m_2) = 1$, that item guarantees that

$$\{m_2 x_2; \ 1 \le x_2 \le m_1\}$$

is a CRS modulo m_1; hence, there is a single $1 \le u_2 \le m_1$ such that $m_2 u_2 \equiv a_1 \ (\mathrm{mod}\, m_1)$. By the same token, there is a single $1 \le u_1 \le m_2$ such that $m_1 u_1 \equiv a_2 \ (\mathrm{mod}\, m_2)$.

The coming example brings a beautiful combinatorial application of the concept of complete residue system, which appeared as the last problem of the 1995 IMO.[1] For what is to come, recall that for $Z \subset \mathbb{R}$ and $r \in \mathbb{R}$ we write $Z + r$ to denote the set $Z + r = \{z + r; \ z \in Z\}$.

Example 11.5 (IMO) Let p be an odd prime. Compute the number of subsets of p elements of the set $\{1, 2, \ldots, 2p\}$ such that the sum of the elements of each such subset is divisible by p.

Solution Let $\mathcal{F}$ be the family of the $\binom{2p}{p} - 2$ subsets of p elements of the set $\{1, 2, \ldots, 2p\}$, different from $X = \{1, 2, \ldots, p\}$ and $Y = \{p + 1, p + 2, \ldots, 2p\}$ (notice that these two sets have sums of elements equal to multiples of p).

Let $\sim$ be the relation defined in $\mathcal{F}$ by setting $A \sim B$ if and only if the following conditions are satisfied:

 (i)   There exists $0 \le r \le p - 1$ such that $(A \cap X) + r = B \cap X \ (\mathrm{mod}\, p)$.
 (ii) $A \cap Y = B \cap Y$.

(Here, according to the paragraph that immediately precedes Example 2.22, for $r \in \mathbb{R}$ and $Z \subset \mathbb{R}$ we let $Z + r = \{z + r; \ z \in Z\}$.)

It is immediate to verify that $\sim$ is an equivalence relation in $\mathcal{F}$. On the other hand, given $A \in \mathcal{F}$ and $B \in \overline{A}$ (the equivalence class of A in $\mathcal{F}$), we have

$$B = (B \cap X) \cup (B \cap Y) = [(A \cap X) + r] \cup (A \cap Y).$$

Therefore, the equivalence class $\overline{A}$ will have as many sets as its elements as there are distinct sets of the form $(A \cap X) + r$, when $r \in \mathbb{Z}$ varies from 0 to $p - 1$.

---

[1] For another approach to this example, see Example 20.7.

In order to address this counting problem, let $A' = A \cap X$ and $S' = \sum_{x \in A'} x$. Since $A \neq X, Y$, we have $\emptyset \neq A' \neq X$, so that $1 \leq |A'| \leq p - 1$. Hence,

$$\sum_{x \in A'+r} x = \sum_{x \in A'} x + r|A'| = S' + r|A'|.$$

On the other hand, since $\gcd(|A'|, p) = 1$, item (a) of Proposition 11.3 guarantees that, as $r \in \mathbb{Z}$ varies from 0 to $p - 1$, the numbers $S' + r|A'|$ thus obtained form a CRS modulo p. In particular, the sums $\sum_{x \in A'+r} x$ are pairwise distinct, so that the sets A', $A' + 1$, ..., $A' + (p - 1)$ are themselves pairwise distinct. Therefore, the class $\overline{A}$ has exactly p sets, and it follows from Proposition 2.18 that there are exactly

$$\frac{1}{p}\left(\binom{2p}{2} - 2\right)$$

distinct equivalence classes.

If we now show that in each of the classes above there is exactly one set whose sum of elements is a multiple of p, it will follow that the number of sets we wish to count is precisely

$$\frac{1}{p}\left(\binom{2p}{2} - 2\right) + 2.$$

(Recall that we cannot forget X and Y!)

For what is left to do, fix $A \in \mathcal{F}$ (and, hence, the class $\overline{A}$) and let $A'' = A \cap Y$. We wish to count how many are the integers $0 \leq r \leq p - 1$ such that the sum of the p elements of the set

$$B = [(A \cap X) + r] \cup (A \cap Y) = (A' + r) \cup A''$$

is a multiple of p (remember that the set B above is a generic element of the class $\overline{A}$). However, for such a B, it follows from what we did above that

$$\sum_{x \in B} x = \sum_{x \in A'+r} x + \sum_{x \in A''} x$$

$$= \sum_{x \in A'} x + r|A'| + \sum_{x \in A''} x$$

$$= \sum_{x \in A} x + r|A'| = S + r|A'|,$$

where $S = \sum_{x \in A} x$. Thus, such a sum will be congruent to 0 modulo p if and only if

$$|A'|r \equiv -S \,(\mathrm{mod}\,p).$$

But since $\gcd(|A'|, p) = 1$, we know from Corollary 10.24 (or from item (a) of Proposition 11.3, with $b = 0$ and $|A'|$ in place of a) that there exists a single $0 \leq r \leq p - 1$ such that the congruence displayed above is satisfied.   □

Back to the development of the theory, given $q, r \in \mathbb{Z}$, with $0 \leq r < n$, recall that $\gcd(nq + r, n) = \gcd(r, n)$. Thus, we can define the gcd between n and a congruence class modulo n by letting

$$\gcd(\overline{r}, n) = \gcd(x, n), \text{ for } x \in \overline{r}; \tag{11.4}$$

in particular,

$$\gcd(\overline{r}, n) = \gcd(r, n).$$

Note that, by Proposition 10.23, the congruence classes $\overline{r}$ for which $\gcd(\overline{r}, n) = 1$ are precisely those formed by the integers invertible modulo n. Therefore, we have the following

Definition 11.6 A **reduced residue system** (we abbreviate **RRS**) modulo n is a set I of integers such that

$$|I \cap \overline{r}| = \begin{cases} 1, & \text{if } \gcd(\overline{r}, n) = 1 \\ 0, & \text{if } \gcd(\overline{r}, n) \neq 1 \end{cases},$$

for every congruence class $\overline{r}$, modulo n.

For a fixed integer $n > 1$, the most important example of RRS modulo n is the set

$$\{x \in \mathbb{Z};\ \gcd(x, n) = 1 \text{ and } 1 \leq x \leq n\}.$$

More generally, if I is a RRS modulo n, then (11.4) easily assures that $|I| = \varphi(n)$, where φ stands for the Euler function. On the other hand, it is clear that every RRS I modulo n can be enlarged to a CRS modulo n, as well as that every CRS modulo n contains a RRS modulo n.

The coming result teaches us how to construct new RRS out of known ones.

Proposition 11.7 *Let $m, n > 1$ be given integers.*

(a) *If an integer a is relatively prime with n and $\{x_1, x_2, \ldots, x_{\varphi(n)}\}$ is a RRS modulo n, then $\{ax_1, ax_2, \ldots, ax_{\varphi(n)}\}$ is also a RRS modulo n.*

(b) If $\{x_1, x_2, \ldots, x_{\varphi(m)}\}$ and $\{y_1, y_2, \ldots, y_{\varphi(n)}\}$ are RRS modulo m and modulo n, respectively, then

$$\{nx_i + my_j; \ 1 \le i \le \varphi(m), \ 1 \le j \le \varphi(n)\} \tag{11.5}$$

is a RRS modulo mn.

Proof

(a) Since every RRS modulo n is part of a CRS, it follows from item (a) of Proposition 11.3 that

$$ax_i \not\equiv ax_j \,(\mathrm{mod}\, n), \ \forall \ 1 \le i < j \le \varphi(n).$$

It now suffices to show that $\gcd(ax_i, n) = 1$, which follows immediately from item (c) of Proposition 6.22.

(b) By invoking again the fact that every RRS modulo n is part of a CRS, item (b) of Proposition 11.3 assures that the set in (11.5) has exactly $\varphi(m)\varphi(n)$ elements, which are pairwise incongruent modulo mn. However, since $\varphi(m)\varphi(n) = \varphi(mn)$ (recall that $\gcd(m, n) = 1$), we conclude that such a set has $\varphi(mn)$ elements. Hence, in order to show that it is indeed a RRS modulo mn, we are left to showing that

$$\gcd(nx_i + my_j, mn) = 1$$

for all $1 \le i \le \varphi(m)$, $1 \le j \le \varphi(n)$. To this end, items (b) and (c) of Proposition 6.22 give

$$\gcd(nx_i + my_j, m) = \gcd(nx_i, m) = \gcd(x_i, m) = 1$$

and, analogously, $\gcd(nx_i + my_j, n) = 1$; therefore, by applying one further time item (c) of that result, we obtain $\gcd(nx_i + my_j, mn) = 1$.   □

With the concept of RRS and item (a) of the former proposition at our disposal, we can give another, conceptually clearer, proof of Euler's theorem.

Theorem 11.8 (Euler)   *If a and n are relatively prime integers, with n > 1, then*

$$a^{\varphi(n)} \equiv 1 \,(\mathrm{mod}\, n). \tag{11.6}$$

Proof Let $k = \varphi(n)$ and $\{x_1, x_2, \ldots, x_k\}$ be a RRS modulo n. Item (a) of the previous proposition assures that $\{ax_1, ax_2, \ldots, ax_k\}$ is also a RRS modulo n; hence, modulo n the integers $ax_1, ax_2, \ldots, ax_k$ are, in some order, congruent to $x_1, x_2, \ldots, x_k$. Termwise multiplication of the corresponding congruences gives

$$a^k x_1 x_2 \ldots x_k = ax_1 \cdot ax_2 \cdot \ldots \cdot ax_k \equiv x_1 x_2 \ldots x_k \,(\mathrm{mod}\, n).$$

Finally, since $\gcd(x_1 x_2 \ldots x_k, n) = 1$, we can invoke item (f) of Proposition 10.6 to cancel $x_1 x_2 \ldots x_k$ out of the congruence above to obtain $a^k \equiv 1 \,(\mathrm{mod}\,n)$.   □

Problems: Sect. 11.1

1. Let $n > 1$ be an integer and $\{a_1, a_2, \ldots, a_n\}$ be a CRS modulo n. Show that, modulo n, we have

$$a_1 + a_2 + \cdots + a_n \equiv \begin{cases} 0, & \text{if } n \text{ is odd} \\ \frac{n}{2}, & \text{if } n \text{ is even} \end{cases}.$$

The following problem generalizes item (b) of Proposition 11.3. The subsequent one generalizes the argument of Example 11.4.

2. Let $m_1, m_2, \ldots, m_k$ be pairwise relatively prime naturals, all greater than 1, and $y_j = \prod_{\substack{1 \le i \le k \\ i \ne j}} m_i$ for $1 \le j \le k$. Show that

$$\{u_1 y_1 + u_2 m_2 + \cdots + u_k m_k; \ 1 \le u_i \le m_i \text{ for } 1 \le i \le k\}$$

is a CRS modulo $m_1 m_2 \ldots m_k$.

3. Use the result of the previous problem to prove the general version of the chinese remainder theorem.

11.2   The Quotient Set $\mathbb{Z}_n$

Let us now examine the quotient set

$$\mathbb{Z}_n = \{\overline{0}, \overline{1}, \ldots, \overline{n-1}\}, \tag{11.7}$$

of $\mathbb{Z}$ with respect to the relation of congruence modulo n. Proposition 10.6 allows us to introduce in $\mathbb{Z}_n$ two operations, to which we shall also refer to as *addition* and *multiplication* and which share properties analogous to those of the usual operations of addition and multiplication of integers. Actually, we have the following important result.

Proposition 11.9 *In $\mathbb{Z}_n$, the operations*

$$\overline{a} \oplus \overline{b} = \overline{a+b} \ \text{ and } \ \overline{a} \odot \overline{b} = \overline{a \cdot b}, \tag{11.8}$$

are well defined operations, commutative and associative. Moreover, $\odot$ is distribu-tive with respect to $\oplus$, and $\overline{0}$ and $\overline{1}$ are identities *(also called* neutral elements*) for $\oplus$ and $\odot$, respectively, in the sense that*

$$\overline{a} \oplus \overline{0} = \overline{a} \ \ and \ \ \overline{a} \odot \overline{1} = \overline{a},$$

for every $\overline{a} \in \mathbb{Z}_n$.

Proof Firstly, we have to show that the operations given by (11.8) are indeed well defined, in the sense that $\overline{a} \oplus \overline{b}$ and $\overline{a} \odot \overline{b}$ are independent of the chosen representatives for the congruence classes $\overline{a}$ and $\overline{b}$.

To this end, if $\overline{a} = \overline{c}$ and $\overline{b} = \overline{d}$, then $a \equiv c \pmod{n}$ and $b \equiv d \pmod{n}$, and it follows from Proposition 10.6 that

$$a + b \equiv c + d \pmod{n} \ \ \text{and} \ \ a \cdot b \equiv c \cdot d \pmod{n}.$$

However, this is the same as

$$\overline{a+b} = \overline{c+d} \ \ \text{and} \ \ \overline{a \cdot b} = \overline{c \cdot d},$$

so that

$$\overline{a} \oplus \overline{b} = \overline{c} \oplus \overline{d} \ \ \text{and} \ \ \overline{a} \odot \overline{b} = \overline{c} \odot \overline{d}.$$

What is left to do is simpler. For instance, for the commutativity of $\oplus$, we get from the commutativity of the addition of integers that

$$\overline{a} \oplus \overline{b} = \overline{a+b} = \overline{b+a} = \overline{b} \oplus \overline{a};$$

analogously, we prove that $\odot$ is commutative and that $\oplus$ and $\odot$ are associative, i.e., that

$$\overline{a} \oplus (\overline{b} \oplus \overline{c}) = (\overline{a} \oplus \overline{b}) \oplus \overline{c}$$

and

$$\overline{a} \odot (\overline{b} \odot \overline{c}) = (\overline{a} \odot \overline{b}) \odot \overline{c}.$$

The checking of the distributivity of $\odot$ with respect to $\oplus$ will be left as an exercise to the reader (cf. Problem 2). Finally,

$$\overline{a} \oplus \overline{0} = \overline{a+0} = \overline{a} \ \ \text{and} \ \ \overline{a} \odot \overline{1} = \overline{a \cdot 1} = \overline{a},$$

as we wished to show. $\qquad\square$

For the sake of exemplifying, we show below the addition and multiplication tables of $\mathbb{Z}_6$:

Addition Table for $\mathbb{Z}_6$

$\oplus$	$\bar{0}$	$\bar{1}$	$\bar{2}$	$\bar{3}$	$\bar{4}$	$\bar{5}$
$\bar{0}$	$\bar{0}$	$\bar{1}$	$\bar{2}$	$\bar{3}$	$\bar{4}$	$\bar{5}$
$\bar{1}$	$\bar{1}$	$\bar{2}$	$\bar{3}$	$\bar{4}$	$\bar{5}$	$\bar{0}$
$\bar{2}$	$\bar{2}$	$\bar{3}$	$\bar{4}$	$\bar{5}$	$\bar{0}$	$\bar{1}$
$\bar{3}$	$\bar{3}$	$\bar{4}$	$\bar{5}$	$\bar{0}$	$\bar{1}$	$\bar{2}$
$\bar{4}$	$\bar{4}$	$\bar{5}$	$\bar{0}$	$\bar{1}$	$\bar{2}$	$\bar{3}$
$\bar{5}$	$\bar{5}$	$\bar{0}$	$\bar{1}$	$\bar{2}$	$\bar{3}$	$\bar{4}$

Multiplication Table for $\mathbb{Z}_6$

$\odot$	$\bar{0}$	$\bar{1}$	$\bar{2}$	$\bar{3}$	$\bar{4}$	$\bar{5}$
$\bar{0}$	$\bar{0}$	$\bar{0}$	$\bar{0}$	$\bar{0}$	$\bar{0}$	$\bar{0}$
$\bar{1}$	$\bar{0}$	$\bar{1}$	$\bar{2}$	$\bar{3}$	$\bar{4}$	$\bar{5}$
$\bar{2}$	$\bar{0}$	$\bar{2}$	$\bar{4}$	$\bar{0}$	$\bar{2}$	$\bar{4}$
$\bar{3}$	$\bar{0}$	$\bar{3}$	$\bar{0}$	$\bar{3}$	$\bar{0}$	$\bar{3}$
$\bar{4}$	$\bar{0}$	$\bar{4}$	$\bar{2}$	$\bar{0}$	$\bar{4}$	$\bar{2}$
$\bar{5}$	$\bar{0}$	$\bar{5}$	$\bar{4}$	$\bar{3}$	$\bar{2}$	$\bar{1}$

Whenever there is no danger of confusion, we shall write simply $+$ and $\cdot$, instead of $\oplus$ and $\odot$, to denote the operations of addition and multiplication in $\mathbb{Z}_n$, respectively. Proceeding this way, we call the reader's attention to the fact that, whenever we write

$$\bar{a} + \bar{b} = \overline{a+b} \text{ and } \bar{a} \cdot \bar{b} = \overline{a \cdot b},$$

each of the $+$ and $\cdot$ signs stand for two different operations. For instance, the first $+$ sign denotes the addition operation in $\mathbb{Z}_n$, whereas the second $+$ sign denotes the usual addition of integers. Moreover, an analogous remark holds for both $\cdot$ signs.

The associativity of the addition of $\mathbb{Z}_n$ immediately gives the *cancellation law*

$$\bar{a} + \bar{b} = \bar{a} + \bar{c} \Rightarrow \bar{b} = \bar{c}.$$

Indeed, if $\bar{a} + \bar{b} = \bar{a} + \bar{c}$, then

$$\overline{-a} + (\bar{a} + \bar{b}) = \overline{-a} + (\bar{a} + \bar{c})$$

and, hence,

$$(\overline{-a + a}) + \overline{b} = (\overline{-a + a}) + \overline{c}.$$

Now, since $\overline{-a} + \overline{a} = \overline{-a + a} = \overline{0}$ and $\overline{0}$ is the neutral element for the addition of $\mathbb{Z}_n$, the equality above gives $\overline{b} = \overline{c}$.

A cancellation law for the multiplication of $\mathbb{Z}_n$ is somewhat more complicated. In order to examine it, we first have the following

Definition 11.10 An element $\overline{a} \in \mathbb{Z}_n$ is a **unit** if there exists $\overline{b} \in \mathbb{Z}_n$ such that $\overline{a} \cdot \overline{b} = \overline{1}$.

As it happens within the ordinary number system of integers, the classes $\pm\overline{1}$ are units in $\mathbb{Z}_n$, for $\pm\overline{1} \cdot \pm\overline{1} = \overline{1}$. On the other hand, the news is that there might be other units; for instance, in $\mathbb{Z}_9$ the classes $\overline{4}$ and $\overline{7}$ are units, since

$$\overline{4} \cdot \overline{7} = \overline{4 \cdot 7} = \overline{1} = \overline{7} \cdot \overline{4}.$$

If $\overline{a} \in \mathbb{Z}_n$ is a unit, then we do have the following cancellation law with respect to multiplication:

$$\overline{a} \cdot \overline{c} = \overline{a} \cdot \overline{d} \Rightarrow \overline{c} = \overline{d}.$$

Indeed, taking $\overline{b} \in \mathbb{Z}_n$ such that $\overline{a} \cdot \overline{b} = \overline{1}$ (and, hence, also $\overline{b} \cdot \overline{a} = \overline{1}$), it follows from $\overline{a} \cdot \overline{c} = \overline{a} \cdot \overline{d}$ that

$$\overline{b} \cdot (\overline{a} \cdot \overline{c}) = \overline{b} \cdot (\overline{a} \cdot \overline{d}).$$

Then, the associativity of multiplication furnishes

$$(\overline{b} \cdot \overline{a}) \cdot \overline{c} = (\overline{b} \cdot \overline{a}) \cdot \overline{d}$$

or, which is the same, $\overline{1} \cdot \overline{c} = \overline{1} \cdot \overline{d}$. This last equality amounts to $\overline{c} = \overline{d}$, as wished.

In particular, if $\overline{a} \in \mathbb{Z}_n$ is a unit, then the element $\overline{b} \in \mathbb{Z}_n$ whose existence is guaranteed by Definition 11.10 is unique, for if

$$\overline{a} \cdot \overline{b} = \overline{1} = \overline{a} \cdot \overline{c},$$

then the cancellation law for multiplication gives $\overline{b} = \overline{c}$. From now on, we shall say that such a $\overline{b} \in \mathbb{Z}_n$ is the **multiplicative inverse** of $\overline{a}$ in $\mathbb{Z}_n$.

The coming proposition characterizes all of the units of $\mathbb{Z}_n$.

Proposition 11.11 *A congruence class $\overline{a} \in \mathbb{Z}_n$ is a unit in $\mathbb{Z}_n$ if and only if* $\gcd(a, n) = 1$. *In particular, $\mathbb{Z}_n$ has exactly $\varphi(n)$ distinct units.*

Proof By definition, $\overline{a} \in \mathbb{Z}_n$ is a unit if and only if there exists $\overline{b} \in \mathbb{Z}_n$ such that $\overline{a} \cdot \overline{b} = \overline{1}$, i.e., $\overline{ab} = \overline{1}$ or, which is the same,

$$ab \equiv 1 \,(\mathrm{mod}\,n).$$

Hence, $\overline{a}$ is a unit in $\mathbb{Z}_n$ if and only if $a \in \mathbb{Z}$ is invertible modulo n. However, by Proposition 10.23, this is the same as asking that $\gcd(a, n) = 1$.

For what is left to do, it suffices to note that, since $\mathbb{Z}_n = \{\overline{1}, \overline{2}, \ldots, \overline{n}\}$, the set of the units of $\mathbb{Z}_n$ coincides with the set of the classes $\overline{a}$ such that $1 \leq a \leq n$ and $\gcd(a, n) = 1$. Since such a set has $\varphi(n)$ elements, we are done. $\qquad\square$

We have now arrived at the most important case.

Corollary 11.12 *If p is prime, then every element of $\mathbb{Z}_p \setminus \{\overline{0}\}$ is a unit of $\mathbb{Z}_p$.*

Proof By the previous proposition, it suffices to show that, if $\overline{a} \neq \overline{0}$ in $\mathbb{Z}_p$, then $\gcd(a, p) = 1$. However, $\overline{a} \neq \overline{0}$ is the same as $a \not\equiv 0 \,(\mathrm{mod}\,p)$, or $p \nmid a$; moreover, the proof of Lemma 6.41 assures that $p \nmid a$ if and only if $\gcd(a, p) = 1$. $\qquad\square$

As far as the arithmetic operations of addition, subtraction, multiplication and division are concerned, the previous corollary allows us to put $\mathbb{Z}_p$, with p prime, on an equal footing with respect to $\mathbb{Q}$ and $\mathbb{R}$ (and, after Chap. 13, also with respect to the set $\mathbb{C}$ of complex numbers). All of these *number systems* are examples of what algebraists call **fields**.

More precisely, in $\mathbb{Z}_n$ (for every $n > 1$, not only n prime) we can define an operation $-$, called *subtraction*, which is analogous to the ordinary subtraction of real numbers. We do so by setting (cf. Problem 3)

$$\overline{a} - \overline{b} = \overline{a - b}.$$

We now restrict to the case in which $n = p$, a prime number. Writing $\overline{a}^{-1}$ to denote the multiplicative inverse of $\overline{a} \in \mathbb{Z}_p \setminus \{\overline{0}\}$ (i.e., letting $\overline{a}^{-1}$ denote the congruence class $\overline{b} \in \mathbb{Z}_p$ whose existence is assured by Definition 11.10), we have

$$\overline{a} \cdot \overline{a}^{-1} = \overline{1};$$

moreover, if $\overline{a} \cdot \overline{b} = \overline{c}$, with $\overline{a}, \overline{b}, \overline{c} \in \mathbb{Z}_p$ and $\overline{b} \neq \overline{0}$, then it is immediate to verify that

$$\overline{a} = \overline{c} \cdot \overline{b}^{-1}.$$

Hence, we can introduce in $\mathbb{Z}_p$ a notion of division which is also similar to the usual division of rational (or real, or complex) numbers, by letting

$$\overline{a} \div \overline{b} = \overline{a} \cdot \overline{b}^{-1},$$

for $\overline{a} \in \mathbb{Z}_p$ and $\overline{b} \in \mathbb{Z}_p \setminus \{\overline{0}\}$.

Problems: Sect. 11.2

1. Write the addition and multiplication tables of $\mathbb{Z}_5$ and $\mathbb{Z}_7$.
2. * Prove that, in $\mathbb{Z}_n$, the multiplication is distributive with respect to addition. More precisely, show that

$$\overline{a} \cdot (\overline{b} + \overline{c}) = (\overline{a} \cdot \overline{b}) + (\overline{a} \cdot \overline{c}),$$

for every $\overline{a}, \overline{b}, \overline{c} \in \mathbb{Z}_n$.
3. * Prove that the operation of subtraction in $\mathbb{Z}_n$ is well defined.
4. Show that, in $\mathbb{Z}_{12}$, all units are equal to their multiplicative inverses.
5. For integers m and n, with $n > 1$, let the **operation of multiplication by m** in $\mathbb{Z}_n$ be given by $m \cdot \overline{a} = \overline{m \cdot a}$. Show that it is well defined and such that

$$m \cdot (\overline{a} + \overline{b}) = m \cdot \overline{a} + m \cdot \overline{b} \text{ and } (m_1 + m_2) \cdot \overline{a} = m_1 \cdot \overline{a} + m_2 \cdot \overline{a},$$

for every $m, m_1, m_2 \in \mathbb{Z}$ and $\overline{a}, \overline{b} \in \mathbb{Z}_n$.
6. If an integer $n > 1$ is not prime, show that $\mathbb{Z}_n$ has *zero divisors*, i.e., that there exist $\overline{a}, \overline{b} \in \mathbb{Z}_n \setminus \{\overline{0}\}$ such that $\overline{a} \cdot \overline{b} = \overline{0}$. Also, show that if p is a prime number, then $\mathbb{Z}_p$ does not possess zero divisors.
7. Let a and n be given integers, with $n > 1$. In $\mathbb{Z}_n$, show that the equation $\overline{a} \cdot x = \overline{b}$ has a solution for every $\overline{b} \in \mathbb{Z}_n$ if and only if $\overline{a}$ is a unit in $\mathbb{Z}_n$. In this case, show that the solution $x \in \mathbb{Z}_n$ is unique, being given by $x = \overline{a}^{-1} \cdot \overline{b}$.

Problems: Sect. II.2

Chapter 12
Primitive Roots and Quadratic Residues

In this last chapter devoted to Number Theory, we return to the analysis of the congruence

$$a^k \equiv 1 \pmod{n}, \tag{12.1}$$

concentrating ourselves in two distinct problems, briefly discussed below.

As a first problem, we know from Euler's theorem that, if $\gcd(a, n) = 1$, then (12.1) is always satisfied for $k = \varphi(n)$, where φ stands for the Euler function. However, for a fixed $a \in \mathbb{Z}$ relatively prime to n, we do not know if such a natural value of k is the least possible one. In this respect, one of the core results of this chapter is the characterization of the integers $n > 1$ for which there exists some integer a, relatively prime with n and such that the least possible natural exponent k in (12.1) is $k = \varphi(n)$; such an a will be called a *primitive root* modulo n.

A second problem we want to address here requires an important change of point of view towards (12.1). Instead of fixing the basis a, prime with n, and looking for the values of $k \in \mathbb{N}$ for which that congruence has a solution, we fix $k \in \mathbb{N}$ and look for the integer value of a which solve it. Note that this amounts to finding the *roots* of the equation $x^k = \bar{1}$ in $\mathbb{Z}_n$. However, due to the elementary character of these notes, we restrict ourselves to the case $k = 2$, when the desired roots $a \in \mathbb{Z}$ will be the *quadratic residues* modulo n.

Along our way through the analysis of the two problems above, several interesting applications will be presented. Among them, we would like to highlight yet another famous theorem of P. de Fermat, characterizing the natural numbers that can be written as the sum of the squares of two integers.

© Springer International Publishing AG, part of Springer Nature 2018
A. Caminha Muniz Neto, *An Excursion through Elementary Mathematics, Volume III*,
Problem Books in Mathematics, https://doi.org/10.1007/978-3-319-77977-5_12

12.1   The Concept of Order Modulo n

Let a and n be integers such that $n > 1$ and $\gcd(a, n) = 1$. We know from Euler's Theorem 10.19 that there always exists a (at least one) natural number k for which $a^k \equiv 1 \pmod{n}$, namely, $k = \varphi(n)$. However, nothing we have done so far guarantees that $\varphi(n)$ is the least such k; for instance $2^3 \equiv 1 \pmod{7}$, even though $\varphi(7) = 6$. These considerations motivate the following

Definition 12.1 Given relatively prime integers a and n, with $n > 1$, the **order** of a, modulo n, denoted $\mathrm{ord}_n(a)$, is the least possible $h \in \mathbb{N}$ for which

$$a^h \equiv 1 \pmod{n}.$$

We obviously have

$$\mathrm{ord}_n(a) \leq \varphi(n).$$

On the other hand, the coming result establishes the most elementary properties of the concept of order modulo n.

Proposition 12.2 *Let $a, n \in \mathbb{Z}$, with $n > 1$ and $\gcd(a, n) = 1$.*

(a) *If $\mathrm{ord}_n(a) = h$, then the integers $1, a, a^2, \ldots, a^{h-1}$ are pairwise incongruent, modulo n. In particular, if $\mathrm{ord}_n(a) = \varphi(n)$, then the set $\{1, a, a^2, \ldots, a^{\varphi(n)-1}\}$ is a RRS modulo n.*

(b) *$a^k \equiv 1 \pmod{n} \Leftrightarrow \mathrm{ord}_n(a) \mid k$. In particular, $\mathrm{ord}_n(a) \mid \varphi(n)$.*

Proof

(a) If there existed $0 \leq k < l < h$ such that $a^l \equiv a^k \pmod{n}$, item (f) of Proposition 10.6 would give us $a^{l-k} \equiv 1 \pmod{n}$, with $0 < l - k < h$. But this would contradict the minimality of h.

For what is left to do, it suffices to see that, if $\mathrm{ord}_n(a) = \varphi(n)$, then the set $\{1, a, a^2, \ldots, a^{\varphi(n)-1}\}$ has $\varphi(n)$ integers relatively prime with n and pairwise incongruent modulo n. Thus, such a set is a RRS modulo n.

(b) Let $\mathrm{ord}_n(a) = h$, so that, in particular, $a^h \equiv 1 \pmod{n}$. If $k = hl$, then

$$a^k = a^{hl} = (a^h)^l \equiv 1^l = 1 \pmod{n}.$$

Conversely, let k be a natural number satisfying $a^k \equiv 1 \pmod{n}$. The division algorithm assures the existence of integers q and r, with $0 \leq r < h$, such that $k = qh + r$. Thus, we have

$$1 \equiv a^k = a^{qh+r} = (a^h)^q \cdot a^r \equiv 1^q \cdot a^r \equiv a^r \pmod{n}.$$

Now, if $r > 0$, then r would be a natural exponent less than h and such that $a^r \equiv 1 \pmod{n}$; in turn, this would contradict the minimality of h. Therefore, $r = 0$ and hence $h \mid k$, i.e., $\mathrm{ord}_n(a) \mid k$.

Finally, $\mathrm{ord}_n(a) \mid \varphi(n)$ follows from the first part of item (b), together with Euler's Theorem 10.19.   □

Example 12.3 Compute the orders of 2 modulo 17 and of 7 modulo 10.

Solution If $h = \mathrm{ord}_{17}(2)$, then $h \mid \varphi(17) = 16$, so that $h = 1, 2, 4, 8$ or 16. It is evident that $h \neq 1, 2$; also, $2^4 \equiv -1 \pmod{17}$, and hence $h \neq 4$. On the other hand, $2^8 = (2^4)^2 \equiv (-1)^2 \equiv 1 \pmod{17}$, so that $h = 8$.

If $h = \mathrm{ord}_{10}(7)$, then $h \mid \varphi(10) = 4$, so that $h = 1, 2$ or 4. Obviously, $h \neq 1$; however, since $7^2 \equiv -1 \pmod{10}$, we also have $h \neq 2$. Thus, $h = 4$.   □

Let us now see how Proposition 12.2 can be applied to an interesting problem.

Example 12.4 If p is an odd prime, prove that all prime factors of $2^p - 1$ are of the form $2kp + 1$, for some $k \in \mathbb{N}$.

Proof If q is a prime factor of $2^p - 1$, then q is odd and $2^p \equiv 1 \pmod{q}$. It follows from item (b) of the previous proposition that $\mathrm{ord}_q(2) \mid p$ and, hence, $\mathrm{ord}_q(2) = 1$ or p. However, if $\mathrm{ord}_q(2) = 1$, then $q = 1$, which is an absurd; therefore, $\mathrm{ord}_q(2) = p$.

On the other hand, it follows from Fermat's little theorem that $2^{q-1} \equiv 1 \pmod{q}$. Then, again by item (b) of the previous proposition, we have

$$p = \mathrm{ord}_q(2) \mid (q - 1).$$

However, since q is odd, there must exist $k \in \mathbb{N}$ such that $q - 1 = 2kp$.   □

The coming result establishes other useful properties of the concept of order modulo n, which will be of paramount importance for further developments.

Proposition 12.5 *Let a and n be relatively prime integers, with $n > 1$.*

(a) $\mathrm{ord}_n(a) = \mathrm{ord}_n(a + n)$.
(b) *If $m > 1$ is a natural number such that $m \mid n$, then $\mathrm{ord}_m(a) \mid \mathrm{ord}_n(a)$.*
(c) *If $\mathrm{ord}_n(a) = h$ and $k \in \mathbb{N}$, then $\mathrm{ord}_n(a^k) = \frac{h}{\gcd(h,k)}$.*
(d) *If $k \in \mathbb{N}$, then $\mathrm{ord}_n(a^k) = \mathrm{ord}_n(a) \Leftrightarrow \gcd(\mathrm{ord}_n(a), k) = 1$.*
(e) *If $\mathrm{ord}_n(a) = h$, then the set $\{a, a^2, \ldots, a^h\}$ has exactly $\varphi(h)$ elements of order h modulo n.*

Proof

(a) It follows from $a \equiv a + n \pmod{n}$ that $a^k \equiv (a + n)^k \pmod{n}$, for every $k \in \mathbb{N}$. In particular, $a^k \equiv 1 \pmod{n}$ if and only if $(a + n)^k \equiv 1 \pmod{n}$, and hence a and $a + n$ have equal orders modulo n.

(b) Since $m \mid n$, item (g) of Proposition 10.6 yields that if $a^k \equiv 1 \pmod{n}$, then $a^k \equiv 1 \pmod{m}$. In particular, since $a^k \equiv 1 \pmod{n}$ when $k = \mathrm{ord}_n(a)$, we have

$$a^{\mathrm{ord}_n(a)} \equiv 1 \pmod{m}.$$

Item (b) of Proposition 12.2 now furnishes $\mathrm{ord}_m(a) \mid \mathrm{ord}_n(a)$.

(c) Let $d = \gcd(h, k)$. By item (b) of Proposition 12.2, we have

$$(a^k)^j \equiv 1 \pmod{n} \Leftrightarrow a^{kj} \equiv 1 \pmod{n} \Leftrightarrow h \mid kj$$

$$\Leftrightarrow \frac{h}{d} \mid \frac{k}{d} \cdot j \Leftrightarrow \frac{h}{d} \mid j,$$

where in the last equivalence we have applied item (a) of Proposition 6.22, together with the fact that $\gcd\left(\frac{h}{d}, \frac{k}{d}\right) = 1$. From this, we immediately get

$$\mathrm{ord}_n(a^k) = \frac{h}{d} = \frac{h}{\gcd(h, k)}.$$

(d) This obviously follows from (c).

(e) By item (d), the number of exponents $1 \leq k \leq h$ such that a^k has order h modulo n equals the number of exponents $1 \leq k \leq h$ which are relatively prime with h. Therefore, it is equal to $\varphi(h)$. $\square$

Problems: Sect. 12.1

1. Compute $\mathrm{ord}_7(2)$, $\mathrm{ord}_{11}(2)$ and $\mathrm{ord}_{15}(7)$.
2. Prove that, for every positive integer n, the number $2^{3^n} + 1$ is not a multiple of 17.
3. Let a and n be relatively prime integers, with $n > 2$. If there exists a natural k such that $a^k \equiv -1 \pmod{n}$, prove that $\mathrm{ord}_n(a)$ is even.
4. (Putnam) Find all $n \in \mathbb{N}$ for which $n \mid (2^n - 1)$.
5. (Turkey) For each $n \in \mathbb{N}$, prove that $n!$ divides the product

$$\prod_{j=0}^{n-1}(2^n - 2^j).$$

6. Given a natural $n > 2$, we label the vertices of a regular $2n$-gon $\mathcal{P}$ as $1, 2, 3, \ldots, n, -n, -(n-1), \ldots, -3, -2, -1$, successively and in the clockwise sense. Then, we choose vertices of $\mathcal{P}$ in the following way: at the first round, we choose vertex 1; on the other hand, if vertex k_i was chosen at the i-th round, then at the

$(i + 1)$-th round we choose the vertex situated $|k_i|$ vertices away from vertex k_i, in the clockwise sense if $k_i > 0$ and in the counterclockwise sense if $k_i < 0$. Such a procedure continues until we choose a vertex which was already chosen in a former round. At the end of this process, let $f(n)$ be the number of non chosen vertices.

(a) If $f(n) = 0$, prove that $2n + 1$ is an odd prime.
(b) Compute $f(1997)$.

7. (IMO—adapted) Given a prime number p, we want to show that there exists a prime number q such that $q \nmid (n^p - p)$, for all $n \in \mathbb{N}$. To this end, do the following items[1]:

(a) Show that it suffices to find a prime q for which $\mathrm{ord}_q(p) = p$ and $q \not\equiv 1 \pmod{p^2}$.
(b) Letting $a = \frac{p^p - 1}{p - 1}$, show that a has at least one prime factor q for which $q \not\equiv 1 \pmod{p^2}$, and that $\mathrm{ord}_q(p) = p$.

12.2 Primitive Roots

Given integers a and n such that $n > 1$ and $\gcd(a, n) = 1$, in this section we will be particularly interested in the case in which $\mathrm{ord}_n(a) = \varphi(n)$. Such a case is so important that is worth the following

Definition 12.6 Let a and n be integers such that $n > 1$ and $\gcd(a, n) = 1$. We say that a is a **primitive root** modulo n if $\mathrm{ord}_n(a) = \varphi(n)$.

Let us take a look at a couple of examples.

Examples 12.7

(a) Since $2^1 \equiv 2 \pmod 3$ and $\varphi(3) = 2$, it follows that $\mathrm{ord}_3(2) = 2 = \varphi(3)$, i.e., 2 is a primitive root modulo 3. Similar computations show that 2 is also a primitive root modulo 5.
(b) Modulo 7 we have $2^1 \equiv 2$, $2^2 \equiv 4$ and $2^3 \equiv 1$. Hence, $\mathrm{ord}_7(2) = 3 < 6 = \varphi(7)$ and it follows that 2 is not a primitive root modulo 7.

The main result of this section is the characterization of all integer values $n > 1$ for which primitive roots modulo n do exist. More precisely, we shall show that an integer $n > 1$ has a primitive root if and only if $n = 2, 4, p^k$ or $2p^k$, for some odd prime p. We start this path by computing the number of pairwise incongruent primitive roots modulo n, provided such a root does exist.

---

[1] The coming items are based on the solution of Professor Samuel B. Feitosa.

Proposition 12.8 *If an integer* $n > 1$ *has a primitive root, say* a, *then every primitive root modulo* n *is congruent to one of the elements of the set*

$$\{a^k;\ 1 \le k \le \varphi(n)\ and\ \gcd(\varphi(n), k) = 1\}.$$

In particular, n *has exactly* $\varphi(\varphi(n))$ *pairwise incongruent primitive roots.*

Proof The second part follows immediately from the first, by the very definition of Euler's function: the number of exponents $1 \le k \le \varphi(n)$ such that k is relatively prime with $\varphi(n)$ is exactly $\varphi(\varphi(n))$.

For the first part, letting a be a primitive root modulo n, we have that $\mathrm{ord}_n(a) = \varphi(n)$ and $A = \{a, a^2 \dots, a^{\varphi(n)}\}$ is a RRS modulo n. Hence, any primitive root modulo n is congruent, modulo n, to one of the elements of A, so that it suffices to see which elements of A have order (modulo n) equal to $\varphi(n)$. However, by item (d) of Proposition 12.5, for $1 \le k \le \varphi(n)$ we have

$$\mathrm{ord}_n(a^k) = \varphi(n) \Leftrightarrow \gcd(\varphi(n), k) = 1.$$

$\square$

Example 12.9 Example 12.7 assures that 2 is a primitive root modulo 5. Since $\varphi(5) = 4$, the previous propositions teaches us that a set of pairwise incongruent primitive roots modulo 5 is

$$\{2^k;\ 1 \le k \le 4\ and\ \gcd(4, k) = 1\} = \{2, 2^3\}.$$

Hence, the pairwise incongruent primitive roots modulo 5 are 2 and 3 (for $2^3 = 8 \equiv 3 \,(\mathrm{mod}\,5)$).

We now go towards characterizing the integers which do possess primitive roots, starting with a necessary condition.

Theorem 12.10 *If and integer* $n > 1$ *has primitive roots, then* $n = 2, 4,\ p^k$ *or* $2p^k$, *where* p *is an odd prime and* k *is a natural number.*

Proof Firstly, note that the integers $n > 1$ which differ from those at the statement of the theorem are either of the form $n = bc$, with $b, c > 2$ being relatively prime integers, or of the form $n = 2^k$, for some integer $k > 2$. It thus suffices to show that such integer values of n do not have primitive roots, and we shall do this by showing that every integer a relatively prime with n satisfies $\mathrm{ord}_n(a) < \varphi(n)$. We look separately at the two possibilities above:

(i) $n = bc$, with $b, c > 2$ relatively prime integers: since $\gcd(b, c) = 1$, we have $\varphi(n) = \varphi(bc) = \varphi(b)\varphi(c)$. On the other hand, condition $b, c > 2$ assures (cf.

Problem 19, page 220) that $\varphi(b)$ and $\varphi(c)$ are both even. If $a \in \mathbb{Z}$ is prime with n, then a is prime with b and c and, from Euler's theorem, we get

$$a^{\varphi(n)/2} = \left(a^{\varphi(b)}\right)^{\varphi(c)/2} \equiv 1^{\varphi(c)/2} \equiv 1 \pmod{b}$$

and

$$a^{\varphi(n)/2} = \left(a^{\varphi(c)}\right)^{\varphi(b)/2} \equiv 1^{\varphi(b)/2} \equiv 1 \pmod{c}.$$

Therefore, item (h) of Proposition 10.6 gives $a^{\varphi(n)/2} \equiv 1 \pmod{n}$; in particular,

$$\operatorname{ord}_n(a) \leq \frac{\varphi(n)}{2} < \varphi(n)$$

and, hence, a is not a primitive root modulo n.

(ii) $n = 2^k$, for some integer $k > 2$: let a be an odd integer (i.e., prime with 2). If we show that

$$a^{2^{k-2}} \equiv 1 \pmod{2^k},$$

we will have

$$\operatorname{ord}_{2^k}(a) \leq 2^{k-2} < 2^{k-1} = \varphi(2^k)$$

and a will not be a primitive root modulo 2^k. For what is left to do, let us make induction on k: the initial case $k = 3$ follows from Proposition 10.9, since $a^2 \equiv 1 \pmod{8}$ for odd a. Assume we had already proved that, for some integer $k \geq 3$, there exists $q \in \mathbb{N}$ such that $a^{2^{k-2}} = 2^k q + 1$. Then,

$$a^{2^{k-1}} = \left(a^{2^{k-2}}\right)^2 = (2^k q + 1)^2 = 2^{2k} q^2 + 2^{k+1} q + 1$$

$$= 2^{k+1}(2^{k-1} q^2 + q) + 1 \equiv 1 \pmod{2^{k+1}},$$

as wished.   □

We are left to establishing the converse of the previous theorem, namely, that all numbers of one of the forms 2, 4, p^k and $2p^k$, for some odd prime p and some $k \in \mathbb{N}$, have primitive roots. It is clear that 1 is a primitive root modulo 2 and 3 is a primitive root modulo 4. The rest of this section is devoted to the analysis of the two remaining cases.

The coming result shows that it suffices to worry about the numbers of the form p^k.

Proposition 12.11 *If p is an odd prime and $a \in \mathbb{Z}$ is a primitive root modulo p^k, then a or $a + p^k$ is a primitive root modulo $2p^k$.*

Proof Let $h = \mathrm{ord}_{2p^k}(a)$. If a is odd, then a is relatively prime with $2p^k$. Then, by using item (b) of Proposition 12.5, together with the fact that $\mathrm{ord}_{p^k}(a) = \varphi(p^k)$, we obtain

$$\varphi(p^k) = \mathrm{ord}_{p^k}(a) \mid h = \mathrm{ord}_{2p^k}(a) \mid \varphi(2p^k) = \varphi(p^k),$$

where, in the last equality above, we used the fact that p is odd. Therefore, $\mathrm{ord}_{2p^k}(a) = \varphi(2p^k)$ and a is a primitive root modulo $2p^k$.

If a is even, change a by $a + p^k$ from the beginning and argue as above.   □

We shall complete the proof of the converse of Theorem 12.10 (i.e., the analysis of the case p^k), in two steps. Nevertheless, before that we need an auxiliary result.

Lemma 12.12 *Let p be an odd prime. If a is a primitive root modulo p^2, then, for every integer $k \geq 1$, we have $a^{\varphi(p^k)} = b_k p^k + 1$, for some $b_k \in \mathbb{Z}$ such that $p \nmid b_k$.*

Proof Let us make induction on $k \geq 1$. Fermat's little theorem gives $a^{p-1} = b_1 p + 1$, for some $b_1 \in \mathbb{Z}$. If $p \mid b_1$, we would have $a^{p-1} \equiv 1 \pmod{p^2}$ and, hence, $\mathrm{ord}_{p^2}(a) \leq p - 1 < \varphi(p^2)$; this would contradict the fact that a is a primitive root modulo p^2.

Assume that, for some integer $k \geq 1$, we have $a^{\varphi(p^k)} = b_k p^k + 1$, with $b_k \in \mathbb{Z}$ such that $p \nmid b_k$. Then, Newton's binomial formula gives

$$a^{\varphi(p^{k+1})} = \left(a^{\varphi(p^k)}\right)^p = (1 + b_k p^k)^p$$

$$= 1 + b_k p^{k+1} + \sum_{j=2}^{p-1} \binom{p}{j} b_k^j p^{jk} + b_k^p p^{pk}.$$

Now, for $1 \leq j \leq p - 1$ Example 6.42 assures the existence of $c_k \in \mathbb{N}$ such that $\binom{p}{j} = pc_k$. Therefore, the last expression above furnishes

$$a^{\varphi(p^{k+1})} = 1 + b_k p^{k+1} + \sum_{j=2}^{p-1} c_k b_k^j p^{jk+1} + b_k^p p^{pk}$$

$$= 1 + b_k p^{k+1} + \left(\sum_{j=2}^{p-1} c_k b_k^j p^{(j-1)k-1} + b_k^p p^{(p-1)k-2}\right) p^{k+2}.$$

Letting t denote the expression within parentheses and noticing that $t \in \mathbb{Z}$, we finally get

$$a^{\varphi(p^{k+1})} = 1 + b_k p^{k+1} + t p^{k+2} = 1 + (b_k + tp) p^{k+1}.$$

However, since $p \nmid b_k$, it suffices to let $b_{k+1} = b_k + tp$ to obtain $a^{\varphi(p^{k+1})} = b_{k+1} p^{k+1} + 1$, with $p \nmid b_{k+1}$. $\qquad\qquad\qquad\qquad\qquad\qquad\qquad\qquad\qquad$ □

We can finally execute the first of the two steps needed to finish the proof of the existence of primitive roots modulo p^k, for an odd prime p.

Theorem 12.13 *Let p be an odd prime and a be relatively prime with p.*

(a) *If a is a primitive root modulo p, then a or $a + p$ is a primitive root modulo p^2.*
(b) *If a is a primitive root modulo p and modulo p^2, then a is a primitive root modulo p^k, for every integer $k \geq 1$.*

Proof

(a) It follows from item (a) of Proposition 12.5 and from our hypotheses that $\operatorname{ord}_p(a + p) = \operatorname{ord}_p(a) = p - 1$. On the other hand, item (b) of Proposition 12.2 assures that

$$\operatorname{ord}_{p^2}(a + p), \ \operatorname{ord}_{p^2}(a) \mid \varphi(p^2) = p(p - 1),$$

whereas item (b) of Proposition 12.5 gives

$$p - 1 = \operatorname{ord}_p(a) \mid \operatorname{ord}_{p^2}(a) \quad \text{and} \quad p - 1 = \operatorname{ord}_p(a + p) \mid \operatorname{ord}_{p^2}(a + p).$$

Hence, $\operatorname{ord}_{p^2}(a) = p - 1$ or $p(p - 1)$, the same being valid for $\operatorname{ord}_{p^2}(a + p)$; thus, we are left to showing that

$$\operatorname{ord}_{p^2}(a) = p - 1 \Rightarrow \operatorname{ord}_{p^2}(a + p) \neq p - 1.$$

To this end, if $a^{p-1} \equiv 1 \pmod{p^2}$, then modulo p^2 we have

$$(a + p)^{p-1} = a^{p-1} + (p - 1)p a^{p-2} + \sum_{j=2}^{p-1} \binom{p-1}{j} p^j a^{p-1-j}$$

$$\equiv a^{p-1} + (p - 1)p a^{p-2}$$

$$\equiv 1 - p a^{p-2} \not\equiv 1,$$

for $p \nmid a$.

(b) Suppose that a is a primitive root modulo p and modulo p^2. We shall use induction on k to prove that a is a primitive root modulo p^k, for every $k \in \mathbb{N}$.

The cases $k = 1$ and $k = 2$ are our hypotheses. Assume, then, that we have already proved that a is a primitive root modulo p^k, for some integer $k \geq 2$.

Items (b) of Propositions 12.2 and 12.5 give us

$$\varphi(p^k) = \mathrm{ord}_{p^k}(a) \mid \mathrm{ord}_{p^{k+1}}(a) \mid \varphi(p^{k+1}) = p\varphi(p^k),$$

so that

$$\mathrm{ord}_{p^{k+1}}(a) = \varphi(p^k) \text{ or } \varphi(p^{k+1}).$$

On the other hand, since a is a primitive root modulo p^2, the previous lemma furnishes $a^{\varphi(p^k)} = b_k p^k + 1$, for some $b_k \in \mathbb{Z}$ such that $p \nmid b_k$. In particular,

$$a^{\varphi(p^k)} \not\equiv 1 \pmod{p^{k+1}},$$

so that $\mathrm{ord}_{p^{k+1}}(a) \neq \varphi(p^k)$. $\square$

Theorem 12.13 puts us in quite a nice position. Indeed, if we show that an odd prime p has a primitive root, say a, then $a + p$ will also be such a root (by item (a) of Proposition 12.5); hence, item (a) of the previous theorem allows us to assume that a is a primitive root modulo p and modulo p^2, and after that item (b) of that same result assures that a is a primitive root modulo p^k, for every $k \in \mathbb{N}$.

We are then left to showing that odd primes have primitive roots. Unfortunately, at this point such a proof is beyond the scope of the material we have at our disposal, and will have to wait until Sect. 19.3 (cf. Theorem 19.20) to be presented. Thus, for the time being we assume the existence of primitive roots modulo p, taking the opportunity to present some instructive examples.

Example 12.14 Prove that 2 is a primitive root modulo 3^k and modulo 5^k, for every $k \in \mathbb{N}$.

Proof From Example 12.7 and the previous theorem, it suffices to show that 2 is a primitive root modulo 9 and modulo 25. Let us check that 2 is a primitive root modulo 9, the case of modulo 25 being entirely analogous: since $\varphi(9) = 6$, we have $\mathrm{ord}_9(2) \mid 6$; however, since none of 2^1, 2^2 or 2^3 is a multiple of 9, we get $\mathrm{ord}_9(2) = 6 = \varphi(9)$. $\square$

The next two examples show how powerful the information on the existence of primitive roots can be.

Example 12.15 If p is prime and $n \in \mathbb{N}$, prove that

$$1^n + 2^n + \cdots + (p-1)^n \equiv \begin{cases} 0 \pmod{p}, & \text{if } (p-1) \nmid n \\ -1 \pmod{p}, & \text{if } (p-1) \mid n \end{cases}.$$

Proof If $(p - 1) \mid n$, say $n = (p - 1)k$, then Fermat's little theorem gives, for $1 \le a \le p - 1$,

$$a^n = a^{(p-1)k} \equiv 1^k = 1 \pmod{p}.$$

Therefore,

$$1^n + 2^n + \cdots + (p - 1)^n \equiv \underbrace{1 + 1 + \cdots + 1}_{p-1} \equiv -1 \pmod{p}.$$

If $(p - 1) \nmid n$, let a be a primitive root modulo p. Since $\{a, a^2, \ldots, a^{p-1}\}$ is a RRS modulo p, we conclude that the numbers $a, a^2, \ldots, a^{p-1}$ are, modulo p and in some order, congruent to $1, 2, \ldots, p - 1$. Hence (also modulo p),

$$1^n + 2^n + \cdots + (p - 1)^n \equiv a^n + a^{2n} + \cdots + a^{(p-1)n} = \frac{a^{pn} - a^n}{a^n - 1}.$$

Now, by invoking Fermat's little theorem once more, we get

$$a^{pn} - a^n = (a^p)^n - a^n \equiv a^n - a^n \equiv 0 \pmod{p}.$$

On the other hand, since $\operatorname{ord}_p(a) = p - 1$ and $(p - 1) \nmid n$, item (b) of Proposition 12.2 guarantees that $p \nmid (a^n - 1)$. Therefore, $p \mid \frac{a^{pn} - a^n}{a^n - 1}$, as we wished to show. $\square$

Example 12.16 (IMO) Find all $n \in \mathbb{N}$ for which n^2 divides $2^n + 1$.

Solution It is clear that such an n must be odd and that $n = 1$ is a possible value. Then, let $n > 1$ be a natural number satisfying the given conditions and write $n = p^k q$, where p is the least prime divisor of n and $k \in \mathbb{N}$ is the exponent of p in the canonical decomposition of n in prime powers. Then, $\gcd(p, q) = 1$ and

$$\frac{2^n + 1}{n^2} \in \mathbb{N} \Rightarrow \frac{2^n + 1}{p^{2k}} \in \mathbb{N}.$$

By Fermat's little theorem, we have $2^p \equiv 2 \pmod{p}$, from where an easy induction gives $2^{p^k} \equiv 2 \pmod{p}$. Thus, modulo p we obtain

$$0 \equiv 2^n + 1 = 2^{p^k q} + 1 \equiv 2^q + 1,$$

so that $2^{2q} \equiv 1 \pmod{p}$. Letting $t = \operatorname{ord}_p(2)$, it follows that $t \mid 2q$ and (again by Fermat's little theorem) $t \mid (p - 1)$; therefore, $t \mid \gcd(2q, p - 1)$. However, since the prime divisors of q are greater than p, this forces us to have $t = 2$ and, consequently, $p = 3$. Hence,

$$\frac{2^n + 1}{n^2} = \frac{2^{3^k q} + 1}{3^{2k} q^2} \in \mathbb{N}, \tag{12.2}$$

which successively implies $2^{3^k q} \equiv -1 \pmod{3^{2k}}$ and

$$2^{2 \cdot 3^k q} \equiv 1 \pmod{3^{2k}}.$$

Since we have already shown (cf. Example 12.14) that 2 is a primitive root modulo 3^{2k}, the above congruence gives

$$2 \cdot 3^{2k-1} = \varphi(3^{2k}) = \mathrm{ord}_{3^{2k}}(2) \mid 2 \cdot 3^k q,$$

so that $3^{k-1} \mid q$. However, since $p = 3$ and $\gcd(3, q) = 1$, we must have $k = 1$. Therefore, $n = 3q$ and we conclude from (12.2) that

$$\frac{8^q + 1}{q^2} \in \mathbb{N}.$$

By the sake of contradiction, assume that $q > 1$ and let w be the least prime divisor of q, say $q = w^l v$, where $v \in \mathbb{N}$ and l is the exponent of w in the canonical decomposition of q as a product of prime powers. Once more from Fermat's little theorem, we have $8^{w^l} \equiv 8 \pmod{w}$ and, hence,

$$0 \equiv 8^q + 1 = 8^{w^l v} + 1 \equiv 8^v + 1 \pmod{w}, \tag{12.3}$$

so that $8^{2v} \equiv 1 \pmod{w}$. Letting $t = \mathrm{ord}_w(8)$, it follows that $t \mid 2v$ and (one more time from Fermat) $t \mid (w - 1)$. However, since the prime factors of v (if they exist) are all greater than w, we conclude that $t = 1$ or 2, and this gives $w \mid (8^1 - 1)$ or $w \mid (8^2 - 1)$. In any case, it follows from $w > p = 3$ that $w = 7$, and (12.3) furnishes

$$0 \equiv 8^v + 1 \equiv 1^v + 1 \equiv 2 \pmod{7},$$

which is an absurd. We finally conclude that $q = 1$ and $n = 3^k q = 3$. $\square$

Problems: Sect. 12.2

1. Show that 2 is a primitive root modulo 29.
2. Let $m, n \in \mathbb{N}$ be such that $m \mid n$, and let $a \in \mathbb{Z}$ be a primitive root modulo n. Prove that a is also a primitive root modulo m.

3. Prove the following generalization of Wilson's theorem: if $n \in \mathbb{N}$ has primitive roots and $1 = a_1 < a_2 < \cdots < a_{\varphi(n)} = n - 1$ are the integers from 1 to n and relatively prime with n, then

$$n \mid (a_1 a_2 \ldots a_{\varphi(n)} + 1).$$

4. (Vietnam) Show that, for every integer $n \geq 1$, there exists an integer $k_n \geq 1$ such that $19^{k_n} - 97$ is a multiple of 2^n.

5. (IMO shortlist) Given $k \in \mathbb{N}$, prove that there are infinitely many natural numbers n for which $n \cdot 2^k - 7$ is a perfect square.

6. (Romania) Find all prime numbers p and q, distinct from 2 and 3 and such that $3pq \mid (a^{3pq-1} - 1)$, for every natural number a relatively prime with $3pq$.

7. (Brazil—adapted) Let $p > 5$ be a prime number for which $\frac{p-1}{2}$ is also prime, and let $f : \mathbb{Z}_+ \to \mathbb{R}$ be given by $f(xy) = f(x)f(y)$ and $f(x + p) = f(x)$, for all $x, y \in \mathbb{Z}_+$.

   (a) Show that $f(0), f(1) \in \{0, 1\}$ and that f is constant if $f(0) = 1$ or $f(1) = 0$.

   (b) From now on, assume that $f(0) = 0$ and $f(1) = 1$. Use Fermat's little theorem to conclude that $f(a) \in \{-1, 1\}$ for every $a \in \mathbb{N}$ such that $p \nmid a$.

   (c) If there exists a primitive root a modulo p such that $f(a) = 1$, then

   $$f(x) = \begin{cases} 0, & \text{if } p \mid x \\ 1, & \text{if } p \nmid x \end{cases}.$$

   (d) If $f(a) = -1$ for every primitive root a modulo p, then

   $$f(x) = \begin{cases} 0, & \text{if } p \mid x \\ x^{p-1} \cdot (-1)^{\mathrm{ord}_p(x)+1} \pmod{p}, & \text{if } p \nmid x \end{cases}.$$

8. (Turkey) Prove that the two following claims on $n \in \mathbb{N}$ are actually equivalent:

   (a) n is square free[2] and, if p is one of its prime divisors n, then $(p-1) \mid (n-1)$.

   (b) For every $a \in \mathbb{N}$, we have that $n \mid (a^n - a)$.

9. (India—adapted) For $n \in \mathbb{N}$, let $s_n = 1 + \sum_{k=1}^{n} k^{n-1}$. The purpose of this problem is to characterize all $n \in \mathbb{N}$ such that $n \mid s_n$. To this end, do the following items:

   (a) If p is prime and $q \in \mathbb{N}$, show that $p \mid \sum_{j=0}^{pq-1} \sum_{l=0}^{p-1} (pj + l)^{n-1}$.

   (b) Conclude that, if $n \mid s_n$, then n is square free.

---

[2] Cf. Problem 6, page 190.

(c) Let $n = p_1 \ldots p_t$, with $p_1 < \cdots < p_t$ being prime numbers, and $q_i = \frac{n}{p_i}$.

    i. Show that $s_n \equiv 1 + q_i \sum_{l=1}^{p_i-1} l^{n-1} \pmod{p_i}$.

    ii. If a is a primitive root modulo p_i, show that

$$s_n \equiv 1 + q_i \left(\frac{a^{p_i(n-1)} - a^{n-1}}{a^{n-1} - 1} \right) \pmod{p_i}.$$

    iii. From ii., successively conclude that, if $n \mid s_n$, then $p_i \mid (a^{n-1} - 1)$, $(p_i - 1) \mid (n - 1)$ and $(p_i - 1) \mid (q_i - 1)$.

    iv. If $n \mid s_n$, use the results of items i. and iii. to conclude that $p_i \mid (q_i - 1)$.

    v. Use the result of the two previous items to prove that $n \mid s_n$ if and only if $p_i(p_i - 1) \mid (q_i - 1)$ for $1 \leq i \leq t$.

10. (Brazil—adapted) Let p be an odd prime, $k \neq p, 2p$ be a natural number such that $1 \leq k < 2(p + 1)$ and $n = 2pk + 1$.

    (a) If n is prime and a is a primitive root modulo n, prove that $\gcd(a^k + 1, n) = 1$.

    (b) Assume that there exists an integer $2 \leq a < n$ such that $a^{kp} \equiv -1 \pmod{n}$ and $\gcd(a^k + 1, n) = 1$.

        i. If $d = \operatorname{ord}_n(a)$, show that $d \mid (n - 1)$ and $d \nmid 2k$. Conclude that $p \mid d$ and, hence, that $p \mid \varphi(2kp + 1)$.

        ii. Use the formula for $\varphi(2kp + 1)$ to infer that there exists an integer $l > 1$ for which $lp + 1$ is a prime divisor of n.

        iii. Use $n = 2kp + 1$ to prove that $n = (lp + 1)(hp + 1)$ for some integer $h \in \{0, 1\}$. Then, conclude that we cannot have $h = 1$, so that n is composite.

11. Show that there exists $n \in \mathbb{N}$ such that the last 1000 digits in the decimal representation of 2^n are all equal to 1 or 2.

    The result of the coming problem is due to the French mathematician of the twentieth century Claude Chevalley, thus being known as **Chevalley's theorem**.

12. If p is an odd prime, we wish to show that, modulo p, the number of distinct solutions of the congruence

$$x_1^4 + x_2^4 + x_3^4 + x_4^4 + x_5^4 \equiv 0 \pmod{p}$$

is a multiple of p. (Here, two solutions $(a_1, \ldots, a_5)$ and $(b_1, \ldots, b_5)$ are considered to be distinct if there exists $1 \leq i \leq 5$ such that $a_i \neq b_i \pmod{p}$.) To this end, do the following items:

(a) If $f(x_1, \ldots, x_5) = 1 - (x_1^4 + x_2^4 + \cdots + x_5^4)^{p-1}$ and m is the number of distinct solutions of the given congruence, then

$$m \equiv \sum_{x_1, \ldots, x_5 \in A} f(x_1, \ldots, x_5) \pmod{p},$$

where $A = \{1, 2, \ldots, p-1\}$.

(b) Modulo p, one has

$$m \equiv p^5 - \sum_{\alpha_1 + \cdots + \alpha_5 = p-1} \sum_{x_1, \ldots, x_5 \in A} x_1^{4\alpha_1} \ldots x_5^{4\alpha_5}$$

$$\equiv - \sum_{\alpha_1 + \cdots + \alpha_5 = p-1} \left(\sum_{x_1 \in A} x_1^{4\alpha_1} \right) \cdots \left(\sum_{x_5 \in A} x_5^{4\alpha_5} \right).$$

(c) If $\alpha_1 + \cdots + \alpha_5 = p - 1$, with $\alpha_i \geq 0$ for $1 \leq i \leq 5$, then there exists $1 \leq i \leq 5$ such that $\alpha_i = 0$ or $(p-1) \nmid 4\alpha_i$.

(d) If $(p-1) \nmid 4\alpha$, use a primitive root a modulo p to conclude that

$$\sum_{x \in A} x^{4\alpha} = \sum_{j=1}^{p-1} a^{4j\alpha} = \frac{a^{4p\alpha} - a^{4\alpha}}{a^{4\alpha} - 1} \equiv 0 \pmod{p}.$$

(e) Finish the proof, showing that $m \equiv 0 \pmod{p}$.

12.3 Quadratic Residues

In this section, we study algebraic congruences of the form

$$x^2 \equiv a \pmod{n}.$$

To this end, we shall need the following

Definition 12.17 Let $a, n \in \mathbb{Z}$, with $n > 1$ and $\gcd(a, n) = 1$. We say that a is a **quadratic residue** modulo n if the congruence

$$x^2 \equiv a \pmod{n}$$

has at least one integer solution x. Otherwise, a is said to be a **quadratic nonresidue** modulo n.

With respect to the above definition, our main task in this section will be to obtain necessary and sufficient conditions for an integer a to be a quadratic residue modulo n. In the text we shall concentrate ourselves in the case of a prime n, leaving the general case for the problems posed at the end of the section.

Proposition 12.18 *Let p be an odd prime.*

(a) *If a is a quadratic residue modulo p, then the congruent $x^2 \equiv a \pmod{p}$ has exactly two incongruent solutions modulo p.*

(b) *Among the integers $1, 2, \ldots, p - 1$ there are exactly $\frac{p-1}{2}$ quadratic residues and $\frac{p-1}{2}$ quadratic nonresidues modulo p.*

Proof

(a) If $x_1, x_2 \in \mathbb{Z}$ are such that $x_1^2 \equiv a \pmod{p}$ and $x_2^2 \equiv a \pmod{p}$, then $x_1^2 \equiv x_2^2 \pmod{p}$, so that $p \mid (x_1^2 - x_2^2)$. However, since p is prime, we conclude that $p \mid (x_1 - x_2)$ or $p \mid (x_1 + x_2)$. Hence, the congruence stated in item (a) has at most two incongruent solutions modulo p, namely, x_1 and $-x_1$ (they are incongruent since p is odd and $\gcd(a, p) = 1 \Rightarrow \gcd(x_1, p) = 1$. On the other hand, letting a be a quadratic residue modulo p, we know that there exists a solution $x_0 \in \mathbb{Z}$ for the congruence $x^2 \equiv a \pmod{p}$. It immediately follows that $-x_0$ is also a solution of such a congruence, with $-x_0 \not\equiv x_0 \pmod{p}$.

(b) Since $\{1, 2, \ldots, p - 1\}$ is a RRS modulo p, in order to count how many of its elements are quadratic residues modulo p, it suffices to compute how many of the numbers $1^2, 2^2, \ldots, (p - 1)^2$ are pairwise incongruent modulo p. To this end, note that if $1 \leq i \leq \frac{p-1}{2}$, then

$$i^2 \equiv (p - i)^2 \pmod{p};$$

on the other hand, if $1 \leq i < j \leq \frac{p-1}{2}$, then

$$i^2 \not\equiv j^2 \pmod{p},$$

since $j^2 - i^2 = (j - i)(j + i)$ and $0 < j - i < j + i < p$.

Therefore, there are exactly $\frac{p-1}{2}$ quadratic residues modulo p and, hence, exactly $(p - 1) - \left(\frac{p-1}{2}\right) = \frac{p-1}{2}$ quadratic nonresidues modulo p. $\square$

The coming proposition is due to Euler, being known in mathematical literature as **Euler's criterion** for quadratic residues.

Proposition 12.19 (Euler) *If p is an odd prime and $a \in \mathbb{Z}$, then a is a quadratic residue modulo p if and only if*

$$a^{\frac{p-1}{2}} \equiv 1 \pmod{p}.$$

Proof Assume first that a is a quadratic residue modulo p. Then, $\gcd(a, p) = 1$ and there exists $x_0 \in \mathbb{Z}$ such that $x_0^2 \equiv a \pmod{p}$. In particular, x_0 is also prime with p, and Fermat's little theorem furnishes

$$a^{\frac{p-1}{2}} \equiv (x_0^2)^{\frac{p-1}{2}} = x_0^{p-1} \equiv 1 \pmod{p}.$$

Conversely, suppose $a^{\frac{p-1}{2}} \equiv 1 \pmod{p}$, so that in particular $\gcd(a, p) = 1$. Then, letting α be a primitive root modulo p, we have that $\{\alpha, \alpha^2, \dots, \alpha^{p-1}\}$ is a RRS modulo p. Therefore, there exists an integer $1 \le k \le p-1$ for which $\alpha^k \equiv a \pmod{p}$. Hence,

$$\alpha^{k\left(\frac{p-1}{2}\right)} \equiv a^{\frac{p-1}{2}} \equiv 1 \pmod{p},$$

and it follows from $\operatorname{ord}_p(\alpha) = p - 1$ that $(p-1) \mid k\left(\frac{p-1}{2}\right)$. We conclude that k is even, say $k = 2l$, and letting $x_0 = \alpha^l$ we get

$$x_0^2 = \alpha^{2l} = \alpha^k \equiv a \pmod{p}.$$

Thus, a is a quadratic residue modulo p.   □

Corollary 12.20 *If p is an odd prime, then an integer a, relatively prime with p, is a quadratic nonresidue modulo p if and only if*

$$a^{\frac{p-1}{2}} \equiv -1 \pmod{p}.$$

Proof Fermat's little theorem gives

$$(a^{\frac{p-1}{2}} - 1)(a^{\frac{p-1}{2}} + 1) = a^{p-1} - 1 \equiv 0 \pmod{p}.$$

However, since p is prime, we get

$$a^{\frac{p-1}{2}} - 1 \equiv 0 \pmod{p} \text{ or } a^{\frac{p-1}{2}} + 1 \equiv 0 \pmod{p}.$$

On the other hand, it follows from Euler's criterion that a is a quadratic nonresidue modulo p if and only if the first of the two congruences above does not hold, i.e., if and only if $a^{\frac{p-1}{2}} + 1 \equiv 0 \pmod{p}$.   □

Remark 12.21 It is possible to prove (cf. Problem 7, page 493) that if $a \in \mathbb{Z}$ is not a perfect square, then there exist infinitely many primes p such that a is a quadratic nonresidue modulo p.

In order to ease computations with quadratic residues and nonresidues, we shall need the following notation, introduced by A-M. Legendre.

Definition 12.22 For integers a and p, with p prime, we let the **Legendre symbol** $\left(\frac{a}{p}\right)$ be given by:

$$\left(\frac{a}{p}\right) = \begin{cases} 1, \text{ if } a \text{ is a quadratic residue modulo } p \\ -1, \text{ if } a \text{ is a quadratic nonresidue modulo } p \\ 0, \text{ if } p \mid a \end{cases}.$$

The convenience of Legendre's symbol is explained by the next two results.

Proposition 12.23 *If p is an odd prime and $a, b \in \mathbb{Z}$, then:*

(a) $a \equiv b \,(\mathrm{mod}\, p) \Rightarrow \left(\frac{a}{p}\right) = \left(\frac{b}{p}\right)$.

(b) *If $p \nmid a$, then $\left(\frac{a^2}{p}\right) = 1$.*

(c) $\left(\frac{a}{p}\right) \equiv a^{\frac{p-1}{2}} \,(\mathrm{mod}\, p)$.

(d) $\left(\frac{a}{p}\right)\left(\frac{b}{p}\right) = \left(\frac{ab}{p}\right)$.

Proof Since the proofs of items (a) and (b) are immediate, let us start by proving item (c): if $p \mid a$, then

$$\left(\frac{a}{p}\right) = 0 \equiv a^{\frac{p-1}{2}} \,(\mathrm{mod}\, p);$$

otherwise, Proposition 12.19 and Corollary 12.20 assure that, modulo p,

$$a^{\frac{p-1}{2}} \equiv \begin{cases} 1 \text{ if } a \text{ is a quadratic residue modulo } p \\ -1 \text{ if } a \text{ is a quadratic nonresidue modulo } p \end{cases} = \left(\frac{a}{p}\right).$$

In what concerns (d), we first note that if $p \mid ab$, then $p \mid a$ or $p \mid b$ and, hence,

$$\left(\frac{a}{p}\right)\left(\frac{b}{p}\right) = 0 = \left(\frac{ab}{p}\right).$$

If $p \nmid ab$, then $p \nmid a$ and $p \nmid b$, and it follows from item (c) that

$$\left(\frac{a}{p}\right)\left(\frac{b}{p}\right) \equiv a^{\frac{p-1}{2}} \cdot b^{\frac{p-1}{2}} = (ab)^{\frac{p-1}{2}} \equiv \left(\frac{ab}{p}\right) \,(\mathrm{mod}\, p).$$

Therefore, p divides the difference $\left(\frac{a}{p}\right)\left(\frac{b}{p}\right) - \left(\frac{ab}{p}\right)$, which assumes an integer value ranging from -2 to 2. However, since p is odd, we conclude that one must have $\left(\frac{a}{p}\right)\left(\frac{b}{p}\right) - \left(\frac{ab}{p}\right) = 0$. $\qquad\square$

The two coming examples bring interesting applications of quadratic residues. In both of them, we shall use the fact that if an integer n satisfies $n \equiv 3 \,(\mathrm{mod}\, 4)$, then

n has a prime divisor p such that $p \equiv 3 \pmod 4$. An argument for the validity of this (easy) fact can be found along the proof of Example 6.40.

Example 12.24 Prove that there does not exist integers x and y for which $y^2 = x^3 + 7$.

Solution By the sake of contradiction, assume that such a pair of integers does exist. Then x is odd, for otherwise we would have $y^2 = x^3 + 7 \equiv 3 \pmod 4$, which in turn contradicts item (c) of Proposition 10.9. Now,

$$y^2 + 1 = x^3 + 8 = (x + 2)[(x - 1)^2 + 3] \tag{12.4}$$

and, since $(x - 1)^2 + 3 \equiv 3 \pmod 4$ (recall that $x - 1$ is even), there exists a prime p such that $p \equiv 3 \pmod 4$ and $p \mid [(x - 1)^2 + 3]$.

Back to (12.4), we conclude that $y^2 + 1 \equiv 0 \pmod p$, and -1 is a quadratic residue modulo p. It then follows from the definition of Legendre's symbol, together with item (c) of the previous proposition, that

$$1 = \left(\frac{-1}{p}\right) \equiv (-1)^{\frac{p-1}{2}} = (-1)^{2k+1} = -1 \pmod p.$$

This contradicts the fact that p is odd. $\qquad\qquad\qquad\qquad\qquad\qquad\qquad$ □

For the next example, we extend the original Fibonacci and Lucas sequences (cf. Problem 2, page 71) to homonymous sequences $(F_n)_{n \in \mathbb{Z}}$ and $(L_n)_{n \in \mathbb{Z}}$, by imposing that the extended sequences satisfy the same recurrence relations as the original ones, namely, $F_1 = F_2 = 1$ and $F_{j+2} = F_{j+1} + F_j$ for every $j \in \mathbb{Z}$, and $L_1 = 1$, $L_2 = 3$ and $L_{j+2} = L_{j+1} + L_j$, for every integer $j \in \mathbb{Z}$ (so that, for instance, $F_0 = 0$, $F_{-1} = 1$, etc., and $L_0 = 2$, $L_{-1} = -1$, etc.). This way, it is a trivial task to prove that $F_{-n} = (-1)^{n-1} F_n$ and $L_{-n} = (-1)^n L_n$ for every $n \in \mathbb{N}$, as well as that the extended sequences continue to satisfy all of the properties listed in Problem 2, page 71, and in Problems 11–13, page 252 (of course, with indices m, n and k now ranging in $\mathbb{Z}$, instead of in $\mathbb{N}$).

Example 12.25 Let $(L_n)_{n \in \mathbb{Z}}$ be the extended Lucas sequence. Prove[3] that L_n is a perfect square if and only if $n = 1$ or 3.

Proof We consider three separate cases:

(i) $n = 2m$, with $m \in \mathbb{Z}$: item (b) of Problem 2, page 71 (with m in place of n), gives

$$L_n = L_{2m} = L_m^2 + 2(-1)^{m-1}.$$

---

[3]The result of this example and those of Problems 6 and 7, as well as their proofs, are due Cohn [10].

Therefore, L_n differs from a perfect square by -2 or 2, so that it cannot be a perfect square itself.

(ii) $n \equiv 1 \pmod 4$: if $n = 1$, then $L_n = L_1 = 1^2$. Otherwise, write $n = 1 + 2 \cdot 3^l \cdot k$, with $l \in \mathbb{Z}_+$ and $k \in \mathbb{Z}$ such that $2 \mid k$ but $3 \nmid k$. Item (d) of Problem 11, page 252 (with k in place of n) guarantees that $L_k \equiv 3 \pmod 4$. Therefore, we can choose a prime $p \equiv 3 \pmod 4$ such that $p \mid L_k$. On the other hand, by applying item (d) of Problem 13, page 253, with $t = 3^l$, we get

$$L_n \equiv -L_1 = -1 \pmod{L_k},$$

so that $L_n \equiv -1 \pmod p$. Finally, using the fact (already established before) that -1 is not a quadratic residue modulo p, we conclude that L_n cannot be a perfect square.

(iii) $n \equiv 3 \pmod 4$: if $n = 3$, then $L_3 = 2^2$, a perfect square. If $n \equiv 3 \pmod 4$ and $n \neq 3$, write $n = 3 + 2 \cdot 3^l \cdot k$, with $l \in \mathbb{Z}_+$ and $k \in \mathbb{Z}$ such that $2 \mid k$ but $3 \nmid k$. Arguing exactly as above, we get

$$L_n \equiv -L_3 = -4 \pmod p,$$

for some prime divisor p of L_k such that $p \equiv 3 \pmod 4$. This time, Proposition 12.23, together with the fact that -1 is not a quadratic residue modulo p, give

$$\left(\frac{-4}{p} \right) = \left(\frac{4}{p} \right) \left(\frac{-1}{p} \right) = -1,$$

and -4 is not a quadratic residue modulo p either. Hence, also as above, L_n is not a perfect square. $\square$

Back to the development of the theory, the next result, due to K. F. Gauss[4] and known as **Gauss' lemma**, furnishes a much simpler procedure than that provided by Euler's criterion to decide whether a certain integer is a quadratic residue or nonresidue modulo p, with p being a given odd prime.

Proposition 12.26 (Gauss) *If p is an odd prime and $a \in \mathbb{Z}$ is prime with p, then $\left(\frac{a}{p} \right) = (-1)^m$, where m is the number of elements of the set*

$$\left\{ 1 \leq j \leq \frac{p-1}{2}; \ ja \equiv -1, -2, \ldots \ or \ -\left(\frac{p-1}{2} \right) \pmod p \right\}. \tag{12.5}$$

---

[4]After Johann Carl Friedrich Gauss, German mathematician of the eighteenth and nineteenth centuries. Gauss is generally considered to be the greatest mathematician of all times. In the several different areas of Mathematics and Physics in which he worked, like Algebra, Analysis, Differential Geometry, Electromagnetism and Number Theory, there are always very important and deep results or methods that bear his name. We refer the reader to [38] for an interesting biography of Gauss.

Proof For $1 \leq j \leq \frac{p-1}{2}$, let $-\frac{p-1}{2} \leq t_j \leq \frac{p-1}{2}$ be the only integer such that

$$ja \equiv t_j \pmod{p}.$$

Let us first prove that

$$1 \leq i < j \leq \frac{p-1}{2} \Rightarrow |t_i| \neq |t_j|.$$

To this end, it suffices to see that, since $\gcd(a, p) = 1$, we have

$$|t_i| = |t_j| \Rightarrow |ia| \equiv |ja| \pmod{p} \Rightarrow |i| \equiv |j| \pmod{p}$$

$$\Rightarrow i \pm j \equiv 0 \pmod{p},$$

which is impossible.

It now follows from what we did above that the numbers $|t_1|, \ldots, |t_{\frac{p-1}{2}}|$ form, in some order, a permutation of $1, 2, \ldots, \frac{p-1}{2}$ (note that none of them equals 0, for $\gcd(a, p) = 1$). Therefore, recalling that m denotes the quantity of indices j for which $t_j < 0$, we get modulo p that

$$(-1)^m a^{\frac{p-1}{2}} \left(\frac{p-1}{2} \right)! \equiv (-1)^m t_1 \ldots t_{\frac{p-1}{2}} = |t_1| \ldots |t_{\frac{p-1}{2}}| = \left(\frac{p-1}{2} \right)!.$$

Hence,

$$a^{\frac{p-1}{2}} \equiv (-1)^m \pmod{p},$$

and the rest follows from item (c) of Proposition 12.23. □

As an application of Gauss' lemma, we next show that 2 is a quadratic residue modulo p (an odd prime) if and only if $p \equiv 0, 1, 2$ or $7 \pmod{8}$.

Example 12.27 If p is an odd prime, then

$$\left(\frac{2}{p} \right) = (-1)^{\lfloor \frac{p+1}{4} \rfloor} = (-1)^{\frac{p^2-1}{8}}.$$

Proof By Gauss' lemma, we have $\left(\frac{2}{p} \right) = (-1)^m$, where

$$m = \# \left\{ 1 \leq k \leq \frac{p-1}{2}; \ 2k \equiv -1, -2, \ldots \text{ or } -\left(\frac{p-1}{2} \right) \pmod{p} \right\}.$$

If $1 \le k \le \lfloor \frac{p-1}{4} \rfloor$, then $2 \le 2k \le 2\lfloor \frac{p-1}{4} \rfloor \le \frac{p-1}{2}$ and, hence,

$$2k \not\equiv -1, -2, \ldots, -\left(\frac{p-1}{2} \right) \pmod{p}.$$

If $\lfloor \frac{p-1}{4} \rfloor + 1 \le k \le \frac{p-1}{2}$, then

$$p - 1 \ge 2k \ge 2\left(\left\lfloor \frac{p-1}{4} \right\rfloor + 1 \right) > 2\left(\frac{p-1}{4} - 1 \right) + 2 = \frac{p-1}{2};$$

in particular, $2k \not\equiv 1, 2, \ldots, \frac{p-1}{2} \pmod{p}$ or, which is the same,

$$2k \equiv -1, -2, \ldots \text{ or } -\left(\frac{p-1}{2} \right) \pmod{p}.$$

Hence,

$$m = \frac{p-1}{2} - \left(\left\lfloor \frac{p-1}{4} \right\rfloor + 1 \right) + 1 = \frac{p-1}{2} - \left\lfloor \frac{p-1}{4} \right\rfloor = \left\lfloor \frac{p+1}{4} \right\rfloor,$$

where in the last equality above we separately considered the cases $p = 4k + 1$ and $p = 4k + 3$.

As for the second equality, it is enough to show that

$$\left\lfloor \frac{p+1}{4} \right\rfloor \equiv \frac{p^2 - 1}{8} \pmod{2}.$$

In turn, to this end it suffices to look at the cases $p = 8k+1$, $p = 8k+3$, $p = 8k+5$ and $p = 8k + 7$. $\square$

Continuing our study of quadratic residues, we are going to prove one of the most famous theorems of Gauss, which establishes a simple relationship between the Legendre symbols $\left(\frac{p}{q} \right)$ and $\left(\frac{q}{p} \right)$, whenever p and q are distinct odd primes. Let us start by rewriting Gauss' lemma in a suitable way.

Lemma 12.28 *If p is an odd prime and $a \in \mathbb{Z}$ is relatively prime with p and also odd, then $\left(\frac{a}{p} \right) = (-1)^M$, with*

$$M = \sum_{j=1}^{\frac{p-1}{2}} \left\lfloor \frac{ja}{p} \right\rfloor.$$

Proof For $1 \leq j \leq \frac{p-1}{2}$, let $0 < r_j < p$ be the remainder upon division of ja by p, so that (cf. Proposition 6.7)

$$ja = \left\lfloor \frac{ja}{p} \right\rfloor p + r_j. \tag{12.6}$$

It is immediate that

$$ja \equiv -1, -2, \ldots, -\left(\frac{p-1}{2}\right) \pmod{p} \Leftrightarrow \frac{p+1}{2} \leq r_j < p.$$

Hence, in the notations of Gauss' lemma, there are exactly m indices $1 \leq j \leq \frac{p-1}{2}$ for which $\frac{p+1}{2} \leq r_j < p$, and it suffices to show that the parity of the number of such indices equals that of $\sum_{j=1}^{\frac{p-1}{2}} \left\lfloor \frac{ja}{p} \right\rfloor$.

Write $s_1, s_2, \ldots, s_m$ (after some reordering, if needed) to denote the remainders r_j such that $\frac{p+1}{2} \leq r_j < p$, and $t_1, \ldots, t_n$ (also after some reordering, if needed) the remainders r_j for which $1 \leq r_j \leq \frac{p-1}{2}$. Then $m + n = \frac{p-1}{2}$, and (as in the proof of Gauss' lemma) for $1 \leq i < j \leq \frac{p-1}{2}$ we have $r_i \neq r_j, p - r_j$. Therefore,

$$\left\{1, 2, \ldots, \frac{p-1}{2}\right\} = \{p - s_1, p - s_2, \ldots, p - s_m\} \cup \{t_1, t_2, \ldots, t_n\},$$

a disjoint union, so that

$$\frac{p^2 - 1}{8} = \sum_{j=1}^{\frac{p-1}{2}} j = mp - \sum_{j=1}^{m} s_j + \sum_{j=1}^{n} t_j.$$

On the other hand, adding equalities (12.6) over $1 \leq j \leq \frac{p-1}{2}$, we obtain

$$a\left(\frac{p^2 - 1}{8}\right) = p \sum_{j=1}^{\frac{p-1}{2}} \left\lfloor \frac{ja}{p} \right\rfloor + \sum_{j=1}^{m} s_j + \sum_{j=1}^{n} t_j.$$

Subtracting both relations above, it then follows that

$$(a - 1)\left(\frac{p^2 - 1}{8}\right) = p \sum_{j=1}^{\frac{p-1}{2}} \left\lfloor \frac{ja}{p} \right\rfloor - mp + 2 \sum_{j=1}^{m} s_j.$$

Finally, recalling that a and p are both odd and looking at the last equality above modulo 2, we get

$$0 \equiv \sum_{j=1}^{\frac{p-1}{2}} \left\lfloor \frac{ja}{p} \right\rfloor - m,$$

as wished.   □

The coming result is known in mathematical literature as the **Quadratic Reciprocity Law** of Gauss. The proof we present is due to the German mathematician of the nineteenth century Ferdinand Eisenstein.

Theorem 12.29 (Gauss) *If p and q are distinct odd primes, then*

$$\left(\frac{p}{q} \right) \left(\frac{q}{p} \right) = (-1)^{\left(\frac{p-1}{2} \right) \left(\frac{q-1}{2} \right)}.$$

Proof By the previous lemma, we have

$$\left(\frac{p}{q} \right) \left(\frac{q}{p} \right) = (-1)^m (-1)^n,$$

with

$$m = \sum_{j=1}^{\frac{q-1}{2}} \left\lfloor \frac{jp}{q} \right\rfloor \quad \text{and} \quad n = \sum_{j=1}^{\frac{p-1}{2}} \left\lfloor \frac{jq}{p} \right\rfloor.$$

It then suffices to show that

$$\sum_{j=1}^{\frac{p-1}{2}} \left\lfloor \frac{jq}{p} \right\rfloor + \sum_{j=1}^{\frac{q-1}{2}} \left\lfloor \frac{jp}{q} \right\rfloor = \left(\frac{p-1}{2} \right) \left(\frac{q-1}{2} \right), \qquad (12.7)$$

for which we use the following double counting argument: since the right hand side of the desired equality counts the number of points of integer coordinates within the closed rectangle

$$\mathcal{R} = \left\{ (x, y) \in \mathbb{R}^2; \ 1 \le x \le \frac{p-1}{2} \text{ and } 1 \le y \le \frac{q-1}{2} \right\},$$

it is enough to count the number of those points in some other way, thus obtaining the left hand side of (12.7).

For what is left to do assume, without any loss of generality, that $p > q$, and consider the straight line $y = \frac{q}{p}x$ (cf. Fig. 12.1). For each $j \ge 1$, the nonnegative

Fig. 12.1 Counting the number of points of $\mathcal{R} \cap \mathbb{Z}^2$

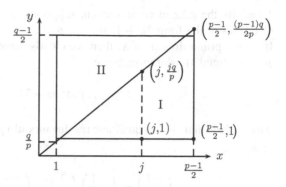

integer $\left\lfloor \frac{jq}{p} \right\rfloor$ counts the number of positive integers less than or equal to $\frac{jq}{p}$; on the other hand, for $1 \leq j \leq \frac{p-1}{2}$ we have $\frac{jq}{p} \notin \mathbb{Z}$, so that $\left\lfloor \frac{jq}{p} \right\rfloor$ counts the number of points of integer coordinates situated on the straight line $x = j$, below the line $y = \frac{q}{p}x$ and above the line $y = 0$. However, since

$$\left\lfloor \frac{jq}{p} \right\rfloor \leq \left\lfloor \frac{(p-1)q}{2p} \right\rfloor = \left\lfloor \frac{q}{2} - \frac{q}{2p} \right\rfloor \leq \left\lfloor \frac{q}{2} \right\rfloor = \frac{q-1}{2},$$

all of the points we are counting actually belong the closed rectangle $\mathcal{R}$. Hence, in the notations of Fig. 12.1, we have

$$\sum_{j=1}^{\frac{p-1}{2}} \left\lfloor \frac{jq}{p} \right\rfloor = \text{\# points of integer coordinates within region I of } \mathcal{R}.$$

Analogously,

$$\sum_{j=1}^{\frac{q-1}{2}} \left\lfloor \frac{jp}{q} \right\rfloor = \text{\# points of integer coordinates within region II of } \mathcal{R},$$

and there is nothing left to do. □

We finish this section by applying the quadratic reciprocity law to the establish the existence of certain prime numbers.[5]

Example 12.30 Prove that there are infinitely many primes of the form $3k + 1$, with $k \in \mathbb{N}$.

---

[5]For a more general result, see Theorem 20.19.

Proof By the sake of contradiction, suppose that there were only a finite number of primes of the form $3k + 1$, say $p_1, p_2, \ldots, p_n$, and let $x = (2p_1 \ldots p_n)^2 + 3$. If p is a prime divisor of x, then we clearly have $p \neq 2, 3, p_1, \ldots, p_n$, so that $p \equiv -1 \pmod 3$. Moreover, since

$$(2p_1 \ldots p_n)^2 \equiv -3 \pmod p,$$

we conclude that -3 is a quadratic residue modulo p. We must then have $\left(\frac{-3}{p}\right) = 1$. However,

$$\left(\frac{-3}{p}\right) = \left(\frac{-1}{p}\right)\left(\frac{3}{p}\right), \quad \left(\frac{-1}{p}\right) = (-1)^{\frac{p-1}{2}}$$

and, by the quadratic reciprocity law,

$$\left(\frac{3}{p}\right)\left(\frac{p}{3}\right) = (-1)^{\left(\frac{p-1}{2}\right)\left(\frac{3-1}{2}\right)} = (-1)^{\frac{p-1}{2}}.$$

Thus,

$$\left(\frac{-3}{p}\right) = (-1)^{\frac{p-1}{2}}\left(\frac{p}{3}\right)^{-1}(-1)^{\frac{p-1}{2}} = \left(\frac{p}{3}\right).$$

Finally, $p \equiv -1 \pmod 3$ implies, by item (a) of Proposition 12.23, that $\left(\frac{p}{3}\right) = \left(\frac{-1}{3}\right) = -1$; in turn, such an equality contradicts the fact that -3 is a quadratic residue modulo p. $\qquad \square$

Problems: Sect. 12.3

1. If p is an odd prime, prove that -1 is a quadratic residue modulo p if and only if $p \equiv 1 \pmod 4$.
2. Prove that there are infinitely many integer values of k for which the equation $x^2 = y^3 + k$ does not admit any integer solutions x, y.
3. Let a, b and c be given integers, not all of which equal to zero, and n be a natural number. If there exist integers x and y, relatively prime with n and such that $ax^2 + bxy + cy^2 = n$, prove that $b^2 - 4ac$ is a quadratic residue modulo n.
4. (Putnam) Prove that there does not exist $x, y \in \mathbb{Z}$ for which $x^2 + 3xy - 2y^2 = 122$.
5. Given $a, b \in \mathbb{Z}$, show that:

   (a) $2b^2 + 3$ has a prime divisor p such that $p \equiv \pm 3 \pmod 8$.
   (b) $(2b^2 + 3) \nmid (a^2 - 2)$.

The two coming problems are also concerned with the extended Fibonacci and Lucas sequences $(F_n)_{n\in\mathbb{Z}}$ and $(L_n)_{n\in\mathbb{Z}}$, as defined in the paragraph immediately before Example 12.25.

6. Let $(L_n)_{n\in\mathbb{Z}}$ be the extended Lucas sequence. Prove that L_n is twice a perfect square if and only if $n = 0$ or ± 6.

7. Let $(F_n)_{n\in\mathbb{Z}}$ be the extended Fibonacci sequence. Prove that F_n is a perfect square if and only if $n = 0, \pm 1, 2$ or 12.

8. If $p \neq 3$ is an odd prime, show that

$$\left(\frac{3}{p}\right) = \begin{cases} 1, & \text{if } p \equiv \pm 1 \ (\text{mod } 12) \\ -1, & \text{if } p \equiv \pm 5 \ (\text{mod } 12) \end{cases} .$$

9. Let m and n be odd naturals, with $n > 1$. Prove that:

   (a) $2^n - 1$ has a prime divisor p such that $p \equiv \pm 5 \ (\text{mod } 12)$.
   (b) $(2^n - 1) \nmid (3^m - 1)$.

10. If $n > 1$ is a natural number for which $p = 2^n + 1$ is prime, do the following items:

   (a) Show that 3 is a quadratic nonresidue modulo p.
   (b) Conclude that 3 is a primitive root modulo p.

11. With respect to quadratic residues modulo a prime power, do the following items:

   (a) Let a and k be given integers, with a being odd. Prove that the congruence $x^2 \equiv a \ (\text{mod } 2^k)$ has an integer solution for every $k > 2$ if and only if the congruence $x^2 \equiv a \ (\text{mod } 8)$ has an integer solution.
   (b) Let p be an odd prime and $a \in \mathbb{Z}$ be relatively prime with p. Prove that the congruence $x^2 \equiv a \ (\text{mod } p^k)$ has an integer solution for every $k \geq 1$ if and only if the congruence $x^2 \equiv a \ (\text{mod } p)$ has an integer solution.

12. * Let a and n be relatively prime integers, with $n > 1$, and $n = 2^k p_1^{k_1} \ldots p_t^{k_t}$ be the canonical factorization of n into prime powers. Prove that the congruence $x^2 \equiv a \ (\text{mod } n)$ has an integer solution if and only if the following conditions are satisfied:

   (i) $a \equiv 1 \ (\text{mod } 2)$ if $k = 1$, $a \equiv 1 \ (\text{mod } 4)$ if $k = 2$ or $a \equiv 1 \ (\text{mod } 8)$ if $k \geq 3$.
   (ii) $a^{\frac{p_i-1}{2}} \equiv 1 \ (\text{mod } p_i)$, for $1 \leq i \leq t$.

13. (APMO) A teacher gathered his n students around a circle. Then, he chose one of them, gave him/her a candy and, proceeding in the counterclockwise sense, distributed candies to the students according to the following rule: he jumped one student and gave a candy to the third one, then jumped two students and gave a candy to the sixth one, jumped three students and gave a candy to the tenth one, and so on, always jumping one student more than the previous time

and giving a candy to the next one. Find all values of n for which every student will have received at least one candy after a finite number of turns of the teacher around the circle.

14. Given a prime p and relatively prime integers a and n, with $n > 1$, we say that a is a **n-th residue** modulo p if the congruence $x^n \equiv a \pmod{p}$ has an integer solution. Generalize Euler's criterion for quadratic residues in the following way: if $d = \gcd(n, p - 1)$, then

$$a \text{ is an } n\text{-th residue modulo } p \Leftrightarrow a^{\frac{p-1}{d}} \equiv 1 \pmod{p}.$$

For the coming problem, the reader will need to use some basic facts on the *elementary symmetric sums* of the roots of a polynomial, for which we refer to Sect. 16.2.

15. (Bulgaria—adapted) The purpose of this problem is to show that, if $p > 5$ is prime and $k > 1$ is integer, then p^3 divides $\binom{kp}{p} - k$. To this end, do the following items:

(a) Until item (f), assume that k is odd. Conclude that

$$\binom{kp}{p} - k = \frac{k}{(p-1)!}(f(\alpha) - f(-\alpha)),$$

where $f(X) = \prod_{j=1}^{p-1}\left(X + \frac{(k-1)p}{2} + j\right)$, is a polynomial with integer coefficients and $\alpha = \frac{(k-1)p}{2}$.

(b) If $f(X) = X^{p-1} + a_{p-2}X^{p-2} + a_{p-3}X^{p-3} + \cdots + a_1X + a_0$, show that $f(\alpha) - f(-\alpha) \equiv 2a_1\alpha \pmod{p^3}$ and, then, deduce that it suffices to show that $a_1 \equiv 0 \pmod{p^2}$.

(c) Use Girard's relation (cf. Proposition 16.6) to show that condition $a_1 \equiv 0 \pmod{p^2}$ is equivalent to the congruence $\sum_{j=1}^{\frac{p-1}{2}} r_j \equiv 0 \pmod{p}$, where $r_j = \prod_{\substack{1 \le i \le \frac{p-1}{2} \\ i \ne j, p-j}} (\alpha + i)$.

(d) Show that, modulo p, equality $(\alpha+j)(\alpha+p-j)r_j = \prod_{i=1}^{\frac{p-1}{2}}(\alpha+i)$ implies the congruence $j^2 r_j \equiv -(p-1)! \pmod{p}$, for $1 \le j \le \frac{p-1}{2}$.

(e) Conclude from item (d) that r_j is a quadratic residue modulo p and $r_i \not\equiv r_j \pmod{p}$, for all $1 \le i < j \le \frac{p-1}{2}$.

(f) Deduce that, modulo p, we have

$$\{r_1, r_2, \ldots, r_{\frac{p-1}{2}}\} = \left\{1^2, 2^2, \ldots, \left(\frac{p-1}{2}\right)^2\right\}.$$

Then, finish the proof for odd k.

(g) If k is even, adapt the steps delineated from items (a) to (f), arguing from the beginning with the polynomial $f(X) = \prod_{j=1}^{p-1} \left(X + \frac{kp}{2} - j \right)$.

12.4 Sums of Two Squares

As an application of the ideas developed so far, in this section we characterize the natural numbers which can be written as the sum of two perfect squares. We start by looking at the case of prime numbers, with the following result of P. de Fermat.

Theorem 12.31 (Fermat) *The following conditions on an odd prime p are equivalent:*

(a) $p \equiv 1 \pmod 4$.
(b) -1 is a quadratic residue modulo p.
(c) p can be written as the sum of two perfect squares.

Proof
(a) $\Rightarrow$ (b): letting $p = 4k + 1$, it follows from item (c) of Proposition 12.23 that

$$\left(\frac{-1}{p} \right) \equiv (-1)^{\frac{p-1}{2}} \equiv (-1)^{2k} \equiv 1 \pmod p$$

and thus -1 is a quadratic residue modulo p.

(b) $\Rightarrow$ (c): let $h \in \mathbb{Z}$ be such that $h^2 + 1 \equiv 0 \pmod p$, and

$$A = \{(x, y);\ x, y \in \mathbb{Z},\ 0 \le x, y < \sqrt{p}\}.$$

By the fundamental principle of counting (cf. Sect. 1.1), we have $|A| = \left(\lfloor \sqrt{p} \rfloor + 1 \right)^2$. Now, since

$$\left(\lfloor \sqrt{p} \rfloor + 1 \right)^2 > \sqrt{p}^2 = p$$

and there are only p possible distinct remainders upon division by p, the pigeonhole principle (cf. Sect. 4.1) implies the existence of distinct ordered pairs $(x_1, y_1), (x_2, y_2) \in A$ for which

$$h x_1 + y_1 \equiv h x_2 + y_2 \pmod p.$$

Letting $a = |x_1 - x_2|$ and $b = |y_1 - y_2|$, we have that both a and b are nonzero and, hence,

$$0 < a^2 + b^2 = |x_1 - x_2|^2 + |y_1 - y_2|^2 < \sqrt{p}^2 + \sqrt{p}^2 = 2p.$$

However, since

$$a^2 + b^2 = |x_1 - x_2|^2 + |y_1 - y_2|^2$$
$$\equiv (x_1 - x_2)^2 + (hx_1 - hx_2)^2 \pmod{p}$$
$$= (h^2 + 1)(x_1 - x_2)^2 \equiv 0 \pmod{p},$$

the only possibility is that $a^2 + b^2 = p$.

(c) $\Rightarrow$ (a): let $p = a^2 + b^2$, with $a, b \in \mathbb{Z}$. Since p is odd, we have a even and b odd, or conversely. Assume, without loss of generality, that a is even and b is odd. Then, Proposition 10.9 gives

$$p = a^2 + b^2 \equiv 0 + 1 \equiv 1 \pmod{4}.$$

$\square$

Before we can proceed, we need an elementary auxiliary result, usually attributed to Euler and known as **Euler's identity**.

Lemma 12.32 *Let m and n be natural numbers, both of which can be written as a sum of two perfect squares. Then, mn can also be written as a sum of two perfect squares.*

Proof Writing $m = a^2 + b^2$ and $n = c^2 + d^2$, we have

$$mn = (a^2 + b^2)(c^2 + d^2)$$
$$= ((ac)^2 + (bd)^2) + ((ad)^2 + (bc)^2) \tag{12.8}$$
$$= (ac + bd)^2 + (ad - bc)^2.$$

$\square$

The coming result also credited to Fermat, provides a necessary and sufficient condition for a natural number to be written as a sum of two perfect squares.

Theorem 12.33 (Fermat) *A natural number $n > 1$ can be written as a sum of two perfect squares if and only if the following condition is satisfied: for each prime p such that $p \equiv 3 \pmod{4}$, the greatest power of p that divides n has even exponent.*

Proof In what follows, write

$$n = 2^a p_1^{a_1} \dots p_k^{a_k} q_1^{b_1} \dots q_l^{b_l}$$

with $a, a_i, b_j \in \mathbb{Z}_+$ and p_i, q_j prime numbers such that $p_i \equiv 1 \pmod{4}$ and $q_j \equiv 3 \pmod{4}$, for all $1 \le i \le k$ and $1 \le j \le l$.

(i) If each b_j is even, then n can be written as a sum of two perfect squares: to prove this, first note that, by Theorem 12.31, each p_i can be written as such; on the other hand, we have $2^a = (2^{a/2})^2 + 0^2$ if a is even $2^a = (2^{(a-1)/2})^2 + (2^{(a-1)/2})^2$ if a is odd; also, writing $b_j = 2c_j$, with $c_j \in \mathbb{Z}$ for $1 \leq j \leq l$, we have $q_j^{b_j} = (q_j^{c_j})^2 + 0^2$. Therefore, repeated applications of Lemma 12.32 allow us to conclude that n can be written as a sum of two squares.

(ii) If n can be written as a sum of two perfect squares, then each b_j is even: it suffices to prove that if n can be written as a sum of two squares and $b_j \geq 1$, then $b_j \geq 2$ and $\frac{n}{q_j^2}$ can also be written as a sum of two squares. To this end, if $n = c^2 + d^2$, with $c, d \in \mathbb{Z}$, then $c^2 + d^2 \equiv 0 \pmod{q_j}$. If $d \equiv 0 \pmod{q_j}$, then $c \equiv 0 \pmod{q_j}$ and hence $n = c^2 + d^2 \equiv 0 \pmod{q_j^2}$; therefore, $b_j \geq 2$ and

$$\frac{n}{q_j^2} = \left(\frac{c}{q_j}\right)^2 + \left(\frac{d}{q_j}\right)^2.$$

By the sake of contradiction, assume that $d \not\equiv 0 \pmod{q_j}$. Then $\gcd(d, q_j) = 1$, so that d is invertible modulo q_j. Letting f denote its inverse modulo q_j, we obtain

$$(cf)^2 + 1 \equiv 0 \pmod{q_j}.$$

This contradicts Theorem 12.31, for $q_j \equiv 3 \pmod 4$.   □

We finish this section by showing that there is essentially only one way of writing a prime congruent to 1 modulo 4 as a sum of two perfect squares.

Proposition 12.34 *If p is a prime number of the form $4k+1$, then there exist unique $x, y \in \mathbb{N}$ such that $x < y$ and $x^2 + y^2 = p$.*

Proof We already know that there exists at least one pair of natural numbers x and y such that $x^2 + y^2 = p$. Assume that a, b is another such pair and note that a, b, x and y are all relatively prime with p and less than $\sqrt{p}$. We may then choose integers $1 \leq c, z < p$ such that $xz \equiv y$ and $ac \equiv b \pmod p$.

We first claim that either $c = z$ or $c + z = p$. Indeed, we have

$$0 \equiv x^2 + y^2 \equiv x^2 + (xz)^2 \equiv x^2(z^2 + 1) \pmod p,$$

so that $z^2 \equiv -1 \pmod p$. Analogously, $c^2 \equiv -1 \pmod p$, and hence p divides $z^2 - c^2 = (z - c)(z + c)$, which in turn is the same as saying that p divides either $z - c$ or $z + c$. However, since $1 \leq c, z < p$, it follows that $-p < z - c < z + c < 2p$, and this (together with the fact that $p \mid (z - c)$ or $p \mid (z + c)$) gives $z - c = 0$ or $z + c = p$.

Now, assume that $c = z$. Then, the choices of c and z guarantee that

$$bxz \equiv acy \equiv ayz \pmod p, \tag{12.9}$$

and cancelling z we obtain $bx \equiv ay \pmod{p}$. However, since $0 < a, b, x, y < \sqrt{p}$, we have $0 < bx, ay < p$, and then $bx = ay$. Therefore,

$$p = x^2 + y^2 = \left(\frac{ay}{b}\right)^2 + y^2 = \left(\frac{y}{b}\right)^2 (a^2 + b^2) = \left(\frac{y}{b}\right)^2 p,$$

which gives $y = b$ and $x = a$.

If $z + c = p$, then, arguing as in (12.9), we arrive at $bx \equiv -ay \pmod{p}$ and, hence, at $bx + ay = p$. Thus, (12.8) furnishes

$$p^2 = (a^2 + b^2)(x^2 + y^2) = (bx + ay)^2 + (by - ax)^2 = p^2 + (by - ax)^2,$$

from which $by = ax$. We now obviously get $x = b$ and $y = a$.   □

Problems: Sect. 12.4

1. The purpose of this problem is to show that there are infinitely many natural numbers that cannot be written as a sum of three squares. More precisely, prove the following result of Euler: there does not exist integers k, l, x, y, z such that $l \geq 0$ and

$$x^2 + y^2 + z^2 = 4^l(8k + 7).$$

2. If $a, b, c \in \mathbb{N}$ are such that $a(a - 1) = b^2 + c^2$, show that $a + b$ is odd.
3. (BMO) Find all pairs of distinct natural numbers x and y such that $\frac{x^2+y^2}{x-y}$ is an integer which divides 1995.
4. Given a natural n, prove that there exist n consecutive naturals such that none of them can be written as the sum of two perfect squares.
5. The purpose of this problem is to prove a theorem, due to Lagrange, which asserts that every natural number can be written as a sum of four perfect squares. To this end, we first observe that the following generalization of Euler's identity holds: if two given naturals m and n can be written as sums of four squares, then mn can also be written this way. Indeed, if $m = a^2 + b^2 + c^2 + d^2$ and $n = w^2 + x^2 + y^2 + z^2$, then a simple (though surely tedious) verification shows that[6]

$$mn = (aw - bx - cy - dz)^2 + (ax + bw + cz - dy)^2$$
$$+ (ay - bz + cw + dx)^2 + (az + by - cx + dw)^2. \qquad (12.10)$$

---

[6]For a natural proof of such an identity, we refer the reader to item (f) of Problem 10, page 327.

Given such a result, do the following items[7]:

(a) Let p be prime, $S = \{x^2;\ x \in \mathbb{Z}_p\}$ and $S' = \{\overline{-1} - y;\ y \in S\}$. Prove that $S \cap S' \neq \emptyset$.
(b) Conclude that, if p is prime, then there exist $x, y, m \in \mathbb{Z}$ such that $1 \leq m < p$ and $x^2 + y^2 + 1 = mp$.
(c) Let p be prime and $1 < m < p$ be such that $mp = x_1^2 + x_2^2 + x_3^2 + x_4^2$, with $x_1, x_2, x_3, x_4 \in \mathbb{Z}$. If $-\frac{m}{2} < y_i < \frac{m}{2}$ is chosen in such a way that $x_i \equiv y_i \pmod{m}$, prove that there exists $1 \leq r < m$ for which $rp = y_1^2 + y_2^2 + y_3^2 + y_4^2$.
(d) Complete the proof of Lagrange's theorem, showing that every natural number can written as the sum of four perfect squares.

6. (IMO shortlist—adapted) The purpose of this problem is to find out all $n \in \mathbb{N}$ for which there exists some $m \in \mathbb{N}$ such that $2^n - 1$ divides $m^2 + 9$. To this end, do the two coming items:

(a) By contradiction, show that if $2^n - 1$ divides $m^2 + 9$ for some $m \in \mathbb{N}$, then n is a power of 2.
(b) If $n = 2^k$, make induction on k to show that there exists $m_k \in \mathbb{N}$ such that $2^{2^k} - 1$ divides $m_k^2 + 9$.

---

[7] We follow the steps delineated in problem XI.14 of the marvelous book [32].

Chapter 13
Complex Numbers

It is an obvious fact that the set of reals is too small to provide a complete description of the set of roots of polynomial functions; for instance, the function $x \mapsto x^2 + 1$, with $x \in \mathbb{R}$, does not have any real root. Historically, the search for such roots strongly motivated the birth of complex numbers and the flowering of complex function theory. In this respect, a major first crowning was the proof, by Gauss, of the famous *Fundamental Theorem of Algebra*, which asserts that every polynomial function with complex coefficients has a complex root.

In this chapter, we concentrate ourselves in the construction and discussion of the elementary properties of the set of complex numbers, postponing to Sect. 15.1 the presentation of one simple proof of the aforementioned theorem of Gauss.

13.1 Basic Definitions and Properties

As was seen in Chapter 1 of [8], one generally looks at the set $\mathbb{R}$ of real numbers as the *number line*: one starts with a straight line (geometric entity), choosing a point in it to correspond to 0; then, one chooses a standard length to correspond to the unit of measure, and two distinct rules to operate with two points along the line (such operations being called addition and multiplication of real numbers), so that one gets a third point as the result of operating (i.e., adding or multiplying) two given points. We then call the points along the line as real numbers, assume that they fulfill the whole line and check that the operations of addition and multiplication satisfy the usual properties of commutativity, associativity etc.

A. Caminha Muniz Neto, *An Excursion through Elementary Mathematics, Volume III*,
Problem Books in Mathematics, https://doi.org/10.1007/978-3-319-77977-5_13

The discussion at the former paragraph hints to the following natural question: would there be some way of defining two operations with the points of an Euclidean plane in a way similar to the ordinary addition and multiplication of real numbers? The answer to this question is *yes* and the resulting set, which we are about to describe, is that of *complex numbers*.

Look at the cartesian plane as the set $\mathbb{R} \times \mathbb{R}$ of ordered pairs (a, b) of reals. Let $\oplus$ and $\odot$ be the operations with the points of such a plane defined by

$$(a, b) \oplus (c, d) = (a+c, b+d) \text{ and } (a, b) \odot (c, d) = (ac-bd, ad+bc), \quad (13.1)$$

where $+$ and $\cdot$ respectively denote the ordinary addition and multiplication of real numbers.

One can straightforwardly verify that $\oplus$ and $\odot$ are *associative* and *commutative* operations, and that $\odot$ is distributive with respect to $\oplus$, i.e., that for all $a, b, c, d, e, f \in \mathbb{R}$, the following properties hold:

i. **Associativity** of $\oplus$ and $\otimes$: $(a, b) \oplus ((c, d) \oplus (e, f)) = ((a, b) \oplus (c, d)) \oplus (e, f)$ and $(a, b) \odot ((c, d) \odot (e, f)) = ((a, b) \odot (c, d)) \odot (e, f)$.
ii. **Commutativity** of $\oplus$ and $\otimes$: $(a, b) \oplus (c, d) = (c, d) \oplus (a, b)$ and $(a, b) \odot (c, d) = (c, d) \odot (a, b)$.
iii. **Distributivity** of $\otimes$ with respect to $\oplus$: $(a, b) \odot ((c, d) \oplus (e, f)) = ((a, b) \odot (c, d)) \oplus ((a, b) \odot (e, f))$

It is also pretty clear (check it!) that $(0, 0)$ and $(1, 0)$ are *identities* for $\oplus$ and $\odot$, respectivamente, i.e., that

$$(a, b) \oplus (0, 0) = (a, b) \text{ and } (a, b) \odot (1, 0) = (a, b),$$

for all $a, b \in \mathbb{R}$. Below, we prove a further important *arithmetic* property of $\odot$, which is known as the **cancellation law**.

Lemma 13.1 *The following cancellation law holds for $\odot$:*

$$(a, b) \odot (c, d) = (0, 0) \Rightarrow (a, b) = (0, 0) \text{ or } (c, d) = (0, 0).$$

Proof Since $(a, b) \odot (c, d) = (ac-bd, ad+bc)$, we get $ac-bd = 0$ and $ad+bc = 0$. Therefore, arguing as in the proof of Euler's identity (cf. Lemma 12.32), we get

$$0 = (ac - bd)^2 + (ad + bc)^2 = (a^2 + b^2)(c^2 + d^2).$$

Hence, either $a^2 + b^2 = 0$ or $c^2 + d^2 = 0$, which is the same as saying that either $(a, b) = 0$ or $(c, d) = 0$. $\qquad\qquad\square$

Now, let us look at the real line as the horizontal axis of the cartesian plane, which is equivalent to identifying each real number x with the ordered pair $(x, 0)$. We have to check whether such an identification is good, in the sense that the results of performing $\oplus$ and $\odot$ with such ordered pairs give (up to identification) the same results we would get if we first performed the usual operations of addition and multiplication of real numbers. This reduces to verifying that

$$(x, 0) \oplus (y, 0) = (x + y, 0) \text{ and } (x, 0) \odot (y, 0) = (xy, 0), \qquad (13.2)$$

which is totally immediate.

Thanks to the above computations, together with the fact that $x \mapsto (x, 0)$ is injective, we can safely consider $\mathbb{R}$, furnished with its usual operations of addition and multiplication, as a subset of $\mathbb{R} \times \mathbb{R}$, furnished with the operations $\oplus$ and $\odot$ defined as in (13.1). This also allows us to call the points of the cartesian plane as *numbers*, more precisely as the **complex numbers**. From now on, we shall denote the set of complex numbers by $\mathbb{C}$.

In what follows, we shall derive an easier way to represent the set of complex numbers and its operations. To this end, let us henceforth denote by i the complex number $(0, 1)$, calling it the **imaginary unit**.[1]

Letting $\approx$ denote the identification of the points along the horizontal axis with real numbers, and taking into account the definition of $\odot$, we are led to conclude that

$$i^2 = (0, 1) \odot (0, 1) = (0 \cdot 0 - 1 \cdot 1, 0 \cdot 1 + 1 \cdot 0) = (-1, 0) \approx -1. \qquad (13.3)$$

Note also that

$$(a, b) = (a, 0) \oplus (0, b) = (a, 0) \oplus ((b, 0) \odot (0, 1)) \approx a + bi, \qquad (13.4)$$

which is said to be the **algebraic form**.

From now on, we shall write simply $+$ and $\cdot$ to respectively denote the operations $\oplus$ and $\odot$, and shall write $a + bi$ to denote the complex number (a, b). We shall also omit the identifications involved, for, as we shall see in a few moments, there is no danger that such a usage induces *wrong arithmetic* inside $\mathbb{C}$. It thus follows from (13.4) that

$$a + bi = 0 \Leftrightarrow (a, b) = (0, 0) \Leftrightarrow a = b = 0.$$

According to (13.3), in the set $\mathbb{C}$ of complex numbers the equation

$$x^2 + 1 = 0$$

---

[1] The names *complex* and *imaginary* are rooted in the historical development of complex numbers. More precisely, when mathematicians started using complex numbers, even without having a precise definition of what they should be, they called them *complex* or *imaginary*, in allusion to the oddness of guessing the existence of "numbers" whose squares could be negative.

has i as a root. As we shall see along this chapter and Chap. 15, this is the core fact related to the importance of complex numbers.

As we anticipated above, a good reason for writing $(a, b) \in \mathbb{C}$ in the algebraic form $a + bi$ is that, upon doing it, we can operate with complex numbers exactly as we do with reals; one just has to take into account that $i^2 = -1$. Indeed, we check below that the computations performed this way lead to the same results as those done by directly using the definitions of $+$ and $\cdot$ (i.e., $\oplus$ and $\odot$) in $\mathbb{C}$:

- Computation of $(a, b) + (c, d)$: by definition, we have $(a, b) + (c, d) = (a + c, b + d)$. On the other hand, operating as we would do with reals, we obtain

$$(a + bi) + (c + di) = (a + c) + (b + d)i.$$

However, since we are writing $(a + c, b + d) = (a + c) + (b + d)i$, both results do coincide.
- Computation of $(a, b) \cdot (c, d)$: by definition, we have $(a, b) \cdot (c, d) = (ac - bd, ad + bc)$. Operating once more as with real numbers, we get

$$(a + bi)(c + di) = ac + adi + bci + bdi^2 = (ac - bd) + (ad + bc)i,$$

since $i^2 = -1$. Here again we are writing $(ac - bd, ad + bc) = (ac - bd) + (ad + bc)i$, so that both results also coincide.

Thanks to the above identifications, we can surely write $\mathbb{R} \subset \mathbb{C}$, with $x \in \mathbb{R}$ written in algebraic form as $x + 0i$. Moreover, by analogy with the addition and multiplication of reals, we shall also call the $+$ and $\cdot$ operations on complex numbers as the **addition** and **multiplication**, respectively.

Proceeding with the development of the theory, let us introduce in $\mathbb{C}$ other two operations, this time extending the subtraction and division of real numbers. Before we go on, we remark that it is common usage to write z, w etc to refer to general complex numbers, so that each one of them has its own algebraic form.

Given $z, w \in \mathbb{C}$, *subtracting* w of z means obtaining a complex number $z - w$ (the *difference* between z and w) such that $z = (z - w) + w$. Writing $z = a + bi$ and $w = c + di$ and operating as with reals, we easily see that the only possible way of defining $z - w$ is by letting

$$z - w = (a + bi) - (c + di) = (a - c) + (b - d)i.$$

Formally, for $z = a + bi$ and $w = c + di$, the complex number $z - w$ is *defined* by

$$z - w = (a - c) + (b - d)i, \tag{13.5}$$

being called the **difference** between z and w, in this order.

Also for $z, w \in \mathbb{C}$, with $w \neq 0$, *dividing* z by w means obtaining a complex number z/w (the *quotient* between z and w) such that $z = (z/w) \cdot w$. Letting

$z = a + bi$ and $w = c + di$, and operating as we do upon rationalising real number, we immediately get

$$
\begin{aligned}
\frac{z}{w} &= \frac{a + bi}{c + di} = \frac{(a + bi)(c - di)}{(c + di)(c - di)} \\
&= \frac{(ac + bd) + (bc - ad)i}{c^2 + d^2} \\
&= \frac{ac + bd}{c^2 + d^2} + \frac{bc - ad}{c^2 + d^2}i.
\end{aligned}
$$

as the only possibility for z/w.

Formally, for $z = a + bi$ and $w = c + di$, with $w \neq 0$, the complex number z/w is *defined* by

$$
\frac{z}{w} = \frac{ac + bd}{c^2 + d^2} + \frac{bc - ad}{c^2 + d^2}i, \tag{13.6}
$$

being called the **quotient** between z and w, in this order.

The particular case of dividing 1 by a nonzero complex number leads us to conclude that every complex number $z \neq 0$ possess an *inverse* with respect to multiplication. Writing $z = a + ib \neq 0$, it follows from (13.6) that such an inverse, which we shall denote by z^{-1} or $1/z$, is given by

$$
z^{-1} = \frac{a}{a^2 + b^2} - \frac{b}{a^2 + b^2}i.
$$

Once again, note that we need not worry about memorizing the above formulas. It just suffices to operate as we do with real numbers.

In order to simplify lots of forthcoming computations, we now introduce the following notation: for $z = a + bi \in \mathbb{C}$, we write $\bar{z}$ to denote the complex number $\bar{z} = a - bi$, and call it the **conjugate** of z. In particular, we have $\bar{\bar{z}} = z$.

Note also that, upon multiplying z by $\bar{z}$, we get as result the real number $a^2 + b^2$, for

$$
z \cdot \bar{z} = (a + bi)(a - bi) = a^2 - (bi)^2 = a^2 + b^2.
$$

We write $|z| = \sqrt{a^2 + b^2}$ and call $|z|$ the **modulus** of z. When $z \in \mathbb{R}$, it is immediate to note that the notion of modulus of a complex number, as defined above, coincides with the usual notion of the modulus of a real number (cf. [8], for instance). In short,

$$
z = a + bi \Rightarrow |z|^2 = z\bar{z} = a^2 + b^2. \tag{13.7}
$$

Fig. 13.1 Conjugation of
complex numbers in the
complex plane

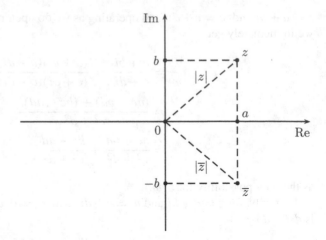

Yet another simple fact concerning the modulus of a complex number is that $|\bar{z}| = |z|$. This will become *geometrically* apparent quite soon (cf. Fig. 13.1), provided we go back to the very definition of $\mathbb{C}$, looking at complex numbers as the points of the cartesian plane. Upon doing this, we obtain a geometric representation of $\mathbb{C}$, which is known as the **complex plane**.[2]

In the complex plane, the horizontal and vertical axes are respectively called the **real** and **imaginary**. The real axis is this formed by the complex numbers which are actually real numbers (i.e., by the ordered pairs $(x, 0) \approx x$), whereas the complex numbers along the imaginary axis are those of the form yi, with $y \in \mathbb{R}$ (this corresponds to the ordered pairs $(0, y) = (y, 0) \cdot (0, 1) \approx yi$); such complex numbers are called **pure imaginary** (so that i is one of them).

Also with respect to the complex plane, since $z = a + bi = (a, b)$, it follows that z and $\bar{z} = a - bi$ are symmetric with respect to the real axis (cf. Fig. 13.1); hence, $|z| = |\bar{z}|$. On the other hand, the real numbers a and b are respectively called the **real** and **imaginary parts** of z, and are denoted

$$a = \text{Re}(z), \quad b = \text{Im}(z).$$

Thus, $z = \text{Re}(z) + i \, \text{Im}(z)$, so that $\bar{z} = \text{Re}(z) - i \, \text{Im}(z)$ and, hence,

$$\text{Re}(z) = \frac{z + \bar{z}}{2} \quad \text{and} \quad \text{Im}(z) = \frac{z - \bar{z}}{2i}. \tag{13.8}$$

---

[2]Also called the *Argand-Gauss plane*, in honor of the amateur Swiss mathematician of the eighteenth century Jean-Robert Argand, and of the great J. C. F. Gauss.

The following lemma collects further useful properties of the conjugation of complex numbers. In order to properly state it, we observe that the associativity of the multiplication of complex numbers guarantees the well definiteness of the $n-th$ power z^n, for $z \in \mathbb{C}$ and $n \in \mathbb{N}$, as $z^1 = z$ and

$$z^n = \underbrace{z \cdot \ldots \cdot z}_{n \text{ times}},$$

for $n > 1$. Moreover, such a definition can be easily extended to integer exponents n by setting, for $z \in \mathbb{C} \setminus \{0\}$, $z^0 = 1$ and, for an integer $n < 0$,

$$z^n = (z^{-n})^{-1} = \frac{1}{z^{-n}}.$$

Then, easy inductive arguments allow us to show that the usual arithmetic rules for powers remain valid, namely:

$$(z^m)^n = z^{mn} \text{ and } (zw)^n = z^n w^n \tag{13.9}$$

for all $z, w \in \mathbb{C} \setminus \{0\}$ and $m, n \in \mathbb{Z}$.

We also take the opportunity to point out (although this would not be needed in the coming lemma) that the usual binomial formula,

$$(z + w)^n = \sum_{k=0}^{n} \binom{n}{k} z^{n-k} w^k$$

(for all $z, w \in \mathbb{C} \setminus \{0\}$ such that $z + w \neq 0$, and every $n \in \mathbb{N}$), also holds true in $\mathbb{C}$, since its proof only depends on properties of addition and multiplication of real numbers that also hold true in $\mathbb{C}$.

We can finally state and prove the aforementioned result.

Lemma 13.2 *If z and w are nonzero complex numbers, then:*

(a) $z \in \mathbb{R} \Leftrightarrow \operatorname{Re}(z) = 0 \Leftrightarrow z = \bar{z}$.
(b) $\overline{z + w} = \bar{z} + \bar{w}$, $\overline{zw} = \bar{z} \cdot \bar{w}$ *and* $\overline{z/w} = \bar{z}/\bar{w}$.
(c) $\overline{z^n} = (\bar{z})^n$, *for every* $n \in \mathbb{Z}$.
(d) $|z| = 1 \Leftrightarrow \bar{z} = 1/z$.

Proof

(a) Letting $z = a + bi$, we have

$$z \in \mathbb{R} \Leftrightarrow b = 0 \Leftrightarrow a + bi = a - bi \Leftrightarrow z = \bar{z}.$$

(b) Writing $z = a + bi$ and $w = c + di$, we compute

$$\overline{z} + \overline{w} = (a - bi) + (c - di)$$
$$= (a + c) - (b + d)i$$
$$= \overline{z + w}$$

and

$$\overline{zw} = \overline{(ac - bd) + (ad + bc)i}$$
$$= (ac - bd) - (ad + bc)i$$
$$= (a - bi)(c - di) = \overline{z} \cdot \overline{w}.$$

From this last one, it follows that

$$\overline{z/w} \cdot \overline{w} = \overline{z/w \cdot w} = \overline{z}$$

and, hence,

$$\overline{z/w} = \overline{z}/\overline{w}.$$

(c) For $n = 0$, the result is immediate and, for $n \in \mathbb{N}$, it easily follows by induction from (b). For an integer $n < 0$, first note that, by item (b), we have

$$1 = \overline{1} = \overline{zz^{-1}} = \overline{z} \cdot \overline{z^{-1}},$$

so that $\overline{z^{-1}} = (\overline{z})^{-1}$; hence, letting $u = z^{-1}$, the first part above, together with (13.9), furnishes

$$\overline{z^n} = \overline{u^{-n}} = (\overline{u})^{-n} = ((\overline{z})^{-1})^{-n} = (\overline{z})^{(-1)(-n)} = (\overline{z})^n.$$

(d) Since (cf. (13.7)) $z \cdot \overline{z} = |z|^2$, we conclude that

$$|z| = 1 \Leftrightarrow z \cdot \overline{z} = 1 \Leftrightarrow \overline{z} = 1/z.$$

□

Example 13.3 (Spain) If z and w are complex numbers of modulus 1 and such that $zw \neq -1$, show that $\frac{z+w}{1+zw}$ is real.

Proof Letting $a = \frac{z+w}{1+zw}$, item (a) of the previous lemma guarantees that it is enough to show that $\overline{a} = a$. To this end, note that, by applying items (b) and (d) of that same result, we get

$$\bar{a} = \frac{\overline{z+w}}{1+zw} = \frac{\bar{z}+\bar{w}}{1+\bar{z}\cdot\bar{w}}$$

$$= \frac{z^{-1}+w^{-1}}{1+z^{-1}w^{-1}} = \frac{w+z}{zw+1} = a.$$

□

Our next result gives quite a useful geometric interpretation of the modulus of the difference of two complex numbers.

Proposition 13.4 *Given* $z, w \in \mathbb{C}$, *the real number* $|z - w|$ *equals the Euclidean distance from* z *to* w *in the cartesian plane subjacent to the corresponding complex plane.*

Proof If $z = a + bi$ and $w = c + di$, then

$$|z - w| = |(a - c) + (b - d)i| = \sqrt{(a - c)^2 + (b - d)^2}.$$

On the other hand (cf. Fig. 13.2), basic Analytic Geometry (cf. Chapter 6 of [9]) assures that the distance from z to w is also given by $\sqrt{(a - c)^2 + (b - d)^2}$. □

The coming inequality (13.10) is known as the **triangle inequality** for complex numbers.

Corollary 13.5 *If* u, v *and* z *are any complex numbers, then*

$$|u - v| \le |u - z| + |z - v|. \tag{13.10}$$

Proof According to Proposition 13.4, the corollary only says that the length of any side of a (possibly degenerate) triangle does not exceed the sum of the lengths of its other two sides. However, we already know this to be true (cf. Theorem 2.26 of [9]). □

Fig. 13.2 Modulus of the difference between two complex numbers

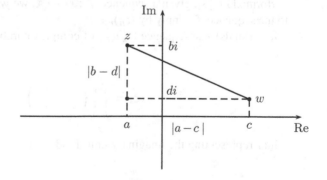

Problems: Sect. 13.1

1. With respect to the operations of addition and multiplication of complex numbers, check that they are associative and commutative, as well as that multiplication is distributive with respect to addition. Check also that $(0, 0)$ and $(1, 0)$ are unit elements for the addition and multiplication, respectively.

2. * Given $z, w \in \mathbb{C}$, prove that:

   (a) $|zw| = |z| \cdot |w|$ and (if $w \neq 0$) $|z/w| = |z|/|w|$.
   (b) $|z + w|^2 = |z|^2 + 2\mathrm{Re}(\overline{z}w) + |w|^2$.

3. * Use item (b) of the previous problem to establish, for all $z, w \in \mathbb{C} \setminus \{0\}$, the validity of the inequality $|z+w| \leq |z|+|w|$, with equality if and only if $z = \lambda w$, for some $\lambda > 0$. This inequality is also known as the **triangle inequality** for complex numbers. Use it to:

   (a) Deduce the validity of (13.10).
   (b) Prove that $||z| - |w|| \leq |z - w|$, for all $z, w \in \mathbb{C}$.

4. For a given $z \in \mathbb{C}$, show that $|z| = 1$ if and only if there exists $x \in \mathbb{R}$ for which $z = \frac{1-ix}{1+ix}$.

5. (OCM) Let a and z be complex numbers such that $|a| < 1$ and $\overline{a}z \neq 1$. If

$$\left| \frac{z - a}{1 - \overline{a}\,z} \right| < 1,$$

   prove that $|z| < 1$.

6. Let $a \in \mathbb{C}$ and $R > 0$ be such that $|a| \neq R$. Let Γ be the circle of center a and radius R in the complex plane (so that $0 \notin \Gamma$), and $\tilde{\Gamma} = \left\{ \frac{1}{z}; \ z \in \Gamma \right\}$. Show that $\tilde{\Gamma}$ is the circle of center $\frac{\overline{a}}{|a|^2 - R^2}$ and radius $\frac{R}{||a|^2 - R^2|}$.

   For the coming problem, we define a **sequence of complex numbers** as a function $f : \mathbb{N} \to \mathbb{C}$. As in our discussion of sequences of real numbers, performed in [8], given a sequence $f : \mathbb{N} \to \mathbb{C}$ we write $z_n = f(n)$ and refer to the sequence f simply by $(z_n)_{n \geq 1}$.

7. (Netherlands) The sequence $(z_k)_{k \geq 1}$ of complex numbers is defined, for $k \in \mathbb{N}$, by

$$z_k = \prod_{j=1}^{k} \left(1 + \frac{i}{\sqrt{j}} \right),$$

   with i representing the imaginary unit. Find all $n \in \mathbb{N}$ for which

$$\sum_{k=1}^{n} |z_{k+1} - z_k| = 1000.$$

8. Given $z, w \in \mathbb{C}$, let $\boldsymbol{u}$ and $\boldsymbol{v}$ be the vectors starting at 0 and having z and w as endpoints, respectively. Prove that the complex numbers $z + w$ and $z - w$ are the endpoints of the vectors $\boldsymbol{u} + \boldsymbol{v}$ and $\boldsymbol{u} - \boldsymbol{v}$, respectively, both of which starting at 0.

   For the coming problem, recall (cf. Sect. 4.3) that a (**partial**) **order relation** in a nonempty set X is a relation $\preceq$ in X which is *reflexive*, *transitive* and *antisymmetric*; moreover, if we have $x \preceq y$ or $y \preceq x$ for all $x, y \in X$, then the order relation $\preceq$ is said to be a **total ordering** of X.

9. Prove that the set of complex numbers cannot be totally ordered. More precisely, prove that there does not exist, on $\mathbb{C}$, a total ordering $\preceq$, which extends the usual order relation in $\mathbb{R}$ and satisfies the conditions

$$0 \preceq z, w \Rightarrow 0 \preceq z + w, zw.$$

   The coming problem generalizes the construction of $\mathbb{C}$, presenting the set $\mathbb{H}$ of **Hamilton's**[3] **quaternions** (also called **quaternionic numbers**).

10. * In $\mathbb{H} = \{(a, b, c, d); a, b, c, d, \in \mathbb{R}\}$, let $\oplus$ and $\odot$ be the operations defined, for $\alpha = (a, b, c, d)$ and $\beta = (w, x, y, z)$, by

$$\alpha \oplus \beta = (a + w, b + x, c + y, d + z)$$

and

$$\alpha \odot \beta = (aw - bx - cy - dz, ax + bw + cz - dy,$$
$$ay - bz + cw + dx, az + by - cx + dw),$$

   where $+$ and $\cdot$ respectively denote the usual operations of addition and multiplication of real numbers. Such operations are called the **addition** and **multiplication** of $\mathbb{H}$. With respect to them, do the following items:

   (a) Show that the function $\iota : \mathbb{R} \to \mathbb{H}$, given by $\iota(x) = (x, 0, 0, 0)$ preserves operations, in the sense that

$$\iota(x) \oplus \iota(y) = \iota(x + y) \ \text{ and } \ \iota(x) \odot \iota(y) = \iota(x + y),$$

   for all $x, y \in \mathbb{R}$.

   (b) Show that the functions $\iota_1, \iota_2, \iota_3 : \mathbb{C} \to \mathbb{H}$, given by $\iota_1(x + yi) = (x, y, 0, 0), \iota_2(x + yi) = (x, 0, y, 0)$ and $\iota_3(x + yi) = (x, 0, 0, y)$, preserve operations, in the sense that

$$\iota_j(z) \oplus \iota_j(w) = \iota_j(z + w) \ \text{ and } \ \iota_j(z) \odot \iota_j(w) = \iota_j(z + w),$$

   for all $z, w \in \mathbb{C}$ and $1 \leq j \leq 3$.

---

[3] After William R. Hamilton, Irish astronomer, mathematician and physicist of the nineteenth century. Quaternions have many important applications in Mathematics and Physics, but unfortunately they lie far beyond the scope of these notes.

(c) In view of items (a) and (b), we hereafter write simply $+$ and $\cdot$ to denote $\oplus$ and $\odot$, respectively, x to denote $(x, 0, 0, 0)$ and $i = (0, 1, 0, 0)$, $j = (0, 0, 1, 0)$ and $k = (0, 0, 0, 1)$. Under these conventions, show that

$$i^2 = j^2 = k^2 = -1$$
$$ij = -ji = k,\, jk = -kj = i,\, ki = -ik = j \tag{13.11}$$

and

$$(a, b, c, d) = a + bi + cj + dk.$$

(d) Show that the results of $(a + bi + cj + dk) + (w + xi + yj + zk)$ and $(a + bi + cj + dk) \cdot (w + xi + yj + zk)$ are the same we would have obtained by using the usual operations of addition and multiplication of real numbers, together with (13.11). From this, conclude that:

   i. The addition of quaternions is commutative, associative and has 0 as identity.
   ii. The multiplication of quaternions is associative, distributive with respect to $+$ and has 1 as identity; nevertheless, it is not commutative.

(e) For $\alpha = a + bi + cj + dk \in \mathbb{H}$, let $\overline{\alpha} = a - bi - cj - dj$ be the **conjugate** of α. For $\alpha, \beta \in \mathbb{H}$, show that $\overline{\alpha\beta} = \overline{\alpha}\,\overline{\beta}$.
(f) For $\alpha = a + bi + cj + dk \in \mathbb{H}$, let $|\alpha| = \sqrt{a^2 + b^2 + c^2 + d^2}$ be the **norm** of α. Show that $\alpha\overline{\alpha} = |\alpha|^2$ and $|\alpha\beta| = |\alpha||\beta|$, for all $\alpha, \beta \in \mathbb{H}$.
(g) Use the result of the previous item to deduce identity (12.10).
(h) Conclude that, given $\alpha \in \mathbb{H} \setminus \{0\}$, there exists a single $\beta \in \mathbb{H}$ for which $\alpha\beta = 1$. Then, show that the following *cancellation law* holds in $\mathbb{H}$: if $\alpha, \beta \in \mathbb{H}$ are such that $\alpha\beta = 0$, then $\alpha = 0$ or $\beta = 0$.

11. (Japan) Let $\mathcal{P}$ be a convex pentagon whose sides and diagonals have, in some order, lengths equal to $l_1, l_2, \ldots, l_{10}$. If all of the numbers $l_1^2, l_2^2, \ldots, l_9^2$ are rational, prove that l_{10}^2 is also rational.

13.2   The Polar Form of a Complex Number

Given $z = a + bi \in \mathbb{C} \setminus \{0\}$, let $\alpha \in [0, 2\pi)$ be the least value, in radians, of the trigonometric angle measured from the positive real axis to the half line joining 0 to z (cf. Fig. 13.3).

   Writing

$$z = |z|\left(\frac{a}{\sqrt{a^2 + b^2}} + \frac{b}{\sqrt{a^2 + b^2}}i\right),$$

Fig. 13.3 The polar form of a complex number

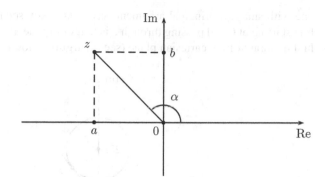

we easily get

$$\cos\alpha = \frac{a}{\sqrt{a^2+b^2}} \quad \text{and} \quad \sin\alpha = \frac{b}{\sqrt{a^2+b^2}},$$

so that

$$z = |z|(\cos\alpha + i\,\sin\alpha). \tag{13.12}$$

Since $\sin(\alpha + 2k\pi) = \sin\alpha$ and $\cos(\alpha + 2k\pi) = \cos\alpha$ for every $k \in \mathbb{Z}$, equality (13.12) continues to hold with $\alpha + 2k\pi$ in place of α. For this reason, from now on we say that the real numbers of the form $\alpha + 2k\pi$, with $k \in \mathbb{Z}$, are the **arguments** of the nonzero complex numbers z, and that α is its **principal argument**.

Letting α be any argument of $z \in \mathbb{C} \setminus \{0\}$, we shall call (13.12) the **polar**, or **trigonometric, form** of z. Note that

$$|\cos\alpha + i\,\sin\alpha| = 1, \tag{13.13}$$

for every $\alpha \in \mathbb{R}$. Hereafter, we shall denote the complex number $\cos\alpha + i\sin\alpha$ simply by $\operatorname{cis}\alpha$. Thus, letting α be an argument of $z \in \mathbb{C} \setminus \{0\}$, it follows from (13.12) that

$$z = |z|\operatorname{cis}\alpha.$$

The coming example presents an interesting, albeit elementary, use of the notion of argument of a nonzero complex number.

Example 13.6 Among all complex numbers z for which $|z - 25i| \le 15$, find the one having smallest principal argument.

Solution The complex numbers satisfying the given condition are those situated on the closed disk bounded by the circle of center $25i$ and radius 15. Among these, the one with smallest principal argument, say z, is easily seen to be such that the half line starting at 0 and passing through z is tangent to the aforementioned circle in the first quadrant of the cartesian plane (see the figure below).

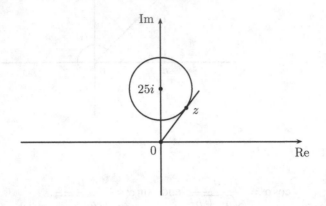

Since the radius of the circle is 15, Pythagoras' Theorem applied to the right triangle with vertices $25i$, z and 0 gives $|z| = 20$. Now, letting $z = a + bi$, we have that a equals the height of such a right triangle relative to its hypotenuse. Therefore, the usual metric relations on right triangles (cf. Chapter 4 of [9], for instance) give $25a = 15 \cdot 20$, so that $a = 12$. From $|z| = 20$ we get

$$20^2 = |z|^2 = a^2 + b^2 = 12^2 + b^2$$

and, since z belongs to the first quadrant, $b = 16$. Hence, $z = 12 + 16i$.   □

Back to the development of the theory, we now present the computational advantages of the polar form of nonzero complex numbers. We start with a simple, albeit very important result.

Proposition 13.7 *Given $\alpha, \beta \in \mathbb{R}$, we have*

$$\operatorname{cis} \alpha \cdot \operatorname{cis} \beta = \operatorname{cis}(\alpha + \beta). \qquad (13.14)$$

Proof It is enough to compute

$$
\begin{aligned}
\operatorname{cis} \alpha \cdot \operatorname{cis} \beta &= (\cos \alpha + i \sin \alpha)(\cos \beta + i \sin \beta) \\
&= (\cos \alpha \cos \beta - \sin \alpha \sin \beta) + i(\cos \alpha \sin \beta + \cos \beta \sin \alpha) \\
&= \cos(\alpha + \beta) + i \sin(\alpha + \beta) \\
&= \operatorname{cis}(\alpha + \beta),
\end{aligned}
$$

where in the third equality above we applied the usual addition formulas of Trigonometry (cf. Chapter 7 of [9], for instance).   □

Fig. 13.4 Geometric interpretation of the multiplication by $\operatorname{cis}\alpha$

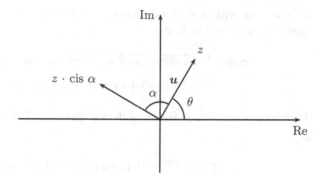

The coming corollary provides the usual geometric interpretation for the multiplication of complex numbers, one of which of modulus 1. For the general case, we refer the reader to Problem 1.

Corollary 13.8 *Let α be a given real number and $z \in \mathbb{C} \setminus \{0\}$. If $\boldsymbol{u}$ is the vector that goes from 0 to z, then the point of the complex plane representing $(\operatorname{cis}\alpha) \cdot z$ is the endpoint of the vector obtained from $\boldsymbol{u}$ by means of a* trigonometric rotation[4] *of center 0 and angle α (cf. Fig. 13.4).*

Proof Letting $z = |z| \operatorname{cis}\theta$, we have from (13.14) that

$$z \cdot \operatorname{cis}\alpha = |z| \operatorname{cis}\theta \cdot \operatorname{cis}\alpha = |z| \operatorname{cis}(\theta + \alpha).$$

Now, the description of the polar form assures that this last complex number is precisely the endpoint of the vector obtained from $\boldsymbol{u}$ by means of a trigonometric rotation of α radians. $\square$

Formula (13.15) below, known as the **first de Moivre's formula**, gives further insight on the computational advantages of the polar representation of nonzero complex numbers.

Proposition 13.9 (de Moivre) *If $z = |z| \operatorname{cis}\alpha$ is a nonzero complex number and $n \in \mathbb{Z}$, then*

$$z^n = |z|^n \operatorname{cis}(n\alpha). \tag{13.15}$$

Proof The case $n = 0$ is trivial. For the time being, assuming we have proved the formula for $n > 0$, let us show how to set its validity also for $n < 0$. To this end,

---

[4]Recall that this means we rotate $\boldsymbol{u}$ through an angle of α radians, in the counterclockwise sense if $\alpha > 0$ and in the clockwise sense if $\alpha < 0$.

let $n = -m$, with $m \in \mathbb{N}$. Given $\theta \in \mathbb{R}$, it follows from item (d) of Lemma 13.2, together with $|\operatorname{cis} \theta| = 1$, that

$$
\begin{aligned}
(\operatorname{cis} \theta)^{-1} &= \overline{\operatorname{cis} \theta} = \overline{\cos \theta + i \sin \theta} = \cos \theta - i \sin \theta \\
&= \cos(-\theta) + i \sin(-\theta) = \operatorname{cis}(-\theta).
\end{aligned}
\tag{13.16}
$$

Therefore, since we are assuming the validity of (13.15) with m in place of n, we get

$$
\begin{aligned}
z^n &= z^{-m} = (|z| \operatorname{cis} \alpha)^{-m} = |z|^{-m} (\operatorname{cis}(m\alpha))^{-1} \\
&= |z|^n \operatorname{cis}(-m\alpha) = |z|^n \operatorname{cis}(n\alpha).
\end{aligned}
$$

For $n > 0$, let us make induction on n, the case $n = 1$ being trivial. Assuming that the result holds true for some $n \in \mathbb{N}$ and invoking (13.14), we obtain

$$
\begin{aligned}
z^{n+1} &= z \cdot z^n = |z| \operatorname{cis} \alpha \cdot |z|^n \operatorname{cis}(n\alpha) \\
&= |z|^{n+1} \operatorname{cis} \alpha \cdot \operatorname{cis}(n\alpha) \\
&= |z|^{n+1} \operatorname{cis}(n+1)\alpha.
\end{aligned}
$$

$\square$

Let us now compute the arguments of the product and the quotient of two nonzero complex numbers.

Corollary 13.10 *If $z = |z| \operatorname{cis} \alpha$ and $w = |w| \operatorname{cis} \beta$ are nonzero complex numbers, then*

$$
zw = |zw| \operatorname{cis}(\alpha + \beta) \quad \text{and} \quad \frac{z}{w} = \frac{|z|}{|w|} \cdot \operatorname{cis}(\alpha - \beta).
$$

In particular, $\alpha + \beta$ (resp. $\alpha - \beta$) is the measure, in radians of an argument for zw (resp. for z/w).

Proof Let us do the proof for $\frac{z}{w}$, the other case being entirely analogous. To this end, it suffices to successively apply (13.16) and (13.14) to get

$$
\frac{z}{w} = |z| \operatorname{cis} \alpha \cdot |w|^{-1} \operatorname{cis}(-\beta) = \frac{|z|}{|w|} \cdot \operatorname{cis}(\alpha - \beta).
$$

$\square$

Given $n \in \mathbb{N}$ and $z \in \mathbb{C} \setminus \{0\}$, we say that a complex number w is an **n-th root** of z provided $w^n = z$. In contrast with what happens with real numbers, we shall see next that the number of n-th roots of each nonzero complex number is exactly n. Nevertheless, whenever there is no danger of confusion, we shall write simply $\sqrt[n]{z}$ to denote any such root. In any case, formula (13.17) below, known as the **second de Moivre's formula**, teaches us how to compute all of them.

Proposition 13.11 (de Moivre) *If $z = |z| \operatorname{cis} \alpha$ is a nonzero complex number and $n \in \mathbb{N}$, then there are exactly n distinct complex values for the n-th root of z. Moreover, such values are given by*

$$\sqrt[n]{|z|} \cdot \operatorname{cis}\left(\frac{\alpha + 2k\pi}{n}\right); \quad 0 \le k < n, \; k \in \mathbb{N}, \qquad (13.17)$$

where $\sqrt[n]{|z|}$ stands for the usual positive real root of $|z|$.

Proof Writing $w = r \operatorname{cis} \theta$, we have

$$w^n = z \Leftrightarrow (r \operatorname{cis} \theta)^n = |z| \operatorname{cis} \alpha$$

$$\Leftrightarrow r^n \operatorname{cis}(n\theta) = |z| \operatorname{cis} \alpha$$

$$\Leftrightarrow r^n = |z| \text{ and } n\theta = \alpha + 2k\pi, \, \exists \, k \in \mathbb{Z}.$$

In turn, these last two equalities take place if and only if $r = \sqrt[n]{|z|}$ and $\theta = \frac{\alpha + 2k\pi}{n}$, for some $k \in \mathbb{Z}$. Hence, there will be as many distinct n-th roots of z as there are distinct complex numbers of the form $\operatorname{cis}\left(\frac{\alpha + 2k\pi}{n}\right)$. Now, it is easy to see that

$$\operatorname{cis}\left(\frac{\alpha + 2k\pi}{n}\right) = \operatorname{cis}\left(\frac{\alpha + 2(k+n)\pi}{n}\right)$$

and

$$\operatorname{cis}\left(\frac{\alpha + 2k\pi}{n}\right) \neq \operatorname{cis}\left(\frac{\alpha + 2l\pi}{n}\right)$$

for $0 \le k < l < n$, so that is suffices to consider the integers k such that $0 \le k < n$.
□

In spite of the above formula, it is worth noticing (and will be clear in a few moments) that its major role is *theoretical*; in other words, it is not always the best way of actually computing the roots of a given complex number. This is due to the fact that not always the polar form of a nonzero complex number is actually useful in computations, as shown by the coming

Example 13.12 Compute the square roots of $7 + 24i$.

Solution If we try to use the second de Moivre's formula, we ought to start by writing $7 + 24i = 25 \operatorname{cis} \alpha$, with $\alpha = \arctan \frac{24}{7}$. Then, we invoke (13.17) to compute $\sqrt{7 + 24i}$ as being equal to either

$$5 \operatorname{cis}\left(\frac{\alpha}{2}\right) \text{ or } 5 \operatorname{cis}\left(\frac{\alpha + 2\pi}{2}\right).$$

We now get to the point: since $\alpha = \arctan \frac{24}{7}$, we shall have to compute $\cos\left(\frac{\alpha}{2}\right)$ and $\sin\left(\frac{\alpha}{2}\right)$ by knowing $\tan\alpha = \frac{24}{7}$. Although this can be done with the adequate trigonometric formulas, it is much simpler to take an alternative path, which we shall now describe.

Write $7 + 24i = (a + bi)^2$, with $a, b \in \mathbb{R}$, so that $a + bi$ is a square root of $7 + 24i$. Expanding $(a + bi)^2$ and equating real and imaginary parts in both sides of the equality thus obtained, we get the system of equations

$$\begin{cases} a^2 - b^2 = 7 \\ ab = 12 \end{cases}.$$

Squaring the second equation and substituting $a^2 = b^2 + 7$ in the result, we arrive at the biquadratic equation $(b^2 + 7)b^2 = 144$, so that $b^2 = 9$. Now, since $ab = 12 > 0$, we conclude that a and b must have equal signs, so that the possible ordered pairs (a, b) are $(a, b) = (3, 4)$ or $(-3, -4)$. Therefore, the square roots we are looking for are $\pm(3 + 4i)$. $\square$

We now present an interesting (and useful) *geometric interpretation* of the n-th roots of a nonzero complex number.

Proposition 13.13 *If z is a nonzero complex number and $n > 2$ is natural, then, in the complex plane, the n-th roots of z are the vertices of a regular n-gon with center at the origin.*

Proof Letting α be an argument of z, it follows from the second formula of de Moivre that the n-th roots of z are the complex numbers $z_0, z_1, \ldots, z_{n-1}$ such that

$$z_k = \sqrt[n]{|z|} \cdot \mathrm{cis}\left(\frac{\alpha + 2k\pi}{n}\right),$$

for $0 \le k < n$.

From (13.13), we obtain

$$|z_k| = \sqrt[n]{|z|} \left| \mathrm{cis}\left(\frac{\alpha + 2k\pi}{n}\right) \right| = \sqrt[n]{|z|},$$

so that all of the points z_k lie on the circle of the complex plane centered at 0 and having radius $\sqrt[n]{|z|}$.

On the other hand, it follows from Corollary 13.10 that

$$\frac{z_{k+1}}{z_k} = \mathrm{cis}\left(\frac{2\pi}{n}\right),$$

for $0 \le k < n$. Therefore, letting u_k denote the vector that goes from 0 to z_k, Corollary 13.8 assures that the angle between u_k and u_{k+1}, in radians and in the counterclockwise sense, is equal to $\frac{2\pi}{n}$.

The result follows immediately from these two facts. $\square$

Fig. 13.5 Geometric configuration of the fourth roots of -1

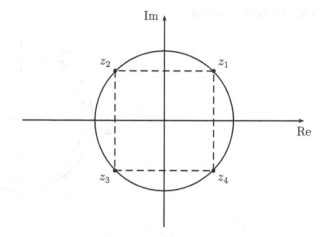

Example 13.14 Figure 13.5 represents the fourth roots of -1. Observe that $z_1 = \sqrt{|-1|}\text{cis}\,\frac{\pi}{4} = \frac{1+i}{\sqrt{2}}$.

As an important particular case of the discussion of roots of complex numbers, we say that the complex number ω is a **root of unity** if there exists a natural number n such that $\omega^n = 1$. In this case, ω is said to be an **n-th root of unity**.

Since $1 = \text{cis}\,0$, the second formula of de Moivre teaches us that the n-th roots of unity are the n complex numbers given by

$$\omega_k = \text{cis}\left(\frac{2k\pi}{n}\right); \ 0 \le k < n, \ k \in \mathbb{Z}. \tag{13.18}$$

Writing $\omega = \text{cis}\,\frac{2\pi}{n}$, it follows from the first formula of de Moivre that

$$\text{cis}\left(\frac{2k\pi}{n}\right) = \text{cis}\left(\frac{2\pi}{n}\right)^k = \omega^k.$$

Therefore, the n-th roots of unity are the complex numbers

$$1, \omega, \ldots, \omega^{n-1}. \tag{13.19}$$

As we shall see along the rest of this book, roots of unity are quite useful in a number of situations, both algebraic and combinatorial. For the time being, a clue of why this is so is provided by the following consequence of the previous proposition. In order to properly state it, from now on we shall refer to the circle of the complex plane with center 0 and radius 1 simply as the **unit circle**.

Corollary 13.15 *If $n > 2$ is a natural number, then the n-th roots of unity are the vertices of the regular n-gon inscribed in the unit circle and having 1 as one of its vertices.*

Fig. 13.6 Sixth roots of unity

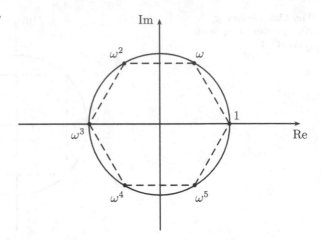

Proof Apply Proposition 13.13 to $z = 1$, noticing that 1 is also one of the roots we are looking at.   □

Example 13.16 In the notations of Fig. 13.6 we have $\omega = \text{cis } \frac{2\pi}{6} = \frac{1+i\sqrt{3}}{2}$, so that the complex numbers $1, \omega, \dots, \omega^5$ are the sixth roots of unity. Note also that $1, \omega^2$ and ω^4 are the cubic roots of unity, whereas $\omega, \omega^3 = -1$ and ω^5 are the cubic roots of -1.

We finish this section by presenting two examples involving roots of unity and which are somewhat more *algebraic* in nature.

Example 13.17 Given $n \in \mathbb{N}$, find all of the solutions of equation $(z - 1)^n = z^n$.

Solution Since $z = 0$ is not a solution, the given equation is equivalent to $\left(1 - \frac{1}{z}\right)^n = 1$. Hence, letting $\omega = \text{cis } \frac{2\pi}{n}$, the above discussion assures that $1 - \frac{1}{z}$ is an n-th root of unity, thus being equal to one of the numbers $1, \omega, \omega^2, \dots, \omega^{n-1}$.

Now, observe that $1 - \frac{1}{z} = 1$ has no solutions. On the other hand, for $1 \le k \le n - 1$ we have $\omega^k \ne 1$, so that $1 - \frac{1}{z} = \omega^k$ if and only if $z = \frac{1}{1-\omega^k}$. Thus, the solutions of the given equation are the $n - 1$ complex numbers

$$\frac{1}{1 - \omega}, \frac{1}{1 - \omega^2}, \dots, \frac{1}{1 - \omega^{n-1}}.$$

□

For the next example, recall (cf. discussion preceding Problem 7, page 326) that **a sequence of complex numbers** is a function $f : \mathbb{N} \to \mathbb{C}$, which, most often, will be simply denoted by $(z_n)_{n \ge 1}$, where $z_n = f(n)$. This being said, we observe that the proof of the formula for the sum of the first n terms of an arithmetic progression with common ratio different from 0 and 1 (cf. item (b) of Proposition 3.12 of [8], for instance) still holds, *ipsis literis*, for a *geometric progression of complex numbers*, i.e., a sequence $(z_n)_{n\geq 1}$ in $\mathbb{C}$ such that $z_{k+1} = qz_k$ for every $k \geq 1$, with $q \in \mathbb{C} \setminus \{0, 1\}$. Hence, we can safely skip the proof of the following auxiliary result.

Lemma 13.18 *For $z \in \mathbb{C} \setminus \{0, 1\}$, we have*

$$1 + z + z^2 + \cdots + z^{n-1} = \frac{z^n - 1}{z - 1}.$$

As a direct corollary of such a result, we note that if $\omega \neq 1$ is an n-th root of unity, then $\omega^n = 1$ and, hence,

$$1 + \omega + \omega^2 + \cdots + \omega^{n-1} = \frac{\omega^n - 1}{\omega - 1} = 0. \tag{13.20}$$

We shall use this formula several times along this text.

Example 13.19 (IMO) Prove that

$$\cos\frac{\pi}{7} - \cos\frac{2\pi}{7} + \cos\frac{3\pi}{7} = \frac{1}{2}.$$

Proof Letting $\omega = \operatorname{cis}\frac{2\pi}{7}$ be a seventh root of unity, we get

$$\operatorname{Re}(\omega + \omega^2 + \omega^3) = \cos\frac{2\pi}{7} + \cos\frac{4\pi}{7} + \cos\frac{6\pi}{7}$$

$$= \cos\frac{2\pi}{7} - \cos\frac{3\pi}{7} - \cos\frac{\pi}{7}.$$

On the other hand, since $\omega^k \cdot \omega^{7-k} = 1$ and $|\omega^k| = 1$, item (d) of Lemma 13.2 furnishes

$$\omega^{7-k} = (\omega^k)^{-1} = \overline{\omega^k}.$$

Therefore, $\overline{\omega + \omega^2 + \omega^3} = \omega^6 + \omega^5 + \omega^4$ and, hence,

$$\operatorname{Re}(\omega + \omega^2 + \omega^3) = \operatorname{Re}(\omega^6 + \omega^5 + \omega^4). \tag{13.21}$$

Now, (13.20) gives

$$\omega + \omega^2 + \cdots + \omega^6 = -1.$$

Taking real parts and using (13.21), we obtain

$$2\operatorname{Re}(\omega + \omega^2 + \omega^3) = \operatorname{Re}(\omega + \omega^2 + \omega^3) + \operatorname{Re}(\omega^6 + \omega^5 + \omega^4) = -1.$$

Thus, it follows from (13.21) that

$$\cos \frac{\pi}{7} - \cos \frac{2\pi}{7} + \cos \frac{3\pi}{7} = -\text{Re}(\omega + \omega^2 + \omega^3) = \frac{1}{2}.$$

$\square$

Problems: Sect. 13.2

1. For $r \in \mathbb{R} \setminus \{0\}$, we define the **homothety** of **center** 0 and **ratio** r as the function $H_r : \mathbb{C} \to \mathbb{C}$ such that $H_r(z) = rz$ for every $z \in \mathbb{C}$. For $\theta \in \mathbb{R} \setminus \{0\}$, we define the **rotation** of **center** 0 and **angle** θ as the function $R_\theta : \mathbb{C} \setminus \{0\} \to \mathbb{C} \setminus \{0\}$ given by $R_\theta(z) = (\text{cis}\,\theta)z$, for every $z \in \mathbb{C} \setminus \{0\}$.

   (a) Let u be the vector starting at 0 and having z as its endpoint. If $w = H_r(z)$, prove that ru is the vector starting at 0 and having endpoint w.
   (b) If $w = r \, \text{cis}\, \theta$ is a nonzero complex number, prove that for every $z \in \mathbb{C} \setminus \{0\}$ we have

$$wz = H_r \circ R_\theta(z).$$

2. (OCM) Let ω be a complex number satisfying $\omega^2 + \omega + 1 = 0$. Compute the value of the product

$$\prod_{k=1}^{27} \left(\omega^k + \frac{1}{\omega^k} \right).$$

3. Let $n \in \mathbb{N}$ be a multiple of 3. Compute the value of $(1 + \sqrt{3}i)^n - (1 - \sqrt{3}i)^n$.
4. Solve, in $\mathbb{C}$, the system of equations

$$\begin{cases} |z_1| = |z_2| = |z_3| = 1 \\ z_1 + z_2 + z_3 = 0 \\ z_1 z_2 z_3 = 1 \end{cases}.$$

5. Let $n \in \mathbb{N}$ and $a \in \mathbb{C} \setminus \{0\}$ be given, with $|a| > 1$. Show that all of the roots of the equation $(z - 1)^n = a(z + 1)^n$ lie on the circle of the complex plane centered at $\frac{1+R^2}{1-R^2}$ and having radius $\frac{2R}{R^2-1}$, where $R = \sqrt[n]{|a|}$.
6. (Ireland) For each $n \in \mathbb{N}$, let $a_n = n^2 + n + 1$. Given $k \in \mathbb{N}$, prove that there exists $m \in \mathbb{N}$ such that $a_{k-1}a_k = a_m$.
7. (Romania) Let $p, q \in \mathbb{C}$, with $q \neq 0$, be such that the second degree equation $x^2 + px + q^2 = 0$ has complex roots of equal modulus. Prove that $\frac{p}{q} \in \mathbb{R}$.

8. Let z_1, z_2 and z_3 be three given complex numbers, not all zero. Prove that z_1, z_2 and z_3 are the vertices of an equilateral triangle in the complex plane if and only if

$$z_1^2 + z_2^2 + z_3^2 = z_1 z_2 + z_2 z_3 + z_3 z_1.$$

9. Given an integer $n > 2$, use complex numbers to prove that

$$\sum_{k=0}^{n-1} \cos \frac{2k\pi}{n} = \sum_{k=0}^{n-1} \sin \frac{2k\pi}{n} = 0.$$

10. Given an integer $n > 2$ and $\alpha \in \mathbb{R}$, use complex numbers to compute each of the sums below in terms of n and α:

   (a) $\sin \alpha + \sin(2\alpha) + \sin(3\alpha) + \cdots + \sin(n\alpha)$.
   (b) $\sin^2 \alpha + \sin^2(2\alpha) + \sin^2(3\alpha) + \cdots + \sin^2(n\alpha)$.

11. (IMO—shortlist) Let $g : \mathbb{C} \to \mathbb{C}$ be a given function and $\omega = \operatorname{cis} \frac{2\pi}{3}$. For each given complex number a, show that there exists only one function $f : \mathbb{C} \to \mathbb{C}$ such that

$$f(z) + f(\omega z + a) = g(z), \forall z \in \mathbb{C}.$$

12. We are given a regular polygon $A_1 A_2 \ldots A_{2n}$ and a point P in the plane. Show that

$$\sum_{k=1}^{n} \overline{A_{2k} P}^2 = \sum_{k=1}^{n} \overline{A_{2k-1} P}^2.$$

13.3 More on Roots of Unity

We begin this section by illustrating the role of roots of unity as a counting tool. Although several applications of algebraic arguments in Combinatorics will appear along the rest of these notes, this is a much broader subject than we will be able to present here. For a more comprehensive introduction we refer the interested reader to the excellent classical book of Van Lint and Wilson [40].

We start by examining, in the coming example, how to use complex numbers as *tags*.

Example 13.20 We put $n > 1$ equally spaced lamps around a circle. At first, exactly one of then is ON. An allowed operation on the states of the lamps is to choose a positive divisor d of n, with $d < n$, and change the status of $\frac{n}{d}$ equally spaced lamps (i.e., from ON to OFF or vice-versa), provided all of them have the same status. Does there exist a sequence of allowed operations that leaves all of the n lamps ON?

Solution No! By contradiction, assume, without loss of generality, that the circle is the unit circle in the complex plane and the lamps are positioned at the points 1, ω, $\ldots$, ω^{n-1}, where $\omega = \operatorname{cis} \frac{2\pi}{n}$. Assume also (again with no loss of generality) that the lamp which is initially ON is the one standing at 1.

For every integer $k \geq 0$, let a_k denote the sum of the complex numbers associated to the lamps which are ON immediately before the $(k+1)$-th operation is performed, so that $a_0 = 1$. Along the $(k + 1)$-th operation, we choose lamps positioned at

$$\omega^l, \ \omega^{l+d}, \ \omega^{l+2d}, \ \ldots, \ \omega^{l+(\frac{n}{d}-1)d},$$

for some pair (l, d) of integers such that $0 \leq l < d$ and $d \mid n$, with $d < n$. Since $\omega^n = 1$, we have

$$
\begin{aligned}
a_{k+1} &= a_k \pm (\omega^l + \omega^{l+d} + \omega^{l+2d} + \cdots + \omega^{l+(\frac{n}{d}-1)d}) \\
&= a_k \pm \frac{\omega^{l+(\frac{n}{d}-1)d} \cdot \omega^d - \omega^l}{\omega^d - 1} = a_k,
\end{aligned}
$$

and it follows that $a_k = 1$ for every $k \geq 0$.

On the other hand, if after m operations all lamps were ON, then we would have

$$a_m = 1 + \omega + \cdots + \omega^n = 0,$$

which is a contradiction.   □

Now, given $m, n, p \in \mathbb{N}$, let us consider the problem of assembling an $m \times n$ rectangle by using rectangular pieces $1 \times p$. Counting 1×1 squares, we obtain an obvious necessary condition for the proposed assembling to be possible: p must divide mn. On the other hand, it is clear that if p divides either m or n, then the assembling is always possible: letting, for instance, $m = pk$, with $k \in \mathbb{N}$, it suffices to juxtapose k rectangles $p \times n$, each of which assembled by piling n pieces $1 \times p$, one on the top of the other. The interesting fact here is that the converse of the above discussion holds true, namely: one can assemble the desired rectangle *only if* p divides m or n. This is the content of the coming theorem, which is due to the American mathematician David Klarner.[5]

Theorem 13.21 (Klarner) *Let m, n and p be given naturals. If we can assemble an $m \times n$ rectangle by using pieces $1 \times p$, then $p \mid m$ or $p \mid n$.*

---

[5]Cf. Klarner and Göbel [24]. Nevertheless, the proof we present here is different from the original one.

Proof Assume we can perform such a task, and note first that $p \leq \max\{m, n\}$. Now, partition the rectangle into m rows $1 \times n$ and n columns $m \times 1$, numbering the rows from 1 to m from top to bottom, and the columns from 1 to n from left to right (as is usual in an $m \times n$ matrix). Then, inside the 1×1 square of intersection of row i and column j, write the complex number ω^{i+j}, where $\omega = \text{cis } \frac{2\pi}{p}$.

Let us now compute the sum of all of the numbers written in the 1×1 squares in two different ways (here we have a double counting!). On the one hand, such a sum is clearly equal to

$$(\omega + \omega^2 + \cdots + \omega^n)(\omega + \omega^2 + \cdots + \omega^m) = \frac{\omega^2(\omega^n - 1)(\omega^m - 1)}{(\omega - 1)^2}, \qquad (13.22)$$

where we have used Lemma 13.18 in the equality above; on the other, since we are assuming that it is possible to assemble the rectangle as wished, the numbers written on it can be grouped together in sets of p numbers each, corresponding to the p horizontal or vertical 1×1 squares forming each one of the $1 \times p$ rectangles used. Such sets of p complex numbers give rise to sums of one of the following types:

$$\omega^{i+j} + \omega^{i+(j+1)} + \cdots + \omega^{i+(j+p-1)}$$

or

$$\omega^{i+j} + \omega^{(i+1)+j} + \cdots + \omega^{(i+p-1)+j},$$

according to whether the $1 \times p$ rectangle is horizontally or vertically positioned, respectively. Again by Lemma 13.18, together with the first de Moivre's formula, each of such sums equals

$$\omega^{i+j}(1 + \omega + \cdots + \omega^{p-1}) = \omega^{i+j} \cdot \frac{\omega^p - 1}{\omega - 1} = 0.$$

Hence, the sum of all of the mn numbers written in the rectangle must also vanish. Therefore, it follows from (13.22) that $(\omega^n - 1)(\omega^m - 1) = 0$ or, which is the same, that

$$\text{cis } \frac{2n\pi}{p} = 1 \quad \text{or} \quad \text{cis } \frac{2m\pi}{p} = 1.$$

Finally, it is clear from these equalities that $p \mid n$ or $p \mid m$. $\qquad \square$

We devote the rest of this section to take a closer look at the structure of the set of n-th roots of unity. We start by letting $U_n(\mathbb{C})$ denote it, so that

$$U_n(\mathbb{C}) = \{z \in \mathbb{C};\ z^n = 1\} = \{\omega_n^k;\ 1 \leq k \leq n\},$$

where $\omega_n = \text{cis } \frac{2\pi}{n}$.

We begin by showing that the intersection of $U_m(\mathbb{C})$ and $U_n(\mathbb{C})$ is again a set of roots of unity.

Proposition 13.22 *Given $m, n \in \mathbb{N}$, one has*

$$U_m(\mathbb{C}) \cap U_n(\mathbb{C}) = U_d(\mathbb{C}),$$

where $d = \gcd(m, n)$. In particular, if m and n are relatively prime, then $U_m(\mathbb{C}) \cap U_n(\mathbb{C}) = \{1\}$.

Proof The second part follows immediately from the first. For the first part, if $z \in U_d(\mathbb{C})$, then $z^d = 1$ and, hence, $z^m = (z^d)^{m/d} = 1$. Analogously, $z^n = 1$, so that $z \in U_m(\mathbb{C}) \cap U_n(\mathbb{C})$. This reasoning shows that $U_d(\mathbb{C}) \subset U_m(\mathbb{C}) \cap U_n(\mathbb{C})$. Conversely, let $z \in U_m(\mathbb{C}) \cap U_n(\mathbb{C})$. Bézout's theorem assures the existence of $x, y \in \mathbb{Z}$ such that $mx + ny = d$. Therefore, since $z^m = z^n = 1$, we get

$$z^d = z^{mx+ny} = (z^m)^x (z^n)^y = 1,$$

so that $z \in U_d(\mathbb{C})$. Hence, $U_m(\mathbb{C}) \cap U_n(\mathbb{C}) \subset U_d(\mathbb{C})$, and we are done. $\square$

We now specialize to a particular type of root of unity, according to the definition below.

Definition 13.23 A complex number $z \in U_n(\mathbb{C})$ is a **primitive n-th root of unity** if

$$U_n(\mathbb{C}) = \{z^m; \ m \in \mathbb{N}\}.$$

It follows immediately from the definition that ω_n is a primitive n-th root of unity. On the other hand, the following result completely characterizes all of such special roots.

Theorem 13.24 *Given $z \in U_n(\mathbb{C})$, the following are equivalent:*

(a) z is primitive.
(b) There exists an integer $1 \le k \le n$ such that $\gcd(k, n) = 1$ and $z = \omega_n^k$.
(c) $\min\{l \in \mathbb{N}; \ z^l = 1\} = n$.

In particular, there are exactly $\varphi(n)$ primitive n-th roots of unity, where $\varphi : \mathbb{N} \to \mathbb{N}$ stands for the Euler function.

Proof

(a) $\Rightarrow$ (b):   let $z = \omega_n^k$, for some integer $1 \le k \le n$. If $\gcd(k, n) = d > 1$, then $n = dn'$ and $k = dk'$, for some $k', n' \in \mathbb{N}$. Also,

$$\{z^j; \ j \in \mathbb{N}\} = \{\omega_n^k, \omega_n^{2k}, \omega_n^{3k}, \ldots\}.$$

Letting $m = n'q + r$, with $q, r \in \mathbb{Z}$ and $0 \le r < n'$, we have

$$\omega_n^{mk} = \omega_n^{(n'q+r)dk'} = \omega_n^{nqk'} \cdot \omega_n^{rk} = \omega_n^{rk} \in \{1, \omega_n^k, \dots, \omega_n^{(n'-1)k}\}.$$

Therefore,

$$\#\{z^j; \ j \in \mathbb{N}\} \le n' = \frac{n}{d} < n.$$

(b) $\Rightarrow$ (c): let $z = \omega_n^k$, where $1 \le k \le n$ is an integer relatively prime with n. Take $l \in \mathbb{N}$ for which $z^l = 1$. Then, $\omega_n^{kl} = 1$ and, thus, $n \mid kl$. However, since $\gcd(k, n) = 1$, item (a) of Proposition 6.22 assures that $n \mid l$; in turn, this gives $l \ge n$. On the other hand, since $z^n = (\omega_n^k)^n = (\omega_n^n)^k = 1$, we conclude that $\min\{l \in \mathbb{N}; \ z^l = 1\} = n$.

(c) $\Rightarrow$ (a): assuming that $\min\{l \in \mathbb{N}; \ z^l = 1\} = n$, we first show that, for integers $1 \le r < s \le n$, we have $z^r \ne z^s$. Indeed, if we had $z^r = z^s$, then we would have $z^{s-r} = 1$, with $1 \le s - r < n$; but this would contradict the assumed minimality of n. It follows from this claim that $z, z^2, \dots, z^n$ are pairwise distinct n-th roots of unity, so that $\{z^j; \ j \in \mathbb{N}\} = U_n(\mathbb{C})$. Then, z is primitive. □

In the sequel, we show that primitive roots of unity provide another way to establish the multiplicative character of the Euler function. To this end, we need a preliminary result which is interesting in itself.

Theorem 13.25 *Let z and w be primitive roots of unity, of orders m and n, respectively. If $\gcd(m, n) = 1$, then zw is a primitive root of unity of order mn.*

Proof Since $z^m = w^n = 1$, we have $(zw)^{mn} = (z^m)^n(w^n)^m = 1$, so that $zw \in U_{mn}(\mathbb{C})$. Now, take $l \in \mathbb{N}$ such that $(zw)^l = 1$. Then,

$$z^l = (w^{-1})^l \in U_m(\mathbb{C}) \cap U_n(\mathbb{C}) = \{1\},$$

where we used Proposition 13.22 in the last passage above. Therefore, $z^l = (w^{-1})^l = 1$.

Now, the previous theorem, together with the fact that z and $w^{-1} = w^{n-1}$ are primitive roots of unity of orders m and n, respectively, guarantees that $l \mid m$ and $l \mid n$. However, since $\gcd(m, n) = 1$, item (e) of Proposition 6.22 gives $mn \mid l$. Hence, $l \ge mn$, so that the equivalences of the previous theorem imply that zw is a primitive root of unity of order mn. □

For our final result, we let $U_n(\mathbb{C})^\times$ denote the subset of $U_n(\mathbb{C})$ formed by the primitive roots of unity of order n:

$$U_n(\mathbb{C})^\times = \{\omega_n^k; \ 1 \le k \le n \text{ and } \gcd(k, n) = 1\}.$$

Theorem 13.26 *If m and n are relatively prime natural numbers, then the map*

$$U_m(\mathbb{C})^\times \times U_n(\mathbb{C})^\times \longrightarrow U_{mn}(\mathbb{C})^\times$$
$$(z, w) \longmapsto zw$$

is a well defined bijection. In particular, $\varphi(mn) = \varphi(m)\varphi(n)$, where $\varphi : \mathbb{N} \to \mathbb{N}$ stands for the Euler function.

Proof The well definiteness of the given map follows immediately from the previous result. On the other hand, if we show that it is a bijection, then two applications of the last part of Theorem 13.24 will furnish

$$\varphi(mn) = \left| U_{mn}(\mathbb{C})^\times \right| = \left| U_m(\mathbb{C})^\times \times U_n(\mathbb{C})^\times \right|$$
$$= \left| U_m(\mathbb{C})^\times \right| \cdot \left| U_n(\mathbb{C})^\times \right| = \varphi(m)\varphi(n).$$

We now show what is left, in two steps:

(i) The map is injective: take $z_1, z_2 \in U_m(\mathbb{C})^\times$ and $w_1, w_2 \in U_n(\mathbb{C})^\times$ such that $z_1 w_1 = z_2 w_2$. Since $\gcd(m, n) = 1$, we have

$$z_2^{-1} z_1 = w_2 w_1^{-1} \in U_m(\mathbb{C}) \cap U_n(\mathbb{C}) = \{1\}.$$

Therefore, $z_2^{-1} z_1 = w_2 w_1^{-1} = 1$, which is the same as $z_1 = z_2$ and $w_1 = w_2$.

(ii) The map is surjective: let $1 \le k \le mn$ be an integer such that $\gcd(k, mn) = 1$. We wish to find nonzero integers x and y such that $\gcd(x, n) = \gcd(y, m) = 1$ and, letting $z = \omega_m^y$, $w = \omega_n^x$, we have $zw = \omega_{mn}^k$ (why does this suffice?). To this end, we must have

$$\text{cis}\, \frac{2(mx + ny)\pi}{mn} = \text{cis}\, \frac{2k\pi}{mn},$$

which happens if $mx+ny = k$. However, since $\gcd(m, n) = 1$, Proposition 6.26 assures that such integers x and y do exist. $\square$

We shall have more to say on the combinatorial and algebraic aspects of roots of unity in Sects. 14.1 and 20.2, respectively.

Problems: Sect. 13.3

1. Give an example of an infinite set $\mathcal{T} \subset \mathbb{C}$ satisfying the following conditions:

   (a) For each $z \in \mathcal{T}$, there exists $n \in \mathbb{N}$ such that $z^n = 1$.
   (b) For all $z, w \in \mathcal{T}$, we have $zw \in \mathcal{T}$.

2. (Croatia) Let n be a given natural number and A be a set of n nonzero complex numbers having the following property: if any two of its elements (not necessarily distinct) are multiplied, we obtain another element of it as result. Find all such possible sets A.

3. * Show that, if $n > 2$, then the product of the primitive n-th roots of unity is equal to 1.

4. A rectangle is assembled by using a combination of pieces $1 \times m$ or $1 \times n$, where m and n are given natural numbers.

   (a) If $d = \gcd(m, n)$, prove that d divides at least one of the dimensions of the rectangle.
   (b) Prove that there exist arbitrarily large rectangles that not necessarily can be assembled by using only pieces $1 \times m$ or only pieces $1 \times n$.

5. We cover an 8×8 chessboard using twenty two pieces, with twenty one of them being 1×3 rectangles and one being an 1×1 square. Find all possible positions of the 1×1 square.

6. Prove the following extension of Klarner's theorem: given natural numbers m, n, p and q such that $1 < q \leq \min\{m, n, p\}$, we can assemble an $m \times n \times p$ parallelepiped by using pieces $1 \times 1 \times q$ if and only if q divides m, n or p.

7. We can further generalize Klarner's theorem as follows: given $m_1, \ldots, m_n \in \mathbb{N}$, let

$$A = I_{m_1} \times \cdots \times I_{m_n},$$

be the set of all sequences $(x_1, \ldots, x_n)$ of positive integers such that $1 \leq x_k \leq m_k$ for $1 \leq k \leq n$. A $(1, \ldots, 1, p)$-*block* in A is a subset of A formed by p sequences $(x_{11}, \ldots, x_{1n}), \ldots, (x_{p1}, \ldots, x_{pn})$ satisfying the following condition: there exists an integer $1 \leq k \leq n$ such that:

   (a) $x_{1j} = x_{2j} = \cdots = x_{pj}$ for every $1 \leq j \leq n$, $j \neq k$.
   (b) $(x_{1k}, x_{2k}, \ldots, x_{pk})$ is a sequence of p consecutive integers.

We say that A can be *partitioned into* $(1, \ldots, 1, p)$-*blocks* if A can be written as the disjoint union of such blocks. Prove that this is possible if and only if p divides one of the numbers $m_1, \ldots, m_n$.

8. (Russia) We wish to partition the set of four-digit natural numbers into subsets of four numbers each, in such a way that the following property is satisfied: the four number that compose each subset have the same digits in some three decimal places, whereas, in the fourth place, their digits are consecutive (for instance, a possibility for the four numbers of a subset could be 1265, 1275, 1285 and 1295). Prove that there does not exist such a partition.

Chapter 14
Polynomials

From a more *algebraic* point of view, real polynomial functions can be seen as *polynomials* with real coefficients. As we shall see from this chapter on, such a change of perspective turns out to be quite fruitful, so much that we shall not restrict ourselves to polynomials with real, or even complex, coefficients; later chapters will deal with the case of polynomials with coefficients in $\mathbb{Z}_p$, for some prime number p. As a result of such generality, we will be able to prove several results on Number Theory which would otherwise remain unaccessible. Our purpose in this chapter is, thus, to start this journey by developing the most elementary algebraic concepts and results on polynomials. To this end, along all that follows we shall write $\mathbb{K}$ to denote one of $\mathbb{Q}$, $\mathbb{R}$ or $\mathbb{C}$, whenever a specific choice of one of these number sets is immaterial.

14.1 Basic Definitions and Properties

In this section we collect the basic definitions and notations on polynomials which will be needed to the further developments of the theory.

Definition 14.1 A sequence $(a_0, a_1, a_2, \ldots)$ of elements of $\mathbb{K}$ is said to *vanish almost everywhere* if there exists an integer $n \geq -1$ such that

$$a_{n+1} = a_{n+2} = a_{n+3} = \cdots = 0.$$

In words, the sequence $(a_0, a_1, a_2, \ldots)$ vanishes almost everywhere if all of its terms, from a certain index on, are all equal to zero. For instance, both sequences

$$(0, 0, 0, 0, 0, 0, \ldots) \quad \text{and} \quad (1, 2, 3, \ldots, n, 0, 0, 0, \ldots)$$

© Springer International Publishing AG, part of Springer Nature 2018
A. Caminha Muniz Neto, *An Excursion through Elementary Mathematics, Volume III*,
Problem Books in Mathematics, https://doi.org/10.1007/978-3-319-77977-5_14

vanish almost everywhere; on the other hand, the sequence $(1, 0, 1, 0, 1, \ldots)$, with
1's and 0's indefinitely alternating themselves, obviously does not vanish almost
everywhere.

Definition 14.2 A **polynomial** on (or with coefficients in) $\mathbb{K}$ is a *formal sum*

$$f(X) = a_0 + a_1 X + a_2 X^2 + a_3 X^3 + \cdots := \sum_{k \geq 0} a_k X^k, \qquad (14.1)$$

where $(a_0, a_1, a_2, \ldots)$ is a sequence of elements of $\mathbb{K}$ vanishing almost everywhere
and we make the convention that $X^0 = 1$ and $X^1 = X$ in the above sum. Whenever
there is no danger of confusion, we shall simply write f, instead of $f(X)$.

Yet with respect to the above definition, it is worth noticing that the choice
of X is completely immaterial; in particular, it does not represent a *variable*, and
could have been substituted by whichever symbol we wanted, say $\square$, for instance.
Also, the correct interpretation of the word *formal* is that two polynomials $f(X) = \sum_{k \geq 0} a_k X^k$ and $g(X) = \sum_{k \geq 0} b_k X^k$ on $\mathbb{K}$ are **equal** if and only if $a_k = b_k$ for
every $k \geq 0$.

Given a polynomial $f(X) = a_0 + a_1 X + a_2 X^2 + a_3 X^3 + \cdots$ on $\mathbb{K}$, we shall
adopt the following conventions:

i. The elements $a_i \in \mathbb{K}$ are called the **coefficients** of f.
ii. If $a_i = 0$, we shall omit the term $a_i X^i$ whenever convenient. In particular,
since the sequence $(a_0, a_1, a_2, \ldots)$ vanishes almost everywhere, there exists an
integer $n \geq 0$ such that we can write

$$f(X) = \sum_{k=0}^{n} a_k X^k.$$

iii. If $a_i = \pm 1$, we shall write $\pm X^i$ in place of $(\pm 1) X^i$, for the corresponding term
of f.
iv. The polynomial $0 + 0X + 0X^2 + \cdots$ is said to be the **identically zero polynomial**
on $\mathbb{K}$. We shall also sometimes say that such a polynomial *vanishes identically*.
Furthermore, whenever there is no danger of confusion with $0 \in \mathbb{K}$, we shall
write 0 to denote the identically zero polynomial on $\mathbb{K}$.
v. More generally (and in consonance with items ii. and iv.), given $\alpha \in \mathbb{K}$ we shall
denote the polynomial $\alpha + 0X + 0X^2 + \cdots$ simply by α, and say that it is the
constant polynomial α. In each case, we shall always rely on the context to
make it clear whether we are referring to the constant polynomial α or to the
element $\alpha \in \mathbb{K}$.

We let $\mathbb{K}[X]$ denote the set of all polynomials on $\mathbb{K}$. In particular, in view of item
v. above, we also make the convention that

$$\mathbb{K} \subset \mathbb{K}[X].$$

On the other hand, the inclusions $\mathbb{Q} \subset \mathbb{R} \subset \mathbb{C}$ furnish the corresponding inclusions

$$\mathbb{Q}[X] \subset \mathbb{R}[X] \subset \mathbb{C}[X].$$

Example 14.3 If $f(X) = 1 + X - X^3 + \sqrt{2}X^7$, then $f \notin \mathbb{Q}[X]$ (for $\sqrt{2} \notin \mathbb{Q}$) but $f \in \mathbb{R}[X]$. On the other hand, the formal sum $g = 1 + X + X^2 + X^3 + X^4 + \cdots$ does not represent a polynomial, for the sequence $(1, 1, 1, \ldots)$ does not vanish almost everywhere.

In what follows, we shall use the ordinary operations of addition and multiplication on $\mathbb{K}$ to define on $\mathbb{K}[X]$ operations

$$\oplus : \mathbb{K}[X] \times \mathbb{K}[X] \to \mathbb{K}[X] \text{ and } \odot : \mathbb{K}[X] \times \mathbb{K}[X] \to \mathbb{K}[X],$$

also called **addition** and **multiplication**, respectively. To this end, we first need the following auxiliary result.

Lemma 14.4 *If* $(a_k)_{k \geq 0}$ *and* $(b_k)_{k \geq 0}$ *are sequences of elements of* $\mathbb{K}$ *that vanish almost everywhere, then so do the sequences* $(a_k \pm b_k)_{k \geq 0}$ *and* $(c_k)_{k \geq 0}$, *where*

$$c_k = \sum_{\substack{i+j=k \\ i,j \geq 0}} a_i b_j = \sum_{i=0}^{k} a_i b_{k-i}.$$

Proof We show that the sequence $(c_k)_{k \geq 0}$ vanishes almost everywhere, leaving the other case as an exercise for the reader (cf. Problem 1). Let $m, n \in \mathbb{Z}_+$ be such that $a_i = 0$ for $i > n$ and $b_j = 0$ for $j > m$. If $k > m + n$ and $i + j = k$, with $i, j \geq 0$, then either $i > n$ or $j > m$, for otherwise we would have $k = i + j \leq n + m$, which is not the case. However, since $i > n \Rightarrow a_i = 0$ and $j > m \Rightarrow b_j = 0$, in any event we have $a_i b_j = 0$, so that

$$c_k = \sum_{\substack{i+j=k \\ i,j \geq 0}} a_i b_j = 0.$$

$\square$

With the previous lemma at our disposal, we can finally formulate the definitions of addition and multiplication for polynomials.

Definition 14.5 Given in $\mathbb{K}[X]$ polynomials

$$f(X) = \sum_{k \geq 0} a_k X^k \text{ and } g(X) = \sum_{k \geq 0} b_k X^k,$$

the **sum** and **product** of f and g, respectively denoted by $f \oplus g$ and $f \odot g$, are the polynomials

$$(f \oplus g)(X) = \sum_{k \geq 0} (a_k + b_k) X^k$$

and

$$(f \odot g)(X) = \sum_{k \geq 0} c_k X^k,$$

with $c_k = \sum_{\substack{i+j=k \\ i,j \geq 0}} a_i b_j$.

Even though the definition of the product of two polynomials may not seen to be a natural one at a first glance, this is not the case. More precisely, the formula for the coefficient c_k of $f \odot g$ is the correct one if we wish $\odot$ to be distributive with respect to $\oplus$, as well as the usual rule for powers, $X^m \odot X^n = X^{m+n}$, to be valid. Indeed, if such properties hold, then computing the product

$$(a_0 + a_1 X + a_2 X^2 + \cdots) \odot (b_0 + b_1 X + b_2 X^2 + \cdots)$$

distributively, we obtain

$$a_0 b_0 = \sum_{\substack{i+j=0 \\ i,j \geq 0}} a_i b_j$$

for the coefficient of X^0,

$$a_0 b_1 + a_1 b_0 = \sum_{\substack{i+j=1 \\ i,j \geq 0}} a_i b_j$$

for the coefficient of X,

$$a_0 b_2 + a_1 b_1 + a_2 b_0 = \sum_{\substack{i+j=2 \\ i,j \geq 0}} a_i b_j$$

for the coefficient of X^2 and so on.

Actually, it is not difficult to convince ourselves (cf. Problem 4) that, as defined, the operations of addition and multiplication of polynomials on $\mathbb{K}$ have the following properties:

  i. **commutativity**: $f \oplus g = g \oplus f$ and $f \odot g = g \odot f$;
 ii. **associativity**: $(f \oplus g) \oplus h = f \oplus (g \oplus h)$ and $(f \odot g) \odot h = f \odot (g \odot h)$;
iii. **distributivity**: $f \odot (g \oplus h) = (f \odot g) \oplus (f \odot h)$,

for all $f, g, h \in \mathbb{K}[X]$.

Example 14.6 Consider the polynomials with real coefficients $f(X) = 1 + X - \sqrt{2}X^2 - 4X^3$ and $g(X) = X + X^2$, where, according to our previous conventions, we omit terms with vanishing coefficients. Then,

$$(f \oplus g)(X) = 1 + 2X + (1 - \sqrt{2})X^2 - 4X^3$$

and

$$
\begin{aligned}
(f \odot g)(X) &= (1 + X - \sqrt{2}X^2 - 4X^3) \odot (X + X^2) \\
&= [1 \odot (X + X^2)] \oplus [X \odot (X + X^2)] \oplus \\
&\quad \oplus [-\sqrt{2}X^2 \odot (X + X^2)] \oplus [-4X^3 \odot (X + X^2)] \\
&= (X + X^2) \oplus (X^2 + X^3) \oplus (-\sqrt{2}X^3 - \sqrt{2}X^4) \oplus \\
&\quad \oplus (-4X^4 - 4X^5) \\
&= X + 2X^2 + (1 - \sqrt{2})X^3 - (\sqrt{2} + 4)X^4 - 4X^5.
\end{aligned}
$$

Our discussion up to this point makes it clear that we can relax notations, henceforth writing simply $+$ and $\cdot$ to denote the operations of addition and multiplication of polynomials. Thus, whenever we add two polynomials, the reader must be aware of the fact that the involved $+$ signs stand for two distinct operations: addition of elements of $\mathbb{K}$, performed on the coefficients of the polynomials, and addition of elements of $\mathbb{K}[X]$. Nevertheless, this will cause no confusion, for the context will always make it clear which operation the $+$ sign refers to. Moreover, an analogous comment holds for the $\cdot$ sign and the two operations of multiplication.

Remark 14.7 All of the above definitions and remarks can be extended, in an obvious way, to polynomials with *integer* coefficients. From now on, whenever necessary, we shall denote the set of such polynomials by $\mathbb{Z}[X]$.

Letting 0 be the identically vanishing polynomial, we have $f + 0 = 0 + f = 0$ for every $f \in \mathbb{K}[X]$, i.e, 0 is the **identity for addition** of polynomials. On the other hand, given $f \in \mathbb{K}[X]$, there exists a single $g \in \mathbb{K}[X]$ for which $f + g = g + f = 0$; indeed, letting $f(X) = a_0 + a_1 X + a_2 X^2 + \cdots$, it is immediate that $g(X) = -a_0 - a_1 X - a_2 X^2 - \cdots$ is such polynomial, which will henceforth be denoted by $-f$. Thus,

$$(-f)(X) = -a_0 - a_1 X - a_2 X^2 - \cdots$$

and, for $f, g \in \mathbb{K}[X]$, we can define the **difference** $f - g$ between f and g by $f - g = f + (-g)$.

For $\alpha \in \mathbb{K}$ and $g(X) = b_0 + b_1 X + b_2 X^2 + \cdots \in \mathbb{K}[X]$, it is immediate to verify that

$$\alpha \cdot g = \alpha b_0 + \alpha b_1 X + \alpha b_2 X^2 + \cdots ;$$

in particular, we have $1 \cdot g = g$ for every $g \in \mathbb{K}[X]$, so that the constant polynomial 1 is the **identity for multiplication** of polynomials.

From now on, whenever there is no danger of confusion, we shall write simply fg, instead of $f \cdot g$, to denote the product of $f, g \in \mathbb{K}[X]$. In particular, if $\alpha \in \mathbb{K}$, then αf will stand for the product of α, seen as a constant polynomial, and f.

The coming definition will play a major role along the rest of the book.

Definition 14.8 Given $f(X) = a_0 + a_1 X + \cdots + a_n X^n \in \mathbb{K}[X] \setminus \{0\}$, with $a_n \neq 0$, the nonnegative integer n is said to be the **degree** of f, and we will write $\partial f = n$ to denote this.

According to the above definition, the notion of degree only applies to nonvanishing polynomials. On the other hand, $\partial f = 0$ for every constant polynomial $f(X) = \alpha$, with $\alpha \in \mathbb{K} \setminus \{0\}$. From now on, whenever we refer to $f \in \mathbb{K}[X] \setminus \{0\}$ by writing

$$f(X) = a_n X^n + \cdots + a_1 X + a_0,$$

unless explicitly stated otherwise we shall tacitly assume that $a_n \neq 0$. In such a case, a_n will be called the **leading coefficient** of f. Also, $f \in \mathbb{K}[X] \setminus \{0\}$ is said to be **monic** if it has leading coefficient equal to 1. Finally, in the above notations a_0 is said to be the **constant term** of f.

The coming proposition establishes two very important properties of the notion of degree of polynomials.

Proposition 14.9 *For $f, g \in \mathbb{K}[X] \setminus \{0\}$, we have:*

(a) $\partial(f + g) \leq \max\{\partial f, \partial g\}$ *if* $f + g \neq 0$.
(b) $fg \neq 0$ *and* $\partial(fg) = \partial f + \partial g$.

Proof Let $\partial f = n$ and $\partial g = m$, with

$$f(X) = a_0 + a_1 X + \cdots + a_n X^n \text{ and } g(X) = b_0 + b_1 X + \cdots + b_m X^m.$$

(a) If $m \neq n$, we can assume, without any loss of generality, that $m > n$. Then

$$(f + g)(X) = (a_0 + b_0) + \cdots + (a_n + b_n)X^n + b_{n+1}X^{n+1} + \cdots + b_m X^m,$$

so that $\partial(f + g) = m = \max\{\partial f, \partial g\}$. If $m = n$ but $f + g \neq 0$, then

$$(f + g)(X) = (a_0 + b_0) + \cdots + (a_n + b_n)X^n,$$

and there are two possibilities: $a_n + b_n = 0$ or $a_n + b_n \neq 0$. In the first case, $\partial(f + g) < n = \max\{\partial f, \partial g\}$. In the second, $\partial(f + g) = n = \max\{\partial f, \partial g\}$. In any case, we have $\partial(f + g) \leq \max\{\partial f, \partial g\}$.

(b) Let $fg = c_0 + c_1 X + c_2 X^2 + \cdots$. If $k > m + n$, we saw in the proof of Lemma 14.4 that $c_k = 0$. Hence, if we show that $c_{m+n} \neq 0$, we will conclude that $fg \neq 0$ and $\partial(fg) = m + n = \partial f + \partial g$. For what is left to do, since $a_i = 0$ for $i > n$ and $b_j = 0$ for $j > m$, it is immediate that

$$c_{m+n} = \sum_{\substack{i+j=m+n \\ i,j \geq 0}} a_i b_j = a_n b_m \neq 0.$$

$\square$

In spite of the fact that the material developed so far is rather elementary, it is enough to present yet another use of roots of unity as tags (cf. Sect. 13.3). To this end, given $f(X) = a_n X^n + \cdots + a_1 X + a_0 \in \mathbb{C}[X]$ and $z \in \mathbb{C}$, it is convenient that we write $f(z)$ to denote the complex number

$$f(z) = a_n z^n + \cdots + a_1 z + a_0. \tag{14.2}$$

(We shall have more to say on this *substitution of X by z* in Sect. 15.1—cf. Definition 15.1.)

Formula (14.3) below is known in mathematical literature as the **multisection formula**.

Theorem 14.10 *For $f \in \mathbb{C}[X] \setminus \{0\}$, let a_k denote the coefficient of X^k in f. If n is a natural number, $\omega = \mathrm{cis}\,\frac{2\pi}{n}$ and $z \in \mathbb{C}$, then*

$$\sum_{n|k} a_k z^k = \frac{1}{n}(f(z) + f(\omega z) + \cdots + f(\omega^{n-1} z)). \tag{14.3}$$

Proof Since $f(X) = \sum_{k \geq 0} a_k X^k$, we have

$$\begin{aligned}
\sum_{j=0}^{n-1} f(\omega^j z) &= \sum_{j=0}^{n-1} \sum_{k \geq 0} a_k \omega^{jk} z^k = \sum_{k \geq 0} \sum_{j=0}^{n-1} a_k \omega^{jk} z^k \\
&= \sum_{k \geq 0} \left(a_k z^k \sum_{j=0}^{n-1} \omega^{jk} \right).
\end{aligned} \tag{14.4}$$

Now, note that if $n \mid k$ then $\omega^{jk} = 1$ for every $j \in \mathbb{Z}$, so that $\sum_{j=0}^{n-1} \omega^{jk} = n$. On the other hand, if $n \nmid k$, then $\omega^k \neq 1$ and, hence,

$$\sum_{j=0}^{n-1} \omega^{jk} = 1 + \omega^k + \cdots + \omega^{(n-1)k} = \frac{\omega^{nk} - 1}{\omega^k - 1} = 0.$$

Therefore, (14.4) reduces to

$$\sum_{j=0}^{n-1} f(\omega^j z) = \sum_{n|k} na_k z^k.$$

□

The coming example relies on the fact that, given $a, b \in \mathbb{K}$, the binomial formula can be used to expand $(aX + b)^n \in \mathbb{K}[X]$ in monomials. The reader is invited to check this.

Example 14.11 (Poland) For each $n \in \mathbb{N}$, compute, in terms of n, the value of the sum

$$\sum_{3|k} \binom{n}{k}.$$

Solution Let

$$f(X) = (X + 1)^n = \sum_{k=0}^{n} \binom{n}{k} X^k.$$

If $\omega = \operatorname{cis} \frac{2\pi}{3}$, then ω is a cubic root of unity and $1 + \omega + \omega^2 = 0$. By applying the multisection formula to f, we get

$$\sum_{3|k} \binom{n}{k} = \frac{1}{3}\left(f(1) + f(\omega) + f(\omega^2)\right)$$

$$= \frac{1}{3}\left(2^n + (\omega + 1)^n + (\omega^2 + 1)^n\right)$$

$$= \frac{1}{3}\left(2^n + (-\omega^2)^n + (-\omega)^n\right)$$

$$= \frac{1}{3}\left(2^n + (-1)^n(\omega^{2n} + \omega^n)\right).$$

Now, note that if $3 \mid n$, then $\omega^n = 1$ and, hence, $\omega^{2n} + \omega^n = 2$; if $3 \nmid n$, then $\omega^n = \omega$ or ω^2, so that $\omega^{2n} + \omega^n = \omega^2 + \omega = -1$. Therefore,

$$\sum_{3|k} \binom{n}{k} = \begin{cases} (2^n + 2(-1)^n)/3, & \text{if } 3 \mid n \\ (2^n + (-1)^{n+1})/3, & \text{if } 3 \nmid n \end{cases}.$$

□

Problems: Sect. 14.1

1. * If $(a_k)_{k \geq 0}$ and $(b_k)_{k \geq 0}$ are almost everywhere vanishing sequences of elements of $\mathbb{K}$, show that the sequences $(a_k \pm b_k)_{k \geq 0}$ also vanish almost everywhere.

2. * Given $n \in \mathbb{N}$ and $a \in \mathbb{K}$, perform the following products of polynomials:

   (a) $(X - a)(X^{n-1} + aX^{n-2} + \cdots + a^{n-2}X + a^{n-1})$.
   (b) $(X + a)(X^{n-1} - aX^{n-2} + \cdots - a^{n-2}X + a^{n-1})$, if n is odd.
   (c) $(X + 1)(X^2 + 1)(X^4 + 1)\ldots(X^{2^n} + 1)$.

3. Compute the coefficient of X^p on both sides of $(X+1)^m(X+1)^n = (X+1)^{m+n}$ to give another proof of Vandermonde's identity[1] (cf. Problem 3, page 48):

$$\sum_k \binom{m}{k}\binom{n}{p-k} = \binom{m+n}{p},$$

   with the sum at the left hand side above ranging through all possible nonnegative integer values of k.

4. * Show that the operations of addition and multiplication of polynomials are commutative and associative, and that multiplication is distributive with respect to addition.

5. * Show that $\mathbb{K}[X]$ does not possess any *zero divisors*. More precisely, show that if $f, g \in \mathbb{K}[X]$ are such that $fg = 0$, then $f = 0$ or $g = 0$.

6. (TT) Find at least one polynomial f, of degree 2001, such that $f(X) + f(1 - X) = 1$.

7. Let $n \in \mathbb{N}$ and, for $0 \leq k \leq 2n$, let a_k be the coefficient of X^k in the expansion of $f(X) = (1 + X + X^2)^n$. Compute, in terms of n, the value of the sum

$$a_3 + a_6 + a_9 + \cdots.$$

8. * Generalize the multisection formula in the following way[2]: given $f \in \mathbb{C}[X] \setminus \{0\}$, a natural number $n > 1$ and an integer $0 \leq r \leq n - 1$, if $\omega = \operatorname{cis} \frac{2\pi}{n}$ and $z \in \mathbb{C}$, then

$$\sum_{k \equiv r \,(\mathrm{mod}\, n)} a_k z^k = \frac{1}{n} \sum_{j=0}^{n-1} \omega^{-rj} f(\omega^j z),$$

   where a_k denotes the coefficient of X^k in f.

---

[1] Alexandre-Theóphile Vandermonde, French mathematician of the eighteenth century.
[2] For another proof of this result, see Example 18.7.

9. Compute, as a function of $m \in \mathbb{N}$, the value of the sum

$$\binom{m}{1} + 4\binom{m}{5} + 16\binom{m}{9} + 64\binom{m}{13} + \cdots.$$

For the coming problem, given $f(X) = a_n X^n + a_{n-1} X^{n-1} + \cdots + a_1 X + a_0 \in \mathbb{K}[X]$ and $m \in \mathbb{N}$, we write $f(X^m)$ to denote the polynomial in $\mathbb{K}[X]$ defined by

$$a_n X^{mn} + a_{n-1} X^{m(n-1)} + \cdots + a_1 X^m + a_0,$$

i.e., obtained by writing X^m in place of X in the expression of f.

10. (Miklós-Schweitzer) A polynomial with real coefficients is called *positively reducible* if it can be written as the product of two nonconstant polynomials whose nonvanishing coefficients are positive. Let $f \in \mathbb{R}[X]$ have nonvanishing constant term. If $f(X^m)$ is positively reducible for some natural number m, prove that f itself is positively reducible.

14.2 The Division Algorithm

We have already seen that it is possible to add, subtract and multiply polynomials, the result being another polynomial. And what about the possibility of *dividing* one polynomial for another? Well, it will not always be possible to do so; in other words, given polynomials $f, g \in \mathbb{K}[X] \setminus \{0\}$, there will not always exist $h \in \mathbb{K}[X]$ such that $f = gh$; for instance, if $\partial f < \partial g$, then such a polynomial h does not exist, for otherwise we would have

$$\partial f = \partial (gh) = \partial g + \partial h \geq \partial g,$$

which is an absurd.

In spite of the above discussion, the following analogue of the division algorithm for integers, called the **division algorithm for polynomials**, still holds.

Theorem 14.12 *If $f, g \in \mathbb{K}[X]$, with $g \neq 0$, then there exist unique $q, r \in \mathbb{K}[X]$ such that*

$$f = gq + r, \text{ with } r = 0 \text{ or } \partial r < \partial g. \tag{14.5}$$

Proof Let us first show that there exists at most one pair of polynomials $q, r \in \mathbb{K}[X]$ satisfying the stated conditions. To this end, let $q_1, q_2, r_1, r_2 \in \mathbb{K}[X]$ be such that

$$f = gq_1 + r_1 = gq_2 + r_2,$$

with $r_i = 0$ or $0 \le \partial r_i < \partial g$, for $i = 1, 2$. Then, $g(q_1 - q_2) = r_2 - r_1$ and, if $q_1 \ne q_2$, Problem 5, page 355, would guarantee that $r_1 \ne r_2$. However, Proposition 14.9 would then give

$$\partial g \le \partial g + \partial(q_1 - q_2) = \partial(g(q_1 - q_2))$$
$$= \partial(r_1 - r_2) \le \max\{\partial r_1, \partial r_2\} < \partial g,$$

which is an absurd. Hence, $q_1 = q_2$ and, then, $r_1 = r_2$.

Let us now establish the existence of polynomials q and r satisfying the stated conditions. To this end, we consider two cases separately:

(i) If $f = 0$, take $q = r = 0$; if g is (a nonzero) constant, say c, take $q = c^{-1}f$ and $r = 0$.
(ii) If $f \ne 0$ and g is nonconstant, we shall make induction on $\partial f \ge 0$.

As initial case, if $\partial f < \partial g$ (such an f does exist, for $\partial g \ge 1$), take $q = 0$ and $r = f$. By induction hypothesis, let $m \ge \partial g$ be an integer and assume that the existence of q and r has been established for any polynomial of degree less than m in place of f. In order to establish the induction step, take $f \in \mathbb{K}[X] \setminus \{0\}$ of degree m, and let $b \ne 0$ denote the leading coefficient and n the degree of g. If a stands for the leading coefficient of f, there are two distinct possibilities:

- $f(X) = ab^{-1}X^{m-n}g(X)$: in this case, $q(X) = ab^{-1}X^{m-n}$ and $r(X) = 0$.
- $f(X) \ne ab^{-1}X^{m-n}g(X)$: letting $h(X) = f(X) - ab^{-1}X^{m-n}g(X)$, we have $h \ne 0$ and

$$h(X) = (aX^m + \cdots) - ab^{-1}X^{m-n}(bX^n + \cdots),$$

so that $\partial h < m$. Therefore, by induction hypothesis, there exists $q_1, r_1 \in \mathbb{K}[X]$ such that $h = gq_1 + r_1$, with $r_1 = 0$ or $\partial r_1 < \partial g$. Thus,

$$f(X) = h(X) + ab^{-1}X^{m-n}g(X)$$
$$= g(X)q_1(X) + r_1(X) + ab^{-1}X^{m-n}g(X)$$
$$= g(X)(q_1(X) + ab^{-1}X^{m-n}) + r_1(X),$$

and it suffices to take $q(X) = q_1(X) + ab^{-1}X^{m-n}$ and $r(X) = r_1(X)$.

$\square$

In the notations of the division algorithm for polynomials (and by analogy with the division algorithm for integers), we say that q is the **quotient** and r is the **remainder** of the division of f by g. Moreover, when $r = 0$ we say that f is **divisible** by g, or that g **divides** f; in this case, as with integers we denote $g \mid f$.

Remark 14.13 As the reader can easily verify by checking the previous proof, the division algorithm still holds for polynomials in $\mathbb{Z}[X]$, provided the leading coefficient of g is equal to ± 1. More precisely, if $f, g \in \mathbb{Z}[X]$, with $g \neq 0$ having leading coefficient equal to ± 1, then there exists unique polynomials $q, r \in \mathbb{Z}[X]$ such that

$$f = gq + r, \text{ with } r = 0 \text{ or } \partial r < \partial g.$$

In all that follows, we shall use this without further comments.

In spite of the name *division algorithm*, the existence part of the presented proof for Theorem 14.12 is actually not *algorithmic*, i.e., it does not provide an actual *algorithm* for finding q and r out of f and g. In order to remedy this, we now revisit the existence part of the proof in a more algorithmic way:

Division Algorithm for Polynomials

1. DO
   - $r \leftarrow f; m \leftarrow \partial f; q \leftarrow 0$;
   - $a \leftarrow$ LEADING COEFFICIENT OF r;

2. WHILE $r \neq 0$ OR $\partial r \geq \partial g$ DO
   - $r(X) \leftarrow r(X) - ab^{-1}X^{m-n}g(X)$;
   - $q(X) \leftarrow q(X) + ab^{-1}X^{m-n}$;
   - $m \leftarrow \partial r$
   - $a \leftarrow$ LEADING COEFFICIENT OF r;

3. READ THE FINAL VALUES OF r AND q.

Let us prove that the above algorithm does stop after a finite number of repetitions of the loop WHILE and give us, at the end, polynomials q and r as wished.

If $r \neq 0$ or $\partial r \geq \partial g$, then the instruction of the WHILE loop substitutes $r(X)$ by the polynomial $r(X) - ab^{-1}X^{m-n}g(X)$. Such a polynomial, if it is not identically zero, has degree less than that of r, for the polynomial $ab^{-1}X^{m-n}g(X)$ has degree m (which is precisely the degree of r immediately before the execution of the loop) and leading coefficient a (which is the leading coefficient of r right before the execution of the loop). Thus, the algorithm stops after a finite number of steps.

After the first attribution, we have

$$f(X) = g(X)q(X) + r(X),$$

for, at the beginning, $q(X) = 0$ and $r(X) = f(X)$. Assume, by induction hypothesis, that after a certain execution of the WHILE loop we have $f(X) = g(X)q(X) + r(X)$ and that, at this moment, $r(X) \neq 0$ and $\partial r \geq \partial g$. Then, the next step does take place and replaces $r(X)$ by $r(X) - ab^{-1}X^{m-n}g(X)$, and $q(X)$ by $q(X) + ab^{-1}X^{m-n}$. Now, since

$$g(X)\left(q(X) + ab^{-1}X^{m-n}\right) + \left(r(X) - ab^{-1}X^{m-n}g(X)\right) \qquad (14.6)$$

equals $g(X)q(X) + r(X)$, it follows from induction hypothesis that, after such additional execution, we will also have (14.6) equal to $f(X)$.

Example 14.14 Follow the algorithm presented above to obtain the quotient and the remainder upon dividing $f(X) = X^4 - 2X^2 + 5X + 7$ by $g(X) = 3X^2 + 1$ in $\mathbb{Q}[X]$.

Solution Following the notations of the previous discussion, we start by storing $b = 3$ and $n = 2$. Then, we have the following steps:

1. We make the attributions $r(X) = X^4 - 2X^2 + 5X + 7$; $m = 4$; $q(X) = 0$; $a = 1$.
2. Since neither $r(X) = 0$ nor $m = 4 < n = 2$, we substitute $r(X)$ by

$$r(X) - ab^{-1}X^{m-n}g(X) =$$

$$= X^4 - 2X^2 + 5X + 7 - \frac{1}{3}X^{4-2}(3X^2 + 1)$$

$$= -\frac{7}{3}X^2 + 5X + 7,$$

m by 2, $q(X)$ by

$$q(X) + ab^{-1}X^{m-n} = 0 + \frac{1}{3}X^{4-2} = \frac{1}{3}X^2$$

and a by $-\frac{7}{3}$.
3. Since we still do not have $r(X) = 0$ or $m = 2 < n = 2$, we replace $r(X)$ by

$$r(X) - ab^{-1}X^{m-n}g(X) =$$

$$= -\frac{7}{3}X^2 + 5X + 7 + \frac{7/3}{3}X^{2-2}(3X^2 + 1)$$

$$= 5X + \frac{70}{9},$$

m by 1, $q(X)$ by

$$q(X) + ab^{-1}X^{m-n} = \frac{1}{3}X^2 + \frac{-7/3}{3}X^{2-2} = \frac{1}{3}X^2 - \frac{7}{9}$$

and a by $\frac{1}{3}$.
4. Now, although $r(X) \neq 0$, we have $m = 1 < n = 2$, so that we do not return to the beginning of the loop. Instead, we simply read

$$r(X) = 5X + \frac{70}{9} \quad \text{and} \quad q(X) = \frac{1}{3}X^2 - \frac{7}{9}.$$

$\square$

The table below looks at the solution of the previous example *schematically*. In order to help the reader in grasping how steps 1. to 4. actually took place, we suggest that he/she tries to identify each of those steps within the table.

Division Algorithm for Polynomials

$$
\begin{array}{rrr|l}
X^4 & -2X^2 +5X & +7 & 3X^2 + 1 \\
\hline
-X^4 & -1/3X^2 & & 1/3X^2 \\
\hline
& -7/3X^2 +5X & +7 & 1/3X^2 - 7/9 \\
\hline
& 7/3X^2 & +7/9 & \\
\hline
& & 5X +70/9 &
\end{array}
$$

In the coming example, given $f(X) = a_n X^n + \cdots + a_1 X + a_0 \in \mathbb{K}[X]$ and $z \in \mathbb{K}$, we stick to the convention set in (14.2), of writing $f(\alpha)$ to denote the element

$$f(\alpha) = a_n \alpha^n + \cdots + a_1 \alpha + a_0 \in \mathbb{K}.$$

As already quoted there, we refer to Sect. 15.1 for more on such *substitution of X by α*.

Example 14.15 Given $f(X) = a_n X^n + \cdots + a_1 X + a_0 \in \mathbb{K}[X]$ and $\alpha \in \mathbb{K}$, show that the remainder upon division of $f(X)$ by $X - \alpha$ is precisely $f(\alpha)$.

Proof We want to show that there exists $q \in \mathbb{K}[X]$ such that

$$f(X) = (X - \alpha)q(X) + f(\alpha).$$

Since this is the same as having $f(X) - f(\alpha) = (X - \alpha)q(X)$, we only need to show that $X - \alpha$ divides $f(X) - f(\alpha)$. We do this by invoking the result of item (a) of Problem 2, page 355, according to which

$$X^k - \alpha^k = (X - \alpha)(X^{k-1} + \alpha X^{k-2} + \cdots + \alpha^{k-2}X + \alpha^{k-1}),$$

for every $k \in \mathbb{N}$. Thus,

$$f(X) - f(\alpha) = \sum_{k=0}^{n} a_k X^k - \sum_{k=0}^{n} a_k \alpha^k = \sum_{k=0}^{n} a_k (X^k - \alpha^k)$$

$$= \sum_{k=0}^{n} a_k (X - \alpha)(X^{k-1} + \alpha X^{k-2} + \cdots + \alpha^{k-2}X + \alpha^{k-1})$$

$$= (X - \alpha) \sum_{k=0}^{n} a_k (X^{k-1} + \alpha X^{k-2} + \cdots + \alpha^{k-2}X + \alpha^{k-1}).$$

$\square$

Problems: Sect. 14.2

1. Given $n \in \mathbb{N}$, find the remainder upon division of $(X^2 + X + 1)^n$ by $X^2 - X + 1$.
2. We are given natural numbers m and n, such that n is a divisor of m. Find the remainder upon division of $X^m + 1$ by $X^n - 1$.
3. Given natural numbers m and n, with $m > n$, find the remainder of the division of $X^{2^m} + 1$ by $X^{2^n} + 1$.
4. Let $a, b \in \mathbb{K}$ be given, with $a \neq 0$. For $f \in \mathbb{K}[X] \setminus \{0\}$, prove that the remainder of the division of f by $aX + b$ is equal to $f\left(-\frac{b}{a}\right)$.
5. Upon dividing $f \in \mathbb{Q}[X]$ by $X + 2$, we obtain -1 as remainder; upon dividing it by $X - 2$, the remainder is 3. Find the remainder of the division of f by $X^2 - 4$.
6. A polynomial $f \in \mathbb{R}[X]$ leaves remainders 4 when divided by $X + 1$ and $2X + 3$ when divided by $X^2 + 1$. Find the remainder of the division of f by $(X+1)(X^2 + 1)$.
7. Let m and n be given natural numbers such that $m = nq + r$, with $q, r \in \mathbb{Z}$ and $0 \leq r < n$. Find the quotient and the remainder upon dividing $X^m + 1$ by $X^n + 1$.

Chapter 15
Roots of Polynomials

In all we have done so far concerning polynomials, at no moment we had any intent of *substituting* X by an element of $\mathbb{K}$. Even in Theorem 14.10 and Example 14.15, the notation $f(z)$, used to denote the complex number obtained by *formally substituting* X by z in the expression of $f \in \mathbb{C}[X]$, was a mere convention. This is no surprise, for we are looking at polynomials as *formal expressions*, rather than as *functions*. In this sense, the **indeterminate** X is a symbol with no arithmetic meaning, and we have even stressed before that we could have used the symbol $\square$, instead.

We remedy this state of things in this chapter, where we introduce the concept of polynomial function associated to a polynomial, as well as that of root of a polynomial. Among the various important results presented, we highlight the presentation of a complete proof of the Fundamental Theorem of Algebra, as well as the searching criterion for rational roots of polynomials with integer coefficients, along with several interesting applications.

15.1 Roots of Polynomials

In order to proceed with our study of polynomials, it is convenient to consider the *polynomial function* associated to a polynomial, according to the following

Definition 15.1 For $f(X) = a_n X^n + \cdots + a_1 X + a_0 \in \mathbb{K}[X]$, the **polynomial function** associated to f is the function $\tilde{f} : \mathbb{K} \to \mathbb{K}$, given for $x \in \mathbb{K}$ by

$$\tilde{f}(x) = a_n x^n + \cdots + a_1 x + a_0.$$

© Springer International Publishing AG, part of Springer Nature 2018
A. Caminha Muniz Neto, *An Excursion through Elementary Mathematics, Volume III*,
Problem Books in Mathematics, https://doi.org/10.1007/978-3-319-77977-5_15

When $f(X) = c$, note that the associated polynomial function $\tilde{f}$ will be the constant function $\tilde{f}(x) = c$ for every $x \in \mathbb{K}$, thus justifying the nomenclature *constant* attributed to a polynomial which is either identically zero or of degree zero.

Definition 15.2 Let $f \in \mathbb{K}[X]$ be a given polynomial, with associated polynomial function $\tilde{f} : \mathbb{K} \to \mathbb{K}$. An element $\alpha \in \mathbb{K}$ is a **root** of f if $\tilde{f}(\alpha) = 0$.

For instance, if $f(X) = X^2 + 1 \in \mathbb{C}[X]$, it is easy to see that $z = \pm i$ are the only roots of f in $\mathbb{C}$. Indeed, the polynomial function associated to f is

$$\tilde{f} : \mathbb{C} \longrightarrow \mathbb{C}$$
$$z \longmapsto z^2 + 1 \, ,$$

so that

$$\tilde{f}(z) = 0 \Leftrightarrow z^2 = -1 \Leftrightarrow z = \pm i.$$

The coming proposition is known as the **root test**. Note that item (a) revisits Example 14.15 in a slightly different way.

Proposition 15.3 *If $f \in \mathbb{K}[X] \setminus \{0\}$ and $\alpha \in \mathbb{K}$, then:*

(a) α *is a root of f if and only if $(X - \alpha) \mid f(X)$ in $\mathbb{K}[X]$.*
(b) *If α is a root of f, then there exists a greatest natural number m such that $(X - \alpha)^m$ divides f. Moreover, writing $f(X) = (X - \alpha)^m q(X)$, with $q \in \mathbb{K}[X]$, one has $\tilde{q}(\alpha) \neq 0$, where $\tilde{q} : \mathbb{K} \to \mathbb{K}$ stands for the polynomial function associated to q.*
(c) *If $\alpha_1, \ldots, \alpha_k \in \mathbb{K}$ are pairwise distinct roots of f, then $f(X)$ is divisible by $(X - \alpha_1) \ldots (X - \alpha_k)$ in $\mathbb{K}[X]$.*

Proof

(a) The division algorithm assures the existence of polynomials $q, r \in \mathbb{K}[X]$ such that

$$f(X) = (X - \alpha)q(X) + r(X),$$

with $r = 0$ or $0 \leq \partial r < \partial(X - \alpha) = 1$. Hence, $r(X) = c$, a constant polynomial. On the other hand, letting $\tilde{f}$ and $\tilde{q}$ be the polynomial functions respectively associated to f and q, it follows from the equality above that

$$\tilde{f}(\alpha) = (\alpha - \alpha)\tilde{q}(\alpha) + c = c,$$

i.e., $f(X) = (X - \alpha)q(X) + \tilde{f}(\alpha)$. Thus,

$$\alpha \text{ is a root of } f \Leftrightarrow \tilde{f}(\alpha) = 0 \Leftrightarrow f(X) = (X - \alpha)q(X).$$

(b) If $m > \partial f$, then $(X - \alpha)^m \nmid f(X)$, for $\partial(X - \alpha)^m = m > \partial f$. Therefore, item
(a) guarantees the existence of a greatest positive integer m for which $(X - \alpha)^m \mid$
$f(X)$, say $f(X) = (X - \alpha)^m q(X)$, with $q \in \mathbb{K}[X]$. In the level of polynomial
functions, we thus obtain

$$\tilde{f}(x) = (x - \alpha)^m \tilde{q}(x), \ \forall x \in \mathbb{K}.$$

Now, if $\tilde{q}(\alpha) = 0$, it would follow from (a) that $q(X) = (X - \alpha)q_1(X)$, for
some $q_1 \in \mathbb{K}[X]$. However, we would then have

$$f(X) = (X - \alpha)^{m+1} q_1(X),$$

a contradiction to the maximality of m.

(c) For the sake of simplicity, let us establish the result of this item for $k = 2$ (the
proof in the general case following by a straightforward inductive argument).
Since α_1 is a root of f, item (a) once more assures the existence of $g \in \mathbb{K}[X]$
such that $f(X) = (X - \alpha_1)g(X)$. Let $\tilde{g}$ denote the corresponding polynomial
function. Since $\alpha_1 \neq \alpha_2$ and α_2 is also a root of f, the last equality above gives

$$0 = \tilde{f}(\alpha_2) = (\alpha_2 - \alpha_1)\tilde{g}(\alpha_2),$$

so that α_2 is also a root of g. Thus, by applying (a) yet another time, we conclude
that there exists $h \in \mathbb{K}[X]$ for which $g(X) = (X - \alpha_2)h(X)$. In turn, this gives

$$f(X) = (X - \alpha_1)(X - \alpha_2)h(X).$$

$\square$

The coming example is an interesting application of the root test.

Example 15.4 (Soviet Union) Label the rows and columns of an $n \times n$ chessboard
from 1 to n, respectively from top to bottom and from left to right. Given $2n$ real
numbers $a_1, \ldots, a_n, b_1, \ldots, b_n$, with $b_1, \ldots, b_n$ being pairwise distinct, write $a_i + b_j$
inside the unit square situated in row i and column j, for $1 \le i, j \le n$. If the
products of the numbers written in the unit squares of each column of the chessboard
are all equal, prove that the products of the numbers written in the unit squares of
each row are all equal too.

Proof Consider the polynomial

$$f(X) = (X + a_1)(X + a_2) \ldots (X + a_n) \in \mathbb{R}[X].$$

Letting $\tilde{f}$ be the polynomial function associated to f, it follows from the statement of the problem the existence of $\alpha \in \mathbb{R}$ such that $\tilde{f}(b_1) = \tilde{f}(b_2) = \ldots = \tilde{f}(b_n) = \alpha$. Then, $\tilde{f}(b_i) - \alpha = 0$ for $1 \leq i \leq n$ and, since $b_1, \ldots, b_n$ are pairwise distinct, item (c) of the root test gives

$$f(X) - \alpha = (X - b_1)(X - b_2) \ldots (X - b_n)$$

(note that both $f(X) - \alpha$ and the polynomial at the right hand side of the equality above are monic and have degree n). Hence,

$$-\alpha = f(-a_i) - \alpha = (-1)^n (a_i + b_1)(a_i + b_2) \ldots (a_i + b_n),$$

and this is the same as saying that the products of the numbers written in the unit squares of each row are all equal to $(-1)^{n-1}\alpha$.   □

As a consequence of the root test, we now describe a rather simple algorithm for finding the quotient of the division of a monic polynomial $f \in \mathbb{K}[X]$ by $X - \alpha$, with $\alpha \in \mathbb{K}$ being a root of f. Such an algorithm is known in mathematical literature as the **Horner-Rufinni algorithm**,[1] and is based on the following homonymous result.

Proposition 15.5 (Horner-Rufinni) *Let*

$$f(X) = X^n + a_{n-1}X^{n-1} + \cdots + a_1 X + a_0$$

be a monic polynomial on $\mathbb{K}$. If $\alpha \in \mathbb{K}$ is a root of f and

$$g(X) = X^{n-1} + b_{n-2}X^{n-2} + \cdots + b_0 \in \mathbb{K}[X]$$

is the quotient upon dividing $f(X)$ by $X - \alpha$, then

$$\begin{cases} b_{n-2} & = \alpha + a_{n-1} \\ b_{n-3} & = \alpha b_{n-2} + a_{n-2} \\ \cdots & \cdots \\ b_{n-i-1} & = \alpha b_{n-i} + a_{n-i} \\ \cdots & \cdots \\ b_0 & = \alpha b_1 + a_1 \end{cases}. \tag{15.1}$$

---

[1]Paolo Ruffini, Italian mathematician, and William G. Horner, English mathematician, both of the eighteenth and nineteenth centuries.

Proof By the sake of notational convenience, let $b_{n-1} = 1$. Firstly, note that

$$f(X) = (X - \alpha)g(X)$$

$$= (X - \alpha)\sum_{i=0}^{n-1} b_{n-1-i} X^{n-1-i}$$

$$= \sum_{i=0}^{n-1} b_{n-1-i} X^{n-i} - \sum_{i=0}^{n-1} \alpha b_{n-1-i} X^{n-1-i}$$

$$= b_{n-1} X^n + \sum_{i=1}^{n-1} b_{n-1-i} X^{n-i} - \sum_{i=0}^{n-2} \alpha b_{n-1-i} X^{n-1-i} - \alpha b_0$$

$$= X^n + \sum_{i=1}^{n-1} b_{n-1-i} X^{n-i} - \sum_{i=1}^{n-1} \alpha b_{n-i} X^{n-i} - \alpha b_0$$

$$= X^n + \sum_{i=1}^{n-1} (b_{n-1-i} - \alpha b_{n-i}) X^{n-i} - \alpha b_0.$$

By comparing the last expression for f with that in the statement of the proposition, we conclude that

$$b_{n-1-i} - \alpha b_{n-i} = a_{n-i}$$

for $1 \leq i \leq n - 1$. This way, we obtain the linear system of equations

$$\begin{cases} b_{n-1} = 1 \\ b_{n-2} - \alpha b_{n-1} = a_{n-1} \\ b_{n-3} - \alpha b_{n-2} = a_{n-2} \\ \quad \cdots \quad \cdots \\ b_{n-i-1} - \alpha b_{n-i} = a_{n-i} \\ \quad \cdots \quad \cdots \\ b_0 - \alpha b_1 = a_1 \\ -\alpha b_0 = a_0. \end{cases}$$

which, in turn, is readily seen to be equivalent to the equalities in (15.1) (note that the last equation of the system can be neglected, for it only represents the necessary compatibility condition for α to be a root of f). $\qquad\square$

The table below, which can be seen as the actual execution of Horner-Ruffini's algorithm, resumes the above discussion. Note that we put all of coefficients of f but the leading one (which equals 1) along the first row; then, in the second row we successively compute, from left to right, the coefficients of g, starting with the leading coefficient $b_{n-1} = 1$. Note that, starting with b_{n-2}, each coefficient of g is equal to the product of the previously computed coefficient (i.e., the one at its left) by α, added to the coefficient of f situated right above this one.

The Horner-Rufinni Algorithm

a_{n-1}	a_{n-2}	a_{n-3}	$\cdots$	a_1	a_0
$b_{n-1} = 1$	$\alpha \cdot 1 + a_{n-1}$	$\alpha \cdot b_{n-2} + a_{n-2}$	$\cdots$	$\alpha \cdot b_2 + a_2$	$\alpha \cdot b_1 + a_1$
	$\underbrace{}_{b_{n-2}}$	$\underbrace{}_{b_{n-3}}$		$\underbrace{}_{b_1}$	$\underbrace{}_{b_0}$

The coming example shows the Horner-Ruffini algorithm in action. It also shows that the algorithm can be used to check whether some given element of $\mathbb{K}$ is actually a root of a given polynomial in $\mathbb{K}[X]$.

Example 15.6 Check that $\sqrt{2}$ is a root of $f(X) = X^5 - 5X^4 + X^3 - 5X^2 - 6X + 30 \in \mathbb{R}[X]$, and use the Horner-Ruffini algorithm to compute the quotient of the division of f by $X - \sqrt{2}$.

Solution We begin assembling the numerical table corresponding to Horner-Rufinni's algorithm by setting $n = 5$ and writing $a_4 = -5$, $a_3 = 1$, $a_2 = -5$, $a_1 = -6$, $a_0 = 30$ along the first row:

$$-5 \mid 1 \mid -5 \mid -6 \mid 30$$

Now starting with $b_4 = 1$ in the leftmost entry of second row, we successively obtain $b_3 = \sqrt{2} \cdot 1 - 5$, $b_2 = \sqrt{2}b_3 + 1 = 3 - 5\sqrt{2}$, $b_1 = \sqrt{2}b_2 - 5 = -15 + 3\sqrt{2}$ and $b_0 = \sqrt{2}b_1 - 6 = -15\sqrt{2}$:

-5	1	-5	-6	30
1	$\sqrt{2} \cdot 1 - 5$	$3 - 5\sqrt{2}$	$-15 + 3\sqrt{2}$	$-15\sqrt{2}$

Finally, since $-\alpha b_0 = -\sqrt{2}(-15\sqrt{2}) = 30 = a_0$, it follows that $\sqrt{2}$ is actually a root of f, and the quotient upon dividing f by $X - \sqrt{2}$ is $X^4 + (\sqrt{2} - 5)X^3 + (3 - 5\sqrt{2})X^2 + (-15 + 3\sqrt{2})X - 15\sqrt{2}$.   □

Back to the development of the theory and in the notations of the root test, if $f(X) = (X - \alpha)^m q(X)$, with $\tilde{q}(\alpha) \neq 0$, we say that m is the **multiplicity** of α as a root of f. In particular, α is a **simple root** of f if $m = 1$, and a **multiple root** of f if $m > 1$.

Example 15.7 Let $f(X) = X^3 - 7X^2 + 16X - 12$ be a polynomial with real coefficients. It is immediate to check that $f(X) = (X - 2)^2(X - 3)$. Therefore, 2 is a double root, whereas 3 is a simple one.

Later in this chapter we shall study in detail the problem of finding whether a given polynomial $f \in \mathbb{K}[X]$ has multiple roots. For the time being, we only need to know what such a root is. Nevertheless, we shall observe that we already have a way of deciding whether an element $\alpha \in \mathbb{K}$ is a multiple root of $f \in \mathbb{K}[X] \setminus \{0\}$: it suffices to check if $f(\alpha) = 0$ and, if this is so, to divide $f(X)$ by $(X - \alpha)^2$ (which can be done by means of two successive applications of Horner-Rufinni's algorithm), verifying whether or not such a division is exact. We shall have more to say about this in a while.

Another interesting consequence of the division algorithm is the fact, due to Lagrange, that the number of roots of a nonzero polynomial cannot exceed its degree. In the coming result, note that we allow multiple roots.

Corollary 15.8 (Lagrange) *If $f \in \mathbb{K}[X] \setminus \{0\}$, then f has at most ∂f roots in $\mathbb{K}$, counted according to their multiplicities.*

Proof Let us make induction on the degree of f. If $\partial f = 0$, then there exists $c \in \mathbb{K} \setminus \{0\}$ such that $f(X) = c$. Hence, the polynomial function $\tilde{f}$ associated to f is the constant function $x \mapsto c$, so that f has $0 = \partial f$ roots.

Now, let f be a polynomial of positive degree, and assume that the result is true for every polynomial in $\mathbb{K}[X] \setminus \{0\}$ whose degree is less than that of f. If f has no roots in $\mathbb{K}$, there is nothing left to do. Otherwise, let $\alpha \in \mathbb{K}$ be a root of f and (according to the root test) let m be the greatest natural number such that $f(X) = (X - \alpha)^m q(X)$, for some $q \in \mathbb{K}[X]$ such that $\tilde{q}(\alpha) \neq 0$. Since

$$\partial q = \partial f - m < \partial f,$$

the induction hypothesis assures that q has at most ∂q roots in $\mathbb{K}$, counted with their multiplicities.

In order to finish the argument, note that if $\beta \neq \alpha$ is a root of f, then

$$0 = \tilde{f}(\beta) = (\beta - \alpha)^m \tilde{q}(\beta)$$

and, hence, β is a root of q. Therefore, the number of roots of f (with multiplicities) equals m (the multiplicity of α) plus that of q. Thus, the induction hypothesis shows that f has at most

$$m + \partial q = \partial f$$

roots in $\mathbb{K}$. $\square$

The previous corollary will be frequently used in one of the two forms below.

Corollary 15.9 *If $f(X) = a_n X^n + \cdots + a_1 X + a_0 \in \mathbb{K}[X]$ admits at least $n + 1$ distinct roots in $\mathbb{K}$, then f vanishes identically, i.e., $a_n = \ldots = a_0 = 0$.*

Proof If $f \in \mathbb{K}[X] \setminus \{0\}$, then the former corollary assures it has at most n roots in $\mathbb{K}$.   □

Corollary 15.10 *Let* $f(X) = a_n X^n + \cdots + a_1 X + a_0$ *and* $g(X) = b_m X^m + \cdots + b_1 X + b_0$ *be polynomials in* $\mathbb{K}[X]$, *with* $m \geq n$. *If* $\tilde{f}(x) = \tilde{g}(x)$ *for at least* $m + 1$ *distinct values of* $x \in \mathbb{K}$, *then* $f = g$, *i.e,* $m = n$ *and* $a_i = b_i$ *for* $0 \leq i \leq n$.

Proof It suffices to apply the previous corollary to the polynomial $f - g$, noticing that the associated polynomial function is $\tilde{f} - \tilde{g}$.   □

If $\mathcal{F}_{\mathbb{K}}$ stands for the set of functions from $\mathbb{K}$ to itself, then, since $\mathbb{K}$ is infinite, the last result above guarantees that the mapping

$$\begin{array}{ccc} \mathbb{K}[X] & \longrightarrow & \mathcal{F}_{\mathbb{K}} \\ f & \longmapsto & \tilde{f} \end{array} , \qquad (15.2)$$

that associates to each $f \in \mathbb{K}[X]$ its polynomial function $\tilde{f} \in \mathcal{F}_{\mathbb{K}}$, is injective. In other words, it says that two polynomials on $\mathbb{K}$ will have equal polynomial functions if they are themselves equal polynomials.[2] Indeed, if $f, g \in \mathbb{K}[X]$ are such that $\tilde{f} = \tilde{g}$, then $\tilde{f}(x) = \tilde{g}(x)$ for infinitely many values of x, and the previous corollary assures that $f = g$.

Thanks to the injectivity of the above map, from now one we shall write f to denote both an element of $\mathbb{K}[X]$ (i.e., a polynomial with coefficients in $\mathbb{K}$) as its associated polynomial function. In particular, whenever we write $f(X)$, we shall be referring ourselves to the *polynomial* f; whenever we write $f(x)$, we shall be looking at the element of $\mathbb{K}$, image of $x \in \mathbb{K}$ through the polynomial function associated to f. Moreover, the context will always clear any possible doubts on this point.

Remark 15.11 Corollary 15.10 assures that, thanks to the fact that $\mathbb{K}$ is infinite, the coefficients of $f \in \mathbb{K}[X]$ are completely determined by the values $f(x)$, with $x \in \mathbb{K}$. Later, in Chap. 18, we shall study in detail the problem of getting a polynomial in $\mathbb{K}[X]$ that assumes prescribed values at prescribed elements of $\mathbb{K}$.

The two coming examples illustrate typical uses of Corollaries 15.9 and 15.10.

Example 15.12 (Moldavia) Find all polynomials $f \in \mathbb{R}[X]$ such that $f(0) = 0$ and

$$f(x^2 + 1) = f(x)^2 + 1, \ \forall x \in \mathbb{R}.$$

---

[2] As we shall see in Sect. 19.3, upon studying polynomials over $\mathbb{Z}_p$, for some prime number p, the fact that $\mathbb{K}$ is infinite here is actually indispensable for the injectivity of (15.2).

Solution Let the sequence $(u_n)_{n \geq 0}$ be defined by $u_0 = 0$ and $u_{n+1} = u_n^2 + 1$, for $n \geq 1$. If $g(X) = X$, then $f(u_0) = 0 = g(u_0)$ and, by assuming $f(u_k) = g(u_k)$, we get

$$f(u_{k+1}) = f(u_k^2 + 1) = f(u_k)^2 + 1$$
$$= g(u_k)^2 + 1 = u_k^2 + 1 = g(u_{k+1}).$$

Therefore, we conclude by induction that $f(u_n) = g(u_n)$ for every nonnegative integer n. However, since the values of the terms u_n are pairwise distinct, we conclude that $f(x)$ and $g(x)$ coincide for an infinite number of distinct values of x. It thus follows from Corollary 15.10 that $f = g$, i.e., $f(X) = X$. $\square$

Example 15.13 (Hong Kong) Let $g(X) = X^5 + X^4 + X^3 + X^2 + X + 1$. Compute the remainder of the division of $g(X^{12})$ by $g(X)$.

Solution The division algorithm assures the existence of $q, r \in \mathbb{R}[X]$ such that

$$g(X^{12}) = g(X)q(X) + r(X),$$

with $r = 0$ or $0 \leq \partial r \leq 4$. Letting $\omega = \text{cis} \frac{2\pi}{6}$ and taking $x = \omega^k$ in the corresponding polynomial functions, with $1 \leq k \leq 5$, we obtain

$$g(\omega^{12k}) = g(\omega^k)q(\omega^k) + r(\omega^k) \tag{15.3}$$

for $1 \leq k \leq 5$.

Now, since $\omega^{12k} = \text{cis}(4k\pi) = 1$, we get $g(\omega^{12k}) = g(1) = 6$. On the other hand, for $1 \leq k \leq 5$, Lemma 13.18 furnishes

$$g(\omega^k) = \omega^{5k} + \omega^{4k} + \omega^{3k} + \omega^{2k} + \omega^k + 1 = \frac{\omega^{6k} - 1}{\omega^k - 1} = 0.$$

Thus, (15.3) reduces to $r(\omega^k) = 6$ for $1 \leq k \leq 5$, so that the polynomial $r - 6$ has at least five distinct roots. However, since $r - 6 = 0$ or $\partial(r - 6) \leq 4$, Corollary 15.8 gives $r - 6 = 0$. $\square$

The coming further consequence of Lagrange's theorem (Corollary 15.8) guarantees that the image of the polynomial function associated to a nonconstant polynomial on $\mathbb{K}$ is an infinite set.

Corollary 15.14 *If $f \in \mathbb{K}[X] \setminus \mathbb{K}$, then the image $\text{Im}(f)$ of the polynomial function associated to f is an infinite subset of $\mathbb{K}$.*

Proof Assume the contrary, i.e., that $\text{Im}(f) = \{\alpha_1, \ldots, \alpha_k\} \subset \mathbb{K}$. Then, for every $x \in \mathbb{K}$ we have $f(x) \in \{\alpha_1, \ldots, \alpha_k\}$. However, since $\mathbb{K}$ is an infinite set, there exists $1 \leq i \leq k$ and pairwise distinct elements $x_1, x_2, \ldots \in \mathbb{K}$ such that

$$f(x_1) = f(x_2) = \ldots = \alpha_i.$$

Hence, $f(X) - \alpha_i$ is a nonvanishing polynomial (for f is nonconstant) with an infinite number of distinct roots, which is a contradiction.   □

Example 15.15 (Canada) Find all nonconstant polynomials $f \in \mathbb{Q}[X]$ such that $f(f(X)) = f(X)^k$ for some given natural number k.

Solution It is easy to see (cf. Problem 1) that for every $x \in \mathbb{Q}$ we have $f(f(x)) = f(x)^k$. Yet in another way, we have $f(y) = y^k$ for every $y \in \text{Im}(f)$. Now, the previous result assures that $\text{Im}(f)$ is an infinite subset of $\mathbb{Q}$, so that Corollary 15.10 gives $f(X) = X^k$.   □

Problems: Sect. 15.1

1. * Show that the usual definition of composite function, when applied to polynomials $f, g \in \mathbb{K}[X] \setminus \{0\}$ of degrees m e n, respectively, produces a polynomial in $\mathbb{K}[X] \setminus \{0\}$ of degree mn. Letting $f \circ g$ denote such a polynomial, show that its associated polynomial function coincides with the composite function $\tilde{f} \circ \tilde{g}$, where $\tilde{f}$ and $\tilde{g}$ denote the polynomial functions associated to f and g, respectively.

2. * Let a, b and r be rational numbers, with $r > 0$ being such that $\sqrt{r}$ is irrational. Do the following items:

   (a) For a fixed $k \in \mathbb{N}$, show that there exist $a_k, b_k \in \mathbb{Q}$ such that $(a \pm b\sqrt{r})^k = a_k \pm b_k\sqrt{r}$.

   (b) If $f \in \mathbb{Q}[X] \setminus \{0\}$, prove that $f(a + b\sqrt{r}) = 0 \Leftrightarrow f(a - b\sqrt{r}) = 0$.

3. One of the roots of the polynomial $X^4 + aX^3 + X^2 + bX - 2$ is $1 - \sqrt{2}$. Find the remaining roots, knowing that a and b are both rational numbers.

4. Let $a, b \in \mathbb{K}$ be given, with $a \neq 0$. For $f \in \mathbb{K}[X] \setminus \{0\}$, prove that the remainder of the division of f by $aX + b$ is equal to $f\left(-\frac{b}{a}\right)$.

5. Does there exist a polynomial $f \in \mathbb{R}[X]$ such that $f(\sin x) = \cos x$ for every x? Justify your answer.

6. Let $\alpha \neq k\pi$ be a complex number. Prove that, in $\mathbb{R}[X]$, the polynomial $X^2 + 1$ divides the polynomial

$$f(X) = (\cos \alpha + X \sin \alpha)^n - \cos(n\alpha) - \sin(n\alpha)X.$$

7. Find all $n \in \mathbb{N}$ for which $X^{n+1} - X^n + 1$ is divisible by $X^2 - X + 1$.

8. (BMO) For $m \in \mathbb{Z}$, prove that the polynomial

$$X^4 - 1994X^3 + (1993 + m)X^2 - 11X + m$$

does not possess two distinct integer roots.

9. (AIME) Find all real numbers a and b such that $X^2 - X - 1$ divides $aX^{17} + bX^{16} + 1$.

10. (Moldavia) Let $n > 4$ be a given natural number and p be a monic polynomial of degree n, having n distinct integer roots, one of which is 0. Compute the number of distinct integer roots of the polynomial $p(p(X))$.

11. (Austrian-Polish) Let $f(x) = ax^3 + bx^2 + cx + d$ be a polynomial with integer coefficients and degree 3. Prove that it is not possible to find four distinct primes p_1, p_2, p_3 and p_4 such that

$$|f(p_1)| = |f(p_2)| = |f(p_3)| = |f(p_4)| = 3.$$

12. (Canada) Let $p(X) = X^n + a_{n-1}X^{n-1} + \cdots + a_1 X + a_0$ be a polynomial with integer coefficients, and assume that there exist distinct integers a, b, c and d for which $p(a) = p(b) = p(c) = p(d) = 5$. Prove that there exists no integer m such that $p(m) = 8$.

13. (IMO shortlist) Find all values of $k \in \mathbb{N}$ such that $X^2 + X + 1$ divides $X^{2k} + 1 + (X + 1)^{2k}$.

   For the coming example, recall (cf. paragraph that precedes Problem 13, page 30) that a *multiset* is a collection $\{\{a_1, a_2, \ldots, a_n\}\}$ of not necessarily distinct elements, with $\{\{a_1, a_2, \ldots, a_n\}\} = \{\{b_1, b_2, \ldots, b_m\}\}$ if and only if $m = n$ and each element appears the same number of times in each multiset.

14. Let $\{\{a_1, a_2, \ldots, a_n\}\}$ and $\{\{b_1, b_2, \ldots, b_n\}\}$ be two distinct multisets, each of which formed by positive integers. If

$$\{\{a_i + a_j; \ 1 \le i < j \le n\}\} = \{\{b_i + b_j; \ 1 \le i < j \le n\}\},$$

prove that n is a power of 2.

15.2   Rational Roots of Polynomials in $\mathbb{Z}[X]$

This short section studies the problem of finding rational roots of polynomials with integer coefficients, presenting several interesting applications. Item (a) of the coming proposition brings the central result, which is known in mathematical literature as the **rational roots test** for polynomials with integer coefficients.

Proposition 15.16 *Let $n > 1$ be an integer, $f(X) = a_n X^n + \cdots + a_1 X + a_0$ be a nonzero polynomial with integer coefficients and p and q be nonzero, relatively prime integers. If $f\left(\frac{p}{q}\right) = 0$, then:*

(a) $p \mid a_0$ and $q \mid a_n$.

(b) If f is monic, then the possible rational roots of f are integers.

(c) $(p - mq) \mid f(m)$ for every $m \in \mathbb{Z}$. In particular, $(p - q) \mid f(1)$ and $(p + q) \mid f(-1)$.

Proof

(a) Starting from $f\left(\frac{p}{q}\right) = 0$, we readily obtain

$$a_n p^n + a_{n-1} p^{n-1} q + \cdots + a_1 pq^{n-1} + a_0 q^n = 0$$

and, hence,

$$\begin{cases} a_0 q^n = p(-a_n p^{n-1} - \cdots - a_1 q^{n-1}) \\ a_n p^n = q(-a_{n-1} p^{n-1} - \cdots - a_0 q^{n-1}) \end{cases}.$$

Therefore, $p \mid a_0 q^n$ and $q \mid a_n p^n$. However, since p and q are relatively prime, item (a) of Proposition 6.22 assures that $p \mid a_0$ and $q \mid a_n$, as we wished to show.

(b) This follows immediately from (a).

(c) Since $f\left(\frac{p}{q}\right) = 0$, we have $f(m) = f(m) - f\left(\frac{p}{q}\right)$ or, which is the same,

$$f(m) = (a_n m^n + \cdots + a_1 m + a_0) - \frac{1}{q^n}(a_n p^n + \cdots + a_1 pq^{n-1} + a_0 q^n).$$

Hence,

$$\begin{aligned} q^n f(m) &= q^n(a_n m^n + \cdots + a_1 m + a_0) - (a_n p^n + \cdots + a_1 pq^{n-1} + a_0 q^n) \\ &= a_n((mq)^n - p^n) + \cdots + a_1 q^{n-1}(mq - p) \\ &= (mq - p)r \end{aligned}$$

for some $r \in \mathbb{Z}$, where we used item (a) of Example 6.3 in the last equality. The above computations show that $(mq - p) \mid q^n f(m)$. Thus, in order to conclude that $(mq - p) \mid f(m)$, it suffices to show (again by item (a) of Proposition 6.22) that $\gcd(mq - p, q^n) = 1$. To this end, by successively applying item (b) of that result, together with Corollary 6.23, we get

$$\gcd(p, q) = 1 \Rightarrow \gcd(mq - p, q) = 1 \Rightarrow \gcd(mq - p, q^n) = 1.$$

The rest follows at once.   □

The previous result can sometimes be applied to establish the irrationality of certain real numbers. We shall now see a couple of examples in this direction.

Example 15.17 Prove that $\sqrt{2 + \sqrt{2 + \sqrt{2}}}$ is irrational.

Proof Letting $\alpha = \sqrt{2 + \sqrt{2 + \sqrt{2}}}$, we have $\alpha^2 - 2 = \sqrt{2 + \sqrt{2}}$ and, then, $(\alpha^2 - 2)^2 = 2 + \sqrt{2}$. Therefore, $((\alpha^2 - 2)^2 - 2)^2 = 2$, so that α is a root of the monic polynomial with integer coefficients

$$f(X) = ((X^2 - 2)^2 - 2)^2 - 2$$
$$= (X^4 - 4X^2 + 2)^2 - 2.$$

Thus, if $\alpha \in \mathbb{Q}$, items (a) and (b) or the former proposition assure that $\alpha \in \mathbb{N}$ and $\alpha \mid f(0) = 2$, so that $\alpha = 1$ ou 2. However, since

$$1 < \alpha = \sqrt{2 + \sqrt{2 + \sqrt{2}}} < \sqrt{2 + \sqrt{2 + 2}} = 2,$$

we have reached a contradiction. □

Example 15.18 Prove that $\tan 10°$ is irrational.

Proof We let $\alpha = \tan 10°$. Applying standard trigonometric formulas (cf. Chapter 7 of [9], for instance), we successively obtain

$$\tan 30° = \frac{\tan 20° + \tan 10°}{1 - \tan 20° \cdot \tan 10°} = \frac{\frac{2 \tan 10°}{1 - \tan^2 10°} + \tan 10°}{1 - \frac{2 \tan^2 10°}{1 - \tan^2 10°}}$$

$$= \frac{\frac{2\alpha}{1 - \alpha^2} + \alpha}{1 - \frac{2\alpha^2}{1 - \alpha^2}} = \frac{3\alpha - \alpha^3}{1 - 3\alpha^2}$$

However, since $\tan 30° = \frac{1}{\sqrt{3}}$, we get

$$\frac{(3\alpha - \alpha^3)^2}{(1 - 3\alpha^2)^2} = \frac{1}{3}$$

or, which is the same,

$$3\alpha^6 - 27\alpha^4 + 33\alpha^2 - 1 = 0.$$

Hence, $\tan 10°$ is a root of the polynomial with integer coefficients $f(X) = 3X^6 - 27X^4 + 33X^2 - 1$, and it suffices to show that it does not possess any rational roots. For what is left to do, first note that, by Proposition 15.16, the possible rational roots of f are ± 1 or $\pm\frac{1}{3}$; nevertheless, direct computations show that $f(\pm 1)$ and $f(\pm\frac{1}{3})$ are all nonzero, as wished.   □

Example 15.19 (Yugoslavia) Find all positive rationals $a \leq b \leq c$ for which all of the numbers

$$a + b + c, \quad \frac{1}{a} + \frac{1}{b} + \frac{1}{c}, \quad abc$$

are integers.

Solution Letting $f(X) = (X - a)(X - b)(X - c)$, we conclude that a, b and c are the roots of f; moreover, they are all rational. On the other hand,

$$f(X) = X^3 - (a + b + c)X^2 + (ab + bc + ca)X - abc,$$

with

$$ab + bc + ca = abc\left(\frac{1}{a} + \frac{1}{b} + \frac{1}{c}\right) \in \mathbb{N}.$$

It thus follows that $f \in \mathbb{Z}[X]$. However, since f is monic, the rational roots test applied to f assures that all of a, b, c must be integers.

Finally, since a, b and c are positive, we conclude that it suffices to find all natural numbers $a \leq b \leq c$ for which $\frac{1}{a} + \frac{1}{b} + \frac{1}{c}$ is also natural. In order to solve this problem, start by noticing that

$$1 \leq \frac{1}{a} + \frac{1}{b} + \frac{1}{c} \leq \frac{3}{a},$$

so that $a \leq 3$. Then, by separately considering the cases $a = 1, 2$ or 3 and iterating the above reasoning, one easily finds

$$(a, b, c) = (1, 1, 1), \ (1, 2, 2), \ (2, 4, 4), \ (2, 3, 6) \text{ or } (3, 3, 3).$$

(For instance, if $a = 3$, then $b, c \geq 3$, and hence

$$1 \leq \frac{1}{a} + \frac{1}{b} + \frac{1}{c} \leq \frac{1}{3} + \frac{1}{3} + \frac{1}{3} = 1,$$

so that $a = b = c = 3$.)   □

We finish this section by presenting an interesting application of item (b) of Proposition 15.16. More precisely we shall characterize all rational numbers α for which $\cos(\alpha\pi)$ is also rational. To this end, we first need to introduce a special class of polynomials.

Let $(f_n)_{n\geq 1}$ be the sequence of polynomials with real coefficients given by $f_1(X) = X$, $f_2(X) = X^2 - 2$ and, for an integer $k \geq 1$,

$$f_{k+2}(X) = Xf_{k+1}(X) - f_k(X). \tag{15.4}$$

The polynomial f_n is called the n-th **Bernstein polynomial**, after the Russian mathematician of the twentieth century Sergei N. Bernstein. The next proposition unfolds some of its properties.

Proposition 15.20 *If $(f_n)_{n\geq 1}$ is the sequence of Bernstein's polynomials, then:*

(a) $f_n(2\cos\theta) = 2\cos(n\theta)$, for every $\theta \in \mathbb{R}$.
(b) f_n is monic, has integer coefficients and degree n.

Proof For item (a), it is obvious that $f_1(2\cos\theta) = 2\cos\theta$, whereas the formula for $\cos(2\theta)$ in terms of $\cos\theta$ gives

$$f_2(2\cos\theta) = (2\cos\theta)^2 - 2 = 2(2\cos^2\theta - 1) = 2\cos(2\theta).$$

Now, let $k \geq 1$ be an integer and assume, by induction hypothesis, that $f_j(2\cos\theta) = 2\cos(j\theta)$ for $1 \leq j \leq k+1$ and every $\theta \in \mathbb{R}$. The product formula $\cos\alpha + \cos\beta = 2\cos\left(\frac{\alpha+\beta}{2}\right)\cos\left(\frac{\alpha-\beta}{2}\right)$ gives

$$2\cos\theta \cdot 2\cos(k+1)\theta - 2\cos(k\theta) = 2\cos(k+2)\theta.$$

Therefore, it follows from (15.4) and the induction hypothesis that

$$\begin{aligned}
f_{k+2}(2\cos\theta) &= 2\cos\theta \cdot f_{k+1}(2\cos\theta) - f_k(2\cos\theta) \\
&= 2\cos\theta \cdot 2\cos(k+1)\theta - 2\cos(k\theta) \\
&= 2\cos(k+2)\theta
\end{aligned}$$

for every $\theta \in \mathbb{R}$.

Item (b) is trivially true for f_1 and f_2. Arguing once more by induction, if f_k and f_{k+1} are monic, with integer coefficients and degrees k and $k+1$, respectively, then the recurrence relation (15.4) readily assures that the same properties hold for f_{k+2}. $\square$

We are finally in position to present the promised result.

Theorem 15.21 *If* $\alpha \in \mathbb{Q}$ *is such that* $\cos(\alpha\pi) \in \mathbb{Q}$, *then* $\cos(\alpha\pi) = \pm 1, \pm\frac{1}{2}$ *or* 0.

Proof Let $m, n, p, q \in \mathbb{Z}$ such that $n, q \neq 0$ and $\cos(\frac{m}{n}\pi) = \frac{p}{q}$. For $\alpha = \frac{m}{n}$, item (a) of the previous proposition gives

$$f_n\left(\frac{2p}{q}\right) = f_n(2\cos(\alpha\pi)) = 2\cos(n\alpha\pi) = 2\cos(m\pi) = 2(-1)^m.$$

Now, item (b) of that result shows that $\frac{2p}{q}$ is a root of the monic polynomial with integer coefficients $f_n(X) - 2(-1)^m$. Hence, item (b) of Proposition 15.16 gives $\frac{2p}{q} \in \mathbb{Z}$. On the other hand, since

$$\frac{2p}{q} = 2\cos\left(\frac{m}{n}\pi\right) \in [-2, 2],$$

we get $\frac{2p}{q} = \pm 2, \pm 1$ or 0. Therefore, $\frac{p}{q} = \pm 1, \pm\frac{1}{2}$ or 0. □

Problems: Sect. 15.2

1. (Canada) Let f be a nonzero polynomial with integer coefficients. If $f(0)$ and $f(1)$ are odd, prove that f does not have any integer roots.
2. (OCM) Prove that there does not exist nonzero integers x, y and z in arithmetic progression, such that $x^5 + y^5 = z^5$.
3. Show that if $(f_n)_{n \geq 1}$ is a sequence of polynomials of real coefficients such that $f_n(2\cos\theta) = 2\cos(n\theta)$ for every $\theta \in \mathbb{R}$, then f_n is the n-th Bernstein polynomial.
4. For $n \in \mathbb{N}$, let f_n be the n-th Bernstein polynomial. Show that

$$z^n + \frac{1}{z^n} = f_n\left(z + \frac{1}{z}\right),$$

for every $z \in \mathbb{C} \setminus \{0\}$.
5. Find the roots of the polynomial $2(2X^2 - 1)^2 - X - 1$ by calling one of them x and:

   (a) Letting $y = 2x^2 - 1$.
   (b) Imposing that $x = \cos\theta$.

   Then, use the previous items to compute $\cos\frac{\pi}{10}$.

6. Let $a, b \in \mathbb{Z}$ be such that $\sqrt[3]{a + \sqrt{b}} + \sqrt[3]{a - \sqrt{b}} \in \mathbb{Q}$.

   (a) Prove that it must be an integer and a divisor of $8a^3$.
   (b) Assuming that $a^2 - b$ is a perfect cube, do the following items:

      i. Prove that the given number is a divisor of $2a$.
      ii. Characterize all such pairs of integers a and b.

7. (Ireland) Prove that $(2 + i)^n \neq (2 - i)^n$ for every $n \in \mathbb{N}$.
   The next problem partially generalizes Example 6.24.

8. (OCM) Do the following items:

   (a) If none of the natural numbers $a_1, a_2, \ldots, a_n$ is a perfect square, show that $\sqrt{a_1} + \sqrt{a_2} + \cdots + \sqrt{a_n}$ is a root of a monic polynomial $f_n \in \mathbb{Z}[X]$, of degree 2^n.
   (b) Show that $\sqrt{2} + \sqrt{3} + \sqrt{5} + \sqrt{7} + \sqrt{11}$ is irrational.

9. (IMO—shortlist) Given $a, b \in \mathbb{Q}$, can the polynomials $X^5 - X - 1$ and $X^2 + aX + b$ have a common complex root?

15.3 The Fundamental Theorem of Algebra

As a particular case of Corollary 15.9, every polynomial of complex coefficients and degree n has at most n complex roots. On the other hand, the polynomial $f(X) = X^2 - 2 \in \mathbb{Q}[X]$ has rational coefficients but does not admit any rational roots; accordingly, $g(X) = X^2 + 1 \in \mathbb{R}[X]$ has real coefficients but no real roots. In both cases, the fact that such polynomials do not admit roots in $\mathbb{Q}$ or $\mathbb{R}$ is due to *deficiencies* of such number sets, in the sense that, in them, one cannot perform certain root extractions.

On the other hand, Bhaskara's formula shows (cf. Problem 1) that every second degree polynomial with complex coefficients has two (possibly equal complex roots), and the coming example looks at the third degree case. The reasoning presented below is essentially the one developed by the Italian mathematicians of the sixteenth century Scipioni del Ferro, Gerolamo Cardano and Niccolò Fontana, known as Tartaglia (in this respect, see also Problem 2).

Example 15.22 A third degree polynomial $f(X) = a_3 X^3 + a_2 X^2 + a_1 X + a_0$ with complex coefficients has a complex root.

Proof We can obviously assume that $a_3 = 1$. Also, for a given $\alpha \in \mathbb{C}$, we have that $z \in \mathbb{C}$ is a root of $f(X)$ if and only if $z - \alpha$ is a root of $f(X + \alpha)$; since

$$f(X + \alpha) = (X + \alpha)^3 + a_2(X + \alpha)^2 + a_1(X + \alpha) + a_0$$

$$= X^3 + (3\alpha + a_2)X^2 + (3\alpha^2 + 2a_2\alpha + a_1)X + f(\alpha),$$

by choosing $\alpha = -\frac{a_2}{3}$ we can assume further that $a_2 = 0$.

We are thus left to showing that a third degree polynomial $f(X) = X^3 + aX + b$, with $a, b \in \mathbb{C}$, has complex roots. To this end, we substitute $z = u + v$ to get

$$
\begin{aligned}
f(z) &= (u + v)^3 + a(u + v) + b \\
&= u^3 + v^3 + 3uv(u + v) + a(u + v) + b \\
&= (u^3 + v^3 + b) + 3(u + v)(uv + a).
\end{aligned}
$$

Since we are trying to find *some* complex root for f, we try to impose that $u^3 + v^3 = -b$ and $uv = -a$, thus getting the system of equations

$$
\begin{cases}
u^3 + v^3 = -b \\
uv = -a
\end{cases}.
$$

In turn, such a system can easily be solved, for u^3 and v^3 are the roots of the second degree equation $X^2 + bX - a^3 = 0$ $\square$

At this point, we could turn to fourth degree polynomials with complex coefficients and, by performing even trickier computations, show that they have complex roots too. Instead, we turn to the general case, which was settled by Gauss in his doctor thesis, in 1799 (nevertheless, see Problem 3 and the subsequent paragraph). This is the content of the coming result, which is know in mathematical literature as the **Fundamental Theorem of Algebra** (we abbreviate **FTA**). The proof we present is due to the French mathematician of the eighteenth century J-B. le R. d'Alembert.

Theorem 15.23 (Gauss) *Every polynomial $f \in \mathbb{C}[X] \setminus \mathbb{C}$ has at least one complex root.*

Proof By the sake of notation, write $f(z) = a_n z^n + \cdots + a_1 z + a_0$ to denote the polynomial function associated to f. Without any loss of generality, we can assume that $a_0 \neq 0$.

For $z \neq 0$, the triangle inequality for complex numbers (cf. Problem 3, page 326) gives

$$
\begin{aligned}
|f(z)| &= |z|^n \left| a_n + \frac{a_{n-1}}{z} + \frac{a_{n-2}}{z^2} + \cdots + \frac{a_0}{z^n} \right| \\
&\geq |z|^n \left(|a_n| - \frac{|a_{n-1}|}{|z|} - \frac{|a_{n-2}|}{|z|^2} - \cdots - \frac{|a_0|}{|z|^n} \right).
\end{aligned}
$$

Therefore, letting

$$
|z| > \max\left\{ \frac{2n|a_{n-1}|}{|a_n|}, \frac{\sqrt{2n|a_{n-2}|}}{\sqrt{|a_n|}}, \ldots, \frac{\sqrt[n]{2n|a_0|}}{\sqrt[n]{|a_n|}} \right\}, \tag{15.5}
$$

we have $\frac{|a_{n-k}|}{|z|^k} < \frac{|a_n|}{2n}$, so that

$$|f(z)| > |z|^n \left(|a_n| - n \frac{|a_n|}{2n} \right) = \frac{|a_n|}{2} |z|^n.$$

However, since $|z| > \sqrt[n]{\frac{2n|a_0|}{|a_n|}}$, we have $|a_n||z|^n > 2n|a_0|$ and, hence, $|f(z)| > n|a_0| \geq |a_0|$.

In short, letting R denote the right hand side of (15.5), we get

$$|z| > R \Rightarrow |f(z)| > |a_0| = |f(0)|.$$

Now, in Sect. 21.2 we will show (cf. Corollary 21.21) that there exists $z_0 \in \mathbb{C}$ such that $|z_0| \leq R$ and

$$|f(z_0)| = \min\{|f(z)|; \ z \in \mathbb{C}, \ |z| \leq R\}.$$

Hence, what we did above assures that

$$|f(z_0)| = \min\{|f(z)|; \ z \in \mathbb{C}\}.$$

By contradiction, let us suppose that $f(z_0) \neq 0$, and show that there exists $h \in \mathbb{C}$ for which $|f(z_0 + h)| < |f(z_0)|$. To this end, we start by noticing that there exist complex numbers $c_0, c_1, \ldots, c_n$, independent of h and such that

$$f(z_0 + h) = a_0 + a_1(z_0 + h) + \cdots + a_n(z_0 + h)^n$$
$$= c_0 + c_1 h + \cdots + c_n h^n;$$

moreover, $c_0 = f(z_0) \neq 0$ and $c_n = a_n \neq 0$. Take the least $1 \leq k \leq n$ such that $c_k \neq 0$. Then, letting $d_j = \frac{c_j}{c_0}$ for $0 \leq j \leq n$, we obtain

$$\frac{|f(z_0 + h)|}{|f(z_0)|} = |1 + d_k h^k + d_{k+1} h^{k+1} + \cdots + d_n h^n|$$

$$\leq |1 + d_k h^k| + |d_{k+1} h^{k+1} + \cdots + d_n h^n|$$

$$= |1 + d_k h^k| + |d_k h^k| \left| \frac{d_{k+1}}{d_k} h + \cdots + \frac{d_n}{d_k} h^{n-k} \right|.$$

Now, estimates analogous to those we did with the aid of (15.5) allow us to choose $r > 0$ such that $|h| \leq r \Rightarrow \left| \frac{d_{k+1}}{d_k} h + \cdots + \frac{d_n}{d_k} h^{n-k} \right| < \frac{1}{2}$. Then,

$$|h| \leq r \Rightarrow \frac{|f(z_0 + h)|}{|f(z_0)|} \leq |1 + d_k h^k| + \frac{1}{2}|d_k h^k|.$$

Let $d_k = s \operatorname{cis} \alpha$ and $h = r \operatorname{cis} \theta$. The first de Moivre formula gives us

$$\frac{|f(z_0 + h)|}{|f(z_0)|} \leq |1 + sr^k \operatorname{cis}(\alpha + k\theta)| + \frac{1}{2} sr^k;$$

therefore, choosing $\theta \in \mathbb{R}$ in such a way that $\alpha + k\theta = \pi$ (i.e., taking $\theta = \frac{\pi - \alpha}{k}$), we get

$$\frac{|f(z_0 + h)|}{|f(z_0)|} \leq |1 + sr^k \operatorname{cis} \pi| + \frac{1}{2} sr^k$$

$$= |1 - sr^k| + \frac{1}{2} sr^k$$

$$= 1 - \frac{1}{2} sr^k < 1,$$

whenever $sr^k < 1$, i.e., $0 < r < \sqrt[k]{\frac{1}{s}}$. With such an r (and with the corresponding h), we have $|f(z_0 + h)| < |f(z_0)|$.   □

An immediate consequence of the FTA is collected in the coming

Corollary 15.24 *If* $f(X) = a_n X^n + \cdots + a_1 X + a_0$ *is a polynomial with complex coefficients and degree* $n \geq 1$, *then there exist complex numbers* $z_1, \ldots, z_n$ *such that*

$$f(X) = a_n(X - z_1) \ldots (X - z_n). \tag{15.6}$$

*The right hand side above is the **factorised form** of* f.

Proof We prove the corollary by induction on n, the case $n = 1$ being immediate. Take an integer $n > 1$, and assume that the corollary is valid for every polynomial with complex coefficients and degree $n - 1$.

If $z_1 \in \mathbb{C}$ is a root of f (whose existence is guaranteed by the FTA), the root test assures the existence of a polynomial g with complex coefficients, such that $f(X) = (X - z_1)g(X)$. Note that g has degree $n - 1$ and leading coefficient a_n; hence, by induction hypothesis, there exist $z_2, \ldots, z_n \in \mathbb{C}$ such that $g(X) = a_n(X - z_2) \ldots (X - z_n)$. Therefore,

$$f(X) = (X - z_1)g(X) = a_n(X - z_1)(X - z_2) \ldots (X - z_n)$$

and there is nothing left to do.   □

A useful variation of the previous corollary is the one given by the next result.

Corollary 15.25 *If* $f(X) = a_n X^n + \cdots + a_1 X + a_0$ *is a polynomial with complex coefficients and degree* $n \geq 1$, *then, given* $\alpha \in \mathbb{C}$, *there exist* $z_1, \ldots, z_n \in \mathbb{C}$ *such that* $f(z_k) = \alpha$ *for* $1 \leq k \leq n$.

Proof Apply the former corollary to $g(X) = f(X) - \alpha$. □

Back to the general case, let $f(X) = a_n X^n + \cdots + a_1 X + a_0$ be a nonzero polynomial with coefficients in $\mathbb{K}$. If there exist elements $\alpha_1, \ldots, \alpha_n \in \mathbb{K}$ for which

$$f(X) = a_n(X - \alpha_1) \ldots (X - \alpha_n),$$

we shall also say that the expression at the right hand side is the **factorised form** of f over $\mathbb{K}$.

Since some of the α_j's may appear several times in the factorised form, if we look only at the distinct ones we conclude, possibly after renumbering them, that there exist $1 \leq m \leq n$ and positive integers $k_1, \ldots, k_m$ such that

$$f(X) = a_n(X - \alpha_1)^{k_1} \cdots (X - \alpha_m)^{k_m},$$

with $k_1 + \cdots + k_m = n$ and $\alpha_1, \ldots, \alpha_m$ pairwise distinct elements of $\mathbb{K}$.

The two coming examples give interesting applications of the factorised form of polynomials.

Example 15.26 Given a natural number $n > 1$, do the following items:

(a) Obtain the factorised form of $f(X) = X^{n-1} + X^{n-2} + \cdots + X + 1$.
(b) Let $A_1 A_2 \ldots A_n$ be a regular polygon with n sides, inscribed in a circle with radius 1. Compute the value of $\overline{A_1 A_2} \cdot \overline{A_1 A_3} \cdot \ldots \cdot \overline{A_1 A_n}$.

Solution

(a) Since

$$(X - 1)f(X) = (X - 1)(X^{n-1} + X^{n-2} + \cdots + X + 1) = X^n - 1,$$

the complex roots of f are precisely the n-th roots of unity distinct from 1. By (13.19), such roots are the complex numbers $\omega, \omega^2, \ldots, \omega^{n-1}$, with $\omega = \operatorname{cis} \frac{2\pi}{n}$. Therefore, the factorised form of f is

$$f(X) = (X - \omega)(X - \omega^2) \ldots (X - \omega^{n-1}). \tag{15.7}$$

(b) Assume, without any loss of generality, that the circle of radius 1 circumscribing $A_1 A_2 \ldots A_n$ is the unit circle centered at the origin of the complex plane, as well as that $A_1 = 1$ and $A_2 = \omega$, with $\omega = \operatorname{cis} \frac{2\pi}{n}$. Then, $A_j = \omega^{j-1}$ for $1 \leq j \leq n$, so that

$$\overline{A_1 A_2} \cdot \overline{A_1 A_3} \cdot \ldots \cdot \overline{A_1 A_n} = |1 - \omega||1 - \omega^2| \ldots |1 - \omega^{n-1}|$$
$$= |(1 - \omega)(1 - \omega^2) \ldots (1 - \omega^{n-1})|$$
$$= |f(1)| = n.$$

□

The next example is concerned with Bernstein polynomials, so that the reader may find it convenient to recall their definition, in the paragraph that precedes Proposition 15.20.

Example 15.27 For $n \in \mathbb{N}$, let f_n be the n-th Bernstein polynomial. Show that

$$f_n(X) = \prod_{k=1}^{n} \left(X - 2\cos \frac{(2k-1)\pi}{2n} \right).$$

Proof Item (a) of Proposition 15.20 gives $f_n(2\cos\theta) = 0$ if $\cos(n\theta) = 0$. However,

$$\cos(n\theta) = 0 \Leftrightarrow n\theta = \frac{\pi}{2} + k\pi, \ \exists\, k \in \mathbb{Z}$$

$$\Leftrightarrow \theta = \frac{\pi}{2n} + \frac{k\pi}{n}, \ \exists\, k \in \mathbb{Z}.$$

Now, for $k = 0, 1, \ldots, n-1$, we get θ respectively equal to $\frac{\pi}{2n}, \frac{3\pi}{2n}, \ldots, \frac{(2n-1)\pi}{2n}$, and all these values of θ give distinct values for $\cos\theta$. Therefore, $2\cos\left(\frac{\pi}{2n}\right)$, $2\cos\left(\frac{3\pi}{2n}\right)$, $\ldots, 2\cos\left(\frac{(2n-1)\pi}{2n}\right)$ are n distinct roots for f_n. Since f_n it is monic and has degree n, the rest follows from Corollary 15.24. $\square$

Problems: Sect. 15.3

1. Prove that the formula giving the roots of a second degree equation still holds to compute the complex roots of $aX^2 + bX + c$, with $a, b, c \in \mathbb{C}$ and $a \neq 0$.
2. Let $a, b \in \mathbb{C}$, with $a \neq 0$. In the notations of the discussion of Example 15.22, if $u_0 \in \mathbb{C}$ is such that u_0^3 is a complex root of $X^2 + bX - a^3$, show that the roots of $X^3 + aX + b$ are given by $u_0 + v_0$, $u_0\omega + v_0\omega^2$ and $u_0\omega^2 + v_0\omega$, with $v_0 = -\frac{a}{u_0}$ and $\omega = \operatorname{cis}\frac{2\pi}{3}$. Moreover, show that such are distinct roots if $4a^3 + b^2 \neq 0$.
3. The purpose of this problem is to delineate the steps followed by the Italian mathematician of the sixteenth century Lodovico Ferrari to find the roots of a fourth degree polynomial $X^4 + aX^3 + bX^2 + cX + d$, with $a, b, c, d \in \mathbb{C}$:

(a) If $z \in \mathbb{C}$ is any complex number, check that

$$z^4 + az^3 + bz^2 + cz + d = \left(z^2 + \frac{az}{2}\right)^2 + \left(b - \frac{a^2}{4}\right)z^2 + cz + d.$$

(b) If $z \in \mathbb{C}$ is a root of the given polynomial and w is any complex number, check that

$$\left(z^2 + \frac{az}{2} + w\right)^2 = \left(2w - b + \frac{a^2}{4}\right)z^2 + (wa - c)z + w^2 - d.$$

(c) Show that if w is a root of the third degree equation

$$(wa - c)^2 = 4(w^2 - d)\left(2w - b + \frac{a^2}{4}\right),$$

then the equality of item (b) reduces to

$$z^2 + \frac{az}{2} + w = \pm\left(\sqrt{2w - b + \frac{a^2}{4}}\,z + \sqrt{w^2 - d}\right),$$

where $\sqrt{2w - b + \frac{a^2}{4}}$ and $\sqrt{w^2 - d}$ stand for any complex square roots of $2w - b + \frac{a^2}{4}$ and $w^2 - d$.

(d) Show, without resorting to the Fundamental Theorem of Algebra, that the given fourth degree polynomial has a complex root.

The reader who went through Example 15.22 may have noticed that we reduced the task of finding a complex root for a third degree polynomial to that of finding a complex root for an associated second degree polynomial; subsequently, in the problem above we reduced the task of finding a complex root for a fourth degree polynomial to that of finding a complex root for an associated third degree polynomial. One is thus naturally tempted to guess whether it would be possible to extend such a recursive reasoning, thus reducing the task of finding a complex root for a general polynomial of degree $n > 1$ to that of finding a complex root for an appropriately associated polynomial of degree $n - 1$. Attempts to implement such a strategy were unsuccessfully pursued by several mathematicians along the sixteenth, seventeenth and eighteenth centuries, until a general consensus started to form around the idea that this would perhaps not be possible. Actually, in the dawn of the nineteenth century, the Norwegian mathematician Niels H. Abel proved that there is no general formula for finding the roots of a fifth degree polynomial in terms of their coefficients; a few years later, the French mathematician Évariste Galois extended such a result for all polynomial of degree $n \geq 5$, obtaining the first major achievement of what is known today as *Galois Theory*. A gentle introduction to the involved ideas, with complete proofs, is the object of the marvelous book [20].

4. * If $f \in \mathbb{R}[X] \setminus \{0\}$ and $z \in \mathbb{C} \setminus \mathbb{R}$, prove that $f(z) = 0 \Leftrightarrow f(\overline{z}) = 0$. Then, conclude that every nonconstant polynomial with real coefficients has an even number of nonreal complex roots, counted with multiplicities.

5. * Prove that every polynomial with real coefficients and odd degree has an odd number of real roots. In particular, such a polynomial always has at least one real root.

6. If $f \in \mathbb{R}[X] \setminus \mathbb{R}$ is monic and has no real roots, show that there exist $g, h \in \mathbb{R}[X]$ such that $f = g^2 + h^2$.

7. * Let $f \in \mathbb{Q}[X] \setminus \mathbb{Q}$ be a polynomial of degree $n \in \mathbb{N}$, and $\alpha \neq 0$ be a complex root of f. Given $m \in \mathbb{Z}$, prove that there exist $b_0, b_1, \ldots, b_{n-1} \in \mathbb{Q}$ such that

$$\alpha^m = b_0 + b_1\alpha + \cdots + b_{n-1}\alpha^{n-1}. \tag{15.8}$$

8. Let $f(X) = a_n X^n + a_{n-1} X^{n-1} + \cdots + a_1 X + a_0$ be a polynomial with complex coefficients and degree $n \geq 1$. If $z \in \mathbb{C}$ is one of its roots f, prove that

$$|z| \leq \max\left\{1, \frac{nA}{|a_n|}\right\},$$

where $A = \max\{|a_0|, |a_1|, \ldots, |a_{n-1}|\}$.

9. * Let $f(X) = a_n X^n + a_{n-1} X^{n-1} + \cdots + a_1 X + a_0$ be a polynomial with integer coefficients, such that $a_n \geq 1$, and $k > 2$ be an integer for which $|a_i| \leq k$ for $0 \leq i < n$. If z is a complex root of f, prove that $\mathrm{Re}(z) < 1 + \frac{k}{2}$.

10. A polynomial f over $\mathbb{C}$ of the form $f(X) = aX^4 + bX^3 + cX^2 + bX + a$, with $a \neq 0$, is said to be a **reciprocal** of degree 4 grau. Compute its roots and, then, formulate and solve the analogous problem for reciprocal polynomials of degree 6.

11. (Singapore)[3] Let f be a polynomial with real coefficients and degree n, such that $f(k) = \frac{k}{k+1}$ for every integer $0 \leq k \leq n$. Compute the possible values of $f(n+1)$.

12. Given an integer $n > 2$, prove that

$$\sin\frac{\pi}{n} \sin\frac{2\pi}{n} \sin\frac{3\pi}{n} \ldots \sin\frac{(n-1)\pi}{n} = \frac{n}{2^{n-1}}.$$

13. For $m \in \mathbb{N}$, show that:

   (a) $\sin\frac{\pi}{2m} \sin\frac{2\pi}{2m} \ldots \sin\frac{(m-1)\pi}{2m} = \frac{\sqrt{m}}{2^{m-1}}$.

   (b) $\sin\frac{\pi}{2m+1} \sin\frac{2\pi}{2m+1} \ldots \sin\frac{m\pi}{2m+1} = \frac{\sqrt{2m+1}}{2^m}$.

14. (Romania) Find all nonconstant polynomials $p \in \mathbb{R}[X]$ such that $p(X^2) = p(X)p(X-1)$.

---

[3] For other approaches to this problem, see Problem 4, page 444, and Problem 3, page 450.

15.4 Multiple Roots

In this section, given $f \in \mathbb{C}[X] \setminus \mathbb{C}$ and $z \in \mathbb{C}$, we derive a useful necessary condition to be satisfied for z to be a multiple root of f. To this end, it is convenient to make the convention of saying that $z \in \mathbb{C}$ a root of f of *multiplicity zero* if z is actually not a root of f. We shall also need the coming

Definition 15.28 Given a polynomial $f(X) = a_n X^n + \cdots + a_1 X + a_0 \in \mathbb{C}[X]$, we define its **derivative**[4] $f' \in \mathbb{C}[X]$ of f as the polynomial

$$f'(X) = n a_n X^{n-1} + (n-1)a_{n-2} X^{n-2} + \cdots + 2a_2 X + a_1$$

if $\partial f > 0$. Otherwise, we let $f' = 0$.

The rule for getting the derivative of a polynomial is rather simple: the derivative of a constant polynomial is the identically zero one, whereas the derivative of a polynomial of degree $n \geq 1$ is obtained by erasing its constant term and performing, for $1 \leq k \leq n$, the *monomial exchange*

$$a_k X^k \mapsto k a_k X^{k-1}.$$

The coming proposition establishes the main properties of derivatives of polynomials.

Proposition 15.29 *For $f_1, \ldots, f_k \in \mathbb{C}[X]$ and $a_1, \ldots, a_k \in \mathbb{C}$, we have:*

(a) $\left(\sum_{i=1}^{k} a_i f_i \right)' = \sum_{i=1}^{k} a_i f_i'.$

(b) $\left(\prod_{i=1}^{k} f_i \right)' = \sum_{i=1}^{k} f_1 \cdots f_i' \cdots f_k.$

Proof

(a) This item follows immediately by induction on $k \geq 1$. Note that the initial cases are $k = 1$ and $k = 2$.

(b) Firstly, let

$$f(X) = a_n X^n + \cdots + a_1 X + a_0 \text{ and } g(X) = b_m X^m + \cdots + b_1 X + b_0.$$

---

[4]The reader acquainted with Calculus has certainly noticed that the definition of f' matches the one presented in Calculus courses by computing limits of Newton's quotients. The point of the present definition and the subsequent proposition is that they will equally apply to polynomials over $\mathbb{Z}_p$, in Chap. 19.

By omitting X whenever convenient, it follows from (a) that

$$(fg)' = \left[\left(\sum_{j=0}^{n} a_j X^j\right) g\right]' = \left(\sum_{j=0}^{n} a_j X^j g\right)' = \sum_{j=0}^{n} (a_j X^j g)'.$$

On the other hand,

$$(a_j X^j g)' = \left(a_j X^j \sum_{i=0}^{m} b_i X^i\right)' = \left(\sum_{i=0}^{m} a_j b_i X^{j+i}\right)' = \sum_{i=0}^{m} (a_j b_i X^{j+i})'$$

$$= \sum_{i=0}^{m} (j+i) a_j b_i X^{j+i-1}$$

$$= \sum_{i=0}^{m} j a_j b_i X^{j+i-1} + \sum_{i=0}^{m} i a_j b_i X^{j+i-1}$$

$$= j a_j X^{j-1} \sum_{i=0}^{m} b_i X^i + a_j X^j \sum_{i=1}^{m} i b_i X^{i-1}$$

$$= (a_j X^j)' g + (a_j X^j) g'.$$

Hence, going back to the previous expression for $(fg)'$, we obtain

$$(fg)' = \sum_{j=0}^{n} \left((a_j X^j)' g + a_j X^j g'\right)$$

$$= \left(\sum_{j=0}^{n} (a_j X^j)'\right) g + \left(\sum_{j=0}^{n} a_j X^j\right) g'$$

$$= \left(\sum_{j=0}^{n} a_j X^j\right)' g + fg'$$

$$= f'g + fg',$$

where we used item (a) once more in the next to last equality.

Finally, the extension to k polynomials $f_1, \ldots, f_k \in \mathbb{C}[X]$ easily follows by induction.

$\square$

Corollary 15.30 *Given $g \in \mathbb{C}[X]$ and $n \in \mathbb{N}$, if $f(X) = g(X)^n$ then $f'(X) = ng(X)^{n-1} g'(X)$. In particular, if $f(X) = (X - a)^n$, then $f'(X) = n(X - a)^{n-1}$.*

Proof Letting $k = n$ and $f_1 = \cdots = f_n = g$ in item (b) of the previous proposition, we obtain

$$f'(X) = \sum_{i=1}^{n} g(X)^{n-1} g'(X) = ng(X)^{n-1} g'(X).$$

The particular case follows from this formula, letting $g(X) = X - a$. $\square$

The coming result relates the multiplicity of a root of a polynomial to its derivative.

Proposition 15.31 *Let $f \in \mathbb{C}[X] \setminus \{0\}$ and $z \in \mathbb{C}$.*

(a) *If z is a root of multiplicity $m \geq 1$ of f, then z is a root of multiplicity $m - 1$ of f'.*
(b) *If z is a root of f and is a root of multiplicity $m - 1$ of f', then z is a root of multiplicity m of f.*

Proof

(a) The root test allows us to write $f(X) = (X - z)^m g(X)$, with $g(z) \neq 0$. On the other hand, item (b) of the previous proposition, together with its corollary, give

$$f'(X) = m(X - z)^{m-1} g(X) + (X - z)^m g'(X)$$

$$= (X - z)^{m-1} [mg(X) + (X - z)g'(X)].$$

Hence, letting $h(X) = mg(X) + (X - z)g'(X)$, we have $h(z) = mg(z) \neq 0$ and $f'(X) = (X - z)^{m-1} h(X)$. By the very definition of multiplicity, we conclude that z is a root of f' of multiplicity $m - 1$.
(b) Again by the root test, we can write $f(X) = (X - z)^k g(X)$, for some integer $k \geq 1$ and $g \in \mathbb{C}[X]$ such that $g(z) \neq 0$. Then, item (a) assures that the multiplicity of α as a root of f' is $k - 1$, so that $k - 1 = m - 1$ and, hence, $k = m$. $\square$

Corollary 15.32 *If $z \in \mathbb{C}$ and $f \in \mathbb{C}[X] \setminus \{0\}$, then z is a multiple root of f if and only if $f(z) = f'(z) = 0$.*

Proof If z is a multiple root of f, this means that it is a root of f with multiplicity is at least 2. Therefore, item (a) of the previous proposition implies that z is a root of f'.

Conversely, if z is a root of both f and f', then item (b) of the previous proposition shows that the multiplicity of z as a root of f is at least 2. In turn, this is the same as saying that z is a multiple root of f. $\square$

Example 15.33 (Sweden) Prove that, for every $n \in \mathbb{N}$, the polynomial

$$1 + X + \frac{1}{2!}X^2 + \cdots + \frac{1}{n!}X^n$$

has no multiple roots.

Proof Let $f(X) = 1 + X + \frac{1}{2!}X^2 + \cdots + \frac{1}{n!}X^n$, and assume that it has a multiple root z. Then, by the former corollary we have $f(z) = f'(z) = 0$. Now, since $f(X) = f'(X) + \frac{1}{n!}X^n$, it would follow that

$$0 = f(z) = f'(z) + \frac{z^n}{n!},$$

i.e., $z = 0$. However, since $f(0) = 1 \neq 0$, we have reached a contradiction. $\square$

In order to refine the conclusion of the previous corollary, we first need to generalize Definition 15.28. We do this now.

Definition 15.34 For $f \in \mathbb{C}[X] \setminus \{0\}$, we define the **$k$-th derivative** of f, denoted $f^{(k)}$, by

$$f^{(k)} = \begin{cases} f, & \text{if } k = 0 \\ \left(f^{(k-1)}\right)', & \text{if } k \geq 1 \end{cases}.$$

It readily follows from this definition that $f^{(1)} = (f^{(0)})' = f'$; thus, $f^{(2)} = (f^{(1)})' = (f')'$, so that we denote $f^{(2)} = f''$. Accordingly, whenever convenient we write $f^{(3)} = f'''$ etc.

If $\partial f = n$ and $0 \leq k \leq n$, an easy induction guarantees that $\partial f^{(k)} \leq n - k$; in particular, $\partial f^{(n)} = 0$ and, hence, $f^{(n+1)} = f^{(n+2)} = \cdots = 0$.

Corollary 15.35 *If $z \in \mathbb{C}$ and $f \in \mathbb{C}[X] \setminus \{0\}$, then z is a root of multiplicity $m \geq 1$ of f if and only if*

$$f(z) = \ldots = f^{(m-1)}(z) = 0 \text{ and } f^{(m)}(z) \neq 0.$$

Proof Suppose first that z is a root of multiplicity m of f. Several applications of item (a) of Proposition 15.31 give us, on the one hand,

$$f(z) = \cdots = f^{(m-1)}(z) = 0$$

and, on the other, that z is a root of multiplicity zero of $f^{(m)}$, i.e., that $f^{(m)}(z) \neq 0$.

Conversely, assume that the stated condition holds and let $k \in \mathbb{N}$ and $g \in \mathbb{C}[X]$ be such that $f(X) = (X - z)^k g(X)$, with $g(z) \neq 0$. A straightforward inductive argument (using again item (a) of Proposition 15.31) assures that z is a root of multiplicity $k - j$ of $f^{(j)}$, for $0 \leq j \leq k$. In particular,

$$f(z) = \ldots = f^{(k-1)}(z) = 0 \text{ and } f^{(k)}(z) \neq 0.$$

Now

$$f(z) = \ldots = f^{(m-1)}(z) = 0 \Rightarrow k \geq m$$

and

$$f^{(m)}(z) \neq 0 \Rightarrow k \leq m.$$

$\square$

In the rest of this section, our purpose is to show that, if $f \in \mathbb{C}[X] \setminus \{0\}$ has degree n and $z \in \mathbb{C}$, then f is completely determined by the values $f^{(k)}(z)$, for $0 \leq k \leq n$. However, before we can do that, we need two further consequences of Proposition 15.31.

Corollary 15.36 *Let $f \in \mathbb{C}[X]$ be such that $f = 0$ or $\partial f \leq n$. If $z \in \mathbb{C}$ satisfies $f(z) = \cdots = f^{(n)}(z) = 0$, then $f = 0$.*

Proof If $f \neq 0$, then the previous corollary, together with the stated conditions, assure that the multiplicity of z as a root of f is at least $n + 1$. However, this would imply that $(X - z)^{n+1}$ would divide f, which in turn contradicts the fact that $\partial f < n$.
$\square$

Corollary 15.37 *Let $f, g \in \mathbb{C}[X] \setminus \{0\}$, with $\partial f, \partial g \leq n$. If there exists $z \in \mathbb{C}$ such that*

$$f(z) = g(z), \quad \ldots, f^{(n)}(z) = g^{(n)}(z),$$

then $f = g$.

Proof An easy induction on $k \geq 1$, with the help of Definition 15.34 and item (a) of Proposition 15.29, assures that

$$(f - g)^{(k)}(X) = f^{(k)}(X) - g^{(k)}(X),$$

for every integer $k \geq 0$. In particular, the stated conditions give

$$(f - g)^{(k)}(z) = f^{(k)}(z) - g^{(k)}(z) = 0,$$

for $0 \leq k \leq n$.

Finally, since $f - g = 0$ or $\partial(f - g) \leq \max\{\partial f, \partial g\} \leq n$, the previous corollary, applied to $f - g$, assures that such a polynomial vanishes identically.
$\square$

The coming result is known in mathematical literature as the **Taylor formula**[5] for polynomials, and is a direct consequence of the last corollary above.

Theorem 15.38 *If $z \in \mathbb{C}$ and $f \in \mathbb{C}[X] \setminus \{0\}$ is of degree n, then*

$$f(X) = f(z) + \frac{f^{(1)}(z)}{1!}(X - z) + \cdots + \frac{f^{(n)}(z)}{n!}(X - z)^n. \tag{15.9}$$

Proof Let

$$g(X) = f(z) + \frac{f'(z)}{1!}(X - z) + \cdots + \frac{f^{(n)}(z)}{n!}(X - z)^n.$$

Since $f \neq 0$, it follows from Corollary 15.36 that at least one of the numbers $f(z)$, $f'(z)$, ..., $f^{(n)}(z)$ is nonzero; therefore, $g \neq 0$. On the other hand, it is immediate to check that, for $0 \leq k \leq n$, we have

$$g^{(k)}(X) = \sum_{j=k}^{n} \frac{f^{(j)}(z)}{(j-k)!}(X - z)^{j-k}$$

and, hence,

$$f(z) = g(z), \ f'(z) = g'(z), \ \ldots, \ f^{(n)}(z) = g^{(n)}(z).$$

However, since f and g both have degrees less than or equal to n, the previous corollary assures that $f = g$. □

Example 15.39 Let $f \in \mathbb{R}[X] \setminus \{0\}$ and $a \in \mathbb{R}$ be such that $f(a) = 0$ and $f^{(k)}(a) \geq 0$ for every $k \geq 1$. Prove that f has no roots along the interval $(a, +\infty)$.

Proof Letting $n = \partial f$, Taylor's formula gives

$$f(X) = \frac{f'(a)}{1!}(X - a) + \cdots + \frac{f^{(n)}(a)}{n!}(X - a)^n.$$

Since $f \neq 0$, at least one of the derivatives $f^{(k)}(a)$, say $f^{(j)}(a)$, is positive. Obviously, $1 \leq j \leq n$. Therefore, for a real number $x > a$, we have

$$f(x) = \frac{f'(a)}{1!}(x - a) + \cdots + \frac{f^{(n)}(a)}{n!}(x - a)^n$$

$$\geq \frac{f^{(j)}(a)}{j!}(x - a)^j > 0.$$

□

---

[5]Brook Taylor, English mathematician of the eighteenth century.

Problems: Sect. 15.4

1. Find all integer values of a for which $f(X) = X^3 - aX^2 + 5X - 2$ has multiple roots.

2. * Establish the following partial generalization of item (a) of Proposition 15.31: if $f, g \in \mathbb{C}[X] \setminus \{0\}$ are such that $g(X)^2 \mid f(X)$, then $g(X) \mid f'(X)$.

3. * For $z_1, \ldots, z_n \in \mathbb{C}$, let $f(X) = (X - z_1) \ldots (X - z_n)$. Prove that, for $z \in \mathbb{C} \setminus \{z_1, \ldots, z_n\}$, we have

$$\frac{f'(z)}{f(z)} = \sum_{j=1}^{n} \frac{1}{z - z_j}.$$

4. * Generalize the previous problem proving that, if $f_1, \ldots, f_k \in \mathbb{C}[X] \setminus \{0\}$, $f = f_1 \ldots f_k$ and $z \in \mathbb{C}$ is not a root of f, then

$$\frac{f'(z)}{f(z)} = \frac{f_1'(z)}{f_1(z)} + \cdots + \frac{f_k'(z)}{f_k(z)}.$$

5. * Let $n > 1$ be an integer and $\omega = \operatorname{cis} \frac{2\pi}{n}$.

   (a) Show that, for every integer $1 \le j \le n$, one has

   $$(\omega^j - 1) \ldots (\omega^j - \omega^{j-1})(\omega^j - \omega^{j+1}) \ldots (\omega^j - \omega^{n-1}) = n\omega^{(n-1)j}.$$

   (b) Compute, in terms of n, the value of $P = \prod_{0 \le k < j < n} (\omega^j - \omega^k)^2$.

6. Given $u, v, w \in \mathbb{C}$, with $w \ne 0$, let $(a_n)_{n \ge 1}$ be such that $a_{k+3} = ua_{k+2} + va_{k+1} + wa_k$ for every $k \ge 1$. Also, let α, β and γ be the complex roots of the polynomial $X^3 - uX^2 - vX - w$.

   (a) If α, β and γ are pairwise distinct, show that $a_n = A\alpha^{n-1} + B\beta^{n-1} + C\gamma^{n-1}$ for every $n \ge 1$, where A, B and C are the solutions of the linear system

   $$\begin{cases} A + B + C = a_1 \\ \alpha A + \beta B + \gamma C = a_2 \\ \alpha^2 A + \beta^2 B + \gamma^2 C = a_3 \end{cases}.$$

   (b) If $\alpha = \beta \ne \gamma$, show that $a_n = (A + Bn)\alpha^{n-1} + C\gamma^{n-1}$ for $n \ge 1$, where A, B and C are the solutions of the linear system

   $$\begin{cases} A + B + C = a_1 \\ \alpha(A + 2B) + \gamma C = a_2 \\ \alpha^2(A + 3B) + \gamma^2 C = a_3 \end{cases}.$$

(c) If $\alpha = \beta = \gamma$, show that $a_n = (A + Bn + Cn^2)\alpha^{n-1}$ for $n \geq 1$, where A, B and C are the solutions of the linear system

$$\begin{cases} A + B + C = a_1 \\ \alpha(A + 2B + 4C) = a_2 \\ \alpha^2(A + 3B + 9C) = a_3 \end{cases}$$

7. * Solve the linear recurrence relation

$$d_{n+3} - 6d_{n+2} + 12d_{n+1} - 8d_n = 0,$$

knowing that $d_1 = 1$, $d_2 = 4$ and $d_3 = 14$.

We now revisit Problem 13, page 22, with the methods we developed in this section.

8. (USA) For each nonempty finite set S of real numbers, let $\sigma(S)$ and $\pi(S)$ respectively denote the sum and product of its elements. Prove that

$$\sum_{\emptyset \neq S \subset I_n} \frac{\sigma(S)}{\pi(S)} = n^2 + 2n - \left(1 + \frac{1}{2} + \cdots + \frac{1}{n}\right)(n + 1).$$

9. Prove the following theorem of Gauss: if $f(X) = a_n X^n + \cdots + a_1 X + a_0$ is a polynomial of degree greater than 1 and with complex coefficients, then the roots of f' are contained in the smallest convex polygon (possibly degenerated) of the complex plane having the roots of f as vertices.

10. * Let f be a nonzero polynomial with integer coefficients and degree n, and $a \in \mathbb{Z}$. Prove that $\frac{f^{(j)}(a)}{j!} \in \mathbb{Z}$ for $0 \leq j \leq n$.

11. Let f be a nonzero polynomial with integer coefficients and p be a prime that does not divide its leading coefficient. If $m \in \mathbb{N}$ is such that

$$f(m) \equiv 0 \,(\text{mod } p) \quad \text{and} \quad f'(m) \not\equiv 0 \,(\text{mod } p),$$

prove that, for every $k \in \mathbb{N}$, there exists $m_k \in \mathbb{N}$ for which

$$f(m_k) \equiv 0 \,(\text{mod } p^k).$$

12. (IMO—shortlist) The real numbers a, b and c are such that there exists exactly one square[6] whose vertices lie on the graph of the polynomial function $f(x) = x^3 + ax^2 + bx + c$. Prove that the length of the sides of such a square is $\sqrt[4]{72}$.

---

[6]One can show (cf. Problem 5, page 444) that, given a simple n-sided polygon, one can always find a polynomial function of degree at most $n - 1$ and whose graph contains the set of vertices of the polygon.

Chapter 16
Relations Between Roots and Coefficients

This chapter is devoted to the proof of some important relations between the coefficients of a polynomial and their complex roots; such results are generically known as the *relations between roots and coefficients* of a polynomial. We also discuss an important theorem of Newton on symmetric polynomials, which will reveal itself to be of central importance for the material of Chap. 20.

In all that follows, recall that $\mathbb{K}$ stands for $\mathbb{Q}$, $\mathbb{R}$ or $\mathbb{C}$.

16.1 Polynomials on Several Indeterminates

For a given $n \in \mathbb{N}$, a **polynomial f in n indeterminates** over $\mathbb{K}$ is a sum of monomials of the form

$$a_{i_1 \ldots i_n} X_1^{i_1} \ldots X_n^{i_n},$$

where $a_{i_1 \ldots i_n} \in \mathbb{K}$ and $i_1, \ldots, i_n$ vary in $\mathbb{Z}_+$, with $a_{i_1 \ldots i_n} = 0$ for almost every (i.e., for all but a finite number of) sequence $(i_1, \ldots, i_n)$ of nonnegative integers. In this case, we write

$$f = f(X_1, X_2, \ldots, X_n) = \sum_{i_1, \ldots, i_n \geq 0} a_{i_1 \ldots i_n} X_1^{i_1} \ldots X_n^{i_n}.$$

The **degree** of a polynomial f as above is the greatest possible value of the sum $i_1 + \cdots + i_n$, such that $a_{i_1 \ldots i_n} \neq 0$.

We let $\mathbb{K}[X_1, \ldots, X_n]$ denote the set of polynomials in n indeterminates over $\mathbb{K}$. On such a set we define, in an obvious way, operations

$$+ : \mathbb{K}[X_1, \ldots, X_n] \times \mathbb{K}[X_1, \ldots, X_n] \to \mathbb{K}[X_1, \ldots, X_n]$$

© Springer International Publishing AG, part of Springer Nature 2018
A. Caminha Muniz Neto, *An Excursion through Elementary Mathematics, Volume III*,
Problem Books in Mathematics, https://doi.org/10.1007/978-3-319-77977-5_16

and

$$\cdot : \mathbb{K}[X_1, \ldots, X_n] \times \mathbb{K}[X_1, \ldots, X_n] \to \mathbb{K}[X_1, \ldots, X_n],$$

respectively denoted **addition** and **multiplication**, which extend the homonymous operations on $\mathbb{K}[X]$ and continue to be commutative, associative etc. For instance, if $f(X_1, X_2) = X_1^2 + X_1 X_2 + X_2^3$ and $g(X_1, X_2) = X_1^3 - \sqrt{2}X_1 X_2$, then

$$f(X_1, X_2) + g(X_1, X_2) = X_1^2 + (1 - \sqrt{2})X_1 X_2 + X_1^3 + X_2^3$$

and

$$f(X_1, X_2) \cdot g(X_1, X_2) = X_1^5 - \sqrt{2}X_1^3 X_2 + X_1^4 X_2 - \sqrt{2}X_1^2 X_2^2$$
$$+ X_1^3 X_2^3 - \sqrt{2}X_1 X_2^4.$$

Given $f \in \mathbb{K}[X_1, \ldots, X_n]$, we can look at it as a polynomial in X_i, with coefficients in $\mathbb{K}[X_1, \ldots, \widehat{X_i}, \ldots, X_n]$, where we have put the *hat* $\wedge$ over X_i to indicate that all of the indeterminates, except for X_i, are present. For example, let

$$f(X_1, X_2, X_3) = X_1^3 X_2 X_3^4 - X_1^3 - 5X_1 X_2 + 10X_1 X_2^5 X_3^2,$$

be a polynomial in $\mathbb{K}[X_1, X_2, X_3]$; writing

$$f(X_1, X_2, X_3) = X_1^3 X_2 \cdot X_3^4 + 10X_1 X_2^5 \cdot X_3^2 - (X_1^3 + 5X_1 X_2),$$

we consider f as a polynomial in X_3, with coefficients in $\mathbb{K}[X_1, X_2]$.

If $f, g \in \mathbb{K}[X_1, \ldots, X_n]$ are given by

$$f(X_1, X_2, \ldots, X_n) = \sum_{i_1, \ldots, i_n \geq 0} a_{i_1 \ldots i_n} X_1^{i_1} \ldots X_n^{i_n}$$

and

$$g(X_1, X_2, \ldots, X_n) = \sum_{i_1, \ldots, i_n \geq 0} b_{i_1 \ldots i_n} X_1^{i_1} \ldots X_n^{i_n},$$

we say that f and g are *equal* provided $a_{i_1 \ldots i_n} = b_{i_1 \ldots i_n}$ for all possible choices of indices $i_1, \ldots, i_n \in \mathbb{Z}_+$.

For fixed $x_1, x_2 \ldots, x_n \in \mathbb{K}$, we write $f(x_1, x_2, \ldots, x_n)$ to denote the element of $\mathbb{K}$ given by

$$f(x_1, x_2, \ldots, x_n) = \sum_{i_1, \ldots, i_n} a_{i_1 \ldots i_n} x_1^{i_1} x_2^{i_2} \ldots x_n^{i_n}.$$

The coming proposition establishes an important relation between such elements of $\mathbb{K}$ and the notion of equality of polynomials in several indeterminates.

Proposition 16.1 *Let $f, g \in \mathbb{K}[X_1, \ldots, X_n]$. If $A_1, \ldots, A_n \subset \mathbb{K}$ are infinite sets, then*

$$f = g \Leftrightarrow f(x_1, x_2, \ldots, x_n) = g(x_1, x_2, \ldots, x_n),$$

for all $x_1 \in A_1, x_2 \in A_2, \ldots, x_n \in A_n$.

Proof If $f = g$, then it is clear that $f(x_1, x_2, \ldots, x_n) = g(x_1, x_2, \ldots, x_n)$ for all $x_1, x_2, \ldots, x_n \in \mathbb{K}$ and, in particular, for $x_1 \in A_1, x_2 \in A_2, \ldots, x_n \in A_n$.

Conversely, suppose that this last condition is satisfied, and let us prove that $f = g$ by using induction on the number n of indeterminates. Corollary 15.10 already gives the validity of this claim for $n = 1$. Assume, by induction hypothesis, that the result holds true for polynomials over $\mathbb{K}$ in $n - 1$ indeterminates, and write

$$\begin{cases} f(X_1, X_2, \ldots, X_n) = \sum_{j=0}^{m} f_j(X_2, \ldots, X_n)X_1^j \\ g(X_1, X_2, \ldots, X_n) = \sum_{j=0}^{p} g_j(X_2, \ldots, X_n)X_1^j \end{cases}, \tag{16.1}$$

with $f_i, g_j \in \mathbb{K}[X_2, \ldots, X_n]$ for all $0 \le i \le m, 0 \le j \le p$.

For arbitrarily fixed $x_2 \in A_2, \ldots, x_n \in A_n$, the hypothesis gives

$$f(x_1, x_2, \ldots, x_n) = g(x_1, x_2, \ldots, x_n)$$

for all $x_1 \in A_1$, i.e.,

$$\sum_{j=0}^{m} f_j(x_2, \ldots, x_n)x_1^j = \sum_{j=0}^{p} g_j(x_2, \ldots, x_n)x_1^j,$$

for all $x_1 \in A_1$. Since A_1 is an infinite set, applying Corollary 15.10 to the polynomials in X_1

$$f(X_1, x_2, \ldots, x_n) = \sum_{j=0}^{m} f_j(x_2, \ldots, x_n)X_1^j$$

and

$$g(X_1, x_2, \ldots, x_n) = \sum_{j=0}^{p} g_j(x_2, \ldots, x_n)X_1^j,$$

we conclude that $m = p$ and

$$f_j(x_2, \ldots, x_n) = g_j(x_2, \ldots, x_n) \tag{16.2}$$

for $0 \le j \le m$. However, since $x_2 \in A_2, \ldots, x_n \in A_n$ were arbitrarily chosen, we conclude that (16.2) holds for all $x_2 \in A_2, \ldots, x_n \in A_n$. Hence, our induction hypothesis guarantees that $f_j = g_j$ for $0 \le j \le m$, so that (16.1) gives $f = g$. $\quad\square$

Remark 16.2 From now on, whenever we deal with polynomials f over $\mathbb{K}$ in two indeterminates, we will generally write $f(X, Y)$ instead of $f(X_1, X_2)$. An analogous remark holds for polynomials f over $\mathbb{K}$ in three indeterminates, which will be generally denoted by $f(X, Y, Z)$, instead of $f(X_1, X_2, X_3)$.

Example 16.3 Write $(X + Y + Z)^3 - (X^3 + Y^3 + Z^3)$ as a product of polynomials of degree 1.

Solution Choose arbitrary $y, z \in \mathbb{R}^*$, with $y \neq z$, and consider the polynomial in X

$$g(X) = f(X, y, z) = (X + y + z)^3 - X^3 - (y^3 + z^3).$$

Since

$$f(-y, y, z) = (-y + y + z)^3 - ((-y)^3 + y^3 + z^3) = 0,$$

the root test assures that the polynomial (in X) $f(X, y, z)$ is divisible by $X - (-y) = X + y$. Analogously, f is also divisible by $X + z$ and, since $\partial g = 2$, item (c) of Proposition 15.3 guarantees that

$$f(X, y, z) = \alpha(X + y)(X + z),$$

for some $\alpha \in \mathbb{R}$ to be found. By evaluating the equality above at 0, we obtain

$$\alpha yz = f(0, y, z) = (y + z)^3 - (y^3 + z^3) = 3yz(y + z),$$

so that $\alpha = 3(y + z)$. Therefore,

$$f(X, y, z) = 3(y + z)(X + y)(X + z)$$

and, then, $f(x, y, z) = 3(y + z)(x + y)(x + z)$ for all $x \in \mathbb{R}$ and $y, z \in \mathbb{R}^*$, with $y \neq z$. Proposition 16.1 now gives

$$f(X, Y, Z) = 3(Y + Z)(X + Y)(X + Z).$$

$\square$

Problems: Sect. 16.1

1. * If $f \in \mathbb{K}[X_1, \ldots, X_n]$ does not vanish identically, prove that there exist infinite sets $A_1, \ldots, A_n \subset \mathbb{K}$ such that $f(x_1, \ldots, x_n) \neq 0$ for all $x_1 \in A_1, \ldots, x_n \in A_n$.

2. Do the following items:

   (a) Prove that the polynomial $(X - Y)^5 + (Y - Z)^5 + (Z - X)^5$ is divisible by $(X - Y)(Y - Z)(Z - X)$.

   (b) Write the polynomial $(X - Y)^5 + (Y - Z)^5 + (Z - X)^5$ as a product of three polynomials of degree 1 and one polynomial of degree 2.

3. Write the polynomial $(X + Y + Z)^5 - X^5 - Y^5 - Z^5$ as a product of three polynomials of degree 1 and one polynomial of degree 2.

16.2   Symmetric Polynomials

For what follows, given $n \in \mathbb{N}$ we recall that $I_n = \{1, 2, \ldots, n\}$; also, a function $\sigma : I_n \to I_n$ is said to be a permutation of I_n if it is a bijection from I_n into itself. The object of study along this section is isolated in the coming definition.

Definition 16.4 A polynomial $f \in \mathbb{K}[X_1, \ldots, X_n]$ is **symmetric** if

$$f(X_1, X_2, \ldots, X_n) = f(X_{\sigma(1)}, X_{\sigma(2)}, \ldots, X_{\sigma(n)}),$$

for every permutation σ of I_n.

For a better understanding of the above definition, let us look at the polynomials $f, g \in \mathbb{K}[X, Y]$ given by

$$f(X, Y) = X^2 + Y^2 - XY + X + Y \text{ and } g(X, Y) = X^3 + Y^3 - X.$$

The first is symmetric, whereas the second one is not, for

$$f(Y, X) = Y^2 + X^2 - YX + Y + X = f(X, Y),$$

and

$$g(Y, X) = Y^3 + X^3 - Y \neq g(X, Y).$$

A particular set of symmetric polynomials, called *elementary*, deserves special attention, and this is the object of the coming definition.

Definition 16.5 For $0 \leq j \leq n$, the j-th **elementary symmetric polynomial** in $X_1, \ldots, X_n$, denoted $s_j = s_j(X_1, \ldots, X_n)$, is defined by

$$s_j(X_1, \ldots, X_n) = \begin{cases} 1, & \text{if } j = 0 \\ \sum_{1 \leq i_1 < \cdots < i_j \leq n} X_{i_1} X_{i_2} \ldots X_{i_j}, & \text{if } 1 \leq j \leq n \end{cases}.$$

In the case $n = 3$, for example, we have

$$s_0 = 1, \quad s_1 = X + Y + Z, \quad s_2 = XY + YZ + XZ, \quad s_3 = XYZ.$$

For an arbitrary natural number n, it is not difficult to prove that s_j is indeed symmetric. On the other hand, the actual importance of the elementary symmetric polynomials lies in the next proposition, which establishes **Girard-Viète formulas**[1] between the coefficients and roots of a given polynomial.

---

[1]François Viète and Albert Girard, French mathematicians of the sixteenth and seventeenth centuries, respectively.

Proposition 16.6 (Girard-Viète) *Let* $f(X) = a_n X^n + \cdots + a_1 X + a_0 \in \mathbb{K}[X] \setminus \mathbb{K}$
have roots $\alpha_1, \ldots, \alpha_n \in \mathbb{K}$, *repeated according to their multiplicities. For* $1 \leq j \leq$
n, *one has*

$$s_j(\alpha_1, \ldots, \alpha_n) = (-1)^j \frac{a_{n-j}}{a_n}. \tag{16.3}$$

Proof By the sake of simplicity, we shall write $s_j(\alpha_i)$ to denote $s_j(\alpha_1, \ldots, \alpha_n)$.
According to our hypotheses, the factorised form of the polynomial f is $f(X) =$
$a_n(X - \alpha_1)(X - \alpha_2) \ldots (X - \alpha_n)$. Expanding the products, we obtain

$$f(X) = a_n X^n - a_n s_1(\alpha_i) X^{n-1} + a_n s_2(\alpha_i) X^{n-2} - \cdots + a_n(-1)^n s_n(\alpha_i).$$

It now suffices to compare such an expression for f with that given in the
statement of the proposition to obtain

$$-a_n s_1(\alpha_i) = a_{n-1}, \ a_n s_2(\alpha_i) = a_{n-2}, \ \ldots, a_n(-1)^n s_n(\alpha_i) = a_0.$$

$\square$

In order to get the right feeling on what the previous proposition says and what
it does not say, let α, β and γ be the complex roots of the polynomial $f(X) =$
$X^3 - 2X^2 + 1$. Girard-Viète relations give

$$\alpha + \beta + \gamma = 2, \ \alpha\beta + \alpha\gamma + \beta\gamma = 0 \text{ and } \alpha\beta\gamma = -1.$$

However, it is worth noticing that such relations do not bring enough information
for us to compute the values of α, β and γ; indeed, if we try to solve the system of
equations formed by them, we will simply obtain the polynomial equations $f(\alpha) =$
0, $f(\beta) = 0$ and $f(\gamma) = 0$. Actually, by multiplying both sides of the second
equation by α, we get

$$\alpha^2(\beta + \gamma) + \alpha\beta\gamma = 0;$$

in turn, substituting $\beta + \gamma$ by $2 - \alpha$ and $\alpha\beta\gamma$ by -1 in this last equality, we arrive at

$$\alpha^2(2 - \alpha) - 1 = 0.$$

In the notations of the former proposition, we shall refer to $s_j(\alpha_1, \ldots, \alpha_n) \in \mathbb{K}$
as the j-th **elementary symmetric sum** of the roots of f. Whenever $\alpha_1, \ldots, \alpha_n$
are understood and there is no danger of confusion with the elementary symmetric
polynomial $s_j = s_j(X_1, \ldots, X_n)$, we shall write simply s_j to denote such a sum.

We now present a series of examples that illustrate a variety of different situations
to which one can profitably apply Girard-Viète relations.

Example 16.7 Let $f(X) = X^n + a_{n-1}X^{n-1} + a_{n-2}X^{n-2} + \cdots + a_1 X + a_0$ be a polynomial of real coefficients, such that $a_{n-1}^2 < 2a_{n-2}$. Prove that f has at least two complex, nonreal roots.

Proof By contradiction, assume that $a_{n-1}^2 < 2a_{n-2}$ but at least $n-1$ of the roots of f are real, say $\alpha_1, \ldots, \alpha_{n-1} \in \mathbb{R}$. Since f and $g(X) = (X - \alpha_1)\ldots(X - \alpha_{n-1})$ have real coefficients, we get from the division algorithm that

$$f(X) = g(X)(X - \alpha_n),$$

for some $\alpha_n \in \mathbb{R}$. Therefore, all roots of f are real.

Now, by applying Girard-Viète relations, we obtain

$$\sum_{i=1}^{n} \alpha_i = -a_{n-1} \text{ and } \sum_{i<j} \alpha_i \alpha_j = a_{n-2}.$$

Hence

$$0 \le \sum_{i=1}^{n} \alpha_i^2 = \left(\sum_{i=1}^{n} \alpha_i\right)^2 - 2\sum_{i<j} \alpha_i \alpha_j = a_{n-1}^2 - 2a_{n-2} < 0,$$

which gives us the desired contradiction. □

Example 16.8 (Croatia) Let a, b and c be nonzero real numbers satisfying $a + b + c = 0$. Prove that

$$\frac{a^5 + b^5 + c^5}{5} = \left(\frac{a^3 + b^3 + c^3}{3}\right)\left(\frac{a^2 + b^2 + c^2}{2}\right).$$

Proof If $f(X) = (X - a)(X - b)(X - c)$, then condition $a + b + c = 0$ gives $f(X) = X^3 + sX - t$, with $s = ab + ac + bc$ and $t = abc$. Now, since $f(a) = 0$, we obtain $a^3 = -sa + t$ and, analogously, $b^3 = -sb + t$ and $c^3 = -sc + t$. By termwise addition of these relations and using condition $a + b + c = 0$ once more, we get

$$a^3 + b^3 + c^3 = -s(a + b + c) + 3t = 3t.$$

In order to deal with the sum $a^5 + b^5 + c^5$, we start by multiplying both sides of the equalities $a^3 = -sa + t$, $b^3 = -sb + t$ and $c^3 = -sc + t$ respectively by a^2, b^2 and c^2; we then add them termwisely to obtain

$$a^5 + b^5 + c^5 = -s(a^3 + b^3 + c^3) + t(a^2 + b^2 + c^2).$$

In the right hand side of the above expression, substitute $a^3 + b^3 + c^3 = 3t$ and

$$a^2 + b^2 + c^2 = (a + b + c)^2 - 2(ab + ac + bc) = -2s$$

to obtain

$$a^5 + b^5 + c^5 = -s \cdot 3t + t(-2s) = -5st.$$

The stated equality is, now, obvious.   □

Example 16.9 (OCM) The roots of the polynomial $f(X) = X^3 - 7X^2 + 14X - 6$ are the lengths of the sides of a triangle. Compute the value of its area.

Solution Let a, b and c be the roots of f and A be the area of the triangle with sides a, b and c. Heron's formula for the area of a triangle (cf. Chapter 7 of [9], for instance) gives

$$A^2 = p(p - a)(p - b)(p - c),$$

where p is the semi-perimeter of the triangle. On the other hand, by Girard-Viète relations, we get

$$p = \frac{1}{2}(a + b + c) = \frac{1}{2}s_1(a, b, c) = \frac{7}{2},$$

so that

$$A^2 = \frac{7}{2}\left(\frac{7}{2} - a\right)\left(\frac{7}{2} - b\right)\left(\frac{7}{2} - c\right).$$

Now, the factorised form of f gives

$$f(X) = (X - a)(X - b)(X - c) = X^3 - 7X^2 + 14X - 6$$

and, hence,

$$f\left(\frac{7}{2}\right) = \left(\frac{7}{2} - a\right)\left(\frac{7}{2} - b\right)\left(\frac{7}{2} - c\right)$$
$$= \left(\frac{7}{2}\right)^3 - 7\left(\frac{7}{2}\right)^2 + 14\left(\frac{7}{2}\right) - 6 = \frac{1}{8}.$$

Thus,

$$A^2 = \frac{7}{2} \cdot f\left(\frac{7}{2}\right) = \frac{7}{2} \cdot \frac{1}{8} = \frac{7}{16},$$

whence $A = \frac{\sqrt{7}}{4}$.   □

Example 16.10 (Romania) Let a, b, c and d be real numbers such that

$$a = \sqrt{4 - \sqrt{5 - a}}, \quad b = \sqrt{4 + \sqrt{5 - b}},$$

$$c = \sqrt{4 - \sqrt{5 + c}} \text{ and } d = \sqrt{4 + \sqrt{5 + d}}.$$

Compute all possible values of the product $abcd$.

Solution Note that $a^2 = 4 - \sqrt{5 - a}$ and, then, $(a^2 - 4)^2 = 5 - a$ or, which is the same, $a^4 - 8a^2 + a + 11 = 0$. Analogously, we have $b^4 - 8b^2 + b + 11 = 0$, so that a and b are roots of the polynomial

$$f(X) = X^4 - 8X^2 + X + 11.$$

Likewise, we conclude that c and d are roots of the polynomial $X^4 - 8X^2 - X + 11$, so that $-c$ and $-d$ are also roots of f.

Therefore, all of a, b, $-c$ and $-d$ are roots of f, and if we know that they are pairwise distinct, we shall conclude that they are all of the roots of the polynomial f; in turn, Girard-Viète relations will give

$$abcd = ab(-c)(-d) = 11.$$

For what is left to do, assume we had $a = b$. This would give $\sqrt{4 - \sqrt{5 - a}} = \sqrt{4 + \sqrt{5 - a}}$ and, hence, $a = 5$. However, since $5 \neq \sqrt{4 - \sqrt{5 - 5}}$, we actually have $a \neq b$. Similarly, we prove that $c \neq d$. Finally, since $-c, -d < 0 < a, b$, we are finished. □

Example 16.11 (Romania) If $x_1, x_2, \ldots, x_n$ are positive real numbers for which $x_1 x_2 \ldots x_n = 1$, prove that

$$\sum_{j=1}^{n} \frac{1}{n - 1 + x_j} \leq 1.$$

Proof Let $p(X) = (X + x_1) \ldots (X + x_n)$. Problem 3, page 393, gives

$$\frac{p'(x)}{p(x)} = \sum_{i=1}^{n} \frac{1}{x + x_i}$$

for $x \neq -x_1, \ldots, -x_n$. Hence, we want to show that

$$\frac{p'(n - 1)}{p(n - 1)} \leq 1.$$

For what is left to do, for $1 \leq i \leq n$ let $a_i = s_i(x_1, \ldots, x_n)$ be the i-th symmetric elementary sum of $x_1, \ldots, x_n$, and set $a_0 = 1$. Then, we have

$$p(X) = \sum_{j=0}^{n} a_j X^{n-j} \quad \text{and} \quad p(X) = \sum_{j=0}^{n-1} (n-j) a_j X^{n-j-1},$$

so that it suffices to show the inequality

$$\sum_{j=0}^{n} a_j (n-1)^{n-j} \geq \sum_{j=0}^{n-1} (n-j) a_j (n-1)^{n-j-1}$$

or, which is the same,

$$\sum_{j=1}^{n-1} a_j (j-1)(n-1)^{n-j-1} + a_n \geq n a_0 (n-1)^{n-1} - a_0 (n-1)^n.$$

Finally, we want to prove that

$$\sum_{j=1}^{n} a_j (j-1)(n-1)^{n-j-1} \geq (n-1)^{n-1}.$$

To this end, the inequality between the arithmetic and geometric means (cf. [8], for instance) furnishes

$$a_j = \sum_{i_1 < \cdots < i_j} x_{i_1} \ldots x_{i_j} \geq \binom{n}{j} \sqrt[\binom{n}{j}]{\prod_{i_1 < \cdots < i_j} x_{i_1} \ldots x_{i_j}} = \binom{n}{j}.$$

Therefore, it is enough to show that

$$\sum_{j=1}^{n} \binom{n}{j} (j-1)(n-1)^{n-j-1} \geq (n-1)^{n-1}.$$

We claim that this last inequality is actually an equality. Indeed, the equality

$$X^n = (X-1+1)^n = \sum_{k=0}^{n} \binom{n}{k} (X-1)^{n-k}$$

give us, upon differentiation,

$$n X^{n-1} = \sum_{k=0}^{n-1} (n-k) \binom{n}{k} (X-1)^{n-k-1}.$$

Now, by evaluating the corresponding polynomial functions at $x = n$, it comes that

$$n^n = \sum_{k=0}^{n-1}(n-k)\binom{n}{k}(n-1)^{n-k-1}$$

$$= \sum_{k=0}^{n-1}[(n-1)-(k-1)]\binom{n}{k}(n-1)^{n-k-1}$$

$$= \sum_{k=0}^{n-1}\binom{n}{k}(n-1)^{n-k} - \sum_{k=0}^{n-1}(k-1)\binom{n}{k}(n-1)^{n-k-1}.$$

On the other hand,

$$n^n = (n-1+1)^n = \sum_{k=0}^{n}\binom{n}{k}(n-1)^{n-k},$$

so that

$$\sum_{k=0}^{n}\binom{n}{k}(n-1)^{n-k} = \sum_{k=0}^{n-1}\binom{n}{k}(n-1)^{n-k} - \sum_{k=0}^{n-1}(k-1)\binom{n}{k}(n-1)^{n-k-1}.$$

After performing the obvious cancellations, we arrive at the equality

$$\binom{n}{n}(n-1)^{n-n} + \sum_{k=0}^{n-1}(k-1)\binom{n}{k}(n-1)^{n-k-1} = 0,$$

which is the same as

$$\sum_{k=0}^{n}(k-1)\binom{n}{k}(n-1)^{n-k-1} = 0.$$

At last, from the above we can write

$$(0-1)\binom{n}{0}(n-1)^{n-0-1} + \sum_{k=1}^{n}(k-1)\binom{n}{k}(n-1)^{n-k-1} = 0,$$

which is precisely the desired equality. $\square$

Problems: Sect. 16.2

1. Let f, g and h be nonzero polynomials in $\mathbb{K}[X_1, \ldots, X_n]$, such that f and g are symmetric and $f = gh$. Prove that h is also symmetric.
2. * A polynomial $f \in \mathbb{K}[X_1, \ldots, X_n]$ is said to be **homogeneous** of degree k if

$$f(tX_1, \ldots, tX_n) = t^k f(X_1, \ldots, X_n)$$

for every $t \in \mathbb{K}$. Find, up to multiplication by a constant, all symmetric and homogeneous polynomials of degree 2 in $\mathbb{R}[X, Y, Z]$.

For the next two problems, S_n stands for the set of all permutations of I_n

3. In $\mathbb{R}[X_1, \ldots, X_n]$, $n > 1$, let be given a nonzero polynomial

$$f(X_1, \ldots, X_n) = \sum_{i_1, \ldots, i_n} a_{i_1 \ldots a_i} X_1^{i_1} \ldots X_n^{i_n}.$$

For $\sigma \in S_n$, define $f^\sigma \in \mathbb{R}[X_1, \ldots, X_n]$ by

$$f^\sigma(X_1, \ldots, X_n) = f(X_{\sigma(1)}, \ldots, X_{\sigma(n)}) = \sum_{i_1, \ldots, i_n} a_{i_1 \ldots a_i} X_{\sigma(1)}^{i_1} \ldots X_{\sigma(n)}^{i_n}.$$

In this respect, do the following items:

(a) Prove that $f^{\sigma \circ \tau} = (f^\sigma)^\tau$, for all $\sigma, \tau \in S_n$.
(b) If $f(X_1, \ldots, X_n) = \prod_{1 \le i < j \le n}(X_i - X_j)$, then $f^\sigma = \pm f$ for every $\sigma \in S_n$.
(c) For f as in (b), write $f^\sigma = \text{sgn}(\sigma)f$, with $\text{sgn}(\sigma) \in \{\pm 1\}$. Show that $\text{sgn}(\sigma \circ \tau) = \text{sgn}(\sigma)\text{sgn}(\tau)$, for all $\sigma, \tau \in S_n$.
(d) Let $A_n = \{\sigma \in S_n; \text{sgn}(\sigma) = 1\}$ (we call the elements of A_n **even permutations**). For $\sigma, \tau \in A_n$, show that $\sigma \circ \tau$ and σ^{-1} also belong to A_n.
(e) Given $1 \le k < l \le n$, we let τ_{kl} denote the **transposition** corresponding to k and l, i.e., the permutation of I_n such that

$$\tau_{kl}(i) = \begin{cases} i, & \text{if } i \ne k, l \\ l, & \text{if } i = k \\ k, & \text{if } i = l \end{cases}.$$

Prove that $\text{sgn}(\tau_{kl}) = -1$, for all $1 \le k < l \le n$.
(f) Show that $|A_n| = \frac{n!}{2}$.

4. For $f \in \mathbb{K}[X_1, \ldots, X_n]$, let $g \in \mathbb{K}[X_1, \ldots, X_n]$ be defined by

$$g = \frac{1}{n!} \sum_{\sigma \in S_n} f^\sigma,$$

where f^σ is given as in the previous problem. Prove that g is a symmetric polynomial, called the **symmetrization** of f, and that $g = f$ if f is itself symmetric.

5. (Brazil) It is known that the polynomial $f(X) = X^3 + pX + q$ has three distinct real roots. Prove that $p < 0$.

6. (Croatia) Let a, b and c are pairwise distinct real numbers satisfying the system of equations

$$\begin{cases} a^3 = 3b^2 + 3c^2 - 25 \\ b^3 = 3c^2 + 3a^2 - 25 \\ c^3 = 3a^2 + 3b^2 - 25 \end{cases},$$

Compute all possible values of the product abc.

7. Given $a, b, c \in \mathbb{C}$ find, as a function of a, b and c, the coefficients of a monic polynomial of degree 3 whose complex roots are the cubes of those of $X^3 + aX^2 + bX + c$.

8. Let $p, q, r \in \mathbb{R}$ and assume that the complex roots of the polynomial $X^3 + pX^2 + qX + r$ are three distinct and positive real numbers. Show that such roots are the lengths of the sides of a triangle if and only if $p^3 - 4pq + 8r > 0$.

9. (Romania) Let a, b and c be nonzero complex numbers, such that

$$a + b + c = \frac{1}{a} + \frac{1}{b} + \frac{1}{c} = 0.$$

If n is a positive integer, show that $a^n + b^n + c^n = 0$ if and only if $3 \nmid n$.

10. (Hong Kong) Let $a_3, a_4, \ldots, a_{100}$ be real numbers, with $a_{100} \neq 0$. Prove that the polynomial

$$f(X) = a_{100}X^{100} + a_{99}X^{99} + \cdots + a_3X^3 + 3X^2 + 2X + 1$$

has at least one nonreal root.

11. Consider all straight lines that meet the graph of the polynomial function $f(x) = 2x^4 + 7x^3 + 3x - 5$ in four distinct points (x_i, y_i), for $1 \leq i \leq 4$. Prove that the number $\frac{1}{4}(x_1 + x_2 + x_3 + x_4)$ does not depend on the chosen line, and compute its value.

12. (Moldavia) In the cartesian plane, a circle intersects the hyperbola of equation $xy = 1$ in four distinct points. Prove that the product of the abscissas of the intersection point is always equal to 1.

13. (Canada—adapted) The complex numbers a, b and c are the roots of $X^3 - X^2 - X - 1$.

(a) Prove that a, b and c are pairwise distinct.
(b) If, for each $n \in \mathbb{N}$, we set

$$S_n = \frac{a^n - b^n}{a - b} + \frac{b^n - c^n}{b - c} + \frac{c^n - a^n}{c - a},$$

show that $S_{k+3} = S_{k+2} + S_{k+1} + S_k$ for every $k \in \mathbb{N}$.
(c) Conclude that $S_n \in \mathbb{Z}$, for every $n \in \mathbb{N}$.

14. (BMO—adapted) For real numbers x and y such that $xy \neq -1$, define

$$x * y = \frac{1 + x + y}{1 + xy}.$$

(a) Given $n \geq 2$ positive reals $x_1, x_2, \ldots, x_n$, show that

$$x_1 * (x_2 * (\cdots * x_n) \cdots) = \frac{s_1 + s_3 + \cdots + s_i}{1 + s_2 + s_4 + \cdots + s_p},$$

where i and p stand for the largest odd and even natural numbers less than or equal to n, respectively, and s_j denotes the j-th elementary symmetric sum of $x_1, x_2, \ldots, x_n$.

(b) If $f(X) = (X + x_1)(X + x_2) \ldots (X + x_n)$, show that

$$x_1 * (x_2 * (\cdots * x_n) \cdots) = \frac{f(1) + (-1)^{n+1} f(-1)}{f(1) + (-1)^n f(-1)}.$$

15. Given an integer $n > 1$, find all real solutions of the system of equations

$$\begin{cases} x_1 + x_2 + \cdots + x_n = n \\ x_1^2 + x_2^2 + \cdots + x_n^2 = n \\ \qquad \cdots \\ x_1^n + x_2^n + \cdots + x_n^n = n \end{cases}.$$

16. (IMO—shortlist) Let $f(X) = X^n + a_{n-1}X^{n-1} + \cdots + a_1 X + 1$ be a polynomial of nonnegative real coefficients and real roots. Prove that $f(x) \geq (x + 1)^n$ for every real number $x \geq 0$.

17. (Ireland) Find all polynomials $f(X) = a_n X^n + \cdots + a_1 X + a_0$ satisfying the following conditions:

(a) $a_j \in \{-1, 1\}$ for $0 \leq j \leq n$.
(b) All roots of f are real.

16.3 Newton's Theorem

Yet with respect to symmetric polynomials, note that $f(X, Y, Z) = X^4 + Y^4 + Z^4$ is symmetric, albeit not one of the symmetric polynomials in X, Y, Z which we called *elementary*. Nevertheless, letting $s_1 = s_1(X, Y, Z)$, $s_2 = s_2(X, Y, Z)$ and $s_3 = s_3(X, Y, Z)$, we can write

$$f(X, Y, Z) = (X^2 + Y^2 + Z^2)^2 - 2(X^2Y^2 + X^2Z^2 + Y^2Z^2)$$

$$= \left[(X + Y + Z)^2 - 2(XY + XZ + YZ) \right]^2$$

$$- 2\left[(XY + XZ + YZ)^2 - 2XYZ(X + Y + Z) \right]$$

$$= (s_1^2 - 2s_2)^2 - 2(s_2^2 - 2s_1s_3)$$

$$= s_1^4 - 4s_1^2 s_2 + 2s_2^2 + 4s_1 s_3$$

$$= g(s_1, s_2, s_3),$$

where

$$g(X, Y, Z) = X^4 - 4X^2Y + 2Y^2 + 4XZ.$$

Thus, at least in this particular case, we were capable of expressing the symmetric polynomial $f(X, Y, Z) = X^4 + Y^4 + Z^4$ as a *polynomial in the elementary symmetric polynomials in X, Y, Z.*

Actually, the coming result, usually attributed to Newton and known in mathematical literature as the **fundamental theorem on symmetric polynomials**, assures that this was no accident. In what follows, $\mathbb{K}$ includes the possibility $\mathbb{K} = \mathbb{Z}$.

Theorem 16.12 (Newton) *If* $f = f(X_1, \ldots, X_n) \in \mathbb{K}[X_1, \ldots, X_n]$ *is symmetric, then there exists* $g \in \mathbb{K}[X_1, \ldots, X_n]$ *such that*

$$f(X_1, \ldots, X_n) = g(s_1, \ldots, s_n),$$

where $s_1, \ldots, s_n \in \mathbb{K}[X_1, \ldots, X_n]$ *are the elementary symmetric polynomials in* $X_1, \ldots, X_n$.

For the proof of this theorem, we shall need to introduce some preliminary concepts. Firstly, let us *order* the monomials in $\mathbb{K}[X_1, \ldots, X_n]$ in the following way: given two distinct n-tuples $(i_1, \ldots, i_n)$ and $(j_1, \ldots, j_n)$ of nonnegative integers, and monomials $a_{(i)}X_1^{i_1} \ldots X_n^{i_n}$ and $b_{(j)}X_1^{j_1} \ldots X_n^{j_n}$ in $\mathbb{K}[X_1, \ldots, X_n]$, we define

$$a_{(i)}X_1^{i_1} \ldots X_n^{i_n} \prec b_{(j)}X_1^{j_1} \ldots X_n^{j_n}$$

$$\Updownarrow$$

$$\exists\, 1 \leq k \leq n; \quad \begin{cases} i_l = j_l \text{ if } l < k \\ i_k < j_k \end{cases}. \tag{16.4}$$

In this case, we say that $b_{(j)}X_1^{j_1} \ldots X_n^{j_n}$ is *greater that* $a_{(i)}X_1^{i_1} \ldots X_n^{i_n}$.

In general, we shall refer to the ordering above as the **lexicographic order** in the monomials of $\mathbb{K}[X_1, \ldots, X_n]$. In particular, for $f \in \mathbb{K}[X_1, \ldots, X_n] \setminus \{0\}$, we define its *leading term* as the maximum monomial with respect to the lexicographic order. Note also that in $\mathbb{K}[X]$ one has $1 \prec X \prec X^2 \prec \cdots$, so that the above notion of leading term of a polynomial in several indeterminates coincides with the usual notion in one indeterminate, which is given by the degree of the monomial.

Example 16.13 If $f = s_1^{k_1} \ldots s_n^{k_n}$, with $s_i \in \mathbb{K}[X_1, \ldots, X_n]$ standing for the i-th elementary symmetric polynomial, then f has leading term

$$X_1^{k_1+k_2+\cdots+k_n} X_2^{k_2+\cdots+k_n} \ldots X_{n-1}^{k_{n-1}+k_n} X_n^{k_n}. \tag{16.5}$$

Indeed, since

$$f = \left(\sum_i X_i\right)^{k_1} \left(\sum_{i<j} X_i X_j\right)^{k_2} \ldots (X_1 \ldots X_n)^{k_n},$$

the definition of lexicographic order guarantees that its leading term is

$$X_1^{k_1} (X_1 X_2)^{k_2} (X_1 X_2 X_3)^{k_3} \ldots (X_1 \ldots X_n)^{k_n},$$

which is exactly (16.5).

With the concept of lexicographic order at our disposal, the key to the proof of Newton's theorem lies in the coming auxiliary result.

Lemma 16.14 *If $f \in \mathbb{K}[X_1, \ldots, X_n]$ is symmetric and $a_{(\alpha)} X_1^{\alpha_1} \ldots X_n^{\alpha_n}$ is its leading term, then*

$$\alpha_1 \geq \cdots \geq \alpha_n.$$

Proof If α_1 is the largest exponent appearing in some monomial of f, then, since f is symmetric, it contains a monomial with $X_1^{\alpha_1}$. If $a_{(i)} X_1^{i_1} \ldots X_n^{i_n}$ is a monomial in f such that $i_1 < \alpha_1$, it follows from the definition of lexicographic order that such a monomial is not the leading one. Hence, the leading term of f must contain $X_1^{\alpha_1}$.

Now, among all of the monomials in f containing $X_1^{\alpha_1}$, choose one with maximum exponent in one of the indeterminates $X_2, \ldots, X_n$, say exponent α_2. By invoking again the symmetry of f, there exists such a monomial containing $X_1^{\alpha_1} X_2^{\alpha_2}$; moreover, from the choice of α_1 we have $\alpha_1 \geq \alpha_2$. In turn, if $a_{(i)} X_1^{\alpha_1} X_2^{i_2} \ldots X_n^{i_n}$ is a monomial in f such that $i_2 < \alpha_2$, then (again from the definition of lexicographic order) such a monomial is not the leading one, which shows that the leading term contains $X_1^{\alpha_1} X_2^{\alpha_2}$. Finally, by repeating the above argument more $n - 2$ times, we obtain the desired result.   □

We can finally present the proof of Newton's theorem.

Proof of Theorem 16.12 Take $f \in \mathbb{K}[X_1, \ldots, X_n]$ symmetric, with leading term $a_{(\alpha)} X_1^{\alpha_1} \ldots X_n^{\alpha_n}$. By the previous lemma, we have $\alpha_1 \geq \cdots \geq \alpha_n$. On the other hand, Example 16.13 guarantees that the symmetric polynomial

$$g(X_1, \ldots, X_n) = a_{(\alpha)} s_1^{\alpha_1 - \alpha_2} s_2^{\alpha_2 - \alpha_3} \ldots s_{n-1}^{\alpha_{n-1} - \alpha_n} s_n^{\alpha_n}$$

also has $a_{(\alpha)} X_1^{\alpha_1} \ldots X_n^{\alpha_n}$ as its leading term, so that the symmetric polynomial $f - g$ has leading term $a_{(\beta)} X_1^{\beta_1} \ldots X_n^{\beta_n}$, with $a_{(\beta)} X_1^{\beta_1} \ldots X_n^{\beta_n} \prec a_{(\alpha)} X_1^{\alpha_1} \ldots X_n^{\alpha_n}$ in the lexicographic order. However, since $\mathbb{K}[X_1, \ldots, X_n]$ contains only a finite number of monomials which are less than $a_{(\alpha)} X_1^{\alpha_1} \ldots X_n^{\alpha_n}$ with respect to the lexicographic order, a finite number of repetitions of the algorithm above gives us $f - g = l(s_1, \ldots, s_n)$, for some polynomial $l \in \mathbb{K}[X_1, \ldots, X_n]$. Thus, the same happens with f. □

It is worth noticing that Newton's theorem is mainly an existence result. Indeed, it is possible to prove that, for generic symmetric polynomials, the algorithm described in the proof of Theorem 16.12 does not finish in polynomial time (i.e., even with the aid of a good computer, we will not be able to use the algorithm to actually express, in finite time, a generic symmetric polynomial in n indeterminates as a polynomial in the corresponding elementary symmetric polynomials). Nevertheless, the coming example shows that the mere existence guaranteed by Newton's theorem can be quite useful. For another interesting application, see Sect. 20.1.

Example 16.15 (Miklós Schweitzer) Let $f \in \mathbb{Z}[X]$ be a nonconstant polynomial and $\omega = \text{cis} \frac{2\pi}{n}$, where $n > 1$ is an integer. Prove that

$$f(\omega) f(\omega^2) \ldots f(\omega^{n-1}) \in \mathbb{Z}.$$

Proof Consider the polynomial $g \in \mathbb{Z}[X_1, \ldots, X_{n-1}]$ given by

$$g(X_1, \ldots, X_{n-1}) = f(X_1) f(X_2) \ldots f(X_{n-1}).$$

If σ is a permutation of I_{n-1}, then

$$\{\sigma(1), \sigma(2), \ldots, \sigma(n-1)\} = \{1, 2, \ldots, n-1\},$$

so that

$$
\begin{aligned}
g(X_{\sigma(1)}, \ldots, X_{\sigma(n-1)}) &= f(X_{\sigma(1)}) f(X_{\sigma(2)}) \ldots f(X_{\sigma(n-1)}) \\
&= f(X_1) f(X_2) \ldots f(X_{n-1}) \\
&= g(X_1, \ldots, X_{n-1}).
\end{aligned}
$$

Thus, g is symmetric and Newton's theorem assures the existence of a polynomial $h \in \mathbb{Z}[X_1, \ldots, X_{n-1}]$ such that $g(X_1, \ldots, X_{n-1}) = h(s_1, \ldots, s_{n-1})$, i.e.,

$$f(X_1) f(X_2) \ldots f(X_{n-1}) = h(s_1, \ldots, s_{n-1}),$$

with s_j standing for the j-th elementary symmetric polynomial in $X_1, \ldots, X_{n-1}$.

Substituting X_j by ω^j in the equality above, we obtain

$$f(\omega)f(\omega^2)\dots f(\omega^{n-1}) = h(s_1(\omega,\dots,\omega^{n-1}),\dots,s_{n-1}(\omega,\dots,\omega^{n-1})).$$

Therefore, in order to show that $f(\omega)f(\omega^2)\dots f(\omega^{n-1}) \in \mathbb{Z}$, it suffices to show that $s_j(\omega,\dots,\omega^{n-1}) \in \mathbb{Z}$ for $1 \le j \le n-1$.

For what is left to do, just note that

$$X^n - 1 = (X-1)(X-\omega)(X-\omega^2)\dots(X-\omega^{n-1})$$
$$= (X-1)(X^{n-1} + X^{n-2} + \dots + X + 1),$$

so that

$$X^{n-1} + X^{n-2} + \dots + X + 1 = (X-\omega)(X-\omega^2)\dots(X-\omega^{n-1})$$
$$= \sum_{j=0}^{n-1} (-1)^j s_j(\omega,\omega^2,\dots,\omega^{n-1})X^{n-1-j}.$$

Hence, we have

$$s_j(\omega,\omega^2,\dots,\omega^{n-1}) = (-1)^j \in \mathbb{Z},$$

as wished.   □

The discussion preceding the former example hints to the following fact: if we wish to effectively express a given symmetric polynomial as a polynomial in the elementary symmetric polynomials in the same number of indeterminates, we shall usually have to rely on *ad hoc* arguments. In this sense, we end this section by discussing a relevant example, for which we shall need the following consequence of Horner-Rufinni's algorithm. Identities (16.6) below are known as **Horner-Ruffini's identities**.

Lemma 16.16 (Horner-Rufinni) *Let*

$$f(X) = X^n - s_1 X^{n-1} + \dots + (-1)^{n-1} s_{n-1} X + (-1)^n s_n$$

be a nonconstant polynomial over $\mathbb{C}$. *If* z *is a complex root of* f *and*

$$g(X) = X^{n-1} + b_1 X^{n-2} + \dots + b_{n-1}$$

is the quotient upon division of $f(X)$ *by* $X - z$, *then*

$$\begin{cases} b_1 = z - s_1 \\ b_2 = z^2 - s_1 z + s_2 \\ \cdots \quad \cdots \\ b_{k+1} = z^{k+1} - s_1 z^k + \cdots + (-1)^{k+1} s_{k+1} \\ \cdots \quad \cdots \\ b_{n-1} = z^{n-1} - s_1 z^{n-2} + \cdots + (-1)^{n-1} s_{n-1} \end{cases} \tag{16.6}$$

Proof In the stated notations, the recurrence formulas (15.1) can be written as

$$\begin{cases} b_1 = z - s_1 \\ b_2 = z b_1 + s_2 \\ \cdots \quad \cdots \\ b_{k+1} = z b_k + (-1)^{k+1} s_{k+1} \\ \cdots \quad \cdots \\ b_{n-1} = z b_{n-2} + (-1)^{n-1} s_{n-1} \end{cases}$$

By solving the linear system above successively for $b_2, \ldots, b_{n-1}$, we obtain, one by one, the stated relations. □

The relations contained in the coming result will give us recurrences that express the symmetric polynomial

$$\sigma_k(X_1, X_2, \ldots, X_n) = X_1^k + X_2^k + \cdots + X_n^k$$

as a polynomial in the elementary symmetric polynomials in $X_1, X_2, \ldots, X_n$. The formulas of items (a) and (b) will be respectively referred to as the first and second **Jacobi's identities**.[2] For a proof of them using generating functions, see Example 21.19.

Proposition 16.17 (Jacobi) *For $z_1, \ldots, z_n \in \mathbb{C}$, if $s_i = s_i(z_1, \ldots, z_n)$ denotes the i-th elementary symmetric sum of $z_1, \ldots, z_n$ and $\sigma_k = z_1^k + \cdots + z_n^k$, then:*

(a) $\sigma_{n+k} = \sum_{j=1}^n (-1)^{j-1} s_j \sigma_{n+k-j}$ *for $k \geq 1$.*

(b) $s_{k+1} = \frac{1}{k+1} \sum_{j=1}^{k+1} (-1)^{j-1} s_{k+1-j} \sigma_j$ *for $1 \leq k \leq n-1$.*

Proof

(a) Let

$$f(X) = (X - z_1) \cdots (X - z_n) = X^n - s_1 X^{n-1} + \cdots + (-1)^n s_n. \tag{16.7}$$

---

[2]Carl Gustav Jakob Jacobi, German mathematician of the nineteenth century.

Since z_i is a root of f, we have

$$z_i^n - s_1 z_i^{n-1} + \cdots + (-1)^{n-1} s_{n-1} z_i + (-1)^n s_n = 0$$

for $1 \leq i \leq n$, and thus

$$z_i^{k+n} - s_1 z_i^{k+n-1} + \cdots + (-1)^{n-1} s_{n-1} z_i^{k+1} + (-1)^n s_n z_i^k = 0.$$

By adding the equalities above for $1 \leq i \leq n$, we obtain

$$\sigma_{n+k} - s_1 \sigma_{n+k-1} + \cdots + (-1)^{n-1} s_{n-1} \sigma_{k+1} + (-1)^n s_n \sigma_k = 0,$$

an equality that is equivalent to the stated one.

(b) It follows from (16.7) that

$$f'(X) = nX^{n-1} - (n-1)s_1 X^{n-2} + \cdots + (-1)^{n-1} s_{n-1}. \tag{16.8}$$

On the other hand, for $z \in \mathbb{C} \setminus \{z_1, \ldots, z_n\}$, Problem 3, page 393 gives

$$f'(z) = \frac{f(z)}{z - z_1} + \cdots + \frac{f(z)}{z - z_n}.$$

If $f_j \in \mathbb{C}[X]$ is such that $f(X) = (X - z_j) f_j(X)$, say

$$f_j(X) = X^{n-1} + b_{1j} X^{n-2} + \cdots + b_{n-1,j},$$

we get

$$f'(z) = \sum_{j=1}^n f_j(z) = \sum_{j=1}^n (z^{n-1} + b_{1j} z^{n-2} + \cdots + b_{n-1,j})$$

$$= nz^{n-1} + \Big(\sum_{j=1}^n b_{1j} \Big) z^{n-2} + \cdots + \Big(\sum_{j=1}^n b_{n-1,j} \Big).$$

However, since the equality above is true for every $z \in \mathbb{C} \setminus \{z_1, \ldots, z_n\}$, Corollary 15.10 gives

$$f'(X) = nX^{n-1} + \Big(\sum_{j=1}^n b_{1j} \Big) X^{n-2} + \cdots + \Big(\sum_{j=1}^n b_{n-1,j} \Big). \tag{16.9}$$

Finally, by equating the corresponding coefficients in (16.8) and (16.9) and substituting identities (16.6), we obtain

$$(-1)^{k+1}(n - k - 1)s_{k+1} = \sum_{j=1}^{n} b_{k+1,j}$$

$$= \sum_{j=1}^{n} (z_j^{k+1} - s_1 z_j^k + \cdots + (-1)^{k+1} s_{k+1})$$

$$= \sigma_{k+1} - s_1 \sigma_k + \cdots + (-1)^{k+1} n s_{k+1},$$

so that

$$(k+1)s_{k+1} = s_k \sigma_1 - s_{k-1}\sigma_2 + \cdots + (-1)^k s_0 \sigma_{k+1}.$$

□

The coming result is also due to Jacobi.

Corollary 16.18 (Jacobi) *For each integer $k \geq 1$, let s_k denote the k-th elementary symmetric polynomial in $X_1, \ldots, X_n$, and $\sigma_k = X_1^k + \cdots + X_n^k$. Then:*

(a) $\sigma_{n+k} = \sum_{j=1}^{n}(-1)^{j-1} s_j \sigma_{n+k-j}$ *for $k \geq 1$.*

(b) $s_{k+1} = \frac{1}{k+1} \sum_{j=1}^{k+1}(-1)^{j-1} s_{k+1-j}\sigma_j$ *for $1 \leq k \leq n$ 1.*

Proof The previous proposition assures that both sides of (a) and (b) are equal upon evaluation in arbitrary $z_1, \ldots, z_n \in \mathbb{C}$. Since $\mathbb{C}$ is an infinite set, Proposition 16.1 guarantees the equality of the corresponding polynomials. □

Problems: Sect. 16.3

1. If $f \in \mathbb{Z}[X]$ is a monic polynomial of degree $n \geq 1$ and complex roots $z_1, \ldots, z_n$, prove that $z_1^k + \cdots + z_n^k \in \mathbb{Z}$ for every integer $k \geq 1$.
2. If $a_1, \ldots, a_n, b_1, \ldots, b_n$ are complex numbers such that

$$a_1^k + \cdots + a_n^k = b_1^k + \cdots + b_n^k$$

for $1 \leq k \leq n$, show that $\{a_1, \ldots, a_n\} = \{b_1, \ldots, b_n\}$.
3. (Japan—adapted) Let $n, k \in \mathbb{N}$, with $2 \leq k \leq n$, and $a_1, \ldots, a_k$ real numbers such that

$$\begin{cases} a_1 + \cdots + a_k = n \\ a_1^2 + \cdots + a_k^2 = n \\ \cdots \\ a_1^k + \cdots + a_k^k = n \end{cases}.$$

If $p(X) = (X + a_1) \ldots (X + a_k)$, prove that $p(X) = \sum_{j=0}^{k} \binom{n}{j} X^{k-j}$.

4. Let $m \in \mathbb{N}$, $\omega = \cos \frac{2\pi}{m} + i \sin \frac{2\pi}{m}$ and $z_1, \ldots, z_n$ be complex numbers such that

$$f(X) = (X - z_1)(X - z_2) \ldots (X - z_n) \in \mathbb{Z}[X].$$

(a) If $g(X) = \prod_{j=0}^{m-1} f(\omega^j X)$, prove that

$$g(X) = (X^m - z_1^m) \ldots (X^m - z_n^m).$$

(b) Show that the polynomial g also has integer coefficients.

(c) Conclude that if $z \in \mathbb{C}$ is a root of a nonzero polynomial with integer coefficients, then so is z^m, for every $m \in \mathbb{N}$.

5. (Brazil—adapted)

(a) Given $n \in \mathbb{N}$, prove that there is at most a finite number of monic polynomials of degree n and integer coefficients, such that all of their complex roots have modulus 1.

(b) Let $f \in \mathbb{Z}[X] \setminus \mathbb{Z}$ be a monic polynomial all of whose complex roots have modulus 1. Prove that all of them are roots of unity.

Chapter 17
Polynomials Over $\mathbb{R}$

This chapter revisits, for real polynomials and departing from the fundamental theorem of Algebra, some classical theorems of Calculus. As applications of them, we shall prove Newton's inequalities, which generalizes the classical inequality between the arithmetic and geometric means of n positive real numbers, and Descartes' rule, which relates the number of positive roots of a real polynomial with the number of changes of sign in the sequence of its nonzero coefficients.

17.1 Some Calculus Theorems

In this section we establish, for real polynomials, some classical results on continuity and differentiability. For a more general approach, we refer the reader to [3] or [8], for instance.

Our first result provides a sufficient way for the existence of real roots in a given interval. For the proof of it, we shall need the following auxiliary result, which appeared as Problem 6, page 386. Nevertheless, this time we give a different proof.

Lemma 17.1 *If $f \in \mathbb{R}[X] \setminus \{0\}$ is monic and has no real roots, then there exist polynomials $g, h \in \mathbb{R}[X]$ such that $f = g^2 + h^2$. In particular, $f(x) > 0$ for every $x \in \mathbb{R}$.*

Proof Problem 4, page 386, assures the existence of complex nonreal numbers $z_1, \ldots, z_k$ such that

$$f(X) = \prod_{j=1}^{k} (X - z_j)(X - \overline{z}_j).$$

Now, if $z_j = a_j + ib_j$, with $a_j, b_j \in \mathbb{R}$, then

$$
\begin{aligned}
(X - z_j)(X - \overline{z}_j) &= (X - a_j - ib_j)(X - a_j + ib_j) \\
&= (X - a_j)^2 - (ib_j)^2 \\
&= (X - a_j)^2 + b_j^2,
\end{aligned}
$$

the sum of the squares of two real polynomials (one of them being constant).

It now suffices to apply several times the analogous of Euler's identity (12.8) for polynomials: for $g_1, g_2, h_1, h_2 \in \mathbb{R}[X]$, we have

$$
(g_1^2 + h_1^2)(g_2^2 + h_2^2) = (g_1 g_2 + h_1 h_2)^2 + (g_1 h_2 - g_2 h_1)^2.
$$

For what is left to do, it follows from the first part that

$$
f(x) = g(x)^2 + h(x)^2 \geq 0,
$$

for every $x \in \mathbb{R}$. However, since f has no real roots, the inequality above must be strict, for every $x \in \mathbb{R}$.   □

The coming result is nothing but the intermediate value theorem for polynomials, being known in mathematical literature as **Bolzano's theorem.**[1]

Theorem 17.2 (Bolzano) *If $f \in \mathbb{R}[X]$ and $a < b$ are real numbers such that $f(a)f(b) < 0$, then there exists $c \in (a, b)$ for which $f(c) = 0$.*

Proof Assuming, without loss of generality, that f is monic and $f(a) < 0 < f(b)$, the last part of the previous lemma assures that f has at least one real root. Let $a_1 \leq \cdots \leq a_k$ be the real roots of f, repeated according to their multiplicities. If $g \in \mathbb{R}[X]$ is such that

$$
f(X) = g(X)(X - a_1)\dots(X - a_k),
$$

then g is monic and has no real roots, so that (again by the previous lemma) $g(x) > 0$ for every $x \in \mathbb{R}$.

By contradiction, suppose that no real root of f belongs to the interval (a, b), and separately consider the three following cases:

(i) $a_k < a$: we have

$$
f(a) = g(a)(a - a_1)\dots(a - a_k) > 0,
$$

which is a contradiction.

---

[1]Bernhard Bolzano, German mathematician of the nineteenth century.

(ii) $b < a_1$: it follows from $f(b) > 0$ that

$$g(b)(b - a_1) \ldots (b - a_k) > 0$$

and, hence, k is even (for $g(b) > 0$ and $b - a_i < 0$ for $1 \leq i \leq k$). On the other hand, since $f(a) < 0$, we get

$$g(a)(a - a_1) \ldots (a - a_k) < 0,$$

so that k is odd (for $g(a) > 0$ and $a - a_i < 0$ for $1 \leq i \leq k$). We have, thus, reached a contradiction.

(iii) $a_l < a < b < a_{l+1}$, for some $1 \leq l < k$: we can arrive at a contradiction in a way similar to that of item (ii). For instance,

$$0 < f(a) = \underbrace{g(a)}_{>0} \underbrace{(a - a_1) \ldots (a - a_l)}_{>0} \underbrace{(a - a_{l+1}) \ldots (a - a_k)}_{<0}$$

implies $k - l$ even.

$\square$

Example 17.3 (Moldavia) Let $f, g \in \mathbb{R}[X]$ be two given polynomials, each of which possessing at least one real root.

(a) Prove that there exists $a \in \mathbb{R}$ such that $f(a)^2 = g(a)^2$.
(b) If $f(1 + X + g(X)^2) = g(1 + X + f(X)^2)$, show that $f = g$.

Proof

(a) Let α and β be real roots of f and g, respectively. We can assume, without any loss of generality, that $\alpha \leq \beta$. If $g(\alpha) = 0$ or $f(\beta) = 0$, there is nothing to do. Otherwise,

$$f(\alpha)^2 - g(\alpha)^2 = -g(\alpha)^2 < 0 \quad \text{and} \quad f(\beta)^2 - g(\beta)^2 = f(\beta)^2 > 0,$$

so that Bolzano's theorem, applied to the polynomial $f(X)^2 - g(X)^2$, guarantees the existence of $a \in (\alpha, \beta)$ such that $f(a)^2 - g(a)^2 = 0$.

(b) Choose $a \in \mathbb{R}$ as in item (a) and define a sequence $(u_n)_{n \geq 1}$ by letting $u_0 = a$ and, for each integer $n \geq 1$, $u_{n+1} = 1 + u_n + g(u_n)^2$. It follows from (a) that $f(u_1) = g(u_1)$. Now assume, as induction hypothesis, that $f(u_k) = g(u_k)$ for some integer $k \geq 1$. Then,

$$f(u_{k+1}) = f(1 + u_k + g(u_k)^2)$$

$$= g(1 + u_k + f(u_k)^2)$$

$$= g(1 + u_k + g(u_k)^2) = g(u_{k+1}).$$

Thus, we have $f(u_n) = g(u_n)$ for every integer $n \geq 1$, with

$$u_{n+1} = 1 + u_n + g(u_n)^2 \geq 1 + u_n > u_n$$

for each such $n \geq 1$. Therefore, Corollary 15.10 gives $f = g$.

□

In what comes next, we establish, for polynomials, Lagrange's **mean value theorem**.

Theorem 17.4 (Lagrange) *Let* $f \in \mathbb{R}[X] \setminus \{0\}$ *and* $a < b$ *be given real numbers. Then, there exists* $a < c < b$ *such that*

$$f'(c) = \frac{f(b) - f(a)}{b - a}.$$

Proof We may assume that $f \neq 0$, and first show that if $f(a) = f(b) = 0$, then there exists $a < c < b$ for which $f'(c) = 0$. Without any loss of generality, we can assume that f has no other roots in the interval (a, b). Indeed, since $f \neq 0$, Corollary 15.9 assures that f has at most a finite number of roots in (a, b); then, if these are $a_1 < a_2 < \cdots < a_k$, it suffices to consider a_1 in place of b.

If there exist $a < c < d < b$ such that $f(c)f(d) < 0$, Bolzano's theorem guarantees the existence of a root of f in the interval (a, b) which is an absurd. Hence, f has constant sign along the interval (a, b). Assume, without loss of generality, that $f(x) > 0$ for $x \in (a, b)$, and let k and l be the multiplicities of a and b as roots of f, respectively. Then, there exists $g \in \mathbb{R}[X]$ such that

$$f(X) = (X - a)^k (X - b)^l g(X),$$

with $g(a), g(b) \neq 0$. Now,

$$a < c < b \Rightarrow f(c) > 0 \Rightarrow (c - a)^k (c - b)^l g(c) > 0 \Rightarrow (-1)^l g(c) > 0.$$

We look at the case of an even l (that of odd l can be dealt with in an analogous way), so that $g(c) > 0$ for every $c \in (a, b)$. If $g(a) < 0$, then Bolzano's theorem would assure the existence of $a < d < \frac{a+b}{2}$ such that $g(d) = 0$; in turn, this would give $f(d) = 0$, which is impossible. Therefore, $g(a) > 0$ and, likewise, $g(b) > 0$. We then have

$$f'(X) = k(X - a)^{k-1}(X - b)^l g(X) + l(X - a)^k (X - b)^{l-1} g(X)$$

$$+ (X - a)^k (X - b)^l g'(X)$$

$$= (X - a)^{k-1}(X - b)^{l-1} h(X),$$

with

$$h(X) = (k(X - b) + l(X - a))g(X) + (X - a)(X - b)g'(X).$$

It now suffices to show that there exists $c \in (a, b)$ such that $h(c) = 0$. However, since

$$h(a)h(b) = -kl(a - b)^2 g(a)g(b) < 0,$$

an application of Bolzano's theorem does the job.

For the general case, let

$$f_1(X) = f(X) - f(a) - \left(\frac{f(b) - f(a)}{b - a} \right)(X - a),$$

which is also a polynomial. Since $f_1(a) = f_1(b) = 0$, what we did above guarantees the existence of $c \in (a, b)$ such that $f_1'(c) = 0$. Finally, it is enough to note that

$$f_1'(X) = f'(X) - \left(\frac{f(b) - f(a)}{b - a} \right).$$

□

The case $f(a) = f(b) = 0$ in the previous result preceded the general version due to Lagrange. For this reason, it is usually referred to as **Rôlle's theorem**, in honor of his discoverer, the French mathematician of the seventeenth century Michel Rôlle.

Corollary 17.5 (Rôlle) *If $f \in \mathbb{R}[X]$ and $a < b$ are real numbers such that $f(a) = f(b) = 0$, then there exists $a < c < b$ for which $f'(c) = 0$.*

The coming example illustrates a typical use of Rôlle's theorem.

Example 17.6 Let $f(X) = a_0 + a_1 X + \cdots + a_{n-1} X^{n-1} + a_n X^n$ be a real polynomial such that

$$\frac{a_0}{1} + \frac{a_1}{2} + \cdots + \frac{a_{n-1}}{n} + \frac{a_n}{n+1} = 0.$$

Prove that f has at least one root in the open interval $(0, 1)$.

Proof It is clear that $f = g'$, where g is the real polynomial

$$g(X) = a_0 X + \frac{a_1}{2} X^2 + \cdots + \frac{a_{n-1}}{n} X^n + \frac{a_n}{n+1} X^{n+1}.$$

Since $g(0) = g(1) = 0$, Rôlle's theorem assures the existence of $a \in (0, 1)$ such that $f(a) = g'(a) = 0$.

□

An important consequence of Lagrange's mean value theorem is the study of the first variation (i.e., of the growth or decay) of real polynomials. This is the subject of the coming corollary.

Corollary 17.7 *Let $f \in \mathbb{R}[X] \setminus \{0\}$ and $I \subset \mathbb{R}$ be an interval.*

(a) If $f'(x) > 0$ for every $x \in I$, then f increases in I.
(b) If $f'(x) < 0$ for every $x \in I$, then f decreases in I.

Proof Let us prove item (a), the proof of (b) being entirely analogous. For $a < b$ in I, the mean value theorem furnishes $c \in (a, b)$ (and, hence, $c \in I$) such that

$$\frac{f(b) - f(a)}{b - a} = f'(c) > 0.$$

In particular, $f(b) > f(a)$.   □

The coming example uses a little more Calculus than we discussed above. As before, we refer the reader to [3] or [8] for the necessary background.

Example 17.8 (Sweden) Prove that, for every natural number n, the real polynomial

$$1 + X + \frac{1}{2!}X^2 + \cdots + \frac{1}{n!}X^n$$

has at most one real root.

Proof Letting $f_n = 1 + X + \frac{X^2}{2!} + \cdots + \frac{X^n}{n!}$, we shall show that (i) f_n has no real roots if n is even, and (ii) f_n has exactly one real root if n is odd.

(i) Assuming n even, we have $\lim_{|x| \to +\infty} f_n(x) = +\infty$. Hence, Weierstrass' theorem (Theorem 8.26 of [8]) assures the existence of $x_0 \in \mathbb{R}$ such that f_n assumes its minimum value at x_0. Suppose, for the sake of contradiction, that $f_n(x_0) \leq 0$. Then, Problem 3 gives $f_n'(x_0) = 0$, and then

$$0 \geq f_n(x_0) = f_n'(x_0) + \frac{x_0^n}{n!} = \frac{x_0^n}{n!}.$$

Since n is even, the inequality above successively gives $x_0 = 0$ and $f_n(0) = 0$, which is an absurd.

(ii) Assuming n odd, Problem 5, page 386, gives an odd number of real roots for f_n. On the other hand, since $f_n' = f_{n-1} > 0$ (for $n - 1$ is even), Example 15.33 shows that f_n has no multiple roots. By the sake of contradiction, suppose that f_n has at least two distinct real roots, say $a < b$. Then, Rôlle's theorem gives $\alpha \in (a, b)$ such that $f_n'(\alpha) = 0$. This way, $f_{n-1}(\alpha) = 0$, and this contradicts (i), for $n - 1$ is even.

                                                                                                                □

Problems: Sect. 17.1

1. If $f \in \mathbb{Z}[X] \setminus \mathbb{Z}$ and $m \in \mathbb{N}$ is such that $m > 1 + \mathrm{Re}(z)$, for every complex root z of f, prove that $|f(m)| > 1$.

2. * Prove, for real polynomials and with the methods of this section, the **sign preserving lemma**: if $f \in \mathbb{R}[X]$ is such that $f(a) > 0$ for some $a \in \mathbb{R}$, then there exists $r > 0$ such that $f(x) > 0$ for every $x \in (a - r, a + r)$.

3. * Given $f \in \mathbb{R}[X] \setminus \mathbb{R}$, let $a \in \mathbb{R}$ and $r > 0$ be such that

$$f(a) = \min\{f(x); \ x \in (a - r, a + r)\}.$$

   Prove that $f'(a) = 0$.

4. Let $f(X) = X^5 - 2X^4 + 2$. Prove that f has exactly three real roots.

5. We are given a real nonconstant polynomial f with positive leading coefficient. Prove that there exists $n_0 \in \mathbb{N}$ for which

$$u > v > n_0 \Rightarrow f(u) > f(v) > 0.$$

6. Prove that there does not exist a polynomial $f \in \mathbb{Z}[X]$ such that $f(n)$ is prime for every nonnegative integer n.

7. (Leningrad) Decide whether there exist four distinct real numbers a, b, c and d such that, for any two of them, say x and y, we have

$$x^{10} + x^9 y + \cdots + xy^9 + y^{10} = 1.$$

8. The positive reals a_1, a_2, a_3, a_4 are such that $a_1 \leq a_2 \leq a_3 \leq a_4$ and $a_1 a_2 a_3 a_4 = 1$. If λ is a real root of

$$X^3 - \left(\sum_{i=1}^{4} a_i \right) X^2 + \left(\sum_{1 \leq i < j \leq 4} a_i a_j \right) X - \left(\sum_{i=1}^{4} \frac{1}{a_i} \right),$$

   prove that $\lambda > a_2$.

9. (Leningrad) Let a, b, c, d and e be real numbers such that the polynomial $aX^2 + (c - b)X + (e - d) = 0$ has a real root greater than 1. Prove that the polynomial $aX^4 + bX^3 + cX^2 + dX + e = 0$ has at least one real root.

10. (Austrian-Polish) Given any real numbers $a_1, \ldots, a_n$, prove that

$$\sum_{i, j=1}^{n} \frac{a_i a_j}{i + j} \geq 0,$$

   with equality if and only if $a_1 = \cdots = a_n = 0$.

11. Let $f(X) = X^3 - 3X + 1$. Compute the number of distinct real roots of the polynomial $f(f(X))$.

12. (Sweden) Let $f \in \mathbb{R}[X]$ be a polynomial of degree n. If $f(x) \geq 0$ for every $x \in \mathbb{R}$, show that

$$f(x) + f'(x) + f''(x) + \cdots + f^{(n)}(x) \geq 0,$$

also for every $x \in \mathbb{R}$.

13. (Soviet Union) Two mathematicians A and B play the following game: it is given a polynomial of even degree greater than 4,

$$f(X) = X^{2n} + a_{2n-1}X^{2n-1} + \cdots + a_1 X + 1,$$

with $a_1, \ldots, a_{2n-1}$ real numbers to be chosen. The players, starting from A, alternate themselves in choosing the coefficients of f, until it is completely determined. At the end, A wins if none of the roots of f is real, and B wins otherwise. Find a winning strategy for B.

14. At the beginning of a class, the teacher wrote a third degree real polynomial in the blackboard. Then, the students alternated themselves, each one performing one of the following operations with the polynomial written in the blackboard (which was then substituted by a new one):

   (a) Add, to the written polynomial, its derivative.
   (b) Subtract, to the written polynomial, its derivative.

   At the end of the class, the polynomial initially written by the teacher reappeared in the blackboard. Prove that at least one of the students made a mistake.

15. (OIM, IMO—adapted) We are given real numbers $a_1, a_2, \ldots, a_n$, with $0 < a_1 < a_2 < \cdots < a_n$. If $f : \mathbb{R} \setminus \{-a_1, \ldots, -a_n\} \to \mathbb{R}$ is given by

$$f(x) = \sum_{j=1}^{n} \frac{a_j}{x + a_j},$$

prove that the set $\{x \in \mathbb{R};\ f(x) \geq 1\}$ consists of the union of n bounded, pairwise disjoint intervals, with sum of lengths equal to $a_1 + a_2 + \cdots + a_n$.

17.2 Newton's Inequalities

In this section, we present a set of inequalities that refine the inequality between the arithmetic and geometric means of a finite set of positive real numbers (cf. Section 5.1 of [8], for instance—actually, there we give four different proofs of such an inequality). To this end, we first need the following auxiliary result.

Lemma 17.9 *If $f \in \mathbb{R}[X] \setminus \{0\}$ has k real roots, counted with multiplicities, then f' has at least $k - 1$ real roots, also counted with multiplicities. In particular, if all roots of f are real, then so are all of the roots of f'.*

Proof For the first part, we can assume that $k > 1$. Let $a_1 < \cdots < a_l$ be the distinct real roots of f, with multiplicities respectively equal to $m_1, \ldots, m_l$, so that $m_1 + \cdots + m_l = k$. We have already seen, in Proposition 15.31 that a_i is a root of f' with multiplicity $m_i - 1$. On the other hand, Rôlle's theorem assures that f' has at least one additional root between a_i and a_{i+1}; since there are $l - 1$ intervals of the form (a_i, a_{i+1}), we thus count at least

$$(m_1 - 1) + \cdots + (m_l - 1) + (l - 1) = k - 1$$

real roots for f'.

For the second part, assume that f has degree n and n real roots, counted with multiplicities. Then, f' has degree $n - 1$ and, by the first part, at least $n - 1$ real roots, also counted with multiplicities. Therefore, all roots of f' are real too. □

For what comes next, given $n > 1$ real numbers $a_1, a_2, \ldots, a_n$ and an index $1 \le j \le n$, we shall write $S_j(a_k)$ to denote the j-th elementary symmetric sum of $a_1, \ldots, a_n$, and $H_j(a_k)$ to denote the arithmetic mean of the $\binom{n}{j}$ summands that compose $S_j(a_k)$, i.e.,

$$H_j(a_k) = \frac{S_j(a_k)}{\binom{n}{j}} = \frac{1}{\binom{n}{j}} \sum_{i_1 < \ldots < i_j} a_{i_1} \ldots a_{i_j}.$$

The coming result is usually credited to Newton, and inequalities (17.1) are usually referred to as **Newton's inequalities**. For the sake of notation, we let $H_0 = 1$.

Theorem 17.10 (Newton) *Let $n > 1$ be an integer and $a_1, a_2, \ldots, a_n$ be given real numbers. If $H_1(a_k), H_2(a_k), \ldots, H_{j+1}(a_k) \ne 0$ for some $1 \le j < n$, then*

$$H_j(a_k)^2 \ge H_{j-1}(a_k) H_{j+1}(a_k), \tag{17.1}$$

with equality if and only if $a_1 = a_2 = \ldots = a_n$.

Proof Let us make induction on n. For $n = 2$, we want to show that $H_1^2(a_1, a_2) \ge H_0(a_1, a_2) H_2(a_1, a_2)$ or, which is the same, $\left(\frac{a_1 + a_2}{2}\right)^2 \ge a_1 a_2$. But this is equivalent to $(a_1 - a_2)^2 \ge 0$, which turns into an equality if and only if $a_1 = a_2$.

Assume, by induction hypothesis, that for any $n - 1$ given real numbers $b_1, \ldots, b_{n-1}$ such that $H_1(b_k), H_2(b_k), \ldots, H_{j+1}(b_k) \ne 0$ for some $1 \le j < n - 1$, we have $H_j(b_k)^2 \ge H_{j-1}(b_k) H_{j+1}(b_k)$, with equality if and only if $b_1 = \ldots = b_{n-1}$.

We now consider $n \ge 3$ real numbers $a_1, a_2, \ldots, a_n$, and let

$$f(X) = (X + a_1) \ldots (X + a_n).$$

The previous lemma assures the existence of $n - 1$ real numbers $b_1, \ldots, b_{n-1}$ such that $b_k \in (a_k, a_{k+1})$ for $1 \le k < n$ and

$$f'(X) = n(X + b_1) \ldots (X + b_{n-1}).$$

In particular,

$$\frac{1}{n} f'(X) = X^{n-1} + \binom{n-1}{1} H_1(b_k) X^{n-2} + \cdots + \binom{n-1}{n-1} H_{n-1}(b_k). \qquad (17.2)$$

On the other hand, the equality

$$f(X) = \sum_{j=0}^{n} \binom{n}{j} H_j(a_k) X^{n-j}$$

furnishes

$$f'(X) = \sum_{j=0}^{n-1} (n - j) \binom{n}{j} H_j(a_k) X^{n-j-1}$$

$$= n \sum_{j=0}^{n-1} \binom{n-1}{j} H_j(a_k) X^{n-1-j},$$

and comparing coefficients with (17.2) we obtain, for $0 \le j \le n - 1$,

$$H_j(a_1, \ldots, a_n) = H_j(b_1, \ldots, b_{n-1}).$$

Now, assume that $H_1(a_k), H_2(a_k), \ldots, H_{j+1}(a_k) \ne 0$ for some $1 \le j \le n - 2$. Then, $H_1(b_k), H_2(b_k), \ldots, H_{j+1}(b_k) \ne 0$ and, by induction hypothesis, for $1 \le j \le n - 2$ we have

$$H_j^2(a_k) = H_j^2(b_k) \ge H_{j-1}(b_k) H_{j+1}(b_k) = H_{j-1}(a_k) H_{j+1}(a_k). \qquad (17.3)$$

For the equality, if $H_j^2(a_k) = H_{j-1}(a_k) H_{j+1}(a_k)$ for some $1 \le j \le n - 2$, then (17.3) gives $H_j^2(b_k) = H_{j-1}(b_k) H_{j+1}(b_k)$. Therefore, the condition for equality in the induction hypothesis furnishes $b_1 = \ldots = b_{n-1}$, and from this it is almost immediate (cf. Problem 1) to show that $a_1 = \ldots = a_n$.

We are left to proving that, if $H_1(a_k), H_2(a_k), \ldots, H_n(a_k) \ne 0$, then

$$H_{n-1}^2(a_k) \ge H_{n-2}(a_k) H_n(a_k),$$

with equality if and only if $a_1 = \ldots = a_n$. We shall prove that this is true provided we merely have $H_n(a_k) \ne 0$. Indeed, in what concerns the desired inequality, it is the same as proving that

$$\left[\binom{n}{n-1}^{-1} \sum_{i=1}^{n} a_1 \ldots \widehat{a_i} \ldots a_n\right]^2 \geq$$

$$\geq \left[\binom{n}{n-2}^{-1} \sum_{i<j}^{n} a_1 \ldots \widehat{a_i} \ldots \widehat{a_j} \ldots a_n\right]\left[\binom{n}{n}^{-1} a_1 \ldots a_n\right].$$

Letting $P = a_1 \ldots a_n$, we observe that to prove the above inequality is the same as to establish that

$$\left(\frac{1}{n} \sum_{i=1}^{n} \frac{P}{a_i}\right)^2 \geq \frac{2P}{n(n-1)} \sum_{i<j} \frac{P}{a_i a_j},$$

or

$$(n-1)\left(\sum_{i=1}^{n} \frac{1}{a_i}\right)^2 \geq 2n \sum_{i<j} \frac{1}{a_i a_j}.$$

If we set $\alpha_i = \frac{1}{a_i}$, we wish to show that

$$(n-1)\left(\sum_{i=1}^{n} \alpha_i\right)^2 \geq 2n \sum_{i<j} \alpha_i \alpha_j.$$

Denoting $S = (n-1)\left(\sum_{i=1}^{n} \alpha_i\right)^2 - 2n \sum_{i<j} \alpha_i \alpha_j$, we compute

$$S = (n-1) \sum_{i=1}^{n} \alpha_i^2 + 2(n-1) \sum_{i<j} \alpha_i \alpha_j - 2n \sum_{i<j} \alpha_i \alpha_j$$

$$= (n-1) \sum_{i=1}^{n} \alpha_i^2 - 2 \sum_{i<j} \alpha_i \alpha_j$$

$$= \sum_{i<j} (\alpha_i - \alpha_j)^2 \geq 0.$$

Now, it is clear that equality holds if and only if $\alpha_1 = \ldots = \alpha_n$, i.e., if and only if $a_1 = \ldots = a_n$. $\qquad\square$

The coming corollary, due to the Scottish mathematician of the eighteenth century Colin McLaurin, embodies the promised refinement of the inequality between the arithmetic and geometric means. Inequalities (17.4) are then known as **McLaurin inequalities**.

Corollary 17.11 (McLaurin) *For $n > 1$ positive real numbers $a_1, a_2, \ldots, a_n$, we have*

$$H_1(a_k) \geq \sqrt{H_2(a_k)} \geq \sqrt[3]{H_3(a_k)} \geq \ldots \geq \sqrt[n]{H_n(a_k)}, \tag{17.4}$$

with equality at each of the above inequalities if and only if $a_1 = \ldots = a_n$.

Proof For the sake of simplicity, write H_j in place of $H_j(a_k)$. Newton's inequalities give us

$$H_1^2 \geq H_0 H_2 = H_2 \quad \text{and} \quad H_2^2 \geq H_1 H_3.$$

Therefore, $H_1^4 \geq H_2^2 \geq H_1 H_3$, so that $H_1^3 \geq H_3$. In turn, this last inequality, together with the second inequality above, gives

$$H_2^2 \geq H_1 H_3 \geq H_3^{1/3} H_3 = H_3^{4/3}$$

and, thus,

$$H_1 \geq H_2^{1/2} \geq H_3^{1/3}.$$

Assume we had already proved that

$$H_1 \geq H_2^{1/2} \geq \cdots \geq H_{k-1}^{1/(k-1)} \geq H_k^{1/k}$$

for some positive integer $k < n$. It follows from Newton's inequalities, together with the last inequality above, that

$$H_k^2 \geq H_{k-1} H_{k+1} \geq H_k^{\frac{k-1}{k}} H_{k+1}, \tag{17.5}$$

which in turn gives us $H_k^{2 - \left(\frac{k-1}{k}\right)} \geq H_{k+1}$ or, which is the same, $H_k^{1/k} \geq H_{k+1}^{1/(k+1)}$.

For the equality, suppose we have $H_k^{1/k} = H_{k+1}^{1/(k+1)}$. Then, $H_k^2 = H_k^{\frac{k-1}{k}} H_{k+1}$, and (17.5) shows that $H_k^2 = H_{k-1} H_{k+1}$. Therefore, it comes from the condition for equality in Newton's inequalities that $a_1 = \ldots = a_n$. $\qquad\square$

Problems: Sect. 17.2

1. * Let $f \in \mathbb{R}[X] \setminus \{0\}$ be a polynomial of degree n with n real roots, counted with their multiplicities. If f' has $n - 1$ equal roots, show that f has n equal roots.
2. We are given nonzero real numbers a, b, c, d such that $4b^2 < 9ac$. Show that the polynomial $X^4 + aX^3 + bX^2 + cX + d$ has at least two nonreal roots.

3. Let a, b, c, d be positive real numbers. Prove that

$$2(a+b+c+d)(ab+ac+ad+bc+bd+cd) \geq 3(abc+abd+acd+bcd),$$

with equality if and only if $a = b = c = d$.

17.3 Descartes' Rule

The main result of this section relates the number of positive roots of a real polynomial to the number of changes of sign along the sequence of its nonvanishing coefficients. Among the results proved so far in this chapter, it depends only on the Bolzano theorem, so that we could have discussed it right after that. Up to details, our presentation follows [26].

We start with the following combinatorial lemma, which can be easily established by induction, for example.

Lemma 17.12 *For each finite sequence v of $+$ and $-$ signs, let $V(v)$ denote the number of pairs of distinct consecutive signs in v. If $\mathcal{A}$ stands for the set of those sequences with initial and final distinct signs, and $\mathcal{B}$ for the set of those sequences with initial and final equal signs, then:*

(a) $v \in \mathcal{A} \Rightarrow V(v) \equiv 1 \pmod{2}$.
(b) $v \in \mathcal{B} \Rightarrow V(v) \equiv 0 \pmod{2}$.

Now, we shall need the following definition.

Definition 17.13 Let

$$f(X) = a_n X^{k_n} + a_{n-1} X^{k_{n-1}} + \cdots + a_1 X^{k_1} + a_0 X^{k_0}$$

be a nonconstant real polynomial, with $k_n > k_{n-1} > \ldots > k_1 > k_0 \geq 0$ and $a_j \neq 0$ for $0 \leq j \leq n$. Define the sequence $v_f = (\alpha_n, \alpha_{n-1}, \ldots, \alpha_1, \alpha_0)$ by setting, for each integer $0 \leq j \leq n$,

$$\alpha_j = \begin{cases} +, & \text{if } a_j > 0 \\ -, & \text{if } a_j < 0 \end{cases}.$$

The **variation** of f, denoted $V(f)$, is the number of pairs of distinct consecutive signs in v_f.

For our purposes, the following one is a crucial result.

Lemma 17.14 *Let g be a nonconstant real polynomial and $c > 0$ be a given real number. If $f(X) = (X - c)g(X)$, then $V(f) - V(g)$ is an odd positive integer.*

Proof We make induction on $\partial g \geq 1$.

(a) $\partial g = 1$: we can surely suppose that g is monic. If $g(X) = X$, then $f(X) = X^2 - cX$, so that $V(f) - V(g) = 1$. Now, assume that $g(X) = X - \alpha$, with $\alpha > 0$. Then, $f(X) = X^2 - (\alpha + c)X + c\alpha$, so that $V(f) - V(g) = 2 - 1 = 1$. The case $g(X) = X + \alpha$, with $\alpha > 0$, is equally easy: $f(X) = X^2 + (\alpha - c)X - c\alpha$ and, since $-c\alpha < 0$, Lemma 17.12 assures that $V(f)$ is odd; hence, $V(f) - V(g) = V(f)$, an odd positive integer.

(b) Assume that the result is valid for every nonconstant real polynomial g such that $\partial g < n$, with $n > 1$ being an integer, and take another such polynomial g, this time of degree n. For the sake of notation, for a given nonconstant real polynomial h, whenever necessary we shall denote by α_h and β_h, respectively, the leading coefficient and the coefficient of the term of lowest degree in h. In particular,

$$h(X) = \alpha_h X^r + \cdots + \beta_h X^s,$$

with $r \geq s$. We distinguish three cases:

(i) All coefficients of g are positive: once again assume (without loss of generality) that g monic, say $g(X) = X^n + \cdots + \beta X^l$, with $l < n$. Then,

$$f(X) = (X - c)g(X) = X^{n+1} + \cdots - c\beta X^l,$$

with $c\beta < 0$, and Lemma 17.12 shows that $V(f)$ is odd and, hence $V(f) - V(g) = V(f)$ is an odd positive integer.

(ii) There exist nonconstant polynomials u and v such that $g = u + v$,

$$u(X) = \alpha_u X^r + \cdots + \beta_u X^s, \quad \text{and} \quad v(X) = \alpha_v X^p + \cdots + \beta_v X^q,$$

with $n = r \geq s$, $p \geq q$ and $p + 1 < s$. Then,

$$\begin{aligned} (X - c)g(X) &= (X - c)u(X) + (X - c)v(X) \\ &= (\alpha_u X^{r+1} + \cdots - c\beta_u X^s) \qquad (17.6) \\ &\quad + (\alpha_v X^{p+1} + \cdots - c\beta_v X^q), \end{aligned}$$

since $p + 1 < s$. Let us now look at two distinct subcases: $\alpha_v \beta_u > 0$ and $\alpha_v \beta_u < 0$.

• $\alpha_v \beta_u > 0$: we obviously have $V(g) = V(u) + V(v)$. However, since $(-c\beta_u)\alpha_v < 0$, the second equality in (17.6) shows that

$$V(f) = V((X - c)g) = V((X - c)u) + V((X - c)v) + 1.$$

By the induction hypothesis (in principle, u has degree $r = n$; nevertheless, $s > 1$ implies that we can apply the induction hypothesis to $w(X) = \alpha_u X^{r-s} + \cdots + \beta_u$, since $u(X) = X^s w(X)$), we obtain

$$V(f) \geq (1 + V(u)) + (1 + V(v)) + 1 > V(g) + 1$$

and, modulo 2,

$$V(f) \equiv (1 + V(u)) + (1 + V(v)) + 1 \equiv V(g) + 1.$$

- $\alpha_v \beta_u < 0$: since $\alpha_v \beta_u < 0$, we have

$$V(g) = V(u) + V(v) + 1$$

and, by (17.6),

$$V(f) = V((X - c)g) = V((X - c)u) + V((X - c)v).$$

Once again from the induction hypothesis, we get

$$V(f) \geq (1 + V(u)) + (1 + V(v)) = V(g) + 1$$

and, modulo 2,

$$V(f) \equiv (1 + V(u)) + (1 + V(v)) \equiv V(g) + 1.$$

(iii) $g(X) = a_n X^n + a_{n-1} X^{n-1} + \cdots + a_k X^k$, with $a_n, \ldots, a_k \neq 0$ and not all of $a_n, a_{n-1}, \ldots, a_k$ positive. Assume, without loss of generality, that $a_n > 0$, and write

$$g = g_0 - g_1 + \cdots + (-1)^t g_t,$$

such that each $g_i - \alpha_i X^{k_i} + \cdots + \beta_i X^{l_i}$ $(k_i \geq l_i)$ has only positive coefficients and $l_i = k_{i+1} + 1$ for $i < t$. Then, $V(g) = t$ and

$$
\begin{aligned}
(X - c)g &= (X - c)g_0 - (X - c)g_1 + \cdots + (-1)^t (X - c)g_t \\
&= (X - c)(\alpha_0 X^{k_0} + \cdots + \beta_0 X^{l_0}) \\
&\quad - (X - c)(\alpha_1 X^{k_1} + \cdots + \beta_1 X^{l_1}) + \cdots \\
&\quad + (-1)^t (X - c)(\alpha_t X^{k_t} + \cdots + \beta_t X^{l_t}) \\
&= \alpha_0 X^{k_0+1} + \cdots - (\alpha_1 + c\beta_0) X^{l_0} + \cdots + (\alpha_2 + c\beta_1) X^{l_1} \\
&\quad + \cdots - (\alpha_3 + c\beta_2) X^{l_2} + \cdots + (-1)^t (\alpha_t + c\beta_{t-1}) X^{l_{t-1}} \\
&\quad + \cdots + (-1)^{t+1} c\beta_t X^{l_t}.
\end{aligned}
$$

Since α_0, $\alpha_1 + c\beta_0$, $\alpha_2 + c\beta_1$, $\ldots$, $\alpha_t + c\beta_{t-1}$ and $c\beta_t$ are all positive, Lemma 17.12 assures that there is an odd number of changes of sign (and, hence, at least one) of consecutive pairs of coefficients in each of the $t + 1$ dotted intervals above. This gives

$$V(f) = V((X - c)g) \geq t + 1 = V(g) + 1.$$

It also follows from the above that, modulo 2,

$$V(f) = V((X - c)g) \equiv t + 1 = V(g) + 1.$$

□

We can finally state and prove the following result, which is known in Mathematics as **Descartes' rule**[2] for polynomials with real coefficients.

Theorem 17.15 (Descartes) *Let* f *be a nonconstant real polynomial. If* $P(f)$ *denotes the number of positive roots of* f*, then* $V(f) - P(f)$ *is a nonnegative even integer.*

Proof We make induction on the degree of f.

If $\partial f = 1$, assume, without loss of generality, that f is monic. If $f(X) = X$ or $f(X) = X + \alpha$, with $\alpha > 0$, then $V(f) - P(f) = 0 - 0 = 0$. If $f(X) = X - \alpha$, with $\alpha > 0$, then $V(f) - P(f) = 1 - 1 = 0$.

Now, let $n > 1$ be a given integer and suppose that the theorem holds true for each nonconstant real polynomial of degree less than n. If

$$f(X) = a_n X^n + a_{n-1} X^{n-1} + \cdots + a_1 X + a_0$$

is a real polynomial of degree, we have two different cases:

- $P(f) = 0$: in this case, it suffices to show that $V(f)$ is even. Since $f(0) = a_0$ and $f(x)$ has the same sign as a_n for every sufficiently large real number x (for $\lim_{x \to +\infty} \frac{f(x)}{x^n} = a_n$), the fact that $P(f) = 0$ assures, via Bolzano's theorem, that $a_n a_0 > 0$. Hence, Lemma 17.12 guarantees that $V(f)$ is even.
- $P(f) > 0$: by taking a positive root c of f, there exists a nonconstant polynomial g such that $f(X) = (X - c)g(X)$. Lemma 17.14 guarantees the existence of an odd positive integer I for which

$$V(f) - P(f) = (V(g) + I) - (P(g) + 1)$$
$$= (V(g) - P(g)) + (I - 1).$$

Since, by induction hypothesis, $V(g) - P(g)$ is nonnegative and even, we conclude that the same holds for $V(f) - P(f)$.

□

---

[2]René Descartes, French mathematician, philosopher and scientist of the seventeenth century. Descartes' legacy to Mathematics and science is a huge one, and came mainly from his landmarking book *Discours de la Méthode* (Discourse on the Method) and its three corresponding appendices. This book marks a turning point on the way of doing science, for, along it, Descartes strongly rejected the scholastic tradition of using speculation, instead of deduction, as the central strategy for the investigation of natural phenomena. On the other hand, its appendix *La Géométrie* layed down the foundations of Analytic Geometry, and nowadays every student is acquainted with cartesian coordinate systems.

The coming corollary examines the case of negative roots.

Corollary 17.16 *If f is a nonconstant real polynomial and $N(f)$ is the number of its negative roots, then $V(f(-X)) - N(f)$ is nonnegative and even.*

Proof Immediate from Descartes' rule, together with the fact that $\alpha \in N(f)$ if and only if $-\alpha \in P(f)$.   □

As a direct consequence of Descartes' rule, if f is a nonconstant real polynomial such that $V(f)$ is odd, then has at least one positive root. More particularly, if $V(f) = 1$, then f has *exactly* one positive root. The coming example takes advantage of this fact.

Example 17.17 Let $a_0, a_1, \ldots, a_{n-1}$ be nonnegative real numbers, at least one of which being nonzero. Compute the number of positive roots of the real polynomial

$$X^n - a_{n-1}X^{n-1} - \cdots - a_1 X - a_0.$$

Solution Denoting by f the given polynomial, Theorem 17.15 assures that $V(f) - P(f)$ is nonnegative and even. However, since $V(f) = 1$, it is clear that $P(f) = 1$.
                                                                                      □

Problems: Sect. 17.3

1. Let $g \in \mathbb{R}[X] \setminus \mathbb{R}$ be a polynomial of degree n and $f(X) = g(X^2)$. If f has exactly n positive roots and n negative roots, show that $V(f) = n$.
2. For each $n \in \mathbb{N}$, show that the real polynomial $X^n - X^{n-1} - X^{n-2} - \cdots - X - 1$ has a positive root a_n satisfying

$$2 - \frac{1}{2^{n-1}} \leq a_n \leq 2 - \frac{1}{2^n}.$$

3. (Leningrad) The third degree real polynomial $aX^3 + bX^2 + cX + d$ has three distinct real roots. Compute the number of real roots of the polynomial

$$4(aX^3 + bX^2 + cX + d)(3aX + b) - (3aX^2 + 2bX + c)^2.$$

Chapter 18
Interpolation of Polynomials

Corollary 15.10 assures that there is at most one polynomial of degree n and assuming preassigned values in $n + 1$ given complex numbers. What we still do not know is whether such a polynomial actually exists. For instance, does there exists a polynomial f with rational coefficients, degree 3 and such that $f(0) = 1$, $f(1) = 2$, $f(2) = 3$ and $f(3) = 0$? In this chapter we study a bunch of techniques that allow us to answer this and alike questions, and which are generically referred to as *interpolation of polynomials*. In particular, we shall study in detail the class of Lagrange interpolating polynomials, which will then be used to solve Vandermonde' linear systems with no Linear Algebra. In turn, the knowledge of the solutions of such linear systems will allow us to study, in Sect. 21.1, an important particular class of linear recurrence relations, thus partially extending the methods of Section 3.2 of [8].

18.1 Bases for Polynomials

For what follows, recall that $\mathbb{K}$ stands for any of the number sets $\mathbb{Q}$, $\mathbb{R}$ or $\mathbb{C}$. We shall firstly need to establish a quite useful notation.

Definition 18.1 For $n \in \mathbb{N}$ and $1 \le i, j \le n$, we let the **Kronecker delta**[1] δ_{ij} be defined by

$$\delta_{ij} = \begin{cases} 1, & \text{if } i = j \\ 0, & \text{if } i \neq j \end{cases}.$$

The advantage of Kronecker's notation comes from the fact that we can use it as a **tag**, in the following sense: given $n \in \mathbb{N}$ and a sequence $(b_1, \ldots, b_n)$ in $\mathbb{K}$, we have

---

[1]Leopold Kronecker, German mathematician of the nineteenth century.

© Springer International Publishing AG, part of Springer Nature 2018
A. Caminha Muniz Neto, *An Excursion through Elementary Mathematics, Volume III*,
Problem Books in Mathematics, https://doi.org/10.1007/978-3-319-77977-5_18

$$\sum_{j=1}^{n} \delta_{ij} b_j = b_i \qquad (18.1)$$

for $1 \leq i \leq n$.

Now, let be given $n \in \mathbb{N}$ and pairwise distinct elements $a_1, a_2, \ldots, a_n$ in $\mathbb{K}$. For $1 \leq i \leq n$, we define $L_i \in \mathbb{K}[X]$ by

$$L_i(X) = \prod_{\substack{1 \leq j \leq n \\ j \neq i}} \frac{X - a_j}{a_i - a_j} = \left(\frac{X - a_1}{a_i - a_1} \right) \cdots \left(\widehat{\frac{X - a_i}{a_i - a_i}} \right) \cdots \left(\frac{X - a_n}{a_i - a_n} \right),$$

with the hat $\overset{\frown}{}$ over a certain factor meaning that it is actually missing from the product. Such polynomials L_i are called the **Lagrange interpolating polynomials** for the set $\{a_1, \ldots, a_n\}$.

It is immediate to verify that

$$L_i(a_j) = \delta_{ij}$$

for all $1 \leq i, j \leq n$. In turn, as we shall see in the coming result, this property allows us to build polynomials attaining preassigned values at a set of pairwise distinct given elements of $\mathbb{K}$. Such a result is usually referred to as the **Lagrange interpolating theorem**.

Theorem 18.2 (Lagrange) *Given $n \in \mathbb{N}$ and $a_1, \ldots, a_n, b_1, \ldots, b_n \in \mathbb{K}$, with $a_1, \ldots, a_n$ being pairwise distinct, there exists exactly one polynomial $f \in \mathbb{K}[X]$, of degree less than n and such that $f(a_i) = b_i$ for $1 \leq i \leq n$. More precisely, such a polynomial is*

$$f(X) = \sum_{j=1}^{n} b_j L_j(X), \qquad (18.2)$$

with the L_j standing for the Lagrange interpolating polynomials for the set $\{a_1, \ldots, a_n\}$.

Proof Uniqueness follows from Corollary 15.10. On the other hand, taking f as in (18.2), it suffices to show that $\partial f < n$ and $f(a_i) = b_i$ for $1 \leq i \leq n$.

Since $\partial L_j = n - 1$ for $1 \leq j \leq n$ and the degree of a sum of polynomials, whenever defined, does not surpass the greatest degree of the summands, it is pretty clear that $\partial f < n$. For what is left to do, just note that

$$f(a_i) = \sum_{j=1}^{n} b_j L_j(a_i) = \sum_{j=1}^{n} b_j \delta_{ji} = b_i,$$

where we have used (18.1) in the last equality above. $\square$

The following corollary provides an equivalent way of formulating the previous result.

Corollary 18.3 *Let* $n \in \mathbb{N}$ *and* $a_1, \ldots, a_n$ *be pairwise distinct elements of* $\mathbb{K}$. *If* $f \in \mathbb{K}[X] \setminus \{0\}$ *satisfies* $\partial f < n$, *then*

$$f(X) = \sum_{i=1}^{n} f(a_i) L_i(X).$$

Proof Indeed, by letting

$$g(X) = \sum_{i=1}^{n} f(a_i) L_i(X),$$

we have $\partial f, \partial g < n$ and, by the proof of Lagrange's theorem, $g(a_i) = f(a_i)$ for $1 \le i \le n$. Therefore, by invoking Corollary 15.10 once more, we get $g = f$.   □

We now collect some interesting applications of Lagrange's interpolation, beginning with the operational aspects involved.

Example 18.4 Find a polynomial $f \in \mathbb{Q}[X]$ such that $f(1) = 12$, $f(2) = 2$, $f(3) = 1$, $f(4) = -6$ and $f(5) = 4$.

Solution Firstly, let us explicitly write Lagrange's interpolating polynomials for the set $\{1, 2, 3, 4, 5\}$:

$$L_1(X) = \prod_{\substack{1 \le j \le 5 \\ j \ne 1}} \frac{X - j}{1 - j} = \left(\frac{X - 2}{1 - 2}\right)\left(\frac{X - 3}{1 - 3}\right)\left(\frac{X - 4}{1 - 4}\right)\left(\frac{X - 5}{1 - 5}\right)$$

$$= \frac{1}{24}(X^4 - 14X^3 + 71X^2 - 154X + 120),$$

$$L_2(X) = \prod_{\substack{1 \le j \le 5 \\ j \ne 2}} \frac{X - j}{2 - j} = \left(\frac{X - 1}{2 - 1}\right)\left(\frac{X - 3}{2 - 3}\right)\left(\frac{X - 4}{2 - 4}\right)\left(\frac{X - 5}{2 - 5}\right)$$

$$= -\frac{1}{6}(X^4 - 13X^3 + 59X^2 - 107X + 60),$$

$$L_3(X) = \prod_{\substack{1 \le j \le 5 \\ j \ne 3}} \frac{X - j}{3 - j} = \left(\frac{X - 1}{3 - 1}\right)\left(\frac{X - 2}{3 - 2}\right)\left(\frac{X - 4}{3 - 4}\right)\left(\frac{X - 5}{3 - 5}\right)$$

$$= \frac{1}{4}(X^4 - 12X^3 + 49X^2 - 78X + 40),$$

$$L_4(X) = \prod_{\substack{1 \le j \le 5 \\ j \ne 4}} \frac{X - j}{4 - j} = \left(\frac{X - 1}{4 - 1}\right)\left(\frac{X - 2}{4 - 2}\right)\left(\frac{X - 3}{4 - 3}\right)\left(\frac{X - 5}{4 - 5}\right)$$

$$= -\frac{1}{6}(X^4 - 11X^3 + 41X^2 - 61X + 30)$$

and

$$L_5(X) = \prod_{\substack{1 \le j \le 5 \\ j \ne 5}} \frac{X - j}{5 - j} = \left(\frac{X - 1}{5 - 1}\right)\left(\frac{X - 2}{5 - 2}\right)\left(\frac{X - 3}{5 - 3}\right)\left(\frac{X - 4}{5 - 4}\right)$$

$$= \frac{1}{24}(X^4 - 10X^3 + 35X^2 - 50X + 24).$$

Hence, according to previous corollary we can take

$$f(X) = 12L_1(X) + 2L_2(X) + L_3(X) - 6L_4(X) + 4L_5(X)$$

$$= \frac{19}{12}X^4 - \frac{55}{3}X^3 + \frac{233}{4}X^2 - \frac{775}{6}X + 86.$$

$\square$

Example 18.5 (IMO Shortlist) Let f be a monic polynomial of real coefficients and degree n, and let $x_1, x_2, \ldots, x_{n+1}$ be pairwise distinct integers. Show that there exists an index $1 \le k \le n + 1$ for which $|f(x_k)| \ge \frac{n!}{2^n}$.

Proof Lagrange's interpolation formula gives

$$f(x) = \sum_{j=1}^{n+1} f(x_j) \prod_{i \ne j} \frac{x - x_i}{x_j - x_i}.$$

Therefore, by comparing the leading coefficients on both sides of the equality above, we obtain

$$1 = \sum_{j=1}^{n+1} \frac{f(x_j)}{\prod_{i \ne j}(x_j - x_i)}.$$

Now, if $M = \max\{|f(x_k)|;\ 1 \le k \le n + 1\}$ and $p_j = \prod_{i \ne j}(x_j - x_i)$, we have

$$1 = \left|\sum_{j=1}^{n+1} \frac{f(x_j)}{p_j}\right| \le \sum_{j=1}^{n+1} \frac{|f(x_j)|}{|p_j|} \le M \sum_{j=1}^{n+1} \frac{1}{|p_j|}.$$

However, since $\sum_{j=1}^{n+1} \frac{1}{|p_j|}$ is symmetric with respect to $x_1, x_2, \ldots, x_{n+1}$, we can assume (after reordering the x_i's, if needed) that $x_1 < x_2 < \cdots < x_{n+1}$. This way,

$$|p_j| = (x_j - x_1) \ldots (x_j - x_{j-1})(x_{j+1} - x_j) \ldots (x_{n+1} - x_j)$$
$$\geq [(j-1) \ldots 2 \cdot 1][1 \cdot 2 \ldots (n+1-j)]$$
$$= (j-1)!(n+1-j)! = \frac{n!}{\binom{n}{j-1}}.$$

In view of the estimates above, we finally get

$$1 \leq M \sum_{j=1}^{n+1} \frac{1}{|p_j|} \leq M \sum_{j=1}^{n+1} \frac{1}{n!} \binom{n}{j-1} = \frac{2^n M}{n!}$$

or, which is the same, $M \geq \frac{n!}{2^n}$. $\qquad \square$

We now show how to use Lagrange's interpolating polynomials to solve certain types of linear systems of equations, known as **Vandermonde systems**. Apart from Example 18.7, Vandermonde systems will also play a role in the discussion of some special linear recurrence relations, in Sect. 21.1.

Proposition 18.6 (Vandermonde) *Given elements $a_1, a_2, \ldots, a_n, \alpha_1, \alpha_2, \ldots, \alpha_n$ of $\mathbb{K}$, with $a_1, a_2, \ldots, a_n$ being pairwise distinct, the linear system of equations*

$$\begin{cases} x_1 + x_2 + \cdots + x_n & = \alpha_1 \\ a_1 x_1 + a_2 x_2 + \cdots + a_n x_n & = \alpha_2 \\ a_1^2 x_1 + a_2^2 x_2 + \cdots + a_n^2 x_n & = \alpha_3 \\ \cdots \\ a_1^{n-1} x_1 + a_2^{n-1} x_2 + \cdots + a_n^{n-1} x_n = \alpha_n \end{cases} \qquad (18.3)$$

admits a unique solution in $\mathbb{K}$. In particular, if $\alpha_1 = \cdots = \alpha_n = 0$, then $x_1 = \cdots = x_n = 0$.

Proof If $f(X) = c_0 + c_1 X + \cdots + c_{n-1} X^{n-1} \in \mathbb{K}[X]$, then, by respectively multiplying the equations of the linear system by $c_0, c_1, \ldots, c_{n-1}$ and adding the results, we obtain

$$f(a_1)x_1 + \cdots + f(a_n)x_n = c_0 \alpha_1 + c_1 \alpha_2 + \cdots + c_{n-1} \alpha_n.$$

Now, if we let $f = L_i$, where L_i is the i-th Lagrange interpolating polynomial for $\{a_1, a_2, \ldots, a_n\}$, then the left hand side of the equality above reduces to x_i, so that

$$x_i = c_0 \alpha_1 + c_1 \alpha_2 + \cdots + c_{n-1} \alpha_n \in \mathbb{K}. \qquad (18.4)$$

$\qquad \square$

The coming example shows how to use the previous proposition to give an alternative proof for the generalized multisection formula (cf. Problem 8, page 355).

Example 18.7 Let $f \in \mathbb{C}[X] \setminus \{0\}$, $n > 1$ be a natural number and $0 \le r \le n - 1$ be an integer. If $\omega = \operatorname{cis} \frac{2\pi}{n}$ and $z \in \mathbb{C}$, show that

$$\sum_{k \equiv r \pmod{n}} a_k z^k = \frac{1}{n} \sum_{j=0}^{n-1} \omega^{-rj} f(\omega^j z),$$

with a_k denoting the coefficient of X^k in f.

Proof Writing

$$f(X) = u_0(X^n) + X u_1(X^n) + \cdots + X^{n-1} u_{n-1}(X^n),$$

with $u_0, u_1, \ldots, u_{n-1} \in \mathbb{C}[X]$, and taking into account that $\omega^n = 1$, we obtain

$$\begin{cases} u_0(z^n) + z u_1(z^n) + \cdots + z^{n-1} u_{n-1}(z^n) & = f(z) \\ u_0(z^n) + \omega z u_1(z^n) + \cdots + (\omega z)^{n-1} u_{n-1}(z^n) & = f(\omega z) \\ u_0(z^n) + (\omega^2 z) u_1(z^n) + \cdots + (\omega^2 z)^{n-1} u_{n-1}(z^n) & = f(\omega^2 z) \\ \cdots \\ u_0(z^n) + (\omega^{n-1} z) u_1(z^n) + \cdots + (\omega^{n-1} z)^{n-1} u_{n-1}(z^n) = f(\omega^{n-1} z) \end{cases}.$$

This is a Vandermonde system in $u_0(z^n)$, $z u_1(z^n)$, $\ldots$, $z^{n-1} u_{n-1}(z^n)$, with

$$z^r u_r(z^n) = \sum_{k \equiv r \pmod{n}} a_k z^k. \tag{18.5}$$

According to (18.4), we have

$$z^r u_r(z^n) = c_0 f(z) + c_1 f(\omega z) + \cdots + c_{n-1} f(\omega^{n-1} z), \tag{18.6}$$

where c_j stands for the coefficient of X^j in the r-th Lagrange interpolating polynomial L_r relative to $\{1, \omega, \omega^2, \ldots, \omega^{n-1}\}$. However,

$$L_r(X) = \frac{(X - 1) \ldots \widehat{(X - \omega^r)} \ldots (X - \omega^{n-1})}{(\omega^r - 1) \ldots \widehat{(\omega^r - \omega^r)} \ldots (\omega^r - \omega^{n-1})}, \tag{18.7}$$

with numerator equal to

$$\frac{X^n - 1}{X - \omega^r} = \frac{X^n - (\omega^r)^n}{X - \omega^r} = X^{n-1} + \omega^r X^{n-2} + \cdots + \omega^{(n-2)r} X + \omega^{(n-1)r}.$$

On the other hand, item (a) of Problem 5, page 393, computes the denominator of (18.7) as being equal to $n\omega^{(n-1)r}$, so that

$$c_j = \frac{\omega^{(n-1-j)r}}{n\omega^{(n-1)r}} = \frac{1}{n}\omega^{-jr}.$$

It finally follows from (18.5), (18.6) and the above expression for c_j that

$$\sum_{k\equiv r \,(\mathrm{mod}\,n)} a_k z^k = \sum_{j=0}^{n-1} c_j f(\omega^j z) = \frac{1}{n}\sum_{j=0}^{n-1} \omega^{-jr} f(\omega^j z).$$

$\square$

A limitation of Lagrange interpolation lies in the fact that it only generates the set of nonzero polynomials having degrees less than or equal to some fixed natural number. We remedy such a situation with the following more general definition.

Definition 18.8 A **basis** for $\mathbb{K}[X]$ is a sequence $(f_0, f_1, f_2, \ldots)$ of elements of $\mathbb{K}[X]$ satisfying the following condition: for every $f \in \mathbb{K}[X]$, there exist unique $n \in \mathbb{Z}_+$ and $a_0, \ldots, a_n \in \mathbb{K}$ such that

$$f(X) = a_0 f_0(X) + \cdots + a_n f_n(X).$$

Example 18.9 A simple way of constructing a basis for $\mathbb{K}[X]$ is to take a sequence $(f_0, f_1, f_2, \ldots)$ in $\mathbb{K}[X]$ such that $\partial f_j = j$ for every $j \geq 0$.

In order to show that such a sequence is indeed a basis for $\mathbb{K}[X]$, we have to establish the following two claims:

(i) If $n \in \mathbb{N}$ and $a_0, a_1, \ldots, a_n$ and $b_0, b_1, \ldots, b_n$ are elements of $\mathbb{K}$ such that

$$a_0 f_0(X) + a_1 f_1(X) + \cdots + a_n f_n(X) = b_0 f_0(X) + b_1 f_1(X) + \cdots + b_n f_n(X),$$

then $a_i = b_i$ for $0 \leq i \leq n$.

(ii) For every $f \in \mathbb{K}[X]$, there exist $n \in \mathbb{N}$ and $a_0, a_1, \ldots, a_n \in \mathbb{K}$ such that

$$f(X) = a_0 f_0(X) + a_1 f_1(X) + \cdots + a_n f_n(X).$$

We leave these checkings for the reader (cf. Problem 1).

As a particular case of the previous example, note that the sequence

$$(1, X, X^2, X^3, \ldots)$$

is a basis for $\mathbb{K}[X]$ (as we already knew).

In what follows, inspired by the binomial numbers, we construct a quite useful basis for $\mathbb{K}[X]$.

Definition 18.10 For $k \in \mathbb{Z}_+$, we define the k-th **binomial polynomial** $\binom{X}{k}$ by setting $\binom{X}{0} = 1$, $\binom{X}{1} = X$ and, for an integer $k > 1$,

$$\binom{X}{k} = \frac{1}{k!}X(X-1)\dots(X-k+1).$$

Since $\partial\binom{X}{k} = k$, para todo $k \in \mathbb{Z}_+$, a direct application of Example 18.9 guarantees that the sequence of binomial polynomials is a basis for $\mathbb{K}[X]$, called the **binomial basis**.

In spite of this, for future use we shall give a direct proof of the validity of condition (ii) of that example. More precisely, let us make induction on $n \geq 0$ to show that there exist $a_0, a_1, \dots, a_n \in \mathbb{K}$ for which

$$X^n = a_0\binom{X}{0} + a_1\binom{X}{1} + \dots + a_n\binom{X}{n}. \tag{18.8}$$

For $n = 0$ and $n = 1$, (18.8) holds due to the very definition of $\binom{X}{0}$ and $\binom{X}{1}$. Now, assume that for some $k \in \mathbb{N}$ there exist $a_0, a_1, \dots, a_k \in \mathbb{K}$ satisfying (18.8) when $n = k$. Then,

$$a_0\binom{X}{0}X + a_1\binom{X}{1}X + \dots + a_k\binom{X}{k}X = X^{k+1} \tag{18.9}$$

with

$$\binom{X}{j}X = \frac{1}{j!}X(X-1)\dots(X-j+1)(X-j+j)$$

$$= \frac{1}{j!}X(X-1)\dots(X-j+1)(X-j)$$

$$+ \frac{j}{j!}X(X-1)\dots(X-j+1)$$

$$= (j+1)\binom{X}{j+1} + j\binom{X}{j}.$$

Therefore, substituting this expression for $\binom{X}{j}X$ (for $0 \leq j \leq k$) into (18.9), we conclude that (18.8) is valid for $n = k + 1$.

Even though the concept of basis for $\mathbb{K}[X]$, as formulated above, does not apply directly to polynomials over $\mathbb{Z}$, when dealing with the binomial basis we have the following result.

Proposition 18.11 *If $f \in \mathbb{C}[X] \setminus \{0\}$ is a polynomial of degree n such that $f(0)$, $f(1), \dots, f(n)$ are all integers, then there exist unique $a_0, a_1, \dots, a_n \in \mathbb{Z}$ such that*

$$f(X) = a_0\binom{X}{0} + a_1\binom{X}{1} + \dots + a_n\binom{X}{n}. \tag{18.10}$$

In particular, $f(x) \in \mathbb{Z}$ for every $x \in \mathbb{Z}$.

Proof Since the binomial polynomials for a basis for $\mathbb{C}[X]$, there exist unique $a_0, a_1, \ldots, a_n \in \mathbb{C}$ such that f can be written as in the statement of the proposition. Thus, it suffices to prove that $a_0, a_1, \ldots, a_n$ are integers.

Arguing once more by induction, start by noticing that $\mathbb{Z} \ni f(0) = a_0$. Now, assume that we have already proved that $a_0, \ldots, a_{k-1} \in \mathbb{Z}$, for some integer $1 \leq k \leq n$. Then, (18.10) gives

$$f(k) - \sum_{j=0}^{k-1} a_j \binom{X}{j}(k) = \sum_{j=k}^{n} a_j \binom{X}{j}(k). \tag{18.11}$$

For $0 \leq j \leq k$, the evaluation $\binom{X}{j}(k)$ coincides with the ordinary binomial number $\binom{k}{j}$. On the other hand, for $k + 1 \leq j \leq n$, we have

$$\binom{X}{j}(k) = \frac{1}{j!} k(k-1) \ldots (k-j+1) = 0$$

Therefore, (18.11) reduces to

$$f(k) - \sum_{j=0}^{k-1} a_j \binom{k}{j} = a_k, \tag{18.12}$$

and the fact that $a_0, a_1, \ldots, a_{k-1} \in \mathbb{Z}$, together with $f(k) \in \mathbb{Z}$, gives $a_k \in \mathbb{Z}$.

For the last part, since $\binom{X}{k}(x) \in \mathbb{Z}$ for every $x \in \mathbb{Z}$ (cf. Problem 2), it follows from (18.10) that $f(x) \in \mathbb{Z}$ for every $x \in \mathbb{Z}$. $\square$

Also with respect to the proof of the proposition above, (18.12) can be written as

$$f(k) = \sum_{j=0}^{k} \binom{k}{j} a_j$$

Therefore, a direct application of Lemma 2.12 furnishes

$$a_k = \sum_{j=0}^{k} (-1)^{k+j} \binom{k}{j} f(j) \in \mathbb{Z}$$

for $0 \leq k \leq n$ and with no inductive argument.

Problems: Sect. 18.1

1. * Complete the discussion of Example 18.9.
2. * Given $k \in \mathbb{Z}_+$, prove that $\binom{X}{k}(x) \in \mathbb{Z}$ for every $x \in \mathbb{Z}$.

3. (USA)[2] Let f be a polynomial of degree n, such that $f(k) = \binom{n+1}{k}^{-1}$ for $0 \le k \le n$. Compute all possible values of $f(n+1)$ in terms of n.

4. (Singapore)[3] We are given a polynomial f of degree n, such that $f(k) = \frac{k}{k+1}$ for every integer $0 \le k \le n$. Compute all possible values of $f(n+1)$ in terms of n.

5. Given an n-sided simple polygon $\mathcal{P}$, show that there exists a polynomial $f \in \mathbb{R}[X]$, of degree at most $n-1$ and such that the graph of f contains the set of vertices of $\mathcal{P}$.

6. Given pairwise distinct real numbers a, b, c and d, solve the linear system of equations

$$\begin{cases} ax_1 + a^2x_2 + a^3x_3 + a^4x_4 = 1 \\ bx_1 + b^2x_2 + b^3x_3 + b^4x_4 = 1 \\ cx_1 + c^2x_2 + c^3x_3 + c^4x_4 = 1 \\ dx_1 + d^2x_2 + d^3x_3 + d^4x_4 = 1 \end{cases}.$$

7. (USA) Let $n > 3$ be an integer and p, p_0, p_1, ..., p_{n-2} be polynomials over $\mathbb{R}$ for which

$$\sum_{j=0}^{n-2} X^j p_j(X^n) = (X^{n-1} + X^{n-2} + \cdots + X + 1)p(X).$$

Prove that $X - 1$ divides $p_j(X)$, for $0 \le i \le n - 2$.

8. Let $n > 1$ be a given natural number.

   (a) Factorise the polynomial

   $$f(X) = \sum_{j=0}^{n}(-1)^j \binom{n}{j} X(X+1)\ldots(\widehat{X+j})\ldots(X+n) - n!.$$

   (b) Conclude that for each $k \in \mathbb{N}$ one has

   $$\sum_{j=0}^{n}(-1)^j \binom{n}{j} \frac{1}{k+j} = \frac{n!}{k(k+1)\ldots(k+n)}.$$

   (c) Compute, in terms of n, the sum of the series

   $$\sum_{k\ge 1} \frac{1}{k(k+1)\ldots(k+n)}.$$

---

[2]For another approach to this problem, see Problem 4, page 450.

[3]For other approaches to this problem, see Problem 11, page 386, and Problem 3, page 450.

9. (Leningrad) A finite sequence $a_1, a_2, \ldots, a_n$ is said to be *p-balanced* if, for $k = 1, 2, \ldots, p$, all sums of the form

$$a_k + a_{k+p} + a_{k+2p} + \cdots$$

are equal. Prove that if $n = 50$ and the sequence $(a_j)_{1 \le j \le 50}$ is *p*-balanced for $p = 3, 5, 7, 11, 13$ and 17, then all of its terms are equal to 0.

18.2 Finite Differences

Another useful, though a little more elaborate, interpolation technique is that of *finite differences*. We shall sketch the rudiments of it along this section.

Definition 18.12 Let h be a real number and $f : \mathbb{R} \to \mathbb{R}$ be a given function. For an integer $k \ge 0$, we define the **k-th finite difference** of f with step h as the function $\Delta_h^k f : \mathbb{R} \to \mathbb{R}$, given by:

(a) $\Delta_h^0 f = f$.
(b) $(\Delta_h^1 f)(x) = (\Delta_h f)(x) = f(x+h) - f(x)$ for each $x \in \mathbb{R}$.
(c) $\Delta_h^k f = \Delta_h(\Delta_h^{k-1} f)$ for $k \ge 2$.

For such a definition to be useful, we need the properties of $\Delta_h^k f$ collected in the coming result.

Proposition 18.13 *In the notations of the previous definition, given $h \in \mathbb{R}$ and functions $f, g : \mathbb{R} \to \mathbb{R}$, we have that:*

(a) *If f is constant, then $\Delta_h f = 0$.*
(b) *If a and b are real constants, then $\Delta_h(af + bg) = a\Delta_h f + b\Delta_h g$.*
(c) *$\Delta_h(fg) = (\Delta_h f)(g + \Delta_h g) + f\Delta_h g$.*
(d) *$\Delta_h^k f = \Delta_h^{k-1}(\Delta_h f)$, for every $k \in \mathbb{N}$.*
(e) *If $k \ge 0$ and $x \in \mathbb{R}$, then*

$$(\Delta_h^k f)(x) = \sum_{j=0}^{k} (-1)^j \binom{k}{j} f(x + (k-j)h).$$

Proof

(a) For $x \in \mathbb{R}$, we have $\Delta_h f(x) = f(x+h) - f(x) = 0$, since f is constant.
(b) For $x \in \mathbb{R}$, we can write

$$(\Delta_h(af + bg))(x) = (af + bg)(x+h) - (af + bg)(x)$$

$$= a(f(x+h) - f(x)) + b(g(x+h) - g(x))$$

$$= a(\Delta_h f)(x) + b(\Delta_h g)(x).$$

(c) Exercise (see Problem 1).

(d) Let us make induction on $k \geq 1$, the case $k = 1$ following directly from Definition 18.12. By induction hypothesis, assume that the result is true when $k = l \geq 1$. For $k = l + 1$, successively applying item (c) of Definition 18.12, the induction hypothesis and once more item (c) of Definition 18.12 (but this time to $\Delta_h f$ in place of f), we obtain

$$\Delta_h^{l+1} f = \Delta_h(\Delta_h^l f) = \Delta_h(\Delta_h^{l-1}(\Delta_h f)) = \Delta_h^l(\Delta_h f).$$

(e) Let us make induction on $k \geq 0$, noticing that if $k = 0$ and $x \in \mathbb{R}$, then

$$\sum_{j=0}^{0} (-1)^j \binom{0}{j} f(x) = (-1)^0 \binom{0}{0} f(x) = f(x) = (\Delta_h^0 f)(x).$$

Now, assume that the formula holds when $k = l \geq 0$, and let us prove its validity for $k = l + 1$. Given $x \in \mathbb{R}$, write $\Delta = (\Delta^{l+1} f)(x)$. By successively applying item (c) of Definition 18.12, the induction hypothesis and Stifel's relation, we obtain

$$\Delta = \Delta_h(\Delta_h^l f)(x) = (\Delta_h^l f)(x + h) - (\Delta_h^l f)(x)$$

$$= \sum_{j=0}^{l} (-1)^j \binom{l}{j} f(x + h + (l - j)h) - \sum_{j=0}^{l} (-1)^j \binom{l}{j} f(x + (l - j)h)$$

$$= f(x + (l + 1)h) + \sum_{j=1}^{l} (-1)^j \left(\binom{l+1}{j} - \binom{l}{j-1} \right) f(x + (l + 1 - j)h)$$

$$- \sum_{j=0}^{l} (-1)^j \binom{l}{j} f(x + (l - j)h)$$

$$= \sum_{j=0}^{l} (-1)^j \binom{l+1}{j} f(x + (l + 1 - j)h)$$

$$- \sum_{j=1}^{l} (-1)^j \binom{l}{j-1} f(x + (l + 1 - j)h) - \sum_{j=0}^{l} (-1)^j \binom{l}{j} f(x + (l - j)h),$$

Hence,

$$(\Delta^{l+1} f)(x) = \sum_{j=0}^{l+1} (-1)^j \binom{l+1}{j} f(x + (l+1-j)h) - (-1)^{l+1} f(x)$$

$$+ \sum_{j=0}^{l-1} (-1)^j \binom{l}{j} f(x + (l-j)h) - \sum_{j=0}^{l} (-1)^j \binom{l}{j} f(x + (l-j)h)$$

$$= \sum_{j=0}^{l+1} (-1)^j \binom{l+1}{j} f(x + (l+1-j)h)$$

$$+ \sum_{j=0}^{l} (-1)^j \binom{l}{j} f(x + (l-j)h) - \sum_{j=0}^{l} (-1)^j \binom{l}{j} f(x + (l-j)h)$$

$$= \sum_{j=0}^{l+1} (-1)^j \binom{l+1}{j} f(x + (l+1-j)h).$$

□

Among the properties of finite differences listed in the previous proposition, the one which is more frequently used in interpolation problems is that of item (e), mostly when combined with the coming simple result. Note that the proposition above refers to finite differences of general real functions, whereas the next one is specifically related to finite differences of polynomials.

We start by extending Definition 18.12 to polynomials, in the obvious way: for fixed $h \in \mathbb{R}$ and $f \in \mathbb{R}[X]$, given an integer $k \geq 0$ we define the **k-th finite difference** of f with step h as the polynomial $\Delta_h^k f \in \mathbb{R}[X]$ given by:

(i) $\Delta_h^0 f = f$.
(ii) $(\Delta_h^1 f)(X) = (\Delta_h f)(X) = f(X + h) - f(X)$.
(iii) $\Delta_h^k f = \Delta_h(\Delta_h^{k-1} f)$, for $k \geq 2$.

In view of the above definition, it is now immediate to check that items from (a) to (d) of the previous proposition still hold for finite differences of polynomials.

Proposition 18.14 *Let $f \in \mathbb{R}[X]$ be a polynomial of degree $n \geq 1$. For an arbitrary fixed step $h \in \mathbb{R} \setminus \{0\}$, the n-th finite differences of f all equal, whereas the l-th ones, for any $l > n$, vanish identically.*

Proof Start by observing that, if $f(X) = a_n X^n + \cdots + a_1 X + a_0$, then

$$(\Delta_h f)(X) = a_n (X + h)^n + \cdots + a_1 (X + h) + a_0$$

$$- (a_n X^n + \cdots + a_1 X + a_0)$$

$$= n a_n h X^{n-1} + g(X),$$

with either $g \equiv 0$ or $\partial g \leq n - 2$. Since $h \neq 0$, we conclude that $\Delta_h f$ has degree $n - 1$. Therefore, arguing by induction on the degree of a polynomial, item (d) of the (extension, for polynomials, of the) previous proposition guarantees that $\Delta_h^n f = \Delta_h^{n-1}(\Delta_h f)$ has degree 0, so that it is constant. Hence, $\Delta_h^l f(X) = 0$ for $l > k$. $\square$

As will be clear from the two coming examples, the main usefulness of the material on finite differences discussed above lies on the fact that, if the interpolation points form an arithmetic progression, then the method of finite differences allow us to compute the value of the corresponding polynomial at any point of such a progression.

Example 18.15 Let $f \in \mathbb{R}[X]$ be a polynomial of degree $m \geq 1$, such that $f(j) = r^j$ for $0 \leq j \leq m$, with r being a given positive real number. Compute all possible values of $f(m + 1)$.

Solution Set $h = 1$ and write $\Delta^k f$ to denote $\Delta_1^k f$ (either the polynomial or its corresponding polynomial function). Item (e) of Proposition 18.13 gives

$$\Delta^k f(x) = \sum_{j=0}^{k} (-1)^j \binom{k}{j} f(x + k - j)$$

for every $x \in \mathbb{R}$. However, since $\partial f = m$, the previous proposition then gives

$$0 = \Delta^{m+1} f(0) = \sum_{j=0}^{m+1} (-1)^j \binom{m + 1}{j} f(m + 1 - j).$$

Therefore,

$$f(m + 1) = \sum_{j=1}^{m+1} (-1)^{j+1} \binom{m + 1}{j} f(m + 1 - j)$$

$$= \sum_{j=1}^{m+1} (-1)^{j+1} \binom{m + 1}{j} r^{m+1-j}$$

$$= r^{m+1} - \sum_{j=0}^{m+1} (-1)^j \binom{m + 1}{j} r^{m+1-j}$$

$$= r^{m+1} - (r - 1)^{m+1},$$

where we used the binomial expansion of $(r - 1)^{m+1}$ in the last equality above. $\square$

Example 18.16 Let f be a polynomial of degree 1992, such that $f(j) = 2^j$ for $1 \leq i \leq 1993$. Compute the remainder upon division of $f(1994)$ by 1994.

Solution Combining item (e) of Proposition 18.13 with Proposition 18.14, and writing again $\Delta^k f$ to denote $\Delta_1^k f$, we have

$$0 = \Delta^{1993} f(1) = \sum_{j=0}^{1993} (-1)^j \binom{1993}{j} f(1994 - j).$$

Hence,

$$f(1994) = \sum_{j=1}^{1993} (-1)^{j+1} \binom{1993}{j} 2^{1994-j}$$

$$= \binom{1993}{0} 2^{1994} - 2 \sum_{j=0}^{1993} (-1)^j \binom{1993}{j} 2^{1993-j}$$

$$= 2^{1994} - 2(2 - 1)^{1993} = 2^{1994} - 2.$$

For what is left to do, note first that 997 is prime (this can be easily checked with the aid of Eratosthenes sieve, exactly as we did in Example 6.37 to show that 641 is prime). Hence, Fermat's Little Theorem 10.15 assures that $2^{996} \equiv 1 \pmod{997}$, so that

$$2^{1992} = (2^{996})^2 \equiv 1 \pmod{997}.$$

From this, it is immediate to get

$$f(1994) = 2^{1994} - 2 = 2^2 - 2 \equiv 2 \pmod{997}.$$

We are then left with the system of linear congruences

$$\begin{cases} f(1994) \equiv 0 \pmod 2 \\ f(1994) \equiv 2 \pmod{997} \end{cases},$$

which has, by the Chinese Remainder Theorem 10.27, a single solution modulus $2 \cdot 997 = 1994$. However, since 2 is clearly a solution, we obtain

$$f(1994) \equiv 2 \pmod{1994}. \qquad \square$$

Problems: Sect. 18.2

1. * Complete the proof of Proposition 18.13.
2. Prove that, for each real function $f : \mathbb{R} \to \mathbb{R}$, we have

$$f(x + kh) = \sum_{j=0}^{k} \binom{k}{j} (\Delta_h^{k-j} f)(x),$$

   for all $h, x \in \mathbb{R}$ and every integer $k \geq 0$.

3. (Singapore)[4] Let f be a polynomial of degree n, such that $f(k) = \frac{k}{k+1}$ for every integer $0 \le k \le n$. Compute $f(n+1)$.

4. (USA)[5] Let f be a real polynomial of degree n, such that $f(k) = \binom{n+1}{k}^{-1}$ for $0 \le k \le n$. Compute all possible values of $f(n+1)$.

5. (IMO—shortlist) A polynomial f, of degree 990, is such that $f(k) = F_k$ for $992 \le k \le 1982$, where F_k stands for the k-th Fibonacci number. Prove that $f(1983) = F_{1983} - 1$.

6. (IMO—shortlist) Let f be a real polynomial such that

   (a) $f(1) > f(0) > 0$.
   (b) $f(2) > 2f(1) - f(0)$.
   (c) $f(3) > 3f(2) - 3f(1) + f(0)$.
   (d) $f(n+4) > 4f(n+3) - 6f(n+2) + 4f(n+1) - f(n)$ for every $n \in \mathbb{N}$.

   Prove that $f(n) > 0$ for every $n \in \mathbb{N}$.

---

[4] For other approaches to this problem, see Problem 11, page 386, and Problem 4, page 444.
[5] For another approach to this problem, see Problem 3, page 444.

Chapter 19
On the Factorisation of Polynomials

The division algorithm for polynomials provides a notion of divisibility in $\mathbb{K}[X]$ when $\mathbb{K} = \mathbb{Q}, \mathbb{R}$ ou $\mathbb{C}$, and such a notion enjoys properties analogous to those of the corresponding concept in $\mathbb{Z}$. It is then natural to ask whether there exists some notion of *primality* in $\mathbb{K}[X]$, which furnishes some sort of *unique factorisation* with properties similar to the unique factorisation of integers. Our purpose in this chapter is to give precise answers to these questions, which shall encompass polynomials with coefficients in $\mathbb{Z}_p$, for some prime integer p.

19.1 Unique Factorisation in $\mathbb{Q}[X]$

Along this section, $\mathbb{K}$ denotes any of $\mathbb{Q}, \mathbb{R}$ or $\mathbb{C}$.

We say that $f, g \in \mathbb{K}[X] \setminus \{0\}$ are **associated** (in $\mathbb{K}[X]$) if there exists $a \in \mathbb{K} \setminus \{0\}$ such that $f = ag$. For instance, the real polynomials

$$f(X) = 4X^2 - 2X + 1 \text{ and } g(X) = 2\sqrt{2}X^2 - \sqrt{2}X + \frac{1}{\sqrt{2}}$$

are associated in $\mathbb{R}[X]$, for $f = \sqrt{2}g$ and $\sqrt{2} \in \mathbb{R} \setminus \{0\}$.

Given $f, g \in \mathbb{K}[X] \setminus \{0\}$, we say that a polynomial $p \in \mathbb{K}[X] \setminus \{0\}$ is a **common divisor** of f and g if $p \mid f, g$. Note that f and g always have common divisors: the constant nonzero polynomials over $\mathbb{K}$, for example.

Definition 19.1 Given $f, g \in \mathbb{K}[X] \setminus \{0\}$, we say that $d \in \mathbb{K}[X] \setminus \{0\}$ is a **greatest common divisor** of f and g, and denote $d = \gcd(f, g)$, if the two following conditions are satisfied:

(a) $d \mid f, g$ in $\mathbb{K}[X]$.
(b) If $d' \in \mathbb{K}[X] \setminus \{0\}$ divides f and g in $\mathbb{K}[X]$, then $d' \mid d$ in $\mathbb{K}[X]$.

© Springer International Publishing AG, part of Springer Nature 2018
A. Caminha Muniz Neto, *An Excursion through Elementary Mathematics, Volume III*,
Problem Books in Mathematics, https://doi.org/10.1007/978-3-319-77977-5_19

The coming result is the analogue, for polynomials, of Theorem 6.13. For this reason, it is also known as **Bézout's theorem**. In order to properly state it, given $f \in \mathbb{K}[X]$ we let $f \mathbb{K}[X]$, or simply (f), if there is no danger of confusion, denote the subset of $\mathbb{K}[X]$ formed by the multiples of f, i.e.,

$$(f) = f \mathbb{K}[X] = \{af;\ a \in \mathbb{K}[X]\}.$$

Theorem 19.2 (Bézout)   *Let* $f, g \in \mathbb{K}[X] \setminus \{0\}$. *If*

$$S = \{af + bg;\ a, b \in \mathbb{K}[X]\},$$

then there exists a polynomial $d \in \mathbb{K}[X] \setminus \{0\}$ *satisfying the following conditions:*

(a) $S = (d)$. *In particular,* $d \mid f, g$ *in* $\mathbb{K}[X]$.
(b) *Every polynomial in* $\mathbb{K}[X] \setminus \{0\}$ *that divides* f *and* g *also divides* d.

Moreover, such a polynomial d *is unique, up to association.*

Proof
(a) Let $d \in S \setminus \{0\}$ be such that

$$\partial d = \min\{\partial h;\ h \in S \setminus \{0\}\}.$$

We first claim that $S = (d)$. Indeed, letting $d = a_0 f + b_0 g$, with $a_0, b_0 \in \mathbb{K}[X]$, and $c \in \mathbb{K}[X]$, we have

$$cd = (ca_0)f + (cb_0)g \in S,$$

i.e., $(d) \subset S$. Conversely, take $h \in S$, say $h = af + bg$, with $a, b \in \mathbb{K}[X]$. By the division algorithm, we have $h = dq + r$, with $q, r \in \mathbb{K}[X]$ and $r = 0$ or $0 \leq \partial r < \partial d$. However, if $r \neq 0$, then $\partial r < \partial d$ and

$$
\begin{aligned}
r &= h - dq \\
&= (af + bg) - (a_0 f + b_0 g)q \\
&= (a - a_0 q)f + (b - b_0 q)g \in S,
\end{aligned}
$$

a contradiction to the minimality of the degree of d in S. Therefore, $r = 0$ and, hence, $h = dq \in (d)$.

For the second part of item (a), it suffices to see that $f, g \in S = (d)$, so that, in particular, f and g are multiples of d in $\mathbb{K}[X]$.

(b) Let $d' \in \mathbb{K}[X] \setminus \{0\}$ be a polynomial dividing f and g in $\mathbb{K}[X]$, say $f = d'f_1$ and $g = d'g_1$, with $f_1, g_1 \in \mathbb{K}[X]$. If $a, b \in \mathbb{K}[X]$, then

$$af + bg = (af_1 + bg_1)d' \in d' \mathbb{K}[X].$$

However, since $af + bg$ is a generic element of S, it follows that

$$d \in (d) = S \subset (d');$$

in particular, d is a multiple of d', as we wished to show.

The last claim is left as an exercise for the reader (cf. Problem 1). $\square$

In the notations of the statement of Bézout's theorem, given $f, g \in \mathbb{K}[X] \setminus \{0\}$ and letting $d \in \mathbb{K}[X] \setminus \{0\}$ be such that $d\,\mathbb{K}[X] = S$, from now on we say that d is a **greatest common divisor** of f and g, and write

$$d = \gcd(f, g).$$

Although a gcd of two nonzero polynomials is not unique, thanks to the last part of Bézout's theorem any two gcd's of them are associated. Thus, whenever convenient we can take d to be monic, and we shall adopt such a usage from now on without further notice.

Bézout's theorem assures that if $d = \gcd(f, g)$, then there exists $a, b \in \mathbb{K}[X]$ such that

$$d = af + bg.$$

Nevertheless, the proof of that result gave no clue on how to *effectively compute* such polynomials a and b. This issue will be dealt with in Problem 3, when we show that Euclid's algorithm remains true for polynomials.

Keeping with the parallelism to the notion of gcd in $\mathbb{Z}$, we say that polynomials $f, g \in \mathbb{K}[X] \setminus \{0\}$ are **relatively prime** if $\gcd(f, g) = 1$. We then have the following important consequence of Bézout's theorem.

Corollary 19.3 *If* $f, g \in \mathbb{K}[X] \setminus \{0\}$, *then* f *and* g *are relatively prime if and only if there exist polynomials* $a, b \in \mathbb{K}[X]$ *for which* $af + bg = 1$.

Proof If $\gcd(f, g) = 1$, then the existence of $a, b \in \mathbb{K}[X]$ as stated above follows directly from Bézout's theorem.

Conversely, if $d = \gcd(f, g)$ and there exist $a, b \in \mathbb{K}[X]$ such that $af + bg = 1$, then, again from Bézout's theorem, we have $1 \in S = d\,\mathbb{K}[X]$, i.e., d is a divisor of the constant polynomial 1. Therefore, $d = 1$ up to association, so that 1 is a gcd of f and g. $\square$

Corollary 19.4 *Let* $f, g \in \mathbb{K}[X] \setminus \{0\}$ *be relatively prime and* $h \in \mathbb{K}[X] \setminus \{0\}$ *be such that* $\partial h < \partial(fg)$. *Then, there exist* $a, b \in \mathbb{K}[X]$ *such that* $a = 0$ *or* $\partial a < \partial g$, $b = 0$ *or* $\partial b < \partial f$ *and* $af + bg = h$.

Proof By the previous corollary, there exist $a_1, b_1 \in \mathbb{K}[X]$ such that $a_1 f + b_1 g = 1$. Then, letting $a_2 = a_1 h$ and $b_2 = b_1 h$, we have $a_2 f + b_2 g = h$. Now, the division algorithm gives $a_2 = gq + a$, with $a = 0$ or $0 \le \partial a < \partial g$. Thus,

$$h = (gq + a)f + b_2 g = af + (qf + b_2)g$$

and, letting $b = qf + b_2$, we have $h = af + bg$, with $a = 0$ or $\partial a < \partial g$. Finally, since $bg = h - af$, if $b \neq 0$ we can compute

$$\partial b + \partial g = \partial(bg) = \partial(h - af) \leq \partial h < \partial(fg) = \partial f + \partial g,$$

so that $\partial b < \partial f$. $\square$

The coming definition is absolutely central for the rest of this chapter.

Definition 19.5 A polynomial $p \in \mathbb{K}[X] \setminus \mathbb{K}$ is **irreducible** over $\mathbb{K}$ if p cannot be written as the product of two nonconstant polynomials in $\mathbb{K}[X]$. A polynomial $p \in \mathbb{K}[X] \setminus \mathbb{K}$ that is not irreducible is said to be **reducible** over $\mathbb{K}$.

In general, it is useful to rephrase the irreducibility condition contrapositively. This way, a polynomial $p \in \mathbb{K}[X] \setminus \mathbb{K}$ is irreducible if and only if the following condition is satisfied:

$$p = gh, \text{ with } g, h \in \mathbb{K}[X] \Rightarrow g \in \mathbb{K} \text{ or } h \in \mathbb{K}. \tag{19.1}$$

Let us now take a look at two simple examples.

Example 19.6 It follows immediately from (19.1) that every polynomial $p \in \mathbb{K}[X]$ of degree 1 is irreducible. Indeed, letting $p = gh$, with $g, h \in \mathbb{K}[X]$, Proposition 14.9 assures that $\partial g + \partial h = \partial p = 1$ and, hence, $\partial g = 0$ or $\partial h = 0$, i.e., g or h is constant. On the other hand, the Fundamental Theorem of Algebra 15.23 guarantees that the irreducible polynomials in $\mathbb{C}[X]$ are precisely those of degree 1.

Example 19.7 If $p \in \mathbb{R}[X]$ is irreducible over $\mathbb{R}$, then $\partial p = 1$ or 2. Indeed, according to Problem 4, page 386, a complex number z is a root of p if and only if so is $\overline{z}$; then, let us consider two cases:

(a) $\partial p \geq 3$ and p has at least one real root, say α: the root test (cf. Proposition 15.3) allows us to write $p(X) = (X - \alpha)h(X)$, for some $h \in \mathbb{R}[X]$ of degree greater than or equal to 2, and p is reducible over $\mathbb{R}$.

(b) $\partial p \geq 3$ and p has two complex nonreal conjugate roots, say z and $\overline{z}$: by invoking the root test again, we have $p(X) = (X - z)(X - \overline{z})h(X)$, for some $h \in \mathbb{C}[X]$ of degree at least 1. However, if $z = a + bi$ and $g(X) = (X - z)(X - \overline{z})$, then

$$g(X) = (X - a - bi)(X - a + bi) = (X - a)^2 + b^2 \in \mathbb{R}[X].$$

Hence, upon dividing p by g, the division algorithm assures that $h \in \mathbb{R}[X]$. Therefore, $p = gh$, with $g, h \in \mathbb{R}[X] \setminus \mathbb{R}$, and p is reducible over $\mathbb{R}$.

Thanks to the last example above, from now on we shall restrict our attention to the irreducibility of polynomials over $\mathbb{Q}$. In this sense, an argument analogous to that of item (a) of the last example (cf. Problem 5) allows us to conclude that if $p \in \mathbb{Q}[X]$ has degree 2 or 3, then p is irreducible over $\mathbb{Q}$ if and only if it does not have a rational root.

Back to the general concept of irreducible polynomial, note that if $p \in \mathbb{Q}[X] \setminus \mathbb{Q}$ is irreducible, then the only divisors of p in $\mathbb{Q}[X]$ are the constant polynomials and those associated with p. This being said, we have the following important auxiliary result, which is the analogue of Lemma 6.41 for polynomials.

Lemma 19.8 *Let $p \in \mathbb{Q}[X] \setminus \mathbb{Q}$ be irreducible. If $f_1, \ldots, f_k \in \mathbb{Q}[X] \setminus \{0\}$ are such that $p \mid f_1 \ldots f_k$, then there exists $1 \leq i \leq k$ such that $p \mid f_i$.*

Proof By induction on k, it suffices to show that if $p \mid fg$, with $f, g \in \mathbb{Q}[X] \setminus \{0\}$, then $p \mid f$ or $p \mid g$. If $p \nmid f$, we first claim that $\gcd(f, p) = 1$; indeed, if $d = \gcd(f, p)$, then $d \mid p$, so that either $d \in \mathbb{Q}$ or d is associated with p in $\mathbb{Q}[X]$. However, if this last possibility holds, it follows from $d \mid f$ that $p \mid f$ too, which is a contradiction. Therefore, $d \in \mathbb{Q}$ and, up to association, we can suppose that $d = 1$.

Now, Corollary 19.3 guarantees the existence of polynomials $a, b \in \mathbb{Q}[X]$ such that $af + bp = 1$, and hence

$$a(fg) + (bg)p = g.$$

Since $p \mid (fg)$, it follows from the equality above and Problem 6 that $p \mid g$, as wished. $\qquad\square$

We shall now establish an analogue of the Fundamental Theorem of Arithmetic for polynomials over $\mathbb{Q}$, thus showing that every polynomial $f \in \mathbb{Q}[X] \setminus \mathbb{Q}$ can be uniquely written, up to association, as a product of a finite number of irreducible polynomials.

Theorem 19.9 *Every polynomial $f \in \mathbb{Q}[X] \setminus \mathbb{Q}$ can be written as a product of a finite number of irreducible polynomials over $\mathbb{Q}$. Moreover, such a decomposition of f is unique in the following sense: if $p_1, \ldots, p_k, q_1, \ldots, q_l \in \mathbb{Q}[X] \setminus \mathbb{Q}$ are irreducible and such that $p_1 \ldots p_k = q_1 \ldots q_l$, then $k = l$ and, up to reordering, p_i and q_i are associated in $\mathbb{Q}[X]$.*

Proof For the existence part, let us make induction on ∂f, the case $\partial f = 1$ being immediate (for we have already seen that, in this case, f is irreducible). As induction hypothesis, assume that the result is true for every nonconstant polynomial with rational coefficients and degree less than n, and take $f \in \mathbb{Q}[X] \setminus \mathbb{Q}$ such that $\partial f = n$. If f is irreducible, there is nothing left to do. Otherwise, we can write $f = gh$, with $g, h \in \mathbb{Q}[X] \setminus \mathbb{Q}$. Thus, $\partial g, \partial h < n$, and the induction hypotheses guarantees that each of g and h can be written as a product of a finite number of irreducible polynomials with rational coefficients, say $g = p_1 \ldots p_j$ and $h = p_{j+1} \ldots p_k$. Then, $f = gh = p_1 \ldots p_j p_{j+1} \ldots p_k$, a product of a finite number of irreducible polynomials over $\mathbb{Q}$.

For the uniqueness part, let $p_1 \ldots p_k = q_1 \ldots q_l$, with $p_1, \ldots, p_k, q_1, \ldots, q_l \in \mathbb{Q}[X] \setminus \mathbb{Q}$ being irreducible over $\mathbb{Q}$. If $k = 1$, we have $p_1 = q_1 \ldots q_l$, and the irreducibility of p_1 assures that $l = 1$. Similarly, $l = 1 \Rightarrow k = 1$. Then, suppose that $k, l > 1$; since $p_k \mid q_1 \ldots q_l$, Lemma 19.8 guarantees the existence of an index

$1 \leq j \leq l$ such that $p_k \mid q_j$. Assume, without loss of generality, that $j = l$. Since q_l is irreducible over $\mathbb{Q}$ and $p_k \notin \mathbb{Q}$, the only possibility is that p_k and q_l be associated in $\mathbb{Q}[X]$, say $p_k = uq_l$, for some $u \in \mathbb{Q} \setminus \{0\}$. Then,

$$p_1 \cdots p_{k-1} = q_1 \cdots q_{l-2}uq_{l-1} = q_1' \cdots q_{l-1}',$$

with $q_i' = q_i$ for $1 \leq i < l - 2$ and $q_{l-1}' = uq_{l-1}$, all irreducible over $\mathbb{Q}$.

Arguing by induction on $\max\{k, l\}$, the induction hypothesis gives $k - 1 = l - 1$ and, up to reordering, p_i associated to q_i' for $1 \leq i \leq l - 1$. Hence, up to reordering p_i and q_i are also associated over $\mathbb{Q}$, for $1 \leq i \leq k$. $\square$

We can summarize the previous result by saying that one has *unique factorisation* in $\mathbb{Q}[X]$. As within $\mathbb{Z}$, if $f \in \mathbb{Q}[X] \setminus \mathbb{Q}$ is factorised as

$$f = p_1 \cdots p_k,$$

with $p_1, \ldots, p_k \in \mathbb{Q}[X] \setminus \mathbb{Q}$ irreducible, then, by gathering together the irreducible factors p_i which are equal up to association, we obtain

$$f = q_1^{\alpha_1} \cdots q_l^{\alpha_1}, \tag{19.2}$$

with $q_1, \ldots, q_l \in \mathbb{Q}[X] \setminus \mathbb{Q}$ irreducible and pairwise non associated, and $\alpha_1, \ldots, \alpha_l \in \mathbb{N}$. Expression (19.2) (which is also unique up to association and reordering) is the **canonical factorisation** of f in $\mathbb{Q}[X]$, and $q_1, \ldots, q_l$ are the **irreducible factors** of f in $\mathbb{Q}[X]$.

Problems: Sect. 19.1

1. * Complete the proof of Theorem 19.2.
2. Let $f, g \in \mathbb{Q}[X] \setminus \mathbb{Q}$ be polynomials with canonical factorisations

$$f = p_1^{\alpha_1} \cdots p_k^{\alpha_k} q_1^{\alpha_1'} \cdots q_l^{\alpha_l'} \quad \text{and} \quad g = p_1^{\beta_1} \cdots p_k^{\beta_k} r_1^{\beta_1'} \cdots r_m^{\beta_m'},$$

with $p_1, \ldots, p_k, q_1, \ldots, q_l$ and $r_1, \ldots, r_m$ being monic, irreducible and pairwise non associated, and $\alpha_i, \alpha_i', \beta_j, \beta_j' \in \mathbb{N}$. Prove that

$$\gcd(f, g) = p_1^{\gamma_1} \cdots p_k^{\gamma_k},$$

with $\gamma_i = \min\{\alpha_i, \beta_i\}$ for $1 \leq i \leq k$.
3. * Prove the following version of Euclid's algorithm for polynomials: given $f, g \in \mathbb{K}[X] \setminus \{0\}$, with $\partial f \geq \partial g$, the division algorithm in $\mathbb{K}[X]$ gives

$$\begin{aligned}
\text{Step 1} \quad &: f \;\; = \;\; gq_1 + r_1 & 0 \le \partial r_1 < \partial g \\
\text{Step 2} \quad &: g \;\; = \;\; r_1 q_2 + r_2 & 0 \le \partial r_2 < \partial r_1 \\
\text{Step 3} \quad &: r_1 \;\; = \;\; r_2 q_3 + r_3 & 0 \le \partial r_3 < \partial r_2 \\
& \quad\;\;\; \cdots & \cdots \\
\text{Step } j \quad &: r_{j-2} = r_{j-1} q_j + r_j & 0 \le \partial r_j < \partial r_{j-1} \\
\text{Step } j+1 \;\; &: r_{j-1} = r_j q_{j+1} + 0 &
\end{aligned}$$

As in $\mathbb{Z}$, note that the execution of the algorithm actually stops after a finite number of steps. Indeed, since $r_1, r_2, \ldots \in \mathbb{K}[X]$ satisfy $\partial g > \partial r_1 > \partial r_2 > \ldots \ge 0$, there must exist a least index j such that $r_j \ne 0$. Show that $r_j = \gcd(f, g)$.

4. Compute $\gcd(X^4 - X^3 - 3X^2 + 2X + 3, X^5 - X^3 + 4X^2 - 6X - 12)$.

5. * If $p \in \mathbb{Q}[X]$ is a polynomial of degree 2 or 3, prove that p is irreducible over $\mathbb{Q}$ if and only if p has no roots in $\mathbb{Q}$.

6. * If $f, g, h \in \mathbb{Q}[X] \setminus \{0\}$ are such that $f \mid g, h$, prove that f divides $ag + bh$, for all $a, b \in \mathbb{Q}[X]$.

7. If $f \in \mathbb{R}[X] \setminus \mathbb{R}$ has degree greater than 1, prove that f can be written, uniquely up to association, as a product of a finite number of irreducible polynomials over $\mathbb{R}$.

8. Prove **Fermat's last theorem for polynomials**: if $f, g, h \in \mathbb{R}[X] \setminus \mathbb{R}$ have no nontrivial common factors, and are such that

$$f(X)^n + g(X)^n = h(X)^n,$$

then $n = 1$ or 2.

9. The purpose of this problem is to prove the **partial fractions decomposition theorem**. To this end, let $\mathbb{K} = \mathbb{Q}$, $\mathbb{R}$ or $\mathbb{C}$, let $f, g \in \mathbb{K}[X] \setminus \{0\}$ be given and think of $\frac{f}{g}$ as the *rational function* that sends each $x \in \mathbb{K} \setminus g^{-1}(0)$ to $\frac{f(x)}{g(x)}$.

   (a) If $\partial f < \partial g$ and $g = g_1^{\alpha_1} \ldots g_k^{\alpha_k}$ is the canonical decomposition of g in irreducible polynomials in $\mathbb{K}[X]$ (established in the previous problem if $\mathbb{K} = \mathbb{R}$), prove that there exist $f_1, \ldots, f_k \in \mathbb{K}[X]$, with $f_j = 0$ or $\partial f_j < \partial(g_j^{\alpha_j})$ for $1 \le j \le k$, and

$$\frac{f}{g} = \sum_{j=1}^{k} \frac{f_j}{g_j^{\alpha_j}}.$$

   (b) If g is irreducible over $\mathbb{K}$ and $k \ge 1$ is an integer, prove that there exist polynomials $q, r_1, \ldots, r_k \in \mathbb{K}[X]$, with $r_j = 0$ or $\partial r_j < \partial g$ for $1 \le j \le k$, and

$$\frac{f}{g^k} = q + \sum_{j=1}^{k} \frac{r_j}{g^j}.$$

19.2   Unique Factorisation in $\mathbb{Z}[X]$

In the previous section, we studied the problem of unique factorisation for polynomials with rational coefficients. In this section, we extend the analysis of this problem to polynomials with integer coefficients. In this respect, the coming definition will be crucial for all that follows.

Definition 19.10 The **content** of $f \in \mathbb{Z}[X] \setminus \{0\}$, denoted $c(f)$, is the gcd of the nonvanishing coefficients of f. If $c(f) = 1$, we say that f is a **primitive polynomial** in $\mathbb{Z}[X]$.

For the sake of notation, if $f = a_n X^n + \cdots + a_1 X + a_0 \in \mathbb{Z}[X] \setminus \{0\}$, we shall write

$$c(f) = \gcd(a_0, \ldots, a_n).$$

The coming lemma, whose proof we leave to the reader as an exercise (cf. Problem 1), establishes two useful, albeit simple, properties of the content of polynomials with integer coefficients.

Lemma 19.11

(a) *If $f \in \mathbb{Z}[X] \setminus \{0\}$ and $a \in \mathbb{Z} \setminus \{0\}$, then $c(af) = |a| \cdot c(f)$. In particular, there exists a primitive $g \in \mathbb{Z}[X] \setminus \{0\}$ such that $f = c(f)g$.*

(b) *If $f \in \mathbb{Q}[X] \setminus \{0\}$, then, up to multiplication by -1, there exist unique relatively prime nonzero integers a, b and $g \in \mathbb{Z}[X]$ primitive such that $f = (a/b)g$.*

The importance of the concept of content for polynomials in $\mathbb{Z}[X]$ lies in the key role it plays on the study of irreducibility of polynomials over $\mathbb{Z}$, as we shall now define.

Definition 19.12 A polynomial $p \in \mathbb{Z}[X] \setminus \mathbb{Z}$ is **irreducible**[1] over $\mathbb{Z}$ if p is primitive and cannot be written as the product of two nonconstant polynomials of $\mathbb{Z}[X]$. A primitive polynomial $p \in \mathbb{Z}[X] \setminus \mathbb{Z}$ which is not irreducible is said to be **reducible** over $\mathbb{Z}$.

As in the previous section, it is useful to rephrase the irreducibility condition for integer coefficient nonconstant polynomials contrapositively, so that a primitive polynomial $p \in \mathbb{Z}[X] \setminus \mathbb{Z}$ is irreducible if and only if the following condition is satisfied:

$$p = gh, \text{ with } g, h \in \mathbb{Z}[X] \Rightarrow g = \pm 1 \text{ or } h = \pm 1. \tag{19.3}$$

The coming result is known in mathematical literature as **Gauss' lemma**.

---

[1]Although this definition is slightly more restrictive than that usually found in most textbooks, it will be sufficient for our purposes.

Proposition 19.13 (Gauss)   *For $f, g \in \mathbb{Z}[X] \setminus \mathbb{Z}$, we have:*

(a) $c(fg) = c(f)c(g)$. In particular, fg is primitive if and only if[2] so are f and g.
(b) If f is primitive, then f is irreducible in $\mathbb{Z}[X]$ if and only if it is irreducible in $\mathbb{Q}[X]$.
(c) If f and g are primitive and associated in $\mathbb{Q}[X]$, then $f = \pm g$.

Proof

(a) Lemma 19.11 shows that if $f = c(f)f_1$ and $g = c(g)g_1$, then $f_1, g_1 \in \mathbb{Z}[X]$
   are primitive and $fg = c(f)c(g)f_1g_1$, so that $c(fg) = c(f)c(g)c(f_1g_1)$.
   Hence, it suffices to show that f_1g_1 is primitive, or, which is the same, that

$$f, g \text{ primitive} \Leftrightarrow fg \text{ primitive}.$$

   If f is not primitive, then there exists a prime $p \in \mathbb{Z}$ such that p divides all
   of the coefficient of f. Then, p divides all of the coefficients of fg, and fg is
   not primitive. Analogously, if g is not primitive, then fg is not primitive too.
   Conversely, assume that $f(X) = a_m X^m + \cdots + a_1 X + a_0$ and $g(X) = b_n X^n + \cdots + b_1 X + b_0$ are both primitive, but fg is not. Take a prime $p \in \mathbb{Z}$
   such that p divides all of the coefficients of fg, and let $k, l \geq 0$ be the smallest
   indices such that $p \nmid a_k, b_l$ (such k and l do exist, for otherwise f or g would
   not be primitive). Letting c_{k+l} denote the coefficient of X^{k+l} in fg, we have

$$c_{k+l} = \cdots + a_{k-2}b_{l+2} + a_{k-1}b_{l+1} + a_k b_l + a_{k+1}b_{l-1} + a_{k+2}b_{l-2} + \cdots .$$

   Now, since $p \mid c_{k+l}$ and $p \mid a_0, \ldots, a_{k-1}, b_0, \ldots, b_{l-1}$, the above equality
   guarantees that $p \mid a_k b_l$, which is a contradiction.
(b) Implication $\Leftarrow$) is clear. For the converse, take $f \in \mathbb{Z}[X] \setminus \mathbb{Z}$ primitive and
   suppose that f is reducible in $\mathbb{Q}[X]$, say $f = gh$, with $g, h \in \mathbb{Q}[X] \setminus \mathbb{Q}$.
   Lemma 19.11 assures that we can take $a, b, c, d \in \mathbb{Z} \setminus \{0\}$ such that $g = (a/b)g_1$
   and $h = (c/d)h_1$, with $g_1, h_1 \in \mathbb{Z}[X]$ primitive. Then,

$$bdf = bg \cdot dh = ag_1 \cdot ch_1 = acg_1h_1.$$

   However, since $g_1 h_1$ is primitive by (a), taking contents in both sides of the
   equality above we obtain $|bd| \cdot c(f) = |ac|$. Therefore, $\frac{ac}{bd} = \pm c(f) \in \mathbb{Z}$ and,
   once again from the equality above, $f = \pm c(f)g_1h_1$, i.e., f is reducible in
   $\mathbb{Z}[X]$.
(c) If $f = (a/b)g$, with $a, b \in \mathbb{Z} \setminus \{0\}$, then $bf = ag$, and taking contents we get
   $|b| \cdot c(f) = |a| \cdot c(g)$. However, since $c(f) = 1$ and $c(g) = 1$, it comes that
   $|a| = |b|$, and hence $a/b = \pm 1$.

$\square$

---

[2]For another proof of this item, see Problem 6, page 469.

We can finally state and prove another theorem of Gauss, which for integer coefficient polynomials is the analogue of Theorem 19.9.

Theorem 19.14 (Gauss) *Every primitive polynomial $f \in \mathbb{Z}[X] \setminus \mathbb{Z}$ can be written as a product of a finite number of irreducible polynomials in $\mathbb{Z}[X]$. Moreover, such a way of writing f is unique up to a reordering of the factors and multiplication of some of them by -1.*

Proof Let us start by establishing the existence of a factorisation of f as a product of irreducible polynomials. To this end, looking at f as a polynomial in $\mathbb{Q}[X]$, Theorem 19.9 assures the existence of irreducible polynomials $p_1, \ldots, p_k \in \mathbb{Q}[X] \setminus \mathbb{Q}$ such that $f = p_1 \ldots p_k$. Write $p_i = (a_i/b_i)q_i$, with $a_i, b_i \in \mathbb{Z} \setminus \{0\}$ relatively prime and $q_i \in \mathbb{Z}[X] \setminus \mathbb{Z}$ primitive. Since q_i is obviously also irreducible in $\mathbb{Q}[X]$, item (b) of Gauss' lemma guarantees that q_i is irreducible in $\mathbb{Z}[X]$. Letting $a = a_1 \ldots a_k$ and $b = b_1 \ldots b_k$, we then have

$$f = (a/b)q_1 \ldots q_k.$$

However, since $q_1, \ldots, q_k$ are all primitive, item (a) of Gauss' lemma guarantees that $q_1 \ldots q_k$ is primitive too. Thus, taking contents in the last equality above, we get

$$|b| = |b| \cdot c(f) = |a| \cdot c(q_1 \ldots q_k) = |a|.$$

It comes that $a/b = \pm 1$, and hence $f = q_1 \ldots q_k$, a product of irreducible polynomials in $\mathbb{Z}[X]$.

The proof of the uniqueness part of the statement parallels that of Theorem 19.9, once we show the following claim: if $p, f, g \in \mathbb{Z}[X] \setminus \mathbb{Z}$ are such that p is irreducible and $p \mid fg$ in $\mathbb{Z}[X]$, then $p \mid f$ or $p \mid g$ in $\mathbb{Z}[X]$. To this end, note firstly (from item (b) of Gauss' lemma) that p is also irreducible in $\mathbb{Q}[X]$. In turn, this being so, we already know that $p \mid f$ or $p \mid g$ in $\mathbb{Q}[X]$. If $p \mid f$ in $\mathbb{Q}[X]$ (the other case is analogous), there exists $f_1 \in \mathbb{Q}[X]$ such that $f = f_1 p$. Choosing $a, b \in \mathbb{Z}$ such that $f_1 = (a/b)f_2$, with $f_2 \in \mathbb{Z}[X] \setminus \mathbb{Z}$ primitive, we have

$$bf = bf_1 p = af_2 p.$$

Now, p and f_2 primitive implies (once more from item (a) of Gauss' lemma) $f_2 p$ primitive; taking contents in the equality above, we obtain $|b| \cdot c(f) = |a| \cdot c(f_2 p) = |a|$. Hence, $a/b = \pm c(f) \in \mathbb{Z}$, so that $f_1 \in \mathbb{Z}[X]$ and $p \mid f$ in $\mathbb{Z}[X]$.   □

Gauss' theorem allows us to define the greatest common divisor of two polynomials $f, g \in \mathbb{Z}[X] \setminus \mathbb{Z}$. To this end, given $f \in \mathbb{Z}[X] \setminus \mathbb{Z}$, start using Gauss' theorem to write

$$f = \pm c(f)f_1^{n_1} \ldots f_k^{n_k},$$

with $n_1, \ldots, n_k \in \mathbb{N}$ and $f_1, \ldots, f_k \in \mathbb{Z}[X]$ being irreducible, pairwise non associated and having positive leading coefficients. This is the **canonical factorisation** of f in $\mathbb{Z}[X]$.

Now, let $g \in \mathbb{Z}[X] \setminus \mathbb{Z}$ have canonical factorisation $g = \pm c(g) g_1^{m_1} \ldots g_l^{m_l}$. If $h \in \mathbb{Z}[X] \setminus \mathbb{Z}$ is irreducible, has positive leading coefficient and is such that $h \mid f, g$, then there exists $1 \leq i \leq k$ and $1 \leq j \leq l$ for which $h = f_i$ and $h = g_j$, so that $f_i = g_j$. In turn, this means that, up to a change of notation, we can write

$$f = \pm c(f) h_1^{n_1} \ldots h_k^{n_k} f_1^{p_1} \ldots f_r^{p_l} \quad \text{and} \quad g = \pm c(g) h_1^{m_1} \ldots h_k^{m_k} g_1^{q_1} \ldots g_s^{q_s},$$

with $h_1, \ldots, h_k, f_1, \ldots, f_r, g_1, \ldots, g_s \in \mathbb{Z}[X] \setminus \mathbb{Z}$ irreducible, with positive leading coefficients and pairwise non associated and $n_1, \ldots, n_k, p_1, \ldots, p_l, q_1, \ldots, q_s \in \mathbb{Z}_+$. We thus let, in $\mathbb{Z}[X]$,

$$\gcd(f, g) = \gcd(c(f), c(g)) h_1^{a_1} \ldots h_k^{a_k},$$

with $a_i = \min\{n_i, m_i\}$, for $1 \leq i \leq k$.

Problems: Sect. 19.2

1. * Prove Lemma 19.11.
2. * Let $f \in \mathbb{Z}[X]$ be a monic and nonconstant polynomial. If f is reducible over $\mathbb{Q}$, prove that there exist monic and nonconstant polynomials $g, h \in \mathbb{Z}[X]$ such that $f = gh$.
3. Let $f, g \in \mathbb{Z}[X] \setminus \mathbb{Z}$ be primitive. For a given common divisor $d \in \mathbb{Z}[X]$ of f and g, show that the following are equivalent:

   (a) $d = \gcd(f, g)$.
   (b) If $d' \in \mathbb{Z}[X]$ divides f and g in $\mathbb{Z}[X]$, then $d' \mid d$.

19.3 Polynomials Over $\mathbb{Z}_p$

We saw in Sect. 11.2 that, for a prime $p \in \mathbb{Z}$, the set $\mathbb{Z}_p$ of congruence classes modulo p can be furnished with operations of addition, subtraction, multiplication and division quite similar to those of $\mathbb{C}$. In turn, thanks to such a resemblance, essentially all of the concepts and results on polynomials studied so far remain true within the set $\mathbb{Z}_p[X]$ of polynomials with coefficients in $\mathbb{Z}_p$.

Our purpose here is to make explicit comments on some similarities and differences between polynomials over $\mathbb{Z}_p$ and over $\mathbb{K}$, with $\mathbb{K} = \mathbb{Q}, \mathbb{R}$ or $\mathbb{C}$. In this sense, we shall leave to the reader the task of checking that all of the other definitions and results presented for $\mathbb{K}[X]$ (except those of Sects. 15.3 and 18.2 and of Chap. 17) remain valid, *ipsis literis*, for $\mathbb{Z}_p[X]$. We also take the opportunity to deduce, with the aid of polynomials over $\mathbb{Z}_p$, some results on Number Theory and Combinatorics inaccessible by other means. In particular, we complete the proof of the existence of primitive roots modulo p, thus completing the discussion of Sect. 12.2.

Given $f(X) = a_n X^n + \cdots + a_1 X + a_0 \in \mathbb{Z}[X]$, we define $\mathbb{Z}_p[X]$ to be the set of formal expressions $\overline{f}$ of the form

$$\overline{f}(X) = \overline{a}_n X^n + \cdots + \overline{a}_1 X + \overline{a}_0, \tag{19.4}$$

where $\overline{a}_0, \overline{a}_1, \ldots, \overline{a}_n$ respectively denote the congruence classes of $a_0, a_1, \ldots, a_n$ modulo p. As before, such an $\overline{f}$ is called a **polynomial** over $\mathbb{Z}_p$.

The correspondence $f \mapsto \overline{f}$ defines a map

$$\pi_p : \mathbb{Z}[X] \longrightarrow \mathbb{Z}_p[X]$$
$$f \longmapsto \overline{f}$$

which is obviously surjective and is called the **canonical projection** of $\mathbb{Z}[X]$ onto $\mathbb{Z}_p[X]$. For $f, g \in \mathbb{Z}[X]$, it is immediate to verify that

$$\overline{f}(X) = \overline{g}(X) \text{ in } \mathbb{Z}_p[X]$$
$$\Updownarrow$$
$$\exists h \in \mathbb{Z}[X]; \ f(X) = g(X) + ph(X) \text{ in } \mathbb{Z}[X].$$

Equivalently, letting

$$p\,\mathbb{Z}[X] = \{ph; \ h \in \mathbb{Z}[X]\},$$

we have

$$\overline{f} = \overline{0} \Leftrightarrow f \in p\mathbb{Z}[X].$$

We extend the operations of addition and multiplication in $\mathbb{Z}_p$ to homonymous operations $+, \cdot : \mathbb{Z}_p[X] \times \mathbb{Z}_p[X] \to \mathbb{Z}_p[X]$ by setting, for $f, g \in \mathbb{Z}[X]$,

$$\overline{f} + \overline{g} = \overline{f + g} \ \text{ and } \ \overline{f} \cdot \overline{g} = \overline{fg}.$$

We leave to the reader the task of verifying the well definiteness of these operations in $\mathbb{Z}_p[X]$, which can be done in a way entirely analogous to the well definiteness of the operations of $\mathbb{Z}_p$ (cf. Sect. 11.2; see, also, Problem 1).

As in $\mathbb{Z}[X]$, we say that a polynomial $\overline{f} \in \mathbb{Z}_p[X] \setminus \{\overline{0}\}$ as in (19.4) has **degree** n if $\overline{a}_n \neq \overline{0}$, i.e., if $p \nmid a_n$. More generally, if $f \in \mathbb{Z}[X] \setminus p\mathbb{Z}[X]$, then $\overline{f} \neq \overline{0}$ and $\partial \overline{f} \leq \partial f$.

The two coming examples use the multiplication of $\mathbb{Z}_p[X]$ to establish interesting properties of binomial numbers.

Example 19.15 If $p \in \mathbb{Z}$ is a prime number and $k \in \mathbb{N}$, prove that $\binom{p^k}{j}$ is a multiple of p, for every integer $1 \leq j < p^k$.

Proof Example 6.42 shows that $(X+\overline{1})^p = X^p + \overline{1}$ in $\mathbb{Z}_p[X]$. Assume, by induction hypothesis, that $(X + \overline{1})^{p^l} = X^{p^l} + \overline{1}$ in $\mathbb{Z}_p[X]$ and for some $l \in \mathbb{N}$. Then, by successively applying the initial case and the inductive hypothesis, we obtain

$$(X + \overline{1})^{p^{l+1}} = \left((X + \overline{1})^p\right)^{p^l} = (X^p + \overline{1})^{p^l} = (X^p)^{p^l} + \overline{1} = X^{p^{l+1}} + \overline{1}.$$

Thus,

$$(X + \overline{1})^{p^k} = X^{p^k} + \overline{1}$$

for every $k \in \mathbb{N}$.

On the other hand, we also have

$$(X + \overline{1})^{p^k} = X^{p^k} + \overline{\binom{p^k}{p^k - 1}} X^{p^k - 1} + \cdots + \overline{\binom{p^k}{1}} X + \overline{1},$$

so that $\overline{\binom{p^k}{j}} = \overline{0}$ for every integer $1 \leq j < p^k$. In other words, p divides $\binom{p^k}{j}$. □

The next example makes use of the binary representation of natural numbers (cf. Example 4.12 of [8] or Problem 11, page 163). For a generalization, see Problem 11.

Example 19.16 (Romania) Prove that the number of odd binomial coefficients in the n-th line of Pascal's triangle is a power of 2.

Proof Let

$$n = 2^{a_k} + 2^{a_{k-1}} + \cdots + 2^{a_1} + 2^{a_0}$$

be the binary representation of n, with $0 \leq a_0 < a_1 < \ldots < a_k$. The previous example allows us to write, in $\mathbb{Z}_2[X]$,

$$\begin{aligned}
(X + \overline{1})^n &= (X + \overline{1})^{2^{a_k}} (X + \overline{1})^{2^{a_{k-1}}} \ldots (X + \overline{1})^{2^{a_0}} \\
&= (X^{2^{a_k}} + \overline{1})(X^{2^{a_{k-1}}} + \overline{1}) \ldots (X^{2^{a_0}} + \overline{1}).
\end{aligned} \tag{19.5}$$

Let S denote the set of natural numbers which can be written as a sum of (at least one of) distinct powers of 2, chosen from $2^{a_k}, 2^{a_{k-1}}, \ldots, 2^{a_1}$ and 2^{a_0}. By the fundamental principle of counting and the uniqueness of binary representation of naturals, we have $|S| = 2^{k+1} - 1$. On the other hand, by expanding the products of the last expression in (19.5), we get

$$(X + \overline{1})^n = \sum_{m \in S} X^m + \overline{1} \qquad (19.6)$$

a sum with exactly 2^{k+1} summands. Now, by invoking once more the formula of binomial expansion, we have

$$(X + \overline{1})^n = X^n + \overline{\binom{n}{1}} X^{n-1} + \cdots + \overline{\binom{n}{n-1}} X + \overline{1}. \qquad (19.7)$$

Therefore, by comparing (19.6) and (19.7), we conclude that exactly 2^{k+1} of the binomial numbers of the form $\binom{n}{j}$ (which compose the n-th line of Pascal's triangle) satisfy $\overline{\binom{n}{j}} \neq \overline{0}$, i.e., are odd.   □

In order to define the polynomial function associated to a polynomial $\overline{f} \in \mathbb{Z}_p[X]$, we have to take some care. Firstly, note that if $f \in \mathbb{Z}[X]$ and $a, b \in \mathbb{Z}$ satisfy $a \equiv b \pmod{p}$, then item (c) of Proposition 10.6 guarantees that

$$f(a) \equiv f(b) \pmod{p};$$

on the other hand, if $\overline{f} = \overline{g}$ in $\mathbb{Z}_p[X]$, we saw above that there exists $h \in \mathbb{Z}[X]$ such that $f(X) = g(X) + ph(X)$. Hence, for $a \in \mathbb{Z}$ we have

$$f(a) = g(a) + ph(a) \equiv g(a) \pmod{p}.$$

Given $\overline{f} \in \mathbb{Z}_p[X]$, the above comments allow us to define the **polynomial function** $\tilde{f} : \mathbb{Z}_p \to \mathbb{Z}_p$ by setting, for $a \in \mathbb{Z}$,

$$\tilde{f}(\overline{a}) = \overline{g(a)}, \qquad (19.8)$$

where $g \in \mathbb{Z}[X]$ is any polynomial for which $\overline{f} = \overline{g}$. Obviously, the image of $\tilde{f}$ is a finite set, for $\mathbb{Z}_p$ is itself finite. From now on, whenever there is no danger of confusion, we shall write (19.8) simply as

$$\overline{f}(\overline{a}) = \overline{f(a)}.$$

The coming example shows that, contrary to what happens with polynomials over $\mathbb{Q}$, $\mathbb{R}$ or $\mathbb{C}$, the polynomial function associated to a nonzero polynomial over $\mathbb{Z}_p[X]$ can vanish identically. In other words, *it is no longer valid* that two distinct polynomials over $\mathbb{Z}_p$ have distinct polynomial functions.

Example 19.17 The polynomial $f(X) = X^p - X \in \mathbb{Z}_p[X]$ is clearly a nonzero element of $\mathbb{Z}_p[X]$. On the other hand, letting $\overline{f} : \mathbb{Z}_p \to \mathbb{Z}_p$ denote its associated polynomial function, Fermat's little theorem gives

$$\overline{f}(\overline{a}) = \overline{a}^p - \overline{a} = \overline{a^p - a} = \overline{0}$$

for every $\overline{a} \in \mathbb{Z}_p$. Thus, $\overline{f}$ vanishes identically.

Let $f \in \mathbb{Z}[X]$ and $a \in \mathbb{Z}$ be given. As in Sect. 15.1, we say that $\overline{a} \in \mathbb{Z}_p$ is a root of $\overline{f}$ provided $\overline{f}(\overline{a}) = \overline{0}$. An easy review of the proof of the root test shows that it continues to hold in $\mathbb{Z}_p[X]$. In particular, from the above example we obtain the following important result.

Proposition 19.18 *In $\mathbb{Z}_p[X]$, we have*

$$X^{p-1} - \overline{1} = (X - \overline{1})(X - \overline{2}) \ldots (X - \overline{(p-1)}).$$

Proof Since $\overline{1}, \overline{2}, \ldots, \overline{p-1}$ are roots of $X^{p-1} - \overline{1}$ in $\mathbb{Z}_p$ (from the last example), item (c) of Proposition 15.3 assures that the polynomial $X^{p-1} - \overline{1}$ is divisible by $(X - \overline{1})(X - \overline{2}) \ldots (X - \overline{(p-1)})$ in $\mathbb{Z}_p[X]$. However, since both such polynomials are monic and have degree $p - 1$, they are actually equal. $\square$

In $\mathbb{Z}_p[X]$, all of the definitions and results of Sect. 19.1 remain true. In particular, we can state the following theorem, whose proof is entirely analogous to that of Theorem 19.9.

Theorem 19.19 *If $p \in \mathbb{Z}$ is prime, then every polynomial $\overline{f} \in \mathbb{Z}_p[X] \setminus \mathbb{Z}_p$ can be written as a product of a finite number of irreducible polynomials over $\mathbb{Z}_p$. Moreover, such a decomposition of $\overline{f}$ is unique up to association and reordering of the irreducible factors.*

As a first application of the previous result, we shall now prove the existence of primitive roots modulo p. For the statement and proof of the coming result, perhaps the reader might want to review the material of Sect. 12.2.

Theorem 19.20 *If $p \in \mathbb{Z}$ is an odd prime and d is a positive divisor of $p - 1$, then the algebraic congruence*

$$x^{p-1} - 1 \equiv 0 \,(\mathrm{mod}\, p) \tag{19.9}$$

has exactly $\varphi(d)$ roots of order d, pairwise incongruent modulo p. In particular, p has exactly $\varphi(p - 1)$ primitive roots pairwise incongruent modulo p.

Proof For a positive divisor d of $p - 1$, let $N(d)$ denote the number of roots of the algebraic congruence (19.9) which are pairwise incongruent modulo p and have order d. Since the roots of (19.9) are the integers $1, 2, \ldots, p - 1$, and since by Proposition 12.2 each of these numbers has order equal to a divisor of $p - 1$, we conclude that

$$\sum_{0 < d \mid (p-1)} N(d) = p - 1.$$

If we show that $N(d) \leq \varphi(d)$, Proposition 8.11 will give

$$p - 1 = \sum_{0<d\mid(p-1)} N(d) \leq \sum_{0<d\mid(p-1)} \varphi(d) = p - 1.$$

Therefore, we will obtain $N(d) = \varphi(d)$ for every positive divisor d of $p - 1$.

Then, let d be a positive divisor of $p - 1$. If $N(d) = 0$, it is clear that $N(d) \leq \varphi(d)$. Otherwise, let a be an integer of order d modulo p; then, the congruence classes $\bar{1}, \bar{a}, \ldots, \bar{a}^{d-1} \in \mathbb{Z}_p$ are pairwise distinct roots of $X^d - \bar{1} \in \mathbb{Z}_p[X]$. On the other hand, Corollary 15.8 guarantees that such a polynomial has at most d distinct roots in $\mathbb{Z}_p$, so that its roots are exactly $\bar{1}, \bar{a}, \ldots, \bar{a}^{d-1}$. Therefore, if $\alpha \in \mathbb{Z}$ is a root of order d of (19.9), then $\bar{\alpha} \in \mathbb{Z}_p$ is a root of $X^d - \bar{1} \in \mathbb{Z}_p[X]$, so that $\bar{\alpha} \in \{\bar{1}, \bar{a}, \ldots, \bar{a}^{d-1}\}$. We conclude that the roots of order d of (19.9) are precisely the elements of order d of the set $\{1, a, \ldots, a^{d-1}\}$ and, hence,

$$N(d) = \#\{0 \leq k \leq d - 1; \ \mathrm{ord}_p(a^k) = d\}.$$

Item (c) of Proposition 12.5 counts the number of elements of the set in the right hand side above: since $\mathrm{ord}_p(a) = d$, we have

$$\mathrm{ord}_p(a^k) = d \Leftrightarrow \frac{d}{\gcd(d, k)} = d \Leftrightarrow \gcd(d, k) = 1.$$

Therefore,

$$\#\{0 \leq k \leq d - 1; \ \mathrm{ord}_p(a^k) = d\} = \#\{0 \leq k \leq d - 1; \ \gcd(d, k) = 1\}$$
$$= \varphi(d).$$

The above argument has shown that either $N(d) = 0$ or $\varphi(d)$, so that, in any case, we have $N(d) \leq \varphi(d)$, as wished.

For what is left to do, it suffices to note that $N(p - 1) = \varphi(p - 1)$, so that there are exactly $\varphi(p - 1)$ integers, pairwise incongruent modulo p and with order $p - 1 = \varphi(p)$ modulo p. In other words, there are precisely $\varphi(p - 1)$ pairwise incongruent primitive roots modulo p. (For the sake of completeness, it is worth noticing that this result agrees with that of Proposition 12.8.) $\square$

We finish this section by exhibiting yet another beautiful application of the theory of polynomials over $\mathbb{Z}_p$ to Number Theory.

Example 19.21 (Miklós-Schweitzer) If $p > 3$ is a prime number satisfying $p \equiv 3 \pmod 4$, prove that

$$\prod_{1 \leq x \neq y \leq \frac{p-1}{2}} (x^2 + y^2) \equiv 1 \pmod p.$$

Proof In all that follows, without further mention, indices in products will vary from 1 to $\frac{p-1}{2}$.

If P stands for the product at the left hand side, then

$$P = \prod_{x \neq 1}(x^2 + 1^2) \cdot \prod_{x \neq 2}(x^2 + 2^2) \cdot \ldots \cdot \prod_{x \neq \frac{p-1}{2}} \left(x^2 + \left(\frac{p-1}{2}\right)^2\right)$$

$$= \frac{\prod(x^2 + 1^2) \cdot \prod(x^2 + 2^2) \cdot \ldots \cdot \prod\left(x^2 + \left(\frac{p-1}{2}\right)^2\right)}{2^{\frac{p-1}{2}} \cdot 1^2 \cdot 2^2 \cdot \ldots \cdot \left(\frac{p-1}{2}\right)^2}.$$

Now, for each integer $1 \leq k \leq \frac{p-1}{2}$, let c_k denote its inverse modulo p. Then, modulo p we get

$$\prod(x^2 + k^2) \equiv k^2 \prod((c_k x)^2 + 1) \equiv k^2 \prod(x^2 + 1).$$

Indeed, the first congruence is immediate and, for the second one, it suffices to observe that $\{(c_k x)^2;\ 1 \leq x \leq \frac{p-1}{2}\}$ is a set of $\frac{p-1}{2}$ pairwise incongruent quadratic residues modulo p; hence, also modulo p, we have

$$\{(c_k x)^2;\ 1 \leq x \leq \frac{p-1}{2}\} = \{x^2;\ 1 \leq x \leq \frac{p-1}{2}\}.$$

It then follows that, modulo p,

$$2^{\frac{p-1}{2}} \cdot 1^2 \cdot 2^2 \cdot \ldots \cdot \left(\frac{p-1}{2}\right)^2 P \equiv 1^2 \cdot 2^2 \cdot \ldots \cdot \left(\frac{p-1}{2}\right)^2 \left(\prod(x^2 + 1)\right)^{\frac{p-1}{2}}$$

or, which is the same,

$$2^{\frac{p-1}{2}} P \equiv \left(\prod(x^2 + 1)\right)^{\frac{p-1}{2}}.$$

Let α be a primitive root modulo p and $Q = \prod(x^2 + 1)$. Since $\{\alpha^{2k};\ 1 \leq k \leq \frac{p-1}{2}\}$ is a set of $\frac{p-1}{2}$ pairwise incongruent quadratic residues modulo p, we have $\{\alpha^{2k};\ 1 \leq k \leq \frac{p-1}{2}\} = \{x^2;\ 1 \leq x \leq \frac{p-1}{2}\}$ modulo p, so that

$$Q = \prod(\alpha^{2k} + 1) \quad \text{and} \quad 2^{\frac{p-1}{2}} P \equiv Q^{\frac{p-1}{2}} \pmod{p}. \tag{19.10}$$

In order to compute the residue of Q, modulo p, let $f \in \mathbb{Z}[X]$ be defined by

$$f(X) = (X - \alpha^2)(X - \alpha^4) \ldots (X - \alpha^{p-1}),$$

so that

$$Q \equiv (-1)^{\frac{p-1}{2}} f(-1) \equiv -f(-1) \,(\mathrm{mod}\, p).$$

(Here we have used the fact that $p \equiv 3 \,(\mathrm{mod}\, 4)$ in the second congruence above.) Now, observe that

$$f(X^2) = (X - \alpha)(X - \alpha^2) \ldots (X - \alpha^{\frac{p-1}{2}})(X + \alpha)(X + \alpha^2) \ldots (X + \alpha^{\frac{p-1}{2}}).$$

For $1 \leq i, j \leq \frac{p-1}{2}$, if $\alpha^i \equiv -\alpha^j \,(\mathrm{mod}\, p)$, then $\alpha^{2i} \equiv \alpha^{2j} \,(\mathrm{mod}\, p)$ and, since α is a primitive root modulo p, we obtain $2i = 2j$, thus $i = j$; however this being the case, we would have $\alpha^i \equiv -\alpha^i \,(\mathrm{mod}\, p)$, which is an absurd. Then, the set $\{\pm\alpha, \pm\alpha^2, \ldots, \pm\alpha^{\frac{p-1}{2}}\}$ is a RRS modulo p, so that it coincides, modulo p, with $\{1, 2, \ldots, p - 1\}$. Therefore, letting $\overline{f} \in \mathbb{Z}_p[X]$ denote the image of f by the projection of $\mathbb{Z}[X]$ onto $\mathbb{Z}_p[X]$ and invoking the result of Proposition 19.18, we arrive at

$$\overline{f}(X^2) = (X - \overline{1})(X - \overline{2}) \ldots (X - \overline{p-1}) = X^{p-1} - \overline{1}$$

and, hence, at

$$\overline{f}(X) = X^{\frac{p-1}{2}} - \overline{1}.$$

However, since $p \equiv 3 \,(\mathrm{mod}\, 4)$, this gives

$$\overline{Q} = -\overline{f}(-\overline{1}) = -(-\overline{1})^{\frac{p-1}{2}} + \overline{1} = \overline{2},$$

i.e., $Q \equiv 2 \,(\mathrm{mod}\, p)$.

Finally, substituting this information in (19.10), we get the congruence $2^{\frac{p-1}{2}} P \equiv 2^{\frac{p-1}{2}} \,(\mathrm{mod}\, p)$, and hence $P \equiv 1 \,(\mathrm{mod}\, p)$.   □

Problems: Sect. 19.3

1. * Given a prime number p, check the well definiteness of the operations of addition and multiplication in $\mathbb{Z}_p[X]$. Also, show that such operations are associative and commutative, have identities respectively equal to the constant polynomials $\overline{0}$ and $\overline{1}$, and that multiplication is distributive with respect to addition.

2. Let $p > 2$ be a given prime. Find, if any, the roots of $X^{p-1} + \overline{1} \in \mathbb{Z}_p[X]$.

3. * Given $f \in \mathbb{Z}[X]$ and an integer root a of f, prove that $\bar{a} \in \mathbb{Z}_p$ is a root of $\bar{f} \in \mathbb{Z}_p[X]$. In particular, conclude that if $\bar{a}_1, \ldots, \bar{a}_k \in \mathbb{Z}_p$ are the roots of $\bar{f}$, then there exists $1 \leq j \leq k$ such that $a \equiv a_j \pmod{p}$.
4. Show that $f(X) = X^3 - 15X^2 + 10X - 84 \in \mathbb{Z}[X]$ has no rational roots.
5. * If $p \in \mathbb{Z}$ is prime and $f \in \mathbb{Z}[X]$, prove that $\bar{f}(X^p) = \bar{f}(X)^p$.
6. * Use the projection $\pi : \mathbb{Z} \to \mathbb{Z}_p$ to prove that if $f, g \in \mathbb{Z}[X] \setminus \mathbb{Z}$ are primitive polynomials (cf. Definition 19.10), then so is fg.
7. Let $p > 2$ be a prime number and $1 \leq d \leq p - 1$ be an integer.

   (a) If $d \nmid (p - 1)$, show that $X^d - \bar{1}$ has no roots in $\mathbb{Z}_p[X]$.
   (b) If $d \mid (p - 1)$, factorise $X^d - \bar{1}$ in $\mathbb{Z}_p[X]$.

8. * Let $p \geq 3$ be a prime number and, for $1 \leq j \leq p - 1$, let $s_j(1, 2, \ldots, p-1)$ denote the j-th elementary symmetric sum of the natural numbers $1, 2, \ldots, p - 1$. Prove that[3]:

   (a) For $1 \leq j \leq p - 2$, we have $s_j(1, 2, \ldots, p - 1) \equiv 0 \pmod{p}$.
   (b) $s_{p-1}(1, 2, \ldots, p - 1) \equiv -1 \pmod{p}$.

9. * If a, b and c are the complex roots of the polynomial $X^3 - 3X^2 + 1$, show that $a^n + b^n + c^n \in \mathbb{Z}$ for every $n \in \mathbb{N}$, and that such a sum is always congruent to 1 modulo 17.
10. (France) For a given $n \in \mathbb{N}$, let I_n denote the number of odd coefficients of the polynomial $(X^2 + X + 1)^n$.

    (a) Compute I_{2^m}, for $m \in \mathbb{Z}_+$.
    (b) Show that, for $m \in \mathbb{N}$, we have

    $$I_{2^m - 1} = \frac{2^{m+1} + (-1)^{m+1}}{3}.$$

    For the coming problem, the reader may find it useful to review the statement of Problem 11, page 163.

11. Prove **Lucas' theorem**: given natural numbers $m \geq n$ and a prime number p, if

    $$m = \sum_{j=0}^{k} m_j p^j \text{ and } \sum_{j=0}^{k} n_j p^j$$

    are the representations of m and n in base p, then

    $$\binom{m}{n} \equiv \prod_{j=0}^{k} \binom{m_j}{n_j} \pmod{p}.$$

---

[3]Note that item (b) provides another proof of part of Wilson's Theorem 10.26.

In particular, conclude that:

(a) $p \mid \binom{m}{n}$ if and only if $m_j < n_j$ for some $0 \le j \le k$.

(b) Exactly $(m_0 + 1)(m_1 + 1) \ldots (m_k + 1)$ binomial numbers of the form $\binom{m}{n}$ are not divisible by p.

(c) No binomial number of the form $\binom{p^{k+1}-1}{n}$ is divisible by p.

12. Let p be a prime number and $k \in \mathbb{N}$. Prove that

$$\binom{p^k(p-1)}{l} \equiv \begin{cases} (-1)^q \ (\mathrm{mod}\ p), & \text{if } l = p^k q;\ 0 \le q \le p - 1,\ q \in \mathbb{Z} \\ 0 \ (\mathrm{mod}\ p), & \text{otherwise.} \end{cases}$$

19.4   More on Irreducible Polynomials

Up to this moment, we do not have at our disposal any method for establishing the irreducibility of a given polynomial. Obviously, in particular cases one can sometimes use the *direct method*, which consists in writing the given polynomial f (say, in $\mathbb{Z}[X]$) as a product of two other polynomials $g, h \in \mathbb{Z}[X]$, and then of solving the resulting system of equations in the coefficients of g and h, thus obtaining a nontrivial factorisation for f or reaching a contradiction. We illustrate such a procedure in the coming example (see, also, Problems 1 and 2).

Example 19.22 Prove that $f(X) = X^4 + 10X^3 + 5X + 1993$ is irreducible over $\mathbb{Q}$.

Proof By Gauss' lemma (cf. Proposition 19.13), it suffices to show that f cannot be written as a product of two nonconstant polynomials with integer coefficients.

We now observe that 1993 is prime, a fact that can be easily established with the aid of Eratosthenes' sieve, for example. Hence, the rational roots test implies that the possible integer roots of f are ± 1 or ± 1993, and direct checking shows that none of these is actually a root.

We are left to discarding the possibility of f having a factorisation of the form

$$f(X) = (X^2 + aX + b)(X^2 + cX + d),$$

with $a, b, c, d \in \mathbb{Z}$. If such happens, then, by expanding the products at the right hand side and comparing coefficients, we obtain the following system of equations:

$$\begin{cases} a + c = 10 \\ ac + b + d = 0 \\ ad + bc = 5 \\ bd = 1993 \end{cases}.$$

The fourth equation, together with the primality of 1993 and the symmetry of the factorisation (two factors of degree 2) give $(b, d) = (1, 1993)$ or $(-1, -1993)$

as the only essentially distinct possibilities. From them, the first and third equations furnish the systems of equations

$$\begin{cases} a + c = 10 \\ a + 1993c = 5 \end{cases} \text{ or } \begin{cases} a + c = 10 \\ a + 1993c = -5 \end{cases}.$$

Finally, it is immediate to check that none of them has integer solutions. □

As testified by the calculations performed in the example above, the use of the direct method to check that a given polynomial $f \in \mathbb{Q}[X]$ is irreducible over $\mathbb{Q}$ is likely to face considerable computational difficulties, even in the case of f having integer coefficients and small degree. Therefore, it would be quite nice to develop a few techniques that could be effectively applied to the investigation of the irreducibility of polynomials. In this sense, we start with the following result.

Proposition 19.23 *Let $f \in \mathbb{Z}[X]$ be monic, and $p \in \mathbb{Z}$ be prime.*

(a) *If $\overline{f} \in \mathbb{Z}_p[X]$ is irreducible in $\mathbb{Z}_p[X]$, then f is irreducible in $\mathbb{Z}[X]$.*
(b) *If $\overline{f}(X) = (X - \overline{a})\overline{g}(X)$, with $\overline{g} \in \mathbb{Z}_p[X]$ irreducible, then either f is irreducible in $\mathbb{Z}[X]$ or it has a root $\alpha \in \mathbb{Z}$ such that $\alpha \equiv a \pmod{p}$.*

Proof

(a) By contraposition, suppose $f(X) = g(X)h(X)$, with $g, h \in \mathbb{Z}[X] \setminus \mathbb{Z}$ monic polynomials. Then, $\overline{f}(X) = \overline{g}(X)\overline{h}(X)$, with $\partial \overline{g} = \partial g$ and $\partial \overline{h} = \partial h$.
(b) Suppose f to be reducible, say $f(X) = h(X)l(X)$, with $h, l \in \mathbb{Z}[X] \setminus \mathbb{Z}$ being monic. Then, $\partial \overline{h} = \partial h$, $\partial \overline{l} = \partial l$ and

$$(X - \overline{a})\overline{g}(X) = \overline{h}(X)\overline{l}(X).$$

However, since $\overline{g}$ is irreducible in $\mathbb{Z}_p[X]$, Theorem 19.19 assures that either $\overline{h}$ or $\overline{l}$ must be associated to $X - \overline{a}$, so that $\partial h = 1$ or $\partial l = 1$.

Assume, without loss of generality, that $\partial h = 1$. Then, $h(X) = X - \alpha$, for some $\alpha \in \mathbb{Z}$, so that α is an integer root of f. Finally, since $\overline{h}(X) = X - \overline{a}$ (for $\overline{h}$ is monic and associated to $X - \overline{a}$), it comes that $\overline{\alpha} = \overline{a}$.

□

Example 19.24 Let $p \in \mathbb{Z}$ be prime, with $p \equiv 5 \pmod{6}$, and $k \in \mathbb{N}$ be such that $kp + 1$ is also prime. Prove that $f(X) = X^3 + pX^2 + pX + kp + 1$ is irreducible in $\mathbb{Z}[X]$.

Proof Projecting f into $\mathbb{Z}_p[X]$, we obtain

$$\overline{f}(X) = X^3 + \overline{1} = (X + \overline{1})(X^2 - X + \overline{1}).$$

We now claim that $-\overline{1}$ is the only root of $\overline{f}$ in $\mathbb{Z}_p$. Indeed, if $\overline{f}(\overline{a}) = \overline{0}$, then $\overline{a}^3 = -\overline{1}$, so that $a^3 \equiv -1 \pmod{p}$. Thus, $a^6 \equiv 1 \pmod{p}$, and hence (by Fermat's little theorem, together with item (b) of Proposition 12.2)

$$\text{ord}_p(a) \mid \gcd(6, p - 1) = 2.$$

Therefore, $\text{ord}_p(a) = 1$ or 2. Since $\text{ord}_p(a) = 1 \Rightarrow \overline{a} = \overline{1}$, which is not a root of $\overline{f}$, we must have $\text{ord}_p(a) = 2$. Then, $a^2 \equiv 1 \pmod{p}$ and $a \not\equiv 1 \pmod{p}$, so that $a \equiv -1 \pmod{p}$ or, which is the same, $\overline{a} = -\overline{1}$.

Since $\overline{3} \neq \overline{0}$ in $\mathbb{Z}_p$, the above claim guarantees that $X^2 - X + \overline{1}$ has no roots in $\mathbb{Z}_p[X]$; thus, it is irreducible in $\mathbb{Z}_p[X]$. The previous proposition then assures that either f is irreducible in $\mathbb{Z}[X]$ or it has a root $\alpha \in \mathbb{Z}$ such that $\alpha \equiv -1 \pmod{p}$. However, the searching criterion for rational roots assures that an integer root α of f must be a divisor of $kp + 1$, which is prime. Then, either $\alpha = -1$ or $\alpha = -(kp+1)$, and testing these possibilities we conclude that none of them is actually a root of f. Thus, f is irreducible. $\square$

Remark 19.25 For a given $p \in \mathbb{N}$, we shall prove in Sect. 20.2 that there exist infinitely many $k \in \mathbb{Z}$ for which $kp + 1$ is prime.

The corollary of the coming result of F. Eisenstein, known in mathematical literature as **Eisenstein's criterion**, will show to be a powerful ally in establishing the irreducibility of certain polynomials of integer coefficients.

Theorem 19.26 (Eisenstein) *Let $f(X) = a_n X^n + \cdots + a_1 X + a_0$ be a polynomial with integer coefficients and degree $n \geq 1$. Let also $p \in \mathbb{Z}$ be prime and $1 \leq k < n$ be an integer such that:*

(a) $p \mid a_0, a_1, \ldots, a_k$.
(b) $p^2 \nmid a_0$ and $p \nmid a_n$.

If $f = gh$, with $g, h \in \mathbb{Z}[X]$, then $\max\{\partial g, \partial h\} \geq k + 1$.

Proof Letting $g(X) = b_r X^r + \cdots + b_0$ and $h(X) = c_s X^s + \cdots + c_0$, we have $a_n = b_r c_s$ and $a_0 = b_0 c_0$. Therefore,

$$p \nmid a_n \Rightarrow p \nmid b_r c_s \Rightarrow p \nmid b_r \text{ and } p \nmid c_s,$$

so that, in $\mathbb{Z}_p[X]$, we have $\overline{f}(X) = \overline{g}(X)\overline{h}(X)$, with $\partial \overline{g} = \partial g = r$ and $\partial \overline{h} = \partial h = s$.

If $X \mid \overline{g}(X)$ and $X \mid \overline{h}(X)$ in $\mathbb{Z}_p[X]$, then $p \mid b_0$ and $p \mid c_0$ in $\mathbb{Z}$, so that $p^2 \mid b_0 c_0 = a_0$, which is not the case; hence, in $\mathbb{Z}_p[X]$, the polynomial X divides at most one of $\overline{g}$ or $\overline{h}$.

Assume, without loss of generality, that X does not divide $\overline{h}$. Since (by hypothesis (b))

$$\overline{g}(X)\overline{h}(X) = \overline{f}(X) = \overline{a}_n X^n + \cdots + \overline{a}_{k+1} X^{k+1}$$
$$= (\overline{a}_n X^{n-k-1} + \cdots + \overline{a}_{k+1})X^{k+1},$$

we thus conclude that $X^{k+1} \mid \overline{g}(X)$. Thus, $r \geq k + 1$. $\square$

Corollary 19.27 (Eisenstein) *Let $f(X) = a_n X^n + \cdots + a_1 X + a_0$ be a polynomial with integer coefficients and degree $n \geq 1$. If there exists $p \in \mathbb{Z}$ prime such that $p \mid a_0, a_1, \ldots, a_{n-1}$, $p^2 \nmid a_0$ and $p \nmid a_n$, then f is irreducible in $\mathbb{Q}[X]$.*

Proof By Eisenstein's theorem, if $f(X) = g(X)h(X)$, with $g, h \in \mathbb{Z}[X]$, then either $\partial g \geq n$ or $\partial h \geq n$, i.e., either g or h is constant. Thus, f cannot be written as a product of two nonconstant polynomials with integer coefficients, and Gauss' lemma guarantees that f is irreducible in $\mathbb{Q}[X]$. $\qquad\square$

The coming example brings a classical application of Eisenstein's criterion, which will be obtained by other means in Sect. 20.2.

Example 19.28 If $p \in \mathbb{Z}$ is prime and $f \in \mathbb{Z}[X]$ is given by

$$f(X) = X^{p-1} + X^{p-2} + \cdots + X + 1, \tag{19.11}$$

then f is irreducible in $\mathbb{Q}$.

Proof We start by noticing that $f(X) = g(X)h(X)$ if and only if $f(X + 1) = g(X + 1)h(X + 1)$, with $g(X)$ and $g(X + 1)$ (as well as $h(X)$ and $h(X + 1)$) having equal degrees. Thus, it suffices to show that $f(X + 1)$ is irreducible in $\mathbb{Q}[X]$.

Since $(X - 1)f(X) = X^p - 1$, we obtain

$$Xf(X + 1) = (X + 1)^p - 1 = X^p + \binom{p}{1}X^{p-1} + \cdots + \binom{p}{p-1}X,$$

and hence

$$f(X + 1) = X^{p-1} + \binom{p}{1}X^{p-2} + \cdots + \binom{p}{p-1}.$$

Now, note that $\binom{p}{p-1} = p$ and recall, from Example 6.42, that $p \mid \binom{p}{k}$ for $1 \leq k \leq p - 1$. Hence, we can apply Eisenstein's criterion with the prime p to conclude that $f(X + 1)$ is irreducible in $\mathbb{Q}[X]$. $\qquad\square$

Our second example uses the full force of Eisenstein's theorem.

Example 19.29 (IMO) Given a natural number $n > 1$, prove that the polynomial $X^n + 5X^{n-1} + 3$ is irreducible in $\mathbb{Z}[X]$.

Proof Let $X^n + 5X^{n-1} + 3 = g(X)h(X)$, with $g, h \in \mathbb{Z}[X]$ monic. With $p = 3$ in Eiseinstein's theorem, we conclude that

$$\max\{\partial g, \partial h\} \geq n - 1,$$

so that either f is irreducible or it has an integer root.

On the other hand, if this last possibility takes place, then the searching criterion for rational roots assures that such a root is ± 1 or ± 3. However, direct computation shows that none of these four numbers is a root of the given polynomial. $\square$

In spite of the theory developed above, we frequently establish the irreducibility of a given polynomial by invoking *ad hoc methods*. Let us take a look at two such examples.

Example 19.30 (Romania) Let $f \in \mathbb{Z}[X]$ be a monic polynomial of degree $n \geq 1$ and such that $f(0) = 1$. If f has at least $n - 1$ complex roots of modulus less than 1, prove that f is irreducible in $\mathbb{Q}[X]$.

Proof By Gauss' lemma, it suffices to show that f is irreducible in $\mathbb{Z}[X]$. To this end, and for the sake of contradiction, assume that $f = gh$, with $g, h \in \mathbb{Z}[X] \setminus \mathbb{Z}$. Without any loss of generality, we can take g and h to be monic; moreover, from $g(0)h(0) = 1$ we get $g(0), h(0) = \pm 1$. On the other hand, Girard-Viète relations (cf. Proposition 16.6) assure that the modulus of the product of the complex roots of f is equal to 1, so that the hypotheses of the example guarantee that f has exactly one complex root of modulus greater than 1. Hence, one of the polynomials g or h, say g, has only complex roots of modulus less than 1. However, since g is monic and $|g(0)| = 1$, by invoking Girard-Viète relations again, we conclude that the modulus of the product of the complex roots of g is equal to 1, which is a contradiction. $\square$

Remark 19.31 Some complex analysis assures that polynomials $f \in \mathbb{Z}[X]$ satisfying the hypotheses of the previous example can be easily constructed. For instance, let $f(X) = a_n X^n + a_{n-1} X^{n-1} + \cdots + a_1 X + a_0 \in \mathbb{Z}[X]$ be such that

$$|a_{n-1}| > |a_n| + |a_{n-2}| + |a_{n-3}| + \cdots + |a_0|. \tag{19.12}$$

If $|z| = 1$, then the triangle inequality for complex numbers gives

$$|f(z) - a_{n-1}z^{n-1}| = |a_n z^n + a_{n-2}z^{n-2} + \cdots + a_1 z + a_0|$$
$$\leq |a_n| + |a_{n-2}| + \cdots + |a_1| + |a_0|$$
$$< |a_{n-1}| = |a_{n-1}z^{n-1}|.$$

In particular, f does not vanish on the unit circle $|z| = 1$ of the complex plane, and Rouché's theorem (cf. [11], page 125) assures that f has exactly $n - 1$ complex roots of modulus less than 1. Of course, the extra requirement that $a_n = a_0 = 1$ does not preclude the validity of (19.12).

Example 19.32 (IMO Shortlist) Let $n > 1$ be an odd integer and $a_1, \ldots, a_n$ be pairwise distinct given integers. Show that

$$f(X) = (X - a_1)(X - a_2) \ldots (X - a_n) - 1$$

is irreducible in $\mathbb{Q}[X]$.

Proof Once more Gauss' lemma reduces the problem to showing that f cannot be written as the product of two nonconstant polynomials with integer coefficients. By contradiction, assume we have $f = gh$, with g and h being nonconstant, monic polynomials with integer coefficients. Since

$$g(a_i)h(a_i) = f(a_i) = -1$$

and $g(a_i), h(a_i) \in \mathbb{Z}$, we must have $g(a_i) = 1$ and $h(a_i) = -1$, or vice-versa; in any event, it happens that $g(a_i) + h(a_i) = 0$. Now, let $f_1 = g + h$. Since g and h are monic, f_1 does not vanish identically. Therefore,

$$\partial f_1 \leq \max\{\partial g, \partial h\} < \partial f = n,$$

so that f_1 has at most $\partial f_1 < n$ complex roots, counted according with their multiplicities. However, since $f_1(a_i) = 0$ for $1 \leq i \leq n$ and $a_1, \ldots, a_n$ are pairwise distinct, we have arrived at a contradiction. □

Problems: Sect. 19.4

1. Use the direct method described in the beginning of this section (and applied in Example 19.22) to prove that $X^4 - X^2 + 1$ is irreducible in $\mathbb{Z}[X]$.
2. * Also with the aid of the direct method, show that $f(X) = X^5 - X^4 - 4X^3 + 4X^2 + 2$ is irreducible over $\mathbb{Q}$.
3. A real polynomial is said to be *positively reducible* if it can be expressed as a product of two nonconstant real polynomials with nonnegative coefficients. Let f be a real polynomial such that, for some $n \in \mathbb{N}$, the polynomial $f(X^n)$ is positively reducible. Show that f is also positively reducible.
4. (IMO shortlist) Let $n > 1$ be an integer and $a_1, \ldots, a_n$ be pairwise distinct positive integers. Prove that

$$f(X) = (X - a_1)(X - a_2)\ldots(X - a_n) + 1$$

is reducible in $\mathbb{Q}[X]$ only in the following cases:

   i. $n = 2$ and $f(X) = (X - a)(X - a - 2) + 1$.
   ii. $n = 4$ and $f(X) = (X - a)(X - a - 1)(X - a - 2)(X - a - 3) + 1$.

5. Let p be an odd prime, $f(X) = 2X^{p-1} - 1$ and $g(X) = (X-1)(X-2)\ldots(X-p+1)$. Prove that at least one of $f(X) - g(X)$ or $f(X) - g(X) + p$ is irreducible in $\mathbb{Z}[X]$.
6. Let $p \in \mathbb{Z}$ be a prime number and $f(X) = a_{2n+1}X^{2n+1} + \cdots + a_n X^n + \cdots + a_1 X + a_0$ be a polynomial with integer coefficients and satisfying the following conditions:

(a) $p^2 \mid a_0, a_1, \ldots, a_n$.
(b) $p \mid a_{n+1}, a_{n+2}, \ldots, a_{2n}$.
(c) $p^3 \mid a_0$ and $p \nmid a_{2n+1}$.

Show that f cannot be written as the product of two nonconstant polynomials with integer coefficients.

7. Let p be a prime number and $k \in \mathbb{N}$. Prove that[4]

$$f(X) = X^{p^{k-1}(p-1)} + X^{p^{k-1}(p-2)} + \cdots + X^{p^{k-1}} + 1$$

is irreducible over $\mathbb{Q}$.

8. (Romania) Let $p \in \mathbb{Z}$ be a prime number. If $f(X) = a_n X^n + a_{n-1} X^{n-1} + \cdots + a_1 X + a_0 \in \mathbb{Z}[X]$ is such that $|a_0| = p$ and $|a_n| + |a_{n-1}| + \cdots + |a_1| < p$, prove that f is irreducible over $\mathbb{Q}$.

9. (Romania) Let $f(X) = a_n X^n + \cdots + a_1 X + a_0$ be a nonconstant polynomial with integer coefficients. If $|a_0| > |a_1| + |a_2| + \cdots + |a_n|$ and $\sqrt{|a_0|} < \sqrt{|a_n|} + 1$, prove that f cannot be written as the product of two nonconstant polynomials with integer coefficients.

10. (Romania) Let $\mathbb{N} = A_1 \cup A_2 \cup \ldots \cup A_k$ be a partition of the set of natural numbers into k nonempty sets. Given $m \in \mathbb{N}$, prove that there exists $1 \leq j \leq k$ for which we can construct infinitely many polynomials f satisfying the following conditions:

    i. $\partial f = m$.
    ii. All of the coefficients of f belong to A_j.
    iii. f cannot be written as the product of two nonconstant polynomials with integer coefficients.

11. Let f be a polynomial with integer coefficients and degree $n \geq 1$, and $z_1, \ldots, z_n$ be the complex roots of f.

    (a) If $\mathrm{Re}(z_j) < 0$ for $1 \leq j \leq n$, prove that all of the coefficient of f have a single sign.
    (b) Prove **Pólya-Szegö's theorem**[5]: assume that there exists $m \in \mathbb{Z}$ such that $f(m)$ is prime, $f(m-1) \neq 0$ and $m > \frac{1}{2} + \mathrm{Re}(z_j)$, for $1 \leq j \leq n$. Then, f cannot be written as a product of two nonconstant polynomials with integer coefficients.

12. (BMO) If the coefficients of a prime number are taken as the coefficients of a polynomial $f \in \mathbb{Z}[X]$, prove that f is irreducible over $\mathbb{Z}$. More precisely, let $n > 1$ be an integer and $a_0, a_1, \ldots, a_{n-1}, a_n$ be natural numbers such that $a_n \neq 0$ and $0 \leq a_i \leq 9$ for $0 \leq i \leq n$. If $f(10)$ is prime, prove that f is irreducible over $\mathbb{Z}$.

---

[4]As Problem 1, page 492, will show, the result of this problem is a particular case of Theorem 20.18.

[5]George Pólya and Gábor Szegö, Hungarian mathematicians of the twentieth century.

Chapter 20
Algebraic and Transcendental Numbers

We start this chapter by inverting the viewpoint of Chap. 15. More precisely, we fix a complex number z and examine the set of polynomials $f \in \mathbb{C}[X]$ for which $f(z) = 0$. As a byproduct of our discussion, we give a (hopefully) more natural proof of the closedness, with respect to the usual arithmetic operations, of the set of complex numbers which are roots of nonzero polynomials of rational coefficients. We then proceed to investigate the special case of roots of unity, which leads us to the study of cyclotomic polynomials and allows us to give a partial proof of a famous theorem of Dirichlet on the infinitude of primes on certain arithmetic progressions. The chapter closes with a few remarks on the set of real numbers which are not roots of nonzero polynomials with rational coefficients.

20.1 Algebraic Numbers Over $\mathbb{Q}$

A complex number α is said to be **algebraic** over $\mathbb{Q}$ if there exists a polynomial $f \in \mathbb{Q}[X] \setminus \{0\}$ such that $f(\alpha) = 0$. A complex number which is not algebraic over $\mathbb{Q}$ is said to be **transcendental** over $\mathbb{Q}$. In this section and the next one we stick to the case of algebraic numbers, leaving a few word on transcendental numbers to be said in Sect. 20.4.

Obviously, every rational number r, being a root of the polynomial $X - r \in \mathbb{Q}[X] \setminus \{0\}$, is algebraic over $\mathbb{Q}$. In turn, the coming example collects less trivial instances of algebraic numbers over $\mathbb{Q}$.

Example 20.1 Let $r \in \mathbb{Q}_+^*$ and $n \in \mathbb{N}$. If ω is an n-th root of unity, then $\sqrt[n]{r}\omega$ is algebraic over $\mathbb{Q}$, for such a number is a root of the nonzero polynomial with rational coefficients $X^n - r$.

We could have also defined, in an obvious way, what one means by a complex number α to be algebraic over $\mathbb{R}$. Nevertheless, such a concept would not be interesting, for every complex number is algebraic over $\mathbb{R}$. Indeed, give a nonreal complex number $\alpha = a + ib$, we have that α is a root of the nonzero polynomial

$$f(X) = (X - (a + bi))(X - (a - bi))$$

$$= X^2 - 2aX + (a^2 + b^2) \in \mathbb{R}[X].$$

For this reason, whenever we consider an algebraic number over $\mathbb{Q}$, we shall simply refer to it as an *algebraic number*.[1]

If a complex number α is algebraic, the set

$$\mathcal{A}_\alpha = \{f \in \mathbb{Q}[X] \setminus \{0\}; \ f(\alpha) = 0\}$$

is nonempty by definition. Then, it is also nonempty the set of nonnegative integers $\{\partial f; \ f \in \mathcal{A}_\alpha\}$, so that there exists $p_\alpha \in \mathcal{A}_\alpha$, monic and of minimum degree. We thus have the following

Definition 20.2 Given a complex number α algebraic over $\mathbb{Q}$, a polynomial $p_\alpha \in \mathbb{Q}[X] \setminus \{0\}$, monic, of minimum degree and having α as a root is called a **minimal polynomial** for α.

The coming proposition and its corollaries collect the most important properties of minimal polynomials of algebraic numbers.

Proposition 20.3 *If $\alpha \in \mathbb{C}$ is algebraic over $\mathbb{Q}$ and p_α is a minimal polynomial of α, then:*

(a) *p_α is irreducible over $\mathbb{Q}$.*
(b) *If $f \in \mathbb{Q}[X]$ is such that $f(\alpha) = 0$, then $p_\alpha \mid f$ in $\mathbb{Q}[X]$.*

In particular, p_α is uniquely determined by α.

Proof

(a) If we had $p_\alpha = fg$, with f and g being nonconstant and of rational coefficients, then the degrees of f and g would be less than that of p_α and at least one of them would have α as a root. In turn, this would contradict the minimality of the degree of p_α. Therefore, p_α is irreducible over $\mathbb{Q}$.
(b) By the division algorithm, there exist polynomials $q, r \in \mathbb{Q}[X]$ such that

$$f(X) = p_\alpha(X)q(X) + r(X),$$

with $r = 0$ or $0 \le \partial r < \partial p_\alpha$. If $r \neq 0$, then

$$r(\alpha) = f(\alpha) - p_\alpha(\alpha)q(\alpha) = 0,$$

with $\partial r < \partial p_\alpha$, and this would again be a contradiction to the minimality of the degree of p_α. Thus, $r = 0$ and, hence, $p_\alpha \mid f$ in $\mathbb{Q}[X]$.

---

[1]Nevertheless, we shall briefly consider algebraic numbers over $\mathbb{Z}_p$ in Sect. 20.3.

Finally, if p_α and q_α were minimal polynomials for α, then item (b) would give $p_\alpha \mid q_\alpha$ in $\mathbb{Q}[X]$. However, since p_α and q_α are both monic and of the same degree, it would come that $p_\alpha = q_\alpha$.

$\square$

Thanks to the former proposition, given $\alpha \in \mathbb{C}$ algebraic, we can refer to p_α as being *the* minimal polynomial of α.

Corollary 20.4 *If $\alpha \in \mathbb{C}$ is algebraic and $f \in \mathbb{Q}[X] \setminus \{0\}$ is a monic, irreducible polynomial such that $f(\alpha) = 0$, then $f = p_\alpha$.*

Proof By the previous result, p_α divides f in $\mathbb{Q}[X]$. However, since f is irreducible, there must exist a nonzero rational number c such that $f = cp_\alpha$. Finally, since f and p_α are both monic, we must have $c = 1$. $\square$

Corollary 20.5 *If $f \in \mathbb{Q}[X] \setminus \mathbb{Q}$ is irreducible, then f has no multiple roots.*

Proof We can assume, without any loss of generality, that f is monic. If some $\alpha \in \mathbb{C}$ is a multiple root of f, then Proposition 15.31 would guarantee that α is also a root of the derivative f' of f. However, since $f \in \mathbb{Q}[X] \setminus \{0\}$ is monic and irreducible, Corollary 20.4 assures that it is the minimal polynomial of α. Therefore, Proposition 20.3 gives $f \mid f'$ in $\mathbb{Q}[X]$, which is a contradiction to the inequality $\partial f > \partial f'$. $\square$

We now collect some interesting applications of Proposition 20.3 and its corollaries.

Example 20.6 (IMO Shortlist) Let f be a nonconstant polynomial of rational coefficients and α a real number such that $\alpha^3 - 3\alpha = f(\alpha)^3 - 3f(\alpha) = -1$. Prove that, for every positive integer n, one has

$$f^{(n)}(\alpha)^3 - 3f^{(n)}(\alpha) = -1,$$

where $f^{(n)}$ stands for the composite of f with itself, n times.

Proof If $g(X) = X^3 - 3X + 1$, then $\partial g = 3$ and, searching rational roots (cf. Proposition 15.16), we conclude that g has none. Hence, Problem 5, page 457, guarantees that g is irreducible over $\mathbb{Q}$, and Corollary 20.4 assures that g is the minimal polynomial of α.

On the other hand, since we are assuming that the polynomial $g \circ f$ also has α as a root, Proposition 20.3 show that g divides $g \circ f$ in $\mathbb{Q}[X]$, say $(g \circ f)(X) = g(X)u(X)$, for some $u(X) \in \mathbb{Q}[X]$.

Finally, note that we have $g(f(\alpha)) = 0$ by assumption. Hence, if we have proved that $g(f^{(k)}(\alpha)) = 0$ for some $k \geq 1$, then

$$g(f^{k+1}(\alpha)) = (g \circ f)(f^k(\alpha)) = g(f^k(\alpha))u(f^k(\alpha)) = 0,$$

and there is nothing left to do. $\square$

Example 19.28, in conjunction with Corollary 20.4, immediately assures that if p is prime and $\omega = \text{cis } \frac{2\pi}{p}$, then the minimal polynomial of ω is

$$p_\omega(X) = X^{p-1} + X^{p-2} + \cdots + X + 1.$$

The coming example uses this fact to give a beautiful alternative proof of Example 11.5, due to the Bulgarian mathematician N. Nikolov.

Example 20.7 (IMO) Let p be an odd prime. Compute how many are the p-element subsets of the set $\{1, 2, \ldots, 2p\}$ such that the sum of its elements is divisible by p.

Solution If $\omega = \text{cis } \frac{2\pi}{p}$, then $\omega^p = 1$ and, hence,

$$(X^p - 1)^2 = \prod_{j=1}^{p}(X - \omega^j) \prod_{j=1}^{p}(X - \omega^j)$$

$$= \prod_{j=1}^{p}(X - \omega^j) \prod_{j=p+1}^{2p}(X - \omega^j)$$

$$= \prod_{j=1}^{2p}(X - \omega^j).$$

Computing the coefficient of X^p at both sides of the equality above and recalling that p is odd, we conclude that

$$2 = \sum_{\{j_1,\ldots,j_p\} \subset I_{2p}} \omega^{j_1 + \cdots + j_p}, \tag{20.1}$$

with the above sum ranging through all p-element subsets $\{j_1, \ldots, j_p\}$ of $I_{2p} = \{1, 2, \ldots, 2p\}$.

On the other hand, if for $0 \le k \le p - 1$ we let c_k denote the number of p-element subsets $\{j_1, \ldots, j_p\} \subset I_{2p}$ for which $j_1 + \cdots + j_p \equiv k \pmod{p}$, then $\omega^p = 1$ assures that the right hand side of (20.1) can be written as $\sum_{k=0}^{p-1} c_k \omega^k$, so that

$$\sum_{k=0}^{p-1} c_k \omega^k = 2.$$

It follows from what we did above that ω is a root of the polynomial with integer coefficients $f(X) = \sum_{k=0}^{p-1} c_k X^k - 2$. Note also that, since

$$\left\{ 1, p-1, 2, p-2, 3, p-3, \ldots, \frac{p-1}{2}, \frac{p+1}{2}, 2p-1 \right\}$$

has p elements and sum of elements congruent to $p - 1$ modulo p, we have $c_{p-1} \neq$
0. Hence, $\partial f = p - 1$.

Now, since the minimal polynomial of ω is $p_\omega(X) = X^{p-1} + \cdots + X + 1$, item
(b) of Proposition 20.3 guarantees that f is an integer multiple of p_ω, say $f = cp_\omega$,
for some $c \in \mathbb{N}$. In particular, by comparing coefficients on both sides of such an
equality, we conclude that

$$c_0 - 2 = c_1 = \cdots = c_{p-1} = c.$$

Finally, in order to compute the value of c, note that $c_0 + c_1 + \cdots + c_{p-1}$ is equal
to the total number of p-element subsets of I_{2p}, so that

$$pc + 2 = c_0 + c_1 + \cdots + c_{p-1} = \binom{2p}{p}.$$

Therefore,

$$c_0 = c + 2 = \frac{1}{p}\left(\binom{2p}{p} - 2\right) + 2.$$

$\square$

Example 20.8 (Romania) Let $f \in \mathbb{Z}[X]$ be a monic polynomial, of odd degree
greater than 1 and irreducible over $\mathbb{Q}$. Suppose also that:

(a) $f(0)$ is square-free.
(b) The complex roots of f have modulus greater than or equal to 1.

Prove that the polynomial $F \in \mathbb{Z}[X]$, given by $F(X) = f(X^3)$, is also irreducible
over $\mathbb{Q}$.

Proof By contradiction, assume that F is reducible over $\mathbb{Q}$. Then, it follows from
Problem 2, page 461, that there exist monic and nonconstant polynomials $g, h \in$
$\mathbb{Z}[X]$ such that $F = gh$. Since $\partial F = 3\partial f$ and ∂f is odd, we conclude that ∂F is
also odd. Hence, Problem 5, page 386, guarantees that F has at least one real root α.

We can assume, without loss of generality, that α is a root of g and that g is
irreducible over $\mathbb{Q}$. Indeed, if α is a root of g but g is reducible over $\mathbb{Q}$, it suffices
to take the monic irreducible factor of g having α as a root, and then to use the
result of Problem 2, page 461 to conclude that such an irreducible factor has integer
coefficients.

In the conditions of the previous paragraph, we know that g is the minimal
polynomial of α. Now, if $\omega \neq 1$ is a cubic root of 1, then

$$0 = F(\alpha) = f(\alpha^3) = f((\alpha\omega)^3) = F(\alpha\omega) = g(\alpha\omega)h(\alpha\omega).$$

We therefore distinguish two cases:

(i) $g(\alpha\omega) = 0$:   since (cf. Problem 4, page 386) the nonreal roots of a real polynomial come in pairs complex-complex conjugate, we have that $\overline{\alpha\omega} = \alpha\omega^2$ is also a root of g. Now, by grouping together in the expression of g all monomials with exponents of each one of the forms $3k$, $3k + 1$ and $3k + 2$, we can write

$$g(X) = a(X^3) + Xb(X^3) + X^2 c(X^3), \tag{20.2}$$

with $a, b, c \in \mathbb{Z}[X]$. Therefore,

$$a(\alpha^3) + \alpha b(\alpha^3) + \alpha^2 c(\alpha^3) = g(\alpha) = 0,$$

$$a(\alpha^3) + \alpha\omega b(\alpha^3) + \alpha^2\omega^2 c(\alpha^3) = g(\alpha\omega) = 0$$

and

$$a(\alpha^3) + \alpha\omega^2 b(\alpha^3) + \alpha^2\omega c(\alpha^3) = g(\alpha\omega^2) = 0.$$

Since the three equalities above for a Vandermond system of equations in $a(\alpha^3)$, $b(\alpha^3)$ and $c(\alpha^3)$, we conclude from Proposition 18.6 (or by direct computation) that $a(\alpha^3) = b(\alpha^3) = c(\alpha^3) = 0$. Thus, letting p denote the minimal polynomial of α^3, it follows from Proposition 20.3 that p divides a, b and c in $\mathbb{Q}[X]$ and, hence (again from Problem 4, page 386), in $\mathbb{Z}[X]$. It follows from (20.2) that $p(X^3)$ divides $g(X)$. However, since g is monic and irreducible, we conclude that $g(X) = p(X^3)$.

Now, with g being a polynomial in X^3, it follows once more from (20.2) that $b, c = 0$ and, hence, that

$$f(X^3) = F(X) = g(X)h(X) = a(X^3)h(X).$$

Such an equality, in turn, implies that h must also be a polynomial in X^3, say $h(X) = l(X^3)$, for some $l \in \mathbb{Z}[X]$.

Finally, we have $f(X^3) = a(X^3)l(X^3)$ and, hence, $f = gl$, with $\partial g, \partial l \geq 1$. But this contradicts the irreducibility of f.

(ii) $h(\alpha\omega) = 0$:   as in item (i), we conclude that $h(\alpha\omega^2) = 0$. Let, also as above,

$$h(X) = a(X^3) + Xb(X^3) + X^2 c(X^3),$$

with $a, b, c \in \mathbb{Z}[X]$. Then,

$$a(\alpha^3) + \alpha\omega b(\alpha^3) + \alpha^2\omega^2 c(\alpha^3) = h(\alpha\omega) = 0$$

and

$$a(\alpha^3) + \alpha\omega^2 b(\alpha^3) + \alpha^2\omega c(\alpha^3) = h(\alpha\omega^2) = 0.$$

Multiplying the first of the equalities above by ω and subtracting the second from the result, we obtain

$$a(\alpha^3) - \alpha^2 c(\alpha^3) = 0.$$

However, since g is the minimal polynomial of α, by invoking Proposition 20.3 once again we conclude that g divides $a(X^3) - X^2 c(X^3)$ in $\mathbb{Q}[X]$ and, hence, in $\mathbb{Z}[X]$ (by Remark 14.13, for $g \in \mathbb{Z}[X]$ and is monic).

Substituting X by 0, we conclude that $g(0)$ must divide $a(0)$ in $\mathbb{Z}$. Letting $a(0) = g(0)m$, with $m \in \mathbb{Z}$, it comes that

$$f(0) = g(0)h(0) = g(0)a(0) = g(0)^2 m.$$

Since $f(0)$ is square-free, it follows that $g(0) = \pm 1$. Hence, letting $z_1, \ldots, z_k$ be the complex roots of g, the relations of Girard-Viète give

$$1 = |g(0)| = |z_1| \ldots |z_k|. \tag{20.3}$$

On the other hand, since

$$f(z_j^3) = F(z_j) = g(z_j)h(z_j) = 0,$$

hypothesis (b) gives $|z_j^3| \geq 1$, so that $|z_j| \geq 1$ for $1 \leq j \leq k$. In turn, such an information, together with (20.3), allows us to conclude that $|z_j| = 1$ for $1 \leq j \leq k$. In particular, $|\alpha| = 1$ and, being a real number, we must have $\alpha = \pm 1$. Thus, $\alpha^3 = \pm 1$ and, since f is irreducible over $\mathbb{Q}$ with $f(\alpha^3) = 0$, we should have $f(X) = X \pm 1$, a contradiction to the fact that $\partial f > 1$. $\square$

The rest of this section is devoted to showing that the set of algebraic numbers is a **subfield** of $\mathbb{C}$, i.e., that it is a nonempty subset of $\mathbb{C}$ which is closed for the operations of addition, subtraction, multiplication and division (by a nonzero algebraic number, of course). We follow [7], and start with the following result.

Theorem 20.9 *If $\alpha, \beta \in \mathbb{C} \setminus \{0\}$ are algebraic, then so is $\alpha + \beta$.*

Proof Let $\alpha = \alpha_1, \ldots, \alpha_m$ be the complex roots of p_α and $\beta = \beta_1, \ldots, \beta_n$ be those of p_β, so that

$$p_\alpha(X) = \prod_{i=1}^{m}(X - \alpha_i) \ \text{ and } \ p_\beta(X) = \prod_{j=1}^{n}(X - \beta_j).$$

Define

$$f(X, X_1, \ldots, X_m) = \prod_{i=1}^{m}\prod_{j=1}^{n}(X - X_i - \beta_j) = \prod_{i=1}^{m} p_\beta(X - X_i)$$

$$= X^{mn} + \sum_{k=0}^{mn-1} f_k(X_1, \ldots, X_m)X^k, \tag{20.4}$$

for certain polynomials $f_0, \ldots, f_{mn-1} \in \mathbb{Q}[X_1, \ldots, X_m]$ (since $p_\beta \in \mathbb{Q}[X]$).

If σ is a permutation of I_m, then

$$f(X, X_{\sigma(1)}, \ldots, X_{\sigma(m)}) = \prod_{i=1}^{m} p_\beta(X - X_{\sigma(i)}) = \prod_{i=1}^{m} p_\beta(X - X_i)$$

$$= f(X, X_1, \ldots, X_m)$$

so that (20.4) furnishes

$$X^{mn} + \sum_{k=0}^{mn-1} f_k(X_{\sigma(1)}, \ldots, X_{\sigma(m)}) X^k = X^{mn} + \sum_{k=0}^{mn-1} f_k(X_1, \ldots, X_m) X^k.$$

Therefore, we have

$$f_k(X_{\sigma(1)}, \ldots, X_{\sigma(m)}) = f_k(X_1, \ldots, X_m)$$

for all $0 \leq k \leq mn - 1$ and σ, and we conclude that f_k is a symmetric polynomial in $X_1, \ldots, X_m$, for $0 \leq k \leq mn - 1$.

Newton's Theorem 16.12 then assures the existence of polynomials $g_k \in \mathbb{Q}[X_1, \ldots, X_m]$ for which

$$f_k(X_1, \ldots, X_m) = g_k(s_1, \ldots, s_m),$$

with $s_1, \ldots, s_m \in \mathbb{Q}[X_1, \ldots, X_m]$ standing for the elementary symmetric polynomials in $X_1, \ldots, X_m$. In particular,

$$f_k(\alpha_1, \ldots, \alpha_m) = g_k(s_1(\alpha_i), \ldots, s_m(\alpha_i)) \in \mathbb{Q},$$

for

$$p_\alpha(X) = X^m - s_1(\alpha_i) X^{m-1} + \cdots + (-1)^m s_m(\alpha_i) \in \mathbb{Q}[X].$$

Thus, if $h(X) = f(X, \alpha_1, \ldots, \alpha_m)$ then, on the one hand,

$$h(X) = X^{mn} + \sum_{k=0}^{mn-1} f_k(\alpha_1, \ldots, \alpha_m) X^k \in \mathbb{Q}[X]$$

while, on the other,

$$h(X) = \prod_{i=1}^{m} p_\beta(X - \alpha_i) = \prod_{i=1}^{m} \prod_{j=1}^{n} (X - \alpha_i - \beta_j).$$

Therefore, h is a nonzero polynomial with rational coefficients and such that $h(\alpha + \beta) = 0$, which guarantees that $\alpha + \beta$ is algebraic. $\square$

Remark 20.10 Suppose that $p_\alpha, p_\beta \in \mathbb{Z}[X]$. Then, in the notations of the above proof, we have $s_1(\alpha_i), \ldots, s_m(\alpha_i) \in \mathbb{Z}$ and $f_k \in \mathbb{Z}[X_1, \ldots, X_m]$ for $0 \leq k \leq mn - 1$. Therefore, once more from Newton's theorem, $g_k \in \mathbb{Z}[X_1, \ldots, X_m]$ and, hence,

$$f_k(\alpha_1, \ldots, \alpha_m) = g_k(s_1(\alpha_i), \ldots, s_m(\alpha_i)) \in \mathbb{Z}$$

for $0 \leq k \leq mn - 1$. We thus conclude that $h \in \mathbb{Z}[X]$, and Problem 4 shows that $p_{\alpha+\beta} \in \mathbb{Z}[X]$.

In order to show that the set of algebraic numbers is closed for products and quotients, we need the following particular case of Problem 4, page 416, for which we present a simpler proof.

Lemma 20.11 *If $\alpha \neq 0$ is algebraic, then so are α^{-1} and α^2.*

Proof Let $f(X) = a_n X^n + a_{n-1} X^{n-1} + \cdots + a_1 X + a_0 \in \mathbb{Q}[X] \setminus \mathbb{Q}$ be such that $a_0 \neq 0$ and $f(\alpha) = 0$. Then, $g(X) = a_0 X^n + a_1 X^{n-1} + \cdots + a_{n-1} X + a_n$ is a nonconstant polynomial with rational coefficients, and it is immediate to check that $g(\alpha^{-1}) = 0$.

For α^2, notice that there exist $u, v \in \mathbb{Q}[X]$, at least one of which nonzero and such that

$$f(X) = u(X^2) + Xv(X^2).$$

Letting $h(X) = u(X)^2 - Xv(X)^2$, we have $h \in \mathbb{Q}[X] \setminus \{0\}$ and

$$
\begin{aligned}
h(\alpha^2) &= u(\alpha^2)^2 - \alpha^2 v(\alpha^2)^2 \\
&= (u(\alpha^2) - \alpha v(\alpha^2))(u(\alpha^2) + \alpha v(\alpha^2)) \\
&= (u(\alpha^2) - \alpha v(\alpha^2)) f(\alpha) = 0.
\end{aligned}
$$

$\square$

Theorem 20.12 *If $\alpha, \beta \in \mathbb{C} \setminus \{0\}$ are algebraic, then so are $\alpha\beta$ and $\frac{\alpha}{\beta}$.*

Proof By the previous lemma, α^2 and β^2 are algebraic. However, since we already know that $\alpha + \beta$ is algebraic, it follows once more from the previous lemma that $(\alpha + \beta)^2$ is algebraic too. Now, since

$$\alpha\beta = \frac{1}{2}((\alpha + \beta)^2 - \alpha^2 - \beta^2),$$

two applications of Theorem 20.9, together with the result of Problem 1, assure that $\alpha\beta$ is also algebraic.

For what is left to do, note that $\frac{\alpha}{\beta} = \alpha \cdot \frac{1}{\beta}$, with $\frac{1}{\beta}$ being algebraic by the previous lemma. Hence, the first part above guarantees that $\frac{\alpha}{\beta}$ is algebraic. $\square$

Problems: Sect. 20.1

1. * If r is a nonzero rational and $\alpha \in \mathbb{C} \setminus \{0\}$ is algebraic, prove directly (i.e., without resorting to Theorem 20.12) that $r\alpha$ is also algebraic.
2. Given $k, n \in \mathbb{N}$, prove that $\cos \frac{2k\pi}{n}$ and $\sin \frac{2k\pi}{n}$ are algebraic.
3. Let p be a prime number and $n \in \mathbb{N}$. Prove that the minimal polynomial of $\sqrt[n]{p}$ is $f(X) = X^n - p$.
4. * Let $\alpha \in \mathbb{C}$ be algebraic. If there exists $f \in \mathbb{Z}[X] \setminus \mathbb{Z}$ monic and such that $f(\alpha) = 0$, prove that $p_\alpha \in \mathbb{Z}[X]$.
5. Let $a_1, a_2, \ldots, a_n$ be natural numbers, and $\alpha = \sqrt{a_1} + \sqrt{a_2} + \cdots + \sqrt{a_n}$. Prove that:

    (a) α is a root of a monic polynomial with integer coefficients.
    (b) If $\alpha \notin \mathbb{Z}$, then α is irrational.

6. (Brazil) Prove that the polynomial $f(X) = X^5 - X^4 - 4X^3 + 4X^2 + 2$ does not admit any roots of the form $\sqrt[n]{r}$, with $r \in \mathbb{Q}$ and $n \in \mathbb{N}$, $n > 1$.
7. Give a proof analogous to that of Theorem 20.9 to show that $\alpha\beta$ is algebraic whenever so are α and β.
8. Let $\alpha \in \mathbb{C}$ be algebraic, with $\partial p_\alpha = n$, and define

$$\mathbb{Q}(\alpha) = \{a_0 + a_1\alpha + \cdots + a_{n-1}\alpha^{n-1}; \ a_0, a_1, \ldots, a_{n-1} \in \mathbb{Q}\}$$
$$= \{f(\alpha); \ f \in \mathbb{Q}[X], \text{ with } f = 0 \text{ or } \partial f \leq n - 1\}.$$

The purpose of this problem is to show that $\mathbb{Q}(\alpha)$ is a subfield of $\mathbb{C}$.[2] To this end, do the following items:

    (a) Show that $\mathbb{Q}(\alpha)$ is closed for addition, subtraction and multiplication.
    (b) Given $\beta \in \mathbb{Q}(\alpha)$, show that there exists a single $f \in \mathbb{Q}[X]$ such that $\beta = f(\alpha)$, with $f = 0$ or $\partial f \leq n - 1$.
    (c) For $\beta = f(\alpha) \in \mathbb{Q}(\alpha) \setminus \{0\}$, with $f \in \mathbb{Q}[X]$ such that $\partial f \leq n - 1$, show that $\gcd(f, p_\alpha) = 1$. Then, conclude that $\frac{1}{\beta} \in \mathbb{Q}(\alpha)$.

9. Given $a, b, c \in \mathbb{Q}$ such that $a + b\sqrt[3]{2} + c\sqrt[3]{4} \neq 0$, show that there exist $x, y, z \in \mathbb{Q}$ for which

$$\frac{1}{a + b\sqrt[3]{2} + c\sqrt[3]{4}} = x + y\sqrt[3]{2} + z\sqrt[3]{4}.$$

Then, find x, y and z if $a = b = 1, c = 2$.

---

[2] For a converse, see Problem 4, page 503.

10. Let $\alpha \in \mathbb{C}$ be algebraic, with $\partial p_\alpha = n$, and $\mathbb{Q}(\alpha)$ be as in Problem 8. Find all functions $\phi : \mathbb{Q}(\alpha) \to \mathbb{C}$ satisfying the following conditions, for all $u, v \in \mathbb{Q}(\alpha)$:

(i) $\phi(u + v) = \phi(u) + \phi(v)$.
(ii) $\phi(uv) = \phi(u)\phi(v)$.

20.2 Cyclotomic Polynomials

The theory of polynomials over $\mathbb{Z}_p$, p prime, allows us to present some of the most elementary properties of the so-called *cyclotomic polynomials*; in particular, we will show that such polynomials are precisely the minimal polynomials of the complex roots of unity. As a byproduct of our study, we will use cyclotomic polynomials to prove a particular case of Dirichlet's theorem on primes in arithmetic progressions.

Given $n \in \mathbb{N}$, recall from Definition 13.23 that the *primitive n-th roots of unity* are the complex numbers of the form ω_n^k, with $\omega_n = \text{cis} \frac{2\pi}{n}$ and $1 \leq k \leq n$ being relatively prime with n. In particular, there are exactly $\varphi(n)$ primitive n-th roots of unity, where $\varphi : \mathbb{N} \to \mathbb{N}$ stands for the Euler function. Given $m, n \in \mathbb{N}$, whenever there is no danger of confusion we shall write simply (m, n) to denote the gcd of m and n.

Definition 20.13 For $n \in \mathbb{N}$, the *n*-th **cyclotomic polynomial** is the polynomial

$$\Phi_n(X) = \prod_{\substack{1 \leq k \leq n \\ (k,n)=1}} (X - \omega_n^k). \tag{20.5}$$

It follows from the above definition that Φ_n is monic with degree $\partial \Phi_n = \varphi(n)$. The coming proposition collects other elementary properties of Φ_n.

Proposition 20.14 *For $n \in \mathbb{N}$, we have:*

(a) $X^n - 1 = \prod_{0<d|n} \Phi_d(X)$.
(b) $\Phi_n \in \mathbb{Z}[X]$.
(c) $\Phi_n(0) = 1$ for $n > 1$.

Proof

(a) First of all, we have

$$\prod_{0<d|n} \Phi_d(X) = \prod_{0<d|n} \Phi_{n/d}(X) = \prod_{0<d|n} \prod_{\substack{1 \leq k \leq n/d \\ (k,n/d)=1}} (X - \omega_{n/d}^k)$$

$$= \prod_{0<d|n} \prod_{\substack{1 \leq k \leq n/d \\ (k,n/d)=1}} (X - \omega_n^{dk}).$$

Now, note that each integer $1 \leq m \leq n$ can be uniquely written as $m = dk$, with $d, k \in \mathbb{N}$ such that $0 < d \mid n$ and $\left(k, \frac{n}{d}\right) = 1$ (d is exactly $d = \gcd(m, n)$). Therefore, the last sum above is clearly equal to

$$\prod_{j=1}^{n} (X - \omega_n^j) = X^n - 1.$$

(b) Let us make induction on $n \in \mathbb{N}$, beginning with $\Phi_1(X) = X - 1 \in \mathbb{Z}[X]$. Given a natural number $n > 1$, assume, by induction hypothesis, that $\Phi_m \in \mathbb{Z}[X]$ for every integer $1 \leq m < n$. Then, if

$$g(X) = \prod_{\substack{1 \leq d < n \\ d \mid n}} \Phi_d(X),$$

we have $g \in \mathbb{Z}[X]$ and, by (a), $X^n - 1 = \Phi_n(X)g(X)$. Since g is monic (for we already know that each Φ_m is monic), Remark 14.13 guarantees that $\Phi_n \in \mathbb{Z}[X]$.

(c) For $n = 2$ this is a direct computation; for $n > 2$, apply the result of Problem 3, page 345. Alternatively, arguing once more by induction, start by noticing that

$$X^2 - 1 = \Phi_1(X)\Phi_2(X) = (X - 1)\Phi_2(X);$$

hence, $\Phi_2(X) = X + 1$ and $\Phi_2(0) = 1$. Let $n > 2$ and suppose, as induction hypothesis, that $\Phi_m(0) = 1$ for every integer $2 \leq m < n$. Then, in the notations of the proof of (b), we have

$$g(0) = \Phi_1(0) \prod_{\substack{1 < d < n \\ d \mid n}} \Phi_d(0) = (-1) \prod_{\substack{1 < d < n \\ d \mid n}} \Phi_d(0) = -1,$$

and it follows from $X^n - 1 = \Phi_n(X)g(X)$ that

$$-1 = \Phi_n(0)g(0) = -\Phi_n(0),$$

as wished. $\square$

Corollary 20.15 *If $p \in \mathbb{Z}$ is prime, then*

$$\Phi_p(X) = X^{p-1} + X^{p-2} + \cdots + X + 1.$$

Proof Item (a) of the previous proposition gives

$$X^p - 1 = \Phi_1(X)\Phi_p(X) = (X - 1)\Phi_p(X),$$

so that

$$\Phi_p(X) = X^{p-1} + X^{p-2} + \cdots + X + 1.$$

□

Example 19.28 has shown that Φ_p is irreducible over $\mathbb{Q}$, so that Φ_p is the minimal polynomial of ω_p. Theorem 20.18 generalizes this fact by proving that, for every $n \in \mathbb{N}$, the minimal polynomial of ω_n coincides with the n-th cyclotomic polynomial Φ_n. We first need to establish a simple auxiliary result.

Lemma 20.16 *Let $f, g \in \mathbb{Z}[X]$ and $p \in \mathbb{Z}$ be a prime number. If $\overline{g} \in \mathbb{Z}_p[X] \setminus \mathbb{Z}_p$ and $\overline{g}^2 \mid \overline{f}$ in $\mathbb{Z}_p[X]$, then $\overline{g} \mid \overline{f'}$ in $\mathbb{Z}_p[X]$.*

Proof If $h \in \mathbb{Z}[X]$ is such that $\overline{f} = \overline{g}^2\overline{h}$ in $\mathbb{Z}_p[X]$, we know that there exists a polynomial $l \in \mathbb{Z}[X]$ such that

$$f(X) = g(X)^2 h(X) + pl(X)$$

in $\mathbb{Z}[X]$. Computing derivatives at both sides of this equality, we obtain

$$f'(X) = 2g(X)g'(X)h(X) + g(X)^2 h'(X) + pl'(X)$$

in $\mathbb{Z}[X]$, and hence

$$\overline{f'}(X) = \overline{g}(X)\big(\overline{2}\,\overline{g'}(X)\overline{h}(X) + \overline{g}(X)\overline{h'}(X)\big)$$

in $\mathbb{Z}_p[X]$. Therefore, $\overline{g} \mid \overline{f'}$ in $\mathbb{Z}_p[X]$.   □

For our next result, recall that if ω is an n-th root of unity, then Proposition 20.3 guarantees that its minimal polynomial p_ω divides $X^n - 1$ in $\mathbb{Q}[X]$. Then, Problem 4, page 486, assures that $p_\omega \in \mathbb{Z}[X]$.

Proposition 20.17 *Let $n, p \in \mathbb{N}$ be such that p is prime and $p \nmid n$. If ω is an n-th root of unity, then $p_\omega(X) = p_{\omega^p}(X)$.*

Proof Let $\zeta = \omega^p$. Since both ω and ζ are roots of $X^n - 1$, item (b) of Proposition 20.3 shows that both p_ω and p_ζ divide $X^n - 1$. By contradiction, assume that $p_\omega \neq p_\zeta$. Then, the irreducibility of these polynomials assures, via Gauss' Theorem 19.14, that $p_\omega p_\zeta$ divides $X^n - 1$ in $\mathbb{Z}[X]$, say

$$X^n - 1 = p_\omega(X)p_\zeta(X)u(X) \qquad (20.6)$$

for some $u \in \mathbb{Z}[X]$.

If $g(X) = p_\zeta(X^p)$, then

$$g(w) = p_\zeta(\omega^p) = p_\zeta(\zeta) = 0$$

so that (once more by Proposition 20.3) p_ω divides g in $\mathbb{Z}[X]$. Let $v \in \mathbb{Z}[X]$ be such that $p_\omega v = g$. In $\mathbb{Z}_p[X]$, Problem 5, page 469 gives

$$\overline{p}_\omega(X)\overline{v}(X) = \overline{g}(X) = \overline{p}_\zeta(X^p) = \left(\overline{p}_\zeta(X)\right)^p,$$

and Theorem 19.19 guarantees the existence of a monic and irreducible polynomial $\overline{h} \in \mathbb{Z}_p[X]$ such that $\overline{h} \mid \overline{p}_\omega, \overline{p}_\zeta$ in $\mathbb{Z}_p[X]$. It follows from (20.6) that $\overline{h}(X)^2 \mid (X^n - \overline{1})$ in $\mathbb{Z}_p[X]$, and the previous lemma gives that $\overline{h}(X) \mid \overline{n}X^{n-1}$ in $\mathbb{Z}_p[X]$. However, since $\overline{h}$ is monic and $\overline{n} \neq \overline{0}$, by applying once again Theorem 19.19 we obtain $1 \leq l \leq n - 1$ such that $\overline{h}(X) = X^l$ in $\mathbb{Z}_p[X]$. Hence, $\overline{h}(X) \nmid (X^n - \overline{1})$ in $\mathbb{Z}_p[X]$, which is a contradiction. $\square$

We can finally state and prove the desired result.

Theorem 20.18 *If $\omega_n = \text{cis} \frac{2\pi}{n}$, then $p_{\omega_n} = \Phi_n$. In particular, $\Phi_n \in \mathbb{Z}[X]$ is irreducible in $\mathbb{Q}[X]$.*

Proof Take $k \in \mathbb{N}$ such that $k > 1$ and $\gcd(k, n) = 1$, and let $k = p_1 \dots p_l$, with $p_1, \dots, p_l$ being primes not dividing n. Repeated applications of the previous proposition give us

$$p_{\omega_n} = p_{\omega_n^{p_1}} = p_{\omega_n^{p_1 p_2}} = \dots = p_{\omega_n^{p_1 \cdots p_l}} = p_{\omega_n^k}.$$

In particular, the $\varphi(n)$ complex numbers ω_n^k, with $1 \leq k \leq n$ and $\gcd(k, n) = 1$, are distinct roots of p_{ω_n}, so that

$$\partial p_{\omega_n} \geq \varphi(n) = \partial \Phi_n.$$

However, since Φ_n is monic, has integer (thus rational) coefficients and ω_n as a root, the definition of minimal polynomial assures that $p_{\omega_n} = \Phi_n$. $\square$

We finish this section by proving a particular case of the famous Dirichlet's theorem on primes in arithmetic progressions. This theorem states that an infinite nonconstant arithmetic progression of natural numbers contains infinitely many prime numbers, provided its first term and common difference are relatively prime. Although Dirichlet's theorem is a natural generalization of Euclid's Theorem 6.38, its known proofs are far beyond the scope of most undergraduate curricula in Mathematics. Nevertheless, the fragments of the theory of cyclotomic polynomials developed so far allow us to present an elementary proof of Dirichlet's theorem, in the particular case of an arithmetic progression whose first term is equal to 1.

Theorem 20.19 (Dirichlet) *If $n \in \mathbb{N}$, then the arithmetic progression $1, 1 + n, 1 + 2n, \dots$ contains infinitely many primes.*

Proof Let $p_1, \dots, p_k$ be any primes and Φ_n denote the n-th cyclotomic polynomial. Since Φ_n is monic, choosing a sufficiently large integer y we have $\Phi_n(ynp_1 \dots p_k) > 1$. Therefore, letting $a = ynp_1 \dots p_k$, we have modulo a that

$$\Phi_n(a) \equiv \Phi_n(0) = 1 \pmod{a}.$$

Hence, $\Phi_n(a) = aq + 1 = ynp_1 \ldots p_k q + 1$, for some $q \in \mathbb{N}$. If p is a prime factor of $\Phi_n(a)$, then $p \neq p_1, \ldots, p_k$ and $\gcd(p, n) = 1$. Therefore, if we prove that $p \equiv 1 \pmod{n}$, it will follow that p is a prime in our arithmetic progression, distinct from $p_1, \ldots, p_k$.

For what is left to do, note that $\gcd(p, a) = 1$, and let $t := \text{ord}_p(a)$. Since $p \mid \Phi_n(a)$ and $\Phi_n(a) \mid (a^n - 1)$ (for $\Phi_n(X) \mid (X^n - 1)$ in $\mathbb{Z}[X]$), we conclude that $a^n \equiv 1 \pmod{p}$; thus, item (c) of Proposition 12.2 shows that $t \mid n$. If we prove that $t = n$ then, again from item (c) of Proposition 12.2, together with $a^{p-1} \equiv 1 \pmod{p}$, we obtain $n \mid (p - 1)$ or, which is the same, $p \equiv 1 \pmod{n}$.

For $c \in \{a, a + p\}$, it follows from $a^t \equiv 1 \pmod{p}$ that $c^t \equiv 1 \pmod{p}$. By the sake of contradiction, assume that $t < n$. Proposition 20.14, together with the fact that $t \mid n$, furnishes

$$c^n - 1 = \prod_{0 < d \mid n} \Phi_d(c) = \Phi_n(c) \prod_{\substack{0 < d < n \\ d \mid n}} \Phi_d(c)$$

$$= \Phi_n(c) h(c) \prod_{0 < d \mid t} \Phi_d(c)$$

$$= \Phi_n(c) h(c)(c^t - 1),$$

for some appropriate polynomial $h \in \mathbb{Z}[X]$. However, since $c \equiv a \pmod{p}$, item (c) of Proposition 10.6 gives

$$\Phi_n(c) \equiv \Phi_n(a) \equiv 0 \pmod{p},$$

so that

$$c^n - 1 = \Phi_n(c) h(c)(c^t - 1) \equiv 0 \pmod{p^2}.$$

On the other hand,

$$(a + p)^n - 1 = a^n - 1 + \sum_{j=1}^{n-1} \binom{n}{j} a^{n-j} p^j$$

and (by the computations above) both $(a + p)^n - 1$ and $a^n - 1$ are multiples of p^2. Thus, looking at the last equality above modulo p^2, we conclude that

$$na^{n-1} p \equiv 0 \pmod{p^2},$$

which is a contradiction. $\qquad\square$

For a self-contained discussion of the general case of Dirichlet's theorem, we refer the interested reader to [27] or [36]. As far as the special case above is concerned, the coming example shows a nontrivial application. At this point, the reader may find it helpful to recall the material of Sects. 12.2 and 12.3.

Example 20.20 For each prime number p, let $g(p) \in \mathbb{N}$ denote the least positive primitive root modulo p. Then, the function $p \mapsto g(p)$ is unbounded.

Proof Recall that if a is a primitive root modulo p then $a^{\frac{p-1}{2}} \equiv -1 \pmod{p}$, and Euler's criterion (cf. Proposition 12.19) assures that a is a quadratic nonresidue modulo p. In turn, given a natural number $n \geq 2$, we need to show that there exists a prime number p such that none of $1, 2, \ldots, n$ is a primitive root modulo p. Then, it suffices to find p such that all of $1, 2, \ldots, n$ are quadratic residues modulo p.

We first claim that it is enough that every prime q satisfying $1 < q \leq n$ is a quadratic residue modulo p. Indeed, if this is so and $1 < m \leq n$ is not a perfect square, then m can be written as $m = k^2 q_1 \ldots q_l$, with $q_1, \ldots, q_l$ being pairwise distinct prime numbers. Letting $q_j \equiv a_j^2 \pmod{p}$, we get

$$m = k^2 q_1 \ldots q_l \equiv k^2 a_1^2 \ldots a_l^2 = (k a_1 \ldots a_l)^2 \pmod{p}.$$

Now, Dirichlet's theorem assures the existence of a prime number p of the form $p = 1 + 2tn!$, for some $t \in \mathbb{N}$. We claim that such a p works for us. Indeed, if $1 < q \leq n$ is prime, then the Quadratic Reciprocity Law gives

$$\left(\frac{p}{q}\right)\left(\frac{q}{p}\right) = (-1)^{\left(\frac{p-1}{2}\right)\left(\frac{q-1}{2}\right)} = (-1)^{tn!\left(\frac{q-1}{2}\right)} = 1,$$

for $n!$ is even. However, since $q \mid n!$, we have $p \equiv 1 \pmod{q}$, and the properties of Legendre's symbol (cf. item (a) of Proposition 12.23) give

$$\left(\frac{p}{q}\right) = \left(\frac{1}{q}\right) = 1.$$

Hence, $\left(\frac{q}{p}\right) = 1$, and q is a quadratic residue modulo p. □

Problems: Sect. 20.2

1. Let $p, k \in \mathbb{N}$ be given, with p being prime. Compute Φ_{p^k} explicitly.
2. If $n > 1$ is an even integer, show that $\Phi_{2n}(X) = \Phi_{2n}(-X)$.
3. If m and n are distinct naturals, prove that Φ_m and Φ_n have no nonconstant common factors in $\mathbb{C}[X]$. In particular, $\Phi_m \neq \Phi_n$.
4. Let $n > 1$ be a natural number and d be the product of the distinct prime factors of n. Show that $\Phi_n(X) = \Phi_d\left(X^{n/d}\right)$.
5. (England) The set $\left\{1, \frac{1}{2}, \frac{1}{3}, \frac{1}{4}, \ldots\right\}$ contains several arithmetic progressions. Given an integer $k > 2$, prove that it contains an arithmetic progression of k terms which is not contained in any arithmetic progression of $k + 1$ terms of the same set.

6. Let $a, n \in \mathbb{N}$, with $a > 1$ and n odd. Prove that the algebraic congruence $x^n \equiv a \pmod{p}$ has a solution for infinitely many primes p.

7. Let a be a natural number which is not a perfect square. Prove that there are infinitely many prime numbers p for which a is a non-quadratic residue modulo p.

8. Let $a, b \in \mathbb{Z}$ be such that for each $n \in \mathbb{N}$ there exists $c \in \mathbb{Z}$ for which $n \mid (c^2 + ac + b)$. Prove that the equation $x^2 + ax + b = 0$ has integer roots.

20.3 Algebraic Numbers Over $\mathbb{Z}_p$

This section is somewhat more abstract than the previous ones, for we extend the concept of algebraic number to consider algebraic numbers over $\mathbb{Z}_p$, for some prime number p. However, the payoff will be worth the effort, for, given $n \in \mathbb{N}$, we will be able to compute the exact number of irreducible polynomials over $\mathbb{Z}_p$ and having degree n.

We depart from a naive though profitable idea, namely, that there exists a *number set* Ω_p containing $\mathbb{Z}_p$ that plays, for $\mathbb{Z}_p$, the same role as $\mathbb{C}$ plays for $\mathbb{Q}$. We start by formalizing the concept of *field*.

Definition 20.21 A **field** is a nonempty set $\mathbb{K}$, furnished with operations $+, \cdot : \mathbb{K} \times \mathbb{K} \to \mathbb{K}$ having the following properties:

(a) $+$ and $\cdot$ are commutative and associative, and $\cdot$ is distributive with respect to $+$.
(b) There exist elements $0, 1 \in \mathbb{K}$, with $0 \neq 1$, such that $a + 0 = a$ and $a \cdot 1 = a$, for every $a \in \mathbb{K}$.
(c) For every $a \in \mathbb{K}$, there exists an element $-a \in \mathbb{K}$ such that $a + (-a) = 0$.
(d) For every $a \in \mathbb{K} \setminus \{0\}$, there exists an element $a^{-1} \in \mathbb{K}$ such that $a \cdot a^{-1} = 1$.

The reader has certainly realized what we mean by *commutative, associative* and *distributive* from his/her previous experience. Nevertheless, let us explain all that from first principles. Commutativity in item (a) means that

$$a + b = b + a \ \text{ and } \ a \cdot b = b \cdot a,$$

whereas associativity stands for

$$(a + b) + c = a + (b + c) \ \text{ and } \ (a \cdot b) \cdot c = a \cdot (b \cdot c),$$

for all $a, b, c \in \mathbb{K}$. In turn, the distributivity of $\cdot$ with respect to $+$ is exactly what one expects:

$$a \cdot (b + c) = a \cdot b + a \cdot c,$$

with the right hand side being a shorthand for the more precise (though somewhat cumbersome) expression $(a \cdot b) + (a \cdot c)$.

Thus, we surely have that $\mathbb{Q}$, $\mathbb{R}$ and $\mathbb{C}$ are fields, and Problem 8, page 486 shows that so is $\mathbb{Q}(\alpha)$, for every $\alpha \in \mathbb{C}$ algebraic over $\mathbb{C}$. Moreover, in all of these cases, the elements 0 and 1 of item (b) in the above definition are the usual complex numbers 0 and 1, and this is the reason why, for a general field $\mathbb{K}$, we also denote them by 0 and 1. A similar remark holds for $-a$ and a^{-1}, i.e., we are simply adopting the same notation we use for the additive (resp. multiplicative) inverses of elements of $\mathbb{C}$ (resp. of $\mathbb{C} \setminus \{0\}$).

Another class of examples of fields we have been dealing with is that of the finite fields $\mathbb{Z}_p$, with $p \in \mathbb{Z}$ prime. In this case, however, we shall stick to the usage of writing $\overline{0}$ and $\overline{1}$ whenever convenient, in order to avoid any possibility of confusion with $0, 1 \in \mathbb{Z}$.

Back to a general field $\mathbb{K}$, the *cancellation laws* for addition and multiplication hold:

$$a + c = b + c \Rightarrow a = b \text{ and } a \cdot c = b \cdot c, \, c \neq 0 \Rightarrow a = b.$$

Indeed,

$$a + c = b + c \Rightarrow (a + c) + (-c) = (b + c) + (-c)$$
$$\Rightarrow a + (c + (-c)) = b + (c + (-c))$$
$$\Rightarrow a + 0 = b + 0 \Rightarrow a = b,$$

and likewise for the multiplication (see Problem 1).

As it happens within $\mathbb{C}$, whenever there is no danger of confusion we shall write ab, instead of $a \cdot b$, to denote the product of elements a and b of a general field $\mathbb{K}$.

We could have developed most of the theory of polynomials by considering the set $\mathbb{K}[X]$ of polynomials over (or with coefficients in) an arbitrary field $\mathbb{K}$, with operations $+, \cdot : \mathbb{K}[X] \times \mathbb{K}[X] \to \mathbb{K}[X]$ extending those of $\mathbb{K}$. Taking for granted the (harmless) assumption that we have done that, we now have at our disposal the following concepts and facts, whose validities the reader can easily check:

1. If $f, g \in \mathbb{K}[X]$ are such that $fg = 0$ (the identically zero polynomial), then $f = 0$ or $g = 0$.
2. The notions of degree (for nonzero polynomials) and roots for polynomials over $\mathbb{K}$ remain true, unchanged. Likewise, $\partial(fg) = \partial f + \partial g$ if $f, g \neq 0$ and $\partial(f + g) \leq \max\{\partial f, \partial g\}$ if $f + g \neq 0$.
3. The division algorithm, the root test and Lagrange's theorem on the number of distinct roots of a nonzero polynomial also continue to hold, with identical proofs.
4. The concept of greatest common divisor for nonzero polynomials over $\mathbb{K}$ is a direct extension of that for polynomials over $\mathbb{Q}$, and Bézout's Theorem 19.2 is also true, with exactly the same proof.

5. Another concept that extends in a likewise manner from $\mathbb{Q}[X]$ is that of irreducible polynomial $f \in \mathbb{K}[X]$. Now, note that we have a version of Theorem 19.9 at our disposal too, again with the same proof.

A major gap on extending the theory for polynomials over arbitrary fields is fulfilled by the coming result, which will be assumed without proof (we refer the interested reader to [6] or [20]).

Theorem 20.22 *Given an arbitrary field* $\mathbb{K}$, *there exists another field* Ω *containing* $\mathbb{K}$, *whose operations extend those of* $\mathbb{K}$ *and such that every* $f \in \Omega[X]$ *has at least one root in* Ω.

The attentive reader has certainly noticed that Ω plays the role of $\mathbb{C}$ for $\mathbb{K}$. We refer to the property of Theorem 20.22 by saying that Ω is an **algebraically closed** field containing $\mathbb{K}$. Also as with $\mathbb{C}$, one now proves that if $f \in \mathbb{K}[X] \setminus \{0\}$ has degree n, then there exists $a \in \mathbb{K}$ (the leading coefficient of f) and $\alpha_1, \ldots, \alpha_n \in \Omega$ such that

$$f(X) = a(X - \alpha_1)(X - \alpha_2) \ldots (X - \alpha_n).$$

The equality above is the **factorised form** of f over Ω.

We can now define, exactly as was done in Sect. 20.1, what one means for an element $\alpha \in \Omega$ to be *algebraic* over $\mathbb{K}$, and consider its *minimal polynomial* $p_\alpha \in \mathbb{K}[X] \setminus \{0\}$ as was done for $\alpha \in \mathbb{C}$ algebraic over $\mathbb{Q}$. This way, Proposition 20.3 and Corollary 20.4 remain true, unchanged.

We now restrict our attention to $\mathbb{K} = \mathbb{Z}_p$, and write Ω_p to denote the field Ω of Theorem 20.22. We first recall the result of Problem 5, page 469, which we write in the following form:

$$f(X^p) = f(X)^p, \ \forall \ f \in \mathbb{Z}_p[X]. \tag{20.7}$$

We shall also need the following auxiliary result.

Lemma 20.23 *Let* $f \in \Omega_p[X] \setminus \{0\}$ *be given. If* $\alpha \in \Omega_p$ *is a root of* f, *then:*

(a) α^p *is also a root of* f.
(b) *There exists a natural number* $m \leq \partial f$ *such that* α *is a root of* $X^{p^m} - X$.

Proof

(a) It follows from (20.7) that $f(\alpha^p) = f(\alpha)^p = \overline{0}^p = \overline{0}$.
(b) Iterating the result of (a), we conclude that $\alpha, \alpha^p, \alpha^{p^2}, \ldots$ are roots of f. Since it has at most ∂f distinct roots, we conclude that there exist integers $0 \leq k < l \leq \partial f$ for which $\alpha^{p^k} = \alpha^{p^l}$. Therefore,

$$\overline{0} = \alpha^{p^l} - \alpha^{p^k} = \left(\alpha^{p^{l-k}}\right)^{p^k} - \alpha^{p^k} = \left(\alpha^{p^{l-k}} - \alpha\right)^{p^k},$$

where we have used the result of Example 19.15 in the last equality above. It comes that $\alpha^{p^{l-k}} - \alpha = 0$, and α is a root of $X^{p^m} - X$, with $m = l - k \leq \partial f$.

$\square$

We are now in position to prove the following

Proposition 20.24 *Let p be prime and $\alpha \in \Omega_p$ be algebraic over $\mathbb{Z}_p$. If $\partial p_\alpha = n$, then:*

(a) *α is a root of $X^{p^n} - X$.*
(b) *α is not a root of $X^{p^m} - X$, for any positive integer $m < n$.*

Proof We already know, from the previous lemma, that α is a root of $X^{p^m} - X$, for some positive integer $m \leq n$. Now, let

$$\Phi : \underbrace{\mathbb{Z}_p \times \mathbb{Z}_p \times \ldots \times \mathbb{Z}_p}_{n \text{ times}} \longrightarrow \Omega_p$$

be defined by

$$\Phi(\overline{a}_0, \overline{a}_1, \ldots, \overline{a}_{n-1}) = \overline{a}_0 + \overline{a}_1 \alpha + \cdots + \overline{a}_{n-1} \alpha^{n-1}.$$

We claim that Φ is injective and each $\beta \in \text{Im}(\Phi)$ is a root of $X^{p^m} - X$. Indeed, if

$$\Phi(\overline{a}_0, \overline{a}_1, \ldots, \overline{a}_{n-1}) = \Phi(\overline{b}_0, \overline{b}_1, \ldots, \overline{b}_{n-1}),$$

for distinct n-tuples $(\overline{a}_0, \overline{a}_1, \ldots, \overline{a}_{n-1})$ and $(\overline{b}_0, \overline{b}_1, \ldots, \overline{b}_{n-1})$ in the domain of Φ, then

$$(\overline{a}_0 - \overline{b}_0) + (\overline{a}_1 - \overline{b}_1)\alpha + \cdots + (\overline{a}_{n-1} - \overline{b}_{n-1})\alpha^{n-1} = \overline{0},$$

so that α would be a root of the nonzero polynomial $(\overline{a}_0 - \overline{b}_0) + (\overline{a}_1 - \overline{b}_1)X + \cdots + (\overline{a}_{n-1} - \overline{b}_{n-1})X^{n-1}$ of $\mathbb{Z}_p[X]$. Since $\partial p_\alpha = n$, this is a contradiction.

For the second part, let $\beta = \overline{a}_0 + \overline{a}_1 \alpha + \cdots + \overline{a}_{n-1} \alpha^{n-1}$. The result of Example 19.15, together with Fermat's little theorem and $\alpha^{p^m} = \alpha$, gives

$$\beta^{p^m} = \left(\overline{a}_0 + \overline{a}_1 \alpha + \cdots + \overline{a}_{n-1} \alpha^{n-1}\right)^{p^m}$$

$$= \overline{a}_0^{p^m} + \overline{a}_1^{p^m} \alpha^{p^m} + \cdots + \overline{a}_{n-1}^{p^m} \alpha^{(n-1)p^m}$$

$$= \overline{a}_0 + \overline{a}_1 \alpha + \cdots + \overline{a}_{n-1} \alpha^{n-1}$$

$$= \beta$$

Let $\mathcal{R}_m$ stand for the set of roots of $X^{p^m} - X$ in Ω_p. The above claims then assure that $|\mathcal{R}_m| \geq p^n$, so that $p^m \geq p^n$ and, hence, $m \geq n$. Since $m \leq n$, we then get $m = n$, and items (a) and (b) follow at once. □

A direct consequence of this proposition is the coming

Corollary 20.25 *Let p be prime and $\alpha \in \Omega_p$ be algebraic over $\mathbb{Z}_p$. If $\partial p_\alpha = n$, then $p_\alpha \mid \left(X^{p^n} - X\right)$ in $\mathbb{Z}_p[X]$.*

Proof This follows from the previous result, together with the analogue of Proposition 20.3 in our setting. □

Another consequence is collected as the next result.

Lemma 20.26 *Let $f \in \mathbb{Z}_p[X] \backslash \mathbb{Z}_p$ be irreducible and of degree d. If $f \mid (X^{p^n} - X)$, then $d \mid n$.*

Proof If $\alpha \in \Omega_p$ is a root of f, then $f = p_\alpha$, and the previous corollary guarantees that $f \mid (X^{p^d} - X)$ and $f \nmid (X^{p^k} - X)$, for every positive integer $k < d$. Since $f \mid (X^{p^n} - X)$, we conclude that $d \leq n$.
Now, let $n = d + t$ and write

$$X^{p^n} - X = X^{p^{d+t}} - X = \left(X^{p^d}\right)^{p^t} - X^{p^t} + X^{p^t} - X$$

$$= \left(X^{p^d} - X\right)^{p^t} + X^{p^t} - X.$$

It readily follows from this equality that

$$\gcd\left(X^{p^n} - X, X^{p^d} - X\right) = \gcd\left(X^{p^t} - X, X^{p^d} - X\right).$$

Assume that $n = dq + r$, with $0 < r < d$. Iterating the gcd equality above, we get

$$\gcd\left(X^{p^n} - X, X^{p^d} - X\right) = \gcd\left(X^{p^r} - X, X^{p^d} - X\right). \tag{20.8}$$

Since f divides the left hand side, it also divides the right hand side. In particular, $f \mid (X^{p^r} - X)$, which is a contradiction. □

We can finally state and prove our main result, for which we let

$a_n = \#\{$monic, pairwise distinct irreducible polynomials of degree n in $\mathbb{Z}_p[X]\}$.

Also, if $a_n > 0$, we write $f_{n1}, f_{n2}, \ldots, f_{na_n}$ to denote such polynomials.

Theorem 20.27 *Let p be prime and $n \in \mathbb{N}$. Then,*

$$X^{p^n} - X = \prod_{0 < d \mid n} f_{d1}(X) \ldots f_{da_d}(X), \tag{20.9}$$

with the product $f_{d1} \ldots f_{da_d}$ taken as 1 if $a_d = 0$.

Proof The analogue of Theorem 19.9 for $\mathbb{Z}_p[X]$ assures that $X^{p^n} - X$ is the product of finitely many irreducible polynomials, which can all be assumed to be monic.

If f is one such polynomial, with $\partial f = d$, the previous lemma shows that $d \mid n$, so that f is one of the polynomials in the right hand side of (20.9). Conversely, if $0 < d \mid n$, $1 \leq j \leq a_d$ and $\alpha \in \Omega_p$ is a root of f_{dj}, then $f_{dj} = p_\alpha$, and Corollary 20.25 guarantees that $f_{dj} \mid \left(X^{p^d} - X \right)$ in $\mathbb{Z}_p[X]$. However, since $d \mid n$, the argument in the proof of the previous lemma leading to (20.8) shows that $\left(X^{p^d} - X \right) \mid \left(X^{p^n} - X \right)$ in $\mathbb{Z}_p[X]$. Therefore, $f_{dj} \mid \left(X^{p^n} - X \right)$ in $\mathbb{Z}_p[X]$.   □

Corollary 20.28 *If p is prime and $n \in \mathbb{N}$, then*

$$a_n = \frac{1}{n} \sum_{0 < d \mid n} \mu\left(\frac{n}{d}\right) p^d > 0, \qquad (20.10)$$

where μ is the Möbius function.

Proof Computing degrees in (20.9), we obtain

$$p^n = \sum_{0 < d \mid n} d a_d.$$

Applying Möbius inversion formula (8.4), we obtain the given expression for a_n. For what is left to do, just note that, since $\text{Im}(\mu) \subset \{-1, 0, 1\}$, we get

$$n a_n = p^n + \sum_{\substack{0 < d < n \\ d \mid n}} \mu\left(\frac{n}{d}\right) p^d \geq p^n - \sum_{\substack{0 < d < n \\ d \mid n}} p^d$$

$$\geq p^n - \sum_{d=1}^{n-1} p^d = p^n - \frac{p^n - p}{p - 1} > 0.$$

                                    □

Corollary 20.29 *Let p be prime and $n \in \mathbb{N}$. If $0 < d \mid n$, then $X^{p^n} - X$ has an irreducible factor of degree d.*

Proof This follows from (20.9), together with the previous corollary.   □

Problems: Sect. 20.3

1. * Prove the cancellation law for the multiplication in a field.
2. Let p be prime and, for $n \in \mathbb{N}$, let $\mathcal{R}_n$ denote the set of roots of $X^{p^n} - X$ in Ω_p. Show that $\mathcal{R}_n$ is a subfield of Ω_p containing $\mathbb{Z}_p$.
3. In the notations of the statement of Theorem 20.27, show that $a_2 = \binom{p}{2}$ and

$$f_{21}(X) \ldots f_{2a_2}(X) = (X^p - X)^{p-1} + \bar{1}.$$

4. Let $\mathbb{K}$ be any field. Prove that $\mathbb{K}[X]$ has infinitely many irreducible polynomials.

20.4 Transcendental Numbers

As was said at the beginning of Sect. 20.1, transcendental numbers are precisely the complex numbers which are not roots of nonzero polynomials with rational coefficients. Nevertheless, up to this point we don't even know whether such numbers do exist.

One possible way of establishing the existence of (real) transcendental numbers is to start by showing that the set $\mathcal{A}$ of (real) algebraic numbers is *countably infinite*, i.e., that its elements can be put into a sequence $(x_n)_{n \geq 1}$, so that $\mathcal{A} = \{x_1, x_2, x_3, \ldots\}$; then, one proceeds to showing that $\mathbb{R}$ is *uncountable*, so that $\mathbb{R} \setminus \mathcal{A}$ is necessarily nonempty. Up to standard facts on countable and uncountable sets, we follow this path in the coming theorem, referring the reader to [8] or [33] for the necessary background.

Theorem 20.30 *The set of real algebraic numbers is countably infinite, whereas that of real transcendental numbers is uncountable.*

Proof Let $\mathcal{A}_n$ denote the set of real numbers which are roots of nonconstant polynomials with rational coefficients and degree n. If we show that each $\mathcal{A}_n$ is countably infinite, so will be the union

$$\mathcal{A} = \bigcup_{n \geq 1} \mathcal{A}_n.$$

However, such a union is precisely the set of real algebraic numbers.

For what is left to do, for each $n \in \mathbb{N}$ let $\mathcal{P}_n$ denote the set of polynomials with rational coefficients and degree n. Since the function

$$F : \mathbb{Q}^* \times \underbrace{\mathbb{Q} \times \cdots \times \mathbb{Q}}_{n-1} \longrightarrow \mathcal{P}_n(\mathbb{Q})$$

given by

$$F(a_n, a_{n-1}, \ldots, a_0) = a_n X^n + a_{n-1} X^{n-1} + \cdots + a_1 X + a_0$$

is a bijection with countably infinite domain, we conclude that $\mathcal{P}_n$ is countably infinite. Now, let $\mathcal{R}_f$ denote the set of real roots of $f \in \mathbb{Q}[X] \setminus \{0\}$. Since each $f \in \mathcal{P}_n$ has a finite number of real roots, it follows that

$$\mathcal{A}_n = \bigcup_{f \in \mathcal{P}_n} \mathcal{R}_f$$

is a countably infinite union of finite sets, hence countably infinite (for it contains $\mathbb{Q}$).

Finally, since $\mathbb{R}$ is uncountable and $\mathcal{A} \subset \mathbb{R}$ is countably infinite, we conclude that $\mathbb{R} \setminus \mathcal{A}$ is uncountable. $\qquad\square$

We now present the classical, explicit construction of the first transcendental number, due to Liouville. Up to details, our discussion follows the marvelous [12].

In the following, we say that an algebraic number α has **degree** n if its minimal polynomial p_α has degree n; in particular, if $n > 1$, then it's clear that α is irrational. We shall also need another piece of terminology: we say that a property $P(k)$, depending on a natural number k, is *true for all sufficiently large k* if there exists $k_0 \in \mathbb{N}$ such that $P(k)$ is true whenever $k > k_0$; moreover, if a specific value for k_0 is irrelevant in the context under discussion, we only say that $P(k)$ is true for all $k \gg 1$.

We are finally in position to state and prove Liouville's theorem. For the sake of comparison, at this point you might want to take a look at the statement of Dirichlet's Lemma 7.7.

Theorem 20.31 (Liouville)   *Let $\alpha \in \mathbb{C}$ be an algebraic number of degree $n > 1$. If $(p_k)_{k \geq 1}$ and $(q_k)_{k \geq 1}$ are sequences of nonzero integers such that $\lim_{k \to +\infty} \frac{p_k}{q_k} = \alpha$, then*

$$\left| \alpha - \frac{p_k}{q_k} \right| > \frac{1}{q_k^{n+1}}, \tag{20.11}$$

for all $k \gg 1$.

Proof Let $\alpha_k = \frac{p_k}{q_k}$. Since $\alpha_k \xrightarrow{k} \alpha$, it follows that $q_k \xrightarrow{k} +\infty$. Since α is algebraic of degree $n > 1$, there exists $f \in \mathbb{Z}[X] \setminus \{0\}$ of degree n and such that $f(\alpha) = 0$, say,

$$f(X) = a_n X^n + \cdots + a_1 X + a_0,$$

with $a_0, a_1, \ldots, a_n \in \mathbb{Z}$.

We now observe that

$$\left| \frac{f(\alpha_k)}{\alpha_k - \alpha} \right| = \frac{1}{|\alpha_k - \alpha|} \cdot |f(\alpha_k) - f(\alpha)|$$

$$= \frac{1}{|\alpha_k - \alpha|} \cdot \left| \sum_{j=1}^{n} a_j (\alpha_k^j - \alpha^j) \right|$$

$$\leq \frac{1}{|\alpha_k - \alpha|} \cdot \sum_{j=1}^{n} |a_j| \cdot |\alpha_k^j - \alpha^j|,$$

where we used the triangle inequality in the last passage above. Since $\alpha_k \xrightarrow{k} \alpha$, we have $|\alpha_k - \alpha| < 1$ for all $k \gg 1$. Under this assumption, we continue to estimate the last expression above by using some elementary algebra, together with the triangle inequality again, to obtain

$$\left| \frac{f(\alpha_k)}{\alpha_k - \alpha} \right| \leq \sum_{j=1}^{n} |a_j| \cdot \frac{|\alpha_k^j - \alpha^j|}{|\alpha_k - \alpha|}$$

$$= \sum_{j=1}^{n} |a_j| \cdot |\alpha_k^{j-1} + \alpha_k^{j-2}\alpha + \cdots + \alpha^{j-1}|$$

$$\leq \sum_{j=1}^{n} |a_j| \left(\left(|\alpha_k - \alpha| + |\alpha|\right)^{j-1} + \left(|\alpha_k - \alpha| + |\alpha|\right)^{j-2}|\alpha| + \cdots + |\alpha|^{j-1} \right)$$

$$\leq \sum_{j=1}^{n} |a_j| \left(\left(1 + |\alpha|\right)^{j-1} + \left(1 + |\alpha|\right)^{j-2}|\alpha| + \cdots + |\alpha|^{j-1} \right)$$

$$< \sum_{j=1}^{n} j|a_j| ((1 + |\alpha|))^{j-1} := C,$$

where C depends only on α, and not on k.

Then, for all $k \gg 1$ (chosen in such a way that $|\alpha_k - \alpha| < 1$), we have

$$\left| \alpha - \frac{p_k}{q_k} \right| = |\alpha_k - \alpha| > \frac{1}{C} |f(\alpha_k)|.$$

Now, since $q_k \xrightarrow{k} +\infty$, we have $|\alpha_k - \alpha| < 1$ and $q_k > C$ for all $k \gg 1$, so that $\frac{1}{C} > \frac{1}{q_k}$ and, hence,

$$\left| \alpha - \frac{p_k}{q_k} \right| > \frac{1}{q_k} |f(\alpha_k)|,$$

for all $k \gg 1$.

As our final step, notice that if $f(\alpha_k) = 0$, then we would have $f(X) = (X - \alpha_k)g(X)$, for some $g \in \mathbb{Z}[X] \setminus \{0\}$ of degree $n-1$. Since $\alpha \neq \alpha_k$ (for, α is irrational), we conclude that α would be a root of g, which contradicts the fact that the minimal polynomial of α has degree n. Therefore, $f(\alpha_k) \neq 0$ and, hence,

$$|f(\alpha_k)| = a_n \left(\frac{p_k}{q_k}\right)^n + \cdots + a_1 \left(\frac{p_k}{q_k}\right) + a_0$$

$$= \frac{1}{q_k^n} |a_n p_k^n + \cdots + a_1 p_k q_k^{n-1} + a_0 q_k^n|$$

$$\geq \frac{1}{q_k^n}.$$

Finally, combining this inequality with the previous one, we get (20.11). □

The coming example brings a clever application of Liouville's theorem to the construction of transcendental numbers. Along its discussion, we shall need some acquaintance with the arithmetic of convergent series, for which we refer the reader to Section 10.3 of [8].

Example 20.32 (Liouville) If $a_1, a_2, a_3, \ldots \in \{1, 2, 3, \ldots, 9\}$, then the real number

$$\alpha = \frac{a_1}{10^{1!}} + \frac{a_2}{10^{2!}} + \frac{a_3}{10^{3!}} + \cdots .$$

is transcendental.

Proof Suppose that α were algebraic, of degree $n > 1$ (it is clearly irrational—see Problem 1). For $k \geq 1$, let

$$\alpha_k = \sum_{j=1}^{k} \frac{a_j}{10^{j!}} = \frac{p_k}{10^{k!}},$$

where $p_k \in \mathbb{N}$. Then, on the one hand, it follows from Liouville's theorem that

$$|\alpha - \alpha_k| > \frac{1}{\left(10^{k!}\right)^{n+1}},$$

for all $k \gg 1$. On the other,

$$|\alpha - \alpha_k| = \left| \sum_{j>k} \frac{a_j}{10^{j!}} \right| < \frac{1}{10^{(k+1)!}} \cdot 9,999 \ldots = \frac{1}{10^{(k+1)!-1}}.$$

Therefore, for $k \gg 1$, we should have

$$\frac{1}{10^{(k+1)!-1}} > \frac{1}{\left(10^{k!}\right)^{n+1}}$$

and, hence, $(k + 1)! - 1 < k!(n + 1)$. However, this last inequality is equivalent to

$$k!(k - n) < 1,$$

which is false for all $k \geq n + 1$.

Finally, since the assumption that α is algebraic leads to a contradiction, we have no alternative but to conclude that α is transcendental. $\qquad \square$

We finish this section by observing that the methods of real Analysis allow one to prove that e and π are transcendental numbers. Both these facts were established already in the nineteenth century, due to the works of the French mathematician

C. Hermite and the German mathematician F. Lindemann, respectively. Proofs are beyond the scope of these notes, and we refer the interested reader to [17] or [28].

Example 20.33 Once we admit the transcendences of e and π, Theorems 20.9 and 20.12 can be applied to establish the transcendence of several other numbers. More generally, if $\alpha \in \mathbb{R}$ is algebraic and β is transcendental, then $\alpha + \beta$ is transcendental. Indeed, if it algebraic we would have

$$\beta = (\alpha + \beta) - \alpha,$$

also algebraic. A similar argument shows that if $\alpha \in \mathbb{R} \setminus \{0\}$ is algebraic and β is transcendental, then $\alpha\beta$ is transcendental.

Problems: Sect. 20.4

1. Prove that the number α of Example 20.32 is irrational.
2. Let $\alpha \in \mathbb{C}$ be algebraic and $\beta \in \mathbb{C}$ be transcendental. Prove directly (i.e., without resorting to Theorem 20.9) that $\alpha + \beta$ is irrational.
3. If $\alpha \in \mathbb{C}$ is transcendental and $n \in \mathbb{N}$, prove that $\sqrt[n]{\alpha}$ and α^n are also transcendental.
4. * Let $\alpha \in \mathbb{C}$ be transcendental, and define

$$\mathbb{Q}[\alpha] = \{a_0 + a_1\alpha + \cdots + a_n\alpha^n; \ n \in \mathbb{N} \text{ and } a_0, a_1, \ldots, a_n \in \mathbb{Q}\}.$$

   Prove that $\mathbb{Q}[\alpha]$ is closed with respect to addition, subtraction and multiplication but is not a subfield of $\mathbb{C}$.

   For the coming problem you might want to use the following fact, which will be stated without proof: if $\alpha \in \mathbb{C}$ is a root of a polynomial

$$f(X) = a_n X^n + a_{n-1} X^{n-1} + \cdots + a_1 X + a_0,$$

   with $a_0, a_1, \ldots, a_n$ algebraic and $a_n \neq 0$, then α is also algebraic. A proof of this fact can be found in any standard text on Field Theory, for instance [6] or [20].
5. (OBMU) In the cartesian plane, let S be the set of circles whose centers have rational coordinates and whose radii have rational lengths. Show that there exists a regular 2016-gon, all of whose vertices do not belong to S.

Chapter 21
Linear Recurrence Relations

In this chapter we complete the work initiated in Section 3.2 of [8] (see also Problems 5, page 79, and 6, page 393), showing how to solve a *linear recurrence relation with constant coefficients and arbitrary order*. We first need to properly define the objects involved, and we do this now.

Definition 21.1 A sequence $(a_n)_{n\geq 1}$ is said to be **linear recurrent** if there exist a positive integer k and complex numbers $u_0, \ldots, u_{k-1}$, not all equal to zero, such that

$$a_{n+k} = u_{k-1}a_{n+k-1} + \cdots + u_0 a_n \qquad (21.1)$$

for every $n \geq 1$.

In the above notations, we sometimes also say that (21.1) is a **linear recurrence relation**; the natural number k is thus said to be the **order** of the linear recurrence relation (21.1).

Given $\alpha_1, \ldots, \alpha_k \in \mathbb{C}$, it is immediate to verify (by induction, for instance) that there is exactly one linear recurrent sequence $(a_n)_{n\geq 1}$ satisfying (21.1) and such that $a_j = \alpha_j$ for $1 \leq j \leq k$. Hence, a linear recurrent sequence of order k falls completely determined upon knowing the linear recurrence relation it satisfies, together with the values of its first k terms.

Definition 21.2 Let $(a_n)_{n\geq 1}$ be a linear recurrent sequence such that

$$a_{n+k} = u_{k-1}a_{n+k-1} + \cdots + u_0 a_n$$

for every $n \geq 1$, with $u_0, \ldots, u_{k-1}$ being given complex numbers, at least one of which is nonzero. The **characteristic polynomial** of $(a_n)_{n\geq 1}$ is the polynomial $f \in \mathbb{C}[X]$ given by

$$f(X) = X^k - u_{k-1}X^{k-1} - \cdots - u_1 X - u_0. \qquad (21.2)$$

© Springer International Publishing AG, part of Springer Nature 2018
A. Caminha Muniz Neto, *An Excursion through Elementary Mathematics, Volume III*,
Problem Books in Mathematics, https://doi.org/10.1007/978-3-319-77977-5_21

21.1 An Important Particular Case

In this section, we shall discuss how to solve a linear recurrence relation for which the corresponding characteristic polynomial has pairwise distinct complex roots. In spite of such a limitation, this case is much simpler to be dealt with than the general one and finds several interesting applications. The result we are interested in is the content of the coming

Theorem 21.3 *Let $(a_n)_{n\geq 1}$ be a sequence satisfying, for every $n \geq 1$, the linear recurrence relation*

$$a_{n+k} = u_{k-1}a_{n+k-1} + \cdots + u_0 a_n,$$

with $u_0, \ldots, u_{k-1}$ being given complex numbers, not all equal to zero. If the complex roots $z_1, z_2, \ldots, z_k$ of the characteristic polynomial of $(a_n)_{n\geq 1}$ are pairwise distinct and $a_j = \alpha_j$ for $1 \leq j \leq k$, then

$$a_n = z_1^{n-1}x_1 + z_2^{n-1}x_2 + \cdots + z_k^{n-1}x_k \; \forall n \geq 1,$$

where $x_1, \ldots, x_k$ is the solution of the Vandermonde system of equations

$$\begin{cases} x_1 + x_2 + \cdots + x_k & = \alpha_1 \\ z_1 x_1 + z_2 x_2 + \cdots + z_k x_k & = \alpha_2 \\ z_1^2 x_1 + z_2^2 x_2 + \cdots + z_k^2 x_k & = \alpha_3 \; . \qquad (21.3) \\ \cdots \\ z_1^{n-1}x_1 + z_2^{n-1}x_2 + \cdots + z_k^{k-1}x_k = \alpha_k \end{cases}$$

Proof Since $z_1, z_2, \ldots, z_k$ are pairwise distinct, Proposition 18.6 assures the existence of a unique solution $x_1, x_2, \ldots, x_k$ for the system of equations (21.3). We can thus let $(b_n)_{n\geq 1}$ be defined by setting

$$b_n = z_1^{n-1}x_1 + z_2^{n-1}x_2 + \cdots + z_k^{n-1}x_k, \; \forall n \geq 1.$$

This way, for $1 \leq j \leq k$, system (21.3) furnishes

$$b_j = z_1^{j-1}x_1 + z_2^{n-1}x_2 + \cdots + z_k^{j-1}x_k = \alpha_j = a_j.$$

On the other hand, since f is the characteristic polynomial of $(a_n)_{n\geq 1}$, it follows from the definition of the b_j's that

$$b_{n+k} - u_{k-1}b_{n+k-1} - \cdots - u_0 b_n =$$

$$= \sum_{j=1}^{k} z_j^{n+k-1} x_j - u_{k-1} \sum_{j=1}^{k} z_j^{n+k-2} x_j - \cdots - u_0 \sum_{j=1}^{k} z_j^{n-1} x_j$$

$$= \sum_{j=1}^{k} z_j^{n-1} x_j (z_j^k - u_{k-1} z_j^{k-1} - \cdots - u_0)$$

$$= \sum_{j=1}^{k} z_j^{n-1} x_j f(z_j) = 0.$$

It follows from the above that $(b_n)_{n \geq 1}$ satisfies the same linear recurrence relation that $(a_n)_{n \geq 1}$, and its first k terms respectively match the first k terms of $(a_n)_{n \geq 1}$. Therefore, a straightforward inductive argument guarantees that $a_n = b_n$ for every $n \geq 1$, as we wished to show. □

The coming example shows that we not necessarily need to actually know the roots of the characteristic polynomial of a linear recurrence relation, in order to apply the result of the previous theorem to obtain interesting information concerning its behavior.

Example 21.4 (Crux) Let $A_1 A_2 A_3 A_4$ be the square of vertices $A_1 = (0, 1)$, $A_2 = (1, 1)$, $A_3 = (1, 0)$ and $A_4 = (0, 0)$. For each integer $n \geq 1$, let A_{n+4} be the midpoint of the line segment $A_n A_{n+1}$. Prove that, when $n \to +\infty$, the sequence of points A_n converges to some fixed point A, and compute the coordinates of such a point.

Proof For each $n \geq 1$, let $A_n(x_n, y_n)$. The stated condition, together with the formula for the coordinates of the midpoint of a line segment (cf. Section 6.2 of [9]), gives

$$2x_{n+4} = x_{n+1} + x_n \ \forall \ n \geq 1,$$

an analogous recurrence relation being valid for the sequence $(y_n)_{n \geq 1}$.

The characteristic polynomial of the above recurrence relation is

$$f(X) = 2X^4 - X - 1 = (X - 1)g(X),$$

with $g(X) = 2X^3 + 2X^2 + 2X + 1$. If a, b and c are the complex roots of g, then Girard-Viète relations give

$$a + b + c = -1 \quad \text{and} \quad ab + ac + bc = 1. \tag{21.4}$$

Hence,

$$a^2 + b^2 + c^2 = (a + b + c)^2 - 2(ab + ac + bc) = (-1)^2 - 2 \cdot 1 = -1 < 0,$$

so that, thanks to Problem 4, page 386, we can assume that a is real and b and c are nonreal complex conjugates. In particular, a, b and c are pairwise distinct and, since 1 is not a root of g, Theorem 21.3 guarantees the existence of complex constants A, B, C and D for which

$$x_n = Aa^{n-1} + Bb^{n-1} + Cc^{n-1} + D, \ \forall \, n \geq 1. \tag{21.5}$$

For us to know what happens when $n \to +\infty$, note firstly that $g(-1)g(0) = -1 < 0$ and, thus, Bolzano's Theorem 17.2 assures that $-1 < a < 0$. Therefore, relations (21.4) give

$$1 = a(b+c) + bc = a(-1-a) + b\overline{b},$$

so that $|b|^2 = a(a+1)+1 < 1$. However, since $|b| = |c|$, it comes that $|b| = |c| < 1$. Then, relation (21.5), together with the result of Example 21.8, gives

$$\lim_{n \to +\infty} x_n = D.$$

In order to compute the value of D, recall that $x_1 = x_4 = 0$ and $x_2 = x_3 = 1$. Hence, by successively letting $n = 1, 2, 3$ and 4 in (21.5), we obtain the linear system of equations

$$\begin{cases} A + B + C + D & = 1 \\ Aa + Bb + Cc + Dd & = 0 \\ Aa^2 + Bb^2 + Cc^2 + Dd^2 & = 0 \\ Aa^3 + Bb^3 + Cc^3 + Dd^3 & = 1 \end{cases}.$$

Multiplying the last three equations by 2 and termwise adding the four equalities thus obtained, we arrive at

$$Ag(a) + Bg(b) + Cg(c) + 7D = 3.$$

However, since a, b and c are roots of g, we conclude that $7D = 3$, and $D = \frac{3}{7}$.

In an analogous way, we prove that $\lim_{n \to +\infty} y_n = \frac{4}{7}$. Thus,

$$\lim_{n \to +\infty} A_n = \left(\frac{3}{7}, \frac{4}{7} \right).$$

$\square$

Problems: Sect. 21.1

1. (OBMU) If $f(x) = e^{-x} \sin x$ and $f^{(n)}$ denotes the n-th derivative of f, compute $f^{(2001)}(0)$.
2. Let $A_1(a_1, b_1), A_2(a_2, b_2), \ldots, A_k(a_k, b_k)$ be $k \geq 2$ given points in the cartesian plane. For each $n \in \mathbb{N}$, let A_{n+k} be the center of mass of $A_n, A_{n+1}, \ldots, A_{n+k-1}$, i.e.,

$$A_{n+k} = \frac{1}{k}\left(A_{n+k-1} + \cdots + A_{n+1} + A_n\right).$$

If $A_m(x_m, y_m)$ for $m > k$, show that

$$x_m \xrightarrow{m} \frac{ka_k + (k-1)a_{k-1} + \cdots + 2a_2 + a_1}{k + (k-1) + \cdots + 2 + 1}$$

and

$$y_m \xrightarrow{m} \frac{kb_k + (k-1)b_{k-1} + \cdots + 2b_2 + b_1}{k + (k-1) + \cdots + 2 + 1}.$$

3. (IMO shortlist) If a is the largest positive root of the polynomial $X^3 - 3X^2 + 1$, show that $\lfloor a^{1788} \rfloor$ and $\lfloor a^{1988} \rfloor$ are both multiples of 17.

21.2   Sequences, Series and Continuity in $\mathbb{C}$

In this section we partially extend the basic facts on convergence of real sequences and series (cf. Chapters 7 and 11 of [8], for instance) to functions with complex values. We observe that the material collected here (more precisely, Theorem 21.20) was used in the proof of the Fundamental Theorem of Algebra, in Sect. 15.3, and will be of fundamental importance for Sect. 21.3. We also show that it can be used to extend the method of generating functions to *complex power series*.

Given $a \in \mathbb{C}$ and $R > 0$, we denote by $D(a; R)$ the **open disk** of center a and radius R, i.e., the subset of $\mathbb{C}$ given by

$$D(a; R) = \{z \in \mathbb{C}; \; |z - a| < R\}.$$

Analogously, the **closed disk** of center a and radius R is the subset $\overline{D(a; R)}$ of $\mathbb{C}$, given by

$$\overline{D(a; R)} = \{z \in \mathbb{C}; \; |z - a| \le R\}.$$

A set $U \subset \mathbb{C}$ is **open** if, for every $a \in U$, there exists $R > 0$ such that $D(a; R) \subset U$. A set $F \subset \mathbb{C}$ is **closed** if $F^c = \mathbb{C} \setminus F$ is open. It is immediate to see that $\emptyset$ and $\mathbb{C}$ are open (in the case of $\emptyset$, there is no way for the openness condition not to be satisfied, since there exists no $z \in \emptyset$). Hence, $\mathbb{C} = \mathbb{C} \setminus \emptyset$ and $\emptyset = \mathbb{C} \setminus \mathbb{C}$ are also closed. We now collect less trivial examples of open and closed sets.

Example 21.5 For all $a \in \mathbb{C}$ and $R > 0$, the open disk $D(a; R)$ is an open set, whereas the closed disk $\overline{D(a; R)}$ is a closed set.

Fig. 21.1 Every open disk is
an open set

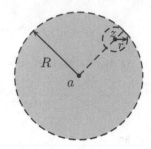

Proof Let $z \in D(a; R)$ be arbitrarily chosen and $r = R - |z - a|$. Then, $r > 0$ and
we claim that $D(z; r) \subset D(a; R)$ (cf. Fig. 21.1), which will be enough to guarantee
that $D(a; R)$ is open.

For what is left to do, take $w \in D(z; r)$. Triangle inequality gives

$$|w - a| = |(w - z) + (z - a)| \leq |w - z| + |z - a|$$

$$\leq r + |z - a| = R,$$

so that $w \in D(a; R)$. However, since this holds with every $w \in D(z; r)$, we
conclude that $D(z; r) \subset D(a; R)$, as wished.

For the second part, it is enough to show that $U = \mathbb{C} \setminus \overline{D(a; R)}$ is open. To this
end, take $z \in U$ and let $r = |z - a| - R$. Then, $r > 0$ and, as in the first part,
we easily show that $w \in D(z; r) \Rightarrow w \in U$, which is the same as $D(z; r) \subset U$.
Therefore, U is indeed open. $\square$

We now turn our attention to sequences of complex numbers, extending to them
the concept of *convergence*.

Definition 21.6 A sequence $(z_n)_{n \geq 1}$ in $\mathbb{C}$ **converges** for a **limit** $z \in \mathbb{C}$, and write
$z_n \to z$ or $\lim_{n \to +\infty} z_n = z$, provided the following condition is satisfied:

$$\forall \epsilon > 0, \ \exists n_0 \in \mathbb{N}; \ n > n_0 \Rightarrow |z_n - z| < \epsilon.$$

Heuristically, the fulfilment of the condition above means that, as $n \to +\infty$, the
terms z_n come closer and closer to z. Indeed, the definition of convergence stipulates
that $(z_n)_{n \geq 1}$ converges to $z \in \mathbb{C}$ if, for an arbitrarily given radius $\epsilon > 0$, there exists
an index $n_0 \in \mathbb{N}$ such that, for $n > n_0$, all of the terms z_n belong to the open disk
$D(z; \epsilon)$.

As in the real case, a **subsequence** $(z_{n_k})_{k \geq 1}$ of a sequence $(z_n)_{n \geq 1}$ is the
restriction of $(z_n)_{n \geq 1}$ to an infinite subset $\{n_1 < n_2 < n_3 < \cdots \}$ of indices; from a
rigorous standpoint, $(z_{n_k})_{k \geq 1}$ is the sequence $f \circ g : \mathbb{N} \to \mathbb{C}$, with $f : \mathbb{N} \to \mathbb{C}$ and
$g : \mathbb{N} \to \mathbb{N}$ given by $f(n) = z_n$ for every $n \in \mathbb{N}$ and $g(k) = n_k$ for every $k \in \mathbb{N}$.

The coming result brings two fundamental properties of the concept of convergence of sequences of complex numbers. In words, item (a) assures that the terms
of a convergent sequence cannot have two distinct limits, and item (b) that every
subsequence of a convergent sequence is also convergent and converges to the same
limit.

Lemma 21.7 *Let $(z_n)_{n\geq 1}$ be a sequence in $\mathbb{C}$.*

(a) If $z_n \to z$ and $z_n \to w$, then $z = w$.
(b) If $(z_{n_k})_{k\geq 1}$ is a subsequence of $(z_n)_{n\geq 1}$ and $z_n \to z$, then $z_{n_k} \to z$.

Proof

(a) If $z \neq w$, then $\epsilon = |z - w| > 0$. However, since $\frac{\epsilon}{2}$ is also positive and $z_n \to z$ and $z_n \to w$, there exist natural numbers n_1 and n_2 such that $|z_n - z| < \frac{\epsilon}{2}$ for $n > n_1$ and $|z_n - w| < \frac{\epsilon}{2}$ for $n > n_2$. Hence, taking an index $n > n_1, n_2$ and applying triangle inequality, we obtain

$$|z - w| = |(z - z_n) + (z_n - w)|$$
$$\leq |z - z_n| + |z_n - w|$$
$$< \frac{\epsilon}{2} + \frac{\epsilon}{2} = \epsilon = |z - w|,$$

which is a contradiction. Therefore, $z = w$.

(b) Given $\epsilon > 0$, the convergence of $(z_n)_{n\geq 1}$ to z assures the existence of $n_0 \in \mathbb{N}$ such that $|z_n - z| < \epsilon$ for $n > n_0$. Now, since $n_1 < n_2 < n_3 < \cdots$, there exists $k_0 \in \mathbb{N}$ with $k > k_0 \Rightarrow n_k > n_0$. Hence, for such values of k, we have $|z_{n_k} - z| < \epsilon$, so that $(z_{n_k})_{k\geq 1}$ also converges to z. $\qquad\square$

Thanks to item (a) of the previous lemma, if a sequence $(z_n)_{n\geq 1}$ in $\mathbb{C}$ converges to $z \in \mathbb{C}$, we shall hereafter say that z is *the* limit of $(z_n)_{n\geq 1}$.

For the purposes of these notes, one of the most important examples of convergent sequence is the one collected below. In this respect, see also Problem 2.

Example 21.8 If $|z| < 1$ and $z_n = z^n$ for every $n \geq 1$, then $(z_n)_{n\geq 1}$ converges to 0.

Proof Firstly, note that $|z_n - 0| = |z^n| = |z|^n$. Now, if $(a_n)_{n\geq 1}$ is the sequence of real numbers given for $n \geq 1$ by $a_n = |z|^n$, then (cf. Example 7.12 of [8], for instance) $a_n \to 0$. Thus, given $\epsilon > 0$, there exists $n_0 \in \mathbb{N}$ such that $0 \leq a_n < \epsilon$ for $n > n_0$. Thus, for $n > n_0$, we have $|z_n - 0| = a_n < \epsilon$, so that $z_n \to 0$. $\qquad\square$

Given a sequence $(z_n)_{n\geq 1}$ of complex numbers, we can write $z_n = x_n + iy_n$ for every $n \geq 1$, with $x_n, y_n \in \mathbb{R}$. Our next result relates the convergence of $(z_n)_{n\geq 1}$ in $\mathbb{C}$ with those of $(x_n)_{n\geq 1}$ and $(y_n)_{n\geq 1}$ in $\mathbb{R}$.

Lemma 21.9 *Let $(z_n)_{n\geq 1}$ be a sequence of complex numbers, with $z_n = x_n + iy_n$ and $x_n, y_n \in \mathbb{R}$ for every $n \geq 1$. Then, $(z_n)_{n\geq 1}$ converges in $\mathbb{C}$ if and only if $(x_n)_{n\geq 1}$ and $(y_n)_{n\geq 1}$ converge in $\mathbb{R}$. Moreover, if $x_n \to a$ and $y_n \to b$, then $z_n \to z$, with $z = a + ib$.*

Proof Firstly, suppose that $z_n \to z$, with $z = a + ib$, $a, b \in \mathbb{R}$, and let $\epsilon > 0$ be given. Since

$$|x_n - a|, |y_n - b| \leq \sqrt{|x_n - a|^2 + |y_n - b|^2} = |z_n - z|,$$

we have $|x_n - a|, |y_n - b| < \epsilon$ whenever $|z_n - z| < \epsilon$. However, since $z_n \to z$, there exists $n_0 \in \mathbb{N}$ such that $n > n_0 \Rightarrow |z_n - z| < \epsilon$. Then, for each such natural n, we do have $|x_n - a|, |y_n - b| < \epsilon$, hence establishing the desired convergences.

Conversely, assume that $x_n \to a$ and $y_n \to b$, and let $z = a + ib$ and $\epsilon > 0$ be given. Since

$$|z_n - z| = \sqrt{|x_n - a|^2 + |y_n - b|^2} \leq |x_n - a| + |y_n - b|,$$

we have $|z_n - z| < \epsilon$ if $|x_n - a|, |y_n - b| < \frac{\epsilon}{2}$. However, since $x_n \to a$ and $y_n \to b$, there exist $n_1, n_2 \in \mathbb{N}$ such that $n > n_1 \Rightarrow |x_n - a| < \frac{\epsilon}{2}$ and $n > n_2 \Rightarrow |y_n - b| < \frac{\epsilon}{2}$. If $n_0 = \max\{n_1, n_2\}$, then, for $n > n_0$, we have $|x_n - a|, |y_n - b| < \frac{\epsilon}{2}$ and, hence, $|z_n - z| < \epsilon$. This establishes the convergence of $(z_n)_{n \geq 1}$ to z. $\square$

We now need to extend to complex sequences another concept related to the convergence of real sequences.

Definition 21.10 A **Cauchy sequence** in $\mathbb{C}$ is a sequence $(z_n)_{n \geq 1}$ satisfying the following condition: given $\epsilon > 0$, there exists $n_0 \in \mathbb{N}$ such that

$$m, n > n_0 \Rightarrow |z_m - z_n| < \epsilon.$$

If $(z_n)_{n \geq 1}$ is a convergent sequence in $\mathbb{C}$, then $(z_n)_{n \geq 1}$ is a Cauchy sequence. Indeed, if $z_n \to z$ and $\epsilon > 0$ is given, then there exists $n_0 \in \mathbb{N}$ such that $|z_n - z| < \frac{\epsilon}{2}$ for every $n > n_0$. Hence, for natural numbers $m, n > n_0$, it follows from the triangle inequality for complex numbers that

$$|z_m - z_n| \leq |z_m - z| + |z - z_n| \leq \frac{\epsilon}{2} + \frac{\epsilon}{2} = \epsilon.$$

Conversely, we have the following important result.

Proposition 21.11 $\mathbb{C}$ *is* **complete**. *More precisely, if* $(z_n)_{n \geq 1}$ *is a Cauchy sequence in* $\mathbb{C}$, *then* $(z_n)_{n \geq 1}$ *converges.*

Proof Let $z_n = x_n + i y_n$, with $(x_n)_{n \geq 1}$ and $(y_n)_{n \geq 1}$ being sequences of real numbers. Given $\epsilon > 0$, take $n_0 \in \mathbb{N}$ as in the definition of Cauchy sequence. Since $|x_m - x_n| \leq |z_m - z_n|$, it follows that $(x_n)_{n \geq 1}$ is a Cauchy sequence in $\mathbb{R}$. Therefore, Theorem 7.27 of [8] guarantees that $(x_n)_{n \geq 1}$ is convergent, say for $x \in \mathbb{R}$. Analogously, there exists $y \in \mathbb{R}$ such that $y_n \to y$. Hence, letting $z = x + iy$, Lemma 21.9 assures that $z_n \to z$. $\square$

As in the real case (cf. [8]), given a sequence $(z_n)_{n \geq 1}$ of complex numbers, we define the **series**[1] $\sum_{k \geq 1} z_k$ as the sequence $(s_n)_{n \geq 1}$, such that $s_n = \sum_{k=1}^{n} z_k$ for every $n \in \mathbb{N}$. Also as in the real case, we say that s_n is the **n-th partial sum** of

---

[1] The reader must pay attention to the fact that we shall sometimes consider sequences $(z_n)_{n \geq 0}$ of complex numbers, so that the corresponding series will be denoted by $\sum_{k \geq 0} z_k$.

the series, and that the series **converges** if there exists the limit $s := \lim_{n \to +\infty} s_n$; moreover, in this case we shall say that s is the **sum** of the series. In symbols, we write

$$\sum_{k \geq 1} z_k = \lim_{n \to +\infty} \sum_{k=1}^{n} z_k,$$

provided the limit at the right hand side does exist.

We now present an example of convergent series which will be of fundamental importance in the next section.

Example 21.12 Given $a \in \mathbb{C} \setminus \{0\}$, show that for $z \in D(0; \frac{1}{|a|})$ we have

$$\frac{1}{1 - az} = \sum_{k \geq 0} a^k z^k.$$

Proof Lemma 13.18 assures that the n-th partial sum of the series is

$$\sum_{k=0}^{n} (az)^k = \frac{1 - (az)^{n+1}}{1 - az} = \frac{1}{1 - az} - \frac{(az)^{n+1}}{1 - az}.$$

Now, for $z \in D(0; \frac{1}{|a|})$ we have $|az| < 1$, so that Example 21.8 gives $(az)^n \to 0$ as $n \to +\infty$. Hence, it follows from the equality above that, for $z \in D(0; \frac{1}{|a|})$,

$$\sum_{k \geq 0} (az)^k = \lim_{n \to +\infty} \sum_{k=0}^{n} (az)^k = \frac{1}{1 - az} - \lim_{n \to +\infty} \frac{(az)^{n+1}}{1 - az} = \frac{1}{1 - az}.$$

□

Another concept of central importance is that of an **absolutely convergent series**, i.e., a series $\sum_{k \geq 0} z_k$ of complex numbers such that the real series $\sum_{k \geq 0} |z_k|$ converges. For $n \in \mathbb{N}$, let $s_n = \sum_{k=1}^{n} z_k$ and $t_n = \sum_{k=1}^{n} |z_k|$, so that $(t_n)_{n \geq 1}$ converges. The triangle inequality for complex numbers furnishes, for integers $m > n$,

$$|s_m - s_n| = \left| \sum_{k=n+1}^{m} z_k \right| \leq \sum_{k=n+1}^{m} |z_k| = t_m - t_n.$$

Now, since $(t_n)_{n \geq 1}$ converges, it is a Cauchy sequence; hence, given $\epsilon > 0$, there exists $k_0 \in \mathbb{N}$ such that $m > n > k_0 \Rightarrow t_m - t_n < \epsilon$. Therefore, we also have $|s_m - s_n| < \epsilon$ for $m > n > k_0$, so that $(s_n)_{n \geq 0}$ is a Cauchy sequence too. Thus, Proposition 21.11 assures its convergence, and we have proved the coming.

Proposition 21.13 *In $\mathbb{C}$, every absolutely convergent series is convergent.*

Before we can go on, we need to extend the concept of continuity for functions defined on a subset of $\mathbb{C}$ and assuming complex values.

Definition 21.14 Given a nonempty subset X of $\mathbb{C}$, we say that a function $f : X \to \mathbb{C}$ is **continuous** if the following condition is satisfied: for every sequence $(z_n)_{n \geq 1}$ of points of X, if $z_n \to z$, with $z \in X$, then $f(z_n) \to f(z)$.

Our next result will allow us to establish the continuity of complex polynomial functions.

Proposition 21.15 *If $f, g : X \to \mathbb{C}$ are continuous functions, then so are $f + g$, $fg : X \to \mathbb{C}$.*

Proof Let a sequence $(z_n)_{n \geq 1}$ in X be given, such that $z_n \to z$, for some $z \in X$. Firstly, the triangle inequality for complex numbers gives

$$|(f + g)(z_n) - (f + g)(z)| = |(f(z_n) - f(z)) + (g(z_n) - g(z)|$$

$$\leq |f(z_n) - f(z)| + |g(z_n) - g(z)|.$$

Now, given $\epsilon > 0$, since $f(z_n) \to f(z)$ and $g(z_n) \to g(z)$, there exists $n_0 \in \mathbb{N}$ such that

$$n > n_0 \Rightarrow |f(z_n) - f(z)|, |g(z_n) - g(z)| < \frac{\epsilon}{2}.$$

Therefore, also for $n > n_0$, the above estimates show that

$$|(f + g)(z_n) - (f + g)(z)| \leq \frac{\epsilon}{2} + \frac{\epsilon}{2} = \epsilon$$

and, hence, that $(f + g)(z_n) \to (f + g)(z)$.

For fg, and applying twice the triangle inequality, we obtain

$$|(fg)(z_n) - (fg)(z)| = |f(z_n)g(z_n) - f(z)g(z)|$$

$$\leq |f(z_n) - f(z)||g(z_n)| + |f(z)||g(z_n) - g(z)|$$

$$\leq |f(z_n) - f(z)||g(z_n) - g(z)|$$

$$+ |f(z_n) - f(z)||g(z)| + |f(z)||g(z_n) - g(z)|.$$

As before, given $\epsilon > 0$, we can take $n_0 \in \mathbb{N}$ such that $n > n_0$ implies

$$|f(z_n) - f(z)| < \min \left\{ \sqrt{\frac{\epsilon}{3}}, \frac{\epsilon}{3(|g(z)| + 1)} \right\}$$

and

$$|g(z_n) - g(z)| < \min \left\{ \sqrt{\frac{\epsilon}{3}}, \frac{\epsilon}{3(|f(z)| + 1)} \right\}.$$

Then, for a natural number $n > n_0$, it follows from the above that

$$|(fg)(z_n) - (fg)(z)| \leq \sqrt{\frac{\epsilon}{3}} \cdot \sqrt{\frac{\epsilon}{3}} + \frac{\epsilon}{3(|g(z)| + 1)} \cdot |g(z)|$$

$$+ |f(z)| \cdot \frac{\epsilon}{3(|f(z)| + 1)}$$

$$< \frac{\epsilon}{3} + \frac{\epsilon}{3} + \frac{\epsilon}{3} = \epsilon,$$

so that $(fg)(z_n) \to (fg)(z)$. $\qquad\qquad\qquad\qquad\qquad\qquad\qquad\qquad\square$

Example 21.16 Given $n \in \mathbb{N}$ and $a_0, a_1, \ldots, a_{n-1}, a_n \in \mathbb{C}$, with $a_n \neq 0$, the complex polynomial function $f : \mathbb{C} \to \mathbb{C}$, given for $z \in \mathbb{C}$ by

$$f(z) = a_n z^n + a_{n-1} z^{n-1} + \cdots + a_1 z + a_0,$$

is continuous.

Proof Obviously, constant functions and the identity function $z \mapsto z$ are continuous (just apply the definition). Thus, by applying several times the second part of the previous proposition, we conclude that, for an integer $0 \leq k \leq n$, the function $z \mapsto a_k z^k$ is also continuous. Now, applying several times the first part of the previous proposition, we conclude that the sum of such functions, when k varies from 0 to n, is also continuous. However, such a sum is precisely the function f. $\qquad\square$

We now turn to the discussion of **power series** in $\mathbb{C}$, namely, a series of the form $\sum_{k \geq 0} a_k z^k$, for some given sequence $(a_n)_{n \geq 0}$ of complex numbers.

Assume that the sequence $(\sqrt[k]{|a_k|})_{k \geq 0}$ is bounded, say with $\sqrt[k]{|a_k|} < A$ for every $k \geq 0$ and some $A > 0$. Then, $|a_k z^k| \leq (Az)^k$ for every $k \geq 0$, so that Example 21.12 guarantees that the series $\sum_{k \geq 0} |a_k z^k|$ converges in $D(0; R)$, where $R = \frac{1}{A}$. In turn Proposition 21.13 then assures that the power series $\sum_{k \geq 0} a_k z^k$ converges in $D(0; R)$. We abbreviate such a situation simply by saying that $D(0; R)$ is a **convergence disk** for the power series under consideration. The next result assures that the function $f : D(0; R) \to \mathbb{C}$ thus defined is continuous.[2] By the sake of notation, we make the convention that $\sum_{k \geq 0} a_k z^k = a_0$ for $z = 0$.

Proposition 21.17 *If $D(0; R)$ is a convergence disk for the power series $\sum_{k \geq 0} a_k z^k$, then the function $f : D(0; R) \to \mathbb{C}$, given for $z \in D(0; R)$ by $f(z) = \sum_{k \geq 0} a_k z^k$, is continuous.*

---

[2]Actually, one can show that f is an example of a *holomorphic* (i.e., *complex differentiable*) function. However, since we shall not need this concept, and in order to maintain the elementary character of these notes, we will only establish the continuity of functions defined by power series. The interested reader can find the relevant results and definitions in [11].

Proof For a fixed $z \in D(0; R)$, take $w \in \mathbb{C}$ such that $|w - z| < \frac{1}{2}(R - |z|)$. Then, $|w| < r$, with $r = \frac{1}{2}(R + |z|) < R$, and for $k \in \mathbb{N}$ we have

$$
\begin{aligned}
|w^k - z^k| &= |w - z||w^{k-1} + w^{k-2}z + \cdots + wz^{k-2} + z^{k-1}| \\
&\leq |w - z|(|w|^{k-1} + |w|^{k-2}|z| + \cdots + |w||z|^{k-2} + |z|^{k-1}) \\
&\leq |w - z|(r^{k-1} + r^{k-2} \cdot r + \cdots + r \cdot r^{k-2} + r^{k-1}) \\
&= kr^{k-1}|w - z|.
\end{aligned}
$$

Hence, under the above restriction on w, Problem 4 yields

$$
\begin{aligned}
|f(w) - f(z)| &= \left| \sum_{k \geq 0} a_k w^k - \sum_{k \geq 0} a_k z^k \right| = \left| \sum_{k \geq 0} a_k(w^k - z^k) \right| \\
&= \left| \sum_{k \geq 1} a_k(w^k - z^k) \right| \leq \sum_{k \geq 1} |a_k||w^k - z^k| \\
&\leq \frac{1}{r} \left(\sum_{k \geq 1} k|a_k|r^k \right) |w - z|.
\end{aligned}
$$

Now, fix a real number c such that $1 < c < \frac{R}{r}$. Since $\sqrt[k]{|a_k|} < \frac{1}{R}$ for $k \geq 0$ and $\sqrt[k]{k} \to 1$ (cf. Example 7.23 of [8], for instance), there exists $k_0 \in \mathbb{N}$ such that $\sqrt[k]{k|a_k|} < \frac{c}{R}$ for $k > k_0$. Hence,

$$
\sum_{k \geq k_0} k|a_k|r^k < \sum_{k \geq k_0} \left(\frac{cr}{R} \right)^k := C < +\infty,
$$

for $0 < \frac{cr}{R} < 1$. Finally, letting $C' = \sum_{k=1}^{k_0} k|a_k|r^k$, it comes that

$$
\begin{aligned}
|f(w) - f(z)| &\leq \frac{1}{r} \left(\sum_{k=1}^{k_0} k|a_k|r^k \right) |w - z| + \frac{1}{r} \left(\sum_{k > k_0} k|a_k|r^k \right) |w - z| \\
&\leq \frac{C'}{r}|w - z| + \frac{C}{r}|w - z|.
\end{aligned}
$$

In short, we have shown that

$$
|w - z| < \frac{1}{2}(R - |z|) \Rightarrow |f(w) - f(z)| < A|w - z|,
$$

for some positive constant A. But this being the case, Problem 8 shows that $z_n \to z \Rightarrow f(z_n) \to f(z)$, and the fact that $z \in D(0; R)$ was arbitrarily chosen guarantees the continuity of f. $\qquad\square$

The coming corollary assures that if a function $f : D(0; R) \to \mathbb{C}$ is given by a power series, then such a series is uniquely determined.

Corollary 21.18 *Let $\sum_{k\geq 0} a_k z^k$ and $\sum_{k\geq 0} b_k z^k$ be convergent power series in the open disk $D(0; R)$. If $\sum_{k\geq 0} a_k z^k = \sum_{k\geq 0} b_k z^k$ for every $z \in D(0; R)$, then $a_k = b_k$ for every $k \geq 0$.*

Proof By evaluating equality $\sum_{k\geq 0} a_k z^k = \sum_{k\geq 0} b_k z^k$ at $z = 0$, we obtain $a_0 = b_0$. Then, cancelling out $a_0 = b_0$ in both sides of $\sum_{k\geq 0} a_k z^k = \sum_{k\geq 0} b_k z^k$, we get $\sum_{k\geq 1} a_k z^k = \sum_{k\geq 1} b_k z^k$ for $z \in D(0; R)$ and, hence, $\sum_{k\geq 1} a_k z^{k-1} = \sum_{k\geq 1} b_k z^{k-1}$ for $z \in D(0; R) \setminus \{0\}$. However, since both sides of this last equality define continuous functions on $D(0; R)$, we conclude that $\sum_{k\geq 1} a_k z^{k-1} = \sum_{k\geq 1} b_k z^{k-1}$ for every $z \in D(0; R)$. Now, looking at such an equality at $z = 0$, we obtain $a_1 = b_1$. By proceeding this way, an easy induction gives $a_k = b_k$ for every $k \geq 0$. $\square$

As a first application of the ideas developed here, we now present a proof of the identities of Proposition 16.17 with the aid of **complex generating functions**.

Example 21.19 For $z_1, \ldots, z_n \in \mathbb{C}$, if $s_i = s_i(z_1, \ldots, z_n)$ denotes the i-th elementary symmetric sum of $z_1, \ldots, z_n$ and $\sigma_k = z_1^k + \cdots + z_n^k$, then:

(a) $\sigma_{n+k} = \sum_{j=1}^{n} (-1)^{j-1} s_j \sigma_{n+k-j}$ for $k \geq 1$.

(b) $s_{k+1} = \frac{1}{k+1} \sum_{j=1}^{k+1} (-1)^{j-1} s_{k+1-j} \sigma_j$ for $1 \leq k \leq n-1$.

Proof Let

$$f(X) = \prod_{j=1}^{n} (1 + z_j X) = \sum_{j=0}^{n} s_j X^j.$$

On the one hand, we have $f'(X) = \sum_{j=1}^{n} j s_j X^{j-1}$; on the other, for $z \in D(0; R)$, with

$$R = \min\{|z_j|^{-1}; \ 1 \leq j \leq n \text{ and } z_j \neq 0\} > 0,$$

Problem 3, page 393, gives

$$f'(z) = \left(\sum_{i=1}^{n} z_i(1 + z_i z)^{-1}\right) f(z) = \sum_{i=1}^{n} \left(\sum_{j\geq 0} (-1)^j z_i^{j+1} z^j\right) f(z)$$

$$= \left(\sum_{j\geq 0} \sum_{i=1}^{n} (-1)^j z_i^{j+1} z^j\right) f(z)$$

$$= \left(\sum_{j\geq 0} (-1)^j \sigma_{j+1} z^j\right) \left(\sum_{l=0}^{n} s_l z^l\right).$$

We now apply the previous corollary in both cases, looking at both expressions for $f'(z)$ as complex power series. In (a), computing the coefficient of z^{k-1} in both expressions for $f'(z)$ we obtain

$$ks_k = \sum_{j+l=k-1} (-1)^j \sigma_{j+1} s_l = \sum_{l=0}^{k-1} (-1)^{k-l-1} \sigma_{k-l} s_l$$

so that

$$(-1)^k k s_k - \sum_{l=0}^{k-1} (-1)^{l+1} \sigma_{k-l} s_l = 0.$$

Doing the same in (b), we get

$$0 = \sum_{j+l=k-1} (-1)^j \sigma_{j+1} s_l = \sum_{l=0}^{n} (-1)^{k-1-l} \sigma_{k-l} s_l$$

and, hence,

$$\sum_{l=0}^{n} (-1)^l \sigma_{k-l} s_l = 0.$$

□

We finish this section by proving a result that extends, for continuous functions $f : \overline{D(a; R)} \to \mathbb{C}$, Theorem 8.26 of [8].

Theorem 21.20 (Weierstrass[3]) *If* $f : \overline{D(a; R)} \to \mathbb{C}$ *is a continuous function, then there exist* z_m *and* z_M *in* $\overline{D(a; R)}$ *such that*

$$|f(z_m)| = \min\{|f(z)|; \ z \in \overline{D(a; R)}\}$$

and

$$|f(z_M)| = \max\{|f(z)|; \ z \in \overline{D(a; R)}\}.$$

Proof Let $(z_n)_{n \geq 1}$ be a sequence in $\overline{D(a; R)}$ such that

$$f(z_n) \to \sup\{|f(z)|; \ z \in \overline{D(a; R)}\}$$

(here, in principle we do not exclude the possibility that such a supremum could be $+\infty$). Write $z_n = x_n + iy_n$, with $x_n, y_n \in \mathbb{R}$. Since $|z_n| \leq R$ for every $n \geq 1$, we have $|x_n|, |y_n| \leq R$ for every $n \geq 1$. Therefore, Theorem 7.25 of [8] guarantees the existence of an infinite set $\mathbb{N}_1 \subset \mathbb{N}$ such that $(x_n)_{n \in \mathbb{N}_1}$ converges to some $x \in \mathbb{R}$. Hence, by the same result we can take a second infinite set $\mathbb{N}_2 \subset \mathbb{N}_1$ such that $(y_n)_{n \in \mathbb{N}_2}$ converges to some $y \in \mathbb{R}$. However, since $(x_n)_{n \in \mathbb{N}_2}$ is a subsequence of $(x_n)_{n \in \mathbb{N}_1}$, we conclude that $(x_n)_{n \in \mathbb{N}_2}$ still converges to x.

Letting $z_M = x + iy$, it follows from Lemma 21.9 that $(z_n)_{n \in \mathbb{N}_2}$ converges to z_M. Hence, Problem 1 gives $|z_M| \leq R$, so that $z_M \in \overline{D(a; R)}$. Now, invoking the

---

[3]Karl Weierstrass, German mathematician of the nineteenth century.

continuity of f, we have

$$f(z_M) = \lim_{\substack{n \to +\infty \\ n \in \mathbb{N}_2}} f(z_n) = \sup\{|f(z)|; \ z \in \overline{D(a; R)}\}.$$

In particular,

$$\sup\{|f(z)|; \ z \in \overline{D(a; R)}\} = \max\{|f(z)|; \ z \in \overline{D(a; R)}\}.$$

The proof of the first part of the theorem is analogous and will be left as an exercise to the reader. □

The result of the following corollary was used in the proof of the Fundamental Theorem of Algebra.

Corollary 21.21 *If $f : \mathbb{C} \to \mathbb{C}$ is a complex polynomial function, then, given $R > 0$, there exist z_m and z_M in $\overline{D(a; R)}$ such that*

$$|f(z_m)| = \min\{|f(z)|; \ z \in \overline{D(a; R)}\}$$

and

$$|f(z_M)| = \max\{|f(z)|; \ z \in \overline{D(a; R)}\}.$$

Proof Example 21.16 and Problem 7 guarantee that $|f| : \mathbb{C} \to [0, +\infty)$ is a continuous function. Thus, so is the function $|f| : \overline{D(a; R)} \to [0, +\infty)$. It now suffices to apply Weierstrass' theorem. □

Problems: Sect. 21.2

1. * Let $(z_n)_{n \geq 1}$ be a sequence in $\mathbb{C}$, converging to $z \in \mathbb{C}$. Prove that $(z_n)_{n \geq 1}$ is **bounded**, i.e., that there exists $M > 0$ such that $|z_n| < M$ for every $n \geq 1$. Moreover, if $|z_n| \leq R$ for every $n \geq 1$, show that $|z| \leq R$.

   For the next problem, we say that a sequence of complex numbers is **divergent** if it is not convergent.

2. If $|z| > 1$ and $z_n = z^n$ for every $n \geq 1$, show that the sequence $(z_n)_{n \geq 1}$ is divergent.

3. * Let $(z_n)_{n \geq 1}$ and $(w_n)_{n \geq 1}$ be sequences in $\mathbb{C}$, converging to z and w, respectively. Prove that:

   (a) If $a \in \mathbb{C}$, then $a z_n \to a z$.
   (b) $z_n \pm w_n \to z \pm w$.
   (c) $z_n w_n \to z w$.
   (d) If $w_n, w \neq 0$, then $z_n / w_n \to z/w$.

4. * If $a, b \in \mathbb{C}$ and $\sum_{k \geq 1} z_k$ and $\sum_{k \geq 1} w_k$ are convergent series of complex numbers, show that $\sum_{k \geq 1} (az_k + bw_k)$ is also convergent, with

$$\sum_{k \geq 1} (az_k + bw_k) = a \sum_{k \geq 1} z_k + b \sum_{k \geq 1} w_k.$$

5. * Extend the comparison test for series to series of complex numbers. More precisely, let $(a_n)_{n \geq 0}$ be a sequence of complex numbers. If there exist positive reals c and M such that $|a_n| \leq cM^n$ for every $n \geq 0$, then the power series $\sum_{n \geq 0} a_n z^n$ converges on the interval $\left(-\frac{1}{M}, \frac{1}{M} \right)$.

6. * Let $\emptyset \neq X \subset Y \subset \mathbb{C}$ and $f : Y \to \mathbb{C}$ be a continuous function. Prove that $f_{|X} : X \to \mathbb{C}$ is also continuous.

7. * Given $\emptyset \neq X \subset \mathbb{C}$ and a continuous function $f : X \to \mathbb{C}$, prove that $|f| : X \to [0, +\infty)$ is also continuous.

8. * Let $\emptyset \neq X \subset \mathbb{C}$ and $f : X \to \mathbb{C}$ be a function satisfying the following condition: given $z \in X$, there exist $A, B > 0$ (at first depending on z), such that

$$w \in X, \ |w - z| < B \Rightarrow |f(w) - f(z)| < A|w - z|.$$

   Show that f is continuous.

9. * If $a \in \mathbb{C} \setminus \{0\}$ and $m \in \mathbb{N}$, show that for $z \in D(0; \frac{1}{|a|})$ we have

$$\frac{1}{(1 - az)^m} = \sum_{n \geq 0} \binom{n + m - 1}{m - 1} a^n z^n.$$

21.3   The General Case

With the material of the previous section at our disposal, we can finally face the discussion of the general case of (21.1), i.e., that in which the complex roots of the characteristic polynomial (21.2) are not necessarily distinct. To this end, we shall use complex generating functions again. The fundamental result is given by the coming

Theorem 21.22 *Let $(a_n)_{n \geq 1}$ be a sequence of complex numbers satisfying, for every $n \geq 1$, the linear recurrence relation*

$$a_{n+k} = u_{k-1} a_{n+k-1} + \cdots + u_0 a_n,$$

with $u_0, \ldots, u_{k-1}$ being given complex numbers and $u_0 \neq 0$. Let $z_1, \ldots, z_l$ be the pairwise distinct roots of the characteristic polynomial (21.2) of $(a_n)_{n \geq 1}$, with multiplicities respectively equal to $m_1, \ldots, m_l$. Then, for every $n \geq 1$, we have

$$a_n = p_1(n - 1)z_1^{n-1} + \cdots + p_l(n - 1)z_l^{n-1},$$

where $p_1, \ldots, p_l \in \mathbb{C}[X]$ are polynomials of degrees less than or equal to $m_1 - 1$, $\ldots$, $m_l - 1$, respectively, and totally determined by the values of $a_1, \ldots, a_k$.

Proof We first claim that there exists a constant $R_0 > 0$ such that $|a_n| \leq R^n$ for every $n \geq 1$ and $R > R_0$. Indeed, if $|a_n| \leq R^n$ for $1 \leq n < m$, with $m > k$, then

$$|a_m| = |u_{k-1}a_{m-1} + \cdots + u_1 a_{m-k+1} + u_0 a_{m-k}|$$

$$\leq |u_{k-1}||a_{m-1}| + \cdots + |u_1||a_{m-k+1}| + |u_0||a_{m-k}|$$

$$\leq |u_{k-1}|R^{m-1} + \cdots + |u_1|R^{m-k+1} + |u_0|R^{m-k}$$

$$= R^{m-k}(|u_{k-1}|R^{k-1} + \cdots + |u_1|R + |u_0|).$$

Hence, if $g(X) = X^k - |u_{k-1}|X^{k-1} - \cdots - |u_1|X - |u_0|$ and $R_0 > 0$ is such that $g(R) > 0$ for $R > R_0$, then, for each one of such R's, the estimates above give

$$|a_m| \leq R^{m-k}(|u_{k-1}|R^{k-1} + \cdots + |u_1|R + |u_0|)$$

$$\leq R^{m-k} \cdot R^k = R^m.$$

It thus suffices to choose, from the very beginning, $R_0 > 0$ such that $|a_1|, \ldots, |a_k| \leq R_0$ and $g(R) > 0$ for every $R > R_0$.

Now, fix $R > R_0$ and let

$$F(z) = \sum_{n \geq 1} a_n z^n.$$

Since $|a_n| \leq R^n$ for every $n \geq 1$, the comparison test for series (cf. Problem 5, page 520) guarantees the convergence of F in the open disk $D(0; \frac{1}{R})$ of the complex plane. Let $f(X) = X^k - u_{k-1}X^{k-1} - \cdots - u_1 X - u_0$ be the characteristic polynomial of $(a_n)_{n \geq 1}$ and $h(X) = -u_0 X^k - u_1 X^{k-1} - \cdots - u_{k-1}X + 1$ its *reciprocal* ($\partial h = k$, since $u_0 \neq 0$). For $z \in D(0; \frac{1}{R})$, we have

$$h(z)F(z) = -\left(\sum_{j=0}^{k-1} u_j z^{k-j}\right)\left(\sum_{n \geq 1} a_n z^n\right) + \sum_{n \geq 1} a_n z^n$$

$$= -\sum_{j=0}^{k-1}\sum_{n \geq 1} u_j a_n z^{n+k-j} + \sum_{n \geq 1} a_n z^n$$

$$= -\sum_{j=0}^{k-1}\sum_{n \geq k-j+1} u_j a_{n-k+j} z^n + \sum_{n \geq 1} a_n z^n.$$

For $j \geq 2$, write

$$\sum_{n \geq k-j+1} u_j a_{n-k+j} z^n = \sum_{n=k-j+1}^{k-1} u_j a_{n-k+j} z^n + \sum_{n \geq k} u_j a_{n-k+j} z^n$$

and analogously for $\sum_{n \geq 1} a_n z^n$. Then, we obtain

$$h(z)F(z) = -\sum_{n \geq k+1} u_0 a_{n-k} z^n - \sum_{n \geq k} u_1 a_{n-k+1} z^n$$

$$- \sum_{j=2}^{k-1} \left(\sum_{n=k-j+1}^{k-1} u_j a_{n-k+j} z^n + \sum_{n \geq k} u_j a_{n-k+j} z^n \right)$$

$$+ \sum_{n=1}^{k-1} a_n z^n + \sum_{n \geq k} a_n z^n$$

$$= -\sum_{n \geq k+1} u_0 a_{n-k} z^n - \sum_{n \geq k} u_1 a_{n-k+1} z^n$$

$$- \sum_{j=2}^{k-1} \sum_{n \geq k} u_j a_{n-k+j} z^n + \sum_{n \geq k} a_n z^n$$

$$- \sum_{j=2}^{k-1} \sum_{n=k-j+1}^{k-1} u_j a_{n-k+j} z^n + \sum_{n=1}^{k-1} a_n z^n.$$

However, since

$$\sum_{j=2}^{k-1} \sum_{n \geq k} u_j a_{n-k+j} z^n = \sum_{n \geq k} \sum_{j=2}^{k-1} u_j a_{n-k+j} z^n,$$

it follows from (21.1) that

$$-\sum_{j=2}^{k-1} \sum_{n \geq k} u_j a_{n-k+j} z^n + \sum_{n \geq k} a_n z^n = \sum_{n \geq k} \left(-\sum_{j=2}^{k-1} u_j a_{n-k+j} + a_n \right) z^n$$

$$= \left(-\sum_{j=2}^{k-1} u_j a_j + a_k \right) z^k + \sum_{n \geq k+1} \left(-\sum_{j=2}^{k-1} u_j a_{n-k+j} + a_n \right) z^n$$

$$= \left(-\sum_{j=2}^{k-1} u_j a_j + a_k \right) z^k + \sum_{n \geq k+1} \left(u_0 a_{n-k} + u_1 a_{n-k+1} \right) z^n.$$

Therefore,

$$h(z)F(z) = -\sum_{n \geq k+1} u_0 a_{n-k} z^n - \sum_{n \geq k} u_1 a_{n-k+1} z^n$$

$$+ \left(-\sum_{j=2}^{k-1} u_j a_j + a_k \right) z^k + \sum_{n \geq k+1} (u_0 a_{n-k} + u_1 a_{n-k+1}) z^n$$

$$- \sum_{j=2}^{k-1} \sum_{n=k-j+1}^{k-1} u_j a_{n-k+j} z^n + \sum_{n=1}^{k-1} a_n z^n$$

$$= \left(a_k - \sum_{j=1}^{k-1} u_j a_j \right) z^k - \sum_{j=2}^{k-1} \sum_{n=k-j+1}^{k-1} u_j a_{n-k+j} z^n + \sum_{n=1}^{k-1} a_n z^n,$$

so that

$$h(z)F(z) = zp(z)$$

for $z \in D(0; \frac{1}{R})$, with $p \in \mathbb{C}[X]$ being a nonzero polynomial of degree $\partial p \leq k - 1$.

Since $h(0) = 1 \neq 0$, by increasing R, if needed, we can assume that $h(z) \neq 0$ for $z \in D(0; \frac{1}{R})$. On the other hand, since

$$f(X) = (X - z_1)^{m_1} \dots (X - z_l)^{m_l},$$

we have

$$h(X) = (1 - z_1 X)^{m_1} \dots (1 - z_l X)^{m_l},$$

so that

$$F(z) = \frac{zp(z)}{(1 - z_1 z)^{m_1} \dots (1 - z_l z)^{m_l}} \tag{21.6}$$

for $z \in D(0; \frac{1}{R})$.

Now, observe that

$$\partial p < k = \partial((1 - z_1 z)^{m_1} \dots (1 - z_l z)^{m_l}).$$

Hence, by applying to (21.6) the partial fractions decomposition formula (cf. Problem 9, page 457), we conclude the existence, for $1 \leq j \leq l$ and $1 \leq n_j \leq m_j$, of constants d_{jn_j}, uniquely determined by the coefficients of p (and, thus, by $a_1, a_2,$ $\dots, a_k$ and $u_0, u_1, \dots, u_{k-1}$), such that

$$F(z) = z \sum_{j=1}^{l} \sum_{n_j=1}^{m_j} \frac{d_{jn_j}}{(1 - z_j z)^{n_j}} \tag{21.7}$$

for $z \in D(0; \frac{1}{R})$.

However, if $r = \min\{\frac{1}{R}, \frac{1}{|z_1|}, \ldots, \frac{1}{|z_l|}\}$, then the result of Problem 9, page 520, assures that

$$\frac{1}{(1 - z_j z)^{n_j}} = \sum_{n \geq 0} \binom{n + n_j - 1}{n_j - 1} z_j^n z^n$$

for $1 \leq j \leq l$, $1 \leq n_j \leq m_j$ and $z \in D(0; r)$. Hence, it follows from (21.7) that, for $|z| < r$,

$$F(z) = z \sum_{j=1}^{l} \sum_{n_j=1}^{m_j} \sum_{n \geq 0} d_{jn_j} \binom{n + n_j - 1}{n_j - 1} z_j^n z^n$$

$$= \sum_{n \geq 0} \left(\sum_{j=1}^{l} \sum_{n_j=1}^{m_j} d_{jn_j} \binom{n + n_j - 1}{n_j - 1} z_j^n \right) z^{n+1}$$

$$= \sum_{n \geq 1} \left(\sum_{j=1}^{l} \sum_{n_j=1}^{m_j} d_{jn_j} \binom{n + n_j - 2}{n_j - 1} z_j^{n-1} \right) z^n.$$

Finally, since a_n is the coefficient of z^n in the defining series for F, the last equality above, together with Corollary 21.18, gives

$$a_n = \sum_{j=1}^{l} \sum_{n_j=1}^{m_j} d_{jn_j} \binom{n + n_j - 2}{n_j - 1} z_j^{n-1}$$

$$= \sum_{j=1}^{l} \sum_{n_j=1}^{m_j} \frac{d_{jn_j}}{(n_j - 1)!} (n + n_j - 2)(n + n_j - 3) \ldots (n + 1) n z_j^{n-1}$$

$$= \sum_{j=1}^{l} p_j(n - 1) z_j^{n-1},$$

where

$$p_j(X) = \sum_{n_j=1}^{m_j} \frac{d_{jn_j}}{(n_j-1)!}(X+n_j-1)(X+n_j-2)\ldots(X+1)z_j^{n-1}$$

$$= c_{j,m_j-1}X^{m_j-1} + \cdots + c_{j1}X + c_{j0}$$

is a polynomial of degree at most $m_j - 1$. □

In order to present a relevant application of the previous result, we first need a definition. Throughout the rest of this section, AP stands for *arithmetic progression*.

Definition 21.23 Given an integer $m > 1$, we call a sequence $(a_n)_{n\geq 1}$ an:

(a) **AP of order 1** if it is an ordinary AP.
(b) **AP of order m** if the sequence $(b_n)_{n\geq 1}$, given for $n \geq 1$ by $b_n = a_{n+1} - a_n$, is an AP of order $m - 1$.

The coming lemma characterizes AP's of order m by means of a linear recurrence relation that generalizes those satisfied by AP's of orders 1 and 2 (cf. Sections 3.1 and 3.3 of [8]).

Lemma 21.24 *A sequence $(a_n)_{n\geq 1}$ is an AP of order m if and only if*

$$\binom{m+1}{0}a_{n+m+1} - \binom{m+1}{1}a_{n+m} + \cdots + (-1)^{m+1}\binom{m+1}{m+1}a_n = 0, \quad (21.8)$$

for every $n \geq 1$.

Proof Firstly, let $(a_n)_{n\geq 1}$ be an AP of order m. If $m = 1$, then $(a_n)_{n\geq 1}$ is an ordinary AP, and (21.8) reduces to $a_{n+2} - 2a_{n+1} + a_n = 0$ for $n \geq 1$, the linear recurrence relation that characterizes ordinary AP's.

By induction hypothesis, assume that (21.8) is valid for $m = k - 1$. For $m = k$, the sequence $(b_n)_{n\geq 1}$ given by $b_n = a_{n+1} - a_n$ is, by definition, an AP of order $k - 1$. Therefore, the induction hypothesis gives

$$\binom{k}{0}b_{n+k} - \binom{k}{1}b_{n+k-1} + \cdots + (-1)^k\binom{k}{k}b_n = 0 \quad (21.9)$$

for every $n \geq 1$, or

$$\binom{k}{0}(a_{n+k+1} - a_{n+k}) - \binom{k}{1}(a_{n+k} - a_{n+k-1}) + \cdots$$

$$\cdots + (-1)^k\binom{k}{k}(a_{n+1} - a_n) = 0 \quad (21.10)$$

for every $n \geq 1$; in turn, this is the same as

$$\binom{k}{0}a_{n+k+1} - \left(\binom{k}{0} + \binom{k}{1}\right)a_{n+k} + \left(\binom{k}{1} + \binom{k}{2}\right)a_{n+k-1} - \cdots$$

$$\cdots + (-1)^k \left(\binom{k}{k-1} + \binom{k}{k}\right)a_{n+1} + (-1)^{k+1}\binom{k}{k}a_n = 0. \tag{21.11}$$

By applying Stifel's relation, together with the facts that $\binom{k}{0} = \binom{k+1}{0}$ and $\binom{k}{k} = \binom{k+1}{k+1}$, we get (21.8) for $m = k$.

Conversely, let $(a_n)_{n \geq 1}$ be a sequence for which (21.8) holds. If $m = 1$, then $a_{n+2} - 2a_{n+1} + a_n = 0$ for $n \geq 1$, and we already know that $(a_n)_{n \geq 1}$ is an AP. By induction hypothesis, suppose that the validity of (21.8) for $m = k - 1$ implies that $(a_n)_{n \geq 1}$ is an AP of order $k - 1$. Then, consider a sequence $(a_n)_{n \geq 1}$ satisfying (21.8) for $m = k$, i.e., such that

$$\binom{k+1}{0}a_{n+k+1} - \binom{k+1}{1}a_{n+k} + \cdots + (-1)^{k+1}\binom{k+1}{k+1}a_n = 0$$

for $n \geq 1$. Using Stifel's relation to write $\binom{k+1}{j} = \binom{k}{j} + \binom{k}{j-1}$ for $1 \leq j \leq k$, together with $\binom{k+1}{0} = \binom{k}{0}$ and $\binom{k+1}{k+1} = \binom{k}{k}$, we successively reobtain (21.11), (21.10) and (21.9), for every $n \geq 1$. Hence, the induction hypothesis assures that $(b_n)_{n \geq 1}$ is an AP of order $k - 1$, so that, by definition, $(a_n)_{n \geq 1}$ is an AP of order k. □

We are finally in position to present the promised application of Theorem 21.22.

Example 21.25 If $(a_n)_{n \geq 1}$ is an AP of order m, then (21.8) holds for every $n \geq 1$, so that the characteristic polynomial of $(a_n)_{n \geq 1}$ is

$$f(X) = \binom{m+1}{0}X^{m+1} - \binom{m+1}{1}X^m + \cdots + (-1)^{m+1}\binom{m+1}{m+1}$$

$$= (X - 1)^{m+1}.$$

Theorem 21.22 thus guarantees the existence of constants $\alpha_0, \alpha_1, \ldots, \alpha_m$ such that

$$a_n = \alpha_0 + \alpha_1(n - 1) + \cdots + \alpha_m(n - 1)^m$$

for every $n \geq 1$. Evaluating the above relation for $n = 1$, we get $\alpha_0 = a_1$; for $n = 2$, $\ldots, m + 1$, we obtain $\alpha_1, \ldots, \alpha_m$ as the solution to the linear system of equations

$$\begin{cases} \alpha_1 + \alpha_2 + \cdots + \alpha_m & = a_2 - a_1 \\ 2\alpha_1 + 2^2\alpha_2 + \cdots + 2^m\alpha_m & = a_3 - a_1 \\ \cdots & \cdots \\ m\alpha_1 + m^2\alpha_2 + \cdots + m^m\alpha_m & = a_{m+1} - a_1 \end{cases}.$$

Notice that the fact that such a linear system always has a single solution is an immediate consequence[4] of the last part of the statement of Theorem 21.22.

Problems: Sect. 21.3

1. Let k be a given natural number and $(a_n)_{n\geq 1}$ be a sequence of complex numbers such that

$$a_n = \frac{1}{2}(a_{n-k} + a_{n+k}),$$

for every natural number $n > k$. Show that

$$a_n = \sum_{j=1}^{k}(A_j + (n-1)B_j)\omega^{j(n-1)}$$

for every $n \geq 1$, where $\omega = \operatorname{cis}\frac{2\pi}{k}$ and $A_1, \ldots, A_k, B_1, \ldots, B_k$ are certain complex constants.

2. Find all real sequences $(a_n)_{n\geq 1}$ such that $a_1 = 1$,

$$a_{n+3} = 5a_{n+2} - 8a_{n+1} + 4a_n$$

for every $n \geq 1$ and $\frac{a_n}{2^n} \xrightarrow{n} 3$.

---

[4]This also follows from basic Linear Algebra, for the transpose of the matrix of the coefficients of the linear system is a Vandermonde matrix of nonvanishing determinant.

Chapter 22
Hints and Solutions

Section 1.1

1. If there exists a bijection $f : I_1 \to I_n$, show that $n = 1$. Now, assume that
   $m, n > 1$ and there exists a bijection $f : I_m \to I_n$. Letting $k = f(m)$, show
   that there exists a bijection $g : I_n \setminus \{k\} \to I_{n-1}$, so that $g \circ f_{|I_{m-1}} : I_{m-1} \to I_{n-1}$
   is also a bijection. Then, use an inductive argument to deduce that $m-1 = n-1$.
2. If $g : I_m \to A$ and $h : I_n \to B$ are bijections, define $f : I_{m+n} \to A \cup B$ by
   letting

$$f(k) = \begin{cases} g(k), & \text{if } 1 \le k \le m \\ h(k-m), & \text{if } m+1 \le k \le m+n \end{cases}.$$

Since g and h are surjective and $m+1 \le k \le m+n$ if and only if $1 \le k-m \le n$,
we conclude that f is also surjective. For the injectivity of f, suppose that
$f(k) = f(l)$, with $1 \le k, l \le m+n$. There are four cases to consider:

- $1 \le k, l \le m$: then $f(k) = f(l) \Rightarrow g(k) = g(l)$, and the injectivity of g
  gives $k = l$.
- $m+1 \le k, l \le m+n$: then $f(k) = f(l) \Rightarrow h(k-m) = h(l-m)$, and the
  injectivity of h gives $k - m = l - m$, thus $k = l$.
- $1 \le k \le m$ and $m+1 \le l \le m+n$: then $f(k) = f(l) \Rightarrow g(k) = h(l-m) \in$
  $A \cap B$, a contradiction to the fact that A and B are disjoint. Therefore, in this
  case it is always the case that $f(k) \ne f(l)$.
- $m+1 \le k \le m+n$ and $1 \le l \le m$: as in the previous case, we always have
  $f(k) \ne f(l)$.

3. By induction, if $k < n$ and $|\cup_{j=1}^{k} A_j| = \sum_{j=1}^{k} |A_j|$, then

$$\left|\cup_{j=1}^{k+1} A_j\right| = \left|\left(\cup_{j=1}^{k} A_j\right) \cup A_{k+1}\right| = \left|\cup_{j=1}^{k} A_j\right| + |A_{k+1}|$$

$$= \sum_{j=1}^{k} |A_j| + |A_{k+1}| = \sum_{j=1}^{k+1} |A_j|.$$

4. Let's make induction on $|A| = |B|$, the case $|A| = |B| = 1$ being trivial. Suppose, by induction hypothesis, that the result is true when $|A| = |B| = n$, for some $n \in \mathbb{N}$, and let A and B be such that $|A| = |B| = n + 1$. Given $x \in A$ and a function $f : A \to B$, let $y = f(x)$. Then $|A \setminus \{x\}| = |B \setminus \{y\}| = n$ and f restricts to a function $g : A \setminus \{x\} \to B \setminus \{y\}$. By the inductive hypothesis g is injective if and only if it's surjective. However, since f is injective (resp. surjective) if and only if g is injective (resp. surjective), it follows that f is injective if and only if it's surjective.

5. If $(a, b) = (c, d)$, then $\{\{a\}, \{a, b\}\} = \{\{c\}, \{c, d\}\}$. If $a = b$, then $\{\{a\}, \{a, b\}\} = \{\{a\}, \{a, a\}\} = \{\{a\}, \{a\}\} = \{\{a\}\}$, so that $\{\{c\}, \{c, d\}\}$ must have just one element. Hence, $c = d$ and $\{\{a\}\} = \{\{c\}\}$, so that $a = c$ and then $a = b = c = d$. The case $a \neq b$ can be treated analogously, and will be left to the reader.

6. For the first part, if $B = B_1 \cup \ldots \cup B_n$, $a \in A$ and $b \in B$, show that the ordered pair (a, b) belongs to the left hand side if and only if it belongs to the right hand side.

7. By induction, if $k < n$ and $|A_1 \times \cdots \times A_k| = \prod_{j=1}^{k} |A_j|$, then

$$|A_1 \times \cdots \times A_k \times A_{k+1}| = |(A_1 \times \cdots \times A_k) \times A_{k+1}|$$

$$= |A_1 \times \cdots \times A_k| \cdot |A_{k+1}|$$

$$= \prod_{j=1}^{k} |A_j| \cdot |A_{k+1}| = \prod_{j=1}^{k+1} |A_j|.$$

Now, observe that Proposition 1.6 deals with the initial case.

8. If $A = \{a_1, \ldots, a_n\}$, then, for a fixed integer $1 \leq i \leq n$, there are exactly k possible choices for an index $1 \leq j \leq k$ such that $a_i \in A_j$. Now, use the fundamental principle of counting.

9. After n moves, the paws will be at a point (a, b) such that $a = x_1 + x_2 + \cdots + x_n$ and $b = y_1 + y_2 + \cdots + y_n$, where $x_i, y_i \in \{-1, 1\}$ for $1 \leq i \leq n$. Hence, we have (i) $-n \leq a, b \leq n$ and (ii) a and b have the same *parity* of n (this is, either a and b are odd, if n is odd, or even, if n is even). Now, first certify yourself that any point (a, b) satisfying conditions (i) and (ii) can be a final destination of the pawn; then, use the fundamental principle of counting to show that there are exactly $(n + 1)^2$ of such points.

10. For item (a), if $(a_1, \ldots, a_k)$ is such a sequence, then $a_1, \ldots, a_k \in \{j, j + 1, \ldots, n\}$, so that there are exactly $n - j + 1$ possibilities for each a_i; then, apply the fundamental principle of counting. For item (b), observe that the set of the wished sequences is the difference between the set of the sequences considered in item (a) and the set of the sequences of k elements of I_n (possibly with repetitions) such that the smallest term of the sequence is greater than or equal to $j + 1$. Finally, by item (b) the sum of item (c) is equal to $\sum_{j=1}^{n} j((n - j + 1)^k - (n - j)^k)$; verify that such a sum equals $1^k + 2^k + \cdots + n^k$.

11. For $k \geq 3$, let I be an odd integer such that $1 \leq I \leq 2^{k-1} - 1$. For each of these I's, let $A_I = \{2^t I \in A\}$, so that

$$A = \bigcup_{I=1}^{2^{k-1}-1} A_I \cup \{2^{k-1} + 1, 2^{k-1} + 3, \ldots, 2^k - 1\},$$

with the index I varying over the set of odd integers from 1 to $2^{k-1} - 1$. From this, conclude that if we wish a subset $X \subset A$ with the stated property to have the greatest possible number of elements, then: (i) $\{2^{k-1} + 1, 2^{k-1} + 3, \ldots, 2^k - 1\} \subset X$; (ii) for each odd integer $1 \leq I \leq 2^{k-1} - 1$, at most half or $\lfloor \text{half} \rfloor + 1$ of the elements of A_I can be in X. Then, refine the analysis in (ii), showing that if $a \in \mathbb{N}$ and $I \geq 3$ is an odd integer satisfying $2^a < I < 2^{a+1}$, then at most $\lfloor \frac{k-a-1}{2} \rfloor + 1$ elements of A_I belong to X. From this, conclude that X has at most

$$S = 2^{k-2} + \left\lfloor \frac{k}{2} \right\rfloor + 1 + \sum_{a=1}^{k-2} 2^{a-1}\left(\left\lfloor \frac{k-a-1}{2} \right\rfloor + 1\right)$$

elements. Finally, separately consider the cases k even and k odd to show that $S = \frac{2^{k+1} + (-1)^k}{3}$.

Section 1.2

1. Exhibit a bijection from the family of the subsets of I_n containing n to the family of subsets of I_{n-1}.

2. For each $x \in A$, show that there are $2^n - 1$ ways of choosing the subsets A_i which contain x.

3. The n straight lines pairwise intersect at $\binom{n}{2} = \frac{n(n-1)}{2}$ distinct points. Choose a line r which is not parallel to any straight line passing through two of these $\binom{n}{2}$ points. There are two kinds of regions: those which possess a highest point with respect to r—which is necessarily one of the $\binom{n}{2}$ points of intersection of two of the n given lines—and those which do not possess a highest point. Show that there is a bijection between the regions of the first kind and the set formed by the

$\binom{n}{2}$ points; then, show that there are exactly $n + 1$ regions of the second kind, so that the total number of regions is equal to $\frac{n(n-1)}{2} + (n + 1)$.

4. For item (a), consider the family of subsets of I_n containing n. For item (b), let $\mathcal{F}$ be an intersecting system in I_n and denote by $\mathcal{F}_c$ the family $\mathcal{F}_c = \{A^c;\ A \in \mathcal{F}\}$. Note that $|\mathcal{F}_c| = |\mathcal{F}|$, $\mathcal{F} \cup \mathcal{F}_c = \mathcal{P}(I_n)$ and show that $\mathcal{F} \cap \mathcal{F}_c = \emptyset$.

5. For item (b), for $1 \leq i \leq n$ let a_i be the number of sets in $\mathcal{I}_n$ which have i as central element. Item (a) gives $a_i = a_{n+1-i}$. Hence,

$$2 \sum_{A \in \mathcal{I}_n} c(A) = \sum_{i=1}^{n} i a_i + \sum_{i=1}^{n} (n + 1 - i) a_{n+1-i}$$

$$= \sum_{i=1}^{n} i a_i + \sum_{i=1}^{n} (n + 1 - i) a_i$$

$$= \sum_{i=1}^{n} (n + 1) a_i = (n + 1) \sum_{i=1}^{n} a_i$$

$$= (n + 1) |\mathcal{I}_n| = (n + 1) \cdot 2^{n-1}.$$

6. For item (a), start by showing that, for and integer $1 \leq k \leq n - 1$, there is a bijection between the partitions of n having k as one of its summands and the partitions of $n - k$. From this, conclude that the contribution of k to $q(n)$ equals $p(n - k)$. For item (b), show that if k is the greatest possible diversity of a partition, then $n \geq 1 + 2 + \cdots + k$, so that $k < \sqrt{2n}$. Then, notice that $q(n) \leq k p(n)$.

Section 1.3

1. Show that there are exactly a_{n+3} ways of writing $n+4$ as a sum of summands 1, 3 or 4, with at least one of these summands being equal to 1; make an analogous reasoning to deal with a_{n+1} and a_n.

2. For an integer $n \geq 3$, let a_n be the number of diagonals of a convex n-gon, so that $a_3 = 0$. Let $A_1 A_2 \ldots A_k A_{k+1}$ be a convex $(k+1)$-gon, so that $A_1 A_2 \ldots A_k$ is a convex k-gon. Counting the number of diagonals of $A_1 A_2 \ldots A_k A_{k+1}$ departing from A_{k+1}, show that $a_{k+1} = a_k + k - 1$.

3. For item (a), consider a tiling of a $2 \times (k + 2)$ checkerboard, and look at the possible ways of filling its upper right 1×1 square with a domino. For item (b), compute $a_1 = 1$, $a_2 = 2$ and, then, apply the results of Chapter 3 of [8].

4. Adapt, to the present case, the solution of Example 1.17.

5. For item (a), fix an element x of A and let $A' = A \setminus \{x\}$. Start by showing that, if a subset B of A has an even number of elements, then there are two possibilities: (i) $x \notin B$, so that B is a subset of A' with an even number of elements; (ii)

$x \in B$, so that $B \setminus \{x\}$ is a subset of A' with an odd number of elements. Then, show that the correspondence $B \mapsto B \setminus \{x\}$ establishes a bijection between the family of subsets of A which contain x and have an even (resp. odd) number of elements and the family of subsets of A' with and odd (resp. even) number of elements. For item (b), recall that $a_n + b_n = 2^n$.

6. More generally, let $f(n, k)$ denote the number of ways of choosing k nonconsecutive chairs out of a row of n chairs. Show that $f(n, k) = f(n-2, k-1) + f(n-1, k)$, then use such a recurrence to conclude that there are exactly 56 ways of choosing the three chairs.

7. For (a), assume we have $k + 1$ disks piled at the leftmost rod. We obviously need at least a_k movements to pile the k upper disks at the central rod; an extra movement puts the larger disk at the rightmost rod; finally, at least more a_k movements move the k disks from the central to the leftmost rod. This amounts to a total of at least $2a_k + 1$ movements, so that $a_{k+1} = 2a_k + 1$. For (b), either make induction on n or note that $a_{k+1} + 1 = 2(a_k + 1)$, taking into account that $a_1 = 1$.

8. For item (b), note that there are two kinds of partitions of I_{n+1} into $k + 1$ nonempty, pairwise disjoint subsets: those in which $\{n + 1\}$ is one of the sets of the partition—there are exactly $S(n, k)$ such partitions—and those in which $n + 1$ belongs to a set of the partition having at least two elements—there are exactly $(k+1)S(n, k+1)$ such partitions. Note that item (c) asks us to compute $S(7, 3)$, for which we can use the recurrence relation of item (b), together with the results of (a).

9. Let a_n denote the number of desired sequences, and b_n the number of sequences of n terms, all of which equal to 0, 1, 2 or 3 and having an odd number of 0's. Argue recursively to show that $a_n = 3a_{n-1} + b_{n-1}$ and $b_n = a_{n-1} + 3b_{n-1}$ for all $n \geq 2$. Then conclude that $a_{k+1} - 6a_k + 8a_{k-1} = 0$ for every $k \geq 2$ and compute $a_1 = 2$, $a_2 = 1 + 3^2 = 10$. Finally, apply the result of Problem 5, page 79.

10. For item (a), use the fundamental principle of counting to show that there are exactly $3 \cdot 2^{n-1}$ flags with n strips, for every integer $n \geq 1$. For item (b), start by observing that $a_1 = 0$ and $a_2 = 6$; then, show that, for $n \geq 1$, there is a bijection between the set of flags with $n + 1$ strips in which the first and last strips have equal colors and the set of flags with n strips in which the first and last strips have distinct colors; finally, use the result of item (a). For item (c), use the result of (b) to show that $a_{n+2} = a_{n+1} + 2a_n$ for every $n \geq 1$; then, solve this recurrence relation.

11. For item (b), let A be a minimal selfish subset of I_n containing n. Note that $1 \notin A$ (for, otherwise, A wouldn't be minimal) and $A \neq \{n\}$, I_n. We claim that the set $A' = (A \setminus \{n\}) - 1$ is a minimal selfish subset of I_{n-2}. To check this, note first that $A' \subset I_{n-2}$; on the other hand, $|A'| = |A| - 1$ and, since $|A| \in A$, we have that $|A| - 1 \in (A - 1) \subset A'$. Thus, A is a selfish subset of I_{n-2}. In order to show that A' is minimal, assume that there would exist $B' \subset A'$ such that $B' \neq A'$ and B' was selfish. Then, prove that $B = (B' + 1) \cup \{n\} \subset A$ would be a selfish subset of I_n containing n. Use the minimality of A to conclude that

$B = A$ and, hence, that $B' = A'$, which is an absurd. Analogously, given a minimal selfish subset B of I_{n-2}, prove that $(B + 1) \cup \{n\}$ is a minimal selfish subset of I_n. For item (c), if a_n denotes the number of minimal selfish subsets of I_n, it follows from items (a) and (b) that $a_n = a_{n-1} + a_{n-2}$ for every integer $n > 2$. Now, verify that $a_1 = a_2 = 1$ and conclude that $a_n = F_n$ for every $n \geq 1$.

12. For item (a), note that there are three distinct types of nonempty subsets of Y: the nonempty subsets of X; the sets $A \cup \{m\}$, where $A \subset X$ is nonempty; the set $\{m\}$. Then, write a_Y as a sum with three summands, each of them corresponding to the subsets of one of these three types; while doing this, note that if $A \subset X$ is nonempty, then $\pi(A \cup \{m\}) = \pi(A)m$. Item (c) follows from (b), together with a straightforward induction.

13. For item (a), adapt to the present case the hint given to the item (a) of the previous problem. To this end, note that if $\emptyset \neq A \subset X$, then $\sigma(A \cup \{m\}) = \sigma(A) + m$ and $\pi(A \cup \{m\}) = \pi(A)m$. Item (c) follows from item (b) by an easy induction.

14. Let $J_0 = \{1\}$ and, for every natural $k \geq 1$, let $J_k = \{2^{k-1} + 1, \ldots, 2^k\}$. Show first that, for every k, there exists a good set $B_k \subset \{1, 2, 3, \ldots, 2^k\}$ such that $|B_k| = b_k$ and $J_k \subset B_k$, i.e., $B_k \cap J_k = J_k$. Since B_k is good and $B_k \cap J_k = J_k$, we must have $B_k \cap J_{k-1} = \emptyset$. Now, let $X_{k-2} = B_k \cap \{1, 2, 3, \ldots, 2^{k-2}\}$. Clearly, every subset of a good set is also a good set; therefore, X_{k-2} is good. Conclude from this fact that $b_k \leq b_{k-2} + 2^{k-1}$. Then, taking a good subset Y_{k-2} of $\{1, 2, 3, \ldots, 2^{k-2}\}$ and looking at $Y_{k-2} \cup J_k$, show that $b_k \geq b_{k-2} + 2^{k-1}$. For item (b), use the recurrence relation of item (a), together with telescoping sums (cf. Chapter 3 of [8]), to get $b_n = \frac{2^{n+1} + (-1)^n}{3}$.

Section 1.4

2. For item (a), expand the products in $(x + y)^n = (x + y) \ldots (x + y)$ (n factors equal to $x + y$) and note that the monomial $x^{n-k} y^k$ appears as many times as the number of ways of choosing a subset of k elements from the set $\{1, 2, \ldots, n\}$. Then, generalize such an argument to get the result of item (b).

3. Let $x_1 = \cdots = x_k = 1$ in (1.12).

4. Notice that each (nonordered) partition of a set A of n elements into k nonempty subsets gives rise to $k!$ ordered partitions of A into k nonempty subsets.

5. If a_j is the only local maximum of a permutation $(a_1, a_2, \ldots, a_n)$ of I_n, then $a_j = n$. Now, show that $a_1 < \cdots < a_{j-1}$ and $a_{j+1} > \cdots > a_n$, so that it suffices to choose the set $\{a_1, \ldots, a_{j-1}\}$. Finally, observe that j may vary from 1 to n.

6. Show that each 6 of the n given points determine a single triangle satisfying the given conditions.

7. Firstly, show that the total number of subsets with an even number of elements equals $\sum_{k\geq 0}\binom{n}{2k}$. Then, use item (b) of Corollary 4.25 of [8].

8. If exactly $2k$ terms of the sequence are 0's, show that there are exactly $\binom{n}{2k}3^{n-2k}$ sequences, so that the total number of them is $\sum_{k\geq 0}\binom{n}{2k}3^{n-2k}$. Now, note that

$$\sum_{k\geq 0}\binom{n}{2k}3^{n-2k} = \frac{1}{2}\sum_{k\geq 0}\left(\binom{n}{2k}3^{n-2k} + \binom{n}{2k+1}3^{n-2k-1}\right)$$

$$+ \frac{1}{2}\sum_{k\geq 0}\left(\binom{n}{2k}3^{n-2k} - \binom{n}{2k+1}3^{n-2k-1}\right)$$

$$= \frac{1}{2}\left((3+1)^n + (3-1)^n\right) = \frac{1}{2}(4^n + 2^n).$$

9. Consider a sequence $(a_1, \ldots, a_{n+k-1})$ of $n+k-1$ terms, with $k-1$ of them equal to 0 and the other n equal to 1, and let $1 \leq i_1 < \ldots < i_{k-1} \leq n-k+1$ be the indices of the former ones. There are $x_1 = i_1 - 1 \geq 0$ terms that come before a_{i_1} (which is the i-th one), $x_2 = i_2 - i_1 - 1 \geq 0$ terms between a_{i_1} and a_{i_2}, ..., $x_k = (n-k+1) - i_{k-1} \geq 0$ terms that come after $a_{i_{k-1}}$. Since all of these are equal to 1, and there are exactly n terms equal to 1, we get $x_1 + \cdots + x_k = n$. Thus, starting from a sequence $(a_1, \ldots, a_{n+k-1})$ as above, we get a solution, in nonnegative integers, of the given equation. Conversely, given a solution $x_1 = j_1, \ldots, x_k = j_k$ in nonnegative integers of the equation, we can form a sequence $(a_1, \ldots, a_{n+k-1})$ as above, letting the terms of indices $j_1+1, j_1+j_2+2, j_1+j_2+\cdots+j_{k-1}+(k-1)$ be equal to 0 and the remaining ones be equal to 1. Therefore, by the bijective principle we only need to count the total number of such sequences, which is rather simple: it suffices to choose a subset of $k-1$ elements of the set of I_{n+k-1} and, then, to impose that all the terms corresponding to the chosen indices are equal to 0, while the remaining ones are equal to 1. Proposition 1.26 gives $\binom{n+k-1}{k-1}$ exactly different choices.

10. If $y_i = x_i - 1$ for $1 \leq i \leq k$, then $y_i \geq 0$ and $y_1 + \cdots + y_k = n - k$. Now, apply the result of the previous problem.

11. For item (a), if $A = \{a_1, \ldots, a_k\} \subset I_n$ is a set without consecutive elements, let $x_1 = a_1 - 1$, $x_{k+1} = n - a_k$ and $x_j = a_j - a_{j-1} - 2$ for $2 \leq j \leq k$. Then, $x_j \geq 0$ for $1 \leq j \leq k+1$, and

$$\sum_{j=1}^{k+1} x_j = (a_1 - 1) + \sum_{j=2}^{k}(a_j - a_{j-1} - 2) + (n - a_k)$$

$$= (a_1 - 1) + (a_k - a_1 - 2(k-1)) + n - a_k$$

$$= n - 2k + 1.$$

For item (b), let $a_1 = x_1 + 1$ and $a_j = \sum_{i=1}^{j} x_i + 2(j-1)$ for $2 \leq j \leq k$. Then, $a_1 \geq 1, a_j - a_{j-1} \geq 2$ for $1 < j \leq k$ and

$$n - a_k = n - \sum_{i=1}^{k} x_i - 2(k-1) = (n - 2k + 1) - \sum_{i=1}^{k} x_i + 1 = x_{k+1} + 1 \geq 1;$$

hence, a solution of equation $\sum_{i=1}^{k+1} x_i = n - 2k + 1$ in nonnegative integers gives rise to a k-element subset of I_n satisfying the conditions of Kaplansky's first lemma. For item (c), the bijective principle, together with the results of items (a) and (b), guarantees that it suffices to count how many are the nonnegative integer solutions of equation $x_1 + \cdots + x_{k+1} = n - 2k + 1$. This follows from Problem 9, which shows that there are exactly $\binom{(n-2k+1)+k}{k} = \binom{n-k+1}{k}$ such solutions.

12. Letting $A \subset I_n$ be such a set, there are two possibilities: (i) $1 \in A$ or (ii) $1 \notin A$. In case (i), $A \setminus \{1\}$ has $k-1$ elements, no consecutive elements and must be contained in $\{3, 4, \ldots, n-1\}$; apply Kaplansky's first lemma to conclude that there are $\binom{n-k-1}{k-1}$ such sets. In (ii), A can be any k-element subset of $\{2, 3, \ldots, n\}$ without consecutive elements; apply Kaplansky's first lemma to conclude that there are $\binom{n-k}{k}$ such sets. Finally, add possibilities.

13. If $A = \{a_1, \ldots, a_k\}$ and $f(a_j) = x_j$, then (1.13) gives $x_1 + \cdots + x_k = n$. Now, apply the result of Problem 10.

14. Let $A = \{1, 3, 5, \ldots, 2n-1\}$ and $B = \{2, 4, 6, \ldots, 2n\}$, so that $I_{2n} = A \cup B$, a disjoint union. Let $\mathcal{F}$ be the family of subsets of I_{2n} with equal numbers of even and odd elements, and let $\mathcal{P}$ be the family of subsets of I_{2n} with n elements. Show that $X \mapsto (X \cap A) \cup (B \setminus X)$ establishes a bijection between $\mathcal{F}$ and $\mathcal{P}$. Then, conclude that $|\mathcal{F}| = |\mathcal{P}| = \binom{2n}{n}$.

15. For $A \subset I_m$, Problem 2, page 13, shows that there are $(2^n - 1)^{|A|}$ sequences $(A_1, \ldots, A_n)$ of subsets of I_m such that $A_1 \cup \ldots \cup A_n = A$. Now, for $0 \leq k \leq m$, we know that I_m has exactly $\binom{m}{k}$ subsets A having k elements. Hence, letting S denote the desired sum, we get

$$S = \sum_{A \subset I_m} \sum_{A_1 \cup \ldots \cup A_n = A} |A_1 \cup \ldots \cup A_n|$$

$$= \sum_{A \subset I_m} (2^n - 1)^{|A|} |A| = \sum_{k=0}^{m} \binom{m}{k} (2^n - 1)^k k$$

$$= m(2^n - 1) \sum_{k=1}^{m} \binom{m-1}{k-1} (2^n - 1)^{k-1}$$

$$= m(2^n - 1)[1 + (2^n - 1)]^{m-1}$$

$$= m(2^n - 1) 2^{n(m-1)}.$$

16. Let (i, j) be an ordered pair of distinct integers chosen from $1, 2, \ldots, n$. Note that there are $(n-2)!$ permutations $(a_1, a_2, \ldots, a_n)$ of $\{1, 2, \ldots, n\}$ in which $a_1 = i$ and $a_2 = j$, and these contribute with $(n-2)!(i-j)^2$ for the sum $\sum_\sigma S_\sigma$. However, the same pair (i, j) gives this same contribution when $a_k = i$ and $a_{k+1} = j$, for some k from 1 to $n-1$. Therefore, $(i-j)^2$ appears $(n-1)(n-2)! = (n-1)!$ times in the sum $\sum_\sigma S_\sigma$. It thus follows that

$$\sum_\sigma S_\sigma = \sum_{(i,j)} (n-1)!(i-j)^2,$$

and the average value of S_σ equals

$$A = \frac{1}{n!} \sum_\sigma S_\sigma = \frac{1}{n!} \sum_{(i,j)} (n-1)!(i-j)^2 = \frac{1}{n} \sum_{(i,j)} (i-j)^2.$$

In order to compute the last sum above, note that

$$\frac{1}{n} \sum_{(i,j)} (i-j)^2 = \frac{1}{n} \sum_{i=1}^{n} \sum_{j=1}^{n} (i-j)^2 = \frac{1}{n} \sum_{l=1}^{n} \sum_{j=1}^{n} (i^2 + j^2 - 2ij)$$

$$= \frac{1}{n} \left[\sum_{i=1}^{n} \sum_{j=1}^{n} i^2 + \sum_{i=1}^{n} \sum_{j=1}^{n} j^2 - 2 \sum_{i=1}^{n} \sum_{j=1}^{n} ij \right]$$

$$= \frac{1}{n} \left[n \sum_{l=1}^{n} i^2 + n \sum_{j=1}^{n} j^2 - 2 \left(\sum_{i=1}^{n} i \right) \left(\sum_{j=1}^{n} j \right) \right]$$

$$= \frac{1}{n} \left[2n \sum_{i=1}^{n} i^2 - 2 \left(\sum_{i=1}^{n} i \right)^2 \right]$$

$$= \frac{1}{n} \left[2n \frac{n(n+1)(2n+1)}{6} - 2 \left(\frac{n(n+1)}{2} \right)^2 \right]$$

$$= \frac{(n-1)n(n+1)}{6} = \binom{n+1}{3}.$$

17. Fix one of the $2n + 1$ given points, draw the diameter passing through it and suppose that such a diameter leaves d of the remaining $2n$ points in one semicircle, and the other $2n - d$ points in the other one. Use combinations to conclude that the number of obtuse triangles having the fixed point as one of their vertices and an acute angle at this vertex equals $d^2 - 2nd + 2n^2 - n$ (therefore, it is always greater than or equal to $n(n-1)$). Adding over the $2n + 1$ given points, conclude that there are at least $(2n+1)\binom{n}{2} = \frac{2n+1)n(n-1)}{2}$

obtuse triangles. Finally, deduce that the total number of acute triangles is at most $\frac{n(n+1)(2n+1)}{6}$.

18. Fixed $1 \leq i \leq n$ and $0 \leq j \leq \min\{i-1, n-i\}$, Proposition 1.26 gives exactly $\binom{i-1}{j}$ ways of choosing j elements of I_n less than i and $\binom{n-i}{j}$ ways of choosing j elements of I_n greater than i. Hence, the additive and multiplicative principles guarantee that i is the central element of

$$\sum_j \binom{i-1}{j}\binom{n-i}{j}$$

sets $A \in \mathcal{I}_n$, with the above sum extending through all indices $0 \leq j \leq \min\{i-1, n-i\}$. For the second equality of item (a), there are at least two possible reasonings:

(i) Computing the coefficient of x^{i-1} in both sides of the equality

$$(1+x)^{i-1}(1+x)^{n-i} = (1+x)^{n-1}$$

we get, with the aid of the binomial formula,

$$\sum_{j \geq 0} \binom{i-1}{j}\binom{n-i}{n-i-j} = \binom{n-1}{i-1}.$$

(ii) If $x_{k+1} = i$ and $A = \{x_1, x_2, \ldots, x_{2k+1}\}$ is as above, write $A = \underline{B} \cup \{x_{k+1}\} \cup \overline{B}$, with $\underline{B} = \{x_1, \ldots, x_k\}$ and $\overline{B} = \{x_{k+2}, \ldots, x_{2k+1}\}$; if $\overline{B}' = \{i+1, i+2, \ldots, n-1, n\} \setminus \overline{B}$, show that the correspondence

$$\underline{B} \cup \{x_{k+1}\} \cup \overline{B} \longmapsto \underline{B} \cup \overline{B}'$$

establishes a bijection between $\mathcal{I}_n$ and the family of subsets of $n-i$ elements of $I_n \setminus \{i\}$. Since $|I_n \setminus \{i\}| = n-1$, use Proposition 1.26 to conclude that i is the central element of $\binom{n-1}{n-i}$ sets in $\mathcal{I}_n$.

The rest of the proof follows immediately from the lines' theorem of Pascal's triangle (cf. item (a) do Corollary 4.25 of [8]):

$$\sum_{A \in \mathcal{I}_n} c(A) = \sum_{i=1}^{n} \binom{n-1}{i-1} i$$

$$= (n-1)\sum_{i=1}^{n} \binom{n-2}{i-2} + \sum_{i=1}^{n} \binom{n-1}{i-1}$$

$$= (n-1) \cdot 2^{n-2} + 2^{n-1} = (n+1) \cdot 2^{n-2}.$$

19. For item (a), if f does satisfy the given conditions, then it is clearly injective, for $f(x_1) = f(x_2) \Rightarrow f(f(x_1)) = f(f(x_2)) \Rightarrow x_1 = x_2$. Now, it follows from the result of Problem 4, page 7, that f is a bijection. Assume that f has k fixed points. Fix $a \in I_n$ such that a is not one of them and let $b = f(a)$. Then, $b \neq a$ and $f(b) = a$. Thus, we can group the remaining $n - k$ elements of I_n (all but the fixed points of f) in pairs (a, b), with $b = f(a)$ and $a = f(b)$, which implies that $n - k$ must be even. For item (b), let $0 \leq k \leq n$ be such that $n - k$ is even. For each choice of k elements $x_1, \ldots, x_k \in I_n$ and for each partition $\{\{a_1, b_1\}, \ldots, \{a_l, b_l\}\}$ of the remaining $n - k$ elements of I_n, there obviously exists a single function $f : I_n \to I_n$ having $x_1, \ldots, x_k$ as its fixed points and satisfying $f(a_i) = b_i$ and $f(b_i) = a_i$ for $1 \leq i \leq l$. Observe that $f(f(x)) = x$ for each $x \in I_n$. For item (c), once fixed $0 \leq k \leq n$ such that $n - k$ is even, there are $\binom{n}{k}$ ways of choosing k elements of I_n to be the fixed points of f. In turn, the number of ways of partitioning the remaining $n - k$ elements of I_n into disjoint 2-element sets is equal to

$$\binom{n - k - 2}{2} \cdots \binom{2}{2} = \frac{(n - k)!}{2^{(n-k)/2}}.$$

Hence, the multiplicative and additive principles assure that the number of functions we wish to count is equal to the given number.

20. Let $|\mathcal{F}| = k$. To each $A = \{a, b, c\} \in \mathcal{F}$, associate its 2-element subsets $\{a, b\}$, $\{b, c\}$ and $\{c, a\}$. The condition given in the statement of the problem assures that each 2-element subset of I_n is associated to at most one of the sets $A \in \mathcal{F}$. Thus, $3k \leq \binom{n}{2}$, which gives us the desired inequality.

21. For each $1 \leq i \leq n$, choose $B_i \subset A_i$ such that $|B_i| = k$. The assumption $|A_i \cap A_j| \leq k - 1$ for $1 \leq i < j \leq n$ guarantees that the correspondence $i \mapsto B_i$ is injective; indeed, if $1 \leq i < j \leq n$ are such that $B_i = B_j = B$, then $|A_i \cap A_j| \geq |B| \geq k$. Now, $\{B_1, \ldots, B_n\}$ is a family of pairwise distinct m-element subsets of A, so that Proposition 1.26 gives $n \leq \binom{m}{k}$.

22. For $1 \leq k \leq n$, we let m_k denote the number of P_i's which contain a_k; conclude that $m_1 + \cdots + m_n = 2n$. If $\mathcal{F}$ is the family of sets $\{P_i, P_j\}$ such that $i \neq j$ and $P_i \cap P_j \neq \emptyset$, use the given conditions to show that $|\mathcal{F}| = \sum_{i=1}^{n} \binom{m_i}{2}$. Now, consider $f : \mathcal{F} \to \{P_1, \ldots, P_n\}$ such that $f = (\{P_i, P_j\}) = P_k$, with $P_k = \{a_i, a_j\}$; conclude that f is injective, so that

$$\sum_{i=1}^{n} \binom{m_i}{2} \leq n$$

and, then $m_1^2 + \cdots + m_n^2 \leq 4n$. Finally, use the inequality between the quadratic and arithmetic means (cf. problem 5.2.3 of [8], for instance) to conclude that $m_1 = \cdots = m_n = 2$.

Section 2.1

1. In each item through (a) to (d), evaluate the corresponding functions at $x \in A$.
2. How many natural numbers from 1 to 1000 have a prime factor less than 10?
3. Use the result of Example 2.6, together with the fact that, since $m < n$, there is no surjective function $f : I_m \to I_n$.
4. Let $C_1, \ldots, C_n$ be the n couples, A be the set of distinct lines of $2n$ people formed with the members of the given couples (and without further restrictions), and A_i be the subset of A formed by those lines in which both members of C_i occupy neighboring positions. We wish to count the number of elements of $A \setminus (A_1 \cup \ldots \cup A_n)$. To this end, apply the inclusion-exclusion principle, noting that $|A| = (2n)!$ and $|A_{i_1} \cap \ldots \cap A_{i_k}| = (2n - k)!(2!)^k$ for $1 \le i_1 < \cdots < i_k \le n$.
5. For $1 \le j \le 5$, let $\mathcal{P}_j$ be the set of permutations $(a_1, a_2, \ldots, a_6)$ of I_6 such that $(a_1, \ldots, a_j)$ is a permutation of I_j. We wish to count $|\mathcal{P} \setminus \bigcup_{j=1}^5 \mathcal{P}_j|$, where $\mathcal{P}$ denotes the set of all permutations of I_6. To this end, if $1 \le i_1 < \cdots < i_k \le 5$, conclude that

$$|\mathcal{P}_{i_1} \cap \ldots \cap \mathcal{P}_{i_k}| = i_1!(i_2 - i_1)! \ldots (i_k - i_{k-1})!(6 - i_k)!$$

Finally, apply the inclusion-exclusion principle and compute the value of each possible summand above.

6. Let $A = \{1 \le m \le n; \gcd(m, n) = 1\}$ and, for $1 \le i \le k$, let $A_i = \{1 \le m \le n; p_i \mid m\}$. Then, $I_n \setminus A = \bigcup_{i=1}^k A_i$ and, in order to compute $|A|$, it suffices to apply inclusion-exclusion. To this end, note that if $1 \le i_1 < \cdots < i_j \le k$ are integers, then $m \in A_{i_1} \cap \ldots \cap A_{i_j}$ if and only if $p_{i_1} \ldots p_{i_j} \mid m$, so that $|A_{i_1} \cap \ldots \cap A_{i_j}| = \frac{m}{p_{i_1} \ldots p_{i_j}}$.
7. Start by showing that, for $q \in \mathbb{N}$, the number of naturals less than or equal to n and divisible by q equals $\lfloor \frac{n}{q} \rfloor$. Then, let $A_j = \{a \in \mathbb{N}; a \le n \text{ and } p_j \mid a\}$, and notice that the number of primes less than or equal to n is given by $n - |A_1 \cup \ldots \cup A_k|$. Finally, apply inclusion-exclusion.
8. Let A denote the set of distinct partitions of n in natural summands and, for $1 \le j \le n$, let A_j denote the subset of A formed by those partitions with at least one summand equal to j. Then, $p(n) = |A|$ and, for $1 \le i_1 < \ldots < i_k \le n$, it is easily seen that $|A_{i_1} \cap \ldots \cap A_{i_k}| = p(n - l)$, where $l = i_1 + \cdots + i_k$. Now, note that $A = A_1 \cup \ldots \cup A_n$ and use inclusion-exclusion to get

$$p(n) = \sum_{k=1}^n (-1)^{k-1} \sum_{l=1}^n \sum_{\substack{1 \le i_1 < \ldots < i_k \le n \\ i_1 + \cdots + i_k = l}} p(n - l).$$

Finally, on the one hand, the last summand has exactly $p_k(l)$ summands; on the other, $1 \le i_1 < \ldots < i_k \le n$ and $i_1 + \cdots + i_k = l$ imply

$$n \geq l \geq 1 + 2 + \cdots + k = \frac{k(k+1)}{2} > \frac{k^2}{2}$$

and, hence, $k < \sqrt{2n}$ and $l > \frac{k^2}{2}$.

9. Let A be the set of solutions $(x_1, \ldots, x_n)$ of the given equation, with $x_t \in \mathbb{Z}_+$ for $1 \leq t \leq n$. For $1 \leq i \leq n$, let A_i be the subset of A formed by those solutions $(x_1, \ldots, x_n)$ in which $x_i > k$. We wish to count $|A \setminus (A_1 \cup \ldots \cup A_n)|$; to this end, apply inclusion-exclusion, recalling (cf. Problem 9, page 29) that $|A| = \binom{m+n-1}{n-1}$ and showing that

$$|A_{i_1} \cap \ldots \cap A_{i_j}| = \binom{m - j(k+1) + n - 1}{n - 1}$$

for $1 \leq i_1 < \cdots < i_j \leq n$. For this last computation, observe that $(x_1, \ldots, x_n) \in A_{i_1} \cap \ldots \cap A_{i_j}$ if and only if $x_{i_1}, \ldots, x_{i_j} \geq k + 1$; hence, letting $y_t = x_t$ for $t \neq i_1, \ldots, i_j$ and $y_{i_l} = x_{i_l} - (k+1)$ for $1 \leq l \leq j$, we have $y_1, \ldots, y_n \geq 0$ and $y_1 + \cdots + y_n = m - j(k+1)$; now, apply once more the result of Problem 9, page 29.

10. For the first part of (a), note that there are 2 ways of choosing n pairwise nonadjacent chairs, and after that there are $n!$ ways of distributing the men in them. For (c), note that if $A_{2i-1} \cap A_{2i}$ was not empty, then w_i should simultaneously sit in chairs $i - 1$ and i; also, if $A_{2i} \cap A_{2i+1}$ was not empty, then w_i and w_{i+1} should both sit in chair i. Item (d) follows from the fact that, for every intersection of the form $A_{i_1} \cap A_{i_2} \cap \ldots \cap A_{i_k}$, exactly k of the n women have their positions entirely determined, whereas the $n - k$ remaining ones can be permuted in any desired way. Finally, for (f) apply the inclusion-exclusion principle.

Section 2.2

1. Explicitly write the summands at both sides.
2. Note that both sides count the number of subsets of a set having n elements.
3. Adapt, to the present case, the idea used in the proof of Example 2.9.
4. Start by choosing k odd elements and k even elements; then, use Lagrange's identity.
5. Let $P_1, P_2, \ldots, P_{11}$ be the patrols and $h_1, h_2, \ldots, h_n$ be the volunteers. Define a function

$$f : \{h_1, \ldots, h_n\} \to \{\{P_i, P_j\};\ 1 \leq i < j \leq 11\}$$

by letting

$$f(h_i) = \{P_j, P_k\} \Leftrightarrow h_i \text{ enters } P_j \text{ and } P_k.$$

The conditions given immediately assure that f is a bijection, so that there are as many volunteers as 2-element subsets of the set of the 11 patrols. Thus, there are $n = \binom{11}{2} = 55$ volunteers. Now, let A be the set of ordered pairs (h_i, P_j) such that $h_i \in P_j$. Since each of the 55 men enters exactly two patrols, A has exactly $2 \cdot 55 = 110$ pairs. On the other hand, if k is the number of members in each patrol, then P_j is the second entry of exactly k ordered pairs in A (those whose first entries are the k members of P_j); however, since there are precisely 11 patrols, it follows that A has $11k$ ordered pairs. Therefore, $11k = 110$ and, hence, $k = 10$.

6. Show that both sides of the desired equality count the total number of fixed points of all permutations of I_n.

7. Let $a_1, a_2, \ldots, a_m$ be the initial terms of the arithmetic progressions, and $n > a_1, \ldots, a_m$ a natural number. Let's count the number of elements of I_n in two different ways, the obvious one being $|I_n| = n$. For the other one, assume that the arithmetic progression of initial term a_i and common ratio d_i has exactly k_i terms in I_n, so that $\sum_{i=1}^{m} d_i = n$ and $a_i + (k_i - 1)d_i \leq n < a_i + k_i d_i$ or, which is the same,

$$\frac{n - a_i}{d_i} < k_i < \frac{n - a_i}{d_i} + 1.$$

Adding all such inequalities for $1 \leq i \leq m$, we conclude that

$$\sum_{i=1}^{m} \frac{n - a_i}{d_i} < n < \sum_{i=1}^{m} \frac{n - a_i}{d_i} + m$$

or, which is the same,

$$\sum_{i=1}^{m} \frac{1}{d_i} - \frac{1}{n} \sum_{i=1}^{m} \frac{a_i}{d_i} < 1 < \sum_{i=1}^{m} \frac{1}{d_i} - \frac{1}{n} \sum_{i=1}^{m} \frac{a_i}{d_i} + \frac{m}{n}.$$

Finally, let $n \to +\infty$ to get $\sum_{i=1}^{m} \frac{1}{d_i} = 1$.

8. Associate 0 or 1 to each line segment which is a side of a smaller triangle, according to whether its vertices have equal or distinct labels, respectively; then, associate to each smaller triangle the sum of the numbers associated to its three sides. Now, observe that the number associated to any smaller triangle is equal to 0, 2 or 3, being equal to 3 if and only if the corresponding triangle is labelled as ABC. Finally, use double counting to show that the sum of the numbers associated to the smaller triangles is always odd.

9. Let x denote the desired number of commissions and y the total number of the remaining ones. Firstly, note that $x + y = 4060$. Now, fix an arbitrary senator and show that he/she enters exactly 268 commissions in which the other two members are both his/her friends or enemies. Finally, use double counting to conclude that $3x + y = 30 \cdot 268$.

10. Let $A_1 A_2 \ldots A_n$ be the n-gon, $A_n A_2 \cap A_1 A_3 = \{X\}$ and $A_2 A_4 \cap A_3 A_5 = \{Y\}$; choose the following $n - 2$ points: a point B in the interior of $A_1 A_2 X$, a point

C in the interior of A_3A_4Y and, for $5 \leq j \leq n$, a point in the interior of $A_2A_3A_j \cap A_{j-1}A_jA_{j+1}$. With the aid of the fundamental principle of counting, show that the total number of triangles formed by three vertices of the n-gon and containing one of the chosen points is equal to

$$2(n-2) + \sum_{j=5}^{n}(j-3)(n-j+2) = \binom{n}{3}.$$

Section 2.3

1. For item (a), we leave to the reader the task of showing that $\sim$ is symmetric. For the reflexivity, let $a \in A$ be given; since $\mathcal{F}$ is a partition of A, there exists $B \in \mathcal{F}$ such that $a \in B$. Therefore, $a, a \in B$ or, which is the same, $a \sim a$. Now, let $a, b, c \in A$ be such that $a \sim b$ and $b \sim c$. There exist $B, C \in \mathcal{F}$ such that $a, b \in B$ and $b, c \in C$. However, since $\mathcal{F}$ is a partition of A and $B \cap C \neq \emptyset$ (for, $b \in B \cap C$), we conclude that $B = C$; therefore, $a \sim c$. Finally, for $a \in A$, take the set $B \in \mathcal{F}$ for which $a \in B$. Then,

$$\bar{a} = \{x \in A; \, x \sim a\} = \{x \in A; \, x, a \in C, \, \exists \, C \in \mathcal{F}\}.$$

Invoking once more the fact that $\mathcal{F}$ is a partition of A, we conclude that B is the only element of $\mathcal{F}$ to which a does belongs. Thus,

$$\bar{a} = \{x \in A; \, x \in B\} = B.$$

The analysis of item (b) is quite similar to that of (a) and will also be left to the reader; we notice that, in this case, if B is a finite set, then there are exactly $|B|$ distinct equivalence classes.

2. $A \sim A$, for $A = A + 0$ modulo 3. If $A \sim B$, with $B = A + j$ modulo 3, then $B \sim A$, with $A = B + (3 - j)$ modulo 3. Finally, if $A \sim B$ and $B \sim C$, with $B = A + i$ modulo 3 and $C = B + j$ modulo 3, show that $C = A + k$ modulo 3, where $0 \leq k \leq 2$ is such that $i + j \equiv k \pmod{3}$.

3. If the involved sets were to be chosen in a specific order, then there would be exactly $\binom{ab}{b,\dots,b} = \frac{(ab)!}{(b!)^a}$ distinct ways of partitioning the set as prescribed. Since the choice is actually unordered, we have to divide by $a!$.

4. For item (b), use the result of (a), together with Proposition 2.16 and the definition of Stirling numbers of second kind. For item (c), use the result of (b), together with that of Example 2.6.

5. Firstly, note that there are exactly $m^p - m$ non monochromatic sequences of p beads, each bead being of one of the m given colors. Apply the equivalence relation (2.16) to such sequences (with p in place of n) to conclude that there are exactly $\frac{m^p - m}{p}$ distinct and non monochromatic necklaces. Since there are exactly

m monochromatic necklaces, there is nothing left to do in order to establish the first part. For the second part, it follows from the first that p divides $m^p - m = m(m^{p-1} - 1)$, for every $m \in \mathbb{N}$. It now suffices to assume that $p \nmid m$, using the remarks that precede the statement of the problem to conclude that $p \mid (m^{p-1} - 1)$.

6. In the set $\mathcal{F}$ of aperiodic sequences of size n, let $\sim$ be the relation defined by setting $(x_1, x_2, \ldots, x_n) \sim (y_1, y_2, \ldots, y_n)$ if and only if there exists $0 \leq k \leq n - 1$ such that $y_i = x_{i+k \,(\text{mod}\,n)}$, for $1 \leq i \leq n$. Show that $\sim$ is an equivalence relation, with respect to which each equivalence class contains exactly n sequences. Then, apply the result of Proposition 2.18.

7. Apply Bollobás' theorem, together with the fact that $\binom{n}{\lfloor n/2 \rfloor}$ is the largest binomial number with numerator equal to n.

8. For the first part of (a), show more generally (by induction, for instance) that if $f^{(m)} = \text{Id}$ for some $m \in \mathbb{N}$, then f is injective, hence bijective (thanks to the result of Problem 4, page 7). The second part of (a) follows from the associativity of composition, together with the fact that $f^{(j)}$ is a bijection for every $j \in \mathbb{Z}$ (so that $a = f^{(j)}(b) \Rightarrow b = \left(f^{(j)}\right)^{-1}(a) = f^{(-j)}(a)$). For the first part of (b), let

$$\bar{a} = \{\ldots, f^{(-2)}(a), f^{(-1)}(a), a, f(a), f^{(2)}(a), \ldots\}$$

be an equivalence class of $\sim$ with more than one element, so that $f(a) \neq a$. If $j > 1$ is the smallest integer such that $f^{(j)}(a) = a$ (by assumption, $f^{(p)}(a) = a$, thus such a j does exist), write $p = jq + r$, with $0 \leq r < p$, and use the minimality of j to conclude that $r = 0$; then, show that $j = p$. Conclude that $\bar{a} = \{a, f(a), f^{(2)}(a), \ldots, f^{(p-1)}(a)\}$. The second part of (b) now follows from the first part, together with the fact that the equivalence classes form a partition of I_n. Item (c) follows from the fundamental principle of counting, together with the fact that, given pairwise distinct elements $a_1, a_2, \ldots a_p \in I_n$, there are exactly $(p - 1)!$ possibilities for the restriction of f to $\{a_1, a_2, \ldots a_p\}$, such that $f(\{a_1, a_2, \ldots a_p\}) = \{a_1, a_2, \ldots a_p\}$ and none of $a_1, a_2, \ldots a_p$ is a fixed point. Indeed, in this case there exists $a \in I_n$ for which $\{a_1, a_2, \ldots a_p\} = \{a, f(a), \ldots, f^{(p-1)}(a)\}$; however, since $f^{(p)}(a) = a$, the number of desired bijections from $\{a_1, a_2, \ldots a_p\}$ to itself amounts to the number of circular permutations of these numbers, i.e., $(p-1)!$. Finally, for (d), once fixed an integer $0 \leq k \leq n$ such that $p \mid (n - k)$, there are $\binom{n}{k}$ ways of choosing k elements of I_n to be the fixed points of f. In turn, the number of ways of partitioning the remaining $n - k$ elements of I_n into disjoint p-element sets is equal to

$$\binom{n-k}{p} \cdot \binom{n-k-p}{p} \cdots \binom{p}{p} = \frac{(n-k)!}{p^{(n-k)/p}}.$$

Hence, the multiplicative and additive principles assure that the number of functions we wish to count equals the given number.

9. For item (a), show that if A, $B \subset I_n$ are distinct subsets with $k > \frac{n}{2}$ elements each, then $A \cap B \neq \emptyset$. For item (b), take $\mathcal{F}$ as the family of subsets of I_n formed by the k-element subsets of I_n which contain n. For item i., fix $A \in \mathcal{F}$ compatible with $\overline{\sigma}$; we can assume, without loss of generality, that $A = \{a_1, \ldots, a_k\}$ and $\sigma = (a_1, \ldots, a_k, a_{k+1}, \ldots, a_n)$. Group the remaining k-element subsets of I_n which intersect A and are compatible with $\overline{\sigma}$ in appropriate pairs. Then, use the fact that $\mathcal{F}$ is an intersecting family to conclude that there are at most other $k - 1$ sets in $\mathcal{F}$ which are still compatible with $\overline{\sigma}$. For ii. assume, without loss of generality, that $A = \{x_1, \ldots, x_k\}$ and $\sigma = (a_1, \ldots, a_k, a_{k+1}, \ldots, a_n)$, with $\{x_1, \ldots, x_k\} = \{a_1, \ldots, a_k\}$. Finally, for iii. use double counting.

Section 2.4

1. Let $\mathcal{S}_n$ be the set of sequences $a = (a_1, \ldots, a_n)$, with $a_i \in \{0, 1\}$ for $1 \leq i \leq n$, and $\mathcal{P}(n)$ be the family of parts of I_n. For $a \in \mathcal{S}_n$, let $f(a) \in \mathcal{P}(n)$ be given by $f(a) = \{1 \leq i \leq n; a_i = 1\}$. Now, check that the function $f : \mathcal{S}_n \to \mathcal{P}(n)$ thus defined is a bijection that *preserves metrics*, in the following sense: if d is the Hamming metric in $\mathcal{S}_n$ and d' is the metric of symmetric difference in $\mathcal{P}(n)$, then $d(a, b) = d'(f(a), f(b))$ for all $a, b \in \mathcal{S}_n$.

2. Use the hypotheses to show that $T = \bigcup_{x \in S} B(x; 3)$, a disjoint union, and $|B(x; 3)| = \binom{n}{0} + \binom{n}{1} + \binom{n}{2} + \binom{n}{3}$ for every $x \in S$. Then, conclude from (b) that $|B(x; 3) \cap S| = 1$ for each $x \in T$ and, letting $k = \frac{n+1}{2}$, show that $3 \cdot 2^{k-2} = k(2k^2 - 3k + 4)$. In order to solve such an equation, start by showing that $3 \mid k$; then, write $k = 3l$ and conclude that $l \leq 4$ and, hence, $l = 4$, $k = 12$ and $n = 23$.

3. Adapt, to the present case, the discussion of Example 2.29 and the hint given to Problem 20, page 32.

4. Let $\mathcal{P}(n)$ be the family of subsets of I_n, furnished with the metric of symmetric difference. Show that, for $1 \leq i \leq k$, the balls $B(A_i; r)$ are pairwise disjoint. Then, use (2.20), together with the fact that $B(A_1; r) \cup \ldots \cup B(A_k; r) \subset \mathcal{P}(n)$.

5. Firstly, if A and B are the sets of members of two of the 1997 subcommittees, note that the set of people who participate of exactly one of these subcommittees is $A \Delta B$, so that $|A \Delta B| \geq 5$. Now, apply the result of the previous problem.

6. Let $x_1, \ldots, x_n$ be the sequences in the dictionary and d the Hamming metric (with 0 standing for *big* and 1 standing for *small*). Show that

$$|B(x_1; 4) \cup \ldots \cup B(x_n; 4)| = 2^{24}.$$

Moreover, for $1 \leq i < j \leq n$, if $x \in B(x_i; 4) \cap B(x_j; 4)$, show that $d(x; x_i) = d(x; x_j) = 4$ and $d(x_i; x_j) = 8$. Use this fact to show that x belongs to at most six of the balls $B(x_i; 4)$ and, then, conclude that

$$|B(x_1; 4) \cup \ldots \cup B(x_n; 4)| \le \left(\binom{24}{0} + \binom{24}{1} + \binom{24}{2} + \binom{24}{3} + \frac{1}{6} \binom{24}{4} \right) n$$

$$= 4096n.$$

Section 3.1

1. Termwise differentiate both sides of (3.2) and, then, substitute $x = 1$. Note that Theorem 3.4 justifies such a procedure.
2. For item (a), argue as was done for the Fibonacci sequence. For items (b) and (c), apply the result of (a), together with (3.4).

Section 3.2

1. Make induction on $k \ge 2$, the initial case $k = 2$ being given by Proposition 3.3.
2. Apply Theorem 3.4 $k - 1$ times, starting from $\frac{1}{1-x} = \sum_{n \ge 0} x^n$. Alternatively, expand $(1 - x)^{-k}$ with the aid of the binomial series.
3. For item (c), it follows from (b) and from (3.5) that $g'(x) = \sum_{n \ge 1} (-x)^{n-1} = \frac{1}{1+x}$ for $|x| < 1$. Hence, $f'(x) = g'(x)$ and $f(0) = g(0)$, so that $f(x) = g(x)$ for $|x| < 1$.
4. For item (b), it follows from the Fundamental Theorem of Calculus that

$$\sum_{n \ge 0} \frac{a_n}{n+1} x^{n+1} = F(x) = \int_0^x F'(t) dt$$

$$= \int_0^x f(t) dt = \int_0^x \sum_{n \ge 0} a_n t^n dt.$$

5. Start by showing that, if $q > |a_1|, \sqrt{|a_2|}, x_0$, where $x_0 = \frac{1}{2} \left(|u| + \sqrt{u^2 + 4|v|} \right)$, then $|a_n| \le q^n$ for every $n \ge 1$. Fix such a q and conclude that the generating function $\sum_{k \ge 1} a_k x^k$ converges in the interval $\left(-\frac{1}{q}, \frac{1}{q} \right)$, thus defining a differentiable function $f : \left(-\frac{1}{q}, \frac{1}{q} \right) \to \mathbb{R}$ by $f(x) = \sum_{k \ge 1} a_k x^k$. Use the recurrence relation satisfied by the sequence $(a_k)_{k \ge 0}$ to write

$$f(x) = \sum_{k \ge 1} a_k x^k = a_1 x + a_2 x^2 + \sum_{k \ge 3} a_k x^k$$

$$= a_1 x + a_2 x^2 + \sum_{k \ge 3} (u a_{k-1} + v a_{k-2}) x^k$$

$$= a_1 x + a_2 x^2 + u x \sum_{k \ge 3} a_{k-1} x^{k-1} + v x^2 \sum_{k \ge 3} a_{k-2} x^{k-2}$$

$$= a_1 x + a_2 x^2 + u x (f(x) - a_1 x) + v x^2 f(x).$$

Conclude that

$$f(x) = \frac{a_1 x + (a_2 - u a_1)x^2}{(1 - \alpha x)(1 - \beta x)}$$

for $|x| < \min\{\frac{1}{q}, \frac{1}{|\alpha|}, \frac{1}{|\beta|}\}$, where α and β are the roots of $x^2 - ux - v = 0$. If $\alpha \neq \beta$, show that there exist unique real constants A and B for which $f(x) = \frac{Ax}{1-\alpha x} + \frac{Bx}{1-\beta x}$. Then, expand $\frac{1}{1-\alpha x}$ and $\frac{1}{1-\beta x}$ in geometric series to get

$$f(x) = \sum_{k \geq 1}(A\alpha^{k-1} + B\beta^{k-1})x^k.$$

Finally, compare this last expression with the fact that $f(x) = \sum_{k \geq 1} a_k x^k$ to obtain $a_k = A\alpha^{k-1} + B\beta^{k-1}$, for every $k \geq 1$. If $\alpha = \beta$, argue similarly. This problem is the content of Theorem 11.45 of [8]; further details can be found there.

6. Use the formula for the number of derangements of I_n, together with (3.8) and the triangle inequality to get

$$\left| d_n - \frac{n!}{e} \right| = n! \left| \sum_{k > n} \frac{(-1)^k}{k!} \right| \leq n! \sum_{k > n} \frac{1}{k!}.$$

Then, estimate

$$\left| d_n - \frac{n!}{e} \right| = \frac{n!}{(n+1)!}\left(1 + \frac{1}{n+2} + \frac{1}{(n+2)(n+3)} + \cdots \right)$$

$$< \frac{1}{n+1}\left(1 + \frac{1}{n+2} + \frac{1}{(n+2)^2} + \cdots \right)$$

$$= \frac{1}{n+1} \cdot \frac{n+2}{n+1} < \frac{1}{n}.$$

Section 3.3

1. Letting $f(x) = \sum_{k \geq 1} a_k x^k$, we get

$$f(x) = \sum_{k \geq 1} a_k x^k = x \sum_{k \geq 1} a_k x^{k-1} = x \sum_{k \geq 0} a_{k+1} x^k$$

$$= 2x + x \sum_{k \geq 1}(a_k + (k+1))x^k$$

$$= 2x + x f(x) + x \sum_{k \geq 1}(k+1)x^k$$

$$= 2x + x f(x) + x g'(x),$$

with $g(x) = \sum_{k \geq 1} x^k = \frac{x}{1-x}$ for $|x| < 1$. Use such a relation to obtain a closed expression for $f(x)$, valid for $|x| < 1$, expand this expression as a power series and obtain the desired formula for a_k as a function of k.

2. For item (b), we have

$$f(x) = 1 + \sum_{n \geq 1} a_n x^n = 1 + \sum_{n \geq 0} a_{n+1} x^{n+1}$$

$$= 1 + \sum_{n \geq 0} (2a_n + n) x^{n+1}$$

$$= 1 + 2xf(x) + \sum_{n \geq 0} (n+2) x^{n+1} - 2 \sum_{n \geq 0} x^{n+1}$$

$$= 1 + 2xf(x) + g'(x) - 2g(x),$$

with $g(x) = \frac{x}{1-x}$. For (c), reduce the sum at the right hand side to a single denominator and, then, compare the coefficients of the numerators at both sides to obtain a linear system of equations in A, B and C. Finally, for item (d), use the result of Problem 2, page 78.

3. We want to compute the coefficient of x^m in the product

$$f(x) = (x + x^2 + \cdots) \ldots (x + x^2 + \cdots),$$

with k factors $x + x^2 + \cdots = \frac{x}{1-x}$ (for $|x| < 1$).

4. We want to compute the coefficient of x^{20} in the product

$$(1 + x + x^2 + \cdots)(x^2 + x^3 + \cdots)(1 + x + x^2 + \cdots)(1 + x + x^2 + \cdots + x^7).$$

Letting $f(x)$, for $|x| < 1$, denote the above expression, we have

$$f(x) = x^2 (1 + x + x^2 + \cdots)^3 (1 + x + x^2 + \cdots + x^7)$$

$$= (x^2 + x^3 + x^4 + \cdots + x^9) \cdot \frac{1}{(1-x)^3}.$$

Now, use the result of Problem (2), page 78.

5. For (a) and the first part of (b), apply the multiplicative and additive principles.

For the coming problem, the reader may find it convenient to read again the paragraph that precedes Example 1.15.

6. For item (a), start by observing that if

$$n = \underbrace{1 + 1 + \cdots + 1}_{k \text{ summands}} + \underbrace{2 + 2 + \cdots + 2}_{l \text{ summands}} + \underbrace{3 + 3 + \cdots + 3}_{m \text{ summands}},$$

then $n = k + 2l + 3m$; conversely, any such expression for n gives rise to a partition of n in which no summand exceeds 3. Now, note that if $n = k+2l+3m$, then $x^n = x^k \cdot x^{2l} \cdot x^{3m}$, and hence (setting $a_0 = 1$)

$$\sum_{n\geq 1} a_n x^n = \left(\sum_{k\geq 0} x^k\right)\left(\sum_{l\geq 0} x^{2l}\right)\left(\sum_{m\geq 0} x^{3m}\right)$$

$$= \frac{1}{1-x} \cdot \frac{1}{1-x^2} \cdot \frac{1}{1-x^3}.$$

if $|x| < 1$. Item (b) can be solved by multiplying both sides by $(1-x)(1-x^2)(1-x^3)$, expanding and comparing coefficients of equal powers of x; the answers are $a = \frac{1}{6}, b = c = \frac{1}{4}$ and $d = \frac{1}{3}$. Finally, for item (c), insert the values of a, b, c and d in the right hand side of the equality of item (b) and use (3.1), together with the result of Problem 2, page 78, to get

$$\frac{1}{1-x^2} = \sum_{k\geq 0} x^{2k}, \quad \frac{1}{1-x^3} = \sum_{k\geq 0} x^{3k},$$

$$\frac{1}{(1-x)^3} = \sum_{n\geq 0} \binom{n+2}{2} x^n, \quad \text{and} \quad \frac{1}{(1-x)^2} = \sum_{n\geq 0} (n+1)x^n.$$

Then, look separately at each of the cases $6 \mid n$, $2 \mid n$ but $3 \nmid n$, $2 \nmid n$ but $3 \mid n$, $2 \nmid n$ and $3 \nmid n$.

6. Letting $d_0 = d_1 = 0$, item (d) follows from

$$f'(x) = \sum_{n\geq 2} \frac{d_n}{(n-1)!} x^{n-1}$$

$$= \sum_{n\geq 2} \frac{(n-1)d_{n-1}}{(n-1)!} x^{n-1} + \sum_{n\geq 2} \frac{(n-1)d_{n-2}}{(n-1)!} x^{n-1}$$

$$= x + x\sum_{n\geq 2} \frac{(n-1)d_{n-1}}{(n-1)!} x^{n-2} + x\sum_{n\geq 2} \frac{d_{n-2}}{(n-2)!} x^{n-2}$$

$$= x + xf'(x) + xf(x).$$

Item (e) follows the fact that $\frac{f'(x)}{f(x)+1} = \frac{d}{dx}\log\left(f(x)+1\right)$. Finally, for (f) integrate the equality in (e).

8. For item (a), the number of ways of choosing the $2a_0$ positions of the 0's, the a_1 positions of the 1's etc is equal to

$$\binom{n}{2a_0}\binom{n-2a_0}{a_1}\binom{n-2a_0-a_1}{a_2}\binom{n-2a_0-a_1-a_2}{a_3} = \frac{n!}{(2a_0)!a_1!a_2!a_3!}.$$

Item (b) amounts to noticing that

$$\sum_{n \geq 0} \frac{s_n}{n!} x^n = \sum_{n \geq 1} \sum \frac{1}{(2a_0)! a_1! a_2! a_3!} x^n$$

$$= \Big(\sum_{a_0 \geq 0} \frac{x^{2a_0}}{(2a_0)!} \Big) \Big(\sum_{a_1 \geq 0} \frac{x^{a_1}}{a_1!} \Big) \Big(\sum_{a_2 \geq 0} \frac{x^{a_2}}{a_2!} \Big) \Big(\sum_{a_3 \geq 0} \frac{x^{a_3}}{a_3!} \Big)$$

$$= \Big(1 + \frac{x^2}{2!} + \frac{x^4}{4!} + \cdots \Big) \Big(1 + \frac{x}{1!} + \frac{x^2}{2!} + \cdots \Big)^3.$$

For (c), start by observing that

$$e^x + e^{-x} = \sum_{k \geq 0} \Big(\frac{x^k}{k!} + \frac{(-x)^k}{k!} \Big) = 2 \sum_{j \geq 0} \frac{x^{2j}}{(2j)!},$$

so that

$$f(x) = \frac{1}{2}(e^x + e^{-x})(e^x)^3 = \frac{1}{2}(e^{4x} + e^{2x}).$$

Finally, expand e^{4x} and e^{2x} in power series and compare the result with $\sum_{n \geq 0} \frac{s_n}{n!}$ to finish the problem.

Section 4.1

1. Divide the given square into four small squares of side 1 each, and let them play the role of pigeonholes. Then, notice that the length of the diagonals of each small square is precisely $\sqrt{2}$.
2. Divide the cube into $9^3 = 729$ small cubes, each of which having edge length equal to $\frac{10}{9}$. Such small cubes will be the pigeonholes, and have diagonals of length $\frac{10\sqrt{3}}{9}$, with $\frac{10\sqrt{3}}{9} < 2$.
3. Let the pigeonholes be the intervals $\left[\frac{l}{n}, \frac{l+1}{n} \right)$, with $0 \leq l < n$, and the pigeons be the fractionary parts[1] of the numbers kx, with $1 \leq k \leq n - 1$. Note that if $\{kx\} \in [0, \frac{1}{n}) \cup [\frac{n-1}{n}, 1)$, for some integer $1 \leq k \leq n - 1$, then there is nothing left to do. Else, $\{kx\} \in \bigcup_{l=1}^{n-2} [\frac{l}{n}, \frac{l+1}{n})$ for $1 \leq k \leq n - 1$, and it suffices to apply the pigeonhole principle.

---

[1] Recall that the **fractionary part** of $x \in \mathbb{R}$ is the real number $\{x\}$, given by $\{x\} = x - \lfloor x \rfloor$. In particular, note that $\{x\} \in [0, 1)$, with $\{x\} = 0$ if and only if $x \in \mathbb{Z}$.

4. By contraposition, if $a_1 \le m, a_2 \le m, \ldots, a_n \le m$, then

$$\frac{a_1 + a_2 + \cdots + a_n}{n} \le \frac{n \cdot m}{n} = m.$$

The other case is entirely analogous.

5. Let us argue as in the previous problem: since each number is a summand of five sums of five consecutive numbers (around the circle), the total sum of all possible sums of five consecutive numbers equals

$$\frac{(1 + 15) \cdot 15}{2} \cdot 5 = 40 \cdot 15.$$

Since there are 15 sums of five consecutive numbers, the average value of each one of the is $\frac{40 \cdot 15}{15} = 40$. Therefore, by the result of the previous problem, at least one of these 15 sums is greater than or equal to 40.

6. Apply the result of Problem 4, counting the total number of color matchings of corresponding sectors in disks A and B, in each of the 200 possible matching positions.

7. Consider the 51 sets $\{0\}$, $\{1, 99\}$, $\{2, 98\}$, ..., $\{49, 51\}$, $\{50\}$. Since there are exactly 100 possible remainders upon division of an integer by 100 (the remainder varies from 0 to 99), the pigeonhole principle guarantees that at least two of the 52 given numbers have remainders belonging to a single one of these 51 sets. Therefore, either their sum or their difference leaves remainder 0 upon division by 100.

8. Letting n be a given natural number, take as pigeons the $n + 1$ numbers 1, 11, 111, ..., 11...1 (this last one with exactly $n + 1$ digits 1). Let the pigeonholes be the remainders upon division by n, i.e., $0, 1, \ldots, n - 1$.

9. Given a natural number n, adapt the hint given to the previous problem to show, with the aid of the pigeonhole principle, that n has multiples of the forms $a = 33\ldots300\ldots0$ and $b = 77\ldots700\ldots0$. Then, show that it is possible to choose $m \in \mathbb{N}$ such that either $a + 10^m b$ or $10^m a + b$ has odd sum of digits.

10. Condition $\gcd(a, n) = 1$ assures that none of the given powers of a is divisible by n. Thus, we can distribute them in boxes labelled from 1 to $n - 1$, according to their remainders upon division by n. If box 1 does contain at least one of the given powers, there is nothing left to do; otherwise, the $n - 1$ given powers of a shall be allocated among $n - 2$ boxes with labels varying from 2 to $n - 1$. By pigeonhole's principle, there exists $2 \le k \le n - 1$ such that the box labelled k contains at least two distinct powers of a, say a^i and a^j, with $1 \le i < j \le n - 1$. Then, we have $a^i = ns + k$ and $a^j = nt + k$, for some $s, t \in \mathbb{Z}$, and it follows that

$$a^i(a^{j-i} - 1) = a^j - a^i = n(t - s).$$

Therefore, $n \mid a^i(a^{j-i} - 1)$ and, since $\gcd(n, a^i) = 1$, we conclude that $n \mid (a^{j-i} - 1)$.

11. If one of the n sums $a_1, a_1 + a_2, a_1 + a_2 + a_3, \ldots, a_1 + a_2 + \cdots + a_n$ leaves remainder 0 when divided by n, there is nothing else to do. Otherwise, distribute them into boxes numbered from 1 to $n - 1$, according to their remainders upon division by n. Since we have n sums and $n - 1$ boxes, pigeonhole's principle guarantees that at least two sums will be in a single box. Letting $a_1 + \cdots + a_k$ and $a_1 + \cdots + a_k + \cdots + a_l$ be these two sums, conclude that $a_k + a_{k+1} + \cdots + a_l$ is a multiple of n.

12. Notice that each element of I_{2n} is of the form $2^k q$, with $k \geq 0$ and $q \in \{1, 3, 5, \ldots, 2n - 1\}$. Then, apply pigeonhole's principle, with the pigeonholes being the n odd numbers $1, 3, 5, \ldots, 2n - 1$ and the pigeons being the $n + 1$ selected elements of I_{2n}.

13. Each element $u \in A$ is of the form $u = 2^{\alpha_1} 3^{\alpha_2} 5^{\alpha_3} 7^{\alpha_4} 11^{\alpha_5}$. Take, as pigeonholes, all possible sequences $(\tilde{\alpha}_1, \tilde{\alpha}_2, \tilde{\alpha}_3, \tilde{\alpha}_4, \tilde{\alpha}_5)$, with $\tilde{\alpha}_i = 0$ or 1 for $1 \leq i \leq 5$. Then, place $u \in A$ into the pigeonhole $(\tilde{\alpha}_1, \tilde{\alpha}_2, \tilde{\alpha}_3, \tilde{\alpha}_4, \tilde{\alpha}_5)$ if and only if α_i and $\tilde{\alpha}_i$ have equal parities, for $1 \leq i \leq 5$.

14. By contraposition, assume that none of the 15 chosen elements is a prime. Then, Corollary 6.36 guarantees that each one of these 15 numbers has a prime divisor which is less than or equal to $\sqrt{1998} \approx 44,69$. Now, notice that there are exactly 14 primes less than or equal to 44, namely: 2, 3, 5, 7, 11, 13, 17, 19, 23, 29, 31, 37, 41 and 43. Since we have chosen 15 numbers, the pigeonhole principle assures that at least two of these will be divisible by a single one of these 14 primes and, this way, will not be relatively prime.

15. On the one hand, A has $2^{10} - 1 = 1023$ nonempty subsets. On the other, the sum of the elements of any such subset of A is at most $90 + 91 + 92 + \cdots + 99 = 945$. Hence, by the pigeonhole principle, at least two nonempty subsets of A, say B and C, have equal sums of elements. Thus, $B \setminus (B \cap C)$ and $C \setminus (B \cap C)$ are the subsets of A we are looking for.

16. If $A \subset \{1, 2, 3, \ldots, 2n + 1\}$ is free of sums, we claim that $|A| \leq n + 1$. Indeed, letting $A = \{1, 3, 5, \ldots, 2n + 1\}$, we have $|A| = n + 1$ and A is free of sums, for the sum of any two (not necessarily distinct) elements of A is an even number, while all elements of A are odd. Now, let $B \subset A$ be such that $|B| = n + 2$; we will show that B is not free of sums. If $B = \{x_1 < x_2 < \cdots < x_{n+2}\}$, then

$$1 \leq x_{n+2} - x_{n+1} < \cdots < x_{n+2} - x_1 \leq 2n.$$

If none of these $n + 1$ differences $x_{n+2} - x_i$ belongs to B, then all of it shall belong to $\{1, 2, 3, \ldots, 2n + 1\} \setminus B$. However, since this last set has only n elements, pigeonhole's principle assures that at least two of the numbers $x_{n+2} - x_i$ are equal. This is a contradiction

17. We first claim that, among the $n + 1$ chosen numbers, there are necessarily two, say x and y, such that $|x - y| \leq 2$. In order to verify such a claim, it suffices to apply the pigeonhole principle, distributing the $n + 1$ chosen elements among the n sets $\{3k - 2, 3k - 1, 3k\}$, with $1 \leq k \leq n$. Now, letting x and y be as above, assume, without loss of generality, that $y > x$. If $y - x = 1$, then $4xy + 1 = 4x(x + 1) + 1 = (2x + 1)^2$. If $y - x = 2$, then $xy + 1 = x(x + 2) + 1 = (x + 1)^2$.

18. We shall show that $f(n) = 2n$. To this end, let

$$\{1, 2, \ldots, 2n\} = C_1 \cup C_2 \cup \cdots \cup C_n$$

be an arbitrary of $I_{f(n)}$ into n sets. Since from n to $2n$ there are $n + 1$ integers, the pigeonhole principle guarantees the existence of $n \leq u < v \leq 2n$ and $1 \leq i \leq n$ such that $u, v \in C_i$. Taking $a = 2u - v \geq 0$ and $x = y = v - u \geq 1$ we have

$$\{a + x, a + y, a + x + y\} = \{u, v\} \subset C_i;$$

therefore, $f(n) \leq 2n$. Now, consider the following partition of I_{2n-1} into n sets:

$$\{1, 2, \ldots, 2n - 1\} = \bigcup_{k=1}^{n-1} \{k, k + n\} \cup \{n\};$$

we shall show that the stated conditions are not fulfilled. Indeed, if

$$\{a + x, a + y, a + x + y\} = \{u, v\} \subset C_k = \{k, k + n\},$$

then $a + x = a + y = k$ and $a + x + y = k + n$, so that $x = n$ and $a = k - x = k - n < 0$, which is an absurd. Finally, it is enough to see that $\{a + x, a + y, a + x + y\}$ has at least two elements and, hence, cannot be equal to $\{n\}$.

19. For item (a), assume that the 21 squares of the chessboard have been painted. Call a column *black* if it has at least two black squares, and *white* if it has at least two white squares. Since we have seven columns, the pigeonhole principle assures that at least four of them are of the same color, say (without loss of generality) black. We can also assume that these are the first four columns of the chessboard, from left to right. If any of these four columns has three black squares, there is nothing left to do: it suffices to take two black squares of another columns to get a rectangle with four black squares in the corners (see Fig. 22.1).

Otherwise, in each of these four black columns there are exactly 2 black squares. Since in a single column there are exactly three ways of choosing these two black squares, the pigeonhole principle assures that at least two of the four

Fig. 22.1 The first case of item (a)

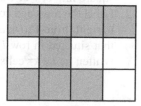

Fig. 22.2 The second case of
item (a)

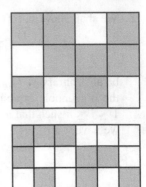

Fig. 22.3 A painting for item
(b)

columns will have two black squares in the same positions. These two pairs
of black squares in the two columns are the corner squares of a rectangle with
sides parallel to those of the chessboard (see Fig. 22.2).

For item (b), the total number of ways of choosing 2 out of 4 squares is $\binom{4}{2} =$
6. Thus, we have to paint black exactly two squares in each of the six columns,
choosing the two black squares at the columns in all six possible distinct ways
(for instance, squares 1 and 2 in the first column, 1 and 3 in the second, 1 and
4 in the third, 2 and 3 in the fourth, 2 and 4 in the fifth and 3 and 4 in the
sixth—see Fig. 22.3).

20. Choose any person, say A; by the pigeonhole principle, A exchange letters on a
single subject with at least $\lfloor \frac{16-1}{3} \rfloor + 1 = 6$ other people. Now, looking at these
six people, there are two cases to consider: either there are two of them which
correspond on the same subject as that of the letters they received/wrote to A—
and in this case there is nothing left to do—or any two of them correspond on
another one of the two left subjects. In this last case, repeat the above reasoning,
applying the pigeonhole principle to these six people, with any two of them
exchanging letters on one of two possible subjects.

Section 4.2

1. Start by observing that if a square can be partitioned into k other squares, then
   it can also be partitioned into $k + 3$ squares. Then, show how to partitionate a
   square into 6, 7 or 8 squares.
2. Label the rows and columns of the chessboard from top to bottom and from left
   to right, from 1 to n, and let a_{ij} denote the number written in the (i, j)-cell (i.e.,
   that situated in row i and column j). If S_k stands for the sum of the numbers
   written in the k^2 cells of a $k \times k$ chessboard, show that

$$S_n = S_{n-2} + a_{nn} - a_{n-1,n-1}.$$

Then, apply an inductive argument.

3. Give a convex polygon $A_1 A_2 \ldots A_k A_{k+1}$, its interior is equal to the union of the interiors of the convex polygons $A_1 A_2 \ldots \widehat{A_i} \ldots A_k A_{k+1}$, where the over A_i means that such a vertex is excluded. Hence, O belongs to one of these convex k-gons, say $A_1 A_2 \ldots A_k$. Now, insert an induction hypothesis and show directly that, for some $1 \le i \le k$, at least one of the angles $\angle A_i O A_{k+1}$ is not acute.

4. Make complete induction on the number k of sides of the polygon. For the induction step, consider a polygon P of n vertices, and draw some of its diagonals in such a way that no two of them intersect in the interior of P. Assume, without loss of generality, that these diagonals partitionate P into triangles. Fix one of the traced diagonals, say AB, and let P_1 and P_2 be the two polygons into which P gets divided by AB. Use the induction hypothesis to show that P_1 and P_2 possess a vertex different from A and B, which is not incident to none of the drawn diagonals.

5. By induction hypothesis, suppose we can paint k towers as described, for a certain $k < n^2$. Place $k+1$ towers in the chessboard, discard the left and upmost tower and apply the induction hypothesis to the k remaining towers. Then, paint the discarded tower with a color (out of the three given ones) distinct from the color(s) used to paint the towers (if any) which are closest to and belong to the same line or column of the discarded one.

6. Suppose that the given property is true for k lines, and draw the $(k + 1)$-th, say r. Choose one of the semiplanes into which r divides the plane and maintain the colors of all regions situated in this semiplane (even those resulting from a division of a former region in two by r); then, change the colors of all regions contained in the opposite semiplane (again, even those coming from a division of a former region in two by r).

7. Make induction on the number of columns of the board (independently of the total number of lines). To this end, say that the chessboard is é blue (resp. red) if the queen can visit all of its blue (resp. red) squares as prescribed in the statement of the problem. For the induction step, apply the induction hypothesis to the $m \times n$ board obtained from an $m \times (n + 1)$ one by the exclusion of its rightmost column. Upon doing this, consider three separate cases: (i) the smaller board is blue but nor red; (ii) the smaller board is red but not blue; (iii) the smaller board is at the same time blue and red. In case (i), if the erased column has any blue square, then at least one of the remaining squares of the line of this blue square is also blue; show that, this being the case, the bigger chessboard is also blue. In case (ii), argue in an analogous way. Finally, in case (iii), start by observing that we can assume that if any square of the erased column is blue (resp. red), then the remaining squares of the same line of this blue (resp. red) square are all red (resp. blue). Let the queen walk along a diagonal of the board to show that it still is blue and red.

8. For item (a), let r be a line passing through two of the n given points, say A
   and B, and which is at a minimum positive distance from one of the remaining
   points, say C. Argue by contradiction to show that, from the n given points, A
   and B are the only ones situated on r. For item (b), use induction. The initial
   case is $n = 3$, while the induction step is taken care of by the result of (a).

9. Make complete induction on n. For the induction step, assume the desired
   property to be true for $1 \leq n \leq k$, with $k \in \mathbb{N}$, and consider $2k + 2$
   points satisfying the given hypotheses. Let P be the smallest convex polygon
   containing such points (so that the vertices of P are some of these points).
   Consider two cases separately:

   (i) There are two consecutive vertices A and B of P such that A is blue and B
       is red: draw AB and apply the induction hypothesis.
   (ii) All vertices of P are monochromatic, say red: take an edge of P and call it
        $B_1 B_2$. Number the blue vertices $A_1, A_2, \ldots, A_k, A_{k+1}$, such that

   $$B_2 \widehat{B_1} A_1 < B_2 \widehat{B_1} A_2 < \cdots < B_2 \widehat{B_1} A_{k+1}.$$

   Define a function $f : \{1, 2, \ldots, k + 1\} \to \mathbb{N}$ by letting

   $$f(i) = \#(\text{pontos vermelhos no interior de } \angle B_2 B_1 A_i) + 2,$$

   so that

   $$2 \leq f(1) \leq f(2) \leq \cdots \leq f(k) \leq f(k + 1) \leq k$$

   (note that the last inequality comes from the fact that P has at least three
   vertices, at least one of which is not interior to $\angle B_2 B_1 A_{k+1}$). Conclude
   that there exists $2 \leq i \leq k$ such that $f(i) = i$, trace $B_1 A_i$ and apply the
   induction hypothesis.

10. Let us make induction on n. For $n = 1$ we can take $k_1 = 3$. Suppose the
    conclusion to be true for $n = l$, and consider a set of m points in the space,
    no four of which being coplanar and each line segment joining two of them
    having one of the numbers $1, 2, \ldots, l, l + 1$ associated to it. Fix one of these m
    points, say A, and look at the $m - 1$ line segments joining A to the remaining
    $m - 1$ points. Since each one of these $m - 1$ segments has associated to it one
    of $l + 1$ given numbers, the pigeonhole principle assures that, out of these $l + 1$
    numbers, a certain number x is associated to at least

    $$M := \left\lfloor \frac{(m - 1) - 1}{l + 1} \right\rfloor - 1$$

    distinct segments. If $M \geq k_l$, we can choose points $A_1, A_2, \ldots, A_{k_l}$ such that
    the line segments AA_i ($1 \leq i \leq k_l$) have x associated to themselves. We now
    look at the line segments joining two of the points $A_1, A_2, \ldots, A_{k_l}$. If one

of them, say $A_i A_j$, also has x associated to itself, there is nothing left to do: $AA_i A_j$ will be a triangle with x associated to all of its sides. Otherwise, points A_1, A_2, ..., A_{k_l} will be a set of k_l points in space, no four of which being coplanar and with one of the l numbers $1, \ldots, x-1, x+1, \ldots, l+1$ associated to each of the line segments joining two of them. By induction hypothesis, we get the existence of a triangle having the desired properties. In order to finish, it suffices to note that $M \geq k_l$ if $m \geq (l+1)(k_l - 1) + 2$.

Section 4.3

1. In each item from (a) to (d) one has to check that the corresponding relation is reflexive, antisymmetric and transitive. For item (a), we already know that $a = a$; $a \leq b$ and $b \leq a$ imply $a = b$; $a \leq b$ and $b \leq c$ imply $a \leq c$. The verification for (b) is quite similar, and we just point out that if Y and Z are subsets of X satisfying $Y \subset Z$ and $Z \subset Y$, then $Y = Z$. We leave item (c) as an equally easy exercise for the reader, referring to items (a) and (b) of Proposition 6.5 (if he/she finds it necessary) for complete arguments. Finally, for item (d), the desired properties are inherited from those of $\preceq$ in A.

2. Apply Mirsky's theorem to A, partially ordered by the divisibility relation.

3. If $(A, \preceq)$ is a partially ordered set such that $|A| = ab + 1$, with $a, b \in \mathbb{N}$, and $c = \min\{a, b\}$, then A contains a chain of $c + 1$ elements. The proof is an easy adaptation of the proof of Theorem 4.25. It starts by letting $n = ab + 1$, $A = \{x_1, \ldots, x_n\}$ and $\tilde{A} = \{(i, x_l); 1 \leq i \leq n\}$, ordered by $\preceq_1$, where

$$(i, x_i) \preceq_1 (j, x_j) \Leftrightarrow (i = j) \text{ or } (i < j \text{ and } x_i \preceq x_j).$$

The rest of the proof goes on almost unchanged.

4. Make induction on m, first noticing that the case $m = 1$ corresponds to the case $a = b$ in the Erdös-Szekeres theorem. For the induction step, write $n^{2^m} + 1$ as $(n^2)^{2^{m-1}} + 1$ and apply the Erdös-Szekeres theorem again.

5. Let $x \in A \cap B$ and assume, by the sake of contradiction, that A and B have distinct lengths, with B being the longest. Let $A_1 = \{y \in A; \ y \preceq x\}$, $A_2 = \{y \in A; \ x \preceq y\}$, $B_1 = \{z \in B; \ z \preceq x\}$, $B_2 = \{z \in B; \ x \preceq z\}$. Show that $A_1 \cup B_2$ and $A_2 \cup B_1$ are chains, and at least one of then is longer than A.

6. For item (a), assume that $A_i \cap B_0 = \emptyset$ for some index $1 \leq i \leq k$, apply the pigeonhole principle and arrive at a contradiction to the fact that B_0 is an antichain. For item (b), start by letting B_i be an antichain of k elements containing x_i. If $j \neq i$, show that $A_j \cap B_i \neq \emptyset$, take $y \in A_j \cap B_i$ and successively conclude that $y \preceq x_j$ and $x_j \npreceq x_i$. Now, reverse the roles of i and j to show that $x_i \npreceq x_j$. For item (c), i., start by observing that $B_i \setminus \{x_i\}$ is an antichain of $k - 1$ elements in $A \setminus C$. On the other hand, if $A \setminus C$ contained an antichain B of k elements, then B would intersect $\{z \in A_i; \ x_i \prec z\}$, which

contradicts the definition of x_i. For (c), ii., the induction hypothesis guarantees that $A \setminus C = C_1 \cup \ldots \cup C_{k-1}$, with $C_1, \ldots, C_{k-1}$ being pairwise disjoint chains; since C is itself a chain, we thus obtain $A = C_1 \cup \ldots \cup C_{k-1} \cup C$, a union of k pairwise disjoint chains. Finally, for (d), i. follows from the hypothesis of (d), together with the fact that a is a maximal element. Subitem ii. is straightforward: $A = \{a\} \cup A_1 \cup \ldots \cup A_k$.

Section 4.4

1. No. On the one hand, one can easily prove that, after any number of operations, each bottle will contain an amount of water expressed in liters by a fraction of the form $\frac{a}{2^k}$, with $a, k \in \mathbb{N}$. On the other hand, if one could reach the situation in which every bottle has the same amount of water, then each one of them would contain $\frac{11}{6}$ l of water.
2. Show that the trinomial's discriminant is an invariant for the allowed operations.
3. No! To see why, first look at the chessboard with its usual alternate black and white painting. Since each 2×1 rectangle covers exactly one black and one white square, after we have distributed any number of rectangular pieces, these will have covered equal numbers of black and white squares (this is the invariant that will solve the problem). Nevertheless, when we cut out two opposite squares, the remaining 62 squares of the chessboard comprise 32 squares of one color and 30 of the other. Hence, the cut out chessboard cannot be covered as asked.
4. If at some moment we replace a and b by $|a-b|$, then the difference between the sums of the numbers written on the blackboard before and after the operation is $(a+b) - |a-b|$, which is always even. Therefore, the parity of the sum of the numbers written on the blackboard at any moment is an invariant with respect to the allowed operations. Now, observe that $1 + 2 + \cdots + 1997 = 999 \cdot 1999$, which is odd.
5. If S and S' denote the sum of the numbers of two consecutive lists of six numbers, check that $S' \equiv S \pmod 7$. Then, note that $1+2+3+4+5+6 = 21$ and $3 + 7 + 2 + 15 + 8 + 8 = 43$, but $21 \not\equiv 43 \pmod 7$.
6. Label the coins consecutively from 1 to 10. Then, associate to each coin a *weight*, which is 0 if the coin is showing *head*, or is equal to the label of the coin, if it shows *tail*. Then, verify that each move does not change the parity of the sum of the weights associated to the coins.
7. Firstly, distribute the members in the two houses at random, and let S denote the sum of the numbers of enemies that each member has in the house he/she is in. Now, choose a member and look at his/her number of enemies within his/her present house; if this is 2, then move him/her to the other house and note that this operation diminishes the value of S. Now, repeat this as long as possible.
8. Note that $a + b + ab + 1 = (a + 1)(b + 1)$. Therefore, if at a certain moment the numbers written on the blackboard are $x_1, x_2, \ldots, x_k$, then the value of the

expression $(x_1 + 1)(x_2 + 1) \ldots (x_k + 1)$ does not change. In the beginning, such a value was $(1 + 1)(2 + 1) \ldots (n + 1) = (n + 1)!$. Hence, letting x denote the number written on the blackboard after $n - 1$ operations, we have $x + 1 = (n + 1)!$, so that $x = (n + 1)! - 1$.

9. No! In order to prove this, let a, b, c, d and a', b', c', d', respectively, denote the quantities of cards at the vertices of the square (in the clockwise sense), immediately before and after some step. If $f(w, x, y, z) = w - x + y - z$, it is immediate to check that $f(a', b', c', d') - f(a, b, c, d) = \pm 3k$, where k is the number of cards removed from one of the vertices at that step. In particular,

$$f(a', b', c', d') \equiv f(a, b, c, d) \, (\text{mod } 3).$$

Thus, the remainder of $f(w, x, y, z)$ upon division by 3 is an invariant associated to the allowed operations. Now, if there were a sequence of operations leading from the first to the second configuration, we should have

$$f(1, 9, 8, 9) \equiv f(1, 0, 0, 0) \quad \text{or} \quad f(1, 9, 8, 9) \equiv f(0, 1, 0, 0) \, (\text{mod } 3),$$

which is an absurd.

10. Start by drawing n line segments satisfying condition (a) and with distinct endpoints. Then, let A_1 and A_2 be blue points and V_1 and V_2 be red ones, such that $A_1 V_1$ and $A_2 V_2$ are two of the drawn line segments. If $A_1 V_1 \cap A_2 V_2 \neq \emptyset$, erase them and draw $A_1 V_2$ and $A_2 V_1$ instead. Now, use the (geometric) triangle inequality to show that the sum of the lengths of the n drawn line segments decreased. Finally, use the finiteness of the set of points, together with the semi-invariant consisting of the sum of the lengths of the n line segments, to show that, at some moment, condition (b) will be satisfied too.

11. A will do. To see why, let us paint the unit cubes black and white, so that adjacent cubes have different colors and each player moves from a unit cube of one color to another one of the other color. We consider two cases: • n is even: group the unit cubes into $n^3/2$ blocks $1 \times 1 \times 2$ and make A start in a white cube, always moving to the black cube of the same $1 \times 1 \times 2$ block. • n is odd: assume, without loss of generality, that the unit cubes placed at the corners of the bigger one are black. Ignore one of these black unit cubes and group the remaining $n^3 - 1$ ones into $(n^3 - 1)/2$ blocks $1 \times 1 \times 2$; again, A should start in the white cube of a block and follow the strategy of the first case.

12. For item (a), number the unit squares of the chessboard, modulo 4, as shown in the table below, noticing that there are 16 squares with each integer from 0 to 3. Any piece of one of the displayed shapes (and this is the invariant!) will occupy four unit squares such that the sum of the numbers written on them is a multiple of 4.

$$1\ 0\ 3\ 2\ 1\ 0\ 3\ 2$$
$$2\ 3\ 0\ 1\ 2\ 3\ 0\ 1$$
$$3\ 2\ 1\ 0\ 3\ 2\ 1\ 0$$
$$0\ 1\ 2\ 3\ 0\ 1\ 2\ 3$$
$$1\ 0\ 3\ 2\ 1\ 0\ 3\ 2$$
$$2\ 3\ 0\ 1\ 2\ 3\ 0\ 1$$
$$3\ 2\ 1\ 0\ 3\ 2\ 1\ 0$$
$$0\ 1\ 2\ 3\ 0\ 1\ 2\ 3$$

For item (b), take five unit squares of the chessboard, placed as shown in the figure below:

If it was possible to number the squares as desired, we would have, modulo 4,

$$a + b + c + d \equiv a + b + d + e \equiv a + b + c + e \equiv d + c + d + e.$$

Hence, it should be the case that $a \equiv c \equiv d \equiv e \pmod{4}$. Check that this would give 30 squares of the chessboard with numbers all congruent modulo 4, which is impossible.

13. Whatever the way of placing the cubes such that the faces touching the table form the board of a 6×6 square, each cube will have 3 visible and 3 invisible faces. For any such placement of the cubes, let v_B and v_W respectively denote the numbers of black and white visible faces. Since

$$|v_B - v_W| \equiv v_B - v_W \equiv v_B + v_W = 60 \equiv 0 \pmod{2},$$

we conclude that $|v_B - v_W|$ is always even (here we have the invariant!). Assume, without loss of generality, that $v_W > v_B$, and hence that $v_W \geq 31$ and $v_B \leq 29$. Show that there is at least one cube with more invisible than visible black faces, and then that it is possible to change the position of this cube in such a way that $|v_B - v_W|$ decreases two units.

14. Note that, after any number of operations, the quantity, say n, of written numbers does not change. On the other hand, thanks to the property we alluded to before the statement of the problem, the product P of the written numbers also does not change. Since there is just a finite number of ways of representing P as a product of n positive integers, there is a largest positive integer m that appears in the blackboard at some moment. We claim that such a number is a multiple of all the other ones written at this moment. For the sake of contradiction, assume that, at this moment, some number k written on the blackboard does not divide m. Then $\operatorname{lcm}(m, k) > m$, so that changing m and

k by $\gcd(m, k)$ and $\operatorname{lcm}(m, k)$, we conclude that m is not the largest possible number to appear on the blackboard. This is an absurd! Thus, by this claim, when m appears on the blackboard, the written numbers are $m_1, m_2, \ldots, m_{n-1}$ and $m_n = m$, with $m_1, m_2, \ldots, m_{n-1}$ dividing m. Now, show that m will always be on the blackboard, so that, for all purposes, the subsequent operations go on as if we were operating only with $m_1, \ldots, m_{n-1}$. Then, we can use the same argument as above with these $n - 1$ numbers, thus concluding that, after a finite number of operations, we shall reach a situation in which the written numbers $x_1, x_2, \ldots, x_n$ are such that $x_1 \mid x_2 \mid \cdots x_{n_1} \mid x_n$. From this time on, nothing changes on the blackboard.

Section 5.1

1. Consider a graph whose vertices are the 100 guests and in which two vertices are adjacent if and only if the corresponding guests know each other. Then, apply the result of Corollary 5.6.
2. Look at the possible degrees of a vertex of a graph with n vertices and, then, apply the pigeonhole principle. You may find it useful to review the discussion of Example 4.2.
3. Assume, without loss of generality, that $V_1 = V_2 = I_n$, for some $n \in \mathbb{N}$. If $f : I_n \to I_n$ is an isomorphism between G_1 and G_2, then f is a permutation of I_n; show that, applying f to the rows of $\operatorname{Adj}(G_1)$, we get $\operatorname{Adj}(G_2)$. Argue similarly for the converse.
4. Let $G_1 = (V_1; E_1)$, $G_2 = (V_2; E_2)$ and $G_3 = (V_3; E_3)$ be given graphs. If $f : V_1 \to V_2$ (resp. $g : V_2 \to V_3$) is a graph isomorphism between G_1 and G_2 (resp. between G_2 and G_3), show that $f^{-1} : V_2 \to V_1$ (resp. $g \circ f : V_1 \to V_3$) is a graph isomorphism between G_2 and G_1 (resp. between G_1 and G_3). For reflexivity, use the identity, $\operatorname{Id} : V_1 \to V_1$.
5. If $V = \{u_1, \ldots, u_n\}$ and $G = (V; E)$ is a labelled graph, then $E \subset \mathcal{P}_2(V)$; conversely, given $E \subset \mathcal{P}_2(V)$, we obtain a labelled graph $G = (V; E)$. Hence, it suffices to count the number of distinct subsets E of $\mathcal{P}_2(V)$, which is $2^{\binom{n}{2}}$.
6. Assume that such a graph, say G, does exist, and let a and b be its vertices of degree 1 and c its vertex of degree 6. Since the graph has seven vertices, c must be adjacent to both a and b. Therefore, if H is the graph obtained from G by the excision of a and b, then H has five vertices, of degrees 2, 3, 4, 5 and 4 (recall that c has lost two of its edges). But this is an absurd, for in H we have a vertex of degree 5 which can have at most four neighbors. Alternatively, apply Havel-Hakimi's algorithm.
7. Try to build the graph by imposing that the vertex of degree 3 is not adjacent to that of degree 6. Then, observe that the vertex of degree 6 must be adjacent to those of degrees 1, 1, 2, 4, 4 and 5. Alternatively, apply Havel-Hakimi's algorithm.

8. Argue by induction on n. For the induction step, if G is a graph of vertices $u_1, v_1, \ldots, u_{k-1}, v_{k-1}$, with $d_G(u_i) = d_G(v_i) = i$ for $1 \leq i \leq k-1$, add a vertex u to G and make it adjacent to $u_1, \ldots, u_{k-1}$; then, add a vertex v and make it adjacent to u.

9. Item (a) follows from the fact that, for given $u, v \in V$, we have $uv \in E$ or $uv \in E^c$; hence, $N_G(u) \cap N_{\overline{G}}(u) = \emptyset$ and $N_G(u) \cup N_{\overline{G}}(u) = V \setminus \{u\}$, so that

$$d_G(u) + d_{\overline{G}}(u) = \#(N_G(u) \cup N_{\overline{G}}(u)) = \#(V \setminus \{u\}) = n - 1.$$

For item (b), it suffices to observe that $E \cap E^c = \emptyset$ and $E \cup E^c = \mathcal{P}_2(V)$, so that

$$|E| + |E^c| = |E \cup E^c| = |\mathcal{P}_2(V)| = \binom{n}{2}.$$

10. Letting $G = (V; E)$ be a self-complementary graph, we have $|E| = |E^c|$. Hence, item (b) of the previous problem gives

$$|E| = \frac{1}{2}\binom{n}{2} = \frac{n(n-1)}{4},$$

so that $4 \mid n(n-1)$. The rest is immediate.

11. $G - B = (V \setminus B; E')$, where E' is the set of edges of G which are not incident to any vertex in B. $G - A = (V; E \setminus A)$.

12. For item (a), note firstly that both $G - u$ and $G_{|V(G)\setminus\{u\}}$ have $V(G) \setminus \{u\}$ as set of vertices. Now, if ϵ is an edge of G, then ϵ is an edge of $G - u$ if and only if ϵ is not incident to u, which is the same as asking that ϵ be an edge of $G_{|V(G)\setminus\{u\}}$. Finally, the proof of (b) is analogous to that of (a), once w apply the discussion of the previous problem.

13. The only vertices of G which change degree in H are u and v, and they do so in such a way that $d_H(u) = d_G(u) + 1$ and $d_H(v) = d_G(v) + 1$. Now, there are three essentially distinct possibilities: (i) u and v have odd degrees in G: then, u and v have even degrees in H, so that H has two odd degree vertices less than G; (ii) u and v have even degrees in G: then, u and v have odd degrees in H, so that H has two vertices of odd degree more than G; (iii) u has even degree and v has odd degree in G: then, u has odd degree and v has even degree in H, so that H has exactly the same number of vertices of odd degree as G. For the second part, let us start with a graph $G = (V; E)$, with $E = \{\epsilon_1, \ldots, \epsilon_m\}$. For $0 \leq k \leq m$, let $G_k = (V; E_k)$, with $E_k = \{\epsilon_{k+1}, \ldots, \epsilon_m\}$ for $0 \leq k < m$, and $E_m = \emptyset$. If v_k denotes the number of vertices of odd degree in G_k, then $v_m = 0$ and, hence,

$$v_0 = \sum_{k=0}^{m-1}(v_k - v_{k+1}) + v_m = \sum_{k=0}^{m-1}(v_k - v_{k+1}).$$

However, by the first part of the problem, we have $v_k - v_{k+1} = -2, 0$ or 2, so that $v_k - v_{k+1}$ is always even. Therefore, v_0, as a sum of even summands, is also even.

14. Item (a) will be left to the reader. For item (b), if $m = 2n$ and $1 \leq i, j \leq 2n$, then

$$\{i, j\} \in E \Leftrightarrow j \equiv i \pm n \pmod{2n}.$$

From this, conclude that vertex $i + n$ (modulo $2n$) is the only neighbor of vertex i; also, the edges of $G(2n, n)$ are those joining vertices i and $i + n$, for $1 \leq i \leq n$, so that there are exactly n edges. For item (c), given $1 \leq i \leq m$, show that vertex i is adjacent to vertices $i + n$ and $i - n$ (modulo m) and that $i + n \neq i - n$ (modulo m), thus giving $d(i) = 2$; for what is left to do, use Euler's theorem to conclude that $|E| = m$. Finally, in what concerns item (d), show that Id : $I_m \to I_m$ is an isomorphism between $G(m, n)$ and $G(m, m - n)$.

15. Verify that the desired graph is the one depicted in Fig. 5.4.

16. Firstly, note that the very definition of bipartite graph assures that, for $u, v \in V$, we have $uv \in E \Rightarrow u \in V_1$ and $v \in V_2$, or vice-versa. Hence, we can identify E with a subset of $V_1 \times V_2$ through the injection $uv \mapsto (u, v)$, if $u \in V_1$ and $v \in V_2$. Therefore, $|E| \leq |V_1 \times V_2| = |V_1| \cdot |V_2|$, with equality if and only if $E = V_1 \times V_2$.

17. If $|V_1| = |W_1|$ and $|V_2| = |W_2|$, take bijections $f_1 : V_1 \to W_1$ and $f_2 : V_2 \to W_2$; then, let $f : V_1 \cup V_2 \to W_1 \cup W_2$ be given by setting $f_{|V_1} = f_1$ and $f_{|V_2} = f_2$, and show that f is an incidence-preserving bijection. Conversely, let $f : V_1 \cup V_2 \to W_1 \cup W_2$ be an incidence-preserving bijection. Since V_1 is an independent set, the same happens to $f(V_1)$. Hence, $f(V_1)$ cannot intersect both W_1 and W_2, so that $f(V_1) \subset W_1$ or $f(V_1) \subset W_2$. Assume that $f(V_1) \subset W_1$ (the other case is entirely analogous). Since every vertex in V_2 is adjacent to every vertex in V_1, we conclude that every vertex in $f(V_2)$ is adjacent to every vertex in $f(V_1) \subset W_1$; then, it follows from the fact that W_1 is an independent set that $f(V_2)$ does not intersect W_1, i.e., $f(V_2) \subset W_2$. Now,

$$|V_1 \cup V_2| = |f(V_1 \cup V_2)| = |f(V_1)| + |f(V_2)|$$
$$\leq |W_1| + |W_2| = |W_1 \cup W_2| = |V_2 \cup V_2|,$$

with the last equality coming from the fact that f is an isomorphism. Therefore, $|f(V_1)| = |W_1|$ and $|f(V_2)| = |W_2|$, so that $f(V_1) = W_1$ and $f(V_2) = W_2$. Finally, this gives $|V_1| = |W_1|$ and $|V_2| = |W_2|$.

19. Assume that G has l vertices of degree 5 in V_1 and k vertices of degree 8, so that $|V_1| = k + l$ and $|V_2| = 16 - l$. Since the graph is bipartite, the edges departing from V_1 are incident to vertices in V_2, and vice-versa. Hence, $8k + 5l = 5(16 - l)$ or, which is the same, $4k + 5l = 40$. It is now immediate to verify that the nonnegative integer solutions of such an equation, with $k > 0$, are $(k, l) = (5, 4)$ or $(10, 0)$. In each one of these cases, exhibit the corresponding bipartite graph.

20. For item (a), apply Euler's theorem. For (b), start by letting the n vertices be those of a regular n-gon, and consider the two following cases: (i) if $r = 2m < n$, join each vertex to the $2m$ closest ones; (ii) if $r = 2m + 1 < n$ (so that n is even), join each vertex to the $2m$ closest ones, as well as to the vertex symmetric with respect to the center of the polygon.

21. Let $P_1, P_2, \ldots, P_{11}$ be the patrols and $h_1, h_2, \ldots, h_n$ be the volunteers, and join h_i to P_j if h_i is a member of P_j. Also, let k be the number of members of each patrol and A_i be the set of men in patrol P_j. Since we have n men altogether, it follows that

$$n = |A_1 \cup A_2 \cup \cdots \cup A_{11}|.$$

Therefore, by inclusion-exclusion,

$$n = |A_1 \cup A_2 \cup \cdots \cup A_{11}|$$
$$= \sum_i |A_i| - \sum_{i<j} |A_i \cap A_j| + \sum_{i<j<k} |A_i \cap A_j \cap A_k| - \cdots.$$

However, since no man belongs to three or more patrols, the only nonzero summands in the last expression above are the two first ones, so that

$$n = \sum_i |A_i| - \sum_{i<j} |A_i \cap A_j| = 11k - \binom{11}{2} = 11k - 55.$$

Now, let us count the number of edges of our (bipartite) graph in two distinct ways: on the one hand, since each man has degree 2, we have a total of $2n$ edges; on the other hand, since each patrol has degree k, the number of edges is equal to $11k$. Thus, $2n = 11k$. Finally, solving the linear system of equations

$$\begin{cases} n = 11k - 55 \\ 2n = 11k. \end{cases},$$

we get $n = 55$ and $k = 10$.

22. For item (a), adapt the reasoning used to solve the previous problem to get $g = m - 1$ and $n = \binom{m}{2}$. For (b), set $V_1 = \{a_1, a_2, \ldots, a_m\}$, $V_2 = \{b_{j,k}; 1 \le j < k \le m\}$ and let a_i to be adjacent to $b_{j,k}$ if and only if $i \in \{j, k\}$.

23. Look at the posed situation as a bipartite graph G, with the elements of the independent sets of vertices being the students and the problems, and such that a student and a problem are adjacent if and only if the student solved the problem. If each problem was solved by k students, apply double counting to conclude that $k = \frac{x}{2}$. Now, let E be the set of students, $e \in E$ and P be the set of the y problems solved by e. If H is the subgraph of G induced by $(E \setminus \{e\}) \cup P$ (cf. Problem 12), then H is also bipartite; apply double counting to H to conclude

that $3(x - 1) = y\left(\frac{x}{2} - 1\right)$. Then, deduce that $(x, y) = (4, 9)$ or $(8, 7)$. Finally, bipartite graphs corresponding to those possibilities can be easily constructed from the respective adjacency matrices A, which we exhibit below:

(i) $(x, y) = (4, 9)$:

$$A = \begin{bmatrix} 1 & 1 & 1 & 1 & 1 & 1 & 1 & 1 & 1 & 0 & 0 & 0 & 0 & 0 & 0 & 0 & 0 & 0 \\ 0 & 0 & 0 & 0 & 0 & 0 & 1 & 1 & 1 & 1 & 1 & 1 & 1 & 1 & 1 & 0 & 0 & 0 \\ 0 & 0 & 0 & 1 & 1 & 1 & 0 & 0 & 0 & 1 & 1 & 1 & 0 & 0 & 0 & 1 & 1 & 1 \\ 1 & 1 & 1 & 0 & 0 & 0 & 0 & 0 & 0 & 0 & 0 & 0 & 1 & 1 & 1 & 1 & 1 & 1 \end{bmatrix}.$$

(ii) $(x, y) = (8, 7)$:

$$A = \begin{bmatrix} 1 & 1 & 1 & 1 & 1 & 1 & 1 & 0 & 0 & 0 & 0 & 0 & 0 & 0 \\ 0 & 0 & 0 & 0 & 1 & 1 & 1 & 1 & 1 & 1 & 0 & 0 & 0 & 1 \\ 0 & 0 & 1 & 1 & 1 & 0 & 0 & 1 & 0 & 0 & 1 & 1 & 0 & 1 \\ 1 & 1 & 0 & 0 & 1 & 0 & 0 & 0 & 1 & 1 & 1 & 1 & 0 & 0 \\ 1 & 0 & 1 & 0 & 0 & 1 & 0 & 0 & 1 & 0 & 1 & 0 & 1 & 1 \\ 1 & 0 & 0 & 1 & 0 & 1 & 0 & 1 & 0 & 1 & 0 & 1 & 1 & 0 \\ 0 & 1 & 1 & 0 & 0 & 0 & 1 & 1 & 0 & 1 & 1 & 0 & 1 & 0 \\ 0 & 1 & 0 & 1 & 0 & 0 & 1 & 0 & 1 & 0 & 0 & 1 & 1 & 1 \end{bmatrix}.$$

24. By the sake of contradiction, assume that there is a way to place the given numbers as stated, and consider the graph $G = (V; E)$ such that $V = \{1, 2, \ldots, 13\}$ and $\{x, y\} \in E$ if and only if x and y are neighbors in the circle. Then, G has 13 vertices and 13 edges, and all vertices have degree 2. Now, look at the sets of vertices $A = \{1, 2, 3, 4, 10, 11, 12, 13\}$ and $B = \{5, 6, 7, 8, 9\}$; we can have at most two edges joining pairs of elements of A: one joining 1 and 4 and another joining 10 and 13. Since we have 13 edges altogether, two of them departing from each vertex, we conclude that at least $4 \cdot 2 + 4 \cdot 1 = 12$ edges join elements of A to elements of B. Since B has five elements, the pigeonhole principle guarantees that at least one of them must receive at least three edges, which is an absurd.

25. Item (a) is immediate and (b) follows from the definitions of chain, antichain, vertex cover and matching. For item (c), note that (from i.) a minimal vertex cover corresponds to a maximal antichain, whereas (from ii.) a maximal matching corresponds to a maximum collection of disjoint chains (thus containing all vertices of degree 0) whose union is $V_1 \cup V_2$.

26. For item (a), show that $\sum_{u \in V} d^+(u)$ counts each subset of two vertices exactly once. Item (b) follows directly from (a). For item (c), show that if (u, v) and (u, w) are edges of G, then there is exactly one transitive K_3 in G involving u, v and w; then, conclude that the number of distinct transitive K_3's in which u enters with out degree 2 is equal to $\binom{d^+(u)}{2}$. For item (d), note that the number of K_3's in G is exactly $\binom{n}{3}$; then, use the results of items (c) and (a). For item

(e), substitute $d^+(u) = (d^+(u) - \overline{d}^+) + \overline{d}^+$ in the formula of (d) and expand, using, whenever convenient, the result of (b).

27. Looking at the described situation in the language of the previous problem, conclude that the number of sets one asks to maximize is at most 112. Then, write down an explicit table of results for the $\binom{14}{2} = 91$ distinct matches, showing that it is indeed possible that one has 112 sets of three players satisfying the stated conditions.

Section 5.2

1.

$$b_{ij} = \sum_{l_1,\ldots,l_k} a_{il_1}^1 a_{l_1 l_2}^2 \cdots a_{l_{k-2}l_{k-1}}^{k-1} a_{l_{k-1}j}^k,$$

For item (a), if $M = (a_{ij})$, use the definition of adjacency matrix to show that a generic summand $a_{ik}a_{kj}$ of the sum defining entry (i, j) of M^2 is equal to 1 if and only if v_k is adjacent to both v_i and v_j. For item (b), make induction on k. Finally, for item (c), use the result of (b) and adapt the argument given to item (a).

2. Show that H coincides with the maximal connected subgraph of G containing u.

3. For the first part assume, by contraposition, that G has at least two connected components, so that G is the union of two disjoint subgraphs H_1 and H_2, respectively with k and l vertices. Use the fact that $k + l = n$, together with *discrete optimization* (cf. Remark 4.32) to get $\binom{k}{2} + \binom{l}{2} \le \binom{n-1}{2}$.

4. Use the result of the first part of the previous problem.

5. Use the comments preceding the statement of the problem to prove that each vertex of $G(m, n)$ belongs to the connected component of one of the vertices 1, 2, ..., d.

6. If G has two connected components H_1 and H_2, then each vertex of H_1 is adjacent to all vertices of H_2 in $\overline{G}$, so that $\overline{G}$ contains the complete bipartite subgraph with independent sets of vertices $V(H_1)$ and $V(H_2)$. The rest is now clear.

7. Create a *walk reduction algorithm* that, as final result, gives a path between two chosen vertices, provided there is a walk from one to the other. Basically, from a given walk such an algorithm should eliminate each portion of it of the form $(v = v_0, v_1, \ldots, v_k = v)$.

8. Execute the algorithm delineated in the proof of Theorem 5.24.

9. Adapt, to the present case, the proof of Theorem 5.24.

10. Let C_1 and C_2 be two paths in G, of largest possible lengths. By the sake of contradiction, suppose that C_1 and C_2 do not have common vertices. Take a

path C in G, joining a vertex of C_1 to a vertex of C_2 and having the smallest possible length. Show that C has only one vertex in common with C_1 and only another one in common with C_2. Now, concatenate a portion of C_1, C and a portion of C_2 to build a walk in G whose length is larger than that of C_1 or of C_2.

11. There are $\binom{n}{k-1}$ ways of choosing $k-1$ vertices out of $u_1, \ldots, u_n$, and each of such choices generates $(k-1)!$ distinct paths with the stated properties. Now, use the additive principle.

12. Adapt, to the present case, the proof of Theorem 5.24 in order to build a cycle in the given graph.

13. Use the result of the previous problem.

14. Let A, B, C, D and E be the mathematicians and A_1, A_2, B_1, B_2, C_1, C_2, D_1, D_2 and E_1, E_2, respectively, be the time intervals during which they slept. Build a graph having these time intervals as vertices, with two such vertices being adjacent if and only if the corresponding time intervals overlap. Then, show that such a graph contains a cycle.

15. Start by attributing the labels $1, 2, \ldots, k$ to a maximal walk (of length k) in G, that does not runs through the same edge twice.

16. We first claim that $G - \epsilon$ has at most two connected components. In order to prove this, let $\epsilon = \{u, v\}$, and let $w \neq u, v$ be another vertex of G. Since G is connected, we have in G a path C joining w and u. Now, there are two possibilities: (i) C does not pass through ϵ: then, we have in $G - \epsilon$ a path joining w and u; (ii) C passes through ϵ: since the vertices of C are pairwise distinct and u is the last one of them, the next to last must necessarily be v; hence, stopping the path C in v, we obtain a path in $G - \epsilon$ joining w to v. Anyhow, every vertex of $G - \epsilon$ can be joined to u or v by a path in $G - \epsilon$, so that $G - \epsilon$ has at most two connected components: one containing u and another containing v. We now claim that $G - \epsilon$ has at least two connected components. Indeed, if $G - \epsilon$ was connected, it would contain a path joining u and v, and such a path would necessarily have length at least 2; adding edge ϵ to such a path would give us a cycle in G, which is an absurd.

17. Let u be the vertex of G with maximum outdegree. If the edge joining u to v is oriented from u to v, there is nothing left to do; otherwise, prove that $d^+(u) + d^-(v) \geq n - 1$, and use this to conclude the existence of $w \in N_G^+(u) \cap N_G^-(v)$.

18. For item (a), number the cities from 1 to n and orient the road joining i and j, with $i < j$, from i to j. For item (b), suppose the wizard executed his intent somehow and note that, if one could always leave every city, then he/she would eventually return to a previously visited city, which is a contradiction. This way, conclude that there exists a city from which one cannot leave, label it as city n and exclude it from the subsequent analysis of the problem. Now, if it was always possible to leave each of the remaining $n - 1$ cities to another one of them, reach an analogous contradiction. Then, conclude that there is one of these remaining $n - 1$ cities such that one cannot leave it to another one of them (i.e., from there one can only reach city n). Label it as city $n - 1$ and exclude it from the subsequent analysis of the problem. Repeat this reasoning as long

as there are cities to consider and conclude that, when just one city is left, then one can go from there to any other city. Finally, for item (c), use the proof of (b) to conclude that every possible way the wizard has to implement his rules is equal, up to isomorphism, to the one described in (a). Then, show that there are exactly $n!$ distinct ways.

19. Given an optimal path $C = (c_0, c_1, \ldots, c_{n^2-1})$, we say that an edge of a unit square is *traversed* if it is a common edge of two consecutive unit squares c_i and c_{i+1} of the path. Consider the graph G having as its vertices all of the $(n+1)^2$ vertices of all of the n^2 unit squares of the chessboard, and as its edges the traversed edges of the unit squares. We classify the vertices of G in three distinct types: (I) the four vertices at the corners of the chessboard; (II) the remaining $4(n-1)$ vertices at the edges of the chess; (III) the $(n-1)^2$ vertices lying in the interior of the chessboard. Letting d_v denote the degree of vertex v, we have $d_v \leq 3$ for every v. Now, note that a path C contains an U if and only if some vertex has degree 3. By the sake of contradiction, assume that an optimal path C does not contain an U and conclude, with the aid of Euler's theorem, that the number m of traversed edges is at most $n^2 - 1$. Then, show that $m = n^2 - 1$ and conclude that the degree of each vertex of type (II) is 1, which is clearly impossible.

20. Let $u_1, u_2, \ldots, u_n$ be the vertices of the graph and $d_1, d_2, \ldots, d_n$ be their respective degrees. Also, let $d_{i1}, d_{i2}, \ldots, d_{id_i}$ be the degrees of the vertices adjacent to u_i. By the sake of contradiction, assume that

$$\frac{d_{i1} + d_{i2} + \cdots + d_{id_i}}{d_i} < \frac{d_1 + d_2 + \cdots + d_n}{n}.$$

Then,

$$\sum_{i=1}^{n} (d_{i1} + d_{i2} + \cdots + d_{id_i}) < \frac{1}{n} \left(\sum_{i=1}^{n} d_i \right) \sum_{i=1}^{n} d_i$$

$$= \frac{(d_1 + d_2 + \cdots + d_n)^2}{n}.$$

However, since each summand d_{ij} at the left hand side is counted exactly d_{ij} times (one time for each of the d_{ij} neighbors of the vertex having degree d_{ij}), it follows that

$$d_1^2 + d_2^2 + \cdots + d_n^2 < \frac{(d_1 + d_2 + \cdots + d_n)^2}{n}.$$

In turn, this contradict the inequality between the quadratic and arithmetic means (cf. problem 5.2.3 of [8], for instance).

22. Start by noticing that such a problem is a natural generalization of Problem 3. Then, adapt to the present setting the discussion of Example 2.29 and the hint given to Problem 20, page 32.

23. Let G be a graph satisfying the stated conditions, fix a vertex u of G and count its neighbors and the neighbors of their neighbors to conclude that $k \geq 5$. If $k = 5$, conclude that such a counting computed exactly one vertex of G twice, and that such vertex is neither u nor a neighbor of u. Then, show that u belongs to a single 4-cycle in G and, by the arbitrariness of u, that G can be partitioned into 4-cycles, which is an absurd. If $k = 6$, construct a graph G of 25 vertices in the following way:

- The set V of vertices of G is the disjoint union $V = \bigcup_{i=1}^{5} A_i$, where each A_i is a set of five vertices and the subgraph G_i of G induced by A_i is a 5-cycle.
- For $1 \leq i < j \leq 5$, establish a bijection $\phi_{ij} : A_i \to A_j$ such that, if $x, y \in A_i$ are adjacent in G_i, then $\phi_{ij}(x)$ and $\phi_{ij}(y)$ are nonadjacent in G_j. Then, impose that x and $\phi_{ij}(x)$ are adjacent in G, for all x, i and j as above.

Finally, show that the graph G thus constructed satisfies the stated condition.

24. Let us interpret the problem in the language of Graph Theory, in the following way: associate a vertex to each room and put an edge between two vertices if and only if there is a door between the corresponding rooms. For two rows of n rooms each, the condition in the statement of the problem amounts to the connectedness of the resulting graph G_n (of $2n$ vertices). Now, let a_n be the number of connected graphs G_n satisfying the given conditions. Firstly, check that $a_1 = 1$ and $a_2 = 5$. Then, use a recursive argument to show that $a_{n+1} = 4a_n + b_n$, where b_n is the number of graphs G_n having exactly two connected components. Finally, show that $b_n = 1 + 2a_1 + 2a_2 + \cdots + 2a_{n-1}$ and conclude, from this last recurrence, that $a_{n+2} = 5a_{n+1} - 2a_n$ for every $n \geq 1$.

Section 5.3

1. For the first part, prove that (a) $\Rightarrow$ (c) $\Rightarrow$ (b) $\Rightarrow$ (a). For the second, the established equivalences assure that the number of distinct paths in a tree with n vertices is equal to the number of distinct ways of choosing two of its vertices.

2. If u is a vertex of T with degree k, show that $T - u$ has k connected components, all of which are themselves trees. Then, apply the result of Lemma 5.32.

3. For item (a), use the result of Proposition 5.33. For (b), show that if $T \cup \{e\}$ did not contain cycles, then T would not be a spanning tree of G. For item (c), assume that G is not a tree and, hence, that it contains some cycle; show that G has at least two spanning trees.

4. Adapt, to the present case, the proof of the correction of Kruskal's algorithm (cf. proof of Proposition 5.36), showing that Dijkstra's algorithm builds a spanning

tree for the given graph and that the weight of such a tree is less than or equal to that of any other spanning tree.

5. Start by following Kruskal's or Dijkstra's algorithm to find one spanning tree with minimal weight.

6. Item (a) simply counts the number of odd integers in the list $1, 2, \ldots, \binom{n}{2}$. For item (b), start by painting a leaf of T, say in blue. Then, if some (but not all) vertices of T have already been painted, choose a vertex v of T which has not been painted yet and is adjacent to an already painted vertex u, and paint v with the color distinct from that of u. Continue this way until there are no unpainted vertices in T. For (c), show that a path has odd weight if and only if its endpoints have distinct colors. Then, use the hint given to the second part of Problem 1. Finally, for (d), use the results of (a) and (c), together with the fact that $x + y = n$.

Section 5.4

1. Given a vertex a of G and changing G by $\overline{G}$, if necessary, we can assume that $d_G(a) \geq 3$. Now, take vertices u, v and w in $N_G(a)$. If two of these vertices, say u and v, are adjacent in G, then a, u and v are the vertices of a K_3 in G. Otherwise, u, v and w are the vertices of a K_3 in $\overline{G}$.

2. Under the assumptions of (a), the pigeonhole principle guarantees that at least one of the remaining $2q - 2$ vertices will receive at least $\lfloor \frac{(2q-2)+1}{2q-1} \rfloor + 1 = 2$ edges from $\{u, v\}$; if w is such a vertex, then uvw is a K_3 in G. If we are as in (b), then $G - \{u, v\}$ has $2q - 2 = 2(q - 1)$ vertices and at least $(q^2 + 1) - ((2q - 2) + 1) = (q - 1)^2 + 1$ edges. By induction hypothesis, $G - \{u, v\}$ (and, hence, G) contains a K_3.

3. Apply the result of the previous problem.

4. Adapt, to the present case, the solution to Problem 2. More precisely, start by taking (from the result of that problem) a K_3 in G, with vertices a, b and c. Now, if one of the remaining seven vertices is adjacent to all of a, b and c, there is nothing to do. Otherwise, assume that k of the remaining seven vertices are adjacent to at least two of a, b and c, whereas the remaining $7 - k$ are adjacent to at most one of a, b and c. Then, we have at least $k + 1$ K_3's and at least $26 - 3 - (2k + (7 - k)) = 16 - k$ edges in $G - \{a, b, c\}$. If $k \geq 4$, there is nothing to do. Otherwise, $k \leq 3$. Separately consider the cases $k = 0, 1, 2$ and 3, arguing in the same way as above.

5. Let $u_1, \ldots, u_{l-1}$ be the vertices of a K_{l-1} in G. If $u_1, \ldots, u_{l-1}$ are all adjacent to at least one of the remaining $n - (l - 1) = (l - 1)(q - 1)$ vertices of G, then we have a K_l in G. Otherwise, at most $(l-2)(n-(l-1)) = (l-1)(l-2)(q-1)$ edges depart from one of $u_1, \ldots, u_{l-1}$ to the remaining vertices. Then, $H :=$ $G - \{u_1, \ldots, u_{l-1}\}$ has $(l - 1)(q - 1)$ vertices and at least

$$\frac{1}{2}(l-1)(l-2)q^2 - (l-1)(l-2)(q-1) - \binom{l-1}{2} = \frac{1}{2}(l-1)(l-2)(q-1)^2$$

edges. By induction hypothesis, H contains a K_l.

6. Construct a graph G having the teams as vertices and such that two of them are adjacent if they have played against each other. The stated conditions assure that $\alpha(G) \leq 2$, and Proposition 5.46 that $\omega(\overline{G}) \leq 2$. Conclude that $\overline{G}$ has at most 100 edges and, then, apply item (b) of Problem 9, page 132, to show that the tournament must have at least 90 matches. In order to build an example with exactly 90 matches, divide the teams in two groups of 10 and, within each group, let each two teams have a match.

7. Model the given situation by means of a graph, having the given people as vertices and with two people being adjacent if and only if they know each other. Then, use the given conditions to obtain upper bounds for $\alpha(G)$ and $\alpha(\overline{G})$.

8. Let two of the given points, say A and B, be joined by an edge if and only if $\angle AOB > 120°$. Elementary geometry guarantees that the resulting graph G does not contain a K_3's. Therefore, Turán's theorem assures that G has at most $T(21;3) = 110$ edges. Now, note that the number of arcs we wish to count is equal to the number of edges of $\overline{G}$, i.e. $\binom{21}{2} - 110 = 100$.

9. Draw the line segment joining two points if and only if its length is greater than $\sqrt{2}$. If G is the resulting graph, use Ptolemy's theorem on cyclic quadrilaterals (cf. Chapter 7 of [9]) to show that G does not contain a K_4. Then, apply Turán's theorem to get the desired upper estimate on the number of edges in G.

10. Consider the graph G having the n people as vertices and such that two people are adjacent if and only if they are friends. Start by observing that G has q edges and, by the stated conditions, does not contain K_3's. Show that, for each vertex u of G, the number of amicable pairs that can be formed from the enemies of u is equal to $q - \sum_{v \in N(u)} d(v)$; from this, conclude that we wish to assure the existence of a vertex u of G for which $\sum_{v \in N(u)} \geq \frac{4q^2}{n^2}$. To this end, note that

$$\sum_{u \in V(G)} \sum_{v \in N(u)} d(v) = \sum_{v \in V(G)} d(v)^2 \geq \frac{1}{n} \left(\sum_{v \in V(G)} d(v) \right)^2 = \frac{4q^2}{n},$$

where the used inequality is that between the arithmetic and quadratic means (cf. (cf. problem 5.2.3 of [8], for instance).

11. Let $m = \lfloor \frac{n}{k-1} \rfloor$ and $p = \lfloor \frac{(k-2)n}{k-1} \rfloor$. Also, let A be the set formed by the vertices u of G such that $d(u) \leq p$, and $B = V(G) \setminus A$, so that $d(v) \geq p+1$ for every $v \in B$. Assume, by contradiction, that $|A| < m$ and, hence, that $|B| > n - m$. Use Turán's theorem to show that $G_{|B}$ contains a K_k which is a contradiction.

12. Let G be the graph having the 1001 people as vertices and such that two people are adjacent if and only if they know each other. Apply the result of the previous problem to $\overline{G}$.

13. Let each element of S correspond to a vertex, and join vertices α and β if and only if the sets $A + \alpha$ and $A + \beta$ are disjoint (here, $A + k := \{x + k;\ x \in A\}$). Thus, letting $A = \{a_1, a_2, \ldots, a_{101}\}$, we conclude that an edge joins two elements α and β of S if $|\alpha - \beta| \neq |a_i - a_j|$, for all distinct $1 \leq i, j \leq 101$. Use the fact that $\#\{|a_i - a_j|;\ 1 \leq i < j \leq 101\} \leq 101 \cdot \frac{100}{2}$ to show that each vertex of the graph thus obtained has degree greater than or equal to $10^6 - 101 \cdot 100$, and note that the problem ends if we find a K_{100} in our graph.

Section 6.1

1. If $n = (a_k a_{k-1} \ldots a_1 a_0)_{10}$ is the decimal representation of the natural number n, then $n = \sum_{j=0}^{k} a_j 10^j$. Now, use the result of Example 6.4, together with Corollary 6.9.

2. Write $10^k = (11 - 1)^k$ and expand the right hand side with the aid of the binomial formula.

3. Write $n = \sum_{j=0}^{k} a_j 10^j$ and use the result of the previous problem, together with Corollary 6.9.

4. If $n = (abc)_{10}$ is the decimal representation of n, use the result of the previous problem to conclude that $a - b + c = 0$ or 11. Then, analyse these two cases separately.

5. If n has k algarisms, use the given conditions to show that $9^k \geq 10^{2(k-1)} - 10^k - 22$ and, then, to conclude that $k = 1$ or 2.

6. For $a \in \mathbb{N}$, start by observing that $(a + 1)(a + 2) \ldots (a + n) = \frac{(a+n)!}{a!}$.

7. Start by showing that a natural number is a multiple of 10 if and only if it is at the same time a multiple of 2 and 5. Thus, since the given number is clearly even, it suffices to show that it is a multiple of 5; to this end, use the result of Example 6.3.

8. Use the result of item (b) of Example 6.3.

9. In item (c), in order to show that $S \subset n\mathbb{Z}$, use the division algorithm and the minimality of n.

10. Start by analysing the first part of the problem. To this end, write $2^{64} + 1 = (2^{64} - 1) + 2$ and factorise $2^{64} - 1$.

11. Start by applying the existence part of the division algorithm to write $m = aq + m_0$, with $0 \leq m_0 \leq a - 1$. Then, arguing by induction, write $q = m_k a^{k-1} + m_{k-1} a^{k-2} + \cdots + m_1$, with $0 \leq m_j \leq a - 1$ for $1 \leq j \leq k$. Now, if

$$m = m_k a^k + m_{k-1} a^{k-1} + \cdots + m_1 a + m_0 = m_l' a^l + m_{l-1}' a^{l-1} + \cdots + m_1' a + m_0',$$

with $0 \leq m_i, m_j' \leq a - 1$ for all i's and j's, use the uniqueness part of the division algorithm to conclude that m_0 and m_0', being remainders upon division of m by a, are equal. Arguing once more by induction, apply the induction hypothesis to $\frac{m - m_0}{a}$ to conclude that $k = l$ and $m_j = m_j'$ for $1 \leq j \leq k$.

13. For items from (a) to (e), it suffices to perform judicious applications of the results of items (a) and (b) of the previous problem. Regarding item (f), start by using the result of item (b).

14. Start by analysing the case in which x belongs to an interval of the form $(n, n + \frac{1}{2})$, for some $n \in \mathbb{Z}$.

15. Start by showing that, under the given conditions, we have $\frac{a^2}{b} + \frac{b^2}{a} + 1 > \frac{a^2+b^2}{ab} - 1 + ab$ and, then, that $1 + (a + b - 1)(a^2 + b^2 - ab) > a^2b^2$. Finally, conclude that $a \geq b \Rightarrow b = 1$.

16. Expanding the given equality we obtain $(x + y)(x + z) = 2xz$. Conclude that one of the sides of this last equality is a multiple of 4, while the other is not.

17. Apply items (a) and (b) of Corollary 6.8.

18. For item (b), use the result of (a); for (a), write $n = 7q + r$, with $r = 0, 1, 2, 3, 4, 5$ or 6, and compute n^3.

19. We certainly have $n > 1$, and can assume $x \leq y < z$; if $x = y$, start by showing that $n = 2$; if $x < y$, write $1 = n^{y-x}(n^{z-y} - 1)$.

20. For $k \geq 5$, show that $k!$ end with 0; then, use the result of Example 6.10.

21. Use the result of Example 6.10.

22. For item (a), write $n = 2^r q$, with r being a nonnegative integer and q an odd natural number. Then, use the given condition to conclude that $q = 1$.

23. By the sake of contradiction, assume this is not the case, and let $N \in \mathbb{N}$ be such that

$$n \geq N \Rightarrow s(3^{n+1}) > s(3^n).$$

Problem 1 assures that $s(3^k)$ is a multiple of 9 for $k \geq 2$, so that

$$s(3^{n+1}) \geq s(3^n) + 9, \quad \forall n \geq N.$$

Adding the above inequality for $n = N, N + 1, \ldots, M - 1$ to get $s(3^M) \geq s(3^N) + 9(M - N) = 9(M - c)$, for some integer $c < M$, which doesn't depend on the chosen $M \geq N$. Now, note that $3^M < 10^{M-c}$ for all sufficiently large M (cf. Section 7.2 of [8], for instance); however, since the greatest possible value for $s(m)$ when $m < 10^{M-c}$ is

$$s(10^{M-c} - 1) = s(\underbrace{99\ldots9}_{M-c}) = 9(M - c),$$

we have arrived at a contradiction.

Section 6.2

1. Apply Euclid's algorithm to $a = 21n + 4$ and $b = 14n + 3$.

2. Firstly, write $2^m = (n - 1)(n + 1)$; secondly, compute all possible values of $\gcd(n - 1, n + 1)$ and, then, use item (b) of Corollary 6.23.

3. First show that, if there exists a natural number which is a term of both progressions, then there will exist infinitely many such natural numbers. Secondly, for the characterization of the existence of a natural number which is a term of both progressions, use Proposition 6.26.

4. Revisit the hint given to Problem 10, page 163.

5. Start by showing that the gcd we wish to compute divides $(n+1)! - n! = n \cdot n!$.

6. Let $d = \gcd(a, b)$ and write $a = du$ and $b = dv$, so that $\gcd(u, v) = 1$. Conclude that $ab \mid (a^2 + b^2)$ if and only if $uv \mid (u^2 + v^2)$. If this is so, since $u, v \mid uv$, we have that $u, v \mid (u^2 + v^2)$ and, hence, $u \mid v^2$ and $v \mid u^2$. Now, use the fact that $\gcd(u, v) = 1$ to obtain $u = v = 1$. Alternatively, write $a^2 + b^2 = abk$, with $k \in \mathbb{N}$, and examine the discriminant of the second degree equation $x^2 - (bk)x + b^2 = 0$, which has a as a natural root.

7. Firstly, analyse the case $\gcd(a, b) = 1$. In this case, let the rectangle has vertices $(0, 0)$, $(a, 0)$, (a, b) and $(0, b)$. Start by showing that if one of its diagonals has a point in common with an 1×1 square, then it crosses the interior of such a square. Subsequently, show that such a diagonal intersects the column of 1×1 squares bounded by the vertical lines $x = j$ and $x = j + 1$ in exactly $\lfloor \frac{b}{a}(j + 1) \rfloor + 1 - \lfloor \frac{b}{a}j \rfloor$ squares.

8. Let d be the desired gcd. It follows from the columns' theorem of Pascal triangle (cf. Section 4.2 of [8]) that d divides the sum $\binom{n+k+1}{k+1}$ of the given numbers. Therefore, d divides $\binom{n+k+1}{k+1} - \binom{n+k}{k} = \binom{n+k}{k+1}$. From this, and with the aid of Stifel's relation (once more according to Section 4.2 of [8]), conclude that d divides $\binom{n+1}{k+1}, \binom{n+2}{k+1}, \ldots \binom{n+k}{k+1}$. Finally, argue inductively.

9. Let d be the gcd we wish to compute. Then, d divides the sum of the given numbers, which (by the lines' theorem of the Pascal triangle—cf. Section 4.2 of [8]) is equal to 2^{2n-1}. Thus, d is a power of 2. Write $n = 2^k q$, with $q \in \mathbb{N}$ odd, and show that

$$\binom{2n}{2t - 1} = 2^{k+1} \cdot \frac{(2n - 1)q}{(2t - 1)(2n - 2t + 1)} \binom{2n - 2}{2t - 2}.$$

From this computation, conclude that 2^{k+1} divides all of the given binomial numbers.

10. If $d = \gcd(n^2 + k, (n+1)^2 + k)$, then $d \mid [((n+1)^2 + k) - (n^2 + k)]$. Therefore, $d \mid [(2n+1)^2 - 4(n^2 + k)]$, and from this it is easy to conclude that $d \mid (4k + 1)$, and hence $d \leq 4k + 1$. Finally, exhibit a value of n for which the corresponding gcd is equal to $4k + 1$.

11. Start by writing $a = du$ and $b = dv$, with $d = \gcd(a, b)$ and $\gcd(u, v) = 1$; then, conclude that $u, v \mid c$ and, hence, that $uv \mid c$.

12. Let $x^2 + y^2 - x = 2xym$, with $m \in \mathbb{N}$. Then, by looking at such an equality as a second degree equation in x, show that the fact that x is an integer guarantees that the discriminant $\Delta = (2ym + 1)^2 - 4y^2 = (2ym + 1 - 2y)(2ym + 1 + 2y)$ must be a perfect square. Finally, show that $\gcd(2ym + 1 - 2y, 2ym + 1 + 2y) = 1$ and use the result of Corollary 6.23.

13. Firstly, use the pigeonhole principle (cf. Sect. 4.1) to show that at least one of the five odd elements of our set is not a multiple of 3, 5 or 7. Now, let a be such an element, and b be any other element of the given set. Use the fact that $\gcd(a, b) \mid (a - b)$ to show that if $\gcd(a, b) > 1$, then it is divisible by 2, 3, 5 or 7. Finally, conclude that a is the element we are looking for.

14. Write $aj + b = m \lfloor \frac{aj+b}{m} \rfloor + r_j$, with $0 \le r_j < m$; then, use the condition $\gcd(a, m) = 1$ to conclude that $r_0, r_1, \ldots, r_{m-1}$ are pairwise distinct.

15. Write $r = du$ and $n = dv$, so that $\gcd(u, v) = 1$; then, let $\{ \frac{ui}{v} \} = \frac{ui}{v} - \lfloor \frac{ui}{v} \rfloor$ and use the result of the previous problem.

16. Use Bézout's theorem to substitute $\gcd(m, n) = mx + ny$, with $x, y \in \mathbb{Z}$, and develop the expression thus obtained.

17. Start by proving that $a_{n+1} = a_1 \ldots a_n + 1$ (cf. item (a) of problem 15 of [8]).

18. Prove item (a) by induction on n and item (b) by induction on k. As for item (c), for subitem (i) use (b) to show, also by induction, that $F_n \mid F_{nq}$; for subitem (ii), use (i) and (a); for (iii), make induction on $q \ge 0$, using (b) and (ii) to perform the induction step. Finally, for item (d), adapt Euclid's algorithm to the situation at hand, with the aid of the results of item (c).

19. Adapt, to the present case, the several items and hints given to the previous problem.

20. For item (a), let u and v be natural numbers such that $au - bv = 1$. The fact that $n > ab$ gives us $nau - nbv = n > ab$ and, hence, $\frac{nu}{b} - \frac{nv}{a} > 1$. Thus, there exists an integer t such that $\frac{nv}{a} < t < \frac{nu}{b}$. Let $x = nu - bt$, $y = at - nv$ and obtain (b) arguing by contradiction; then, reduce (c) to (a) and (d) to (b). Concerning (e), let $S = ab - a - b$ and show that, for $0 \le m < S$, exactly one of the numbers m and $S - m$ can be written as prescribed.

21. For (a), arguing by contradiction, let $x, y, z \in \mathbb{Z}_+$ such that $2abc - ab - bc - ca = xbc + yac + zab$. Then,

$$2abc = (x + 1)bc + (y + 1)ac + (z + 1)ab,$$

so that $a \mid (x + 1)bc$. Since $\gcd(a, bc) = 1$, we conclude that $a \mid (x + 1)$ and, hence, that $x + 1 \ge a$. Similarly, $y + 1 \ge b$ and $z + 1 \ge c$. But these give

$$(x + 1)bc + (y + 1)ac + (z + 1)ab \ge a \cdot bc + b \cdot ac + c \cdot ab = 3abc,$$

which is a contradiction. For item (b), if $n > 2abc - ab - bc - ca$, then $n > a \cdot bc - a - bc$ and the previous problem guarantees (since a and bc are relatively prime) the existence of integers $x, t \in \mathbb{Z}_+$ such that $n = xbc + ta$. Without loss of generality, we can assume that $0 \le x < a$; indeed, if $x \ge a$, write $x = aq + x'$, with $0 \le x' < a$, thus obtaining $n = x'bc + (t + qbc)a$. For (c), letting $x \le a - 1$, we have

$$ta = n - xbc \ge (2abc - ab - bc - ca) - (a - 1)bc = abc - ab - ac,$$

whence $t > bc - b - c$. However, since b and c are relatively prime, by invoking once more the result of the previous problem we conclude that there exist $y, z \in \mathbb{Z}_+$ such that $t = bz + cy$. Finally, for item (d), in the notations of the previous items we get

$$n = xbc + ta = xbc + (bz + cy)a = xbc + yac + zab.$$

22. Adapt the items and hints given to the previous problem and make induction on $n \geq 2$.
23. For the first part of item (a), use induction. For item (b), show that if $0 < a_1 < a_2 < a_3 < \cdots < a_{n-1}$ are the remainders obtained in the execution of Euclid's algorithm when $a = a_{n+1}$ and $b = a_n$, then $a_j \geq F_{j-1}$ for $1 \leq j \leq n$. Finally, show that (c) follows from (a) and (b).
24. Let k be the only positive integer such that $2^k \leq n < 2^{k+1}$, and let $M = \text{lcm}(1, 2, \ldots, n)$. Show that $2^k \mid M$ and

$$\sum_{j=1}^{n} \frac{1}{j} = \frac{1}{M} \sum_{j=1}^{n} \frac{M}{j},$$

with $\frac{M}{j}$ being odd if and only if $j = 2^k$.

Section 6.3

1. Firstly, $p + q$ is even. Also, if $p < q$, then $p < \frac{p+q}{2} < q$, so that $(p+q)/2$ is not prime.
2. Start by writing $\frac{m}{n} = \sum_{j=1}^{2k+1} \frac{1}{j} - 2 \sum_{j=1}^{k} \frac{1}{2j} = \sum_{j=k+1}^{2k+1} \frac{1}{j}$. Then, group the summands of this last sum in pairs to write it as a sum of fractions with all numerators equal to $3k + 2 = p$. Finally, use the fact that p is prime.
3. We can assume that $x \neq y$. If $x < y$, prove that $x < p < y$ and, hence, that $\gcd(p, x) = 1$. Write $2xy = p(x + y)$ to conclude that $p \mid y$. Let $y = pz$, with $z \in \mathbb{N}$, to get $z = \frac{x}{2x-p}$, thus showing that $(2x - p) \mid x$ and, hence, that $(2x - p) \mid p$. Finally, this gives $2x - p = 1$.
4. Start by separately considering the cases $n = 6k, 6k+1, \ldots, 6k+5$. For $n = 6k$, for instance, use the fact that $k, 2k+1, 3k+1$ and $6k+1$ are pairwise relatively prime to conclude that $k = 1$.
5. Start by using the given condition to show that $n \mid a_n$. Then, if p is a prime number which does not divide n, show that $p \nmid a_n$.
6. For item (a), use the FTA (for whose proof we did not use the infiniteness of prime numbers). For the first part of (b) use, beside the result of item (a), the fundamental principle of counting. Finally, show that the result of (b) generates the inequality $n \leq 2^k \sqrt{n}$, which, in turn, generates a contradiction.

7. In each case, adapt the idea of the proof of Example 6.40. To this end, observe that every prime $p \neq 3$ is of the form $3k \pm 1$ and every prime $p \neq 2, 3$ is of the form $6k \pm 1$.

8. Factorise $2^{2^n} - 1$ and use the result of Problem 4, page 177.

9. Use Lagrange's identity—cf. Example 2.9—to write $\binom{2p}{p}$; then, use the result of Example 6.42.

10. For item (a), show that if n is not a power of 2, then it is possible to factorise $a^n + 1$. For item (b), argue similarly.

11. Consider the cases n even and n odd separately. In the case $n = 4k$, compute $\gcd\left(\frac{n}{2} - 1, n\right)$; in the case $n = 4k + 2$, compute $\gcd\left(\frac{n}{2} - 2, n\right)$; finally, in the case of an odd n, compute $\gcd\left(n, \frac{n+1}{2}\right)$.

12. Factorise $c_{m+j} - c_k$ and, then, apply the result of Problem 6, page 163.

13. If $n = p_1^{\alpha_1} \ldots p_k^{\alpha_k}$ is the canonical decomposition of n, write $u = p_1^{\beta_1} \ldots p_k^{\beta_k}$ and $v = p_1^{\gamma_1} \ldots p_k^{\gamma_k}$, with $\beta_i, \gamma_i \geq 0$, and compute how many are the ordered pairs (β_i, γ_i) for which $\max\{\beta_i, \gamma_i\} = \alpha_i$. Then, apply the result of Corollary 6.48.

14. Write $\frac{a}{b^2} = \frac{c}{d}$, with $\gcd(c, d) = 1$, and deduce that $b = x^d$ and $a = x^c$, for some $x \in \mathbb{N}$. Then, conclude that $x = 1$ or $cx^{2d} = dx^c$. In the second case, analyse the subcases $c < 2d, c = 2d$ and $c > 2d$.

15. If n is odd, then there are no solutions, for the left and right hand sides of the given equality have distinct parities. If n is even, then $n = 5 + d_3^2 + d_4^2$, with d_3 and d_4 having distinct parities. If n is a multiple of 4, show that $n = p^2 + 21$, where p is the smallest odd prime that divides n; then, show that $p \mid 21$ and conclude that there are no solutions. If n is of the form $4k + 2$, show that $n = 5(p^2 + 1)$, where, as above, p is the smallest odd prime divisor of n. Conclude that $p = 5$ and, then, that $n = 130$.

16. Write $n = d_{13}q_{13} = d_{14}q_{14} = d_{15}q_{15}$, with $q_{13}, q_{14}, q_{15} \in \mathbb{N}$. Condition (b) is, thus, equivalent to $\frac{1}{q_{13}} + \frac{1}{q_{14}} + \frac{1}{q_{15}} = 1$. Therefore, start by getting all natural numbers $a < b < c$ for which $\frac{1}{a} + \frac{1}{b} + \frac{1}{c} = 1$.

17. Examine the cases $1 \leq n \leq 30$ to conjecture that all naturals of the form $2^k - 2$ satisfy the stated property. Then, prove this fact by writing, for $n = 2^k - 2$,

$$I(n) = \sum_{j=1}^{2^{k-1}-1} d(2j - 1) + \sum_{i=1}^{k-1} \left(d(2^i) + d(2^i \cdot 3) + \cdots + d(2^k - 2^i) \right)$$

$$= \left(2^{k-1} - 1\right)^2 + \sum_{i=1}^{k-1} \left(1 + 3 + 5 + \cdots + (2^{k-i} - 1)\right)$$

$$= \left(2^{k-1} - 1\right)^2 + \sum_{i=1}^{k-1} 2^{2k-2i-2}$$

$$= \left(2^{k-1} - 1\right)^2 + \frac{2^{2k-2} - 1}{3} = \frac{(2^k - 2)(2^k - 1)}{3}.$$

18. First note that $3 \mid n$. Now, write $n = 2^\alpha \cdot 3^\beta p_1^{k_1} \cdots p_t^{k_t}$, with $3 < p_1 < \cdots < p_t$ being primes and $\alpha \geq 0, \beta \geq 1$ being integers. Then, note that the given relation reduces to $3(\alpha + 1)(\beta + 1)(k_1 + 1) \cdots (k_t + 1) = 2^\alpha 3^\beta p_1^{k_1} \cdots p_t^{k_t}$. Conclude that $\beta = 1$ or 2 and n does not possess other prime factors apart from 2 or 3. Finally, analyse these two cases separately.

19. For an arbitrary prime number p, it suffices to show that $e_p(2m) + e_p(2n) \geq e_p(m) + e_p(n) + e_p(m+n)$. To this end, use Legendre's formula, together with item (f) of Problem 13, page 164.

20. For item (a), use Legendre's formula to show that $e_2(2n) \geq 2e_2(n) + 1$ for every $n \in \mathbb{N}$. Alternatively, use Lagrange's identity—cf. Example 2.9—and induction. For item (b), refine the computations of the first hint to (a) to show that $e_2(2n) \geq 2e_2(n) + 2$ if and only if n is not a power of 2.

21. By contradiction, let $d_1 < d_2 < \sqrt{n}$ be divisors of n, so that $d_1 < d_2 < \frac{n}{d_2} < \frac{n}{d_1}$. It suffices to show that there exists no perfect square between d_1 and $\frac{n}{d_1}$. Letting $d_1 = a, d_2 = a+u, \frac{n}{d_2} = a+v$ and $\frac{n}{d_1} = a+w$, with $0 < u < v < w$, we have $a(a+w) = n = (a+u)(a+v)$. If $k \geq 1$ is such that $(k-1)^2 \leq a < k^2$, we claim that $k^2 < a + w$. Indeed, since $\sqrt{a} < k \leq \sqrt{a} + 1$, it suffices to show that $(\sqrt{a} + 1)^2 < a + w$ or which is the same, $2\sqrt{a} < w - 1$. From $a(a + w) = (a + u)(a + v)$ we obtain $aw = au + av + uv$, so that $a \mid uv$ and $a \leq uv$. It comes that $aw = au + av + uv \geq au + av + a$, whence $w \geq u + v + 1$. Then, $w - 1 \geq u + v \geq 2\sqrt{uv} \geq 2\sqrt{a}$. Finally, we cannot have equality in the last inequality, for otherwise $u = v \Rightarrow a + u = a + v \Rightarrow d_2 = \frac{n}{d_2}$, which is impossible.

22. Write $n = a_0 + a_1 \cdot 2 + a_2 \cdot 2^2 + \cdots + a_m 2^m$, with $a_j \in \{0, 1\}$ for $0 \leq j \leq m$, so that $a_0 + a_1 + \cdots + a_m = k$. Then, compute

$$\left\lfloor \frac{n}{2^j} \right\rfloor = a_j + a_{j+1} \cdot 2 + \cdots + a_m \cdot 2^{m-j}$$

and, finally, use Legendre's formula to show that $e_2(n) = n - k$.

23. Let p be prime and $t \in \mathbb{N}$ be such that $(a + kb)(b + ka) = p^t$. Assuming, without loss of generality, that $a \leq b$, we have $a + kb \geq b + ka > 1$, so that there exist $r, s \in \mathbb{Z}$ for which $1 \leq r \leq s, r + s = t$ and $a + kb = p^s$, $b + ka = p^r$. Conclude that $(b+ka) \mid (a+kb)$ and, writing $\frac{a+kb}{b+ka} = k - \frac{(k^2-1)a}{b+ka}$, that $(b + ka) \mid (k^2 - 1)$. Now, consider two separate cases: (i) if $p > 2$, use the fact that $b+ka$ is a power of p and $\gcd(k-1, k+1) \leq 2$ to conclude that $b+ka$ divides $k - 1$ or $k + 1$, so that $b + ka \leq k + 1$ and, hence, $b = a = 1$. Then, show that the solutions are the 3-tuples $(a, b, k) = (1, 1, p^r - 1)$, for every integer $r \geq 1$ and every prime $p > 2$. (ii) If $p = 2$, start by observing that at least one of a and b is odd, so that k is also odd. Now, since $(b + ka) \mid (k^2 - 1)$ and $b + ka = 2^r$, show that $(b + ka) \mid 2(k + 1)$ or $2(k - 1)$; in particular, $b + ka \leq 2k + 2$ and, hence, $a = 1$ or 2. If $a = 2$, arrive at a contradiction; if $a = 1$, we have that $(b + k) \mid 2(k + 1)$ or $k - 1$, which splits the analysis into that of two subcases: $(b + k) \mid (2k + 2)$ or $(b + k) \mid (2k - 2)$. In the first

subcase, show that $b = k + 2$ or $b = 1$. If $b = 1$, conclude that $k = 2^r - 1$; if $b = k + 2$, write $s = 2u$ to get $(a, b, k) = (1, 2^u + 1, 2^u - 1)$, for some $u \in \mathbb{N}$. In the second subcase, show that $b = k - 2$ and, then, let $s = 2u$ to get $(a, b, k) = (1, 2^u - 1, 2^u + 1)$, for some $u \in \mathbb{N}$.

24. Choose a prime p greater than $1 + 2 + \cdots + 1000$ and show that the set $A = \{p, 2p, 3p, \ldots, 1000p\}$ satisfies the required conditions.

25. For item (a), use the result of Problem 4, page 177, together with the fact that $p(q^{2^k}) = q$ for every $k \geq 0$. For item (b), factorise $q^{2^k} - 1$ and use the minimality of k to conclude that one cannot have $p(q^{2^{k-1}} - 1) > p(q^{2^{k-1}})$.

26. If t is the product of all primes less than or equal to m, let $y_1 = x_{t_1}$, where p_{t_1} is a prime greater than t. Letting p be the greatest prime divisor of y_1 and $r \in \mathbb{N}$ be such that $p = p_r$, define $y_2 = x_{t_2}$, with $t_2 > r, t_1$, and convince yourself that $\gcd(y_1, y_2) = 1$. Once you have chosen elements $y_1 = x_{t_1} < \cdots < y_j = x_{t_j}$ of A satisfying item (b), with $j < m$, let p be the greatest prime factor of y_j and $r \in \mathbb{N}$ such that $p = p_r$; define $y_{j+1} = x_{t_{j+1}}$, with $t_{j+1} > r, t_j$. Continue in this way until $j = m$. Then, show that the set B thus obtained does satisfy condition (c).

Section 7.1

1. If (x_0, y_0, z_0) is a solution, then $(x_0 a^{n+1}, y_0 a^{n+1}, z_0 a^n)$ is also a solution, for any $a \in \mathbb{N}$. Now, get a solution by letting $x = y = k + 1$.

2. Use Proposition 7.1 to write $x + y$, $y + z$ and $x + z$ in terms of the parameters u, v and d, as prescribed by that result.

3. Imitate the proof of Proposition 7.1. More precisely, if $x, y, z > 0$, start by observing that $z - x$ and $z + x$ are both even and, if $d = \gcd\left(\frac{z-x}{2}, \frac{z+x}{2}\right)$, the equality

$$\left(\frac{z-x}{2d}\right)\left(\frac{z+x}{2d}\right) = 2\left(\frac{y}{d}\right)^2$$

guarantees the existence of $u, v \in \mathbb{Z}$ such that $\frac{z-x}{2d} = 2v^2$ and $\frac{z+x}{2d} = u^2$. Now, show that $\gcd(u, 2v) = 1$.

4. In each case, use Fermat's descent method. Specifically for item (d), start by showing that x and y leave equal remainders upon division by 3; then, by computing $(3k + r)^3$, conclude that if x and y are not divisible by 3, then $x^3 + 5y^3$ is not divisible by 9.

5. Start by writing $x = u - v$ and $y = u + v$, with $u, v \in \mathbb{Q}$, to get $x^2 + xy + y^2 = 3u^2 + v^2$. Then, let $u = \frac{a}{c}$ and $v = \frac{b}{c}$, with $a, b, c \in \mathbb{Z}$ and $\gcd(b, c) = 1$, to get $3u^2 + v^2 = \frac{3a^2 + b^2}{c^2}$. Now, for item (a) use the result of Example 7.3; for item (b), adapt the proof of Proposition 7.1 to solve the equation $3a^2 + b^2 = c^2$.

6. If z is odd, use the result of item (c) of Corollary 6.8 to show that there are no solutions. If z is even, use item (b) of that same result to conclude that w, x and y are also even; then, apply Fermat's descent method.

7. Multiply both sides of the given equality by 4, complete squares and write $2x + 3$, $2y + 3$ and $2z + 3$ as fractions with a common denominator to reduce this problem to the previous one.

8. Start by multiplying out to show that it suffices to obtain rational solutions to the linear system of equations

$$\begin{cases} ax + 2cy + 2bz = 1 \\ bx + ay + 2cz = 0 \\ cx + by + az = 0 \end{cases}.$$

Then, use Cramer's rule to reduce the problem to showing that the only solution of the diophantine equation $a^3 + 2b^3 + 4c^3 = 6abc$ is $a = b = c = 0$. Finally, for this last step, reduce to the case of $a, b, c \in \mathbb{Z}$ and apply Fermat's descent method.

9. Start by letting $A_1 A_2$ be a diameter of the given circle. Then, use (7.1) to choose points $A_3, \ldots, A_n$ such that $\overline{A_1 A_i}$ and $\overline{A_2 A_i}$ are rationals, for $3 \leq i \leq n$. Finally, apply Ptolemy's theorem (cf. Theorem 4.18 of [9], for instance) to show that $\overline{A_i A_j}$ is also rational.

10. The solution we present to this problem uses more the ideas presented along this section that any specific result we have derived. Anyhow, since it is much more difficult than the previous ones, we present a complete solution: start by simplifying the stated relation to get $a^2 + c^2 - ac = b^2 + d^2 + bd$ or, which is the same,

$$(2a - c)^2 + 3c^2 = (b + 2d)^2 + 3b^2. \tag{22.1}$$

By contradiction, assume that $ab + cd = p$, with p prime. Condition $a > b > c > d > 0$ gives $p = ab + cd \geq 4 \cdot 3 + 2 \cdot 1 = 14$, so that $p \geq 17$. On the other hand,

$$2p = 2ab + 2cd = (2a - c)b + (b + 2d)c$$

and, hence, $\gcd(2a - c, b + 2d)$ divides $2p$. For $\gcd(2a - c, b + 2d)$ to be even, we should have b and c both even, and this would give $p = ab + cd$ also even, which is impossible. Therefore,

$$\gcd(2a - c, b + 2d) = 1 \text{ or } p.$$

We claim that $\gcd(2a - c, b + 2d) = 1$. On the contrary, suppose that $\gcd(2a - c, b + 2d) = p$. Then, (22.1) would give us $p^2 \mid 3(b^2 - c^2)$; in turn, since $p \neq 3$, this would assure that $p^2 \mid (b^2 - c^2)$. However, $p = ab + cd > b$ forces $0 < b^2 - c^2 < p^2$, which is an absurd.

Now, let $x = 2a - c$ and $y = b + 2d$. Then, $\gcd(x, y) = 1$, and it follows from (22.1) that $x^2 - y^2 = 3(b^2 - c^2)$ or, which is the same,

$$(x - y)(x + y) = 3(b - c)(b + c).$$

We consider two cases separately:

(i) b and c have distinct parities: the fact that $p = ab + cd$ gives $\gcd(b, c) = 1$. Since $b + c$ and $b - c$ are both odd, this implies $\gcd(b + c, b - c) = 1$. An analogous reasoning also gives $\gcd(x + y, x - y) = 1$.
    If $3 \mid (x + y)$, then

$$x - y = \gcd(x - y, (b + c)(b - c)) = \gcd(x - y, b + c) \gcd(x - y, b - c)$$

and

$$x + y = 3 \gcd(x + y, (b + c)(b - c)) = 3 \gcd(x + y, b + c) \gcd(x + y, b - c).$$

Writing $\alpha = \gcd(x - y, b - c)$, $\beta = \gcd(x - y, b + c)$, $\gamma = \gcd(x + y, b - c)$ and $\delta = \gcd(x + y, b + c)$, we get $x - y = \alpha\beta$ and $x + y = 3\gamma\delta$. On the other hand,

$$b - c = \gcd(b - c, x - y) \gcd(b - c, x + y) = \alpha\gamma$$

and, analogously, $b + c = \beta\delta$. Solving for a, b, c and d, we obtain

$$4a = \alpha\beta + \beta\delta + 3\gamma\delta - \alpha\gamma, \quad 2b = \alpha\gamma + \beta\delta,$$

$$2c = -\alpha\gamma + \beta\delta \quad \text{and} \quad 4d = -\alpha\beta - \beta\delta + 3\gamma\delta - \alpha\gamma.$$

Thus,

$$8p = 8(ab + cd) = \beta\delta(2\alpha^2 + 3\delta^2).$$

However, since $b + c$ is odd, we conclude that both β and δ are odd and, hence, $\beta\delta(2\alpha^2 + 3\delta^2)$ is odd too, which is impossible.
    If $3 \mid (x - y)$, we reach a contradiction in a likewise manner.

(ii) b and c have equal parities: in this case, b and c must both be odd, for, otherwise, 2 would divide $ab + cd = p$. Thus, in the above notations, x and y must also be odd. It follows that

$$\gcd(b + c, b - c) = \gcd(x + y, x - y) = 2,$$

for we already have $\gcd(x, y) = 1$.

If $3 \mid (x + y)$ (the other case is, once more, entirely analogous), then writing

$$\left(\frac{x-y}{2}\right)\left(\frac{x+y}{2}\right) = 3\left(\frac{b-c}{2}\right)\left(\frac{b+c}{2}\right)$$

and letting

$$\alpha = \gcd\left(\frac{x-y}{2}, \frac{b-c}{2}\right), \quad \beta = \gcd\left(\frac{x-y}{2}, \frac{b+c}{2}\right),$$

$$\gamma = \gcd\left(\frac{x+y}{2}, \frac{b-c}{2}\right), \quad \delta = \gcd\left(\frac{x+y}{2}, \frac{b+c}{2}\right)$$

we arrive, as in the previous case, at $2a = \alpha\beta + \beta\delta + 3\gamma\delta - \alpha\gamma$, $b = \alpha\gamma + \beta\delta$, $c = -\alpha\gamma + \beta\delta$ and $2d = -\alpha\beta - \beta\delta + 3\gamma\delta - \alpha\gamma$. Hence,

$$2p = 2(ab + cd) = \beta\delta(2\alpha^2 + 3\delta^2).$$

Neither β nor δ are equal to p, for, otherwise, we would have $b > p$, which contradicts the fact that $ab + cd = p$. Therefore, $\beta\delta \leq 2$, and this gives $0 < d < c \leq 2 - \alpha\gamma \leq 1$, which is an absurd.

Section 7.2

1. Substitute $d = 4k + 3$, $m = 4l + 3$ and use Corollary 6.8.
2. Just note that $x_{n+1} + y_{n+1}\sqrt{2} = (x_1 + y_1\sqrt{2})(x_n + y_n\sqrt{2})$.
3. Imitate the discussion of Example 7.6, observing that $x = y = 1$ is a solution of the stated equation.
4. If $x^2 - dy^2 = m$ and $a^2 - db^2 = 1$, with $a, b \in \mathbb{N}$, then

$$(a - b\sqrt{d})(x_0 + y_0\sqrt{d}) \cdot (a + b\sqrt{d})(x_0 + y_0\sqrt{d}) = m$$

or, which is the same,

$$[(ax_0 + bdy_0) - (ay_0 + bx_0)\sqrt{d}][(ax_0 + bdy_0) + (ay_0 + bx_0)\sqrt{d}] = m.$$

Therefore, $x_1 = ax_0 + bdy_0$ and $y_1 = ay_0 + bx_0$ also solve the given equation. Now, observe that

$$x_1^2 + dy_1^2 = (a^2 + db^2)(x_0^2 + dy_0^2) + 4abx_0y_0d > x_0^2 + dy_0^2,$$

so that $(x_1, y_1) \neq (x_0, y_0)$.

5. Multiply the equality $n^2 + (n + 1)^2 = m^2$ by 2, complete squares and apply the result of the previous problem.

6. By completing squares, conclude that the given equation is equivalent to $(2x + y)^2 - 5y^2 = 4$. Now, use the result of Problem 4, noticing that $x = y = 1$ is a solution of the stated equation.

7. By adapting the hint given to Problem 4, show that if $u, v \in \mathbb{N}$ are such that $u^2 - \Delta v^2 = 1$, then $\alpha := ux_0 - \Delta v y_0$ and $\beta := uy_0 - vx_0$ are an integer solution of $x^2 - \Delta y^2 = 4an$. Now, note that

$$ax^2 + bxy + cy^2 = n \Rightarrow (2ax + by)^2 - \Delta y^2 = 4an,$$

so that, in order to generate an integer solution of equation $ax^2 + bxy + cy^2 = n$, it suffices to show that it is possible to solve, in $\mathbb{Z}$, the linear system of equations

$$\begin{cases} 2ax + by = \alpha \\ y = \beta \end{cases}.$$

In turn, such a task is equivalent to showing that $2a \mid (\alpha - d\beta)$, which can be done by writing

$$\alpha - d\beta = (ux_0 - \Delta v y_0) - b(uy_0 - vx_0)$$
$$= u(x_0 - by_0) + bv(x_0 - by_0) + 4acvy_0,$$

for the last expression is a sum of multiples of $2a$.

Chapter 8

2. Prove first that $\prod_{0<d|n} d = \prod_{0<d|n} \frac{n}{d}$. Then, let P denote this product and use such an equality to show that $P^2 = n^{d(n)}$.

3. For $0 < d \mid n$, we have $d + \frac{n}{d} \geq 2\sqrt{n}$. Now, add all such inequalities over all positive divisors of n.

4. For the second equation, show that an ordered pair (x, y) is a solution if and only if $x = n - \frac{n^2}{n+y}$, so that $n + y$ must be a divisor of n^2 greater than n. Now, use the result of Example 6.47 and (8.3) to conclude that there are exactly $\frac{1}{2}(d(n^2) - 1)$ distinct possibilities for y and, hence, for (x, y). In what concerns the first equation, adapt the hint given to the second one to conclude that there are $d(n^2)$ distinct solutions. Finally, conclude that we ought to have $3d(n^2) = 157$, which is an absurd.

5. For item (a), use the result of Lemma 8.3. For item (b), if $\gcd(m, n) = 1$, use the result of item (a) to compute $|D_1(mn)| - |D_3(mn)|$ in terms of $|D_1(m)| - |D_3(m)|$ and $|D_1(n)| - |D_3(n)|$; then, apply an inductive argument.

6. For item (a), multiply both sides of the desired equality by n. For item (b), make a direct computation for the sum of the positive divisors of $2^{p-1}(2^p - 1)$.

7. The second part of item (b) uses item (b) of Problem 10, page 190.

8. Show that $s(ab) \geq s(a)b$, for all $a, b \in \mathbb{N}$. To this end, exhibit, in terms of the positive divisors of a, a set of positive divisors of ab whose sum is equal to $s(a)b$.

9. Apply the result of Proposition 8.4.

10. Apply the result of Proposition 8.4.

11. Use Proposition 8.4 and the result of Problem 1 to conclude that both sides of the desired equality can be seen as arithmetic multiplicative functions, say F and G. Then, if p is prime and $\alpha \in \mathbb{N}$, show that $F(p^\alpha) = G(p^\alpha)$.

12. Apply double counting—cf. Sect. 2.2—first showing that, for $1 \leq j \leq n$, the left hand side counts $f(j)$ exactly $\lfloor \frac{n}{j} \rfloor$ times.

13. Start by applying the result of the previous problem; then, apply Proposition 8.4 to show that $F(m) = 1$ if m is a perfect square and $F(m) = 0$ otherwise.

14. If $f(j) = 1$ for every $j \in \mathbb{N}$ and $F(n) = \sum_{0 < d \mid n} f(d)$, then $F(n) = d(n)$ for every $n \in \mathbb{N}$. Now, apply the result of Problem 12.

15. Use the Möbius inversion formula together with the result of Problem 1.

17. Use the result of Proposition 8.4.

18. Start by proving that, for a fixed positive divisor d of m, there are exactly $\frac{m}{d} \cdot \varphi(d)$ ordered pairs (d, n) as stated. Then, use the result of Corollary 8.13.

20. For item (a), use item (a) of Problem 19. For (b), write $S_m(n) = \sum_{i=1}^{k}(n - a_i)^m$ and expand the binomial $(n - a_i)^m$. For the first part of (c), use (b); as for the second part, separately consider the cases n even and n odd and apply the conclusion of item (a) in the case of an even n.

21. For item (a), choose d in such a way that $\gcd(m, n) = \frac{n}{d}$ and a such that $m = \frac{n}{d} \cdot a$; then, conclude that $\gcd(a, d) = 1$. For item (b), use the result of (a) to show that $\sum_{0 < d \mid n} \left(\frac{n}{d}\right)^k S_k(d) = 1^k + 2^k + \cdots + n^k$. Item (c) follows from Möbius inversion formula, applied to the result of (b). Finally, for items (d) and (e), use the formula of (c).

Section 9.1

1. Use the Prime Number Theorem.

2. Use the divergence of the harmonic series (cf. Section 7.4 of [8], for instance), in conjunction with Lemma 3.15.

3. Use again the divergence of the harmonic series, together with Theorem 9.5.

4. For the first part of item (a) use the fact that, if p is prime and $n < p \leq 2n$, then $\gcd(n!, p) = 1$; for the second part, use the binomial formula. For item (b), use (a), together with the fact that the number of primes p satisfying $n < p \leq 2n$ is exactly $\pi(2n) - \pi(n)$. For the second part of item (c), use the fact that $\frac{x}{2} \geq x^{2/3}$

and $x + 2 \leq \frac{5x}{4}$ for $x \geq 8$. For item (d), make a direct checking of the cases $2 \leq n \leq 8$; then, prove the remaining cases by induction on $n \geq 8$. Finally, for item (e), use items (c) and (d), together with the (obvious!) inequality $\pi(x) \leq 2\lfloor \frac{x}{2} \rfloor + 2$.

5. Note that

$$I(n) = (\tau(1) + \tau(3) + \cdots + \tau(i(n))) + (\tau(2) + \tau(4) + \cdots + \tau(2\lfloor n/2 \rfloor))$$

$$= (1 + 3 + \cdots + i(n)) + (\tau(1) + \tau(2) + \cdots + \tau(\lfloor n/2 \rfloor))$$

$$= \frac{1}{4}(i(n) + 1)^2 + I\lfloor n/2 \rfloor.$$

This gives (a), and iterating such a recurrence relation, we get (b). For (c), we have

$$I(n) \leq \frac{1}{4} \sum_{k=0}^{t} \left(\frac{n}{2^k} + 1 \right)^2 = \frac{1}{4} \left(n^2 \sum_{k=0}^{t} \frac{1}{4^k} + n \sum_{k=0}^{t} \frac{1}{2^{k-1}} + t + 1 \right)$$

$$= \frac{1}{4} \left(\frac{n^2}{3} \left(4 - \frac{1}{4^t} \right) + n \left(4 - \frac{1}{2^{t-1}} \right) + t + 1 \right).$$

Now, since $t = \lfloor \log_2 n \rfloor \leq n$, we obtain (d) by estimating

$$I(n) \leq \frac{1}{4} \left(\frac{4n^2}{3} + 4n + t + 1 - \frac{n^2}{3 \cdot 4^{\lfloor \log_2 n \rfloor}} - \frac{2n}{2^{\lfloor \log_2 n \rfloor}} \right)$$

$$\leq \frac{1}{4} \left(\frac{4n^2}{3} + 4n + t + 1 \right) = \frac{1}{4} \left(\frac{4n^2}{3} + 4n + \lfloor \log_2 n \rfloor + 1 \right)$$

$$\leq \frac{1}{4} \left(\frac{4n^2}{3} + 4n + +n + 1 \right) = \frac{4n^2 + 15n + 3}{12}.$$

The estimates leading to (e) are quite analogous and will be left to the reader. The first part of (f) now follows immediately from (d) and (e), together with $T(n) = \frac{n(n+1)}{2}$. Hence,

$$\frac{4 - \frac{12}{n} - \frac{1}{n^2}}{6 + \frac{6}{n}} \leq \frac{I(n)}{T(n)} \leq \frac{4 + \frac{15}{n} + \frac{3}{n^2}}{6 + \frac{6}{n}}$$

for every $n > 1$, from where we get $\lim_{n \to +\infty} \frac{I(n)}{T(n)} = \frac{2}{3}$. Thus, if $r \neq \frac{2}{3}$, then there exists $n_0 \in \mathbb{N}$ such that

$$n > n_0 \Rightarrow \left| \frac{I(n)}{T(n)} - \frac{2}{3} \right| < \left| r - \frac{2}{3} \right|.$$

In particular, $n > n_0 \Rightarrow \frac{I(n)}{T(n)} \neq \frac{2}{3}$.

Section 9.2

1. From Chebyschev's theorem, there is at least one prime between p_n and $2p_n$.
2. Letting p be the largest prime which does not exceed n, apply Chebyschev's theorem to show that $2p > n$.
3. If $n = 2k$, write $1!2!\ldots n! = 2^k k!(3!5!\ldots,(2k-1)!)^2$ and, then, apply the result of the previous problem. The case $n = 2k + 1$ can be dealt with in an analogous way.
4. Let p and q denote the largest primes that do not exceed m and n, respectively. Use the result of Problem 2 to compare the greatest powers of p and q in both sides of the given equation.
5. Start by using the result of Problem 2 to show that $n = p$, a prime number; to this end, follow a path analogous to the hint given to the previous problem. Then, if q is the greatest integer which is less than or equal to $p - 1$, conclude that $m \geq 2q$. Finally, use the remark that precedes the statement of this problem to conclude that $q \leq 5$ and, hence, that $p - 1 \leq 6$. To solutions are $m = n = 2$ or $m = 10$, $n = 7$.
6. Let $n > 1$ be a natural number satisfying the stated conditions, and $p < \sqrt{n}$ be a prime number. Then $p \mid n$, for otherwise we would have $p^2 < n$ and $\gcd(n, p^2) = 1$. Thus, n is divisible by $p_1, \ldots, p_k$, where $p_1 < \cdots < p_k$ are those prime numbers less than $\sqrt{n}$. Now, let $l \in \mathbb{N}$ be such that $2^l \leq \sqrt{n} < 2^{l+1}$. Since (invoking again the stronger version of Chebychev's theorem, alluded to above) for every integer $k > 1$ we have at least two integers between 2^k and 2^{k+1}, we obtain

$$p_1 \ldots p_k \geq 2(2^2 \ldots 2^{l-1})^2 = 2^{l^2}.$$

This way,

$$2^{l^2} \leq p_1 \ldots p_k \leq n < 2^{2l+2},$$

so that $l \leq 2$. Hence, there are three cases to consider:

(i) $l = 0$: in this case, we have $1 \leq n < 4$, and it is immediate to check that $n = 2$ or 3 are the possible solutions.

(ii) $l = 1$: we have $4 \leq n < 16$ and $2 \mid n$, so that $n \in \{4, 6, 8, 10, 12, 14\}$. A simple inspection show that the possible solutions are $n = 4, 6, 8$ or 12.

(iii) $l = 2$: in this last case we have that $16 \leq n < 64$ and $2, 3 \mid n$. Therefore, $n \in \{18, 24, 30, 36, 42, 48, 54, 60\}$ and, once more by inspection, we obtain the solutions: $n = 18, 24$ or 30.

Section 10.1

1. Make induction on $m \geq 2$.
2. Use the elementary properties of congruences, together with the fact that $10 \equiv -1 \pmod{11}$.
4. Use congruences to compute the remainder of the given number upon division by 4. Then, apply the result of Proposition 10.9.
5. For $k \in \mathbb{Z}_+$, show that $7^{4k} \equiv 1 \pmod{10}$, $7^{4k+1} \equiv 7 \pmod{10}$, $7^{4k+2} \equiv -1 \pmod{10}$ and $7^{4k+3} \equiv 3 \pmod{10}$. Then, compute the remainder of the division of 3^{10} by 4.
6. If all prime divisors of n were congruent to 1, modulo 4, use the elementary properties of congruences to conclude that we were to have $n \equiv 1 \pmod 4$.
7. The first equality gives us $2^{32} + 2^{28} \cdot 5^4 \equiv 0 \pmod{641}$; the second, $2^{28} \cdot 5^4 \equiv (-1)^4 \pmod{641}$.
8. Assume that the decimal representation of n has $k + 1$ digits, and let m be the natural number formed by the first k digits of n. Show that the stated condition is equivalent to $6 \cdot 10^k + m = 4(10m + 6)$ or, also, to $3 \cdot 10^k = 13m + 12$. Then, conclude that it suffices to find all natural numbers k for which $10^k \equiv 4 \pmod{13}$.
9. Start by observing that $2^n + 3^n \equiv 0 \pmod 7$ if and only if $6^n + 9^n \equiv 0 \pmod 7$, which, in turn, is equivalent to $(-1)^n + 2^n \equiv 0 \pmod 7$. Then, compute the possible congruences of 2^n, modulo 7.
10. Apply congruence modulo 3 to show that either p or q must be a multiple of 3, hence equal to 3. Then, assuming $q = 3$, write $p^2 + 9p + 9 = n^2$ and solve such an equality for p.
11. For item (a), start by reducing the terms of the Fibonacci sequence modulo 2 and modulo 4, and note that the two reduced sequences thus obtained are periodic. Items (b), (c), (d) and (e) can be dealt with analogously.
12. Start by using the result of item (c) of Problem 2, page 71, to conclude that if $d = \gcd(F_n, L_n)$, then $d \mid 4$. Now, distinguish the cases $3 \nmid n$ and $3 \mid n$, and apply the results of the previous problem.
13. For items (a) and (b), use induction on $n \in \mathbb{N}$. For item (c), start by using the result of item (a) to write $2F_{m+2k} = F_m L_{2k} + F_{2k} L_m$. Then use the result of the item (b) of Problem 2, page 71, as well as (a) again, to get

$$2F_{m+2k} = F_m(L_k^2 + 2(-1)^{k-1}) + F_k L_k L_m \equiv 2(-1)^{k-1} F_m \pmod{L_k}.$$

Now, apply the result of item (b) of Problem 11, taking into account that $2 \mid k$ and $3 \nmid k$. For item (d), argue in an analogous way, writing

$$2L_{m+2k} = 5F_m F_{2k} + L_m L_{2k} = 5F_m F_k L_k + L_m(L_k^2 + 2(-1)^{k-1}) \equiv -2L_m \pmod{L_k}$$

and, hence, $L_{m+2k} \equiv -L_m \pmod{L_k}$.
14. Apply item (d) of Proposition 10.9.

15. Look at the equation modulo 3 to conclude that m is even. Then, factorize $k^2 - 2^m$ and argue as in Example 10.11.

16. If $n \geq 5$, show that $m^p \equiv 3 \pmod{10}$ and, hence, that $p > 2$. Then, use modulo 3 to conclude that $3 \mid m$, so that $27 \mid m^p$. Finally, look at the equation modulo 27 to conclude that there is no solution when $n \geq 9$.

17. Use modulo 3 to show that z is even, say $z = 2t$. Then, factor $5^{2t} - 4^y$ and argue as in Example 10.11.

18. If $y = 1$, then $p = 2$ and $x = 1$. If $y > 1$, write $p^x = (y + 1)(y^2 - y + 1)$ and show that, letting $d = \gcd(y + 1, y^2 - y + 1)$, one has $d = 1$ or 3. Finally, consider separately the cases $d = 1$ and $d = 3$.

19. Write $a = 2^k \alpha$ and $b = 2^l \beta$, with $k, l \in \mathbb{N}$ and α, β being odd naturals. From this, show that $kb = la$ and $\alpha^b + \beta^a = 2^{c-kb}$. Then, use modulo 4 to conclude that $\alpha = \beta = 1$ and $\frac{k}{2^k} = \frac{l}{2^l}$. Finally, analysing the function $f(x) = \frac{x}{2^x}$, show that, if $k, l \geq 4$, then $k = l$.

20. Firstly, show that it suffices to consider the case $b = a^n$. Then, in order to analyse it, make induction on $n \geq 0$, using congruence modulo a at the induction step.

21. Start by looking at the given equation modulo 3 to show that b is even. Then, let $b = 2c$ and conclude that $2^{c+1} = 15^a + 1$ or $2^{c+1} = 3^a + 5^a$. Let $d = c + 1$ and, in the first case, use modulo 3 to conclude that d is even and, hence, that $2^{d/2} - 1 = 3^a$ and $2^{d/2} + 1 = 5^a$. In the second case, use modulo 4 to show that a is odd, so that, if $c - 2 > 0$, we have

$$2^{c-2} = 3^{a-1} - 3^{a-2} \cdot 5 + \cdots + 5^{a-1} \equiv a \pmod 2.$$

Thus, conclude that $c - 2 = 0$.

22. Using the fact that $1992 = 24 \cdot 83$, show that the division of 10^{1992} by $10^{83} + 7$ leaves remainder 7^{24}. Then, if $q \in \mathbb{N}$ is such that $10^{1992} = (10^{83} + 7)q + 7^{24}$, use congruence modulo 10 to compute the last digit of q.

23. For item (a), if x is even, say $x = 2z$, we get $(7^z - 2)(7^z + 2) = 3^y$. Conclude that both $7^z - 2$ and $7^z + 2$ must be powers of 3, and use the fact that $(7^z + 2) - (7^z - 2) = 4$ to get a contradiction. For item (b), assuming that if $y > 1$, show that $7^x = 3^y + 4 \equiv 4 \pmod 9$, and that this gives $x \equiv 2 \pmod 3$. Hence, deduce that $x \equiv 2 \pmod 3$ and $x \equiv 1 \pmod 2$ implies $x \equiv 5 \pmod 6$. Item (c) now follows from letting $x = 6q + 5$ and computing

$$7^x = 7^{6q+5} = (7^6)^q \cdot 7^5 \equiv (-1)^q \cdot (-2) \equiv \pm 2 \pmod{13}.$$

Finally, for (d) and (e), compute $3^y \equiv 1, 3$ or $9 \pmod{13}$ and, hence, $3^y + 4 \equiv 5, 7$ or $0 \pmod{13}$; then, show that this contradicts (c).

25. Looking at the equation modulo $x + 1$, we obtain $(-1)^{z+1} \equiv 1 \pmod{x + 1}$. Therefore, from $x > 1$ show that z is odd. Now, write

$$(x + 1)^y = x^z + 1 = (x + 1)(x^{z-1} - x^{z-2} + \cdots - x + 1),$$

to get $(x+1)^{y-1} = x^{z-1} - x^{z-2} + \cdots - x + 1$. If x is odd, show that both sides of this last equation have different parities, which cannot occur. Now, use the fact that x is odd, together with

$$x^{z-1} = \frac{(x+1)^y - 1}{x} = (x+1)^{y-1} + \cdots + (x+1) + 1,$$

to show that y is also even. Then, letting $x = 2s$ and $y = 2t$, we have

$$\left[(x+1)^t - 1\right]\left[(x+1)^t + 1\right] = (2s)^z. \tag{22.2}$$

Note now that $(x+1)^t - 1$ and $(x+1)^t + 1$ are consecutive even numbers, so that they have gcd equal to 2. Use modulo x to get $(x+1)^t - 1 \equiv 0 \pmod{x}$ and $(x+1)^t + 1 \equiv 2 \pmod{x}$, so that the first factor is a multiple of x, whereas the second one is of the form $xq + 2$, with 2 being its only common factor with x. Use this to conclude from (22.2) that

$$\begin{cases} (x+1)^t - 1 = 2s^z \\ (x+1)^t + 1 = 2^{z-1} \end{cases}.$$

Then, finish the solution.

Section 10.2

1. Apply Fermat's little theorem with $p = 7$.
2. Start by considering the cases $p = 2, 3, 5$. If $p \neq 2, 3, 5$, write the given difference as

$$(a \cdot 10^{8p} - 10^8) + 2(a \cdot 10^{7p} - 10^7) + \cdots + 9(a - 1), \tag{22.3}$$

where $a = \underbrace{11 \ldots 1}_{p}$. Then, apply Fermat's little theorem to conclude that $a \equiv$
1 (mod p) and to analyse the remainder of the division by p of each summand of (22.3).
3. Show that, after performing $k \geq 1$ operations, the j-th term of the sequence goes to position $2^k j \pmod{2n+1}$. Then, use Fermat's little theorem.
4. Firstly, use modulo 11 to conclude that there are no integer solutions such that $x \equiv 0 \pmod{11}$. Then, if $x \not\equiv 0 \pmod{11}$, use Fermat's little theorem to conclude that 11 divides $x^5 + 1$ or $x^5 - 1$ and, hence, that $y^2 \equiv -5$ or $-3 \pmod{11}$. Finally, show that such congruences do not have any solutions.
5. We may assume $p > 2$. In this case, if $n = pq + r$, Fermat's little theorem assures that it suffices to find an integer $0 \leq r < p$ such that $2^{q+r} \equiv r \pmod{p}$ for infinitely many naturals q. For what is left to do, choose $r = 1$ and, then, infinitely many q's.

6. For the sake of contradiction, assume that there exists an integer $n > 1$ such that $n \mid (3^n - 2^n)$. If p is the smallest prime factor of n, with $n = mp$, use Fermat's little theorem to conclude that $3^m \equiv 2^m \pmod p$. Then, use the minimality of p to show that $\gcd(m, p - 1) = 1$. Finally, write $mx = (p - 1)y + 1$, with $x, y \in \mathbb{N}$, and use Fermat's little theorem once more to reach a contradiction.

7. For item (b), start by using the fact that $\gcd(p, q - 1) = 1$ to write $px = (q - 1)y + 1$, with $x, y \in \mathbb{N}$. Then, apply Fermat's little theorem.

8. If k stands for the number of digits of 5^n, show that the problem is equivalent to finding $m > n$ for which $5^m - 5^n \equiv 0 \pmod{10^k}$. Prove that $n \geq k$ and conclude that it suffices for us to have $5^{m-n} \equiv 1 \pmod{2^k}$. Then, use Euler's theorem.

9. With the aid of Euler's theorem, prove first that if l is odd, then $\gcd(2^{\varphi(l)m} - 3, l) = 1$, for every $m \in \mathbb{N}$. Then, choose $k_1, \ldots, k_n$ inductively.

10. Use Möbius inversion formula to get $a_n = \sum_{0<d\mid n} \mu(d)2^{\frac{n}{d}}$. Now, use Euler's theorem to show that, if p is prime and p^α is the greatest power of p that divides n, then $p^\alpha \mid a_n$. To this end, write $n = p^\alpha k$, with $\gcd(k, p) = 1$, and, from this,

$$
a_n = \sum_{0<d\mid p^\alpha k} \mu(d)2^{\frac{p^\alpha k}{d}} = \sum_{0<d\mid k} \mu(d)2^{\frac{p^\alpha k}{d}} + \sum_{0<d\mid k} \mu(pd)2^{\frac{p^\alpha k}{pd}}
$$

$$
= \sum_{0<d\mid k} \mu(d)2^{\frac{p^\alpha k}{d}} - \sum_{0<d\mid k} \mu(d)2^{\frac{p^{\alpha-1}k}{d}}
$$

$$
= \sum_{0<d\mid k} \mu(d)\left\{\left(2^{\frac{p^{\alpha-1}k}{d}}\right)^p - 2^{\frac{p^{\alpha-1}k}{d}}\right\}
$$

$$
= \sum_{0<d\mid k} \mu(d)2^{p^{\alpha-1}\cdot\frac{k}{d}}\left(2^{\frac{p^{\alpha-1}(p-1)k}{d}} - 1\right)
$$

$$
= \sum_{0<d\mid k} \mu(d)2^{p^{\alpha-1}\cdot\frac{k}{d}}\left(2^{\varphi(p^\alpha)\cdot\frac{k}{d}} - 1\right).
$$

11. For item (a), argue by contradiction. More precisely, if $a = (p_1-1)\ldots(p_k-1)$, with $k \geq 1$ minimum as in (a), use the fact that $\gcd(2^{a/2} - 1, 2^{a/2} + 1) = 1$ to conclude that there exist $l < k$ primes among $p_1, \ldots, p_k$, say $p_1, \ldots, p_l$, such that $2^{a/2} - 1 = p_1^{\alpha_1} \ldots p_l^{\alpha_l}$. Then, letting $b = (p_1 - 1)\ldots(p_l - 1)$, use the fact that $(2^b - 1) \mid (2^{a/2} - 1)$ to contradict the minimality of k.

12. For item (b), use Fermat's little theorem. For (f), if $q \nmid x$, show that $x^{(q-1)/2} \equiv \pm 1 \pmod q$, and analogously for $y^{(q-1)/2}$ and $z^{(q-1)/2}$; then, use the equality $x^{(q-1)/2} + y^{(q-1)/2} + z^{(q-1)/2} = 0$, together with $q > 5$, to get a contradiction. Finally, for (g) ii., observe that if $q \mid a$ and $\gcd(a, d) = 1$, then $q \nmid d$.

13. For item (e), you will need to use the result of Example 8.14. For item (h), first show that $q_n = 2^{n-1}q_1 + (2^{n-1} - 1)$, for every integer $n \geq 1$; then, use Fermat's little theorem to show that, if q_1 is an odd prime, then it is possible to choose $n \geq 2$ such that $q_1 \mid q_n$.

Section 10.3

1. We prove both items simultaneously. If x is an integer such that $ax \equiv b \pmod{n}$, then we have $ax = nq + b$ for some $q \in \mathbb{Z}$. Therefore, $b = xa + (-q)n$, a linear combination of a and n, so that Bézout's theorem furnishes $\gcd(a, n) \mid b$. Now, let $\gcd(a, n) = d$, with $d \mid b$. Then,

$$ax \equiv b \pmod{n} \Leftrightarrow \frac{a}{d}x \equiv \frac{b}{d} \left(\bmod \frac{n}{d}\right).$$

However, since $\gcd\left(\frac{a}{d}, \frac{n}{d}\right) = 1$, the existence of solution follows from Corollary 10.25. For what is left to do, start by noticing that

$$ax \equiv ax_0 \pmod{n} \Leftrightarrow x \equiv x_0 \left(\bmod \frac{n}{d}\right) \Leftrightarrow x = x_0 + \frac{n}{d}t, \ \exists t \in \mathbb{Z}.$$

On the other hand, it is immediate that

$$x_0 + \frac{n}{d}t_1 \equiv x_0 + \frac{n}{d}t_2 \pmod{n} \Leftrightarrow t_1 \equiv t_2 \pmod{d},$$

so that there are as many pairwise incongruent solutions for the original equation as there are integer values for t pairwise incongruent modulo d. Thus, the original equation has exactly $d = \gcd(a, n)$ pairwise incongruent solutions modulo n.

2. Assuming that the given congruence has any solution at all, use Bézout's theorem, together with item (d) of Proposition 10.6, to show that the gcd between $\gcd(a_1, a_2, \ldots, a_k)$ and n divides b. Then, observe that

$$\gcd(\gcd(a_1, a_2, \ldots, a_k), n) = \gcd(a_1, a_2, \ldots, a_k, n). \tag{22.4}$$

For the converse, apply Bézout's theorem once more, together with (22.4).

3. Letting x denote the total of soldiers, show that it satisfies a system of two linear congruences with relatively prime moduli. Then, use the proof of the Chinese remainder theorem to show that

$$x \equiv 7 \cdot 12 \cdot 12 + 5 \cdot 1 \cdot 13 \pmod{12 \cdot 13}$$

and, hence, that $x = 132q + 17$ for some $q \in \mathbb{Z}$. Finally, use the fact that $600 < x < 700$.

4. For item (a), start by observing that, if u, v is a solution of the linear diophantine equation and $x = m_1u + a_1 = m_2v + a_2$, then x solves (10.7) for $k = 2$.

5. Apply Euler's theorem.

6. If p and q are distinct primes, Example 6.29 assures that $2^p - 1$ and $2^q - 1$ are relatively prime. In view of this fact, show that an integer x satisfies the stated conditions if and only if it solves a certain system of linear congruences. Then, apply the Chinese remainder theorem.

7. Adapt, to the present case, the idea of the proof of Example 10.28.

8. For implication $\Leftarrow$ of item (a), let $a_i \in \mathbb{Z}$ be an integer solution of $f(x) \equiv 0 \,(\text{mod } p_i^{\alpha_i})$. If x is any solution of the system of linear congruences

$$\begin{cases} x \equiv a_1 \,(\text{mod } p_1^{\alpha_1}) \\ x \equiv a_2 \,(\text{mod } p_2^{\alpha_2}) \\ \quad \cdots \\ x \equiv a_k \,(\text{mod } p_k^{\alpha_k}) \end{cases}$$

whose existence is guaranteed by the Chinese remainder theorem, use item (c) of Proposition refprop:propriedades da congruencia to conclude that $f(x) \equiv 0 \,(\text{mod } p_i^{\alpha_i})$ for $1 \leq i \leq k$ and, hence, that $f(x) \equiv 0 \,(\text{mod } m)$. For item (b), let $S(t)$ be a set of $N(t)$ pairwise incongruent solutions, modulo t, for the congruence $f(x) \equiv 0 \,(\text{mod } t)$. Use the result of (a) to infer the existence of a bijection $f : S(m) \to S(p_1^{\alpha_1}) \times \cdots \times S(p_k^{\alpha_k})$; then, apply the fundamental principle of counting.

9. If a is the initial term and r the common ration of the progression, the given hypotheses are equivalent to the existence of $x, y \in \mathbb{N}$ such that $x^2 \equiv a \,(\text{mod } r)$ and $y^3 \equiv a \,(\text{mod } r)$. By the same token, the thesis is equivalent to the existence of $z \in \mathbb{N}$ such that $z^6 \equiv a \,(\text{mod } r)$. Make complete induction on r, the case $r = 1$ being trivial. For the induction step, separately consider the cases $r = p^\alpha$, for some prime p and some $\alpha \in \mathbb{N}$, and $r = st$, with $s, t > 1$ relatively prime. In the first case, if x and y are as above and y' is the inverse of y modulo p^α, show that $z = xy'$ solves the problem. In the second case, let u and v be terms of the arithmetic progressions $(a + ks)_{k \geq 0}$ and $(a + kt)_{k \geq 0}$, respectively, such that $u^6 \equiv a \,(\text{mod } s)$ and $v^6 \equiv a \,(\text{mod } t)$; use the Chinese remainder theorem to find $z \in \mathbb{N}$ such that $z^6 \equiv a \,(\text{mod } st)$.

Section 11.1

1. Note that $a_1 + a_2 + \cdots + a_n \equiv 1 + 2 + \cdots + n \,(\text{mod } n)$.

2. Adapt the proof of item (b) of Proposition 11.3 to the present case.

3. Adapt the discussion of Example 11.4 to this case, this time using the result of the previous problem.

Section 11.2

2. This amounts to noticing that $\overline{a} \cdot (\overline{b + c}) = \overline{a} \cdot \overline{b + c} = \overline{a \cdot (b + c)} = \overline{a \cdot b + a \cdot c} = \overline{a \cdot b} + \overline{a \cdot c} = \overline{a} \cdot \overline{b} + \overline{a} \cdot \overline{c}$.

3. If $\bar{a} = \bar{c}$ and $\bar{b} = \bar{d}$, then $a \equiv c \pmod{n}$ and $b \equiv d \pmod{n}$, so that $a - b \equiv c - d \pmod{n}$. But this is the same as saying that $\bar{a} = \bar{c}$ and $\bar{b} = \bar{d}$ imply $\overline{a - b} = \overline{c - d}$, which is exactly what we wanted to verify.

4. The units of $\mathbb{Z}_{12}$ are $\bar{1}, \bar{5}, \bar{7}$ and $\overline{11}$ (note that $\varphi(12) = 4$). Since $\bar{1}^2 = \bar{5}^2 = \bar{7}^2 = \overline{11}^2 = \bar{1}$, there is nothing left to do.

5. Adapt, to the present case, the discussion of the well definiteness of the operations of addition and multiplication in $\mathbb{Z}_n$.

6. For the first part, if $n > 1$ is not prime, write $n = ab$, with $1 < a, b < n$. For the second, note that if $\bar{a} \cdot \bar{b} = \bar{0}$ in $\mathbb{Z}_p$, then $p \mid ab$ and, hence, $p \mid a$ or $p \mid b$.

7. For the first part, look at the particular case $\bar{a} \cdot x = \bar{1}$. For the second, multiply both sides of $\bar{a} \cdot x = \bar{b}$ by $\bar{a}^{-1}$ and use the associativity of multiplication.

Section 12.1

2. By contradiction, if $2^{3^n} \equiv -1 \pmod{17}$, then $2^{2 \cdot 3^n} \equiv 1 \pmod{17}$. Now, use the congruence $2^{16} \equiv 1 \pmod{17}$ to conclude that $\mathrm{ord}_{17}(2) = 1$ or 2, which is a contradiction.

3. If $m = \mathrm{ord}_n(a)$, then $m \mid 2k$ and, by Euler's theorem, $m \mid \varphi(n)$. Hence, $m \mid \gcd(2k, \varphi(n)) = 2d$, where $d = \gcd(k, \varphi(n)/2)$ (recall that $n > 2 \Rightarrow \varphi(n)$ even—cf. Problem 19, page 220). If $m \mid d$, then $m \mid k$, which contradicts the given congruence. Therefore, $m = 2d'$ for some divisor d' of d, and hence m is even.

4. If $n \mid (2^n - 1)$, then n is odd. If $n > 1$ and $k = \mathrm{ord}_n(2)$, then $1 < k \mid n, \varphi(n)$. Now choose, from the very beginning, the least possible natural number $n > 1$ such that $n \mid (2^n - 1)$. Arrive at a contradiction by showing that $k \mid (2^k - 1)$.

5. Let S denote the given product. Legendre's formula (6.10) assures that the greatest power of 2 that divides $n!$ has exponent

$$e_2(n) = \sum_{j=1}^{+\infty} \left\lfloor \frac{n}{2^j} \right\rfloor < \sum_{j=1}^{+\infty} \frac{n}{2^j} = n;$$

therefore, $e_2(n) \leq n - 1$ and, hence, $2^{e_2(n)} \mid S$. Now, let $2 < p \leq n$ be prime and l be the exponent of the greatest power of p that divides S. Use the fact that $2^n - 2^t \equiv 0 \pmod{p^j}$ if and only if $\mathrm{ord}_{p^j}(2) \mid (n - t)$, together with the idea of the proof of Legendre's formula, to conclude that

$$l = \sum_{j \geq 1} \left\lfloor \frac{n}{\mathrm{ord}_{p^j}(2)} \right\rfloor > \sum_{j \geq 1} \left\lfloor \frac{n}{p^j} \right\rfloor = e_p(n)).$$

6. Show that $k_{i+1} \equiv 2k_i \pmod{2n + 1}$. Then, for item (a) conclude that $f(n) = 0$ if and only if the set $\{1, 2, 2^2, 2^3, \ldots\}$ contains a RRS modulo $2n + 1$. For item (b), show that

$$f(1997) = 2 \cdot 1997 - \mathrm{ord}_{2 \cdot 1997 + 1}(2)$$

and compute $\mathrm{ord}_{2 \cdot 1997+1}(2) = 8 \cdot 23 = 184$ by using the equalities $2 \cdot 1997 + 1 = 5 \cdot 17 \cdot 47$, $\mathrm{ord}_{17}(2) = 8$ and $\mathrm{ord}_{47}(2) = 23$.

7. Assume that we have found q as in (a), so that $p \mid (q - 1)$. Arguing by contradiction, suppose further that there exists $n \in \mathbb{N}$ such that $n^p \equiv p \pmod{q}$. Since $p \neq q$, we have $\gcd(n, q) = 1$, and Fermat's little theorem gives $1 \equiv n^{q-1} \equiv p^{\frac{q-1}{p}} \pmod{q}$, which is impossible. For (b), note that $a \equiv p^{p-1} + p^{p-2} + \cdots + p + 1 \equiv p + 1 \pmod{p^2}$, so that a has at least one prime factor q for which $q \not\equiv 1 \pmod{p^2}$. In turn, if $k = \mathrm{ord}_q(p)$, then the fact that $a \mid (p^p - 1)$ gives $k \mid p$; if $k = 1$, we would have $0 \equiv a \equiv p^{p-1} + p^{p-2} + \cdots + p + 1 \pmod{q}$, so that $p = q$, a contradiction.

Section 12.2

1. Compute $2^{14} \equiv -1 \pmod{29}$ and, then, conclude that $\mathrm{ord}_{29}(2) = 28$.
2. Separately consider each one of the cases $n = 2, 4, p^k$ and $2p^k$, with p being an odd prime and $k \in \mathbb{N}$.
3. Letting a be a primitive root modulo n, note that the a_i's are congruent modulo n, is some order, to $a, a^2, \ldots, a^{\varphi(n)}$, so that

$$a_1 a_2 \ldots a_{\varphi(n)} \equiv a^{\frac{\varphi(n)(\varphi(n)+1)}{2}} \pmod{n}.$$

   Now, if $n = 2$ or 4, give a direct argument to show that $a^{\frac{\varphi(n)(\varphi(n)+1)}{2}} \equiv -1 \pmod{n}$; if $n = p^k$ or $2p^k$, for some odd prime p and some $k \in \mathbb{N}$, use the fact that $\mathrm{ord}_n(a) = \varphi(n)$ to conclude that $a^{\frac{\varphi(n)}{2}} \equiv -1 \pmod{n}$—be careful: $n \mid (a^{\frac{\varphi(n)}{2}} + 1)(a^{\frac{\varphi(n)}{2}} - 1)$ does not necessarily imply $n \mid (a^{\frac{\varphi(n)}{2}} + 1)$ or $n \mid (a^{\frac{\varphi(n)}{2}} - 1)$.
4. Firstly, adapt the proof of Lemma 12.12 to the present case, showing (by induction) that for every integer $n \geq 3$ there exists an odd integer q such that $19^{2^{n-2}} = 2^q + 1$. Back to the posed problem, start by taking $k_1 = 1$, and $k_2 = 2$. Then if $19^{k_n} - 97 = 2^n q$ and q is even, take $k_{n+1} = k_n$; if q is odd, take $k_{n+1} = k_n + 2^{n-2}$.
5. Ultimately, the solution of this problem also relies in a variation of the idea of the proof of Lemma 12.12. Indeed, one wants to find infinitely many $m \in \mathbb{N}$ such that $m^2 + 7 \equiv 0 \pmod{2^k}$. Start by showing that it suffices to establish the existence, for every $l \in \mathbb{N}$, of $m_l \in \mathbb{N}$ such that $m_l^2 + 7 \equiv 0 \pmod{2^l}$; then, prove this last statement by induction on l. For the induction step, if $m_l^2 + 7 = 2^l n_l$, with $l \geq 3$ and n_l odd, try $m_{l+1} = a m_l$ and show that it is enough to have $a^2 = 2^l q + 1$, with q odd; then, prove that $a = 5^{2^{l-3}}$ works.

6. If $a^{3pq-1} \equiv 1 \pmod{3pq}$, then $a^{3pq-1} \equiv 1 \pmod{p}$; in particular, letting a be a primitive root modulo p, we must have $(p-1) \mid (3pq-1)$ and, analogously, $(q-1) \mid (3pq-1)$. From this, conclude that $(p-1) \mid (3q-1)$ and $(q-1) \mid (3p-1)$, from where $p = 11$ and $q = 17$ or vice-versa. Finally, show that $a^{3 \cdot 11 \cdot 17 - 1} \equiv 1 \pmod{3 \cdot 11 \cdot 17}$ for every natural number a relatively prime with $3 \cdot 11 \cdot 17$. To show that $p = 11$ and $q = 17$ if $p \le q$, one can argue in the following manner: since $p \le q$, we have

$$\frac{3p-1}{q-1} \le \frac{3q-1}{q-1} = 3 + \frac{2}{q-1} \le 3 + \frac{2}{4},$$

so that $\frac{3p-1}{q-1} = 1$, 2 or 3; analysing each one of these cases separately, we conclude that $3p - 1 = 2(q-1)$ and, hence, $\frac{3q-1}{p-1} = \frac{9q-3}{2q-4}$, a number greater than 4 and less than 8; thus, $\frac{9q-3}{2q-4} = 5$, 6 or 7, and these possibilities lead to $q = 17$ as the only actual one.

7. For item (d), take a separate look at each one of the cases $\mathrm{ord}_p(x) = 1, 2, \frac{p-1}{2}$ and $p - 1$—here we used the fact that $\frac{p-1}{2}$ is also prime.

8. For (a) $\Rightarrow$ (b), write $n = p_1 \ldots p_k$, with $p_1 < \cdots < p_k$ being primes such that $(p_i - 1) \mid (n - 1)$; then, apply Fermat's little theorem to conclude that $a^n \equiv a \pmod{p_i}$, for every $1 \le i \le k$. For (b) $\Rightarrow$ (a), take a prime divisor p of n and let a be a primitive root modulo p^2 to obtain a contradiction if $p^2 \mid n$; then, let a be a primitive root modulo p, to show that $(p-1) \mid (n-1)$.

9.

$$s_n \equiv 1 + q_i \left(\frac{u^{p_i(n-1)} - a^{n-1}}{a^{n-1} - 1} \right) \pmod{p_i}.$$

For item (a), observe that

$$\sum_{j=0}^{pq-1} \sum_{l=0}^{p-1} (pj+l)^{n-1} \equiv \sum_{j=0}^{pq-1} \sum_{l=0}^{p-1} l^{n-1} \equiv pq \cdot \sum_{l=0}^{p-1} l^{n-1} \equiv 0 \pmod{p}.$$

For item (b), use (a) to conclude that, if $n = p^2 q$, with p prime and q natural, then $s_n \equiv 1 \pmod{p}$. Item i. goes through with a computation analogous to that of the hint given to item (a). Item ii. follows immediately from i., together with the fact that, modulo p_i, we have $\{a, a^2, \ldots, a^{p_i-1}\} = \{1, 2, \ldots, p_i - 1\}$. Fpr iii., if $n \mid s_n$, then $s_n \equiv 0 \pmod{p_i}$. However, since $p_i \mid (a^{p_i(n-1)} - a^{n-1})$, the formula in ii. guarantees that the only way of having $s_n \equiv 0 \pmod{p_i}$ is to have $p_i \mid (a^{n-1} - 1)$. Then, $p_i - 1 = \mathrm{ord}_{p_i}(a) \mid (n-1)$ and, hence, $(p_i - 1) \mid (q_i - 1)$ (since $n - 1 = (p_i - 1)q_i + (q_i - 1)$). For item iv., if $n \mid s_n$, then $p_i \mid s_n$, and it follows from i. and iii. (more precisely from $(p_i - 1) \mid (n - 1)$) that

$$0 \equiv s_n = 1 + q_i \sum_{l=1}^{p_i-1} l^{n-1} \equiv 1 + q_i \sum_{l=1}^{p_i-1} 1 \equiv 1 + q_i(p_i - 1) \,(\mathrm{mod}\,p_i).$$

Hence, $q_i(p_i - 1) \equiv -1 \equiv p_i - 1 \,(\mathrm{mod}\,p_i)$, so that $q_i \equiv 1 \,(\mathrm{mod}\,p_i)$. Finally in what concerns item v., it is enough to show that if $p_i(p_i - 1) \mid (q_i - 1)$ for $1 \le i \le t$, then $p_i \mid s_n$ for $1 \le i \le t$; to this end, it suffices to use the congruence of item i., together with the fact (easily deducible from our hypotheses) that $(p_i - 1) \mid (n - 1)$.

10. For item (a), if $\gcd(a^k + 1, n) > 1$ then $n \mid (a^k + 1)$, and hence $a^{2k} \equiv 1 \,(\mathrm{mod}\,n)$; this contradicts the fact that $\mathrm{ord}_n(a) = 2pk$. For i., it follows from $a^{kp} \equiv -1 \,(\mathrm{mod}\,n)$ that $a^{2kp} \equiv 1 \,(\mathrm{mod}\,n)$, and thus $d \mid 2kp = n-1$. On the other hand, if $d \mid 2k$, then $a^{2k} \equiv 1 \,(\mathrm{mod}\,n)$ and, this way, $n \mid (a^k - 1)(a^k + 1)$; however, since $\gcd(a^k + 1, n) = 1$, we get $n \mid (a^k - 1)$ and then $a^{kp} \equiv 1 \,(\mathrm{mod}\,n)$, an absurd. Now, if $d \mid 2kp$ and $d \nmid 2k$, it is immediate to see that $\gcd(d, p) > 1$, and hence that $p \mid d$; therefore, the last part of i. follows from Euler's theorem. For ii., it is pretty clear that $p \nmid (2kp + 1)$; therefore, the formula for $\varphi(2kp+1)$ assures that $p \mid (q - 1)$, for some prime divisor q of $2kp + 1$; thus, it suffices to let $q - 1 = lp$, noticing that $l > 1$, for otherwise both p and $q = p + 1$ would be odd primes, which is impossible. Finally, for iii., if $2kp + 1 = (lp + 1)u$ then $u \equiv 1 \,(\mathrm{mod}\,p)$, so that $u = hp + 1$ for some $h \ge 1$. If $h \ge 2$, then $n = (lp + 1)(hp + 1) \ge (2p + 1)^2 > 2(2p + 1)p + 1 \ge 2kp + 1 = n$, which is impossible. Therefore, $h = 0$ or 1, and if $h = 1$ then $(p + 1) \mid (2kp + 1)$, an absurd (for $p + 1$ is even and $2kp + 1$ is odd).

11. Show[2] first (by induction) that for every $k \in \mathbb{N}$ there exists a natural number m_k of k digits, all of which equal to 1 or 2, such that $2^k \mid m_k$. Now, note that $5 \nmid q_{1000}$, and use the fact that 2 is a primitive root modulo 5^{1000} to find $k \in \mathbb{N}$ with $2^k \equiv q_{1000} \,(\mathrm{mod}\,5^{1000})$. Finally, writing $2^k = a \cdot 5^{1000} + q_{1000}$, for some $a \in \mathbb{N}$, show that 2^{k+1000} has the desired properties.

12. Item (a) follows from Fermat's little theorem, together with the fact that $x_1^4 + x_2^4 + x_3^4 + x_4^4 + x_5^4 \equiv 0 \,(\mathrm{mod}\,p)$ if and only if $f(x_1, \ldots, x_5) \equiv 1 \,(\mathrm{mod}\,p)$. Item (b) follows from the multinomial formula desenvolvimento multinomial (cf. Problem 2, page 28) and from the result of (a).

Section 12.3

1. Use Euler's criterion.
2. Adapt the proof of Example 12.24 to the present case to obtain infinitely many integer values of k having the desired property.

---

[2]This is based on the solution given by professor Carlos Gustavo Tamm de A. Moreira.

3. Start by multiplying equality $ax^2 + bxy + cy^2 = n$ by $4a$ and completing squares.

4. First show that if such x and y did exist, then they would be relatively prime with 122. Then, use the result of the previous problem to conclude that 17 should be a quadratic residue modulo 61. Finally, apply the quadratic reciprocity law and Proposition 12.23 to reach a contradiction.

5. For item (a), if every prime divisor of $2b^2 + 3$ was congruent to ± 1 modulo 8, we would have $2b^2 + 3 \equiv \pm 1 \pmod 8$; show that this is impossible. For (b), if $(2b^2 + 3) \mid (a^2 - 2)$ and p is as in (a), then $a^2 \equiv 2 \pmod p$, so that 2 would be a quadratic residue modulo p; use the result of Example 12.27 to reach a contradiction.

6. Let $L_n = 2x^2$. If n is odd, use item (b) of Problem 11, page 252, together with the fact that L_n is even, to show that $n \equiv 3$ or $9 \pmod{12}$. Then, apply several times item (e) of this same problem to get $L_n \equiv L_3$ or $L_9 \equiv 4 \pmod 8$, and show that this gives $x^2 \equiv 2 \pmod 4$, which is impossible. Now, assuming that n is even, consider the following three separate cases: $n \equiv 0 \pmod 4$, $n \equiv 6 \pmod 8$ or $n \equiv 2 \pmod 8$. If $n \equiv 0 \pmod 4$, write $n = 2 \cdot 3^l \cdot k$, with $k \in \mathbb{Z}$ such that $2 \mid k$, $3 \nmid k$. Argue as in the proof of Example 12.25 to obtain $L_n \equiv -L_0 = -2 \pmod{L_k}$ and, hence, $x^2 \equiv -1 \pmod{L_k}$. Show that this furnishes a contradiction, exactly as the one obtained in the proof of Example 12.25. If $n \equiv 6 \pmod 8$ but $n = 6$ (note that $L_6 = 18 = 2 \cdot 3^2$), write $n = 6 + 2 \cdot 3^l \cdot k$, with $k \in \mathbb{Z}$ such that $4 \mid k$ and $3 \nmid k$. Use item (d) of Problem 13, page 253, to obtain $L_n \equiv -L_6 = -18 \pmod{L_k}$, and hence $x^2 \equiv -9 \pmod{L_k}$. Also as in the proof of Example 12.25, conclude that one can take a prime divisor p of L_k such that $p \equiv 3 \pmod 4$, so that $x^2 \equiv -9 \pmod p$. Then, show that $3 \nmid L_k$, so that $\gcd(p, 9) = 1$ and, thus, $\left(\frac{-9}{p}\right) = \left(\frac{9}{p}\right)\left(\frac{-1}{p}\right) = \left(\frac{-1}{p}\right) = -1$. Finally, in the case $n \equiv 2 \pmod 8$, just note that $L_{-n} = L_n$ and $-n \equiv 6 \pmod 8$.

7. Let $F_n = x^2$. If $n \equiv 1 \pmod 4$ but $n \neq 1$, write $n = 1 + 2 \cdot 3^l \cdot k$, with $k \in \mathbb{Z}$ such that $2 \mid k$, $3 \nmid k$. Use item (c) of Problem 13, page 253, to get $F_n \equiv -F_1 = -1 \pmod{L_k}$; then, arrive at a contradiction, exactly as in the proof of Example 12.25. If $n \equiv 3 \pmod 4$, use the fact that $F_{-n} = F_n$ and $-n \equiv 1 \pmod 4$. If n is even, write $n = 2m$ and apply item (a) of Problem 13, page 253 to get $F_n = F_m L_m$. Then, distinguish two distinct subcases: if $\gcd(n, 3) = 1$, use Problem 12, page 252 to conclude that L_m is also a perfect square; then, apply the result of Example 12.25. If $\gcd(n, 3) = 3$, use Problem 12, page 252, again to conclude that L_m is twice a perfect square; then, apply the result of the previous problem.

8. Apply Gauss' lemma, as was done in Example 12.27, or the quadratic reciprocity law.

9. For item (a), if every prime divisor of $2^n - 1$ was congruent to ± 1 modulo 12, we would have $2^n - 1 \equiv \pm 1 \pmod{12}$; show that this is impossible. For (b), if $(2^n - 1) \nmid (3^m - 1)$ and p is as in (a), then $3^m \equiv 1 \pmod p$ or, which is the same, $3^{m+1} \equiv 3 \pmod p$; since $m + 1$ is even, it would follow that 3 would be

a quadratic residue modulo p; now, use the result of the previous problem to get a contradiction.

10. For item (a), use the quadratic reciprocity law. For item (b), use the result of Corollary 12.20, together with the fact that $p - 1$ is a power of 2.

11. For item (a) make induction on $k \geq 3$, showing that if $x_k^2 \equiv a \pmod{2^k}$, then $x_{k+1}^2 \equiv a \pmod{2^{k+1}}$, with $x_{k+1} = x_k$ or $x_k + 2^{k-1}$. For (b) make an analogous induction; more precisely, if $x_k^2 = p^k q + a$, let $x_{k+1} = x_k + p^k t$, with $t \in \mathbb{Z}$, and impose the validity of the congruence $x_{k+1}^2 \equiv a \pmod{p^{k+1}}$ to deduce that t must satisfy the linear congruence $2x_k t \equiv -q \pmod{p}$; finally, show that it is always possible to choose such a t.

12. If $x^2 \equiv a \pmod{n}$ for some integer x, then $x^2 \equiv a \pmod{2^k}$ and $x^2 \equiv a \pmod{p_i^{k_i}}$, for $1 \leq i \leq t$; it then suffices to apply the results of the previous problem, together with Euler's criterion. Conversely, if conditions (i) and (ii) are satisfied, apply once more the results of the previous problem and Euler's criterion to assure the existence of integers $x_0, x_1, \ldots, x_t$ such that $x_0^2 \equiv a \pmod{2^k}$ and $x_i^2 \equiv a \pmod{p_i^{k_i}}$, for $1 \leq i \leq t$; then, invoke the chinese remainder theorem to get, out of the x_i's, an integer x such that $x^2 \equiv a \pmod{n}$.

13. Number the children in the counterclockwise sense and in a *continuous* way, so that the child at positions $1, n + 1, 2n + 1, 3n + 1, \ldots$ is the same one, and analogously for the remaining children. Show that the teacher will give candies to the children situated at positions $\frac{k(k+1)}{2}$, for $k \in \mathbb{N}$, so that it suffices to find all natural values of n for which the congruence $\frac{x(x+1)}{2} \equiv a \pmod{n}$ has a solution for every integer $1 \leq a \leq n$. Now, notice that if such a congruence has a solution, then the same holds with the congruence $(2x+1)^2 \equiv 8a+1 \pmod{n}$. Finally, use the results of Problem 12 to show that n must be a power of 2.

14. For $\Rightarrow$, use Fermat's little theorem. For $\Leftarrow$, let α be a primitive root modulo p, and $j \in \mathbb{N}$ be such that $\frac{n}{d} j \equiv -1 \left(\bmod \frac{p-1}{d} \right)$; then, take $k \in \mathbb{N}$ such that $\alpha^k \equiv a^j \pmod{p}$ and show that $k \mid d$ and $x_0 = \alpha^{k/d}$ solves the congruence $x^n \equiv a \pmod{p}$.

15. Items (a) to (d) only require simple algebraic manipulations and elementary properties of congruences. For the first part of item (e), use Wilson's theorem, together with Euler's criterion; for the second part assume, by the sake of contradiction, that $r_i \equiv r_j \pmod{p}$ and use the fact that $p \nmid (j^2 - i^2)$. For the second part of (f), conclude that

$$\sum_{j=1}^{\frac{p-1}{2}} r_j \equiv \sum_{j=1}^{\frac{p-1}{2}} j^2 \equiv 0 \pmod{p},$$

where, in the last congruence, we used the fact that $p \geq 5$.

Section 12.3

1. Start by showing, with the aid of Proposition 10.9, that if such integers k, l, x, y and z did exist, with $l \geq 1$, then x, y and z would all be even. Then, use this to reduce the analysis to the case $l = 0$, and use again the results of Proposition 10.9.

2. By contradiction, suppose that $a + b$ is even. Then, $a + c^2 = a^2 - b^2 \equiv 0 \pmod 4$, and hence $a \equiv -c^2 \pmod 4$. From this, conclude that $a(a-1)$ has a prime factor congruent to 3 modulo 4 and which appears in its canonical decomposition with odd exponent.

3. Observe that $1995 = 3 \cdot 5 \cdot 7 \cdot 19$, with $3, 7, 19 \equiv 3 \pmod 4$. On the other hand, if d is a positive divisor of 1995 (the case $d < 0$ can be handled in a totally analogous way), analyse separately the following four possibilities: $d = 1$, $d = 5$, $1 < d \mid (3 \cdot 7 \cdot 19)$ and $d = 5d'$, with $1 < d' < 3 \cdot 7 \cdot 19$. For instance: (i) if $d = 5$, then $x^2 + y^2 = 5(x - y)$ if and only if $(2x - 5)^2 + (2y + 5)^2 = 50$, so that $2x - 5 = \pm 1, \pm 7$; (ii) if $1 < d \mid (3 \cdot 7 \cdot 19)$, then, since $x^2 + y^2 \equiv 0 \pmod d$, the argument sketched in item (ii) of the proof of Theorem 12.33 guarantees that $d \mid x$ and $d \mid y$; setting $x = da$ and $y = db$, we obtain $a^2 + b^2 = a - b$, which is equivalent to the first of the four possibilities above.

4. Example 6.40 guarantees the existence of infinitely many primes of the form $4k + 3$; then, choose distinct primes $q_1, \ldots, q_n$, all of which congruent to 3 modulo 4, and apply the chinese remainder theorem, together with Theorem 12.33, to the system of linear congruences $x \equiv -i + q_i \pmod{q_i^2}$, for $1 \leq i \leq n$.

5. For item (a), note that $|S| = \frac{p+1}{2}$. For (b), use (a). Item (c) follows from (b), together with the fact that $y_1^2 + y_2^2 + y_3^2 + y_4^2 \equiv mp \pmod p$. Finally, for (d), start by using a descent argument to show that p can be written as a sum of four squares; then, apply several times identity (12.10) to show that any $n \in \mathbb{N}$ can be written as a sum of four squares.

6. For item (a), assume that n has an odd divisor $d > 1$. Then, since $(2^d - 1) \mid (2^n - 1)$, we have that $(2^d - 1) \mid (m^2 + 9)$, which is a sum of two perfect squares. However, since $2^d - 1$ is of the form $4k - 1$, he must have some prime divisor p which is also of this form. It now follows from a slight modification of the proof of Theorem 12.31 that $p \mid m$ and $p \mid 3$, and hence $p = 3$. Therefore, $3 \mid (2^d - 1)$, which is an absurd, for $2^d - 1 \equiv (-1)^d - 1 \equiv -2 \pmod 3$ (recall that d is odd). For item (b), start by observing that the case $k = 1$ is trivial. Now, assume that $(2^{2^{k-1}} - 1) \mid (m_{k-1}^2 + 9)$ and notice that $(2^{2^{k-1}} + 1) \mid ((3 \cdot 2^{2^{k-2}})^2 + 9)$. Then, with the aid of the chinese remainder theorem, choose $m_k \in \mathbb{N}$ such that

$$m_k \equiv m_{k-1} \pmod{2^{2^{k-1}} - 1} \quad \text{and} \quad m_k \equiv 3 \cdot 2^{2^{k-2}} \pmod{2^{2^{k-1}} + 1}.$$

It is immediate to verify that $2^{2^{k-1}} - 1$ and $2^{2^{k-1}} + 1$ divide $m_k^2 + 9$; however, since such numbers are relatively prime, we conclude that their product also divides $m_k^2 + 9$.

Section 13.1

2. For item (b), write $|z + w|^2 = (z + w)(\overline{z + w})$ and use the result of item (b) of Lemma 13.2, together with (13.8).

3. For item (a), apply inequality $|z + w| \le |z| + |w|$, with $u - z$ in place of z and $z - v$ in place of w. For item (b), observe that the desired inequality is equivalent to $-|z - w| \le |z| - |w| \le |z - w|$; then, in order to show that $|z| - |w| \le |z - w|$, apply inequality $|z + w| \le |z| + |w|$, writing $z - w$ in place of z.

4. If z can be written as in the statement, use the second part of item (a) of Problem 2 to conclude that $|z| = 1$. Conversely, suppose that $|z| = 1$. Impose that $z = \frac{1-iw}{1+iw}$, with $w \in \mathbb{C}$, to obtain $w = i\frac{1-z}{1+z}$; then, use items (b) and (d) of Lemma 13.2, together with the fact that $|z| = 1$, to conclude that $w \in \mathbb{R}$.

5. Expand both sides of the inequality $|z - a|^2 < |1 - \overline{a}z|^2$, with the aid of the result of item (b) of Problem 2.

6. First note that

$$z \in \Gamma \Leftrightarrow |z - a|^2 = R^2 \Leftrightarrow |z|^2 - 2\mathrm{Re}(\overline{a}z) + |a|^2 = R^2.$$

Then, set $w = \frac{1}{z}$ to successively obtain $\left|\frac{1}{w} - a\right|^2 = R^2$, hence $|a|^2|a^{-1} - w|^2 = R^2|w|^2$, then $|w|^2(|a|^2 - R^2) - 2\mathrm{Re}(aw) + 1 = 0$. Finally, write this last equality as $|w|^2 - 2\mathrm{Re}\left(\frac{aw}{|a|^2 - R^2}\right) + \frac{1}{|a|^2 - R^2} = 0$ and *complete squares*.

7. First note that $z_{k+1} = z_k\left(1 + \frac{i}{\sqrt{k+1}}\right)$. Then, compute $|z_k|$ in terms of k by using telescoping products (cf. [8], for instance). Finally, note that $z_{k+1} - z_k = \frac{z_k i}{\sqrt{k+1}}$.

8. Review the definitions of addition and subtraction of vectors, in Chapter 8 of [9].

9. Assume such a total ordering does exist and arrive at a contradiction. You may wish to use that $i^2 = -1$.

10. Checking items (a) to (e) is somewhat tedious, though elementary. In what concerns (f), the first part is immediate from the definition of multiplication in $\mathbb{H}$; for the second part, it suffices to use the identity $|\alpha\beta|^2 = \alpha\beta\overline{\alpha}\overline{\beta}$, together with the result of item (e). Item (g) follows from the second part of (f). Finally, if $\alpha \in \mathbb{H} \setminus \{0\}$, then $\alpha \cdot \frac{\overline{\alpha}}{|\alpha|^2} = 1$; therefore, if $\alpha\beta = 0$ and $\alpha \ne 0$, then

$$0 = \frac{\overline{\alpha}}{|\alpha|^2}(\alpha\beta) = \left(\frac{\overline{\alpha}}{|\alpha|^2}\alpha\right)\beta = 1 \cdot \beta = \beta.$$

On the other hand, if $\alpha\beta = 1$, then $\alpha\beta - \alpha\frac{\overline{\alpha}}{|\alpha|^2} = 0$ or, which is the same, $\alpha\left(\beta - \frac{\overline{\alpha}}{|\alpha|^2}\right) = 0$; however, since $\alpha \ne 0$, it follows that $\beta = \frac{\overline{\alpha}}{|\alpha|^2}$ (here, we wrote $\alpha - \beta$ to denote $\alpha + (-\beta)$, where $-\beta = (-w) + (-x)i + (-y)j + (-z)k$ if $\beta = w + xi + yj + zk$).

11. In the complex plane, let a, b, c, d and e be the complex numbers associated to the vertices of $\mathcal{P}$ and assume, without any loss of generality, that $l_{10} = |b - c|$. Use a geometric argumento to show that we can suppose $d = 0$ and $e = 1$. Then, observe that the stated hypotheses are equivalent to $|a|^2$, $|b|^2$, $|c|^2$, $|a - 1|^2$, $|b - 1|^2$, $|c - 1|^2$, $|a - b|^2$ and $|a - c|^2$ being all rationals. Conclude, with the aid of the result of Problem 2, that $\operatorname{Re}(a)$, $\operatorname{Re}(b)$, $\operatorname{Re}(c)$, $\operatorname{Re}(a\bar{b})$ and $\operatorname{Re}(a\bar{c})$ are rationals and, hence, that $\operatorname{Im}(a)^2 = |a|^2 - \operatorname{Re}(a)^2$ is rational. Since

$$\operatorname{Re}(a\bar{b}) = \operatorname{Re}(a)\operatorname{Re}(\bar{b}) - \operatorname{Im}(a)\operatorname{Im}(\bar{b})$$
$$= \operatorname{Re}(a)\operatorname{Re}(b) + \operatorname{Im}(a)\operatorname{Im}(b),$$

deduce that $\operatorname{Im}(a)\operatorname{Im}(b)$ is rational; analogously, show that $\operatorname{Im}(a)\operatorname{Im}(c)$ is also rational. Now, since (once more from Problem 2) $|b - c|^2 = |b|^2 + |c|^2 - 2\operatorname{Re}(b\bar{c})$, it suffices to show that $\operatorname{Re}(b\bar{c})$ is rational. To this end, note that

$$\operatorname{Re}(b\bar{c}) = \operatorname{Re}(b)\operatorname{Re}(c) + \operatorname{Im}(b)\operatorname{Im}(c),$$

and then it is enough to show that $\operatorname{Im}(b)\operatorname{Im}(c)$ is rational. For what is left to do, write

$$\operatorname{Im}(b)\operatorname{Im}(c) = \frac{\operatorname{Im}(a)\operatorname{Im}(b) \cdot \operatorname{Im}(a)\operatorname{Im}(c)}{\operatorname{Im}(a)^2}.$$

Section 13.2

1. Adapt, to the present case, the proof of Corollary 13.8.
2. Start by observing that $\omega + \frac{1}{\omega} = -1$, $\omega^{3k} = 1$, $\omega^{3k+1} = \omega$ and $\omega^{3k+2} = \omega^2$ for every $k \in \mathbb{Z}$.
3. Put $1 \pm \sqrt{3}i$ in polar form and use the first de Moivre's formula.
4. Take conjugates in the second equation to obtain $\bar{z}_1 + \bar{z}_2 + \bar{z}_3 = 0$. Then, use the first equation and item (d) of Lemma 13.2 to conclude that $\frac{1}{z_1} + \frac{1}{z_2} + \frac{1}{z_3} = 0$ and, hence, that $z_1 z_2 + z_1 z_3 + z_2 z_3 = 0$. Finally, observe that $0 = z_1(z_1 z_2 + z_1 z_3 + z_2 z_3) = z_1^2(z_2 + z_3) + 1 = -z_1^3 + 1$, with identical relations being true for z_2 and z_3.
5. Adapting to the present case the solution given to Example 13.17, write $z - 1 = R\omega(z + 1)$, where ω is an n-th root of unity, so that $z = \frac{1+R\omega}{1-R\omega}$. Now, compute

$$z - \left(\frac{1 + R^2}{1 - R^2} \right) = -\frac{2R}{R^2 - 1}\left(\frac{\omega - R}{1 - R\omega} \right) = -\frac{2R}{R^2 - 1}\left(\frac{\omega - R}{\omega(\bar{\omega} - R)} \right)$$

and take modulus on both sides.
6. Write $a_n = (n - \omega)(n - \omega^2)$, with $\omega = \operatorname{cis}\frac{2\pi}{3}$. Then, recall that $1 + \omega + \omega^2 = 0$, from which we obtain $(k - 1 - \omega)(k - \omega^2) = (k + \omega^2)(k - \omega^2) = k^2 - \omega$ and, analogously, $(k - 1 - \omega^2)(k - \omega) = (k + \omega)(k - \omega) = k^2 - \omega^2$.
7. Letting $w^2 = p^2 - 4q^2$, use Bhaskara's formula to show that the roots of the given equation have the same modulus if and only if $\operatorname{Re}(\bar{p}w) = 0$. Then,

substitute $p = r \operatorname{cis} \alpha$, $q = s \operatorname{cis} \beta$ and $w = t \operatorname{cis} \gamma$ into $\omega^2 = p^2 - 4q^2$, with $r, s, t \in \mathbb{R}_+$, and conclude that $|\alpha - \beta|$ must be an integer multiple of π.

8. Write $z_2 = z_2' + z_1$ and $z_3 = z_3' + z_1$ to reduce to the case $z_1 = 0$; then, write $z_2'' = w z_2'$ and $z_3'' = w z_3'$, for an appropriate complex number w of modulus 1, to reduce to the case $z_2' = 1$.

9. Setting $\omega = \operatorname{cis} \frac{2\pi}{n}$, use the first formula of de Moivre to compute the real and imaginary parts of the left hand side of (13.20).

10. For item (a), adapt the hint given to the previous problem. Do the same with item (b); you may find it helpful to use that $\sin^2 x = \frac{1}{2}(1 - \cos(2x))$.

11. Perform twice the substitution $z \mapsto \omega z + a$ in the stated equality.

12. Let O be the center and R the radius of the circle circumscribed to the polygon. By looking to the plane as the complex plane, we can assume that O corresponds to 0, A_1 to R, A_2 to $R\omega$, with $\omega = \operatorname{cis} \frac{2\pi}{2n}$, and P to z. Hence, $A_k = R\omega^{k-1}$ for $1 \le k \le 2n$, and

$$\overline{A_{2k}P}^2 = |R\omega^{2k-1} - z|^2$$

$$= R^2 + |z|^2 - 2\operatorname{Re}(R\omega^{2k-1}\overline{z}).$$

Analogously, $\overline{A_{2k-1}P}^2 = R^2 + |z|^2 - 2\operatorname{Re}(R\omega^{2k-2}\overline{z})$, so that

$$\sum_{k=1}^{n} \overline{A_{2k}P}^2 = \sum_{k=1}^{n} \overline{A_{2k-1}P}^2$$

if and only if

$$\sum_{k=1}^{n} \operatorname{Re}(R\omega^{2k-1}\overline{z}) = \sum_{k=1}^{n} \operatorname{Re}(R\omega^{2k-2}\overline{z}).$$

In turn, this is the same as asking that

$$\operatorname{Re}\left(R\overline{z}\sum_{k=1}^{n}\omega^{2k-1}\right) = \operatorname{Re}\left(R\overline{z}\sum_{k=1}^{n}\omega^{2k-2}\right).$$

Finally, show that $\sum_{k=1}^{n}\omega^{2k-1} = \sum_{k=1}^{n}\omega^{2k-2} = 0$.

Section 13.3

1. For a given $n \in \mathbb{N}$, $U_n(\mathbb{C})$ is a *finite* subset of $\mathbb{C}$ satisfying both conditions.

2. Start by using the finiteness of A to show that all of its elements are roots of unity. Then, if $z \in A$ and $z^k = 1$ for some natural k, show that $k \mid n$.

3. You may use item (d) of Problem 21, page 220, together with the fact that $\varphi(n)$ is even for $n > 2$ (cf. Problem 19, page 220).

4. For item (a), since d divides both m and n, substituting each piece $1 \times m$ by $\frac{m}{d}$ pieces $1 \times d$ and each piece $1 \times n$ by $\frac{n}{d}$ pieces $1 \times d$, we conclude that the rectangle can be assembled by using only pieces $1 \times d$. The rest follows from Klarner's theorem. For item (b), let q be a natural number such that $q > mn$. According to item (a) of Problem 20, page 179, q can be written as $q = mx + ny$, with $x, y \in \mathbb{N}$. Therefore, if $p, q \in \mathbb{N}$ are such that $q > mn$, then we can assemble a $p \times q$ rectangle by forming p rows $1 \times q$, with each row composed by x pieces $1 \times m$ and y pieces $1 \times n$ (x and y as above). However, if $m, n \nmid p, q$, then Klarner's theorem assures that one cannot assemble a $p \times q$ rectangle by using only pieces $1 \times m$ or only pieces $1 \times n$.

5. Arguing as in the proof of Klarner's theorem, write $x^i y^j$ inside the 1×1 square of the chessboard situated at row i and column j, for $1 \le i, j \le 8$. After we cover the board, assume that the 1×1 square is the one situated at row k and column l. Use double counting to get

$$(1 + x + x^2)E(x, y) + (1 + y + y^2)F(x, y) + x^k y^l$$
$$= (x + x^2 + \cdots + x^8)(y + y^2 + \cdots + y^8),$$

for some polynomial expressions E and F in x and y. Let $\omega = \text{cis}\,\frac{2\pi}{3}$. Substitute $x = y = \omega$ to get $\omega^{k+l} = 1$ and, hence $k + l \equiv 0 \,(\text{mod}\,3)$. Substitute $x = \omega^2$, $y = \omega$ to get $\omega^{2k+l} = 1$ and, hence $2k + l \equiv 0 \,(\text{mod}\,3)$. Finally, conclude that $k, l \equiv 0 \,(\text{mod}\,3)$, and show that any one of the choices $(k, l) = (3, 3)$, $(3, 6)$, $(6, 3)$ or $(6, 6)$ do correspond to actual possible positions of the 1×1 square (i.e., show that the remaining 63 squares of the chessboard can be filled with the 21 given rectangles 1×3.

6. Adapt, to the present case, the proof of Klarner's theorem.

7. Once more, adapt the proof of Klarner's theorem to the present case.

8. Use the result of the previous problem.

Section 14.1

2. The results of items (a) and (b) are respectively $X^n - a^n$ and $X^n + a^n$. As for item (c) recall that, according to Example 4.12 of [8], every natural number can be uniquely written, up to reordering, as a sum of powers of 2 with nonnegative, pairwise distinct integer exponents.

3. For $(X + 1)^m (X + 1)^n$, note that each product that contributes to the coefficient of X^p comes from a product of the form $aX^k \cdot bX^{p-k}$; now, note that $a = \binom{m}{k}$ and $b = \binom{n}{p-k}$.

4. Start by establishing the commutativity and distributivity. For the associativity of multiplication, start by using the distributivity to show that, if $f(gh_1) = (fg)h_1$ and $f(gh_2) = (fg)h_2$, then $f(g(h_1 + h_2)) = (fg)(h_1 + h_2)$. Then, iterate such a reasoning to show that it suffices to establish the (obvious) equality $aX^m(bX^n \cdot cX^p) = (aX^m \cdot bX^n) \cdot cX^p$, for all $a, b, c \in \mathbb{K}$ and $m, n, p \in \mathbb{Z}_+$.

5. By contraposition, show that if $f, g \in \mathbb{K}[X] \setminus \{0\}$, then $fg \neq 0$. To this end, write $f(X) = \sum_{i=k}^{m} a_i X^i$ and $g(X) = \sum_{j=l}^{n} b_j X^j$, with $a_k, b_l \neq 0$, and examine the coefficient of X^{k+l} in fg.

6. Letting $g(X) = f(X) + \frac{1}{2}$, we are left to finding g, of degree 2001, such that $g(X) + g(1-X) = 0$ or, which is the same, $g(X) = -g(1-X)$. Setting $g(X) = (X-a)^{2001}$, we must have $(X-a)^{2001} = -(1-X-a)^{2001} = (X+a-1)^{2001}$.

7. Apply the multisection formula.

8. Adapt the proof of Theorem 14.10 to the present case.

9. Apply the result of the previous problem to

$$f(X) = (X+1)^m = \sum_{k=0}^{m} \binom{m}{k} X^k,$$

$n = 4, r = 1$ and $z = \sqrt{2}$.

10. Let a_0 be the constant term of f and write $f(X^m) = g(X)h(X)$, with $g, h \in \mathbb{R}[X] \setminus \mathbb{R}$ being such that their nonvanishing coefficients are positive. Let $g(X) = \sum_{i=0}^{k} b_i X^i$ and $h(X) = \sum_{j=0}^{l} c_j X^j$. Since $a_0 \neq 0$ and $a_0 = b_0 c_0$, we have $b_0, c_0 \neq 0$. Therefore, $b_0, c_0 > 0$, and hence $a_0 > 0$. Suppose, for the sake of contradiction, that g has a term of the form $b_t X^t$, with $b_t > 0$ and $m \nmid t$. Then, the coefficient of X^t in $f(X^m)$ is

$$\sum_{\substack{i+j=t \\ i,j \geq 0}} b_i c_j = \sum_{\substack{i+j=t \\ i>0, j \geq 0}} b_i c_j + b_t c_0 > 0.$$

Now, reach a contradiction and finish the proof.

Section 14.2

1. Compute $(X^2 + X + 1)^n$ by writing $X^2 + X + 1 = (X^2 - X + 1) + 2X$. Then, use the uniqueness part of the division algorithm.

2. Write $X^m + 1 = X^m - 1 + 2 = (X^n)^{m/n} - 1 + 2$. Then, use the result of item (a) of Problem 2, page 355, to show that $X^n - 1$ divides $(X^n)^{m/n} - 1$.

3. Start by noticing that

$$X^{2^m} - 1 = (X^{2^{m-1}} + 1)(X^{2^{m-1}} - 1)$$

$$= (X^{2^{m-1}} + 1)(X^{2^{m-2}} + 1)(X^{2^{m-2}} - 1) = \cdots,$$

so that $X^{2^n} + 1$ divides $X^{2^m} - 1$.

4. With $\alpha = -\frac{b}{a}$ in Example 14.15, we get

$$f(X) = \left(X + \frac{b}{a}\right)q(X) + f\left(-\frac{b}{a}\right) = (aX+b)a^{-1}q(X) + f\left(-\frac{b}{a}\right).$$

5. Write $f(X) = (X + 2)q_1(X) - 1 = (X - 2)q_2(X) + 3$, with $q_1, q_2 \in \mathbb{Q}[X]$. Then, write $q_1(X) = (X - 2)q(X) + r$, with $r \in \mathbb{Q}$, to get $f(X) = (X^2 - 4)q(X) + (X+2)r - 1$. Finally, use the uniqueness part of the division algorithm to conclude that $r = 1$.

6. Write $f(X) = (X + 1)(X^2 + 1)q(X) + (aX^2 + bX + c)$, for some $a, b, c \in \mathbb{R}$. Then, show that the remainders of the divisions of $aX^2 + bX + c$ by $X + 1$ and $X^2 + 1$ are respectively equal to $a - b + c$ and $bX + (c - a)$. Finally, apply the uniqueness part of the division algorithm twice to obtain a linear system of equations in a, b and c.

7. Assume, without loss of generality, that $m \geq n$. Follow the division algorithm to obtain the quotient $X^{m-n} - X^{m-2n} + \cdots + (-1)^{q-1}X^{n+r}$ and the remainder $(-1)^q X^r + 1$. Alternatively, if q is odd write

$$X^m + 1 = X^{nq+r} + 1 = \left(X^{nq} + 1\right)X^r + (-X^r + 1)$$

and factor $X^{nq} + 1 = (X^n)^q + 1$; if q is even, write

$$X^m + 1 = X^{nq+r} + 1 = \left(X^{nq} - 1\right)X^r + (X^r + 1)$$

and factor $X^{nq} - 1 = (X^{2n})^{q/2} - 1$.

Section 15.1

1. Firstly, show that it suffices to consider the case in which f is of the form $f(X) = aX^n$, for some $a \in \mathbb{K} \setminus \{0\}$, and g is nonconstant. Now, in order to show that $\widetilde{f \circ g} = \tilde{f} \circ \tilde{g}$, use Corollary 15.14.

2. For item (a), use the binomial formula; for (b), write $f(X) = c_n X^n + \cdots + c_1 X + c_0$ and use (a) to infer the existence of rational numbers A and B such that $f(a \pm b\sqrt{r}) = A \pm B\sqrt{r}$. Then, apply the result of problem 1.3.3 of [8].

3. Use the result of the previous problem to conclude that the given polynomial is divisible by $X^2 - 2X - 1$.

4. Use the division algorithm to write $f(X) = (aX + b)q(X) + r$, for some constant r. Then, perform the substitution $X \mapsto -\frac{b}{a}$.

5. Start by observing that, if such an f does exist, then $f(y)^2 = 1 - y^2$ for every $0 \leq y \leq 1$ and, hence, for every $y \in \mathbb{R}$. Then, conclude that $\partial f = 1$ and arrive at a contradiction.

6. Use the root test for $\pm i$, together with the first de Moivre formula.

7. If $n > 1$, write $X^{n+1} - X^n + 1 = X^{n-1}(X^2 - X + 1) - (X^{n-1} - 1)$ to conclude that it is necessary and sufficient to have $X^{n-1} - 1$ divisible by $X^2 - X + 1$. Since $X^2 - X + 1 = (X + \omega)(X + \omega^2)$, with $\omega \neq -1$ being a cubic root of 1, we are searching for the naturals $n > 1$ such that $(-\omega)^{n-1} = 1$ and $(-\omega^2)^{n-1} = 1$.

8. Assume that the given polynomial has an integer root, say r, so that

$$r^4 - 1994r^3 + (1993 + m)r^2 - 11r + m = 0.$$

Look at such an equality modulo 2 to conclude that both m and r are even. Now, suppose that a and b are distinct integer roots of the polynomial. Termwise subtract the equalities

$$a^4 - 1994a^3 + (1993 + m)a^2 - 11a + m = 0$$

and

$$b^4 - 1994b^3 + (1993 + m)b^2 - 11b + m = 0,$$

and factor $a - b$ out of the result to obtain

$$(a + b)(a^2 + b^2) - 1994(a^2 + ab + b^2) + (1993 + m)(a + b) = 11.$$

Then, reach a contradiction.

9. Write $X^2 - X - 1 = (X - u)(X - v)$, with $u = \frac{1+\sqrt{5}}{2}$ and $v = \frac{1-\sqrt{5}}{2}$. Invoke the root test to conclude that it suffices to find all $a, b \in \mathbb{R}$ such that $au^{17} + bu^{16} + 1 = 0$ and $av^{17} + bv^{16} + 1 = 0$. Multiply the first relation by u^{16}, the second by v^{16} and subtract the results to obtain $a = \frac{u^{16} - v^{16}}{u - v}$. Now, letting $x_n = \frac{u^n - v^n}{u - v}$, use (3.4) to conclude that $(x_n)_{n \geq 1}$ is the Fibonacci sequence. Then, conclude that $a = 987$ and compute the value of b in an analogous way.

10. Write $p(X) = X(X - \alpha_2) \ldots (X - \alpha_n)$, with $\alpha_2, \ldots, \alpha_n$ being pairwise distinct nonzero integers. For $d \in \mathbb{Z}$, show that $p(p(d)) = 0$ if and only if $d = \alpha_i$ for some $1 \leq i \leq n$ or $p(d) = \alpha_i$ for some $2 \leq i \leq n$. In this last case, we have $d \neq 0$; moreover, writing $\alpha = \alpha_i$, show that there exists $q \in \mathbb{Z}^*$ such that $d(d - \alpha)q = \alpha$. Conclude that $dq + 1 = -1$ and, hence, that $d = \frac{\alpha}{2}$. Finally, use this to deduce that $(d - \alpha_2) \ldots (d - \alpha_n) = 2$, and use the fact that $n > 4$ to reach a contradiction.

11. Up to reordering, there are two distinct cases: (i) $f(p_1) = f(p_2) = f(p_3) = 3$ and $f(p_4) = -3$; (ii) $f(p_1) = f(p_2) = 3$ and $f(p_3) = f(p_4) = -3$. In case (i), show that $f(X) = a(X - p_1)(X - p_2)(X - p_3) + 3$ and, then, use the condition $f(p_4) = -3$, together with the fact that the p_i's are distinct primes, to reach a contradiction. In case (ii), show that $f(X) = a(X - p_1)(X - p_2)(X - q) + 3$ and compute $f(0)$ to conclude that $q = \frac{3 - d}{ap_1 p_2}$; then, use the conditions $f(p_3) = f(p_4) = -3$ to reach a contradiction.

12. Use the root test to conclude that $p(X) = 5 + (X - a)(X - b)(X - c)(X - d)q(X)$ for some $q \in \mathbb{Z}[X]$. Then, assume that there exists $m \in \mathbb{Z}$ as in the statement and let $x = m$ in the corresponding polynomial functions to reach a contradiction.

13. If $\omega = \operatorname{cis} \frac{2\pi}{3}$, then ω and ω^2 are cubic roots of unity and are the roots of $X^2 + X + 1$. Use these two facts to find out the natural numbers k for which $\omega^{2k} + 1 + (\omega + 1)^{2k} = 0$ and $(\omega^2)^{2k} + 1 + (\omega^2 + 1)^{2k} = 0$.

14. Start by defining $f(X) = \sum_{i=1}^{n} X^{a_i}$ and $g(X) = \sum_{i=1}^{n} X^{b_i}$, and computing

$$f(X)^2 = f(X^2) + 2 \sum_{i<j}^{n} X^{a_i + a_j} \quad \text{and} \quad g(X)^2 = g(X^2) + 2 \sum_{i<j}^{n} X^{b_i + b_j}.$$

Then, use the stated conditions to show that 1 is a root of $f - g$ and $(f(X) + g(X))(f(X)-g(X)) = f(X^2)-g(X^2)$. Writing $f(X)-g(X) = (X-1)^k h(X)$, with $k \in \mathbb{N}$ and $h \in \mathbb{Z}[X]$ such that $h(1) \neq 0$, obtain $f(X^2) - g(X^2) = (X^2 - 1)^k h(X^2)$ and, hence,

$$(f(X) + g(X))h(X) = (X + 1)^k h(X^2).$$

Finally, evaluate both sides at $x = 1$ to get $n = 2^{k-1}$.

Section 15.2

1. Were a an integer root of f, show that the fact that $f(0)$ is odd, together with the rational roots test, would give a to be odd. Then, show that if a is odd, then $f(a)$ and $f(1)$ have distinct parities, so that $f(a)$ would be odd and, hence, we could not have $f(a) = 0$.

2. Letting r be the common difference of the arithmetic progression, write $x = y - r, z = y + r$ and conclude that the polynomial $X^5 - 10X^4 - 20X^2 - 2$ has the rational root x/r. Then, apply the test for rational roots to get a contradiction.

3. Since $\{2\cos\theta;\ \theta \in \mathbb{R}\} = [-2, 2]$, which is an infinite set, Corollary 15.10 guarantees that there is at most one polynomial $f_n \in \mathbb{R}[X]$ satisfying the stated condition.

4. Start by observing that the desired property holds if and only if $z^n f_n(z + \frac{1}{z}) = z^{2n} + 1$ for every $z \in \mathbb{C} \setminus \{0\}$. However, letting $f_n(X) = \sum_{j=0}^{n} a_j X^j$, the equality above is equivalent to the polynomial equality $\sum_{j=0}^{n} a_j X^{n-j}(X^2+1)^j = X^{2n} + 1$. Thus, it suffices to establish the desired equality for at least $n + 1$ distinct complex values of z. To this end, letting $z = \operatorname{cis}\theta$, we get $\frac{1}{z} = \operatorname{cis}(-\theta)$ and, hence,

$$f_n(z + \frac{1}{z}) = f_n(2\cos\theta) = 2\cos(n\theta) = \operatorname{cis}(n\theta) + \operatorname{cis}(-n\theta) = z^n + \frac{1}{z^n}.$$

5. For item (a), note that one also has $x = 2y^2 - 1$, and hence $2(y^2 - x^2) = x - y$. Thus, either $x = y$ or $2(y+x) = -1$, whence $2x^2-x-1 = 0$ or $4x^2+2x-1 = 0$. For item (b), if $x = \cos\theta$, then $2x^2-1 = \cos(2\theta)$ and $2(2x^2-1)^2-1 = \cos(4\theta)$. Hence, we are left to solving $\cos(4\theta) = \cos\theta$, and basic Trigonometry gives $x = \cos\theta = \cos 0, \cos\left(\frac{2\pi}{3}\right), \cos\left(\frac{\pi}{10}\right)$ or $\cos\left(\frac{3\pi}{10}\right)$. Finally, compare these value with those found in item (a) to get $\cos\left(\frac{\pi}{10}\right) = \frac{-1+\sqrt{5}}{4}$.

6. Let $\alpha = \sqrt[3]{a + \sqrt{b}} + \sqrt[3]{a - \sqrt{b}}$. Use the fact that $(u+v)^3 = u^3 + v^3 + 3uv(u+v)$ to compute $\alpha^3 = 2a + 3\sqrt[3]{a^2 - b}\,\alpha$. For item (a), write $(\alpha^3 - 2a)^3 = 27(a^2-b)\alpha^3$ and apply the rational roots test to the polynomial $X^9 - 6aX^6 - (15a^2 - 27b)X^3 - 8a^3$. For item (b), write $a^2 - b = n^3$, with $n \in \mathbb{Z}$, to get $\alpha^3 - 3n\alpha - 2a = 0$. In order to get the result of i., search rational roots in the polynomial $X^3 - $

$3nX - 2a$. For ii., first write $\alpha = 2m$, with $m \in \mathbb{Z}$, to get $a = 4m^3 - 3mn$ and $b = a^2 - n^3 = (4m^3 - 3mn)^2 - n^3$. Now, write $\alpha = 2m + 1$ to get $2a = (2m + 1)(4m^2 + 4m + 1 - 3n)$; hence, n must be even, say $n = 2n' + 1$, with $n' \in \mathbb{Z}$, so that $a = (2m + 1)(2m^2 + 2m - 3n')$ and $b = a^2 - n^3 = [(2m + 1)(2m^2 + 2m - 3n')]^2 - 8(n')^3$.

7. If $(2 + i)^n = (2 - i)^n$, show that $\left(\frac{3}{5} + \frac{4}{5}i\right)^n = 1$, so that $\frac{3}{5} + \frac{4}{5}i = \omega^k$ for some integer $1 \le k < n$, with $\omega = \operatorname{cis} \frac{2\pi}{n}$. Conclude that $\cos \frac{2k\pi}{n} = \frac{3}{5}$ and apply the result of Theorem 15.21.

8. For item (a), let $f_n(X)$ be the product of all 2^n factors of the form

$$X - \left(\pm \sqrt{a_1} \pm \sqrt{a_2} \pm \cdots \pm \sqrt{a_n}\right).$$

Then, f_n is certainly monic and has the given number as one of its roots. We make induction on n to show that $f_n \in \mathbb{Z}[X]$, noting that this is a straightforward computation for $n = 1$. Assuming its validity for $n = k$, let $a_1, a_2, \ldots, a_k, a_{k+1}$ be given as in the statement of the problem. Then,

$$f_{k+1}(X) = \prod \left(X + \sqrt{a_{k+1}} \pm \sqrt{a_1} \pm \cdots \pm \sqrt{a_k}\right)$$
$$\prod \left(X - \sqrt{a_{k+1}} \pm \sqrt{a_1} \pm \cdots \pm \sqrt{a_k}\right)$$
$$= f_k\left(X + \sqrt{a_{k+1}}\right) f_k\left(X - \sqrt{a_{k+1}}\right),$$

with the above products ranging over all 2^k possible choices of $+$ and $-$ signs. Now, direct computation gives

$$f_k\left(X \pm \sqrt{a_{k+1}}\right) = g_k(X) \pm \sqrt{a_{k+1}} h_k(X),$$

for some $g_k, h_k \in \mathbb{Z}[X]$, with g_k monic, $\partial g_k = 2^k$ and $\partial h_k \le 2^k - 1$. Thus, we obtain

$$f_{k+1}(X) = \left(g_k(X) + \sqrt{a_{k+1}} h_k(X)\right)\left(g_k(X) - \sqrt{a_{k+1}} h_k(X)\right)$$
$$= g_k(X)^2 - a_{k+1} h_k(X)^2,$$

and it is now clear that f_{k+1} has integer coefficients. For item (c), the rational roots test, together with (a), shows that the given number is either an integer or an irrational. However, the estimates $1.4 < \sqrt{2} < 1.5$, $1.7 < \sqrt{3} < 1.8$, $2.2 < \sqrt{5} < 2.3$, $2.6 < \sqrt{7} < 2.7$ and $3.3 < \sqrt{11} < 3.4$ give $11.2 < \sqrt{2} + \sqrt{3} + \sqrt{5} + \sqrt{7} + \sqrt{11} < 11.7$, so that it is not an integer.

9. On the contrary, suppose the two polynomials have a common root z, which is not rational (for $X^5 - X - 1$ has no rational root). Then,

$$z + 1 = z(z^2)^2 = z(-az - b)^2 = z(a^2z^2 + 2abz + b^2)$$
$$= (-az - b)a^2z + 2ab(-az - b) + b^2z$$
$$= -a^3(-az - b) - 3a^2bz - 2ab^2 + b^2z,$$

so that $(1 - a^4 + 3a^2b - b^2)z = a^3b - 2ab^2 - 1$. However, since $a, b \in \mathbb{Q}$ but $z \notin \mathbb{Q}$, this gives

$$a^4 - 3a^2b + b^2 = a^3b - 2ab^2 = 1. \tag{22.5}$$

The second equality gives $a \neq 0$, hence $b^2 = \frac{a^3b-1}{2a}$. Substituting such a relation in the first equality, one gets $a^4 - 3a^2b + \frac{a^3b-1}{2a} = 1$, then $b = \frac{2a^5-2a-1}{5a^3}$. Putting this back into the second equality of (22.5), we finally obtain

$$a^3\left(\frac{2a^5 - 2a - 1}{5a^3}\right) - 2a\left(\frac{2a^5 - 2a - 1}{5a^3}\right)^2 = 1.$$

This is equivalent to $a^{10} + 3a^6 - 11a^5 - 4a^2 - 4a - 1 = 0$, which is an absurd, for the rational root test assures that the polynomial $X^{10} + 3X^6 - 11X^5 - 4X^2 - 4X - 1$ has no rational root.

Section 15.3

1. Show that the deduction of Bhaskara's formula (cf. Section 2.3 of [8], for instance) still holds in the realm of complex numbers.
2. For the first part, the roots of $u^3 = u_0^3$ are u_0, $u_0\omega$ and $u_0\omega^2$; now, review the discussion of Example 15.22. For the second part, if $u_0 + v_0 = u_0\omega + v_0\omega^2$, show that $u_0 + v_0(\omega + 1) = 0$ and, hence, that $u_0^2 = a(\omega + 1)$. Then, use the fact that $u_0^6 + bu_0^3 - a^3 = 0$ to obtain the relation $bu_0(\omega + 1) = 2a^2$. Finally, square it and substitute $u_0^2 = a(\omega + 1)$ to arrive at $4a^3 + b^2 = 0$. Argue in an analogous way to investigate the equalities $u_0 + v_0 = u_0\omega^2 + v_0\omega$ and $u_0\omega + v_0\omega^2 = u_0\omega^2 + v_0\omega$.
3. Item (b) follows from applying elementary algebra to the equality of item (a), together with the fact that $z^4 + az^3 + bz^2 + cz + d = 0$. For (d), Example 15.22 shows that one can always find w as in the first part of item (c), and hence solve the quadratic equations displayed in the second part of item (c). Upon doing this, show that z will be a complex root of $X^4 + aX^3 + bX^2 + cX + d$.
4. For the first part, write $f(z) = a_nz^n + \cdots + a_1z + a_0$, with $a_0, a_1, \ldots, a_n \in \mathbb{R}$ and use items (b) and (c) of Lemma 13.2 to conjugate both sides of the equality $f(z) = 0$. For the second, make induction on the degree of f; for the induction step, note that, for $z \in \mathbb{C} \setminus \mathbb{R}$, one has $(X - z)(X - \overline{z}) = X^2 - 2\mathrm{Re}(z)X + |z|^2 \in \mathbb{R}[X]$.

5. Use the result of the previous Problem 4.
6. Use the proof of the previous problem to write $f(X) = u(X)v(X)$, with $u(X) = \prod_{j=1}^{k}(X - z_j)$ and $v(X) = \prod_{j=1}^{k}(X - \bar{z}_j)$ for some $z_1, \ldots, z_k \in \mathbb{C} \setminus \mathbb{R}$. Now, note that if $u(X) = g(X) - ih(X)$ for some $g, h \in \mathbb{R}[X]$, then $v(X) = g(X) + ih(X)$.
7. Writing $f(X) = a_n X^n + a_{n-1} X^{n-1} + \cdots + a_1 X + a_0$, with $a_n \neq 0$, it follows from $f(\alpha) = 0$ that

$$\alpha^n = -\frac{a_{n-1}}{a_n} \cdot \alpha^{n-1} - \cdots - \frac{a_1}{a_n}\alpha - \frac{a_0}{a_n}.$$

   Now, argue by induction to prove (15.8) for $m \geq n$. For the case $m = -1$, expand the equality $\alpha^{-1} f(\alpha) = 0$ in terms of the coefficients of f; for the general case $m < 0$, use induction again.
8. Apply the triangle inequality to get, for $|z| \geq 1$,

$$|a_n||z|^n \leq |a_{n-1}||z|^{n-1} + \cdots + |a_1||z| + |a_0| \leq nA|z|^{n-1} \leq nA|z|^{n-1}.$$

9. Letting $z = \alpha + \beta i$ be a complex root of f such that $\alpha > 2$, adapt the hint given to the previous problem (i.e., apply the triangle inequality) to obtain $|a_n z^n + a_{n-1} z^{n-1}| < \frac{k|z|^{n-1}}{|z|-1}$. Subsequently, use this to show that $|z| < 1 + \frac{k}{|a_n z + a_{n-1}|}$ and, then, substitute $z = \alpha + i\beta$.
10. For the case of degree 4, if $z \in \mathbb{C}$ is a root of f, then $z \neq 0$ and $a\left(z^2 + \frac{1}{z^2}\right) + b\left(z + \frac{1}{z}\right) + c = 0$; then, let $w = z + \frac{1}{z}$ and apply the result of Problem 4, page 378, for $n = 2$. For the case of degree 6, argue analogously, starting with $f(X) = aX^6 + bX^5 + cX^4 + dX^3 + cX^2 + bX + a$, with $a \neq 0$.
11. Use the stated conditions to get $(X + 1)f(X) - X = aX(X - 1)\ldots(X - n)$. Then, compute a by evaluating both sides at -1.
12. As in Example 15.26, substitute X by 1 in the factorised form of $X^{n-1} + X^{n-2} + \cdots + X + 1$; then, use some Trigonometry to show that, if $\omega = \operatorname{cis}\frac{2\pi}{n}$, then $|1 - \omega^k| = 2\sin\frac{k\pi}{n}$ for $1 \leq k \leq n$.
13. Use the result of the previous problem, together with the relation $\sin(\pi - x) = \sin x$.
14. If $z \in \mathbb{C}$ is a root of p, show that z^2 and $z - 1$ are also roots. Then, use the finiteness of the number of roots to successively conclude that $|z| = 1$, $|z-1| = 1$ and $z = \omega$ or $z = \bar{\omega}$, where $\omega = \operatorname{cis}\frac{2\pi}{3}$. Finally, apply the result of Problem 4.

Section 15.4

1. Use the result of Corollary 15.32, pretty much along the same lines of Example 15.33, to show that $8a^3 - 25a^2 - 180a + 608 = 0$. Then, conclude that $a = 4$.

2. Write $f(X) = g(X)^2 h(X)$ and, then, compute f'.

3. Use item (b) of Proposition 15.29 to show that $f'(z) = \sum_{j=1}^{n} \frac{f(z)}{z - z_j}$.

4. Adapt, to the present case, the hint given to the previous problem.

5. For item (a), let $f(X) = X^n - 1 = \prod_{j=0}^{n-1}(X - \omega^j)$ and check that

$$(\omega^j - 1)\dots(\omega^j - \omega^{j-1})(\omega^j - \omega^{j+1})\dots(\omega^j - \omega^{n-1}) = f'(\omega^j) = n\omega^{(n-1)j}.$$

For item (b), note first that the substitution $X \mapsto 0$ in (15.7) gives $\omega\omega^2\dots\omega^{n-1} = (-1)^{n-1}n$. Then,

$$P = \prod_{k=0}^{n-1}(\omega^k - 1)\dots(\omega^k - \omega^{k-1})(\omega^{k+1} - \omega^k)\dots(\omega^{n-1} - \omega^k)$$

$$= \prod_{k=0}^{n-1}(-1)^{n-1-k}(\omega^k - 1)\dots(\omega^k - \omega^{k-1})(\omega^k - \omega^{k+1})\dots(\omega^k - \omega^{n-1})$$

$$= (-1)^{\frac{n(n-1)}{2}}\prod_{k=0}^{n-1}n\omega^{(n-1)k} = (-1)^{\frac{(n-1)(n-2)}{2}}n^n.$$

6. For item (a), set $b_n = A\alpha^{n-1} + B\beta^{n-1} + C\gamma^{n-1}$ for every $n \geq 1$, with A, B and C being the solutions of the linear system

$$\begin{cases} A + B + C = a_1 \\ \alpha A + \beta B + \gamma C = a_2 \\ \alpha^2 A + \beta^2 B + \gamma^2 C = a_3 \end{cases},$$

show that $(b_n)_{n \geq 1}$ satisfies $b_{k+3} = ub_{k+2} + vb_{k+1} + wb_k$ for every $k \geq 1$, and that $a_k = b_k$ for $1 \leq k \leq 3$. Then, show that $a_n = b_n$ for every $n \geq 1$. Items (b) and (c) can be dealt with in similar ways, except for the fact that, in order to establish the equality $b_{k+3} = ub_{k+2} + vb_{k+1} + wb_k$, you will have to use the fact that $X^3 - uX^2 - vX - w$ has multiple roots.

7. First of all, find the roots of $X^3 - 6X^2 + 12X - 8$ (look for integer roots). Then, apply the result of the previous problem.

8. Letting $f(X) = \prod_{k=1}^{n}\left(1 + \frac{1}{k}X^k\right)$, show that $f(1) = n + 1$ and

$$f(X) = 1 + \sum_{\emptyset \neq S \subset I_n} \frac{1}{\pi(S)}X^{\sigma(S)}.$$

Now, compute $f'(1)$ and, with the aid of the result of Problem 4, $\frac{f'(1)}{f(1)}$.

9. It suffices to show that, if some half-plane of the complex plane contains the roots of f, then it also contains those of f'. By the sake of contradiction,

suppose that there exists a line r such that one of the half-planes it determines contains a root w of f', whereas the other contains all of the roots of f. Let $u \in \mathbb{C}$ have modulus 1 and such that the vector $\boldsymbol{u}$ is perpendicular to r; if $f(X) = a(X - z_1) \ldots (X - z_n)$ and θ and θ_j respectively denote arguments of u and $z_j - w$, use the result of Problem 3 to show that

$$\mathrm{Re}\left(\sum_{j=1}^{n} |z_j - w|^{-1} \cos(\theta_j - \theta)\right) = 0.$$

Then, note that $(z_j - w)/u$ has an argument belonging to the interval $(-\frac{\pi}{2}, \frac{\pi}{2})$, and use the equality above to reach a contradiction.

10. If $f(X) = \sum_{k=0}^{n} c_k X^k$, with $c_k \in \mathbb{Z}$, then $f(X) = \sum_{k=0}^{n} c_k (a + (X - a))^k$. Expand all of the binomials $(a + (X - a))^k$ and compare the result with (15.9), for $z = a$.

11. Make induction on $k \geq 1$. For the induction step, if $m_k \in \mathbb{N}$ is such that $f(m_k) \equiv 0 \,(\mathrm{mod}\, p^k)$ and $f'(m_k) \not\equiv 0 \,(\mathrm{mod}\, p)$, let $m_{k+1} = m_k + xp^k$, with $x \in \mathbb{Z}$ to be found. Then, use Taylor's formula (15.9), together with the result of the previous problem, to show that

$$f(m_{k+1}) \equiv f(m_k) + f'(m_k)xp^k \,(\mathrm{mod}\, p^{k+1}).$$

Subsequently, show that it is possible to choose x so that $f(m_{k+1}) \equiv 0 \,(\mathrm{mod}\, p^{k+1})$. Finally, since $m_{k+1} \equiv m_k \,(\mathrm{mod}\, p)$ and $f' \in \mathbb{Z}[X]$, we have $f'(m_{k+1}) \equiv f'(m_k) \not\equiv 0 \,(\mathrm{mod}\, p)$.

12. Start by showing that one can assume that the center of the square is the origin of the cartesian plane. Then, if (u, v) is one of its vertices, note that the other ones are $(-v, u)$, $(-u, -v)$ and $(v, -u)$. Substitute these points in the defining equation for the graph, i.e., $y = f(x)$, to obtain $au^2 + c = 0$, $v = u^3 + bu$ and $av^2 + c = 0$, $u = -v^3 - bv$. Conclude that $a(u - v)(u + v) = 0$ and, hence, that either $a = c = 0$ or $u = \pm v$ and $au^2 + c = 0$. Discard the possibility $u = \pm v$ and successively obtain $a = c = 0$ and $u^2(u^2 + b)^3 + b(u^2 + b) + 1 = 0$. Conclude that the polynomial $X^2(X^2 + b)^3 + b(X^2 + b) + 1 = 0$ has only two real roots, so that $X(X + b)^3 + b(X + b) + 1 = 0$ has only one real root $\alpha = u^2$ (apart from multiplicities). Thus, show that such a root must be a multiple one, and therefore that $(X + b)^3 + 3X(X + b)^2 + b = 0$. Letting $A = \alpha + b$, perform some simple algebraic manipulations to arrive at $(16 - 2b^2)bA = 2b^2 - 16$, so that $b^2 = 8$ or $bA = -1$. Discard the possibility $bA = -1$ and conclude that, if $b^2 = 8$, then $(bA)^2 - 4(bA) - 8 = 0$. Finally, if ℓ stands for the length of the sides of the square, show that

$$\ell^2 = 2\big(u^2 + u^2(u^2 + b)^2\big) = 2(A - b)(A^2 + 1)$$

$$= \frac{2(bA - b^2)\big((bA)^2 + b^2\big)}{b^3} = \frac{2(bA - 8)\big((bA)^2 + 8\big)}{8b}$$

$$= \frac{(bA)^2 - 4(bA) - 32}{b} = \sqrt{72}.$$

Section 16.1

1. Adapt, to the present case, the proof of Proposition 16.1.
2. For item (a), adapt the argument of the solution of Example 16.3. For item (b), start by writing

$$(X - Y)^5 + (Y - Z)^5 + (Z - X)^5 = (X - Y)(Y - Z)(Z - X)f(X, Y, Z),$$

with f of degree 2. Then, use the equality $-y^5 + (y - z)^5 + z^5 = -yz(y - z)f(0, y, z)$ to show that $f(0, Y, Z) = 5(Y^2 - YZ + Z^2)$ and, analogously, that $f(X, 0, Z) = 5(X^2 - XZ + Z^2)$ and $f(X, Y, 0) = 5(X^2 - XY + Y^2)$. Finally, show that

$$f(X, Y, Z) = 5(X^2 + Y^2 + Z^2 - XY - XZ - YZ).$$

3. Letting $f = f(X, Y, Z)$ stand for the given polynomial, adapt the reasoning used in the solution of Example 16.3 to obtain

$$f(X, Y, Z) = (X + Y)(X + Z)(Y + Z)g(X, Y, Z),$$

with $g \in \mathbb{R}[X, Y, Z]$ of degree 2. Now, use the equality $f(0, y, z) = yzg(0, y, z)$ to show that $g(0, Y, Z) = Y^2 + YZ + Z^2$ and, likewise, $g(X, 0, Z) = X^2 + XZ + Z^2$ and $g(X, Y, 0) = X^2 + XY + Y^2$. Finally, obtain

$$g(X, Y, Z) = X^2 + Y^2 + Z^2 + XY + XZ + YZ.$$

Section 16.2

1. Use the symmetry of f and g to show that, for a fixed permutation σ of I_n, we have $h(x_1, \ldots, x_n) = h(x_{\sigma(1)}, \ldots, x_{\sigma(n)})$ for infinitely many elements $x_1, \ldots, x_n \in \mathbb{K}$. Then, use the result of Problem 1, page 398, as well as that of Proposition 16.1.
3. For item (a), apply the definition twice. For (b), just note that all factors $X_i - X_j$, for $1 \leq i < j \leq n$, will appear in f^σ as $\pm(X_i - X_j)$. For (c), use the result of

(a). Item (d) follows straightforwardly from (c). For item (e), and writing τ in place of τ_{kl} for the sake of simplicity, we have

$$f^\tau = \prod_{1 \le i < j \le n} \left(X_{\tau(i)} - X_{\tau(j)} \right)$$

$$= \prod_{\substack{1 \le i < j \le n \\ i,j \ne k,l}} \left(X_{\tau(i)} - X_{\tau(j)} \right) \cdot \prod_{k < j \ne l} \left(X_{\tau(k)} - X_{\tau(j)} \right) \cdot$$

$$\cdot \prod_{l < j} \left(X_{\tau(l)} - X_{\tau(j)} \right) \cdot \prod_{i < k} \left(X_{\tau(i)} - X_{\tau(k)} \right) \cdot$$

$$\cdot \prod_{k \ne i < l} \left(X_{\tau(i)} - X_{\tau(l)} \right) \cdot \left(X_{\tau(k)} - X_{\tau(l)} \right)$$

Finally, for (f), show that the map $\sigma \mapsto \sigma \circ \tau_{12}$ is a bijection from A_n to $S_n \setminus A_n$.

4. If S_n stands for the set of all permutations of I_n and $\tau \in S_n$, use the fact that $\sigma \mapsto \tau\sigma$ defines a bijection on S_n.

5. Adapt the argument given to the proof of Example 16.7.

6. Letting $a^2 + b^2 + c^2 = k$, we have $a^3 + 3a^2 = 3k - 25$, so that a is a root of $f(X) = X^3 + 3X^2 + (25 - 3k)$; analogously, b and c are roots of such a polynomial. Now, factorise f and use Girard-Viète relations to conclude that $a + b + c = -3$ and $ab + ac + bc = 0$, and hence that $a^2 + b^2 + c^2 = 9$.

7. Letting z_1, z_2, z_3 be the complex roots of the given polynomial, we want to find the coefficients of $f(X) = (X - z_1^3)(X - z_2^3)(X - z_3^3)$ in terms of a, b and c. To this end, use Girard-Viète relations, together with the result of Example 16.3. For instance, letting $g(X) = X^3 + aX^2 + bX + c$, we have $g(X) = (X - z_1)(X - z_2)(X - z_3)$ and, hence,

$$z_1^3 + z_2^3 + z_3^3 = (z_1 + z_2 + z_3)^3 - 3(z_1 + z_2)(z_1 + z_3)(z_2 + z_3)$$

$$= (-a)^3 - 3(-a - z_3)(-a - z_2)(-a - z_1)$$

$$= -a^3 - 3g(-a) = -a^3 + 3ab - 3c.$$

8. First of all, show that if a, b and c are such roots, then the hypothesis of a, b and c being positive assures that a, b and c are the lengths of the sides of a triangle if and only if $(a+b-c)(a+c-b)(b+c-a) > 0$. Then, substitute $a+b+c = -p$ at the left hand side, expand it and use the fact that $ab + ac + bc = q$ and $abc = -r$.

9. If $f(X) = (X - a)(X - b)(X - c)$, then $f(X) = X^3 - p$, with $p = abc \ne 0$. Now, argue as in the hint given to Problem 7, page 386.

10. Apply the result of Example 16.7 to the polynomial $g(X) = X^{100} + 2X^{99} + 3X^{98} + \cdots + a_{98}X^2 + a_{99}X + a_{100}$. Then, show that its roots are the inverses of the roots of f.

11. Letting $ax + by = c$ denote the equation of a straightline satisfying the stated conditions, substitute $y = -\frac{b}{a}x - \frac{c}{a}$ in the equation defining the graph of f and, then, use Girard-Viète relations.

12. Let $(x - a)^2 + (y - b)^2 = R^2$ be the equation of the circle. Substitute $y = \frac{1}{x}$ and, then, use Girard-Viète relations.

13. For item (a), apply Corollary 15.32. For (b), use the fact that $a^3 = a^2 + a + 1$, and analogously for b and c. Finally, for (c) use (b) and Girard-Viète relations.

14. For item (a), use induction. For item (b), apply the result of (a), noticing that

$$f(X) = X^n + s_1 X^{n-1} + s_2 X^{n-2} + \cdots + s_{n-1} X + s_n.$$

15. Make induction on $n \geq 2$ to conclude that $x_1 = x_2 = \ldots = x_n = 1$. For the induction step, if

$$f(X) = (X - x_1)(X - x_2) \cdots (X - x_n)$$

$$= X^n + a_{n-1} X^{n-1} + \cdots + a_1 X + a_0,$$

note that

$$0 = f(x_1) + f(x_2) + \cdots + f(x_n)$$

$$= n + a_{n-1}n + \cdots + a_1 n + a_0 n = nf(1).$$

16. Firstly, show that the roots of f are negative. Then, apply Girard-Viète relations, together with the inequality between the arithmetic and geometric means, to deduce that $a_k \geq \binom{n}{k}$ for $1 \leq k \leq n - 1$.

17. Letting $x_1, \ldots, x_n$ denote the roots of f, use Girard-Viète relations to conclude that

$$\sum_{i=1}^{n} x_i^2 = \left(\sum_{i=1}^{n} x_i\right)^2 - 2 \sum_{1 \leq i < j \leq n} x_i x_j = 3$$

and $\prod_{i=1}^{n} x_i^2 = 1$. Then, apply the inequality between the arithmetic and geometric means to show that $n \leq 3$. Finally, separately consider each of the cases $n = 1, 2$ and 3; in the case $n = 3$, you may wish to use some Calculus (namely, the study of the first variation and/or the intermediate value theorem) to find out which of the possible polynomials have only real roots. In this respect, see [8] or the material of Sect. 17.1.

Section 16.3

1. For $1 \leq k \leq n$, use Newton's theorem and Girard-Viète relations. For $k \geq n+1$, use item (a) of Proposition 16.17.

2. Use item (b) of Proposition 16.17 to prove by induction on j that, for $1 \le j \le n$, the j-th elementary symmetric sums of $a_1, \ldots, a_n$ and $b_1, \ldots, b_n$ coincide. Then, compare the coefficients of the polynomials $\prod_{j=1}^{n}(X - a_j)$ and $\prod_{j=1}^{n}(X - b_j)$.

3. Apply item (b) of Proposition 16.17 to prove, by induction on j, that $s_j = \binom{n}{j}$ for $1 \le j \le k$.

4. For item (a), factorise $X^m - z^m$ over $\mathbb{C}$ and use the result to factorise g over $\mathbb{C}$. For (b), use Newton's theorem to show that the elementary symmetric sums of $z_1^m, \ldots, z_n^m$ are integers. Finally, for (c), just note that $g(X) = h(X^m)$, for some $h \in \mathbb{Z}[X]$.

5. For item (a), let $f(X) = X^n + a_{n-1}X^{n-1} + \cdots + a_1X + a_0$ be a polynomial with integer coefficients and such that all of its complex roots have modulus 1; use Girard-Viète relations to show that $|a_{n-k}| \le \binom{n}{k}$ for $0 < k \le n$. For item (b), let $\alpha_1, \ldots, \alpha_n$ be the complex roots of f. Use Newton's theorem to show that, for each integer $k \ge 1$, the polynomial $f_k(X) = (X - \alpha_1^{2^k}) \ldots (X - \alpha_n^{2^k})$ has integer coefficients. Then, apply the result of (a) to assure the existence of natural numbers $k < l$ such that $f_k = f_l$. Finally, use such an equality to show that α_j is a root of unity, for $1 \le j \le n$.

Section 17.1

1. For each nonreal complex root $z = a + bi$ of f, write the factor $(X - z)(X - \bar{z})$ of f as in the proof of Lemma 17.1. Letting c be the leading coefficient of f, $x_1, \ldots, x_k$ its real roots and $z_1, \bar{z}_1, \ldots, z_l, \bar{z}_l$ the nonreal ones, with $z_j = a_j + ib_j$, we have

$$f(X) = c \prod_{j=1}^{k}(X - x_j) \prod_{i=1}^{l}(X - z_i)(X - \bar{z}_i)$$

$$= c \prod_{j=1}^{k}(X - x_j) \prod_{i=1}^{l}\left((X - a_i)^2 + b_i^2\right).$$

Use this way of writing f to estimate $|f(m)|$, and notice that $|f(m)| \ne 0, 1$.

2. If $\alpha < a < \beta$ are the roots of f closest to a, apply Bolzano's theorem to intervals of the form (a, b) or (b, a) contained in the interval (α, β).

3. Arguing by contradiction, assume that $f'(a) > 0$ (the case $f'(a) < 0$ is completely analogous) and apply item (a) of Corollary 17.7, together with the result of the previous problem.

4. Apply Bolzano's theorem, Corollary 17.7 and the result of the previous problem, noticing that $f'(x) = 0 \Leftrightarrow x = 0$ or $\frac{8}{5}$, and $f(-1), f(\frac{8}{5}) < 0 < f(0)$.

5. Take $n_0 \in \mathbb{N}$ greater that the largest real roots of f and f'. Use Bolzano's theorem, together with the fact that $f(x) > 0$ for sufficiently large x (cf. estimates analogous to those preceding (15.5)) to show that $f(x) > 0$ for $x > n_0$. Then, use Corollary 17.7 to show that $f(u) > f(v)$ for $u > v > n_0$.

6. Assume, without loss of generality, that f has positive leading coefficient. According to the previous problem, take $n_0 \in \mathbb{N}$ such that $u > v > n_0 \Rightarrow f(u) > f(v) > 0$. Let $m > n_0$ be an integer for which $f(m) = p$, a prime number. If $f(X) = a_n X^n + a_{n-1} X^{n-1} + \cdots + a_1 X + a_0$, then

$$f(m + p^2) = a_n(p^2 + m)^n + \cdots + a_1(p^2 + m) + a_0$$
$$= p^2 l + (a_k m^k + \cdots + a_1 m + a_0)$$
$$= p^2 l + f(m) = f(m)(f(m)l + 1).$$

Now, check that the choice of m gives $f(m + p^2) > f(m)$, and conclude that $f(m + p^2)$ is composite.

7. Firstly, show that the given condition is equivalent to $x^{11} - x = y^{11} - y$. Then, prove that, for any $c \in \mathbb{R}$ the polynomial $f(x) = X^{11} - X - c$ has at most three distinct real roots; to this end, you shall need to use Corollary 17.7 and Problem 3, in a way analogous to that of the hint to Problem 4.

8. Since $\lambda \neq 0$, show that $f(\lambda) = 1$, where $f(X) = (X - a_1)(X - a_2)(X - a_3)(X - a_4)$. Then, show that f decreases in $(-\infty, a_1)$, so that $f(0) = 1$ gives $\lambda \geq a_1$. Finally, note that if $a_1 \leq \lambda \leq a_2$, then $f(\lambda) \leq 0$.

9. Let $f(X) = aX^4 + bX^3 + cX^2 + dX + e$, and take $t > 1$ such that t^2 is a real root of $aX^2 + (c-b)X + (e-d)$. Show that $f(t)f(-t) = (bt^2 + d)(1 - t^2) < 0$ and, then, apply the theorem of Bolzano.

10. Set $f(X) = \sum_{i=1}^m a_i X^i$, $g(X) = \sum_{i,j=1}^n \frac{a_i a_j}{i+j} X^{i+j}$ and conclude that $xg'(x) = f(x)^2 \geq 0$, for every $x \geq 0$. Then, use Corollary 17.7 to obtain $g(1) \geq g(0) = 0$, with equality if and only if g is constant.

11. Start by showing that f has three distinct real roots $\alpha < \beta < \gamma$, with $-2 < \alpha < -1, 0 < \beta < 1$ and $1 < \gamma < 2$. Then, note that $f(f(x)) = 0$ if and only if $f(x) = \alpha, \beta$ or γ. Finally, find out the numbers of distinct real roots of each one of the polynomials $f(X) - \alpha$, $f(X) - \beta$ and $f(X) - \gamma$.

12. Set $g(x) = f(x) + f'(x) + f''(x) + \cdots + f^{(n)}(x)$ and assume, by the sake of contradiction, that g attains negative values. Then, $f \neq 0$ and the condition $f(x) \geq 0$ for $x \in \mathbb{R}$ assures that n is even and f has positive leading coefficient. Hence, $\lim_{|x| \to +\infty} f(x) = +\infty$, and Weierstrass' theorem (Teorema 8.26 of [8]) guarantees the existence of $x_0 \in \mathbb{R}$ such that g attains its minimum value at x_0, with $g(x_0) < 0$. It follows from Problem 3 that $g'(x_0) = 0$. However, since $g'(x_0) = f'(x_0) + f''(x_0) + \cdots + f^{(n)}(x_0)$, we have $0 > g(x_0) = f(x_0) + g'(x_0) = f(x_0) \geq 0$, which is a contradiction.

13. Firstly, show that B can play in such a way that, when exactly three coefficients are left to be chosen, at least two of them are coefficients of powers X^r, with an odd r. Then, after A plays, we shall have $f(X) = g(X) + a_k X^k + a_k X^l$, with g being a completely determined real polynomial, $1 \leq k, l \leq 2n - 1$ being

distinct integers such that a_k and a_l are still left to be chosen and l odd. Since $f(2) = g(2) + 2^k a_k + 2^l a_l$ and $f(-1) = g(-1) + (-1)^k a_k - a_l$, we shall have $f(2) + 2^l f(-1) = g(2) + 2^l g(-1) + (2^k + (-1)^k 2^l) a_k$, so that B can play by choosing $a_k = -\frac{g(2) + 2^l g(-1)}{2^k + (-1)^k 2^l}$. Finally, conclude that Bolzano's theorem assures the existence of at least one real root for f, no matter how A plays next.

14. Without loss of generality, we can assume that the leading coefficient of the polynomial f originally written in the blackboard is positive. Let α be the smallest and β be the greatest real root of f (possibly with $\alpha = \beta$). If $\alpha < \beta$, show that $f'(\alpha) > 0$ or $f'(\beta) > 0$; if $f'(\beta) > 0$ (the case $f'(\alpha) > 0$ can be dealt with analogously), conclude that the next polynomial to be written either has no root in the interval $[\beta, +\infty)$ or has a root in the interval $(\beta, +\infty)$. If $\alpha = \beta$, conclude that $f(X) = (X - \alpha)^3$ and show that both $f + f'$ and $f - f'$ fall into the first case.

15. Setting $b_i = -a_i$ for $1 \le i \le n$, show that the condition $f(x) \ge 1$ is equivalent to $\frac{p(x)}{q(x)} \le 0$, with $q(X) = \prod_{i=1}^n (X - b_i)$ and

$$p(X) = \prod_{i=1}^n (X - b_i) + \sum_{i=1}^n \left(b_i \prod_{j \ne i} (X - b_j) \right).$$

Show that p has a root x_1 in the interval $(b_1, +\infty)$; then, use Bolzano's theorem to show that p also has a root x_i in each of the intervals (b_i, b_{i-1}), for $2 \le i \le n$. By distinguishing the cases n even and n odd, show that the sum of the lengths of the solution intervals of the inequality $\frac{p(x)}{q(x)} \le 0$ is $\left| \sum_{i=1}^n x_i - \sum_{i=1}^n b_i \right|$. Finally, apply Girard-Viète relations to show that $\sum_{i=1}^n x_i = 0$.

Section 17.2

1. Firstly, use the definition of the derivative of a polynomial to show that if $g, h \in \mathbb{R}[X]$ are such that $g' = h'$, then there exists $c \in \mathbb{R}$ for which $f = g + c$. Now, letting $f'(X) = a(X + b)^{n-1}$, show that $f(X) = \frac{a}{n}(X + b)^n + c$, for some $c \in \mathbb{R}$. Finally, use the hypothesis on f to conclude that $f(X) = (X + \alpha)^n$, for some $\alpha \in \mathbb{R}$.

2. For the sake of contradiction, assume that the given polynomial has only real roots. Then, use Newton's inequalities, together with Girard-Viète relations, to conclude that $3a^2 \ge 8b$, $4b^2 \ge 9ac$ and $3c^2 \ge 8bd$.

3. Writing $H_j = H_j(a, b, c, d)$, the given inequality is equivalent to $H_1 H_2 \ge H_3$. Now, it follows from Newton's inequalities that $H_1 \ge \frac{H_2}{H_1} \ge \frac{H_3}{H_2}$.

Section 17.3

1. First of all, note that $V(f) = V(f(-X)) \leq n$. Then, apply Descartes' rule twice to conclude that $V(f) + V(f(-X)) = 2n$.

2. Example 17.17 assures that f has precisely one positive root. It now suffices to apply Bolzano's theorem to show that $f\left(2 - \frac{1}{2^{n-1}}\right)$ and $f\left(2 - \frac{1}{2^n}\right)$ have distinct signs. To this end, letting f denote the given polynomial and $g(X) = (X-1)f(X)$, note that $g(X) = X^{n+1} - 2X^n + 1$; then, use that $1 < 2 - \frac{1}{2^{n-1}} < 2 - \frac{1}{2^n}$ to show that it suffices to prove that $g\left(2 - \frac{1}{2^{n-1}}\right)$ and $g\left(2 - \frac{1}{2^n}\right)$ have distinct signs.

3. If $f(X) = aX^3 + bX^2 + cX + d$, we want to compute the number of real roots of $g = 2ff'' - (f')^2$. Assume, without loss of generality, that $a = 1$. Moreover, if $\alpha < \beta < \gamma$ are the roots of f, show that changing $g(X)$ by $h(X) = g(X+\beta)$, we can assume (also without loss of generality) that $\beta = 0$ and, hence, that $c < 0$ and $d = 0$. Under such simplifications, an immediate computation furnishes $g(X) = 3X^4 + 4bX^3 + 6cX^2 - c^2$. It now suffices to apply Descartes' rule to conclude that g has exactly one positive root and exactly one negative root.

Section 18.1

1. For item (i), use the fact that $\partial f_j = j$ to conclude that $a_n = b_n$; then, argue by induction. For (ii), take $n = \partial f$ and argue by induction on n; for the induction step, start by choosing a_n to be equal to the leading coefficient of f.

2. For $k = 0$ and $k = 1$ the result is obvious. For $k \geq 2$, show that $\binom{X}{k}(x) = 0$ if $0 \leq x \leq k - 1$, $\binom{X}{k}(x) = \binom{x}{k}$ if $x \geq k$ and $\binom{X}{k}(x) = (-1)^k \binom{-x-1+k}{k}$ if $x < 0$.

3. Apply the Lagrange interpolation theorem.

4. Apply the Lagrange interpolation theorem.

5. Adapt, to the present case, the proof of Proposition 18.6.

6. Let $\omega = \text{cis}\frac{2\pi}{n}$. For $1 \leq k \leq n - 1$, substitute $x = \omega^k$ into the polynomial functions corresponding to p and to the p_j's, and use the result of Proposition 18.6.

7. For item (a), look at the sum at the right hand side as the result of a Lagrange interpolation. Item (b) follows directly from the factorisation obtained in (a). Finally, for item (c), start by writing

$$\sum_{k \geq 1} \frac{1}{k(k+1)\dots(k+n)} = \frac{1}{n!}\sum_{k \geq 1}\sum_{j=0}^{n}(-1)^j\binom{n}{j}\frac{1}{k+j}.$$

Then, for a fixed $l \geq n$, show that $\frac{1}{l}$ appears in the last sum above with coefficient $\sum_{j=0}^{n}(-1)^j\binom{n}{j} = 0$.

8. For each prime p in the set $A = \{3, 5, 7, 11, 13, 17\}$, let α_p denote the common value of the sums $a_k + a_{k+p} + a_{k+2p} + \cdots$ when k varies from 1 to p, and

$\omega_p = \text{cis} \frac{2\pi}{p}$. If $f(X) = a_{50}X^{50} + \cdots + a_2X^2 + a_1X$, apply the general version of the multisection formula (cf. Problem 8, page 355, or Example 18.7) for $r = 0, 1, \ldots, p - 1$ in order to obtain a Vandermonde system of equations in the p unknowns $f(\omega_p^j)$, for $0 \le j \le p - 1$. Then, use the result of Proposition 18.6 to conclude that the solution of such a system is $f(1) = p\alpha_p$ and $f(\omega_p) = \cdots = f(\omega_p^{p-1}) = 0$, thus obtaining $p - 1$ distinct and nonzero roots for f. Finally, notice that $\sum_{p \in A}(p - 1) = 50$ and that 0 is also a root of f to obtain $f = 0$.

Section 18.2

2. Let us make induction on $k \ge 1$, the case $k = 1$ being immediate. By induction hypothesis, assume that the formula holds for a certain $k \in \mathbb{N}$. For $k + 1$, we have:

$$f(x + (k + 1)h) = f(x + kh) + (\Delta_h^1 f)(x + kh)$$

$$= f(x + kh) + \sum_{j=0}^{k} \binom{k}{j}(\Delta_h^{k-j}(\Delta_h^1 f))(x)$$

$$= \sum_{j=0}^{k} \binom{k}{j}(\Delta_h^{k-j} f)(x) + \sum_{j=0}^{k} \binom{k}{j}(\Delta_h^{k+1-j} f)(x)$$

$$= \sum_{j=0}^{k} \binom{k}{j}(\Delta_h^{k-j} f)(x) + (\Delta_h^{k+1} f)(x)$$

$$+ \sum_{j=1}^{k} \binom{k}{j}(\Delta_h^{k-(j-1)} f)(x).$$

Now, by performing a change of indices in the last sum above and applying Stifel's relation, we obtain $f(x + (k + 1)h)$ successively equal to

$$\sum_{j=0}^{k} \binom{k}{j}(\Delta_h^{k-j} f)(x) + (\Delta_h^{k+1} f)(x) + \sum_{j=0}^{k-1} \binom{k}{j+1}(\Delta_h^{k-j} f)(x)$$

$$= (\Delta_h^{k+1} f)(x) + \sum_{j=0}^{k-1} \left(\binom{k}{j} + \binom{k}{j+1}\right)(\Delta_h^{(k+1)-(j+1)} f)(x) + (\Delta_h^0 f)(x)$$

$$= \sum_{j=0}^{k+1} \binom{k+1}{j}(\Delta_h^{k+1-j} f)(x).$$

3. Adapt, to the present case, the solutions of Examples 18.15 and 18.15.

4. Since $\partial f = n$, Proposition 18.14 assures that $\Delta^{n+1} f = 0$. Hence, it follows from item (e) of Proposition 18.13 that

$$0 = (\Delta^{n+1} f)(0) = \sum_{j=0}^{n+1} (-1)^j \binom{n+1}{j} f(n+1-j)$$

$$= \sum_{j=1}^{n+1} (-1)^j \binom{n+1}{j} \cdot \frac{1}{\binom{n+1}{n+1-j}} + f(n+1)$$

$$= \sum_{j=1}^{n+1} (-1)^j + f(n+1),$$

so that

$$f(n+1) = -\sum_{j=1}^{n+1} (-1)^j = \begin{cases} 0, & \text{if } n \equiv 1 \,(\mathrm{mod}\,2) \\ 1, & \text{if } n \equiv 0 \,(\mathrm{mod}\,2) \end{cases}.$$

5. Use item (e) of Proposition 18.13 to get the equality

$$0 = (\Delta_1^{991} f)(992) = \sum_{j=0}^{991} (-1)^j \binom{991}{j} f(1983 - j).$$

Then, use the fact that, for every $n \in \mathbb{N}$, we have

$$\sum_{j=1}^{n} \binom{n}{j} F_{2n+1-j} = F_{2n+1} - 1$$

(cf. problem 18 of Section 4.3 of [8], for instance).

6. The given conditions assure that $(\Delta_1^k f)(0) > 0$ for $0 \le k \le 3$ and $(\Delta_1^4 f)(n) > 0$ for each $n \in \mathbb{N}$. Now show that, for every $m \in \mathbb{N}$, we have

$$(\Delta_1^{m-1} f)(n) = \sum_{k=0}^{n-1} (\Delta_1^m f)(k) + (\Delta_1^{m-1} f)(0).$$

Finally, use this formula to successively show that $(\Delta_1^3 f)(n) > 0$, $(\Delta_1^2 f)(n) > 0$, $(\Delta_1 f)(n) > 0$ and $f(n) = (\Delta_1^0 f)(n) > 0$, for every $n \in \mathbb{N}$.

Section 19.1

2. Firstly, show that $p_1^{\gamma_1} \ldots p_k^{\gamma_k}$ divides both f and g in $\mathbb{Q}[X]$. Secondly, show that if $h \in \mathbb{Q}[X] \setminus \{0\}$ is monic and divides f in $\mathbb{Q}[X]$, then $h = p_1^{\delta_1} \ldots p_k^{\delta_k} q_1^{\delta_1'} \ldots q_l^{\delta_l'}$, with $0 \le \delta_i \le \alpha_i$ for $1 \le i \le k$, and $0 \le \delta_i' \le \alpha_i'$ for $1 \le i \le l$.
3. Adapt, to the present case, the proof of Euclid's algorithm for integers.
4. Apply Euclid's algorithm to find $X^2 - 3$ as the answer.
5. Argue by contraposition.
7. Start by using Corollary 15.24, together with the result of Problem 4, page 386, along the same lines of item (b) of Example 19.7.
8. Compute derivatives to show that $f^{n-1} f' + g^{n-1} g' = h^{n-1} h'$; then, multiply by h on both sides to obtain $(f'h - fh')f^{n-1} = (gh' - g'h)g^{n-1}$. Use the fact that f and h have no nontrivial common factors, together with Theorem 19.2 and the result of the previous problem to show that $f'h - fh' \ne 0$; accordingly, show that $gh' - g'h \ne 0$. Assume, without loss of generality, that $\partial f \ge \partial g$. Then, apply again the result of the previous problem to the equality $(f'h - fh')f^{n-1} = (gh' - g'h)g^{n-1}$, conclude that $f^{n-1} \mid (gh' - g'h)$ in $\mathbb{R}[X]$. Finally, use this last relation, together with $f^n + g^n = h^n$, to obtain the following degree estimates

$$(n - 1)\partial f \le \partial g + \partial h - 1 \le \partial f + 2\partial g - 1$$

and conclude the solution of the problem.
9. For item (a), make induction on k, using the result of Corollary 19.4 to show that there exist $\tilde{f}_1, f_k \in \mathbb{K}[X]$ such that

$$\frac{f}{g} = \frac{\tilde{f}_1}{g_1^{\alpha_1} \ldots g_{k-1}^{\alpha_{k-1}}} + \frac{f_k}{g_k^{\alpha_k}},$$

with $\tilde{f}_1 = 0$ or $\partial \tilde{f}_1 < \partial(g_1^{\alpha_1} \ldots g_{k-1}^{\alpha_{k-1}})$ and $f_k = 0$ or $\partial f_k < \partial(g_k^{\alpha_k})$. For item (b), start by dividing f by g^k, thus getting $f = g^k q + r$, with $r = 0$ or $0 \le \partial r < \partial(g^k)$; then, divide r by g^{k-1} and proceed inductively.

Section 19.2

1. For the uniqueness part of item (b), let a, b, c, d be nonzero integers with $\gcd(a, b) = \gcd(c, d) = 1$, and $g, h \in \mathbb{Z}[X] \setminus \mathbb{Z}$ be primitives, with $(a/b)g = (c/d)h$. Calculate contents in both sides of $(ad)g = (bc)h$.
2. Since f is reducible over $\mathbb{Q}$, there exist monic and nonconstant polynomials $g_1, h_1 \in \mathbb{Q}[X]$ such that $f = g_1 h_1$. Take $a, b, c, d \in \mathbb{Z} \setminus \{0\}$ such that $\gcd(a, b) = \gcd(c, d) = 1$ and $g_1 = \frac{a}{b}g$, $h_1 = \frac{c}{d}g$, with $g, h \in \mathbb{Z}[X]$ monic and nonconstant. Take contents and apply Gauss' lemma to the equality $bd f(X) = acg(X)h(X)$ to conclude that $bd = ac$.

3. First of all, note that d is primitive. Now, for $(a) \Rightarrow (b)$, let f, g and d have canonical factorisations $f = \pm f_1^{n_1} \ldots f_k^{n_k}$, $g = \pm g_1^{m_1} \ldots g_l^{m_l}$ and $d = h_1^{a_1} \ldots h_k^{a_k}$. Show that if $d' \mid f, g$ in $\mathbb{Z}[X]$, then the canonical factorisation of d' is $d' = h_1^{a'_1} \ldots h_k^{a'_k}$, with $a'_i \leq a_i$ for $1 \leq i \leq k$. For $(b) \Rightarrow (a)$, argue in an analogous way.

Section 19.3

1. For the verification of the listed properties, follow the corresponding verification for polynomials in $\mathbb{K}[X]$, with $\mathbb{K} = \mathbb{Q}$, $\mathbb{R}$ or $\mathbb{C}$.
2. Apply Fermat's little theorem.
3. For the first claim, note that

$$f(a) = 0 \Rightarrow \overline{f(a)} = \overline{0} \Rightarrow \overline{f}(\overline{a}) = \overline{0}.$$

For the second, use the root test to write

$$\overline{f}(X) = (X - \overline{a}_1) \ldots (X - \overline{a}_k) \overline{g}(X),$$

for some nonzero $\overline{g} \in \mathbb{Z}_p[X]$ with no roots in $\mathbb{Z}_p$. Then, perform the substitution $X \mapsto \overline{a}$ and apply the second part of Problem 6, page 281 to conclude that, for some index $1 < j \leq k$, we have $\overline{a} - \overline{a}_j = \overline{0}$.

4. If $a \in \mathbb{Q}$ is a rational root of f, the rational roots test shows that $a \in \mathbb{Z}$ and $a \mid 84$. In order to eliminate some of the left possibilities, project f into $\mathbb{Z}_3[X]$ and $\mathbb{Z}_5[X]$ to conclude, with the aid of the previous problem, that $a \equiv 0 \pmod 3$ and $a \equiv 4 \pmod 5$, and hence that $a = -6, -21$ ou 84.

5. Make induction on ∂f; for the induction step, apply the result of Example 6.42.

6. Assume that fg is not primitive and choose a prime p that divides all of its coefficients. From the equality $\overline{fg} = \overline{f}\,\overline{g} = \overline{0}$ in $\mathbb{Z}_p[X]$, conclude that $\overline{f} = \overline{0}$ or $\overline{g} = \overline{0}$ and reach a contradiction.

7. Let a be a primitive root modulo p. For item (a), show that $(a^k)^d \not\equiv 1 \pmod p$, for every integer $1 \leq k \leq p - 1$. For (b), show that $(a^{\frac{k(p-1)}{d}})^d \equiv 1 \pmod p$ for every integer $1 \leq k \leq d$, and then conclude that

$$X^d - \overline{1} = \prod_{k=1}^{d} (X - \overline{a}^{\frac{k(p-1)}{d}}).$$

8. Use the result of Proposition 19.18, together with Girard-Viète relations.

9. Show that, modulo 17, we have

$$X^3 - \bar{3}X^2 + \bar{1} = (X - \bar{4})(X - \bar{5})(X + \bar{6}).$$

Then, let $s_k = a^k + b^k + c^k$ and $t_k = 4^k + 5^k + (-6)^k$ and conclude, with the aid of the methods of Sect. 16.2, that

$$\begin{cases} s_{k+3} - 3s_{k+2} + s_k = 0 \\ s_1 = 3, s_2 = 0, s_3 = -1 \end{cases} \quad \text{and} \quad \begin{cases} t_{k+3} - 3t_{k+2} + t_k = 0 \\ t_1 = 3, t_2 = 0, t_3 = -1 \end{cases}.$$

Finally, use the above to show that $s_n \in \mathbb{Z}$ for every $n \in \mathbb{N}$ and $s_n \equiv t_n \pmod{17}$.

10. For item (a), project on $\mathbb{Z}_2[X]$ and apply the result of Problem 5 with $p = 2$. For (b), start by writing the binary representation of $2^m - 1$; then, project on $\mathbb{Z}_2[X]$ and apply the result of Problem 5; finally, argue by induction on m.

For the coming problem, the reader may find it useful to review the statement of Problem 11, page 163.

11. For the proof of Lucas' theorem, compute in $\mathbb{Z}_p[X]$, with the aid of Example 19.15:

$$(X + \bar{1})^m = (X + \bar{1})^{m_k p^k} \ldots (X + \bar{1})^{m_1 p} (X + \bar{1})^{m_0}$$
$$= (X^{p^k} + \bar{1})^{m_k} \ldots (X^p + \bar{1})^{m_1} (X + \bar{1})^{m_0}$$

Now, use the uniqueness of representation in base p to conclude that the coefficient of X^n in the right hand side is exactly

$$\overline{\binom{m_k}{n_k}} \ldots \overline{\binom{m_1}{n_1}} \overline{\binom{m_0}{n_0}}.$$

For (a), if $p \mid \binom{m}{n}$ then $p \mid \binom{m_j}{n_j}$ for some $0 \le j \le k$; in turn, if $n_j \le m_j$, then $n_j \le m_j < p$, and the fact that p is prime avoids $\binom{m_j}{n_j} = \frac{m_j!}{n_j!(m_j - n_j)!}$ to have a prime factor p. Item, (b) follows from (a) and the uniqueness of representation in base p, since we ought to have $0 \le n_j \le m_j$ for $0 \le j \le k$, which gives exactly $m_j + 1$ possibilities for n_j. Finally, for item (c) note that $p^{k+1} - 1 = (p-1)p^k + \cdots + (p-1)p + (p-1)$, so that $m_0 = m_1 = \ldots = m_k = p - 1$. Hence, from (b) there are p^{k+1} binomial numbers of the form $\binom{p^{k+1}-1}{n}$ which are not divisible by p; thus, no binomial number is left to be divisible by p.

12. First of all, note that the equality $X^p - 1 = (X-1)(X^{p-1} + X^{p-2} + \cdots + X + 1)$ gives, in $\mathbb{Z}_p[X]$,

$$(X - \bar{1})^p = (X - \bar{1})(X^{p-1} + X^{p-2} + \cdots + X + \bar{1}),$$

so that $(X - \bar{1})^{p-1} = X^{p-1} + X^{p-2} + \cdots + X + \bar{1}$. Therefore, bi expanding the binomial at the left hand side and comparing coefficients, we get $(-1)^l \binom{p-1}{l} \equiv 1 \pmod{p}$. Now, if $l = p^k(p-1)$, then Lucas' theorem gives

$$\binom{p^k(p-1)}{l} = \binom{p^k(p-1)}{p^k q} \equiv \binom{p-1}{q} \equiv (-1)^q \pmod{p}.$$

For what is left to do, note that item (b) of the previous problem assures that exactly p binomial numbers of the form $\binom{p^k(p-1)}{l}$ are not divisible by p; since we have already found p such numbers (for q can vary from 0 to $p-1$), the remaining ones are multiples of p.

Section 19.4

1. Firstly, show that the given polynomial has no integer roots. Then, invoke Gauss' lemma and examine the possibility $X^4 - X^2 + 1 = (X^2 + aX + b)(X^2 + cX + d)$, with $a, b, c, d \in \mathbb{Z}$.

2. Gauss' lemma assures that it suffices to show that f cannot be written as the product of two nonconstant polynomials with integer coefficients. To this end, first note that f has no integer roots; hence, if it can be written as the product of two nonconstant polynomials with integer coefficients, we must have

$$f(X) = (X^2 + aX + b)(X^3 + cX^2 + dX + e),$$

for some $a, b, c, d, e \in \mathbb{Z}$. Expanding the products at the right hand side and comparing coefficients, conclude that $be = 2$, so that $(b, e) = (1, 2)$, $(-1, -2)$, $(2, 1)$ or $(-2, -1)$. Then, check that each of these possibilities leads to a contradiction when compared with the remaining relations involving a, b, c, d and e.

3. Write $f(X^n) = g(X)h(X)$, with $g, h \in \mathbb{R}[X] \setminus \mathbb{R}$ as in the statement of the problem and $n > 1$ (the case $n = 1$ is trivial). If $g(X) = b_k X^k + b_{k+1} X^{k+1} + \cdots$ and $h(X) = c_l X^l + c_{l+1} X^{l+1} + \cdots$, with $b_k, c_l \neq 0$, and changing g and h respectively by $g(X) = b_k + b_{k+1} X + \cdots$ and $h(X) = c_l X^{k+l} + c_{l+1} X^{k+l+1} + \cdots$, we can suppose that $g(0), h(0) \neq 0$. Therefore, from the beginning let $g(X) = b_0 + b_1 X + \cdots$ and $h(X) = c_l X^l + c_{l+1} X^{l+1} + \cdots$, with $b_0, c_l \neq 0$. Use the fact that $f(X^n) = g(X)h(X)$ to show that $n \mid l$. Then, cancel the terms of minimal degree at both sides of the equality $f(X^n) = g(X)h(X)$ and argue by induction on ∂f to show that $g(X) = g_1(X^n)$ and $h(X) = h_1(X^n)$, for some $g_1, h_1 \in \mathbb{R}[X] \setminus \mathbb{R}$ as prescribed in the statement of the problem

4. Once more from Gauss' lemma, we examine when it does happen that $f = gh$, with $g, h \in \mathbb{Z}$ monic and nonconstant. Adapting the idea of the proof of Example 19.32, we have $1 = f(0) = g(0)h(0)$, so that $g(0) = h(0) = \pm 1$;

hence, $a_1, a_2, \ldots, a_n$ are pairwise distinct roots of $g - h$. If $g - h \neq 0$, then $\partial(g - h) \leq \max\{\partial g, \partial h\} < \partial f = n$, so that $g - h$ must have less than n distinct roots. Therefore, $g - h = 0$, and hence $f = g^2$, or

$$g(X)^2 - 1 = (X - a_1) \ldots (X - a_n).$$

In particular, $n = 2k$ for some $k \in \mathbb{N}$, and the unique factorisation in $\mathbb{Z}[X]$ gives, without loss of generality,

$$g(X) - 1 = (X - a_1) \ldots (X - a_k) \text{ and } g(X) + 1 = (X - a_{k+1}) \ldots (X - a_{2k}),$$

with $a_1 < a_2 < \cdots < a_k$ and $a_{k+1} < a_{k+2} < \cdots < a_{2k}$. Therefore,

$$2 = (X - a_{k+1}) \ldots (X - a_{2k}) - (X - a_1) \ldots (X - a_k)$$

and, by evaluating the equality above at $a_1, a_2, \ldots, a_k$, respectively, we obtain

$$(a_i - a_{k+1}) \ldots (a_i - a_{2k}) = 2$$

for $1 \leq i \leq k$. From this point, conclude that if $k \geq 3$ then at least two of the numbers $a_{k+1}, a_{k+2}, \ldots, a_{2k}$ would be equal, which is not the case. Finally, look separately at the cases $k = 1$ and $k = 2$ to get the listed polynomials.

5. Apply Eisenstein's criterion in conjunction with the result of item (a) of Problem 8, page 469.

6. For the sake of contradiction, assume that $f = gh$, with $g, h \in \mathbb{Z}[X] \setminus \mathbb{Z}$, and examine the equality $\overline{f}(X) = \overline{g}(X)\overline{h}(X)$ in $\mathbb{Z}_p[X]$.

7. As in Example 19.28, it suffices to establish the irreducibility of $g(X) = f(X + 1)$. To this end, start by applying several times the result of Problem 5, page 469, to show that, in $\mathbb{Z}_p[X]$, one has

$$\overline{g}(X) = \left(X^{p^{k-1}} + \overline{1}\right)^{p-1} + \left(X^{p^{k-1}} + \overline{1}\right)^{p-2} + \cdots + \left(X^{p^{k-1}} + \overline{1}\right) + \overline{1}.$$

Then, also as in Example 19.28, use the fact that

$$\sum_{j=1}^{p}(Y + \overline{1})^{p-j} = \sum_{j=1}^{p}\overline{\binom{p}{j-1}}Y^{p-j} = Y^{p-1}$$

to get $\overline{g}(X) = X^{p^{k-1}(p-1)}$. Then, look at the constant coefficient of $g(X) = f(X + 1)$ to conclude that

$$g(X) = X^m + a_{m-1}X^{m-1} + \cdots + a_1 X + p,$$

with $m = p^{k-1}(p-1)$ and $a_1, a_2, \ldots, a_{m-1} \in \mathbb{Z}$ being multiples of p. Finally, apply Eisenstein's criterion.

8. Thanks to Gauss' lemma, we only need to show that f is irreducible in $\mathbb{Z}[X]$. By contradiction, assume we have $f = gh$, with $g, h \in \mathbb{Z}[X] \setminus \mathbb{Z}$ and (with no

loss of generality, since $f(0) = p) g(0) = \pm 1$. Use Girard-Viète relations to conclude that g, and thus f, has a complex root z of modulus less than 1. Then, substitute the expression for f in the equality $f(z) = 0$ and use the hypotheses on the coefficients of f to reach a contradiction.

9. Let $z_1, \ldots, z_n$ be the complex roots of f. If $|z_j| \leq 1$ for some $1 \leq j \leq n$, then

$$|a_0| = |a_1 z_j + \cdots + a_n z_j^n|$$

$$\leq |a_1||z_j| + \cdots + |a_n||z_j|^n$$

$$\leq |a_1| + \cdots + |a_n|,$$

which is an absurd; therefore, $|z_j| > 1$ for $1 \leq j \leq n$. Now, suppose that $f = gh$, with $g, h \in \mathbb{Z}[X] \setminus \mathbb{Z}$, say $g(X) = b_0 + b_1 X + \cdots + b_r X^r$ and $h(X) = c_0 + c_1 X + \cdots + c_s X^s$, with $r, s \geq 1$ and $b_r, c_s \neq 0$. By renumbering the z_j's, if necessary, we can assume that the roots of g are $z_1, \ldots, z_r$. Then, Girard-Viète relations give

$$|b_0| = |b_r||z_1| \ldots |z_r| > |b_r| \quad \text{and} \quad |c_0| = |c_s||z_{r+1}| \ldots |z_n| > |c_s|.$$

Hence, $|b_0| \geq |b_r| + 1$ and $|c_0| \leq |c_s| + 1$, so that

$$|a_0| = |b_0||c_0| \geq (|b_r| + 1)(|c_s| + 1)$$

$$= |b_r||c_s| + |b_r| + |c_s| + 1$$

$$\geq |b_r||c_s| + 2\sqrt{|b_r||c_s|} + 1$$

$$= (\sqrt{|b_r||c_s|} + 1)^2$$

$$= (\sqrt{|a_n|} + 1)^2.$$

This way, $\sqrt{|a_0|} \geq \sqrt{|a_n|} + 1$, which is an absurd.

11. For item (a), use an argument analogous to that of the proof of Lemma 17.1. For item (b), suppose that $f = gh$, with $g, h \in \mathbb{Z}[X] \setminus \mathbb{Z}$. Then, it follows from (a) that all of the coefficients of $g_1(X) = g\left(X + m - \frac{1}{2}\right)$ have a single sign. Hence, $g_2(X) := g_1(-X)$ has nonzero coefficients of alternating signs, so that $|g_2(x)| < |g_1(x)|$ for every $x > 0$ and, thus,

$$g\left(-x + m - \frac{1}{2}\right) < g\left(x + m - \frac{1}{2}\right)$$

for every $x > 0$; in particular, $|g(m-1)| < |g(m)|$. Conclude that $|g(m)| \geq 2$ and, analogously, $|h(m)| \geq 2$. Finally, use the fact that $f(m)$ is prime to reach a contradiction.

12. Combine Pólya-Szegö's theorem (cf. previous problem) with the result of Problem 9, page 386.

Section 20.1

1. If $f(X) = a_n X^n + \cdots + a_1 X + a_0 \in \mathbb{Q}[X] \setminus \mathbb{Q}$ has α as one of its roots, look at

$$g(X) = \frac{a_n}{r^n} X^n + \frac{a_n}{r^{n-1}} X^{n-1} + \cdots + \frac{a_1}{r} X + a_0 \in \mathbb{Q}[X] \setminus \mathbb{Q}.$$

2. The first de Moivre's formula gives $(\cos \frac{2k\pi}{n} + i \sin \frac{2k\pi}{n})^n = 1$. Expand the binomial at the left hand side and apply identity $\sin^2 \frac{2k\pi}{n} + \cos^2 \frac{2k\pi}{n} = 1$ to find nonzero polynomials with rational coefficients having $\cos \frac{2k\pi}{n}$ and $\sin \frac{2k\pi}{n}$ as roots. Alternatively, for $\cos \frac{2k\pi}{n}$ use the result of Example 15.27, together with $\cos \frac{2k\pi}{n} = 2\cos^2 \frac{k\pi}{n} - 1$ and Theorems 20.9 and 20.12; for $\sin \frac{2k\pi}{n}$, use Trigonometry, together with the fact that $\cos \frac{2k\pi}{n}$ is algebraic.
3. Apply Eisenstein's criterion (cf. Corollary 19.27).
4. First use Gauss' Theorem 19.14 to show that there exists $g \in \mathbb{Z}[X] \setminus \mathbb{Z}$, irreducible over $\mathbb{Z}$ and such that $g(\alpha) = 0$. The, use Gauss' Lemma 19.13 to conclude that g is also irreducible over $\mathbb{Q}$, so that $g = p_\alpha$.
5. For item (a), apply Remark 20.10 several times, together with the fact that $p_{\sqrt{a_j}}(X) = X - \sqrt{a_j}$ or $X^2 - a_j$, according to whether $\sqrt{a_j} \in \mathbb{N}$ or $\sqrt{a_j} \notin \mathbb{N}$. For (b), use (a) and the rational roots test (cf. Proposition 15.16).
6. Problem 2, page 475, shows that f is irreducible over $\mathbb{Q}$. Now, suppose that there exist relatively prime integers a and b and a natural number $n > 1$ such that $f(\alpha) = 0$, with $\alpha = \sqrt[n]{|\frac{a}{b}|}$. Let

$$g(X) = bX^n - a = b(X - \alpha)(X - \alpha\omega)\ldots(X - \alpha\omega^{n-1}),$$

with $\omega = \operatorname{cis} \frac{2\pi}{n}$. Corollary 20.4 shows that $f = p_\alpha$; hence, item (b) of Proposition 20.3 guarantees that f divides g in $\mathbb{Q}[X]$ and, hence, in $\mathbb{Z}[X]$ (for $f \in \mathbb{Z}[X]$ is monic). Thus, it follows from Problem 4, page 386, the existence of integers $1 \le k < l \le n - 1$ such that

$$f(X) = (X - \alpha)(X - \alpha\omega^k)(X - \alpha\overline{\omega}^k)(X - \alpha\omega^l)(X - \alpha\overline{\omega}^l).$$

By examining the independent term of f, conclude that $-\alpha^5 = 2$ and, then, $\alpha = -\sqrt[5]{2}$. How, arrive at a contradiction.
7. If $\alpha = \alpha_1, \ldots, \alpha_m$ are the roots of p_α and $\beta = \beta_1, \ldots, \beta_n$ are those of p_β, the natural candidate to be looked at is

$$h(X) = \prod_{\substack{1 \le i \le m \\ 1 \le j \le n}} (X - \alpha_i \beta_j) = (\alpha_1 \ldots \alpha_m)^n \prod_{i=1}^m p_\beta \left(\frac{1}{\alpha_i} X \right).$$

Now, note that $\alpha_1 \ldots \alpha_m = \pm p_\alpha(0) \in \mathbb{Q}$.

8. Item (a) is almost immediate. For multiplication, start by taking $f, g \in \mathbb{Q}[X] \setminus \{0\}$ and writing $fq = p_\alpha q + r$, with $r = 0$ or $\partial r < \partial p_\alpha = n$; then, show that $f(\alpha)g(\alpha) = r(\alpha)$, so that $f(\alpha)g(\alpha) \in \mathbb{Q}(\alpha)$. For (b), let $\beta = f(\alpha) = g(\alpha)$, with $f, g \in \mathbb{Q}[X]$ such that $f = 0$ or $\partial f \leq n - 1$ and likewise for g; note that $(f - g)(\alpha) = 0$, and successively conclude that $p_\alpha \mid (f - g)$ and $f - g = 0$. Finally, for (c), show that $\gcd(f, p_\alpha) = 1$ follows from the irreducibility of p_α. Therefore, Corollary 19.4 gives $fg + p_\alpha q = 1$, for some $g, q \in \mathbb{Q}[X]$ satisfying $g = 0$ or $\partial g < \partial p_\alpha = n$ and $q = 0$ or $\partial q < \partial f$. Then, evaluate this equality at α to show that $\frac{1}{\beta} \in \mathbb{Q}(\alpha)$.

9. For the first part, note that $p_{\sqrt[3]{2}}(X) = X^3 - 2$, so that $\mathbb{Q}(\sqrt[3]{2}) = \{a + b\sqrt[3]{2} + c\sqrt[3]{4}; a, b, c \in \mathbb{Q}\}$. The second part can be solved following the steps in the hint given to Problem 8, page 200. Alternatively, follow the steps listed in the hint given to item (c) of the previous problem; more precisely, letting $f(X) = 1 + X + 2X^2$ (so that $1 + \sqrt[3]{2} + 2\sqrt[3]{4} = f(\sqrt[3]{2})$), use Euclid's algorithm to find $g, q \in \mathbb{Q}[X]$ such that $fg + p_{\sqrt[3]{2}} q = 1$ and then note that $f(\sqrt[3]{2})g(\sqrt[3]{2}) = 1$.

10. Start by showing that $\phi(0) = 0$ for any such ϕ. Then, use (ii) to show that $\phi(1) = 0$ or $\phi(1) = 1$. If $\phi(1) = 0$, conclude that $\phi \equiv 0$; if $\phi(1) = 1$, show successively that $\phi(m) = m$ for every $m \in \mathbb{Z}$ and $\phi(r) = r$ for every $r \in \mathbb{Q}$. Now, letting $p_\alpha(X) = X^n + a_{n-1}X^{n-1} + \cdots + a_1 X + a_0$, note that

$$0 = \phi(0) = \phi(p_\alpha(\alpha)) = \phi(\alpha)^n + a_{n-1}\phi(\alpha)^{n-1} + \cdots + a_1\phi(\alpha) + a_0,$$

so that $\beta = \phi(\alpha)$ is also a root of p_α. In turn, show that, for any $f \in \mathbb{Q}[X] \setminus \{0\}$ satisfying $\partial f \leq n - 1$, we have $\phi(f(\alpha)) = f(\phi(\alpha)) = f(\beta)$, so that ϕ is completely determined by the fact that $\phi(\alpha) = \beta$. Conversely, for any such root β of p_α, let $\phi : \mathbb{Q}(\alpha) \to$ be given by $\phi(f(\alpha)) = f(\beta)$, for any $f \in \mathbb{Q}[X]$ satisfying $f \equiv 0$ or $\partial f \leq n - 1$. It is straightforward to show that (i) holds. For (ii), let $f, g \in \mathbb{Q}[X] \setminus \{0\}$ be given and write $fg = p_\alpha q + r$, with $q, r \in \mathbb{Q}[X]$ such that $r \equiv 0$ or $\partial r \leq n - 1$. Then, show that

$$\phi(f(\alpha)g(\alpha)) = \phi((fg)(\alpha)) = \phi(r(\alpha)) = r(\beta)$$
$$= f(\beta)g(\beta) = \phi(f(\alpha))\phi(g(\alpha)).$$

Section 20.2

1. Use item (a) of Proposition 20.14 for $n = p^{k-1}$ and $n = p^k$.
2. A typical root of Φ_{2n} is $\omega = \operatorname{cis} \frac{2k\pi}{2n}$, for some integer $1 \leq k \leq 2n$ such that $\gcd(k, 2n) = 1$. Now, use the fact that $-\omega$ is also of this form to conclude that $-\omega$ is a root of Φ_{2n} too. Then, show that Φ_{2n} is a product of factors of the form $(X - \omega)(X + \omega)$.

3. If Φ_m and Φ_n have a nonconstant common factor in $\mathbb{C}[X]$, then, being the minimal polynomials of any of its common roots, they are identical. Then, letting $\Phi_m = \Phi_n = f$, item (a) of Proposition 20.14 shows that f^2 divides $X^{mn} - 1$, which is impossible.

4. Firstly, with n and d as in the statement, note that $\varphi(n) = \varphi(d) \cdot \frac{n}{d}$, so that $\partial\Phi_n = \partial f$, where $f(X) = \Phi_d(X^{n/d})$. Now, if $\omega = \text{cis}\,\frac{2k\pi}{n}$, with $1 \le k \le d$ and $\gcd(k, d) = 1$, and $\eta \in \mathbb{C}$ is an $\frac{n}{d}$-th root of ω, show that η is an n-th primitive root of unity. Finally, conclude that Φ_n and f have the same roots.

5. Dirichlet's theorem allows us to choose $d \in \mathbb{N}$ tal such that $1 + kd = p$, a prime number. Now, look at the arithmetic progression $\frac{1}{(p-1)!}, \frac{1+d}{(p-1)!}, \ldots, \frac{1+(k-1)d}{(p-1)!}$.

6. The general version of Dirichlet's theorem assures that the arithmetic progression $(n + 2, 2n + 2, 3n + 2, \ldots)$ contains infinitely many primes. Letting $p = kn + 2 > a$ be one such prime and g be a primitive root modulo p, the fact that $\{g, g^2, \ldots, g^{p-1}\}$ is a RRS modulo p assures the existence of an integer $1 \le t \le p - 1$ such that $g^t \equiv a \pmod{p}$. Now, apply Bézout's theorem to guarantee the existence of natural numbers u and v such that $nu = t + (p - 1)v$, so that $(g^u)^n \equiv a \pmod{p}$.

7. Firstly, consider the case of an odd a. Since a is not a perfect square, there exist natural numbers b and t, and distinct odd primes $q_1, \ldots, q_t$ such that $a = b^2 q_1 \ldots q_t$. Then, Proposition 12.23 guarantees that it suffices to find infinitely many prime numbers p for which $\left(\frac{q_1}{p}\right) \ldots \left(\frac{q_t}{p}\right) = -1$. In turn, Quadratic Reciprocity Law assures that such a relation is equivalent to

$$\prod_{j=1}^{t} \left(\frac{p}{q_j}\right) = (-1)^{1+\left(\frac{p-1}{2}\right)\left(\frac{s-t}{2}\right)},$$

with $s = \sum_{j=1}^{t} q_j$. Now, apply Dirichlet's theorem, choosing $p = 4kq_1 \ldots q_t + 1$.

8. Let $\Delta = a^2 - 4b$. Multiplying both sides of the given equation by 4 and completing squares, show that the hypotheses of the problem guarantee that the quadratic congruence $x^2 \equiv \Delta \pmod{n}$ has a solution for every $n \in \mathbb{N}$. If Δ is not a perfect square, write $\Delta = \alpha\beta^2$, with $\alpha, \beta \in \mathbb{Z}$ and $|\alpha| > 1$ being square free. Then, use the result of the previous problem to reach a contradiction.

Section 20.3

1. We parallel the proof of the cancellation law for addition:

$$a \cdot c = b \cdot c \Rightarrow (a \cdot c) \cdot c^{-1} = (b \cdot c) \cdot c^{-1}$$

$$\Rightarrow a \cdot (c \cdot c^{-1}) = b \cdot (c \cdot c^{-1})$$

$$\Rightarrow a \cdot 1 = b \cdot 1 \Rightarrow a = b,$$

2. If $\alpha^{p^n} = \alpha$ and $\beta^{p^n} = \beta$, then

$$(\alpha + \beta)^{p^n} = \alpha^{p^n} + \beta^{p^n} = \alpha + \beta,$$

so that $\alpha + \beta \in \mathcal{R}_n$. Argue in the same way to show that $\alpha\beta \in \mathcal{R}_n$ and (if $\alpha \neq 0$) $\alpha^{-1} \in \mathcal{R}_n$. Then, show that $\mathcal{R}_n$ is a field.

3. For the first part, use (20.10). For the second, use (20.9), together with the fact that $f_{11}(X) \ldots f_{1a_1}(X) = X^p - X$ and

$$X^{p^2} - X = (X^p)^p - X^p + X^p - X$$

$$= (X^p - X)^p + X^p - X$$

$$= (X^p - X)\big((X^p - X)^{p-1} + \bar{1}\big).$$

4. Certainly every $f \in \mathbb{K}[X]$ of degree 1 is irreducible. Now, if $f_1, f_2, \ldots, f_k \in \mathbb{K}[X]$ are irreducible (there are at least two such polynomials, X and $X - 1$), let $f = f_1 f_2 \ldots f_k + 1$, choose an irreducible factor $g \in \mathbb{K}[X]$ of f and show that $g \neq f_1, f_2, \ldots, f_k$.

Section 20.4

1. Note that the decimal expansion of α is not periodic.
2. For the sake of contradiction, if $\alpha + \beta = r$, with $r \in \mathbb{Q}$, and $f(\alpha) = 0$ for $f(X) = a_n X^n + \cdots + a_1 X + a_0 \in \mathbb{Q}[X] \setminus \{0\}$, then

$$a_n(r - \beta)^n + \cdots + a_1(r - \beta) + a_0 = 0.$$

Use this equality to show that β would be algebraic.

3. If $\sqrt[n]{\alpha}$ was algebraic, then several applications of Theorem 20.12 assure that the same would be true of $(\sqrt[n]{\alpha}) = \alpha$. If α^n were algebraic, and $f \in \mathbb{Q}[X] \setminus \{0\}$ satisfies $f(\alpha^n) = 0$, find $g \in \mathbb{Q}[X] \setminus \{0\}$ such that $g(\alpha) = 0$.

4. For the second part, if $\mathbb{Q}[\alpha]$ was a subfield of $\mathbb{C}$ there would exist $n \in \mathbb{N}$ and $a_0, a_1, \ldots, a_n \in \mathbb{Q}$ such that $a_n \neq 0$ and $\frac{1}{\alpha} = a_0 + a_1\alpha + \cdots + a_n\alpha^n$. Show that this would force α to be algebraic.

5. Let $\alpha \in \mathbb{R}$ and, for $1 \leq k \leq 2016$, let A_k be the point of the cartesian plane whose coordinates are the real and imaginary parts of $(\alpha + i\alpha^2)\omega^{k-1}$, with $\omega = \cos\frac{2\pi}{2016} + i\sin\frac{2\pi}{2016}$. Then $A_1 A_2 \ldots A_{2016}$ is a regular 2016-gon, and it suffices to show that it is possible to choose α in such a way that $A_k \notin S$ for every k. Let Γ be one of the circles in S, with center $O = (a, b)$ and radius r, so that $a, b, r \in \mathbb{Q}$. Letting $\cos\frac{2(k-1)\pi}{2016} = c$ and $\sin\frac{2(k-1)\pi}{2016} = s$, we have

$$(\alpha + \alpha^2 i)(c + si) = (c\alpha - s\alpha^2) + (s\alpha + c\alpha^2)i.$$

Then, $A_k \in \Gamma \Leftrightarrow \overline{A_k O} = r$ or, which is the same, if and only if

$$\left((c\alpha - s\alpha^2) - a\right)^2 + \left((s\alpha + c\alpha^2) - b\right)^2 - r^2 = 0.$$

In turn, this last equality is equivalent to

$$\alpha^4 + (1 - 2bc + 2as)\alpha^2 + (2ac - 2bs)\alpha + (a^2 + b^2 - r) = 0.$$

However, since c and s are algebraic over $\mathbb{Q}$ (cf. Problem 2, page 486), we conclude from the above remark that α should be algebraic over $\mathbb{Q}$. Therefore, by choosing $\alpha \in \mathbb{R}$ to be transcendental, we conclude that $A_k \notin S$ for every k.

Section 21.1

1. For every integer $k \geq 0$, set $f^{(k)}(x) = a_k e^{-x} \sin x + b_k e^{-x} \cos x$, so that

$$f^{(k+1)}(x) = -(a_k + b_k)e^{-x} \sin x + (a_k - b_k)e^{-x} \cos x$$

and, hence, $a_{k+1} = -(a_k + b_k)$, $b_{k+1} = a_k - b_k$. Then, conclude that $b_{k+2} + 2b_{k+1} + 2b_k = 0$ for $k \geq 0$.

2. The characteristic polynomial of the involved recurrence relations is $f(X) = kX^k - X^{k-1} - \cdots - X - 1$. If $\alpha \in \mathbb{C}$ is a root of f such that $|\alpha| \neq 1$, we have

$$k|\alpha|^k = |\alpha^{k-1} + \cdots + \alpha + 1|$$
$$\leq |\alpha|^{k-1} + \cdots + |\alpha| + 1$$
$$= \frac{|\alpha|^k - 1}{|\alpha| - 1},$$

so that $k|\alpha|^{k+1} - (k+1)|\alpha|^k + 1 \leq 0$. Study the first variation of $g(X) = (X-1)f(X) = kX^{k+1} - (k+1)X^k + 1$ to conclude that $|\alpha| < 1$. If $\alpha \in \mathbb{C}$ is a root of f such that $|\alpha| = 1$, review the above estimates to obtain

$$k = k|\alpha|^k = |\alpha^{k-1} + \cdots + \alpha + 1|$$
$$\leq |\alpha|^{k-1} + \cdots + |\alpha| + 1 = k$$

and, hence, to conclude (from the condition for equality in the triangle inequality for complex numbers, cf. Problem 3, page 326) that $\alpha^j = \lambda_j \cdot 1$, for some $\lambda_j > 0$. Up to this point, we have shown that, apart from 1, all of the other complex roots of f have modulus less than 1. Moreover, they are all simple roots, for

$g'(X) = k(k+1)X^{k-1}(X-1)$ implies that the only multiple root of g is 1, with multiplicity 2. Now, use Theorem 21.3 to obtain, for $m > k$,

$$x_m = u_1 + u_2\alpha_2^{m-1} + \cdots + u_k\alpha_k^{m-1},$$

with $\alpha_1 = 1, \alpha_2, \ldots, \alpha_k$ being the complex roots of f. Then, conclude that $x_m \xrightarrow{m} u_1$. In order to compute u_1, note that

$$\begin{cases} u_1 + u_2 + \cdots + u_k = a_1 \\ u_1 + u_2\alpha_2 + \cdots + u_k\alpha_k = a_2 \\ \cdots \\ u_1 + u_2\alpha_2^{k-1} + \cdots + u_k\alpha_k^{k-1} = a_k \end{cases}$$

and $f(X) = (X-1)h(X)$, with $f(X) = (X-1)(kX^{k-1} + (k-1)X^{k-2} + \cdots + 2X + 1)$. Finally, in the above system, multiply the j-th equation (from top to bottom) by j and add the results to reach

$$\frac{k(k+1)}{2}u_1 + \sum_{j=2}^{k} u_j h(\alpha_j) = ka_k + (k-1)a_{k-1} + \cdots + 2a_2 + a_1.$$

Argue analogously to y_m.

3. If $f(X) = X^3 - 3X^2 + 1$, start by showing that f has real roots $a > b > c$, such that $-\frac{6}{10} < c < -\frac{5}{10}, \frac{6}{10} < b < \frac{7}{10}$ and $0 < b^n + c^n < 1$ for every integer $n \geq 2$. If $a_n = a^n + b^n + c^n$ for $n \geq 1$, show that $a_1, a_2, a_3 \in \mathbb{Z}$ and $a_{k+3} = 3a_{k+2} - a_k$ for $k \geq 1$; then, conclude from what we did above that $\lfloor a^n \rfloor = a_n - 1$. Finally, use the linear recurrence relation satisfied by the sequence $(a_n)_{n \geq 1}$ to show that $a_{k+17} \equiv a_k \pmod{17}$; alternatively, invoke the result of Problem 9, page 469.

Section 21.2

1. For the second part, suppose that $|z| > R$ and let $\epsilon = |z| - R$. Use the definition of convergence to find an $n \in \mathbb{N}$ such that $|z_n - z| < \epsilon$, and deduce that $|z_n| > R$, which is a contradiction.

2. Use the result of the previous problem.

3. Adapt the reasoning presented in the proof of Proposition 21.15.

4. Apply the result of items (a) and (b) of the previous problem to the sequence of partial sums of the series $\sum_{k \geq 1}(az_k + bw_k)$.

5. Adapt, to the present case, the proof of Proposition 3.7.

7. Start by observing that, if $(z_n)_{n \geq 1}$ is a sequence in X such that $z_n \to z$, with $z \in X$, then the triangle inequality gives $\big||f(z_n)| - |f(z)|\big| \leq |f(z_n) - f(z)|$.

8. Start by observing that, if $(z_n)_{n \geq 1}$ is a sequence in X such that $z_n \to z$, then $|z_n - z| < B$ for every sufficiently large index n; hence, $|f(z_n) - f(z)| \leq$

$A|z_n - z|$, also for every sufficiently large index n. Now, given $\epsilon > 0$, note that $|f(z_n) - f(z)| < \epsilon$ if $|z_n - z| < \frac{\epsilon}{A}$; then, use the convergence of $(z_n)_{n \geq 1}$ to z.

9. Case $m = 1$ is the content of Example 21.12. For $m = 2$ and $|z| < \frac{1}{|a|}$, we get from the initial case that

$$\frac{1}{(1 - az)^2} = \left(\sum_{k \geq 0} a^k z^k\right) \left(\sum_{l \geq 0} a^l z^l\right) = \sum_{k,l \geq 0} a^{k+l} z^{k+l}$$

$$= \sum_{n \geq 0} (n + 1) a^n z^n.$$

By induction, if

$$\frac{1}{(1 - az)^{m-1}} = \sum_{n \geq 0} \binom{n + m - 2}{m - 2} a^n z^n,$$

then

$$\frac{1}{(1 - az)^m} = \frac{1}{1 - az} \cdot \frac{1}{(1 - az)^{m-1}}$$

$$= \sum_{k \geq 0} a^k z^k \cdot \sum_{l \geq 0} \binom{l + m - 2}{m - 2} a^l z^l$$

$$= \sum_{k,l \geq 0} \binom{l + m - 2}{m - 2} a^{k+l} z^{k+l}$$

$$= \sum_{n \geq 0} \binom{n + m - 1}{m - 1} a^n z^n,$$

where, in the last equality above, we have used the columns' theorem of Pascal triangle (cf. Proposition 4.17 of [8]).

Section 21.3

1. Use Theorem 21.22, together with the fact that the characteristic polynomial of the given sequence is

$$X^{2k} - 2X^k + 1 = (X^k - 1)^2 = \prod_{j=1}^{k} (X - \omega^j)^2.$$

2. An easy inspection shows that the characteristic polynomial of $(a_n)_{n \geq 1}$ is $(X - 1)(X - 2)^2$. Thus, Theorem 21.22 gives $a_n = A + (B + C(n - 1))2^{n-1}$, for some real constants A, B and C. Since $\frac{a_n}{2^n} = \frac{1}{2}(B + C(n - 1))$, we must have $C = 0$ and $\frac{B}{2} = 3$. Finally, $a_1 = 1$ gives $A + B = 1$.

Glossary

Problems tagged with a country's name refer to any round of the corresponding national mathematical olympiad. For example, a problem tagged "Brazil" means that it appeared in some round of some edition of the Brazilian Mathematical Olympiad. Problems proposed in other mathematical competitions, or which appeared in mathematical journals, are tagged with a specific set of initials, as listed below:

AIME	American Invitational Mathematics Examination
APMO	Asian-Pacific Mathematical Olympiad
Austrian-Polish	Austrian-Polish Mathematical Olympiad
BMO	Balkan Mathematical Olympiad
Crux	Crux Mathematicorum, a mathematical journal of the Canadian Mathematical Society
EKMC	Eötvös-Kürschák Mathematics Competition (Hungary)
IMO	International Mathematical Olympiad
IMO shortlist	Problem proposed to the IMO, though not used
Miklós-Schweitzer	The Miklós-Schweitzer Mathematics Competition (Hungary)
OCM	State of Ceará Mathematical Olympiad
OCS	South Cone Mathematical Olympiad
OBMU	Brazilian Mathematical Olympiad for University Students
OIM	Iberoamerican Mathematical Olympiad
ORM	Rioplatense Mathematical Olympiad
Putnam	The William Lowell Mathematics Competition
Saint Petersburg	Mathematical competition of the city of Saint Petersburg, Russia
TT	The Tournament of the Towns

© Springer International Publishing AG, part of Springer Nature 2018   635
A. Caminha Muniz Neto, *An Excursion through Elementary Mathematics, Volume III*,
Problem Books in Mathematics, https://doi.org/10.1007/978-3-319-77977-5

Bibliography

1. M. Aigner, G. Ziegler, *Proofs from THE BOOK* (Springer, Heidelberg, 2010)
2. G. Andrews, *Number Theory* (Dover, Mineola, 1994)
3. T. Apostol, *Calculus*, Vol. 1 (Wiley, New York, 1967)
4. T. Apostol, *Calculus*, Vol. 2 (Wiley, New York, 1967)
5. T. Apostol, *Introduction to Analytic Number Theory* (Springer, New York, 1976)
6. R. Ash, *Basic Abstract Algebra: for Graduate Students and Advances Undergraduates* (Dover, Mineola, 2006)
7. A. Caminha, Uma prova elementar de que os números complexos algébricos sobre $\mathbb{Q}$ formam um corpo. Matemática Universitária **52/53**, 14–17 (2015) (in Portuguese)
8. A. Caminha, *An Excursion Through Elementary Mathematics I - Real Numbers and Functions* (Springer, New York, 2017)
9. A. Caminha, *An Excursion Through Elementary Mathematics II - Euclidean Geometry* (Springer, New York, 2018)
10. J.H.E. Cohn, Square Fibonacci numbers, etc. Fibon. Quart. **2**, 109–113 (1964)
11. J.B. Conway, *Functions of One Complex Variable I* (Springer, New York, 1978)
12. R. Courant, H. Robbins, *What Is Mathematics* (Oxford University Press, Oxford, 1996)
13. R. Diestel, *Graph Theory* (Springer, New York, 2000)
14. R. Dilworth, A decomposition theorem for partially ordered sets. Ann. Math. **51**, 161–166 (1950)
15. P. Erdös, E. Szekeres, A combinatorial problem in geometry. Compos. Math. **2**, 463–470 (1935)
16. S.B. Feitosa, *Turán's Theorem* (in Portuguese) (Classnotes, 2006)
17. D.G. de Figueiredo, *Números Irracionais e Transcendentes* (in Portuguese) (SBM, Rio de Janeiro, 2002)
18. W. Fulton, *Algebraic Curves*. Freely available at http://www.math.lsa.umich.edu/ wfulton
19. F. Galvin, A proof of Dilworth's chain decomposition theorem. Am. Math. Monthly **101**, 352–353 (1994)
20. C.R. Hadlock, *Field Theory and its Classical Problems* (Washington, MAA, 2000)
21. P. Halmos, *Naive Set Theory* (Springer, New York, 1974)
22. N. Hartsfield, G. Ringel, *Pearls in Graph Theory* (Academic Press, San Diego, 1990)
23. R. Honsberger, *Mathematical Gems III* (Washington, MAA, 1985)
24. D.A. Klarner, F. Göbel, Packing boxes with congruent figures. Indag. Math. **31**, 465–472 (1969)
25. Y. Kohayakawa, C.G.T. de A. Moreira, *Tópicos em Combinatória Contemporânea* (in Portuguese) (IMPA, Rio de Janeiro, 2001)
26. A.G. Kurosch, *Curso de Algebra Superior* (in Spanish) (MIR, Moscow, 1968)

27. E. Landau, *Elementary Number Theory* (AMS, Providence, 1999)
28. S. Lang, *Algebra* (Springer, New York, 2002)
29. H.N. Lima, *Limites e Funções Aritméticas* (in Portuguese). Preprint
30. L. Mirsky, A dual of Dilworth's decomposition theorem. Am. Math. Monthly. **78**, 876–877 (1971)
31. M. Reid, *Undergraduate Algebraic Geometry* (Cambridge University Press, Cambridge, 1988)
32. J. Roberts, *Elementary Number Theory: a Problem Oriented Approach*. (MIT Press, Cambridge, 1977)
33. W. Rudin, *Principles of Mathematical Analysis*, 3rd edn. (McGraw-Hill, Inc., New York, 1976)
34. E. Scheinerman, *Mathematics: A Discrete Introduction*, 3rd edn. (Brooks Cole, Boston, 2012)
35. S. Singh, *Fermat's Enigma: The Epic Quest to Solve the World's Greatest Mathematical Problem* (Anchor Books, New York, 1998)
36. E. Stein, R. Shakarchi, *Fourier Analysis: An Introduction* (Princeton University Press, Princeton, 2003)
37. I. Stewart, D. Tall, *Algebraic Number Theory and Fermat's Last Theorem*, 4th edn. (CRC Press, Boca Raton, 2015)
38. M.B.W. Tent, *Prince of Mathematics: Carl Friedrich Gauss* (A.K. Peters Ltd, Wellesley, 2006)
39. P. Turán, An extremal problem in graph theory. Mat. Fiz. Lapok **41**, 435–452 (1941)
40. J.H. Van Lint, R.M. Wilson, *Combinatorics* (Cambridge University Press, Cambridge, 2001)
41. H. Wilf, *Generatingfunctionology* (Academic, San Diego, 1990)

Index

© Springer International Publishing AG, part of Springer Nature 2018   639
A. Caminha Muniz Neto, *An Excursion through Elementary Mathematics, Volume III*,
Problem Books in Mathematics, https://doi.org/10.1007/978-3-319-77977-5

Printed in the United States
By Bookmasters